Canadian Mathematical Society
Société Mathématique du Canada

CONFERENCE PROCEEDINGS · Volume 8

OSCILLATION, BIFURCATION AND CHAOS

Proceedings of the
1986 Annual Seminar
held July 13–25, 1986

F. V. Atkinson, W. F. Langford, and A. B. Mingarelli, Editors

Published by the American Mathematical Society
for the Canadian Mathematical Society
PROVIDENCE, RHODE ISLAND

Volume 8
1986 Annual Seminar on Oscillation, Bifurcation and Chaos
held at the University of Toronto, Toronto, Ontario
July 13–25, 1986

Proceedings of a conference sponsored by the Natural Sciences and Engineering Research Council of Canada, University of Toronto, University of Guelph, and University of Ottawa

Accepted for publication by the Canadian Mathematical Society

Published by the American Mathematical Society

1980 *Mathematics Subject Classification* (1985 *Revision*). Primary 34B25, 34CXX, 34K05, 58FXX, 70HXX, 92A15, 35B05, 35B22, 35J25.

Library of Congress Cataloging-in-Publication Data

Seminar on Oscillation, Bifurcation, and Chaos (1986: University of Toronto)
Proceedings of the 1986 Annual Seminar on Oscillation, Bifurcation, and Chaos held July 13–25, 1986.
(Conference proceedings / Canadian Mathematical Society, ISSN 0731-1036; v. 8)
Sponsored by the Natural Sciences and Engineering Research Council of Canada ... et. al.
Includes Bibliographies.
1. Oscillations—Congresses. 2. Bifurcation theory—Congresses. 3. Chaotic behavior in systems—Congresses. I. Atkinson, F. V. II. Langford, W. F. (William F.) III. Mingarelli, Angelo B. (Angelo Bernardo), 1952– . IV. Canadian Mathematical Society. V. Natural Sciences and Engineering Research Council Canada. VI. Title. VII. Series: CMS conference proceedings; v. 8.
QA865.S46 1986 531′.322 87-11402
ISBN 0-8218-6013-5

Printed in the United States of America.
Reprinted 1988

DEDICATION

This volume is dedicated

to the memory of

GEOFFREY JAMES BUTLER

1944 - 1986.

CONTENTS

PREFACE

The year 1986 marked the sesquicentennial of the publication in 1836 of the classic memoir by J. Sturm on boundary value problems for second order equations. The 1986 Annual Seminar of the Canadian Mathematical Society, an International Conference on Oscillation, Bifurcation and Chaos, was held at the University of Toronto, from July 13 to 25, 1986. In honour of Sturm, the theme of the first week of the Conference was Sturmian theory and oscillation theory for second order equations, and modern generalizations. The second week was devoted to current research in bifurcation theory and chaos. The Organizing Committee consisted of F.V. Atkinson of Toronto, V. Jurdjevic of Toronto, W.F. Langford of Guelph and A.B. Mingarelli of Ottawa.

In all, 92 lectures were presented, by speakers from 5 continents. Plenary Lectures, each of 50 minutes duration, were given by D.G. Aronson, F.V. Atkinson, F. Brauer, P.A. Deift, W.N. Everitt, M. Golubitsky, G. Iooss, H.K. Langer (3 lectures), J. Lützen (2 lectures), H.G. Kaper, K. Kirchgässner, A.M. Krall, I. Kupka, M.K. Kwong, J. Mallet-Paret, J.E. Marsden, J. Mawhin, R.D. Nussbaum, D. Race, J.A. Sanders, J. Sotomayor, H. Troger, C.C. Tung, and J.S.W. Wong. The remaining talks consisted of Special Session Lectures (30 minutes) and Contributed Papers (20 minutes). Approximately half of the lectures are represented by papers in these Proceedings. All papers contained herein are complete refereed papers in final form, and will not be submitted for publication elsewhere.

The volume as a whole is dedicated to the memory of Geoffrey James Butler (1944-1986), of the University of Alberta, who died of cancer the evening prior to the start of the Conference. Geof had intended to participate in the Conference. One of Canada's most promising young researchers, he also chaired the Canadian Mathematical Olympiad Committee and lead the Canadian team in the International Mathematical Olympiad. His kindness, determination, and courage in the face of despair can only serve as a glimpse of the greatness of this human spirit.

We acknowledge, with thanks, the support of the following: The Natural Sciences and Engineering Research Council of Canada for a major Conference Grant, and the Universities of Toronto, Guelph and Ottawa for their financial as well as secretarial assistance.

F.V. Atkinson, Toronto

W.F. Langford, Guelph

A.B. Mingarelli, Ottawa

LIST OF SPEAKERS

Plenary Lectures.

DONALD G. ARONSON, University of Minnesota.
Analytic and Numerical Studies of Forced and Coupled Oscillators.

FREDERICK V. ATKINSON, University of Toronto.
Sturmian Theory for the Half-linear Equation.

FRED BRAUER, University of Wisconsin.
Harvesting in Delayed-Recruitment Population Models.

PERCY A. DEIFT, Courant Institute.
On the Existence of Eigenvalues of the Schrödinger Operator $H - \lambda W$ in a Gap of the Spectrum of H.

WILLIAM N. EVERITT, University of Birmingham.
Linear Control Theory and Quasi-differential Equations.

MARTIN GOLUBITSKY, University of Houston.
Bifurcation and Symmetry.

GÉRARD IOOSS, Université de Nice.
A Codimension-2 Bifurcation in the Presence of $SO(2)xO(2)$ Symmetry.

HEINZ LANGER, Technical University, Dresden.
Indefinite Inner Product Spaces in the Spectral Theory of Ordinary Differential Operators. (3 lectures)

JESPER LÜTZEN, University of Copenhagen, Denmark.
The Emergence of Sturm-Liouville Theory.
The Theorems Named after Liouville: The Career of A French Mathematician in the Middle of the 19th Century.

HANS G. KAPER, Argonne National Laboratory.
Oscillation Theory for Linear Second-order Differential Systems.

KLAUS KIRCHGÄSSNER, Universität Stuttgart.
Nonlinearly Resonant Waves and Chaotic Behaviour of Fluids.

ALLAN M. KRALL, Pennsylvania State University.
Weak and Distributional Solutions to some Differential Equations Connected with Special Functions.

IVAN KUPKA, Université de Grenoble.
Observers and Observability for Control Systems.

MAN KAM KWONG, Northern Illinois University.
Uniqueness of the Positive Solution of a Nonlinear Differential Equation.

JOHN MALLET-PARET, Brown University.
A Bifurcation Gap for a Singularly Perturbed Delay Equation.

JERROLD E. MARSDEN, University of California, Berkley.
Bifurcation and Stability in the Dynamics of a Liquid Drop.

JEAN MAWHIN, Université de Louvain.
Semilinear Boundary Value Problems on Unbounded Domains.

ROGER D. NUSSBAUM, Rutgers University.
Global Continuation and Asymptotic Behaviour for Periodic Solutions of a Differential-delay Equation.

DAVID RACE, University of Surrey, England.
Some Dirichlet and Conditional Dirichlet Criteria for Second Order Differential Expressions.

JAN A. SANDERS, Free University Amsterdam.
Normal Form Theory and Invariant Theory.

JORGE SOTOMAYOR, Instituto de Matemâtica Pura e Aplicada, Brazil.
Bifurcations of Umbilical Points and Lines of Principal Curvature.

HANS TROGER, Technical University Vienna.
Applications of Bifurcation Theory in Mechanical Engineering.

CHIN-CHU TUNG, Graduate School of Academica Sinica, Beijing.
Dynamics of Integers Illustrated by Collatz's Conjecture.

JAMES S. W. WONG, China Dyeing Works, Hong Kong.
Classifications of Second Order Nonlinear Oscillations.

Special Session and Contributed Lectures.

CALVIN D. AHLBRANDT, University of Columbia.
Sturmian Theory for Matrix Difference Equations.

W. ALLEGRETTO, University of Alberta.
Criticality and the λ-property for Elliptic Equations.

JACQUES BELAIR, Université de Montréal.
A Functional Differential Equation with Two Time Delays.

CLAUDE E. BILLIGHEIMER, McMaster University.
Spectral Theory of Differential and Difference Operators.

EDWARD A. CELARIER, University of Toronto.
The Structure of Basins of Attraction for Rössler's Equations.

JOHN CHADAM, McMaster University.
Bifurcation Problems Arising in Geochemistry.

SHAOZHU CHEN, Shandong University, China.
On the Non-oscillation of a Nonlinear Emden-Fowler Equation.

RANDALL S. DUMONT, University of Toronto.
Asymptotic Statistical Properties of Chaotic Power Spectra.

URI ELIAS, Technion, Israel.
Minors of the Wronskian and Oscillation of an Ordinary Differential Equation.

LYNN H. ERBE, University of Alberta.
Oscillation Results for Second Order Differential Systems.

WILLIAM W. FARR, University of Minnesota.
Degenerate Hopf Bifurcations in a Continuous Stirred Tank Reactor with Sequential Reactions.

DIETRICH FLOCKERZI, Universität Wurzburg.
Interaction of Two Pairs of Pure Imaginary Eigenvalues in the Normal Space of an Invariant N-Torus.

K. GOPALSAMY, Flinders University of South Australia.
Oscillation and Nonoscillation in Vector Systems of Delay Differential Equations.

EDWARD A. GROVE, University of Rhode Island.
Oscillation and Asymptotic Behaviour of Second Order Neutral Differential Equations With Deviating Arguments.

JOSEPH GRUENDLER, North Carolina A&T State University.
Homoclinic Bifurcations in Autonomous Systems.

EVANS M. HARRELL II, Georgia Institute of Technology.
Some Connections and Contrasts Between the Spectral Theories of Schrödinger and Laplace-Beltrami Operators.

B. J. HARRIS, Northern Illinois University.
Zero Counting Results.

BRIAN D. HASSARD, State University of New York at Buffalo.
Hopf Bifurcation for Autonomous Delay-Differential Systems.

JOSEF HOFBAUER, University of Vienna.
Averaging for Difference Equations.

JOHN A. HOLBROOK, University of Guelph.
Quaternionic Chaos.

JOHN W. HOOKER, Southern Illinois University.
Recessive Solutions of Disconjugate Matrix Difference Equations.

KE-LEI HUANG, Beijing Institute of Aeronautics and Astronautics.
Bifurcation of Quasiperiodic Solutions of Nonlinear Vibration Systems with Quasicyclic Coordinates.

KONCAY HUSEYIN, University of Waterloo.
On Certain Perturbation Techniques for Bifurcation Analyses.

GARY D. JONES, Murray State University.
Asymptotic Behavior of Solutions of a Fourth Order Linear Differential Equation.

PIERRE JOYAL, Université de Montréal.
Coefficients Related to the Number of Limit Cycles in the Generalized Homoclinic Bifurcation.

ROBERT M. KAUFFMAN, University of Alabama.
On the Supression of Singularities at Infinity in Heat Equations.

MARTIN KLAUS, Virginia Polytechnic Institute and State University.
On the Variation-Diminishing Property of Schrödinger Operators.

MARK A. KON, Columbia University.
Smoothing by Elliptic Semigroups.

MARTIN KRUPA, University of Waterloo.
On Resonant Double Imaginary Pair Bifurcation.

MUSTAFA KULENOVIC, University of Rhode Island.
Linearized Oscillation Theory for Second Order Differential Equations.

GERASIMOS LADAS, University of Rhode Island.
Sturm Comparison Theorems for Neutral Differential Equations.

BIKKAR S. LALLI, University of Saskatchewan.
Integral Averages and Second Order Superlinear Oscillation.

WILLIAM F. LANGFORD, University of Guelph.
Subharmonic Bifurcations in Nonlinear Mathieu Equations.

ROGER T. LEWIS, University of Alabama at Birmingham.
Eigenvalues of the Schrödinger Operators Below the Essential Spectrum.

LANCE L. LITTLEJOHN, Utah State University.
Sturm-Liouville Operators and Orthogonal Polynomials.

JACK W. MACKI, University of Alberta.
On Regular Growth and Asymptotic Stability.

JOYCE R. MCLAUGHLIN, Rensselaer Polytechnic Institute.
Exact Solutions for the Inverse Spectral Problem for the Timoshenko Beam.

KENNETH R. MEYER, University of Cincinnati.
Homoclinic Phenomena in Almost Periodic Systems.

KONSTANTIN MISCHAIKOW, Brown University.
A Bifurcation into Pathology for Hamiltonian Systems with two Degrees of Freedom.

MARTIN E. MULDOON, York University.
Variation of Zeros with Respect to a Parameter.

JAMES S. MULDOWNEY, University of Alberta.
Generalized Disconjugacy.

K. WAYNE NAGATA, University of Guelph.
Symmetry-breaking Perturbations of a Symmetric Hopf Bifurcation.

DANIEL C. OFFIN, University of Missouri.
The Maslow-Arnold Index in Classical Mechanics.

KENNETH PALMER, University of Miami.
Transversal Heteroclinic Points and Cherry's Example of a Nonintegrable Hamiltonian System.

L. A. PELETIER, University of Leiden.
Bound States for the Prescribed Mean Curvature Equation.

V. SREE HARI RAO, Osmania University, Hyderabad.
Criteria for Oscillating Periodic Solutions for Systems.

LOUISE A. RAPHAEL, Howard University.
On Relating Generalized Expansions to Fourier Integrals.

A. RAY, Ottawa.
Damped Oscillations and their Implications.

DAVID L. ROD, University of Calgary.
On a Theorem of Ziglin in Hamiltonian Systems.

THOMAS ROGERS, University of Alberta.
The Number of Bifurcations of Periodic Orbits of Any Period for the Quadratic Family.

FRANZ ROTHE, University of North Caroline at Chapel Hill.
Periods of Oscillations of Hamiltonian Systems in the Plane.

CHRISTIANE ROUSSEAU, Université de Montréal.
Bifurcations Methods in Quadratic Systems.

STEPHEN SCHECTER, North Carolina State University.
Stable Manifolds in the Method of Averaging.

DIETER SCHMIDT, Universität Gesamthochschule Essen.
Global Representations for the Solutions of Singular Second-Order Linear Differential Equations.

ALAN SELBY, University of Saskatchewan.
On the Flow and Structure of Invariant Vector Fields.

WILLIAM F. SHADWICK, University of Waterloo.
Cartan's Method of Equivalence and the Transcendents of Painlevé.

JOHN KENNETH SHAW, Virgina Tech University.
Hamitonian Systems with Parameter in Boundary Values.

MICHAEL W. SMILEY, Iowa State University.
Time-Periodic Solutions of Nonlinear Wave Equations in Balls.

LAWRENCE TURYN, Wright State University.
Bifurcation Without Mixed Mode Solutions.

ANDRE L. VANDERBAUWHEDE, State University of Gent, Belgium.
Secondary Bifurcation of Periodic Solutions in Autonomous Systems.

YUANMING WANG, Nanjing Institute of Technology, China.
Global Existence and Nonexistence for a Class of Nonlinear Degenerate Reaction-diffusion Systems.

EWA B. WEINMÜLLER, Technical University, Vienna.
Stability of Singular Boundary Value Problems.

M. W. WONG, York University, Toronto.
On Bound States of a One Dimensional Relativistic Hamiltonian.

ZHUO-CHENG YANG, University of Alberta.
Complete Chaos and Sensitive Dependence in a Simple Epidemiological Model.

ANTON ZETTL, Northern Illinois University.
Norm Inequalities for Derivatives and Differences.

Part 1
Sturm Symposium

Canadian Mathematical Society
Conference Proceedings
Volume 8 (1987)

RECESSIVE SOLUTIONS OF SYMMETRIC THREE TERM RECURRENCE RELATIONS

Calvin D. Ahlbrandt
Department of Mathematics
University of Missouri
Columbia, Missouri 65211

John W. Hooker
Department of Mathematics
Southern Illinois University
Carbondale, Illinois 62901

ABSTRACT. This paper studies symmetric linear recurrence relations with matrix coefficients. It is shown that a hypothesis of disconjugacy implies existence of recessive (or principal) matrix solutions. These results extend those of Patula [1979] for nonoscillatory scalar equations and relax certain sign conditions on the coefficients which were required in a previous study of principal solutions by the authors [1984]. This brings the theory for matrix difference equations closer to the continuous theory as developed by Hartman and Reid. The fundamental tool employed here is a discrete version of Morse's Sturmian separation theorem for matrix equations. Applications are given to discrete matrix Riccati equations. Examples include equations with periodic and constant coefficients. Continued fraction representations are given for the dominant solution of certain algebraic matrix quadratic equations.

1. INTRODUCTION.

The key step in extending the concept of principal (or recessive) solutions to disconjugate systems is a discrete version of a Sturm separation theorem. A form of that result which suffices for the present purposes is presented in Section 3. Those results depend upon a previous "variational" formulation of disconjugacy. A discrete version of Reid's

0731-1036/87 $1.00 + $.25 per page

famous "roundabout theorem" has recently been obtained by the authors [1986b]. That result is presented in Section 2. Recessive (principal) solutions of disconjugate systems are developed in Section 4. Associated with the recessive solution at ∞ is the so called "distinguished" solution of the discrete Riccati equation. In contrast to the differential equation case where the distinguished solution is the minimal right extensible hermitian solution, solutions of discrete Riccati equations can jump across asymptotes. However, a version of the minimality property of the distinguished solution does hold. That result and the dual at $-\infty$ are given in Sections 5 and 6. Section 7 considers cases where the coefficients are periodic in n and, as a special case, Section 8 studies Riccati equations and characterizes recessive solutions when the coefficients are constant. As in Riccati differential equations, periodicity of the coefficients assures periodicity of the distinguished solution at ∞ (with the same period as the coefficients). Hence constant coefficients imply that the distinguished solution at ∞ is constant, i.e., it is a solution of the "discrete algebraic Riccati equation." Section 9 gives some connections with other subjects and literature.

We now fix the notation and list blanket hypotheses on the coefficients. Consider operators l and L defined by

$$l[x]_n = -\Delta(C_{n-1}\Delta x_{n-1}) + A_n x_n = 0 \tag{1.1}$$

and

$$L[X]_n = -\Delta(C_{n-1}\Delta X_{n-1}) + A_n X_n = 0, \tag{1.2}$$

where Δ denotes the forward difference operator, $\Delta u_n = u_{n+1} - u_n$. These equations have the three term forms

$$l[x]_n = -C_n x_{n+1} - C_{n-1}x_{n-1} + B_n x_n = 0 \tag{1.3}$$

and

$$L[X]_n = -C_n X_{n+1} - C_{n-1}X_{n-1} + B_n X_n = 0, \tag{1.4}$$

respectively, where B_n is defined by

(1.5) $$B_n = C_n + C_{n-1} + A_n.$$

Assume the following blanket hypothesis for all applicable values of n :

(H) *A_n and C_n are real symmetric or complex hermitian matrices of order r with C_n nonsingular and B_n positive definite.*

The notation $H \geq K$ will mean that $H-K$ is positive semi-definite.

2. <u>DEFINITIONS AND PRELIMINARY RESULTS</u>. Let $(u,v) = v'u$ denote the inner product of R^r (or C^r) where $'$ denotes the transpose or conjugate transpose for vectors in R^r or C^r, respectively. The following properties of the forward difference operator Δ are listed for future reference:

(2.1)
$$\begin{aligned}\Delta(a_n, b_n) &= (a_n, \Delta b_n) + (\Delta a_n, b_{n+1}) \\ &= (a_{n+1}, \Delta b_n) + (\Delta a_n, b_n),\end{aligned}$$

(2.2) $$\Sigma_M^p \Delta C_n = C_{p+1} - C_M,$$

(2.3) $$\Sigma_M^p (t_n, \Delta s_{n-1}) = (t_{p+1}, s_p) - (t_M, s_{M-1}) - \Sigma_M^p (\Delta t_n, s_n)$$

A "bracket function" is defined by

(2.4)
$$\begin{aligned}\{u,v\}_n &= (C_{n-1}\Delta v_{n-1}, u_n) - (v_n, C_{n-1}\Delta u_{n-1}) \\ &= (v_n, C_{n-1}u_{n-1}) - (C_{n-1}v_{n-1}, u_n).\end{aligned}$$

Note that

(2.5) $$\{u,v\}_n = -\{v,u\}_n'.$$

For matrix sequences U_n, V_n, define

$$\{U,V\}_n = U_n' C_{n-1} \Delta V_{n-1} - (C_{n-1}\Delta U_{n-1})' V_n$$

$$= U_{n-1}' C_{n-1} V_n - U_n' C_{n-1} V_{n-1}.$$

Proposition 2.1. *Assume hypothesis* (H). *If* u *and* v *are solutions of* (1.1) *on* [M,N], *then* $\{u,v\}_n$ *is constant. If* U *and* V *are matrix solutions of* (1.2), *then* $\{U,V\}_n$ *is constant.*

Definition: Vector sequences u and v are *conjoined* [Reid, 1971] if $\{u,v\}_n \equiv 0$. A vector sequence u is *self-conjoined* if $\{u,u\}_n \equiv 0$, i.e., if

(2.6) $$u_{n-1}' C_{n-1} u_n \equiv u_n' C_{n-1} u_{n-1}.$$

Similarly, a matrix sequence X is called *prepared* [Hartman, 1957], or isotropic [Coppel, 1971] if $\{X,X\}_n \equiv 0$, or equivalently, if $X_{n-1}' C_{n-1} X_n$ is hermitian for every n.

Proposition 2.2. *Assume hypothesis* (H) *and let* U_n *be any matrix solution of* (1.2). *Then each of the* 2rxr *matrices*

$$\mathfrak{U}_n = \begin{bmatrix} U_n \\ C_{n-1}\Delta U_{n-1} \end{bmatrix}$$

has the same rank. In particular if $\mathfrak{U}_n$ *has full column rank* r *at one point* n, *then* $\mathfrak{U}_n$ *has rank* r *for all* n.

For positive integers M and N with M < N, equation (1.1) is called ***disconjugate*** on [M-1, N] if the real interval [M-1, N] contains no pair of ***conjugate intervals***, where distinct real intervals [p,p+1) and [q, q+1) are called *conjugate intervals* if there exists a self-conjoined vector solution of (1.1) such that

(2.7) $$x_p' C_p x_{p+1} \le 0 \text{ and } x_q' C_q x_{q+1} \le 0,$$

with $x_{p+1} \neq 0$, $x_{q+1} \neq 0$, and x_n is not identically zero for n in the interior of the smallest interval containing $[p, p+1)$ and $[q, q+1)$. The following result was established in Ahlbrandt & Hooker [1986b].

Theorem 2.1. *Assume hypothesis* (H) *and* M *and* N *are integers with* $M < N$. *Then the following conditions are equivalent:*

(*i*) *If* u *is a self-conjoined vector solution of* (1.1) *on* $[M, N]$ *with* $u'_{M-1}C_{M-1}u_M \leq 0$ *and* $u_M \neq 0$, *then*

$$u'_n C_n u_{n+1} > 0 \quad \textit{for} \quad n = M, \dots, N-1.$$

(*ii*) *If* v *is a self-conjoined vector solution of* (1.1) *on* $[M, N]$ *with* $v'_{N-1}C_{N-1}v_N \leq 0$ *and* $v_{N-1} \neq 0$, *then*

$$v'_n C_n v_{n+1} > 0 \quad \textit{for} \quad n = M-1, \dots, N-2.$$

(*iii*) *Equation* (1.1) *is disconjugate on* $[M-1, N]$.

(*iv*) *There exists a prepared matrix solution* X *of* (1.2) *on* $[M, N]$ *with* $X'_{n-1}C_{n-1}X_n > 0$, *for* $n = M, \dots, N$.

(*v*) *There exists a sequence of hermitian matrices* W_n, *defined for* $n = M, \dots, N+1$, *with*

$$W_n + C_{n-1} > 0, \quad \textit{for} \quad n = M, \dots, N,$$

satisfying the Riccati matrix equation

$$W_{n+1} = A_n + W_n(W_n + C_{n-1})^{-1}C_{n-1}, \quad \textit{for} \quad n = M, \dots, N.$$

(*vi*) *The quadratic form defined by*

$$J_2[\eta] = \Sigma_M^N (\Delta\eta_{n-1})'C_{n-1}\Delta\eta_{n-1} + \eta'_n A_n \eta_n$$

is positive definite on the class of vector sequences η *with*

$$\eta_{M-1} = 0 = \eta_N.$$

3. A STURM SEPARATION THEOREM. Consider a matrix solution U_n of (1.2). Proposition 2.2 assures that under hypothesis (H), the 2rxr matrix $\mathfrak{U}_n$ has constant rank. In case U_n is prepared and $\mathfrak{U}_n$ has full rank, the solution U_n will be referred to as a *conjoined basis*. The following separation theorem is a discrete version of Lemma 7.1 of Reid [1971].

Theorem 3.1. *Suppose that hypothesis* (H) *holds, equation* (1.1) *is disconjugate on* [M-1, N] *and* U_n *is a conjoined basis. Then*

(*i*) *there exist at most* r *points* m *in* [M, N] *such that* U_m *is singular*

and

(*ii*) *there exist at most* r *points* m *in* [M, N) *for which there exist unit vectors* $\pi(m)$ *such that*

$$(\pi(m))'U_m'C_mU_{m+1}\pi(m) < 0.$$

Points m such that U_m is singular will be called *focal points* and if m is as in (ii), then [m, m+1) will be called a *focal interval* of the solution U_n.

Corollary: *Suppose* (1.1) *is disconjugate on some neighborhood of* ∞, *say* $[M, \infty)$, *and* U_n *is any conjoined basis. Then there exists an integer* P *in* $[M, \infty)$ *such that*

$$D_n \equiv U_n'C_nU_{n+1} > 0 \quad on \quad [P, \infty), \tag{3.1}$$

i.e., D_n *is eventually positive definite. In particular,* U_n *is eventually nonsingular.*

A specific example where the hypotheses of Theorem 3.1 are satisfied is $A_n > 0$, $C_n > 0$, $U_{M-1} = 0$, and $U_M = I$.

For fixed integers M, N with M < N, define a sesquilinear form $\mathcal{S}$ by

$$(3.2)\qquad \mathfrak{S}[\varphi,\ \theta] = \sum_M^N (C_{n-1}\Delta\varphi_{n-1},\ \Delta\theta_{n-1}) + (A_n\varphi_n,\ \theta_n).$$

Lemma 3.1. *Suppose that* u *and* v *are conjoined solutions of* (1.1) *on* $[M, N]$. *For* $p, q \in [M, N]$ *and sequences* φ, θ *defined by*

$$(3.3)\qquad \varphi_n = \begin{cases} u_n, & M-1 \le n \le p \\ 0, & \textit{otherwise}, \end{cases}$$

$$\theta_n = \begin{cases} v_n, & M-1 \le n \le q \\ 0, & \textit{otherwise}, \end{cases}$$

then

$$(3.4)\qquad \mathfrak{S}[\varphi,\ \theta] = \begin{cases} -(C_{M-1}\Delta u_{M-1}, v_{M-1}), & \textit{if } p \ne q. \\ -(C_{M-1}\Delta u_{M-1}, v_{M-1}) + (u_{p+1}, C_p v_p), & \textit{if } p = q. \end{cases}$$

Before establishing this Lemma, let us give its application to the proof of Theorem 3.1.

Proof of Theorem 3.1(i). In order to avoid the notation of subscripts on subscripts, define an index set K by

$$K = \{m \mid m \in [M, N] \text{ such that } U_m \text{ is singular}\}.$$

Suppose K contains more than r points. For each $m \in K$, let $\pi(m)$ be a unit vector such that

$$(3.5)\quad U_m\ \pi(m) = 0$$

Define solutions $u_n(m)$ of (1.1) by

$$(3.6)\qquad u_n(m) = U_n\pi(m).$$

Then $u_m(m) = 0$, $u_{m-1}(m) \ne 0$, $u_{m+1}(m) \ne 0$, since if $u_n(m) \equiv 0$, then $\pi(m)$ would be in the null space of the full rank matrix $\mathfrak{U}_n$. Define variations $\eta_n(m)$ by

$$(3.7)\qquad \eta_n(m) = \begin{cases} u_n(m), & n \le m \\ 0, & n \ge m \end{cases}.$$

Note that $\eta_N(m) = 0$ for all $m \in K$. However, the set $\eta_{M-1}(m)$, $m \in K$ is linearly dependent and there exist scalars $d(m)$, $m \in K$, not all zero, such that

$$\sum_{m \in K} \eta_{M-1}(m)\, d(m) = 0.$$

Set

$$\eta_n = \sum_{m \in K} \eta_n(m)\, d(m).$$

Then

$$\eta_{M-1} = 0 = \eta_N, \quad \eta_n \not\equiv 0 \text{ on } [M-1, N].$$

Indeed, if $\eta_n \equiv 0$ and k is the largest member of K such that $d(k) \neq 0$, then $0 = \eta_{k-1} = u_{k-1}(k)d(k) \neq 0$. (In fact, this type of proof implies that the set of functions $\eta_n(m)$, $m \in K$, are linearly independent on $[M-1, N]$.) We wish to show that $J[\eta] = 0$, which is contrary to disconjugacy. Expand $J[\eta]$ in terms of the sesquilinear form $\mathcal{S}$ defined in (3.2). Then

$$J[\eta] = \mathcal{S}[\eta, \eta] = \mathcal{S}[\sum_{p \in K} \eta_n(p)d(p), \sum_{q \in K} \eta_n(q)d(q)]$$

$$= \sum_{p \in K} \sum_{q \in K} d(p)d'(q)\mathcal{S}[\eta_n(p), \eta_n(q)].$$

Since relations (3.5) and (3.6) imply $u_p(p) = 0$, it follows from Lemma 3.1 with

$$u_n = u_n(p), \quad v_n = u_n(q), \quad \eta_n(p) = \varphi_n, \quad \eta_n(q) = \vartheta_n,$$

that

$$\mathcal{S}[\eta_n(p), \eta_n(q)] = -(C_{M-1}\Delta U_{M-1}\pi(p), U_{M-1}\pi(q))$$

$$= -\pi'(q)U'_{M-1}C_{M-1}\Delta U_{M-1}\pi(p)$$

and

$$J[\eta] = -\sum_{p \in K} d(p)(\sum_{q \in K} d'(q)\pi'(q)U'_{M-1})C_{M-1}\Delta U_{M-1}\pi(p) = 0$$

because

$$\sum_{q \in K} d(q)\eta_{M-1}(q) = 0.$$

But, this is contrary to J positive definite on the class of η with $\eta_{M-1} = 0 = \eta_N$ and the linear independence of the $\eta_n(m)$.

Proof of Theorem 3.1(ii). The only modifications required are as follows: Set

$$K = \{m \mid M \le m < N \text{ for which there exists a unit vector } \pi(m) \text{ such that } (\pi(m))'U_m'C_mU_{m+1}\pi(m) < 0\}.$$

Note that $m < N$ makes (for u_n defined by (3.6)) φ_n and ϑ_n of Lemma 3.1 satisfy $\varphi_N = 0 = \vartheta_N$. Also, note that in case $p = q$ the terms

$$\sum_{p\in K} d(p)d'(p)(U_{p+1}\pi(p),\ C_pU_p\pi(p))$$

are all nonpositive, since

$$\pi'(p)U_p'C_pU_{p+1}\pi(p) < 0,\ d(p)d'(p) \ge 0.$$

Hence, $J[\eta]$, for η defined as before, gives

$$J[\eta] \le 0 \quad \text{and hence} \quad \eta \equiv 0.$$

Since $u_{m+1}(m) \ne 0$, $u_m(m) \ne 0$, the functions $\eta_n(m)$ are linearly independent contrary to $\eta \equiv 0$ and not all $d(m)$ being 0. Thus, given Lemma 3.1, we have Theorem 3.1.

Proof of Lemma 3.1. Case 1: $p < q$

The upper limit in the summation may be replaced by $p + 1$. Then break off the $p + 1$ term to obtain

$$\mathfrak{S}[\varphi,\ \vartheta] = (C_p(-u_p), \Delta v_p) + \sum_M^p (C_{n-1}\Delta u_{n-1},\ \Delta v_{n-1}) + (A_nU_n, V_n).$$

Use (2.3) with $t_n = C_{n-1}\Delta u_{n-1}$, $s_{n-1} = v_{n-1}$ for

$$\mathfrak{S}[\varphi,\ \vartheta] = -(C_pu_p,\ \Delta v_p) + (C_p\Delta u_p,\ v_p)$$
$$-(C_{M-1}\Delta u_{M-1},\ v_{M-1}) + \sum_M^p (l[u]_n,\ v_n)$$

The Lemma follows in Case 1 upon noting that

$$(u_{p+1}, C_pv_p) = (C_pv_p,\ u_{p+1})' = (v_{p+1}, C_pu_p)' = (C_pu_p, v_{p+1}).$$

Case 2: $p > q$. Use Case 1 as follows:

$$\mathfrak{S}[\varphi, \vartheta] = (\mathfrak{S}[\vartheta, \varphi])' = -(C_{M-1}\Delta v_{M-1}, u_{M-1})'$$

$$= -(u_{M-1}, C_{M-1}v_M - C_{M-1}v_{M-1})$$

$$= -(u_{M-1}, C_{M-1}v_M) + (u_{M-1}, C_{M-1}v_{M-1})$$

$$= -(C_{M-1}u_{M-1}, v_M) + (u_{M-1}, C_{M-1}v_{M-1}).$$

Since $\{u, v\} = 0 \Rightarrow \{v, u\} = 0 \Rightarrow (C_{M-1}u_{M-1}, v_M) = (u_M, C_{M-1}v_{M-1})$ we have the result

$$\mathfrak{S}[\varphi, \vartheta] = -(\Delta u_{M-1}, C_{M-1}v_{M-1}) = -(C_{M-1}\Delta u_{M-1}, v_{M-1}).$$

Case 3: $p = q$. As in Case 1, isolate the $p+1$ term, and sum by parts for

$$\mathfrak{S}[\varphi, \vartheta] = (C_p u_p, v_p) + \sum_M^p (t_n, \Delta s_{n-1}) + (A_n u_n, v_n)$$

$$= (C_p u_p, v_p) + (C_p \Delta u_p, v_p) - (C_{M-1}\Delta u_{M-1}, v_{M-1})$$

$$= (C_p u_{p+1}, v_p) - (C_{M-1}\Delta u_{M-1}, v_{M-1}),$$

where s_n and t_n are as in Case 1. Hence Lemma 3.1 holds.

4. RECESSIVE SOLUTIONS. We now make use of Theorem 2.1 and the Corollary to Theorem 3.1 in order to establish existence of a recessive (principal) solution at ∞ under hypothesis (H) and eventual disconjugacy. In Ahlbrandt and Hooker [1984], we used the hypotheses $A_n > 0$ and $C_n > 0$, which imply disconjugacy on every interval (see parts (iii) and (vi) of the Theorem 2.1 above). Even in the scalar case, examples occur where the solution x_n is oscillatory in the sense of having sign changes on every neighborhood of ∞, but recessive solutions exist. For example, if $b_n = 3$ and $c_n = (-1)^n$, then J is positive definite due to diagonal dominance of the matrix S of the associated quadratic form (see, for example, Ahlbrandt and

Hooker [1985], p. 14). Hence, (1.1) is disconjugate on every interval, but every nontrivial real solution eventually oscillates by the Corollary to Theorem 3.1. However, due to the following study a recessive solution exists. This is in contrast to the behavior of differential equations where recessive solutions at ∞ exist if and only if the equation is eventually nonoscillatory. Because Patula [1979] assumed $C_n > 0$, his study did not establish existence of recessive solutions for all eventually disconjugate equations. The reason for this contrast between differential and difference equations is the form of the Legendre condition in the calculus of variations in continuous and discrete cases (see Ahlbrandt & Hooker [1985], p. 14). Note that $B_n > 0$ in Hypothesis (H) is the strengthened Legendre condition. However, even the following definition of recessive solutions is not as general as that of Ahlbrandt [1972] where conclusion (iii) of Theorem 4.1 below is taken as a definition.

The definition of a recessive solution of difference equations which we now present is analogous to that of Hartman [1957] for Sturm-Liouville operators. It stems from the property $\int^{\infty}(cu^2)^{-1}dt = \infty$, which characterizes principal solutions of self-adjoint differential equations (cf. Hartman [1964]). Principal solutions for *n*th-order scalar difference equations have been discussed by Hartman [1978]. For computation of recessive (principal) solutions of scalar difference equations, see Gautschi [1967] or Olver and Sookne [1972]. Our construction parallels that of Hartman [1957]. Our proof of essential uniqueness employs Reid's construction of principal solution [1958], but does not require any variational proofs beyond that of the above Theorems 2.1 and 3.1.

<u>Definition</u>: A prepared solution X of (1.2) is called "*recessive*" (or principal) at ∞ if X is such that there exists an integer M for which

(4.1) $X_n'C_nX_{n+1}$ is positive definite for $n \geq M$

and

(4.2) $$\sum_{i=M}^{n} k'(X_i'C_iX_{i+1})^{-1}k \to \infty, \text{ as } n \to \infty,$$

for every unit vector k. A prepared solution satisfying (4.1) is called *dominant* if

(4.3) $$\sum_{i=M}^{\infty} k'(X_i'C_iX_{i+1})^{-1}k$$

converges for every unit vector k.

Theorem 4.1. *Assume hypothesis* (H) *and* (1.1) *is eventually disconjugate. Then*

(*i*) *every conjoined basis* X *satisfies* (4.1) *for some* M;

(*ii*) *there exists a solution* Z *which is recessive at* ∞;

(*iii*) *if* Y *is any prepared solution with* $\{Y, Z\}$ *nonsingular, then* Y *is dominant and*

(4.4) $$Y_n^{-1}Z_n \to 0 \quad \text{as} \quad n \to \infty;$$

(*iv*) Z *is essentially unique, i.e., if* X *and* Z *are recessive at* ∞, *then there exists a nonsingular matrix* H *such that* $X_n = Z_nH$ *for all* n;

(*v*) *if* M *is such that* (1.1) *is disconjugate on* $[M-1, \infty)$ *and* $X_n(M,N)$ *is defined as the solution of the two point boundary value problem*

$$X_{M-1}(M,N) = I, \quad X_N(M,N) = 0$$

then the recessive solution at ∞ *is determined* up *to nonsingular constant multiples as*

$$\lim_{N\to\infty} X_n(M,N).$$

Proposition 4.1. *If* X_n *is a solution of the homogeneous matrix difference equation* $\Delta X_{n-1} + A_nX_n = 0$, *and* X_n *is nonsingular for all* $n \geq M-1$ *for some* $M > 0$, *then the solution of* $\Delta Y_{n-1} + A_nY_n = B_n$, $n \geq M$, *with initial value* Y_{M-1}

is

$$Y_n = X_n(X_{M-1}^{-1}Y_{M-1} + \sum_{i=M}^{n} X_{i-1}^{-1}B_i), \quad n \geq M-1,$$

where the summation equals 0 *if* $n = M-1$.

In order to express one solution in terms of another we first introduce some notation which parallels the concept of a definite integral. If $X = X_n$ is a prepared solution of $L[X_n] = 0$, which is nonsingular for $n \geq M-1$, for brevity of notation we define

$$(4.5) \qquad S_n(X) = \begin{cases} 0, & \text{if } n = M-1 \\ \sum_{i=M}^{n} (X_{i-1}'C_{i-1}X_i)^{-1} & \text{if } n \geq M \end{cases}$$

Now we give the following analogue of the usual reduction of order result expressing all solutions in terms of one solution.

Proposition 4.2. *Assume Hypothesis* (H) *and let* X_n *be a prepared solution of* (1.2) *which is nonsingular for all* $n \geq M-1$ *for some* $M > 0$. *Then every solution* Y_n *of* (1.2) *for* $n \geq M-1$ *is given by*

$$(4.6) \qquad Y_n = X_n(P + S_n(X)Q), \quad n \geq M-1,$$

for constant matrices

$$(4.7) \qquad P = X_{M-1}^{-1}Y_{M-1} \text{ and } Q = \{X_{M-1}, Y_{M-1}\} \equiv \{X_n, Y_n\}, \quad n \geq M-1.$$

Conversely, if P *and* Q *are given constant matrices,* (4.6) *defines a solution of* $L[Y_n] = 0$ *for* $n \geq M-1$.

Proof. Given X_n and Y_n as stated, we know

$$\{X_n, Y_n\} = X_n'C_{n-1}\Delta Y_{n-1} - [C_{n-1}\Delta X_{n-1}]'Y_n = Q,$$

a constant matrix (i.e., Q is independent of n). Since X_n is non-singular for $n \geq M-1$, we can write

$$C_{n-1}\Delta Y_{n-1} - X_n'^{-1}[C_{n-1}\Delta X_{n-1}]'Y_n = X_n'^{-1}Q.$$

Hence

$$\Delta Y_{n-1} - C_{n-1}^{-1}[(C_{n-1}\Delta X_{n-1})X_n^{-1}]'Y_n = C_{n-1}^{-1}X_n'^{-1}Q. \tag{4.8}$$

Since X_n is a prepared solution of $L[X_n] = 0$, X_n is a solution of the homogeneous equation related to (4.8) (i.e., substituting X_n for Y_n makes the left-hand side of (4.8) equal 0). Then by variation of parameters, Proposition 4.1, the solution Y_n of the non-homogeneous equation (4.8) can be written as

$$Y_n = X_n(X_{M-1}^{-1}Y_{M-1} + \sum_{i=M}^{n} X_{i-1}^{-1}C_{i-1}^{-1}X_i'^{-1}Q),$$

which is of the form (4.6) with $P = X_{M-1}^{-1}Y_{M-1}$. Conversely, given P and Q, one may verify that (4.6) is a solution of $L[Y_n] = 0$, which completes the proof.

<u>Proposition 4.3</u>. *The solution* Y_n of $L[Y_n] = 0$ *given by* (4.6) *is a prepared solution if and only if*

$$P'Q = Q'P. \tag{4.9}$$

<u>Proof</u>. We need to show that $Y_{n-1}'C_{n-1}Y_n$ is hermitian for $n \geq M$. By Proposition 2.1 it suffices to show this for $n = M$. From (4.6) with $n = M$, we have

$$Y_{M-1}'C_{M-1}Y_M = Y_{M-1}'C_{M-1}X_MP + Y_{M-1}'C_{M-1}X_M(X_{M-1}'C_{M-1}X_M)^{-1}Q. \tag{4.10}$$

Since $Y_{M-1}' = P'X_{M-1}'$ (from (4.6) with $n = M-1$), (4.10) can be written

$$Y_{M-1}'C_{M-1}Y_M = P'X_{M-1}'C_{M-1}X_MP+P'X_{M-1}'C_{M-1}X_M(X_{M-1}'C_{M-1}X_M)^{-1}Q. \tag{4.11}$$

Since X_n is a prepared solution, the first term on the right in (4.11) is hermitian and the second term equals $P'Q$. Thus $Y_{M-1}'C_{M-1}Y_M$ is hermitian if and only if $P'Q$ is hermitian, which completes the proof.

We now proceed to establish the existence of recessive and

dominant solutions of $L[X_n] = 0$, assuming Hypothesis (H) and eventual disconjugacy.

Proof of Theorem 4.1: Conclusion (i) follows from the Corollary to Theorem 3.1. Establish conclusion (ii) as follows:

Let M be such that (1.1) is disconjugate on $[M-2, \infty)$. Theorem 2.1 assures that if a solution X is determined by

(4.12) $$X_{M-2} = 0, \; X_{M-1} = I,$$

then $X_n'C_nX_{n+1}$ is positive definite for $n \geq M-1$. Choose $P = Q = I$ and use the final statement of Proposition 4.2 and Proposition 4.3 to conclude that Y_n defined by

(4.13) $$Y_n = X_n(I + S_n(X)), \; n \geq M-1,$$

is prepared, and since $I + S_n(X)$ is bounded below by I and increasing, Y_n is nonsingular for $n \geq M-1$. Thus, we may reverse the roles of X_n and Y_n in Proposition 4.2, which then tells us that

(4.14) $$X_n = Y_n(\tilde{P} + S_n(Y)\tilde{Q}), \; n \geq M-1,$$

where $\tilde{P} = Y_{M-1}^{-1}X_{M-1} = I$ and $\tilde{Q} = \{Y_n, X_n\}$, a constant matrix. From

$$\tilde{Q} = \{Y_n, X_n\} = -\{X_n, Y_n\}' = -Q' = -I,$$

we conclude that (4.14) reduces to

(4.15) $$X_n = Y_n(I - S_n(Y)).$$

Hence

(4.16) $$I = X_n^{-1}X_n = X_n^{-1}Y_n(I - S_n(Y)).$$

Substituting (4.13) into (4.16) yields

(4.17) $$I = (I + S_n(X))(I - S_n(Y)).$$

Thus $I - S_n(Y)$ is the sequence of inverses of the increasing, positive- definite sequence $I + S_n(X)$, and $I + S_n(X) > I$ for $n \geq M$. It follows that $I - S_n(Y)$ is a decreasing, positive-definite sequence satisfying $0 < I - S_n(Y) < I$ for all $n \geq M$. Hence $S_n(Y)$ is an increasing sequence with a positive-definite limit $S_\infty(Y) \leq I$ as $n \to \infty$. Indeed, $S_\infty(Y)$ is the unique matrix such that for every constant unit vector k, $k'S_n(Y)k \to k'S_\infty(Y)k$, as $n \to \infty$. Now define the sequence

$$Z_n = Y_n(S_\infty(Y) - S_n(Y)), \quad n \geq M-1. \tag{4.18}$$

This is of the form (4.6) with $P = S_\infty(Y)$ and $Q = \{Y_n, Z_n\} = -I$, so by Proposition 4.2, Z_n is a solution of $L[Z_n] = 0$. We will show that Z_n is prepared and recessive.

Since $S_n(Y)$ is hermitian for all $n \geq M$, $S_\infty(Y)$ is hermitian; hence, by Proposition 4.3, Z_n is prepared. Also, $S_\infty(Y) - S_n(Y)$ is positive definite and Y_n is nonsingular for each $n \geq M$, so Z_n is nonsingular. Thus Z_n satisfies the hypotheses of Proposition 4.2, and we may write Y_n in terms of Z_n as

$$Y_n = Z_n(\hat{P} + S_n(Z)\hat{Q}), \quad n \geq M-1, \tag{4.19}$$

where, by (4.7), $\hat{P} = Z_{M-1}^{-1}Y_{M-1} = S_\infty(Y)^{-1}$ and $\hat{Q} = \{Z_n, Y_n\}$. Since $\hat{Q} = \{Z_n, Y_n\} = -\{Y_n, Z_n\}' = I$, (4.19) becomes

$$Y_n = Z_n(S_\infty(Y)^{-1} + S_n(Z)), \quad n \geq M-1.$$

Proceeding as in steps (4.16) and (4.17), we obtain

$$I = (S_\infty(Y) - S_n(Y)(S_\infty^{-1}(Y) + S_n(Z)). \tag{4.20}$$

But $S_\infty(Y) - S_n(Y)$ is positive definite and decreasing, and tends to 0 as $n \to \infty$. It then follows from (4.20) that $S_n(Z)$ is increasing for $n \geq M-1$, and for every unit $r \times 1$ vector k, $k'S_n(Z)k$ is positive and increasing as n increases, and

$\lim_{n\to\infty} k'S_n(Z)k = \infty$. Therefore Z_n is a recessive solution of $L[Z_n] = 0$, by definition.

Similarly, from the discussion above, we know that $S_n(Y)$ is a positive-definite, increasing sequence for $n \geq M$, with a positive-definite limit $S_\infty(Y) \leq I$, and it follows that Y_n is a dominant solution.

Establish condition (iii) of Theorem 4.1 by use of (4.6) with Z playing the role of X, i.e., if Y is any prepared solution, then there exist P, Q such that

$$(4.21) \qquad Y_n = Z_n(P + S_n(Z)Q).$$

By hypothesis $Q = \{Z_n, Y_n\} = -\{Y_n, Z_n\}'$ is nonsingular and $P'Q = Q'P$ since Y is prepared. From (4.21) we have

$$(4.22) \qquad Z_n^{-1}Y_nQ^{-1} = PQ^{-1} + S_n(Z).$$

Since Z is recessive at ∞, we conclude that (note PQ^{-1} is hermitian)

$$k'Z_n^{-1}Y_nQ^{-1}k \to \infty \quad \text{as} \quad n \to \infty$$

for every unit vector k and $Z_n^{-1}Y_nQ^{-1}$ is increasing in n. Hence the least eigenvalue of $Z_n^{-1}Y_nQ^{-1}$ tends to ∞ as $n \to \infty$, Y_n is eventually nonsingular, and $QY_n^{-1}Z_n \to 0$ as $n \to \infty$. Since Q is nonsingular we have $Y_n^{-1}Z_n \to 0$. Because Y is a conjoined basis, the Corollary to Theorem 3.1 assures that there exists an integer p in $[M, \infty)$ such that

$$Y'_{i-1}C_{i-1}Y_i > 0 \quad \text{for} \quad i \geq p.$$

Replace M by p in $S_n(Y)$ and in Proposition 4.2 for the representation

$$Z_n = Y_n(\tilde{P} + S_n(Y)\tilde{Q}), \quad n \geq p-1.$$

But

$$Y_n^{-1}Z_n = \tilde{P} + S_n(Y)\tilde{Q} \to 0 \quad \text{as} \quad n \to \infty.$$

Therefore

$$S_n(Y)\tilde{Q} \to -\tilde{P} = -Y_{p-1}^{-1}Z_{p-1}.$$

Thus $\tilde{Q}$ is nonsingular and

$$S_n(Y) \to -\tilde{P}\tilde{Q}^{-1} \quad \text{as} \quad n \to \infty,$$

i.e., Y is dominant and conclusion (iii) of Theorem 4.1 has been established.

We now establish (iv) by the construction of (v).

In the matrix differential equation case, Reid [1958] gave a construction of the principal solution at ∞ by means of a limiting case of a solution of a two-point boundary value problem. Reid's construction for the system

$$U' = AU + BV, \ V' = CU - A'V$$

started with a solution $U_{st}(x)$, $V_{st}(t)$ defined by

$$U_{st}(s) = I, \ U_{st}(t) = 0.$$

He showed that $V_{st}(s)$ had a limit $V_{s\infty}(s)$ as t became infinite. Then $(U_{s\infty}(x), V_{s\infty}(x))$ was a principal solution at ∞. Constructions of this type have been used in the construction of recessive solutions of three-term recurrence relations [Gautschi, 1967; Olver and Sookne, 1972]. In particular, such methods have been employed in numerical evaluation of Bessel functions $J_n(x)$ from their recurrence relation $-J_{n+1}(x) - J_{n-1}(x) + (2n/x)J_n(x) = 0$.

We now use the same idea to construct a recessive solution at ∞. Of course, the previously constructed Z_n is recessive, but for any nonsingular matrix K, Z_nK is also recessive.

Let M and N be integers with $M < N$. For Z_n constructed as above, define $S_n(Z)$ as before. Then $S_n(Z)$ is increasing in n, positive definite for $n \geq M$ and $(S_n(Z))^{-1} \to 0$

as $n \to \infty$. We define $X_n(M,N)$ as the solution defined by the two-point boundary value problem of condition (v). If such a solution exists, it must have the form

$$X_n(M,N) = Z_n(P + S_n(Z)Q), \quad n \geq M-1.$$

But the choice

$$I = Z_{M-1}P, \quad Q = -(S_N(Z))^{-1}P$$

uniquely determines $X_n(M,N)$ as

$$X_n(M,N) = Z_n(I - S_n(Z)[S_N(Z)]^{-1})Z_{M-1}^{-1}$$

with the consequence

$$X_n(M,N) \to Z_n Z_{M-1}^{-1} \text{ as } N \to \infty.$$

Since $P'Q = -P'[S_N(Z)]^{-1}P$ is hermitian, $X_n(M,N)$ is prepared for each N and since

$$X_n(M,\infty) \equiv Z_nK \quad \text{for} \quad K = Z_{M-1}^{-1},$$

$X_n(M,\infty)$ is prepared and recessive at ∞. We now show that the recessive solution at ∞ is essentially unique. Let Y_n be any other recessive solution at ∞. Without loss of generality, assume M is sufficiently large that Y_n is nonsingular for $n \geq M-1$. Replace Z_n in the previous derivation of $X_n(M,\infty)$ by Y_n for the conclusion

$$X_n(M,\infty) \equiv Y_nY_{M-1}^{-1} = Z_nZ_{M-1}^{-1}$$

i.e.,

$$Y_n = Z_nH \quad \text{for} \quad H = Z_{M-1}^{-1}Y_{M-1}.$$

This completes the proof of Theorem 4.1. It is of interest to note that examples are known where a solution Z_n exists as in (4.4) for

$$x_{n+1} + x_{n-1} = b_nx_n$$

in cases where the equation is oscillatory (see Section 9), i.e., since $c_n = 1$, the equation fails to be disconjugate on every neighborhood of ∞. In those cases the terms of $S_n(Z)$ can have various signs or be undefined. The investigation of usage of relation (4.4) of Theorem 4.1 as a definition of the recessive solution is beyond the scope of the present study. See Ahlbrandt [1972] for usage of that definition in differential equations.

5. EVENTUALLY MINIMAL SOLUTIONS OF DISCRETE RICCATI EQUATIONS. Under hypothesis (H) and eventual disconjugacy, we have shown that every conjoined basis X_n is eventually nonsingular. Associated with any such X_n is a hermitian solution W_n of the Riccati difference equation

$$(5.1)\qquad R[W_n] \equiv -W_{n+1} + W_n(W_n + C_{n-1})^{-1}C_{n-1} + A_n = 0$$

which is defined by

$$(5.2)\qquad W_n = (C_{n-1}\Delta X_{n-1})X_{n-1}^{-1}.$$

Because of the identity

$$(5.3)\qquad X_n'L[X_n] = X_n'R[W_n]X_n,$$

for X and W related by (5.2), it follows that solutions W of the Riccati equation generate solutions X of $L[X] = 0$ by means of the first order recurrence relation

$$(5.4)\qquad X_n = C_{n-1}^{-1}(W_n + C_{n-1})X_{n-1}.$$

Furthermore, the identity

$$(5.5)\qquad W_n + C_{n-1} = (X_{n-1}^{-1})'(X_{n-1}'C_{n-1}X_n)X_{n-1}^{-1}$$

and the Corollary to Theorem 3.1 immediately yield the following Proposition.

<u>Proposition 5.1</u>. *Suppose hypothesis* (H) *holds and* (1.1) *is eventually disconjugate. If* W *is any right extensible hermitian solution of* (5.1), *then there exists an integer* M *such that*

(5.6) $W_n + C_{n-1}$ *is positive definite on* $[M, \infty)$.

For Z a recessive solution of (1.2) at ∞, let W_n^- denote the associated solution of the Riccati equation (5.1). Note that W^- is right extensible and any other recessive solution at ∞, say $X = ZC$, C n.s., generates the same solution, W^-. We know from the above proposition that W_n^- is eventually bounded below by $-C_{n-1}$.

In the continuous case, Reid [1971] called the solution of the Riccati equation associated with the principal solution at ∞ the *distinguished solution at* ∞. He showed that a hermitian solution W was right extensible if and only if it satisfied $W \geq W^-$ on the entire domain of W. However, Riccati difference equations can have hermitian solutions which are initially "below" W^- but are right extensible. Hence the minimality property can be, at best, *eventual minimality*, in the sense that every right extensible hermitian solution W must satisfy

(5.7) $W_n \geq W_n^-$, for large n.

<u>Theorem 5.1</u>. *Suppose hypothesis* (H) *holds and* (1.1) *is eventually disconjugate. If* W^- *is the distinguished solution at* ∞ *associated with the recessive solution at* ∞ *of* (1.2), *then* W^- *is the unique solution which has the eventual minimality property.*

<u>Proof</u>. Let W be any hermitian solution of (5.1) in some neighborhood of ∞ and let M be sufficiently large that W and W^- are defined on $[M, \infty)$ and satisfy (5.6). Let X be the solution of (1.2) defined by $X_{M-1} = I$ and the two term recurrence relation (5.4). Select Z so that $Z_{M-1} = I$. Then

Z can be represented in terms of X as

(5.8) $$Z_n = X_n(I + S_n(X)Q), \text{ for } n \geq M-1.$$

Let $U_n(M,N)$ be the solution of the two point boundary value problem

(5.9) $$U_{M-1}(M,N) = I, \; U_N(M,N) = 0.$$

Then, by Theorem 4.1,

(5.10) $$\lim_{N\to\infty} U_n(M,N) = Z_n, \; n \geq N-1.$$

But there exists Γ_N, namely

$$\Gamma_N = \{X_{M-1}, U_{M-1}(M,N)\},$$

such that

(5.11) $$U_n(M,N) = X_n(I + S_n(X)\Gamma_N)$$

Choice of $n = N$ gives $I + S_N(X)\Gamma_N = 0$. Relations (5.8) and (5.11) imply

$$X_n^{-1}Z_n - I = S_n(X)Q$$

and

$$X_n^{-1}U_n(M,N) - I = S_n(X)\Gamma_N.$$

Since (5.10) holds, we conclude that

$$\lim_{N\to\infty} S_n(X)\Gamma_N = S_n(X)Q.$$

But, for $n > M-1$, $S_n(X)$ is positive definite. Thus

(5.12) $$\lim_{N\to\infty} (S_N(X))^{-1} = \lim_{N\to\infty} (-\Gamma_N) = -Q$$

Set $\Gamma = -Q$. Then Γ is hermitian, nonnegative, and we have

(5.13) $$X_n = Z_n(I + S_n(Z)\Gamma)$$

But

$$\Gamma = \{Z_M, X_M\} = Z'_{M-1}C_{M-1}X_M - Z'_M C_{M-1}X_{M-1}$$

$$= C_{M-1}X_M - Z'_M C_{M-1}$$

$$= (W_M + C_{M-1}) - (W_M^- + C_{M-1}).$$

Consequently,

(5.14) $$W_M - W_M^- = \Gamma = \lim_{N\to\infty} (S_N(X))^{-1} \geq 0$$

with equality if and only if X is recessive.

6. MAXIMAL LEFT EXTENSIBLE SOLUTIONS OF THE REVERSE RICCATI EQUATION.

The usual time reversal $t = -s$ for differential equations allows behavior near $-\infty$ to be read off from known results for solutions defined near ∞. If n in (1.4) is replaced by $-m$ and terms are reordered, then we have

(6.1) $$-C_{-m-1}X_{-m-1} - C_{-m}X_{-m+1} + B_{-m}X_{-m} = 0.$$

Set

(6.2) $$E_m = B_{-m}, \; F_m = C_{-m-1}, U_m = X_{-m}$$

and define D by

(6.3) $$E_m = F_m + F_{m-1} + D_m.$$

Then $D_m = A_{-m}$. Also, (6.1) becomes

(6.4) $$-F_m U_{m+1} - F_{m-1}U_{m-1} + E_m U_m = 0,$$

which is of the form (1.4). Behavior as $n \to -\infty$ of (1.4) can now be read off from behavior of (6.4) as $m \to \infty$. As in Section 5, set

(6.5) $$V_m = (F_{m-1}\Delta U_{m-1})U_{m-1}^{-1},$$

i.e.,

(6.6) $$U_m = F_{m-1}^{-1}(V_m + F_{m-1})U_{m-1}.$$

Furthermore, identity (5.5) becomes

$$(6.7) \qquad V_m + F_{m-1} = (U_{m-1}^{-1})'(U'_{m-1}F_{m-1}U_m)U_{m-1}^{-1}.$$

But, if we define R_n by

$$(6.8) \qquad \begin{aligned} V_m &= C_n(U_m - U_{m-1})U_{m-1}^{-1} = C_n(X_n - X_{n+1})X_{n+1}^{-1} \\ &= -C_n(X_{n+1} - X_n)X_{n+1}^{-1} = -C_n(\Delta X_n)X_{n+1}^{-1} = -R_{n+1}, \end{aligned}$$

i.e., for R defined by

$$(6.9) \qquad R_n = C_{n-1}(\Delta X_{n-1})X_n^{-1},$$

then R is a solution of the *reverse Riccati equation*

$$(6.10) \qquad R_n = -A_n + R_{n+1}[C_n - R_{n+1}]^{-1}C_n.$$

Theorem 6.1. *Suppose hypothesis* (H) *holds and* (1.4) *is disconjugate on some interval* $(-\infty, N]$. *If* R *is any left extensible hermitian solution of the reverse Riccati equation* (6.10) *then*

$$(6.11) \qquad R_n - C_{n-1} \textit{ is negative definite in some neighborhood of } -\infty.$$

Also, there exists a left extensible hermitian solution R_n^+ *of* (6.10) *which has the property that for each left extensible hermitian solution* R_n *of* (6.10), *there exists a neighborhood of* $-\infty$ *on which*

$$(6.12) \qquad R_n \le R_n^+.$$

Proof. Note that Theorem 2.1(iv) assures that for every $M < N$, there exists a prepared solution X_n with

$$0 < X'_{n-1}C_{n-1}X_n = U'_{m+1}F_mU_m$$

for $n = M, \ldots, N$, i.e., for $m = -N, \ldots, -M$. Thus (6.4) is disconjugate in some neighborhood of ∞ and the previous results may be applied. In particular V_m^-, the eventually minimal solution, exists and $R^+_{n+1} = -V^-_m$ will be the dual solution of (6.10). That is, if R_n is any hermitian solution of (6.10) which is left extensible, then on some neighborhood of $-\infty$, we will have (from $V^-_m \le V_m$, $V_m + F_{m-1} > 0$), the conclusions

$$R^+_{n+1} \ge R_{n+1} \quad \text{and} \quad -R_{n+1} + C_n > 0.$$

7. PERIODIC COEFFICIENT RICCATI EQUATIONS.

Suppose that A_n and C_n are *periodic of period* p, i.e.,

(7.1) $$A_{n+p} = A_n, \ C_{n+p} = C_n \quad \text{for all } n.$$

Theorem 7.1. *Assume hypothesis* (H) *holds, the coefficients are periodic of period* p *and* (1.1) *is disconjugate on* $(-\infty, \infty)$. *Then* W^-_n *and* R^+_n *exist on* $(-\infty, \infty)$ *and are periodic of period* p.

Proof. The construction of the recessive solution Z at ∞ given in Section 4 can be based at any point M. Hence the distinguished solution at ∞ extends to $(-\infty, \infty)$. In order to establish periodicity, introduce auxiliary solutions $W^{(1)}$ and $W^{(2)}$ of (5.1) by

(7.2) $$W^{(1)}_n = W^-_{n+p}, \ W^{(2)}_n = W^-_{n-p}.$$

Then, in some neighborhood of ∞,

(7.3) $$W^-_n \le W^{(1)}_n = W^-_{n+p}$$

and

(7.4) $$W^-_m \le W^{(2)}_m \le W^-_{m-p}.$$

Set $m = n+p$ for

(7.5) $$W^-_{n+p} \le W^-_n,$$

which, when combined with (7.3), implies periodicity of W^-.

8. CONSTANT COEFFICIENTS; the Discrete Algebraic Riccati Equation, the Characteristic Equation and a Representation of the Recessive Solutions at $\pm\infty$.

We now restrict our study to the case where A, B, C satisfy the following hypothesis.

(HC) A, B, *and* C *are constant hermitian matrices with* $A > 0$ *and* $C > 0$.

Then $B = 2C + A$ and (1.1) is disconjugate on $(-\infty, \infty)$ by Theorem 2.1. Theorem 7.1 then implies that W^- and R^+ are of period 1 and, hence, are constant solutions of the steady state equations

$$W = A + W(W + C)^{-1}C \tag{8.1}$$

and

$$R = -A + R(C - R)^{-1}C, \tag{8.2}$$

respectively.

Proposition 8.1. *Under hypothesis* (HC), *a matrix* W *is a solution of* (8.1) *if and only if* W *is a solution of the "discrete algebraic Riccati equation"*

$$WC^{-1}W - AC^{-1}W - A = 0. \tag{8.3}$$

In particular, if W *is a solution of* (8.3), *then* $W + C$ *is nonsingular and, if* W *is hermitian, then* $W + C > 0$.

Proof. If W is a solution of (8.1), then $W + C$ is nonsingular and satisfies

$$(W-A)C^{-1}(W+C) = W, \tag{8.4}$$

which reduces to (8.3). Conversely, (8.3) may be written as (8.4) and if $(W+C)k = 0$, then

$$0 = (W-A)C^{-1}(W+C)k = Wk = -Ck.$$

Consequently, any solution of (8.3) has $W + C$ nonsingular and is a solution of (8.1). The conclusion in case W is hermition follows from Proposition 5.1.

<u>Proposition 8.2.</u> *Assume hypothesis* (HC) *and matrices* W *and* Λ *are related by*

(8.5) $$C\Lambda = W + C.$$

Then W *is a solution of* (8.3) *if and only if* Λ *is a solution of the "characteristic equation"*

(8.6) $$-C\Lambda^2 + B\Lambda - C = 0.$$

Furthermore, if W *is a hermitian solution, then* Λ *is similar to a diagonal matrix* $D = diag(\lambda_i)$, *with* λ_i *real. If* W *is negative [positive] definite, then the* λ_i *lie in* (0,1) [(1, ∞)].

<u>Proof.</u> Suppose λ is an eigenvalue of Λ and x is a corresponding eigenvector. Then λ, x is an eigenvalue, eigenvector pair for the "generalized eigenvalue problem" (see Golub & Van Loan [1983], Section 7.7)

(8.7) $$(W+C)x = \lambda Cx.$$

Hence, from the generalized Rayleigh quotient

(8.8) $$\lambda = x'(W+C)x/(x'Cx)$$

we conclude that λ is real and positive. If $W < 0$, then $x'(W+C)x' < x'Cx$ and $\lambda < 1$. If $W > 0$, then $x'(W+C)x' > x'Cx'$ and $\lambda > 1$.

Furthermore, suppose P is a unitary matrix such that (use the Schur canonical form; see Golub & Van Loan, pp. 192-193, Theorem 7.1-2 and Corollary 7.1-3)

$$P'[C^{-1/2}(W+C)C^{-1/2}]P = D = diag(\lambda_i).$$

Then

$$C\Lambda = W + C = C^{1/2}PDP'C^{1/2}$$

and

(8.9) $$\Lambda = C^{-1/2}PDP'C^{1/2} = Q^{-1}DQ$$

for $Q = P'C^{1/2}$.

Proposition 8.3. *Assume hypothesis* (HC) *and* Λ *is a solution of* (8.6). *Then* Λ *is nonsingular and if* X_n *is defined for all* n *by the recurrence relation*

$$X_n = \Lambda X_{n-1}, \quad X_0 = I, \tag{8.10}$$

then X *is a solution of* (1.4) *on* $(-\infty, \infty)$ *and* X_n *has the form*

$$X_n = \Lambda^n, \quad \text{for all } n. \tag{8.11}$$

Note that in (8.11) we use the conventions $\Lambda^0 = I$ and $\Lambda^{-n} = (\Lambda^{-1})^n$, for Λ nonsingular.

Theorem 8.1. *Under hypothesis* (HC), *the discrete algebraic Riccati equation* (8.3) *has exactly one negative definite solution* W^- *and one positive definite solution* W^+. *If* W *is any hermitian solution of* (8.3), *then*

$$W^- \leq W \leq W^+ \tag{8.12}$$

Furthermore, $W^- > -C$, $A < W^+ < A + C$, *and* $W^- + W^+ = A$.

Theorem 8.2. *Assume* (HC). *If* P *and* Q *are defined by*

$$P = C^{-1}W^- + I, \quad Q = C^{-1}W^+ + I, \tag{8.13}$$

then $\lambda(P) \in (0,1)$, $\lambda(Q) \in (1,\infty)$, $P = Q^{-1}$, *and the recessive solutions at* $-\infty$ *and* ∞, *which take on the value* I *at zero, are given, respectively, by*

$$Y_n = Q^n \quad \text{and} \quad Z_n = P^n. \tag{8.14}$$

Hence, $Z_n \to 0$ *as* $n \to \infty$, $Y_n \to 0$ *as* $n \to -\infty$, Y_n *is dominant at* ∞ *and* $Y_n^{-1}Z_n \to 0$ *as* $n \to \infty$.

Theorem 8.3. *Assume hypothesis* (HC) *and* W_n *is a hermitian solution of*

(8.15) $$W_{n+1} = A + W_n(W_n+C)^{-1}C$$

with $W_1 > W^-$. *Then* W_n *extends to the right and converges to* W^+ *as* $n \to \infty$. *If* R_n *is any hermitian solution of*

(8.16) $$R_n = -A + R_{n+1}[C - R_{n+1}]^{-1}C,$$

with $R_0 < W^+ - A$, *then* R_n *extends to the left and converges to* $W^- - A$ *as* $n \to -\infty$.

<u>Proof of Theorem 8.1</u>. Suppose that W_n is the solution of (8.15) with $W_1 = 0$. Then (5.1) implies W_n is right extensible, positive definite and therefore, $W_n \geq W^-$. (Use $W_n(W_n+C_{n-1})^{-1}C_{n-1} = (W_n^{-1}+C_{n-1}^{-1})^{-1}$, $n > 1$.) Hence $W^- \leq W_1 = 0$. But W^- is nonsingular by (8.4). For uniqueness, let $W < 0$ be any solution of (8.3) with $W \neq W^-$. Then relation (5.14) implies that not all eigenvalues of $S_N(X)$ go to ∞, i.e., there exists a unit vector k such that

$$\lim_{N\to\infty} k'S_N(X)k \quad \text{exists.}$$

Hence the series

$$\sum_{i=1}^{\infty} k'(X'_{i-1}CX_i)^{-1}k$$

converges, and, for $CG = W + C$, we have $X_i = G^i$, and

$$k'(G^i)^{-1}(W+C)^{-1}(G^i)'^{-1}k \to 0, \text{ as } i \to \infty.$$

Since there exists a positive α such that $(W+C)^{-1} \geq \alpha I$, then

$$|(G^i)'^{-1}k|_2 \to 0, \text{ as } i \to \infty.$$

But G is of the form $G = Q^{-1}DQ$, for $D = \mathrm{diag}(\lambda_i)$, and we conclude that $(D^{-1})^i u \to 0$, for $u = (Q')^{-1}k \neq 0$, contrary to

$\lambda_i \in (0,1)$. Hence, W^- is the only negative definite solution.

The results on W^+ in Theorem 8.1 are consequences of R^+ being a maximal hermitian solution of

$$(8.17) \qquad RC^{-1}R + AC^{-1}R - A = 0.$$

For R and W hermitian matrices related by $W = -R$, W is a solution of (8.3) if and only if R is a solution of (8.17). Since R^+ is the maximal solution of (8.17), $W = -R^+$ is the minimal solution W^- of (8.3). Thus R^+ is the unique positive definite solution of (8.17). Furthermore, if W and R are related by $W = R + A$ then W is a solution of (8.3) if and only if R is a solution of (8.17). Hence $W^+ = R^+ + A = -W^- + A$. Since $0 < R^+ < C$, we have $A < W^+ < A + C$ and the proof of Theorem 8.1 is complete.

<u>Proof of Theorem 8.2</u>. Postmultiply (8.6) by Λ^{-2} to conclude that $\Lambda = P^{-1}$ is a solution of (8.6). Define W by $\Lambda = C^{-1}W + I$. There exists a unitary matrix U such that

$$P = C^{-1/2}\, U\, DU'C^{1/2}$$

for $D = \mathrm{diag}(\lambda_i)$, $\lambda_i \in (0,1)$. Hence $D^{-1} > I$ and

$$C\Lambda = CP^{-1} = C^{1/2}\, U\, D^{-1}U'C^{1/2} > C.$$

Hence $W = C\Lambda - C$ is positive definite and by uniqueness $W = W^+$; hence $\Lambda = Q = P^{-1}$. Relation (5.4) gives

$$\{Y,Z\} = Y_0'CZ_1 - Y_1'CZ_0 = (W^- + C) - (W^+ + C)' = W^- - W^+ < 0$$

from which Theorem 4.1(iii) applies.

<u>Proof of Theorem 8.3</u>. Let Y and Z be as in Theorem 8.2 and let X be the solution determined by

$$X_n = C^{-1}(W_n + C)X_{n-1}, \quad X_0 = I.$$

Then there exist matrices H and K such that

$$X_n = P^n H + Q^n K;$$

indeed, H and K are given by

$$H = (W^+ - W^-)^{-1}(W^+ - W_1), \quad K = (W^+ - W^-)^{-1}(W_1 - W^-).$$

The assumption $W_1 > W^-$ implies that K is nonsingular. Since

$$W_n = C(X_n X_{n-1}^{-1} - I),$$

we observe that W_n will converge to W^+ if and only if

$$X_n X_{n-1}^{-1} \to Q \quad \text{as} \quad n \to \infty.$$

But,

$$X_{n+1}X_n^{-1} - Q =$$

$$[P^{n+1}H + Q^{n+1}K - Q(P^n H + Q^n K)](P^n H + Q^n K)^{-1} =$$

$$(P - Q)P^n H[Q^{-n}P^n H + K]^{-1}Q^{-n} \to 0.$$

If V_m is any hermitian solution of the Riccati equation associated with (6.4) such that $V_1 > V^-$, then $V_m \to V^+$ as $m \to \infty$. Hence

$$R_0 = -V_1 < -V^- = R^+ = -W^-, \text{ i.e., } R_0 < W^+ - A \text{ implies}$$

$$R_{n+1} = -V_m \to -V^+ = V^- - A = -R^+ - A = W^- - A, \quad \text{as } n \to -\infty.$$

9. CONCLUDING REMARKS. One might expect that the existence of extremizing solutions of (8.3) might be obtained as a corollary to the work of Coppel [1974] (see also Gohberg,

Lancaster and Rodman [1983], chapter 4) on symmetric equations

$$(9.1) \qquad \hat{A} + W\hat{B} + \hat{B}^*W + W\hat{C}W = 0.$$

However, that is not immediate since (8.3) is not symmetric. If W is any hermitian solution of (8.3), then W must be a solution of a symmetric equation (9.1) since $AC^{-1}W$ must be symmetric. Indeed, set

$$\hat{A} = -A, \quad \hat{B} = -(\frac{1}{2})(AC^{-1}), \quad \hat{C} = C^{-1}.$$

Knowledge of W^+ with the properties established above will imply that

$$\hat{B} + \hat{C}W^+ = \frac{1}{2}[Q-P] = \frac{1}{2}[Q-Q^{-1}]$$

and $\hat{B} + \hat{C}W^+$ is similar to $D - D^{-1}$ where D is diagonal with entries in $(1, \infty)$. Since Coppel's W^+ is the only solution with eigenvalues of $\hat{B} + \hat{C}W^+$ in the right half plane, our W^+ must agree with Coppel's, even though the above results are required to see this, i.e., it is not apparent how Section 8 might be read off from Coppel's theorem. On the other hand, certain special cases of (9.1) might be solved by the present methods.

The existence of recessive solutions for three term scalar recurrence relations

$$(9.2) \qquad x_n = b_n x_{n-1} + a_n x_{n-2}$$

is intimately related to convergence of the continued fraction $K(a_n/b_n)$

$$(9.3) \qquad b_0 + \frac{a_1}{b_1} + \frac{a_2}{b_2} + \frac{a_3}{b_3} + \cdots$$

(Note that this b_n is the previous b_{n-1} and if $a_n = -1$, then $c_n = 1$. Lange points out that (see Kalton and Lange [1986], Section 5) the choice $a_n = -1$ and $b_{2k} = 0$, $b_{2k+1} = 1$, gives the continued fraction $-K(-1/b_n)$, i.e.,

$$(9.4)\qquad \frac{1}{1} - \frac{1}{0} - \frac{1}{1} - \frac{1}{0} - \cdots$$

which converges to 0 (also see Wall [1967] for definition of convergence). This example of (9.2) has linearly independent solutions f_n and g_n determined by $f_0 = 0$, $f_1 = 1$, $g_0 = 1$, $g_1 = 0$. It follows readily that $f_{2k} = 0$, $f_{2k+1} = (-1)^k$ and f is recessive in the sense that $f_n/g_n \to 0$ as $n \to \infty$. This is contrary to the behavior of differential equations because the recessive solution f is oscillatory and the equation fails to be disconjugate on any neighborhood of ∞. The basic result equating convergence of continued fractions (9.3) to existence of a recessive (minimal, distinguished) solution of the linear recurrence relation (9.2) is due to Pincherele [1892]. Gautschi [1967], Theorem 1.1, apparently incorrectly states Pincherele's theorem because the above example has $f_0 = 0$, contrary to his assertion that if the continued fraction converges, then the recurrence relation has a minimal solution with $f_0 \neq 0$. Jones and Thron [1980] have a statement of Pincherele's theorem which does not have this difficulty. Unfortunately from a computational viewpoint, much of the literature in the analytic theory of continued fractions uses the term "converges" to include "converges to ∞."

We will not attempt to pursue a matrix version of Pincherele's theorem here. However, if in Theorem 8.3, one selects $W_1 = 0$, then $W_2 = A$,

$$(9.5)\qquad W_{n+1} = A + [C^{-1} + W_n^{-1}]^{-1},\ n \geq 2$$

and $W_n \to W^+$ as $n \to \infty$. Successive substitutions leads to the matrix continued fraction, (set $D = C^{-1}$)

$$(9.6)\qquad A + I/D + I/A + I/D + I/A + \ldots,$$

(using reverse polish arithmetic) which converges to W^+ in the sense that (9.5) converges, if A and C are positive definite. A better starting value would be $W_1 = A + \theta C$, for θ any value in $(0,1)$, for example, $\theta = 1/2$. Numerical experiments with Cleve Moler's MATLAB suggest that convergence is fairly slow, but stable. The inverses of the positive definite matrices could be calculated by using Cholesky factors (see Golub and Van Loan [1983]).

The fact that $Q = P^{-1}$ leads as follows to another possible recurrence relation which seems to converge to W^+.

$$I = PQ = (C^{-1}W^- + I)(C^{-1}W^+ + I)$$

and $$W^- + W^+ = A \quad \text{imply} \quad W^-C^{-1}W^+ = -A,$$

$$W^+ = A - W^- = A + A(W^+)^{-1}C,$$

and $$W^+ = A[I + (W^+)^{-1}C]$$

for the suggested iteration

$$W_1 = A + C/2$$

$$W_{n+1} = A[I + W_n\backslash C],\ n = 1,\ 2,\ \ldots$$

which seems to require symmetrizing W_{n+1} at each step, i.e. replace W_{n+1} by $(W_{n+1} + W_{n+1}')/2$, before iterating. In the scalar case, the corresponding continued fraction is

$$a + \frac{ac}{a} + \frac{ac}{a} + \frac{ac}{a} + \ldots \ .$$

Note that this continued fraction is of period 1 whereas the previous continued fraction was of period 2. Merkes and Scott [1960] studied periodic and reverse periodic continued fractions in terms of fixed points of the associated linear fractional transformation. The above identity $W^+ + W^- = A$ seems related to their results. Postmultiplication of (8.6) by Λ^{-1} suggests the iteration for Q of

$$\Lambda_1 = I, \ \Lambda_{n+1} = C^{-1}B - \Lambda_n^{-1}, \ n = 1, 2, \ldots .$$

Indeed, if $C\Lambda_n = W_n + C$ defines W_n, then $W_1 = 0$, $W_2 = A$ and W_n satisfies (9.5) and converges to W^+. Hence, Λ_n is nonsingular and converges to Q. Also, $\lambda(\Lambda_n) > 1$, for n = 2, 3, ... Related studies on matrix continued fractions include those of Rasmussen, Stokes, and Fair.

Recognition of the Riccati equation as a generalization of the characteristic equation was noted by Smith. Background work for this material includes the references below to Brockett, Fort, Morse, Mingarelli, and the various papers and books of Reid. Higher order difference equations have been considered by Hartman [1978], [1964] and Peterson. Extensive work has been done on numerically computing W^+ for certain equations (9.1). The reader is referred to Vaughan [1970], Dorato and Lewis [1971], Hewer [1971], and Pappas, Lamb, & Sandell [1980].

Symmetric three term recurrence relations with scalar C_n and matrix B_n arise in the Lanczos method for estimating eigenvalues of real symmetric matrices (see Golub & Van Loan [1983] and Parlette [1980]).

Lancaster and Rodman [1980] consider a "unilateral equation"

$$Z^2 + K_1 Z + K_0 = 0$$

with hermitian coefficients, which is related to symmetric equations (9.1). Equation (8.6), when premultiplied by $-C^{-1}$, is of this form with $K_1 = -C^{-1}B$, $K_0 = I$, except K_1 need not be hermitian.

Since in Theorem 8.2 $Q = P^{-1}$ and $P^{2n} \to 0$, it follows that $P - Q$ is nonsingular and P, Q provides a "complete set of solvents" (in the terminology of Lancaster and Tismenetsky [1985]) for the matrix polynomial

$$L[\lambda] = \lambda^2 I - C^{-1}B\lambda + I.$$

Professor Heinz Langer has pointed out that the methods of Gohberg, Lancaster, and Rodman, [1982] and [1983], and Lancaster and Tismenetsky [1985] (Theorem 2, pg 524) provide information about the eigenvalues of all solutions Λ of (8.6); namely, they are contained in the union of the sets of eigenvalues of P and eigenvalues of Q. To see this directly, suppose λ is an eigenvalue of such a solution Λ with eigenvector x. Then

$$0 = (\Lambda^2 - C^{-1}B\Lambda + I)x = (\lambda^2 I - C^{-1}B\lambda + I)x = (\lambda I - Q)(\lambda I - P)x$$

and λ is either an eigenvalue of P or of Q. To relate this to monic polynomials with hermitian coefficients, set $x = C^{-1/2}y$ and premultiply the above by $C^{1/2}$ for the monic polynomial

$$\lambda^2 I - \lambda C^{-1/2}BC^{-1/2} + I.$$

Acknowledgement: The authors wish to express their appreciation to the referee for his constructive comments.

REFERENCES

1972 Ahlbrandt, C.D., Principal solutions of selfadjoint differential systems and their reciprocals, Rocky Mountain J. Math 2, (1972), 169-182.

1984 Ahlbrandt, C.D. and Hooker, J.W., Riccati transformations and principal solutions of discrete linear systems, in Proc. 1984 Workshop "Spectral Theory of Sturm-Liouville Differential Operators," H.G. Kaper and A. Zettl, eds., ANL-84-87, Argonne National Lab., Argonne, Illinois, 1984.

1985 Ahlbrandt, C.D. and Hooker, J.W., "A variational view of nonoscillation theory for linear difference equations," Proceedings of Thirteenth Midwest Differential Equations Conference, J.L. Henderson, ed., Univ. of Missouri-Rolla.

1986a Ahlbrandt, C.D. and Hooker, J.W., Disconjugacy criteria for second order linear difference equations, Proc. of International Conference on Qualitative Theory of Differential Equations, Edmonton, Alberta, Canada, 1984, (to appear).

1986b Ahlbrandt, C.D. and Hooker, J.W., Riccati matrix difference equations and disconjugacy of discrete linear systems. (Preprint).

1970 Brockett, Roger W., Finite Dimensionsl Linear Systems, Wiley, New York, 1970.

1971 Coppel, W.A., Disconjugacy, Lecture Notes in Mathematics 220, Springer-Verlag, New York, 1971.

1974 Coppel, W.A., Matrix quadratic equations, Bulletin of the Australian Math. Soc. 10(1974), 377-401.

1971 Dorato, Peter and Levis, Alexander H., Optimal linear regulators: the discrete-time case, IEEE Transactions on Automatic Control, Vol. AC-16, #6(December 1971), 613-620.

1972 Fair, Wyman, A convergence theorem for noncommutative continued fractions, J. Approximation Theory 5(1972), 74-76.

1948 Fort, T., Finite Differences and Difference Equations in the Real Domain, Oxford University Press, London, 1948.

1967 Gautschi, Walter, Computational aspects of three-term recurrence relations, SIAM Review, 9(1967), 24-82.

1983 Golub, Gene H. and Van Loan, Charles F., Matrix Computations, Johns Hopkins University Press, 1983, Baltimore.

1982 Gohberg, I., Lancaster, P., and Rodman, L., Matrix Polynomials, Academic Press, New York (1983).

1983 Gohberg, I., Lancaster, P., and Rodman, L., "Matrices and Indefinite Scalar Products" Operator Theory: Advances and Applications, Vol. 8, Birkhauser Verlag, Boston (1983).

1957 Hartman, P., Self-adjoint, non-oscillatory systems of ordinary, second order, linear differential equations, Duke Math. J. 24(1957), 25-36.

1964 Hartman, P., Ordinary Differential Equations, John Wiley and Sons, New York 1964.

1978 Hartman, P., Difference equations: disconjugacy, principal solutions, Green's functions, complete monotonicity, Trans. Amer. Math. Soc. 246 (1978), 1-30.

1971 Hewer, G.A., An iterative technique for the computation of the steady-state gains for the discrete optimal regulator, IEEE Trans. Automatic Control, vol. AC-16, 382-383, Aug. 1971.

1981 Hooker, J.W., and Patula, W.T., Riccati type transformations for second- order linear difference equations, J. Math. Anal. Appl. 82(1981), 451- 462.

1985 Hooker, J.W., Kwong, M.K., and Patula, W.T., Riccati type transformations for second-order linear difference equations II, J. Math. Anal. Appl. 107(1985), 182-196.

1986 Hooker, J.W., Kwong, M.K., and Patula, W.T., "Oscillatory second order linear difference equations and Riccati equations," SIAM J. Math. Anal. (to appear).

1980 Jones, W.B. and Thron, W.J., "Continued fractions: analytic theory and applications," Addison-Wesley Pub. Co., Reading, Mass., 1980.

1986 Kalton, N.J., and Lange, L.J., Equimodular limit periodic continued fractions in Analytic Theory of Continued Fractions II, Lecture Notes in Mathematics, Vol. 1199, Springer-Verlag, New York, 1986.

1980 Lancaster, P. and Rodman, L., Existence and uniqueness theorems for the algebraic Riccati equation, Internat. J. Control 32, 285-309 (1980).

1985 Lancaster, P., and Tismenetsky, M., "The Theory of Matrices," second edition, 1985, Academic Press, New York.

1960 Merkes, E.P. and Scott, W.T., Periodic and Reverse Periodic Continued Fractions, Michigan Math. J. 7(1960), 23-29.

1983 Mingarelli, A.B., Volterra-Stieltjes Integral Equations and Generalized Ordinary Differential Expressions, Lecture Notes in Mathematics 989, Springer-Verlag, New York, 1983.

1981 Moler, Cleve, MATLAB Users Guide, Dept. of Computer Science, Univ. of New Mexico, 1981.

1930 Morse, M., A generalization of the Sturm separation and comparison theorems in n-space, Mathematische Annalen 103(1930), 72-91.

1972 Olver, F.W.J., and Sookne, D.J., Note on backward recurrence algorithms, Math. Comput. 26(1972), 941-947.

1980 Pappas, T., Laub, A.J. and Sandell, N.R., On the numerical solution of the discrete-time algebraic Riccati equation, IEEE Trans. on Automatic Control, vol. AC-25, no. 4, August 1980, 631-641.

1980 Parlett, B.N., "The Symmetric Eigenvalue Problem," Prentice-Hall, Englewood Cliffs, N.J., 1980.

1979 Patula, W.T., Growth, oscillation, and comparison theorems for second order linear difference equations, SIAM J. Math. Anal. 10(1979), 1272- 1279.

1984 Peterson, A.C., Boundary value problems for an n-th order linear difference equation, SIAM J. Math. Anal. 15(1984), 124-132.

1892 Pincherle, S., Su la génération de systèmes récurrents au moyen d'une équation linéaire différentielle, Acta Math., 16(1892), pp. 341-363.

1977 Rasmussen, Carl H., Oscillatory and asymptotic behavior of systems of ordinary linear differential equations, Ph.D. thesis, Univ. of Connecticut, Xerox University Microfilms, Ann Arbor Michigan #77-31,215.

1956 Reid, W.T., Oscillation criteria for linear differential systems with complex coefficients, Pacific J. Math. 6(1956), 733-751.

1958 Reid, W.T., Principal solutions of non-oscillatory, self-adjoint linear differential systems, Pacific J. Math 8(1958), 147-169.

1963 Reid, W.T., Riccati matrix differential equations and non-oscillation criteria for associated linear differential systems, Pacific J. Math. 10(1963), 665-685.

1971 Reid, W.T., Ordinary Differential Equations, Wiley, New York, 1971.

1972 Reid, W.T., Riccati Differential Equations, Mathematics in Science and Engineering, 86. Academic Press, New York, 1972.

1980 Rodman, L., On extremal solutions of the algebraic Riccati equation, in Algebraic and Geometric Methods in Linear Systems Theory, AMS Lectures on Applied Mathematics 18 (1980), Eds. C.I. Brynes and C.F. Martin, 311-327.

1986 Smith, Donald R., Decoupling and order reduction via the Riccati transformation, SIAM Review, (to appear).

1973 Stokes, A.N., Differential inequalities and the matrix Riccati equation Ph.D. thesis, Australian National University, Canberra, 1972, See also, abstract, Bull. Australian Math. Soc. 9(1973), 315-317.

1970 Vaughan, D.R., A nonrecursive algebraic solution for the discrete Riccati equation, IEEE Trans. Automatic Control 15, 597-599, 1970.

1967 Wall, H.S., Analytic theory of continued fractions, Chelsea Pub. Co. (1967), New York.

Canadian Mathematical Society
Conference Proceedings
Volume 8 (1987)

ON THE NONOSCILLATION OF AN EMDEN-FOWLER EQUATION

F.V.Atkinson and Shaozhu Chen

ABSTRACT. This paper deals with the nonoscillation of the equation $y'' + y\exp(|y|^m - t) = 0$ for $0 < m < 2$, and with the asymptotic behaviour: the derivatives of all solutions tend to zero as $t \to \infty$ for all $m > 1$. Our results prove a part of the conjecture which states that nonoscillation holds for $0 < m < 2$ and oscillatory solutions may exist for $m \geq 2$.

1. INTRODUCTION. The equation

$$(1) \qquad y'' + y\exp(|y|^m - t) = 0, \qquad m > 0,$$

is of interest in connection with radial solutions of the equation

$$(2) \qquad \Delta u + f(u) = 0$$

in R^2; the case that $f(u)$ behaves as $\exp(u^2)$ is well-known to be of a borderline nature [1,2].

It is easy to see that every bounded solution $y(t)$ is nonoscillatory and satisfies

$$\lim_{t\to\infty} y'(t) = 0, \qquad \lim_{t\to\infty} y(t) = \text{const} \neq 0.$$

In other words, every oscillatory solution , if it exists, must be unbounded. Conversely, the equation (1) certainly has bounded solutions, namely those which tend to any assigned non-zero limit at infinity.

If $0 < m < 1$, then for any real constants a and b, $a^2 + b^2 \neq 0$, the equation (1) has a solution $y(t)$ such that

$$y(t) = a + bt + o(1) \quad \text{as} \quad t\to\infty,$$

since

$$\int^{\infty} t|a + bt| \exp(|a + bt|^m - t)dt < \infty.$$

This is also true for $m = 1$ if $|b| < 1$. For details see [3,5].

1980 Mathematics Subject Classification. 34C10, 15.

The question here is whether all solutions are nonoscillatory, i.e. whether the equation (1) is nonoscillatory. Recently, F.V.Atkinson conjectured that nonoscillation holds for $0 < m < 2$, but that for $m \geqq 2$ oscillatory solutions may exist. We shall prove the nonoscillation in section 3 for $0 < m \leqq 1$ and in section 5 for $1 < m < 2$.

The main approach to dealing with the case $m = 1$ is to consider the "energy function"

$$E(t) = \frac{1}{2}y'^2(t) + e^{-t}F(y(t)), \tag{3}$$

where $y(t)$ is any solution of (1) and

$$F(y) = \int_0^y s\exp(|s|^m)ds = \int_0^{|y|} s\exp(s^m)ds. \tag{4}$$

Here, for simplicity, we have written $E(t)$ instead of $E(t,y(t))$.

Since

$$E'(t) = -e^{-t}F(y(t)) \leqq 0,$$

the function $E(t)$ is nonincreasing and tends to a nonnegative number h (h may depend on the solution $y(t)$). Then we shall show that $h < \frac{1}{2}$ for all solutions which implies that all solutions are nonoscillatory.

This approach will be sharpened in section 4 to obtain the interesting result that $h = 0$ for any solution of (1) and for any $m > 1$. In view of this result we are able to employ a method used by Coffman and Wong in [4] to prove the nonoscillation for $1 < m < 2$.

Since $h = 0$ for $m > 1$, the derivatives of all solutions tend to zero as $t\to\infty$. Therefore, the part of Atkinson's conjecture concerning the existence of oscillatory solutions is that if $m \geqq 2$ an oscillatory solution may exist, but its amplitude would grow unboundedly and its frequency would tend to zero as $t\to\infty$. Thus, we should call it an extremely slow oscillation.

2. PRELIMINARIES. Since $E(t)$ is a nonincreasing function of t, we know that the derivative of any solution $y(t)$ is bounded. Thus, every solution is bounded by a linear function and can be continued to infinity. In the case $m>1$, we can estimate the bound of solutions more exactly. To do this, we first investigate the function $F(y)$.

If $|y| > 1$, then

$$\begin{aligned} F(y) &= \int_0^y s\exp(|s|^m)ds \\ &= \int_0^1 s\exp(s^m)ds + \int_1^{|y|} s\exp(s^m)ds \\ &= \int_0^1 s\exp(s^m)ds + \frac{1}{m}[|y|^{2-m}\exp(|y|^m) - e] - \frac{2-m}{m}\int_1^{|y|} s^{1-m}\exp(s^m)ds. \end{aligned}$$

Because

$$\int_1^{|y|} s^{1-m}\exp(s^m)\,ds \sim \frac{1}{m}|y|^{2-2m}\exp(|y|^m) \quad \text{as} \quad |y|\to\infty,$$

we get the following important estimate

$$(5) \qquad F(y) = \frac{1}{m}|y|^{2-m}\exp(|y|^m)[1 + O(|y|^{-m})] \quad \text{as} \quad |y|\to\infty.$$

Suppose $y(t)$ is any nontrivial solution of (1); then

$$F(y(t)) \leqq E(t_0)\, e^t \quad \text{for} \quad t \geqq t_0.$$

From (5) we have for large $|y(t)|$ that

$$(6) \qquad |y(t)| \leqq (t + (m-2)\log|y(t)| + \log mE(t_0))^{1/m}.$$

Using (6) in the case $1 < m \leqq 2$, iterating (6) once in the case $m > 2$ and taking $\max(\log|y(t)|,1)/t = o(1)$ into account, we obtain

LEMMA 1. *Let* $y(t)$ *be any solution of the equation* (1). *Then for all sufficiently large* $t \geqq t_0$

$$(7) \qquad |y(t)| \leqq \begin{cases} \text{a linear function}, & \text{if } 0 < m \leqq 1, \\ (t + \log mE(t_0))^{1/m}, & \text{if } 1 < m \leqq 2, \\ (t + \frac{m-2}{m}\log t + \log mE(t_0) + o(1))^{1/m}, & \text{if } m > 2. \end{cases}$$

Next, comparing the equation (1) with the nonoscillatory linear equation

$$z'' + \frac{1}{4t^2} z = 0$$

and the oscillatory linear equation

$$z'' + (\frac{1}{4} + \varepsilon)z/t^2 = 0, \quad \varepsilon > 0,$$

respectively, we have

LEMMA 2. *A solution* $y(t)$ *of the equation* (1) *satisfying*

$$(8) \qquad |y(t)| \leqq (t - b\log t + a)^{1/m}$$

is nonoscillatory if $b > 2$, *or if* $b = 2$ *and* $a \leqq -\log 4$.

LEMMA 3. *If* $y(t)$ *is a nonoscillatory solution of the equation* (1), *then for any* $\varepsilon > 0$, *there is a sequence* t_n *with* $t_n \to \infty$ *as* $n \to \infty$, *such that*

$$|y(t_n)| < (t_n - 2 \log t_n + \log(\tfrac{1}{4} + \varepsilon))^{1/m}, \quad n = 1,2,\cdots.$$

The following lemma will be needed in section 4.

LEMMA 4. *Suppose* p *and* k *are two constants*, $0 < p < 1$, $k > 0$. *If two positive numbers* a *and* b *satisfy* $b < a$ *and*

$$a^p + b^p = k(a - b),$$

then

$$a^{1-p} - b^{1-p} = \frac{2(1-p)}{k}(1 + o(1)) \quad \text{as} \quad b \to \infty.$$

Set $x = a/b$; then $x > 1$ and

$$kb^{1-p} = \frac{x^p + 1}{x - 1}.$$

Since the function $(x^p+1)/(x-1)$ is decreasing for $x > 1$, we have $x \to 1+0$ as $b \to \infty$.

On the other hand, we have

$$a - b = (a^{1-p} - b^{1-p})(a^p + b^p) + a^p b^{1-p} - a^{1-p} b^p$$

and

$$(a^p b^{1-p} - a^{1-p} b^p)/(a - b) = [x^{1-p}(x^{2p-1} - 1)]/(x - 1)$$
$$= 2p - 1 + o(1) \quad \text{as} \quad b \to \infty.$$

Then Lemma 4 follows immediately.

3. THE NONOSCILLATION FOR $0 < m < 2$. The main result in this section is

THEOREM 1. *The equation* (1) *does not have any solutions which are oscillatory as* $t \to \infty$ *if* $0 < m < 2$.

Suppose $y(t)$ is any nontrivial solution of (1). If $0 < m < 1$, then by Lemma 1 we have

$$|y(t)|^m - t < -t/2$$

for all sufficiently large t. The solution $y(t)$ is therefore seen to be nonoscillatory by comparing (1) with the nonoscillatory equation

$$z'' + z \exp(-t/2) = 0.$$

We now assume $m = 1$, i.e. we consider the equation

$$y'' + y \exp(|y| - t) = 0 \tag{9}$$

and the function

$$(10)\qquad E(t) = \frac{1}{2}y'^2(t) + e^{-t}[(|y(t)| - 1)\exp y(t) + 1].$$

Let h be the limit of $E(t)$ at infinity. By Lemma 2, we know that $y(t)$ is nonoscillatory if $h < \frac{1}{2}$. For if $h < \frac{1}{2}$ we can choose a t_0 so large that $E(t_0) < \frac{1}{2}$; then (8) is eventually satisfied since

$$|y'(t)| \leqq (2E(t_0))^{1/2} < 1, \quad t \geqq t_0,$$

and $|y(t)|$ is bounded by a linear function with a slope less than 1. On this account it suffices to show that any solution $y(t)$ with $h \geqq \frac{1}{2}$ is nonoscillatory.

Assume $y(t)$ is an oscillatory solution with $h \geqq 1/2$. Let t_n and s_n be two sequences such that

$$(11)\qquad \begin{aligned} &t_n < s_n < t_{n+1}, \quad y(t_n) = y'(s_n) = 0, \\ &y(t) \neq 0 \quad \text{for} \quad t_n < t < t_{n+1}, \quad n = 1,2,\cdots. \end{aligned}$$

From (10) we have

$$(12)\qquad e^{-s_n}[(|y(s_n)| - 1)\exp|y(s_n)| + 1] > \frac{1}{2},$$

and hence

$$(13)\qquad |y(s_n)| \to \infty \quad \text{as} \quad n \to \infty.$$

Since $|y'(t)|$ is bounded, i.e.

$$(14)\qquad |y'(t)| \leqq (2E(t_1))^{1/2} = A \quad \text{for} \quad t \geqq t_1,$$

it follows that

$$(15)\qquad s_n - t_n \to \infty \quad \text{and} \quad t_{n+1} - s_n \to \infty \quad \text{as} \quad n \to \infty.$$

Therefore, for all large n, the intervals (s_n-1, s_n+1) are non-overlapping.

We claim that there is a $k > 0$ such that for all large n

$$(16)\qquad F(y(t))e^{-t} > kF(y(s_n))e^{-s_n}, \quad |t - s_n| \leqq 1.$$

In fact, from (14) we have

$$(17)\qquad |y(t) - y(s_n)| \leqq A \quad \text{for} \quad |t - s_n| \leqq 1.$$

From (13) and (17) we can find a constant $L > 0$ such that

$$(18)\qquad F(y(t)) > LF(y(s_n))$$

for $|t - s_n| \leqq 1$ and for all large n. Thus, (16) holds for $k=L/e$.

If we write

$$I_n = E(s_n - 1) - E(s_n + 1) = \int_{s_n+1}^{s_n-1} E'(t)dt = \int_{s_n-1}^{s_n+1} F(y(t))e^{-t}dt,$$

then (12) and (16) imply that $I_n \geqq k$ for all large n. We get then $E(t) \to -\infty$ as $t \to \infty$ which is impossible. This contradiction proves the nonoscillation for $m = 1$.

To deal with the case $1 < m < 2$ we have to know more about the asymptotic behaviour of solutions. We leave the proof to section 5.

4. AN ASYMPTOTIC PROPERTY OF SOLUTIONS FOR $m > 1$. Our main result in this section is

THEOREM 2. *If* $m > 1$, *then for any solution* $y(t)$ *of the equation* (1) *the function* $E(t)$ *defined by* (3) *tends to zero as* $t \to \infty$.

By the definition of $E(t)$, the following corollary is immediate.

COROLLARY. *If* $m > 1$, *then the derivative of every solution of the equation* (1) *vanishes at infinity*.

Theorem 2 is obviously true if $y(t)$ is a bounded solution.

Suppose $y(t)$ is an unbounded nonoscillatory solution of (1). Then we have $y'(t) \to 0$ as $t \to \infty$. For if $y'(t)$ tended to a non-zero number as $t\to\infty$, (7) would be violated. So we have

$$F(y(t)) = (h + o(1))e^t \quad \text{as} \quad t \to \infty.$$

Assume $h > 0$. Since $|y(t)| \to \infty$, in virtue of (5) we get

$$|y(t)|^{2-m}\exp(|y(t)|^m) = (mh + o(1))e^t$$

and

$$|y(t)| = (t + (m-2)\log|y(t)| + \log(mh) + o(1))^{1/m}.$$

By iterating and noticing that $t^{-1}\log|y(t)| = o(1)$, we finally obtain

$$|y(t)| = (t + \frac{m-2}{m}\log t + \log(mh) + o(1))^{1/m} \quad \text{as} \quad t \to \infty, \tag{19}$$

which contradicts Lemma 3 since $(m-2)/m > -1$. Therefore, Theorem 2 holds for any nonoscillatory solution.

It suffices to show that $h = 0$ for any oscillatory solution $y(t)$. Assume that $h > 0$ and that t_n and s_n, $n=1,2,\cdots$, are two sequences satisfying (11) with t_1 suitably large.

We certainly have $|y(s_n)| \to \infty$ and

$$F(y(s_n)) = (h + o(1))e^{s_n} \quad \text{as} \quad n \to \infty.$$

From this we can derive as above that

(20) $$|y(s_n)| = (s_n + \frac{m-2}{m}\log s_n + \log(mh) + o(1))^{1/m} \quad \text{as} \quad n \to \infty.$$

We observe from (20) that the solution $y(t)$ oscillates, roughly speaking, between the following curves

$$\pm(t + \frac{m-2}{m}\log t + \text{const})^{1/m}$$

and has almost the same slope at t_n for large n, i.e. $|y'(t_n)| \sim (2h)^{1/2}$. We can then expect that $s_n = O(n^{m/(m-1)})$. Indeed we can prove that

(21) $$s_n \sim Kn^M \quad \text{as} \quad n \to \infty,$$

writing here and in the sequel $M = m/(m-1)$ and $K = (2/M^2h)^{M/2}$, although we shall see later that the proof of (21) is not trivial.

Once we prove (21), we will be able to obtain a contradiction. In fact, let

(22) $$I_n = E(s_n - \frac{1}{n}) - E(s_n) = \int_{s_n - \frac{1}{n}}^{s_n} F(y(t))e^{-t}dt \geqq \frac{1}{n} F(y(s_n - \frac{1}{n}))e^{-s_n}.$$

Since $|y'(t)|$ is bounded, we have $|y(s_n - \frac{1}{n})| = |y(s_n)| - \eta_n$ with $\eta_n = O(\frac{1}{n})$. Moreover, (20) and (21) imply

(23) $$|y(s_n - \frac{1}{n})| \sim |y(s_n)| \sim s_n^{1/m} \sim K^{1/m} n^{1/(m-1)}$$

and

(24) $$\eta_n|y(s_n)|^{m-1} = O(1) \quad \text{as} \quad n \to \infty.$$

On the other hand, from (5) we get as $n \to \infty$

$$\begin{aligned} F(y(s_n - \tfrac{1}{n})) &= F(|y(s_n)| - \eta_n) \\ &\sim \tfrac{1}{m}(|y(s_n)| - \eta_n)^{2-m} \exp((|y(s_n)| - \eta_n)^m) \\ &= \tfrac{1}{m}|y(s_n)|^{2-m} \exp(|y(s_n)|^m)(1 - \eta_n/|y(s_n)|)^{2-m} \cdot \\ &\quad \cdot \exp[-m\eta_n|y(s_n)|^{m-1} + O(\eta_n{}^2|y(s_n)|^{m-2})]. \end{aligned}$$

By (24) there is a constant $k > 0$ such that

(25) $$F(y(s_n - \frac{1}{n})) \geqq kF(y(s_n)) \quad \text{for all large} \quad n.$$

Using (22), (25), (23) and (20) we obtain

$$I_n \geqq \frac{k}{n} F(y(s_n))e^{-s_n} \sim \frac{k}{mn}|y(s_n)|^{2-m} \exp(|y(s_n)|^m - s_n)$$

$$\sim \frac{kh}{n} s_n^{(2-m)/m} s_n^{(m-2)/m} = \frac{kh}{n},$$

which implies that $\Sigma I_n = \infty$ and $E(t) \to -\infty$ as $t \to \infty$. This contradiction completes the proof of Theorem 2.

It remains to show (21). To this end, we let

$$f(t) = (t - 2\log t)^{1/m}, \quad g(t) = (t - \frac{1}{m} \log t)^{1/m}.$$

By assuming t_1 is suitably large, using (20) and the fact that

$$(2-m)/m < 1/m < 2,$$

we have four uniquely determined sequences a_n, b_n, c_n and d_n such that

$$t_n < a_n < b_n < s_n < c_n < d_n < t_{n+1},$$

$$(26) \qquad |y(a_n)| = f(a_n), \quad |y(b_n)| = g(b_n),$$

$$|y(c_n)| = g(c_n), \quad |y(d_n)| = f(d_n), \quad n=1,2,\cdots.$$

We shall then prove the following statements one by one.

$$(27) \qquad |y'(a_n)| = (2h)^{1/2} + o(1), \quad |y'(d_n)| = (2h)^{1/2} + o(1) \quad \text{as } n \to \infty.$$

$$(28) \qquad s_n - b_n = O(1), \quad c_n - s_n = O(1) \quad \text{as } n \to \infty.$$

$$(29) \qquad b_n - a_n = o(a_n^{(1-m)/2m}), \quad d_n - c_n = o(d_n^{(1-m)/2m}) \quad \text{as } n \to \infty.$$

$$(30) \qquad a_n^{1/M} = d_{n-1}^{1/M} + 2(2h)^{-1/2}M^{-1}(1 + o(1)) \quad \text{as } n \to \infty.$$

(i) PROOF OF (27). Since $|y'(a_n)| < |y'(t_n)| = (2h)^{1/2} + o(1)$ and $|y(t)| \leqq f(t)$ for $t_n \leqq t \leqq a_n$ and

$$|y'(a_n)| = |y'(t_n) - \int_{t_n}^{a_n} y(s)\exp(|y(s)|^m - s)ds|$$

$$\geqq |y'(t_n)| - \int_{t_n}^{a_n} s^{(1/m)-2} ds > |y'(t_n)| - \frac{m}{m-1} t_n^{(1-m)/m},$$

we have

$$|y'(a_n)| = (2h)^{1/2} + o(1) \quad \text{as } n \to \infty.$$

The second relation in (27) can be obtained similarly.

(ii) PROOF OF (28). Notice that $y(a_n)$, $y(b_n)$, $y(c_n)$ and $y(d_n)$ have the same sign as $y(s_n)$. For the sake of simplicity, we assume $y(s_n) > 0$. The case of $y(s_n) < 0$ can be handled similarly. Because $y(t) \geqq (t - m^{-1}\log t)^{1/m}$ for

$b_n \leqq t \leqq s_n$, we have

$$y'(b_n) = \int_{b_n}^{s_n} y(s)\ \exp(y^m(s)-s)ds$$
$$\geqq \int_{b_n}^{s_n}(1-\ \log\ s/(ms))^{1/m}ds > P(s_n-b_n),$$

where P is a positive number independent of n. Then, s_n-b_n is bounded for all n since $|y'(t)|$ is bounded. The proof of the second relation in (28) is similar.

(iii) PROOF OF (29). Without any loss of generality, we still assume $y(s_n) > 0$. Set $x_n(t) = t-a_n$. Then

$$y'(t) = y'(a_n) - \int_{a_n}^{t} y(s)\exp(y^m(s)-s)ds$$
$$= y'(a_n) - y(u_n(t))x_n(t)\exp(y^m(u_n(t))-u_n(t)),$$

where $a_n < u_n(t) < t$ for $t \in (a_n,b_n]$. Since y(t) is increasing on $[a_n,b_n]$, there is a unique $p_n(t) \in (1/m,2)$ such that

$$y(t) = (t - p_n(t)\ \log\ t)^{1/m}.$$

Set $q_n(t) = p_n(u_n(t))$ and $U_n(t) = u_n^{-q_n}(u_n-q_n\ \log\ u_n)^{1/m}$. Then

$$y'(a_n) - y'(t) = U_n(t)x_n(t).$$

From this we get

$$(31)\qquad y(t) - y(a_n) = \int_{a_n}^{t} y'(s)ds \geqq y'(t)x_n(t)$$
$$= (y'(a_n)-U_n(t)x_n(t))x_n(t).$$

On the other hand, we have

$$(32)\qquad y(t)-y(a_n) = (t-p_n(t)\ \log\ t)^{1/m} - (a_n-2\ \log\ a_n)^{1/m}$$
$$= (t-p_n(t)\ \log\ t)^{1/m} - (a_n-p_n(t)\ \log\ a_n)^{1/m}$$
$$+ (a_n-p_n(t)\ \log\ a_n)^{1/m} - (a_n-2\ \log\ a_n)^{1/m}$$
$$= V_n(t)x_n(t) + W_n(t),$$

where

$$V_n(t) = \frac{1-(p_n(t)/v_n(t))}{m[v_n(t)-p_n(t)\ \log\ v_n(t)]^{1-(1/m)}}$$

with $a_n < v_n(t) < t$, and

$$W_n(t) = \log a_n \, [a_n - w_n(t) \log a_n]^{(1/m)-1}[2-p_n(t)]$$

with $p_n(t) < w_n(t) < 2$. Consequently, we can establish the following estimates which are independent of t,

$$V_n(t) = O(a_n^{(1/m)-1}),$$

$$W_n(t) = O(a_n^{(1/m)-1} \log a_n)$$

and

$$U_n(t) \leqq A = \text{const.}$$

Comparing (32) with (31) produces an inequality with respect to $x_n(t)$:

$$(33) \qquad U_n x_n^2 - (y'(a_n) - V_n)x_n + W_n \geqq 0.$$

The left-hand side of (33) has two positive roots for all large n. The smaller root is

$$(34) \qquad \begin{aligned} \alpha_n(t) &= \frac{1}{2U_n}\{y'(a_n) - V_n - [(y'(a_n) - V_n)^2 - 4U_n W_n]^{1/2}\} \\ &= W_n(y'(a_n) - V_n)^{-1} + o(W_n) = O(a_n^{(1/m)-1} \log a_n) \end{aligned}$$

and the bigger one $\beta_n(t)$ satisfies

$$\beta_n(t) = (y'(a_n) - V_n)U_n^{-1} - \alpha_n \geqq y'(a_n)A^{-1} + o(1).$$

Therefore, (33) implies that

$$(35) \qquad x_n(t) \leqq \alpha_n(t)$$

holds if $x_n(t)$ is very small. In view of the monotonity and convexity of $y(t)$ on $[a_n, b_n]$, all the functions $U_n(t)$, $V_n(t)$, $W_n(t)$ and $\alpha_n(t)$ are continuous. Therefore, as t continuously grows from a_n to b_n, $x_n(t)$ can never exceed $\alpha_n(t)$. Taking $t = b_n$ in (35) yields

$$b_n - a_n = O(a_n^{(1-m)/m} \log a_n) = o(a_n^{(1-m)/2m}) \quad \text{as } n \to \infty.$$

The second relation in (29) can be shown in the same way.

(iv) PROOF OF (30). By (27) we have as $n \to \infty$

$$\frac{(a_n - 2 \log a_n)^{1/m} + (d_{n-1} - 2 \log d_{n-1})^{1/m}}{a_n - d_{n-1}} = (2h)^{1/2}(1+o(1)).$$

Since

$$(a_n - 2\log a_n)^{1/m} + (d_{n-1} - 2\log d_{n-1})^{1/m}$$
$$= a_n^{1/m}(1 - 2a_n^{-1}\log a_n)^{1/m} + d_{n-1}^{1/m}(1 - 2d_{n-1}^{-1}\log d_{n-1})^{1/m}$$
$$= (a_n^{1/m} + d_{n-1}^{1/m})(1 + o(1)),$$

we get

$$(a_n^{1/m} + d_{n-1}^{1/m})/(a_n - d_{n-1}) = (2h)^{1/2}(1+o(1)) \quad \text{as} \quad n \to \infty.$$

Because $(2h)^{1/2}(1+o(1))$ is greater than a fixed positive number for all large n, Lemma 4 may clearly be applied to give (30).

We are now able to prove (21). By (28) and (29) we have $s_n \sim a_n$, $s_{n-1} \sim d_{n-1}$, then we have from (30)

$$s_n^{1/M} = s_{n-1}^{1/M} + \left(\frac{2}{h}\right)^{1/2} M^{-1}(1 + o(1)).$$

Hence,

$$s_n^{1/M} \sim \left(\frac{2}{h}\right)^{1/2} M^{-1} n \quad \text{as} \quad n \to \infty,$$

and (21) follows immediately.

5. THE PROOF OF THEOREM 1 FOR $1 < m < 2$. According to Theorem 2, we are able to prove Theorem 1 in the case $1 < m < 2$. Set $z = y/\psi$ and $u = z'$, where $\psi = t^{1/m}$ and $y(t)$ is any nontrivial solution of (1). Following Coffman and Wong [4] we take a function

$$\Sigma = \psi^k u^2 + \psi^{k-1}\psi'' z^2 + 2\psi^{k-2} e^{-t} F(\psi z), \tag{36}$$

where k is a constant. A straightforward differentiation gives

$$\Sigma' = -(4-k)\psi^{k-1}\psi' u^2 + (\psi^{k-1}\psi'')' z^2 + Q(t),$$

where

$$Q(t) = 2(k-2)\psi^{k-3}\psi' e^{-t}\int_0^{|\psi z|} s\exp(s^m)\,ds + 2\psi^{k-1}\psi' z^2 \exp(|\psi z|^m - t)$$
$$-2\psi^{k-2} e^{-t}\int_0^{|\psi z|} s\exp(s^m)\,ds$$
$$= 2k\psi^{k-3}\psi' e^{-t}\int_0^{|\psi z|} s\exp(s^m)\,ds + 2m\psi^{k-3}\psi' e^{-t}\int_0^{|\psi z|} s^{m+1}\exp(s^m)\,ds$$
$$- 2\psi^{k-2} e^{-t}\int_0^{|\psi z|} s\exp(s^m)\,ds.$$

Let $k = 4$ and $s = \psi(t)r$; then we get

$$\Sigma' = -2(m-1)(2-m)m^{-3}t^{(4/m)-3}z^2 + Q(t) \tag{37}$$

and

$$Q(t) = -2t^{4/m}e^{-t}\int_0^{|z(t)|} r(1-r^m - \frac{4}{mt})\exp(tr^m)\,dr. \tag{38}$$

According to Theorem 2, for any solution $y(t)$ there is a t_0 so large that

$$E(t_0) < \exp(-4/m).$$

By Lemma 1, we have for large t

$$|z(t)| = \frac{|y(t)|}{\psi(t)} \leqq (1 + t^{-1}\log E(t_0))^{1/m} < (1 - 4/(mt))^{1/m}$$

and hence the integrand on the right-hand side of (38) is positive. So we have $Q(t) \leqq 0$ for all large t, which implies $\Sigma(t)$ is non-increasing for large t. Then we can prove that for any oscillatory solution $y(t)$, there is a constant B such that

$$|y(t)| \leqq Bt^{1/2}. \tag{39}$$

Thus Lemma 2 implies that the solution $y(t)$ is nonoscillatory. This contradiction shows that all solutions of (1) are nonoscillatory for $1 < m < 2$.

The proof of (39) is omitted because it is exactly the same as that of Theorem 6 in [4].

REMARK. As in [4] we rewrite

$$Q(t) = (\partial/\partial\sigma)(2\psi^2(\sigma)e^{-\sigma}F(\psi(\sigma)z(t))\Big|_{\sigma=t}.$$

One of the conditions of Theorem 6 in [4], that $\psi^2(t)e^{-t}F(\psi(t)\alpha)$ must be a non-increasing function of t for *every* $\alpha > 0$, is violated for large α, say, $\alpha = 1$. So, we cannot simply use that result to get the nonoscillation for $1 < m < 2$. But, fortunately, for that theorem to hold we need only that $Q(t) \leqq 0$ for any *solution* $y(t)$ of (1), which turns out to be true because every solution of (1) has the property described in Theorem 2.

ACKNOWLEDGEMENT. The authors are grateful to the referee for correcting errors in the typed manuscript and for several helpful suggestions. One of the authors, S.Chen, wishes to express his gratitude for the hospitality of the Department of Mathematics at the University of Toronto where this work was accomplished.

BIBLIOGRAPHY

1. F.V.Atkinson and L.A.Peletier, Ground states of $-\Delta u = f(u)$ and the Emden-Fowler equation, Arch.Rational Mech.Anal. 93(1986),103-127.
2. F.V.Atkinson and L.A.Peletier, Ground states and Dirichlet problems for $-\Delta u = f(u)$ in R^2, ibid. 96(1986),147-165.
3. S.Chen, Asymptotic linearity of the solutions of nonlinear differential equations, to appear in Bull.Austral.Math.Soc.
4. C.V.Coffman and J.S.W.Wong, Oscillation and nonoscillation of solutions of generalized Emden-Fowler equations, Trans.Amer.Math.Soc. 167(1972),399-434.
5. W.F.Trench, Asymptotic behavior of solutions of an nth order differential equation, Rocky Mountain J.Math. 14(1984),441-450.

DEPARTMENT OF MATHEMATICS
UNIVERSITY OF TORONTO
TORONTO, ONTARIO M5S 1A1
CANADA

DEPARTMENT OF MATHEMATICS
SHANDONG UNIVERSITY
JINAN, SHANDONG
PEOPLE'S REPUBLIC OF CHINA

Canadian Mathematical Society
Conference Proceedings
Volume 8 (1987)

Singular Self-adjoint Fourth Order Difference Operators

C.E. Billigheimer

§1. Introduction

We discuss in this paper singular self-adjoint difference operators in a Hilbert space derivable from formally self-adjoint fourth order linear difference operators which are related to a formally self-adjoint five-term recurrence relation.

The formal difference operators P acting on sequences of complex numbers $y = \{y_n\}_{-2}^{\infty}$ have the form

$$(Py)_n = d_n y_{n+2} + c_n y_{y+1} + b_n y_n + c_{n-1} y_{n-1} + d_{n-2} y_{n-2} \quad (n \geq 0) \tag{1.1}$$

where the coefficients b_n, c_n, d_n are real and $d_n > 0$.

The associated recurrence relations are of the form

$$d_n y_{n+2} + c_n y_{n+1} b_n + c_{n-1} y_{n-1} + d_{n-2} y_{n-2} = \lambda\, a_n y_n \quad (n \geq 0) \tag{1.2}$$

where a_n, b_n, c_n, d_n are real and a_n, $d_n > 0$, and λ is a complex parameter.

We may take the coefficients a_n in (1.2) to be equal to 1 by setting $y_n^+ = a_n^{1/2} y_n, b_n^+ = a_n^{-1} b_n, c_n^+ = a_n^{-1/2} c_n a_{n+1}^{-1/2}, d_n^+ = a_n^{-1/2} d_n a_{n+2}^{-1/2}$ and omitting the superscripts. We then have instead of (1.2) the recurrence relations

$$d_n y_{n+2} + c_n y_{n+1} + b_n y_n + c_{n-1} y_{n-1} + d_{n-2} y_{n-2} = \lambda y_n \quad (n \geq 0) \tag{1.3}$$

The theory is analogous to that of self-adjoint linear differential operators of fourth order. Of special interest is the singular case in which the classical Weyl (27) limit-point, limit-circle distinction for second order differential operators extended to general even-order differential operators by Kodaira (19) plays a fundamental role.

The spectral theory, self-adjoint extensions of symmetric operators, resolutions of the identity and the corresponding eigen- function expansions and resolvent operators have been discussed for second order self-adjoint linear differential operators by Weyl (27), Stone (25), Titchmarsh (26), Krein (20), Levitan (23), Yosida (29), Levinson (21, 22), Coddington (10), Hartman(16), Hartman and Wintner (17), Wintner (28), and others, and for general even order self-adjoint linear differential operators by Birkhoff (9), Kodaira (19), Glazman (15), Naimark (24), and Everitt (13, 14). Comprehensive accounts of the theory are found in the books by Stone (25), Titchmarsh (26), Coddington and Levinson (11), Akhiezer and Glazman (10), Naimark (24), Atkinson (2), and especially in the book of Dunford and Schwartz (12), which also contains an extensive bibliography.

The relation between the continuous differential operator and discrete difference operator case has been discussed extensively in the book by Atkinson (2). Hellinger (18), Stone (25), Berezanskii (3) and others have discussed the second order difference operator case and the related three-term recurrence relations which are of importance in the theory of the classical orthogonal polynomials. We have previously discussed regular and singular boundary problems for five-term recurrence relations (Billigheimer (4) and (5)) and the theory of regular self-adjoint fourth order difference operators (Billigheimer (6)). We now consider the theory of singular self-adjoint fourth order difference operators. Methods and results are based especially on the parallel treatment of differential operators in the work of Akhiezer and Glazman (1) and Naimark (24). The work can be generalized to the case of a linear difference operator with real coefficients of even order.

We consider in §2 singular difference operators in a suitable Hilbert space derived from the formal difference operator P defined in (1.1). We obtain characterizations of certain symmetric difference operators in terms of their deficiency indices, and in §3 we discuss their self-adjoint extensions. In a sequel to this paper we consider resolvents of these self-adjoint extensions and corresponding classical Green's functions, the associated resolutions of the identity and corresponding classical eigenfunction expansions and related results.

§2. Singular Difference Operators

We consider the Hilbert space $H = \ell^2$ of sequence of vectors of complex numbers $y = \{y_n\}_{-2}^{\infty}$ of summable square such that $\sum_{n=-2}^{\infty} |y_n|^2 < \infty$ with the scalar product for two vectors $x, y \in \ell^2$ defined by

$$(x,y) = \sum_{n=-2}^{\infty} x_n \bar{y}_n \tag{2.1}$$

and the corresponding norm $\|x\|$ for a vector $x \in \ell^2$ given by

$$\|x\| = (x,x)^{1/2} \tag{2.2}$$

The difference operators considered in this and the subsequent sections are defined in terms of the formal difference operator P acting on sequences $y = \{y_n\}_{-2}^{\infty}$ in ℓ^2 given by $z = Py$ where

$$z_n = (Py)_n = d_n y_{n+2} + c_n y_{n+1} + b_n y_n + c_{n-1} y_{n-1} + d_{n-2} y_{n-2} \quad (n \geq 0) \tag{2.3}$$

with

$$z_{-2} = (Py)_{-2} = z_{-1} = (Py)_{-1} = 0 \tag{2.4}$$

where b_n, c_n, d_n are real numbers and $d_n > 0$.

We have for two sequences $x, y \in \ell^2$ satisfying also $Px, Py \in \ell^2$ the Lagrange identity given in the following lemma:

<u>Lemma 2.1:</u> For $x, y \in \ell^2$ satisfying $Px, Py \in \ell^2$ we have the identity

$$(Px, y) - (x, Py) = [xy]_{\infty} - [xy]_{-1} \tag{2.5}$$

where $[xy]_m$ is defined for $m \geq -1$ by

$$[xy]_m = d_m(x_{m+2}\bar{y}_m - x_m\bar{y}_{m+2}) + c_m(x_{m+1}\bar{y}_m - x_m\bar{y}_{m+1}) + d_{m-1}(x_{m+1}\bar{y}_{m-1} - x_{m-1}\bar{y}_{m+1}) \tag{2.6}$$

and $[xy]_\infty = \lim_{m\to\infty} [xy]_m$.

__Proof:__ We define $(xy)_{m(0)} = \sum_{n=-2}^{m} x_n\bar{y}_n$. Then from (2.3) we obtain

$$(Px,y)_{m(0)} - (x,Py)_{m(0)} = [xy]_m - [xy]_{-1}.$$

Since $x, y \in \ell^2$ and Px, $Py \in \ell^2$, we can let $m \to \infty$ in the above equation and obtain that $[xy]_\infty = \lim_{m\to\infty} [xy]_m$ exists and we have result (2.5).

We have the following lemma:

__Lemma 2.2:__ For y_{-2}, y_{-1}, y_0, y_1 prescribed arbitrarily, there exists a unique solution vector $y = \{y_n\}_{-2}^{\infty}$ of the inhomogeneous recurrence relations

$$(Py)_n - \lambda y_n = f_n \qquad (0 \le n < \infty)$$

where $f = \{f_n\}_{-2}^{\infty}$ is a given vector with complex components f_n, and $f_{-2} = f_{-1} = 0$ say, and λ is a complex parameter.

__Proof__: The result is obvious from the form of difference operator P given in (2.3).

We also require the following lemma proved in §2 (Billigheimer (6)):

__Lemma 2.3:__ The inhomogeneous recurrence relations

$$(Py)_n - \lambda y_n = f_n \qquad (0 \le n \le m)$$

have a solution satisfying $y_{-2} = y_{-1} = y_0 = y_1 = 0$, $y_{m-1} = y_m = y_{m+1} = y_{m+2} = 0$ if and only if $f = \{f_n\}_{-2}^{m+2}$ with complex components f_n and $f_n = 0$ $(n = -2, -1, m+1, m+2)$ is orthogonal to all solutions $z = \{z_n\}_{-2}^{m+2}$ of the homogeneous recurrence relations

$$(Pz)_n - \lambda z_n = 0 \qquad (0 \le n \le m)$$

in the sense that $(f,z)_m = \sum_{n=-2}^{m+2} f_n\bar{z}_n = 0$.

Let $H_{(2)}$ be the projection subspace of H of vectors $y = \{y_n\}_{-2}^{\infty}$ such that

$$y_{-2} = y_{-1} = y_0 = y_1 = 0. \qquad (2.7)$$

We denote the operator P differently according to the domain on which it acts.

By A' we denote the operator with domain $D_{A'}$ consisting of sequences $y = \{y_n\}_{-2}^{\infty}$ in ℓ^2 satisfying (2.7) and containing at most a finite number of non-zero elements. We see immediately from (2.5), (2.6) that A' is symmetric with $D_{A'}$ dense in $H_{(2)}$.

By B we denote the operator with domain D_B consisting of sequences $y \in \ell^2$ such that $Py \in \ell^2$. D_B is the maximal linear manifold in ℓ^2 on which P can be considered as an operator in ℓ^2. Since D_B includes all sequences with at most a finite number of non-zero components which are dense in H, we have that D_B is dense in H. We define the adjoint B^* of B as the operator in H with domain D_{B*} such that for $z \in D_{B*}$

$$z_{-2} = z_{-1} = 0, \ (B^*z)_{-2} = (B^*z)_{-1} = 0 \qquad (2.8)$$

and for all $y \in D_B$

$$(By,z) = (y,B^*z).$$

This uniquely defines B^*.

Let A be the closure in $H_{(2)}$ of A'. Then $y \in D_A$ satisfies (2.7) and A is a closed symmetric operator with domain dense in $H_{(2)}$. We define the adjoint A^* of A as the operator with domain D_{A*} such that for $z \in D_{A*}$

$$(A^*z)_{-2} = (A^*z)_{-1} = 0, \ (A^*z)_0 = (Pz)_0, \ (A^*z)_1 = (Pz)_1 \qquad (2.9)$$

and for all $y \in D_A$

$$(Ay, z) = (y, A^* z).$$

We see that this uniquely defines A^*. We define A'^* with conditions corresponding to (2.9), and A^{**} with conditions corresponding to (2.8). Then we have the following theorem:

Theorem 2.1: $A^* = B.$ (2.10)

Proof: Since A is the closure of A' we have

$$A^* = A'^*.$$

We see that, since $y \in D_{A'}$ satisfies (2.7) and contains at most a finite number of non-zero components, for $y \in D_{A'}$, $z \in D_B$ we have from (2.5), (2.6)

$$(A'y,z) = (y,Bz).$$

From (2.4), (2.9) we now deduce $z \in D_{A'*}$ and $A'*z = Bz$. Hence $B \subset A'^*$ and thus

$$B \subset A^* \tag{2.11}$$

We now show that $A^* \subset B$. For $z \in D_{A^*}$ set

$$(A^*z)_n = w_n \qquad (n \geq -2) \tag{2.12}$$

From (2.9) we have $w_{-2} = w_{-1} = 0$, $w_0 = (Pz)_0$, $w_1 = (Pz)_1$.

Denote by $y = |y_n|_{-2}^{\infty}$ any solution of

$$(Py)_n = w_n \qquad (n \geq -2) \tag{2.13}$$

Let $g = \{g_n\}_{-2}^{\infty}$ be a sequence in ℓ^2 with $g_{-2} = g_{-1} = 0$ and with $g_n = 0$ for $n > m$, which has $\{g_n\}_{-2}^{m+2}$ orthogonal in H_m to all solutions $y = \{y_n\}_{-2}^{m+2}$ of

$$(Py)_n = 0 \qquad (0 \leq n \leq m).$$

Then by Lemma 2.3 there exists a sequence $x = \{x_n\}_{-2}^{m+2}$ which satisfies $x_{-2} = x_{-1} = x_0 = x_1 = 0$ and $x_{m-1} = x_m = x_{m+1} = x_{m+2} = 0$ and

$$(Px)_n = g_n \qquad (0 \leq n \leq m) \tag{2.14}$$

Take also $x_n = 0$ for $n > m + 2$. Then $x = \{x_n\}_{-2}^{\infty}$ satisfies $x \in D_{A'} \subset D_A$. Now since $z \in D_{A^*}$ we have by (2.14), (2.13)

$$(g,z) = (Ax,z) = (x,A^*z) = (x,w) \tag{2.15}$$

Also using the conditions satisfied by x and the Lagrange identity (2.5) we have

$$(g,y) = (Px,y) = (x,Py) = (x,w) \qquad (2.16)$$

From (2.15) (2.16) we deduce that for all g satisfying the conditions specified above we have

$$(g,\ z-y) = 0.$$

Since also m is arbitary, we deduce from the definition of g that

$$\{(z-y)_n\}_0^{\infty} = \{u_n\}_0^{\infty}$$

where $u = \{u_n\}_{-2}^{\infty}$ is a solution of

$$(Pu)_n = 0 \qquad (n \geq 0).$$

Hence using (2.13), (2.12) we have that

$$(Pz)_n = (Py)_n = w_n = (A^*z)_n \qquad (n \geq 2).$$

Also by definitions (2.9), (2.4) we have

$$(A^*z)_{-2} = (Pz)_{-2} = 0, (A^*z)_{-1} = (Pz)_{-1} = 0, (A^*z)_0 = (Pz)_0, (A^*z)_1 = (Py)_1.$$

Hence we have

$$(Pz)_n = (A^*z)_n \qquad (n \geq -2),$$

and

$$A^* \subset B \qquad (2.17)$$

From (2.11), (2.17) we deduce

$$A^* = B.$$

From Theorem 2.1 we deduce that B is closed, and we have the corollary:

<u>Corollary 2.1:</u> $$B^* = A \qquad (2.18)$$

<u>Proof.</u> From (2.10) we have $A^* = B$.

For $z \in D_{A'}$ and all $y \in D_B$ we have

$$(By,z) = (y,A'z)+[yz]_{\infty}-[yz]_{-1} = (y,A'z) = (y,B^*z).$$

Hence $A' \subset B^*$. But B^* is closed. Hence $A = \overline{A'}$ is contained in B^*

and we have $A \subset B^*$. On the other hand, for all $y \in D_A$, for $z \in D_{B*}$ we have

$$(Ay,z) = (By,z) = (y,B^*z) = (y,A^*z).$$

Hence using (2.7) for D_A we see from (2.10) that

$$(B^*z)_n = (A^*z)_n = (Bz)_n \qquad ((n \geq 2).$$

Now for $x \in D_{B*}$ and for all $y \in D_B$ we have

$$(B^*z,y) - (z,By) = 0.$$

We may write this as

$$\sum_2^{\infty} (Bz)_n \bar{y}_n - \sum_2^{\infty} z(\overline{By})_n + \sum_{-2}^{1} (B^*z)_n \bar{y}_n - \sum_{-2}^{\infty} z_n(\overline{By})_n = 0.$$

Using assumptions (2.8) for B^* we obtain from the Lagrange identity (2.5)

$$[zy]_\infty - [zy]_1 + (B^*z)_1\bar{y}_1 + (B^*z)_0\bar{y}_0 - z_1(\overline{By})_1 - z_0(\overline{By})_0 = 0$$

or

$$[zy]_\infty - d_1(z_3\bar{y}_1 - z_1\bar{y}_3) - c_1(z_2\bar{y}_1 - z_1\bar{y}_2) - d_0(z_2\bar{y}_0 - z_0\bar{y}_2) + (B^*z)_1\bar{y}_1$$

$$+ (B^*z)_0\bar{y}_0 - z_1(d_1\bar{y}_3 + c_1\bar{y}_2 + b_1\bar{y}_1 + c_0\bar{y}_0 + d_{-1}\bar{y}_{-1})$$

$$- z_0(d_0\bar{y}_2 + c_0\bar{y}_1 + b_0\bar{y}_0 + c_{-1}\bar{y}_{-1} + d_{-2}\bar{y}_{-2}) = 0.$$

We take for y a sequence with a finite number of non-zero terms.

Taking also $y_{-1} = y_0 = y_1 = 0 \neq y_{-2}$ we obtain $z_0 = 0$.

Taking now $y_0 = y_1 = 0 \neq y_{-1}$ we obtain $z_1 = 0$.

Taking now $y_1 = 0 \neq y_0$ we obtain $(B^*z)_0 = d_0 z_2$.

Taking now $y_0 = 0 \neq y_1$ we obtain $(B^*z)_1 = d_1 z_3 + c_1 z_2$.

With (2.8) we now may write

$$(B^*z)_n = (Bz)_n \qquad (n \geq -2)$$

where for $z \in D_{B*}$ we have $z_n = 0 \quad (n = -2,-1,0,1)$.

Hence we have that $B^* \subset A$ and we deduce from the previous that $B^* = A$.

If we define A^{**} imposing parallel conditions to (2.8) on A^{**} as on B^*, then we have for the closed operator A the relations

$$A^{**} = B^* = A.$$

Incidentally we have shown above that for $z \in D_{B*}$ and for all $y \in D_B$ we have

$$[yz]_\infty = 0.$$

We deduce the following characterization of the domain of operator A:

<u>Lemma 2.4:</u> The operator A has domain D_A consisting of sequences $y \in D_B$ determined by the requirement that $y \in D_A$ if and only if for all $z \in D_B$

$$[yz]_\infty - [yz]_{-1} = 0. \tag{2.19}$$

<u>Proof.</u> Since by (2.18) of Corollary 2.1 we have $B^* = A$, we have $y \in D_B$ satisfies $y \in D_A$ if and only if

$$(y,Bz) - (By,z) = 0 \tag{2.20}$$

for all $z \in D_B$. By the Lagrange identity (2.5) we have that (2.20) is equivalent to (2.19).

We have an alternative characterization of the domain of the operator A:

<u>Lemma 2.5</u> The operator A has domain D_A consisting of sequences $y \in D_B$ such that

$$y_{-2} = y_{-1} = y_0 = y_1 = 0 \tag{2.21}$$

and for all $z \in D_B$

$$[yz]_{-1} = 0, \ [yz]_\infty = 0. \tag{2.22}$$

<u>Proof.</u> Since $y \in D_A$ as stated previously by its definition satisfies

(2.21), we derive from (2.6) that $[yz]_{-1} = 0$. Hence condition (2.19) of Lemma 2.4 becomes (2.22).

We introduce the projection subspace $H_{(0)}$ of H consisting of vectors $y = \{y_n\}_{-2}^{\infty}$ such that

$$y_{-2} = y_{-1} = 0, \qquad (2.23)$$

The deficiency indices of the closed symmetric operator A are defined as the dimensions of the deficiency subspaces in $H_{(0)}$ of A, denoted by $N_{\lambda(0)}$, which are the orthogonal complements in $H_{(0)}$ of the ranges R_λ of the operators $A - \lambda E$ where E is the identity operator for $\mathrm{Im}\lambda < 0$ and $\mathrm{Im}\lambda > 0$ respectively. $R_\lambda \subset H_{(0)}$ by (2.4). We may write

$$H_{(0)} = R_\lambda \oplus H_{\lambda(0)}. \qquad (2.24)$$

Thus if $y = \{y_n\}_{-2}^{\infty}$, then the projection vector $y_{(0)}$ belongs to $H_{\lambda(0)}$ if and only if for all $z \in D_A$ we have

$$(y, (A-\lambda E)z) = 0. \qquad (2.25)$$

Since from (2.21) $z_{-2} = z_{-1} = z_0 = z_1 = 0$, and from (2.4) $(Az)_{-2} = (Az)_{-1} = 0$, condition (2.25) determines the dimension of the space $N_{\lambda(0)}$ of vectors $y_{(0)}$ independently of y_{-2}, y_{-1} which are left undefined.

Now in view of (2.10) $A^* = B$, and hence condition (2.25) may be interpreted as follows:

$y_0 \in N_{\overline{\lambda}(0)}$ if and only if for all $z \in D_A$

$$((A^*-\overline{\lambda}E)y, z) = 0 \qquad (2.26)$$

Using condition (2.21) for $z \in D_A$ and (2.10) $A^* = B$, we see that the deficiency indices (m,n) of A are the number of linearly independent solutions of summable square $y_{(o)} = \{y_n\}_{-2}^{\infty}$ where $y_{(o)} \in \ell^2$, $y_{-2} = y_{-1} =$

0 of the equations

$$(By)_n = \bar{\lambda} y_n \qquad (n \geq 2) \qquad (2.27)$$

for $\mathrm{Im}\lambda < 0$ and $\mathrm{Im}\lambda > 0$ respectively.

We may now choose y_{-2}, y_{-1} arbitrarily, say from the equations

$$(By)_n = \bar{\lambda} y_n \qquad (n = 0,1) \qquad (2.28)$$

and denote the space of the resulting vectors $y = \{y_n\}_{-2}^{\infty}$ by N_λ. This is consistent with the notation $N_{\lambda(0)}$ used above. Obviously we have then

$$\dim N_\lambda = \dim N_{\lambda(0)}.$$

We term the $N_\lambda(\mathrm{Im}\lambda < 0, \mathrm{Im}\lambda > 0)$ deficiency subspaces of the operator A. We can write equations (2.27, (2.28 in view of (2.9), (2.10) as follows: $y \in N_\lambda$ if and only if

$$A^* y = \bar{\lambda} y_{(0)} \qquad (2.29)$$

The deficiency indices (m,n) are determined by the dimensions of N_λ at a single non-real value of λ with $\mathrm{Imd}\lambda < 0$ and $\mathrm{Imd}\lambda > 0$ respectively. (cf. Akhiezer and Glazman §78(1)).

We now have the following theorem:

Theorem 2.2: The deficiency indices of the operator A are of the form (m,m) where

$$2 \leq m \leq 4. \qquad (2.30)$$

Proof: From (2.10) the form of B given by (2.3) and the result of Lemma 2.2 we see that $0 \leq m, n \leq 4$.

Also since the coefficients b_n, c_n, d_n in (2.3) are real, to each summable square solution y of $(By)_n = \bar{\lambda} y_n (h \geq 0)$ corresponds a summable square solution $\bar{y}$ of $(B\bar{y})_n = \lambda\, \bar{y}_n (n \geq 0)$. Hence m = n.

To obtain the result that $m \geq 2$ we make use of the Neumann decomposition formula for the domain of the adjoint operator A^* of the symmetric operator A

$$D_{A*} = D_A \oplus N_{\bar{\lambda}} \oplus N_\lambda \qquad (2.31)$$

where $\operatorname{Im}\lambda > 0$. The proof of the above relation follows the lines of the proof given by Akhiezer and Glazman §80 (1). We see immediately, since by (2.10) $A^* = B$, that

$$D_{A^*} \supset D_A \oplus N_{\bar{\lambda}} \oplus N_\lambda .$$

Conversely, using the orthogonal decomposition formula (2.24) for the vector $A^*f - \lambda f$ where $f \in D_{A*}$ we obtain

$$(A^*f)_{(0)} - \lambda f_{(0)} = (A - \lambda E)h + (\bar{\lambda} - \lambda)g_{\bar{\lambda}(0)} \qquad (2.32)$$

where $h \in D_A$, $g_{\bar{\lambda}} \in N\lambda$. Now by (2.29) we have

$$A^* g_{\bar{\lambda}} = \bar{\lambda} g_{\bar{\lambda}(0)} .$$

Hence since $A \subset A^*$ in view of (2.10) and by (2.9) $(A^*f)_{(0)} = A * f$ we may write (2.32) as

$$A^*(f - g_{\bar{\lambda}} - h) = \lambda(f - g_{\bar{\lambda}} - h)_{(0)} .$$

Hence by (2.29) we have that

$$f - g_{\bar{\lambda}} - h = g_\lambda$$

where $g_\lambda \in N_{\bar{\lambda}}$, and we have for any vector $f \in D_{A*}$ the decomposition

$$f = h + g_\lambda + g_{\bar{\lambda}}$$

and

$$D_{A*} \subset D_A \oplus N_{\bar{\lambda}} \oplus N_\lambda .$$

We thus obtain (2.31).

To obtain the uniqueness of the decomposition we assume for $h \in D_A$, $g_\lambda \in N_{\bar{\lambda}}$, $g_{\bar{\lambda}} \in N_\lambda$ that

$$h + g_\lambda + g_{\bar{\lambda}} = 0. \qquad (2.33)$$

Then operating on (2.33) by A^* we obtain by (2.29)

$$Ah + \lambda g_{\lambda(0)} + \bar{\lambda} g_{\bar{\lambda}(0)} = 0. \qquad (2.34)$$

Subtracting (2.33) multplied by λ we obtain

$$(A - \lambda)h + \lambda g_{\lambda(0)} - \lambda g_{\lambda} + \overline{\lambda} g_{\lambda(0)} - \lambda g_{\overline{\lambda}} = 0.$$

We deduce

$$(A - \lambda)h + (\overline{\lambda} - \lambda)g_{\overline{\lambda}(0)} = 0 \qquad (2.35)$$

and

$$-\lambda(g_{\lambda n} + g_{\overline{\lambda}n}) = 0 \qquad (n = -2, -1). \qquad (2.36)$$

From (2.35) and the orthogonality relation (2.24) we deduce $g_{\overline{\lambda}(0)} = 0$. Then according to (2.28) we deduce $g_{\overline{\lambda}n} = 0 (n = -2, -1)$. Thus $g_{\overline{\lambda}} = 0$. Proceeding analogously we obtain also $g_{\lambda} = 0$. From (2.33) we now obtain h = 0 and the uniqueness of the decomposition (2.31) is proved.

To show now that the deficiency number $m \geq 2$, since we have shown above that $\dim N_{\overline{\lambda}} = \dim N_{\lambda} = m$, it is sufficient in view of (2.31) to prove that

$$\dim D_{A*} \geq 4 \ (\mathrm{mod}\ D_A) \qquad (2.37)$$

We show the existence of a set of four linearly independent vectors, which with their linear combinations lie in D_{A*} but with the exception of the zero vector do not lie in D_A. We choose the linearly independent vectors ϕ_i $(1 \leq i \leq 4)$ defined by

$$\phi_{1n} = p_n(\lambda), \phi_{2n} = q_n(\lambda), \ \phi_{3n} = r_n(\lambda), \phi_{4n} = s_n(\lambda) \quad \text{for} \quad -2 \leq n \leq k$$
$$\text{and} \quad \phi_{in} = 0 \quad \text{for} \quad n > k, \qquad (2.38)$$

for some integer $k \geq 1$, where the vectors $p_n(\lambda)$, $q_n(\lambda)$, $r_n(\lambda)$, $s_n(\lambda)$ $(n \geq -2)$ are solutions of the recurrence relations

$$(Py)_n = \lambda y_n \qquad (n \geq 0) \qquad (2.39)$$

with

$$(Py)_{-2} = (Py)_{-1} = 0 \qquad (2.40)$$

for some complex number λ, satisfying the initial conditions

$$p_{-2}=0, p_{-1}=0, p_0=1, p_1=0, q_{-2}=0, q_{-1}=0, q_0=0, q_1=1,$$
$$r_{-2}=1, r_{-1}=0, r_0=0, r_1=0, s_{-2}=0, s_{-1}=1, s_0=0, s_1=0 \tag{2.41}$$

The $\phi_i (1 \leq i \leq 4)$ are four linearly independent vectors modulo D_A since the assumption $y_n = \sum_{i=1}^{4} \alpha_i \phi_{in} (n \geq -2)$, $y \in D_A$, since y satisfies (2.7) leads to $\alpha_i = 0$ $(1 \leq i \leq 4)$. Hence we have shown (2.37) and $m \geq 2$. Thus we have the result of the lemma.

From Theorems 2.1 and 2.2 we obtain the following corollary:

Corollary 2.2: The number m of linearly independent solutions of summable square $y \in \ell^2$ of the recurrence relations (2.39), (2.40) for $\mathrm{Im}\lambda \neq 0$ is independent of $\lambda(\mathrm{Im}\lambda \neq 0)$ and satisfies

$$2 \leq m \leq 4. \tag{2.42}$$

We also have the following theorem:

Theorem 2.3: If the deficiency indices of the symmetric operator A are (m,m), $2 \leq m \leq 4$, then for real λ the number $m(\lambda)$ of linearly independent solutions of summable square $y \in \ell^2$ of the recurrence relations (2.39), (2.40), in general depending on λ, satisfies

$$m(\lambda) \leq m. \tag{2.43}$$

Proof: Let λ be a real number. Now $y \in D_A$ satisfies (2.21) and thus obviously, from the form of the operator A given by (2.3), we see that A cannot have any eigenvalues. Hence in particular the real number λ cannot be an eigenvalue of A. We deduce that if N_λ is the eigenmanifold of dimension $m(\lambda)$ of A^* corresponding to the eigenvalue λ, then D_A and N_λ are linearly independent since otherwise λ would be an eigenvalue of A. Set $D = D_A \oplus N_\lambda$.

Let $\bar{A}$ be an operator coinciding with A^* on $D = D_{\bar{A}}$. Then it can be readily verified that $\bar{A}$ is a symmetric extension of A which has deficiency indices (m,m) and hence

$$\dim D_{\bar{A}} \leq m \pmod{D_A}.$$

Hence we deduce $m(\lambda) \leq m$.

We now obtain the following corollary:

Corollary 2.3: If for one value of λ, real or complex, the homogeneous recurrence relations (2.39), (2.40) have four linearly independent solutions of summable square $y \in \ell^2$, then the deficiency indices of the operator A are (4,4).

Proof: For complex λ ($\operatorname{Im} \lambda \neq 0$) the result is immediate from the definition of deficiency indices and Corollary 2.2. For real λ the result follows from (2.43).

§3. Self-adjoint Extensions

Since by (2.10) $A^* = B$ and B is the singular difference operator with maximal domain generated by P, we call A the closed symmetric singular difference operator with minimal domain generated by P, since any closed symmetric operator generated by P is an extension of the operator A. Let $\bar{A}$ be a symmetric extension of A, $A \subset \bar{A}$, such that, in accordance with (2.10),

$$A \subset \bar{A} \subset A^* = B \tag{3.1}$$

Hence for $z \in D_A$ we have

$$(\bar{A}z)_n = (Pz)_n \quad (0 \leq n \leq m), \; (Az)_{-2} = (Az)_{-1} = 0 \tag{3.2}$$

We define the adjoint $\bar{A}^*$ of $\bar{A}$ uniquely by specifying for $y \in D_{\bar{A}^*}$ that

$$((\bar{A}^*y)_n = 0 \; (n = -2,-1), \; (\bar{A}^*y)_n = (Py)_n \; (n = 0,1) \tag{3.3}$$

and then we have also

$$(\bar{A}^*y)_n = (Py)_n \quad (n \geq 2). \tag{3.4}$$

To determine $\bar{A}$ we need merely to specify the domain of $\bar{A}$. Using the

Lagrange identity (2.5) we readily obtain the following theorem:

<u>Theorem 3.1:</u> A self-adjoint extension $\bar{A}$ of A satisfies

$$A \subset \bar{A} \subset B \tag{3.5}$$

and is defined by the two conditions

(i) For y, $z \in D_{\bar{A}}$ we have

$$[yz]_{\infty} - [yz]_{-1} = 0 \tag{3.6}$$

(ii) If $z \in D_B$ satisfies

$$[yz]_{\infty} - [yz]_{-1} = 0 \tag{3.7}$$

for all $y \in D_{\bar{A}}$, then $z \in D_{\bar{A}}$.

We can use the theory of self-adjoint extensions of symmetric operators (cf. Akhiezer and Glazman 79-80 (1)) to obtain the following characterization:

<u>Theorem 3.2:</u> Let the deficiency indices of A be (m,m) where $2 \le m \le 4$ and let the vectors $u_k = \{u_{kn}(\lambda)\}_{n=-2}^{\infty}$ $(k = 1,2,\ldots,m)$ form an orthonormal system of solutions of summable square $u_k \in \ell^2$ of the equations

$$(Pu)_n = \lambda u_n \qquad (n \ge 0) \tag{3.8}$$

for some fixed non-real λ, where the u_k are polynomials in λ with real coefficients specified by real values $u_{kn}(n = -2,-1,0,1)$ not all zero for each fixed k $(k = 1,\ldots,m)$ say. Then there exists a one-one correspondence between the class of all self-adjoint extensions $\bar{A}_\theta$ of the operator A and the class of all unitary matrices $\theta = \{\theta_{ik}\}$ $(1 \le i, k \le m)$ of order m defined by

$$A \subset \bar{A}_\theta \subset B \tag{3.9}$$

where $y \in D_{\bar{A}_\theta}$ is given by

$$y = v + x \tag{3.10}$$

where $v \in D_A$ and x is a linear combination of the vectors where $v \in D_A$, then we obtain

$$\{(A^* - \bar{\lambda}) \sum_{i=1}^{m} \beta_i w_i\}_n = \{(\lambda - \bar{\lambda}) \sum_{i=1}^{m} \beta_i u_i\}_n \qquad (n \geq 0)$$

$$= \{A - \bar{\lambda})v\}_n \qquad (n \geq 0)$$

But $\{\sum_{i=1}^{m} \beta_i u_i\}_{n=-2}^{\infty}$ is orthogonal to $\{A-\bar{\lambda})v\}_{n=-2}^{\infty}$ by (2.24).

Hence $\{\sum_{i=1}^{m} \beta_i u_i\}_n = 0 \qquad (n \geq 0)$

and the components $n = -2, -1$ determined by relation (3.8) are also zero. since the $u_i (1 \leq i \leq m)$ were chosen orthonormal we deduce $\beta_i = 0$ $(1 \leq i \leq m)$ and the $w_i (1 \leq i \leq m)$ are linearly independent modulo D_A as asserted.

From the second Neumann formula (cf. Akhiezer and Glazman (1) 80) for self-adjoint extensions of symmetric operator A we now have that $y \in D_{\bar{A}_\theta}$ is given by

$$y = v + g_{\lambda} + Ug_{\lambda} \tag{3.12}$$

where $v \in D_A$, $g_\lambda \in N_{\bar{\lambda}}$ and U is a unitary operator from $N_{\bar{\lambda}}$ into N_λ. This becomes (3.10), (3.11) in our case.

We can characterize self-adjoint extensions in an alternative manner in terms of generalized boundary conditions as follows:

<u>Theorem 3.3:</u> The self-adjoint extension $\bar{A}_\theta$ of the operator A with deficiency indices (m,m) defined in Theorem 3.2 is determined by

$$A \subset \bar{A}_\theta \subset B \tag{3.13}$$

and the boundary conditions that $y \in D_B$ satisfies $y \in D_{\bar{A}_\theta}$ if and only if

$$[yw_i]_\infty - [yw_i]_{-1} = 0 (i = 1,2,\dots,m) \tag{3.14}$$

where the vectors $w_i = w_i^{(\theta)}$ determined by the operator $\bar{A}_\theta$ are given by equations (3.11), are linearly independent modulo D_A, and satisfy deficiency indices (m,m) defined in Theorem 3.2 is determined by

$$A \subset \bar{A}_\theta \subset B \qquad (3.13)$$

and the boundary conditions that $y \in D_B$ satisfies $y \in D_{\bar{A}_\theta}$ if and only if

$$[yw_i]_\infty - [yw_i]_{-1} = 0 (i = 1,2,\ldots,m) \qquad (3.14)$$

where the vectors $w_i = w_i^{(\theta)}$ determined by the operator $\bar{A}_\theta$ are given by equations (3.11), are linearly independent modulo D_A, and satisfy

$$[w_i w_j]_\infty - [w_i w_j]_{-1} = 0 \ (i, j = 1,2,\ldots,m) \qquad (3.15)$$

Proof: Since by (3.10) we have $w_i \in D_{\bar{A}_\theta}$ using the Lagrange identity (2.5) for $y \in D_{\bar{A}_\theta}$ and w_i we have

$$(\bar{A}_\theta y, w_i) - (y, \bar{A}_\theta w_i) = [yw_i]_\infty - [yw_i]_{-1}$$

and since $\bar{A}_\theta$ is self-adjoint we obtain condition (3.14). We obtain (3.15) directly from (3.11) or since by (3.10) $w_i \in D_{\bar{A}_\theta}$, (3.15) follows from (3.14). The linear independence modulo D_A of the vectors w_i follows as in the proof of Theorem 3.2.

Conversely, if $y \in D_B$ satisfies (3.14) where w_i is given by (3.11), then using conditions (2.22) on $v \in D_A$ we obtain from (3.10)

$$[yz]_\infty - [yz]_{-1} = 0 \qquad \text{for all} \quad z \in D_{\bar{A}_\theta}$$

and from Theorem 3.1 we deduce that $y \in D_{\bar{A}_\theta}$.

Alternatively we can characterize the self-adjoint extensions of the operator A as follows:

Theorem 3.4: A self-adjoint extension $\bar{A}$ of the operator A with deficiency indices (m,m) is defined by

$$A \subset \bar{A} \subset B \qquad (3.16)$$

and the boundary conditions that $y \in D_B$ satisfies $y \in D_{\bar{A}}$ if and only if

$$[yw_i]_\infty - [yw_i]_{-1} = 0 \quad (i - 1,2,\ldots,m) \qquad (3.17)$$

where the vectors w_i are certain vectors determined by the operator $\bar{A}$ which satisfy $w_i \in D_B$, are linearly independent modulo D_A, and satisfy

$$[w_i w_j]_\infty - [w_i w_j]_{-1} = 0 \quad (i, j = 1,2,\ldots,m) \qquad (3.18)$$

Conversely conditions (3.17), (3.18) determine a self-adjoint extension $\bar{A}$ of A.

<u>Proof:</u> Since the deficiency indices of A are (m,m) we have from (3.12) that $\dim D_{\bar{A}} = m \pmod{D_A}$. Hence there exist in $D_{\bar{A}}$ m vectors $w_i (1 \leq i \leq m)$ which are linearly independent modulo D_A. $y \in D_{\bar{A}}$ if and only if

$$y = v + \sum_{i=1}^{m} c_i w_i \qquad (3.19)$$

where $y \in D_A$. As in the proof of Theorem.(3.3) using the Lagrange identity (2.5) and conditions (2.19) on $v \in D_A$ we find that condition (3.17) and hence (3.18) are necessary and sufficient for y to satisfy $y \in D_{\bar{A}}$.

To prove the converse statement we show that for arbitrary $w_i \in D_B (1 \leq i \leq m)$, linearly independent modulo D_A and satisfying (3.18) we obtain from boundary conditions (3.17) a self-adjoint extension $\bar{A}$ of A.

Let $D_{\bar{A}}$ denote the set of vectors $y \in D_B$ satisfying (3.17). Then

$$\dim D_{\bar{A}} = m \pmod{D_A} \qquad (3.20)$$

This follows from the fact that the dimension of D_B modulo D_A is equal to 2m by the Neumann formula (2.31) since by (2.10) $B = A^*$.

We have imposed m linearly independent conditions (3.17) on D_B to obtain $D_{\bar{A}}$ in the sense that the equations

$$\sum_{j=1}^{m} d_j [yw_j]_\infty - \sum_{j=1}^{m} d_j [yw_j]_{-1} = 0 \qquad (3.21)$$

for all $y \in D_B$, have only the trivial solution, $d_j = 0$ $(1 \le i \le m)$, since otherwise we would have

$$[yz]_\infty - [yz]_{-1} = 0$$

for all $y \in D_B$ where $z = \sum_{j=1}^{m} \bar{d}_j w_j$, and hence by (2.19) of Lemma 2.4 for $y \in D_B$, $z \in D_B$ that $\sum_{j=1}^{m} \bar{d}_j w_j \in D_A$, contradicting the assumption that the w_j are linearly independent modulo D_A.

To see that dim $D_{\bar{A}} = m$ (mod D_A), let $y \in D_B$ be represented according to (2.31) by

$$y = v + \sum_{i=1}^{2m} c_i u_i \tag{3.22}$$

where $v \in D_A$, and the vectors $u_i \in D_B$ $(1 \le i \le 2m)$ are linearly independent modulo D_A. By (2.22) we have

$$[vu_i]_\infty - [vu_i]_{-1} = 0, \ (i = 1,\dots,2m). \tag{3.23}$$

For $y \in D_{\bar{A}}$ we hence require in view of (2.22) from (3.17)

$$\sum_{i=1}^{2m} c_i\{[u_i w_j]_\infty - [u_i w_j]_{-1}\} = 0 \quad (j = 1,\dots,m). \tag{3.24}$$

Assume the rank of the matrix W given by

$$W = \{W_{ij}\} = \{[u_i w_j]_\infty - [u_i w_j]_{-1}\} \quad (1 \le i \le 2m),\ 1 \le j \le m \tag{3.25}$$

is r. Then we have in (3.24) 2m-r independent coefficients c_i which can be chosen arbitrarily while the other r coefficients c_i are given in terms of these. Thus there are in $D_{\bar{A}}$ 2m-4 vectors linearly independent modulo D_A.

We show r=m. This follows from the linear independence of the conditions (3.17) expressed as above. We require using (3.22) in view of (2.19) that in conformity with (3.21) the equations

$$\sum_{j=1}^{m} d_j\{[u_i w_j]_\infty - [u_i w_j]_{-1}\} = 0 \quad (i = 1,\dots,2m) \tag{3.26}$$

have only the trivial solution $d_j = 0$ ($j = 1,\dots,m$). Since the equations (3.26) have m-r linearly independent solutions for the d_j where r is the rank of the matrix (3.25) defined above, we see that we must have r=m. Hence we have shown that

$$\dim \bar{A} = m \pmod{D_A}. \qquad (3.20)$$

We now let $D_{A'}$ denote the set of all vectors y of the form

$$y = v + \sum_{j=1}^{m} c_j w_j \qquad (3.27)$$

where the c_j are constants and $v \in D_A$. We show $D_{A'} = D_{\bar{A}}$.

From (3.18) and (2.19) we see that every vectors $y \in D_{A'}$ satisfies conditions (3.17). Hence $D_{A'} \subset D_{\bar{A}}$. but, obviously from (3.27) we have

$$\dim A' = m \pmod{D_A}. \qquad (3.28)$$

From (3.20), (3.28) and $D_{A'} \subset D_{\bar{A}}$ we deduce that $D_{A'} = D_{\bar{A}}$.

We now show that $D_{\bar{A}} = D_{A'}$ determines a self-adjoint extension of A. Condition (3.7) of Theorem (3.1) is satisfied for y, $z \in D_{A'}$ given by (3.27) in view of (3.18) and (2.19). Also every vector $z \in D_B$ which satisifes (3.7) for all $y \in D_{A'}$ and hence in particular for $y = w_j$ ($j = 1,\dots,2$) must by definition belong to $D_{\bar{A}} = D_{A'}$. Hence by Theorem 3.1 $D_{\bar{A}}$ is the domain of a self-adjoint extension of A.

From the proof of Theorem 3.4 we easily derive the following corollary:

<u>Corollary 3.1:</u> If a self-adjoint extension $\bar{A}$ of the operator A is defined by (3.16) together with the m linearly independent boundary conditions given by (3.17), (3.18) then the deficiency indices of A are (m,m).

<u>Proof</u>: Assume that the deficiency indices of A are (n,n). Then

using the Neumann formula (2.31) we find that $m \leq 2n$ and as in the proof of Theorem (3.4) we have

$$\dim D_{\bar{A}} = 2n - m \pmod{D_A}.$$

Now the m vectors $w_i (1 \leq i \leq m)$ are linearly independent modulo D_A and satisfy $w_i \in D_{\bar{A}}$. Hence $2n-m \geq m$ and we deduce $n \geq m$.

Assume $n > m$. then by Theorem 3.4 a self-adjoint extension $\bar{A}'$ of the operator A is determined by (3.18) together with n boundary conditions of the form (3.17), (3.18) given by n vectors in D_B linearly independent modulo D_A consisting of the m vectors $w_i (1 \leq i \leq m)$ used in determining $\bar{A}$ together with $n-m$ additional vectors. Then

$$\dim D_{\bar{A}'} = n \pmod{D_A}.$$

Now obviously $\bar{A}' \subseteq \bar{A}$. By the maximality of self-adjoint extensions we require however $\bar{A}' = \bar{A}$, and we have a contradiction in the dimensions of the corresponding domains. Hence we deduce that $n=m$ as stated in the corollary.

We term the operator A quasi-regular or of Category II (cf. Billigheimer (5)) in the case that the deficiency indices of A are (4,4). this case is in some aspects analogous to the limit-circle category of Weyl (25) in the case of second-order differential equations. Here the number of boundary conditions determining self- adjoint extensions of A according to Theorem 3.4 is equal to 4, as in the case of the regular operator considered in Section 3 (Billigheimer (6)).

We now discuss the special case where the deficiency indices of the operator A are (2,2). In this case we term the operator A of Category I (cf. Billigheimer (5)). This case in some aspects analogous to the limit-point category of Weyl (25) in the case of second-order differential operators. Here the self-adjoint extensions of A can be characterized by explicit boundary conditions independent of the form of the difference operator P. We first prove the following lemma characterizing the domain of the operator B:

<u>Lemma 3.1:</u> If the deficiency indices of the operator A are (2,2), then for arbitrary elements $y, z \in D_B$ we have

$$[yz]_\infty = 0 \qquad (3.29)$$

<u>Proof:</u> We consider the vectors ϕ_i, $1 \leq i \leq 4$, defined in (2.38). Then the $\phi_i \in D_B$ but are linearly independent modulo D_A. Now since we have assumed the deficiency indices of A are (2,2) we deduce from the Neumann formula (2.31), and the fact (2.10) that $A^* = B$, that the dimension of D_B modulo D_A is 4. Hence every vector $y \in D_B$ is expressible in the form

$$y = v + \sum_{i=1}^{4} \alpha_i \phi_i \qquad (3.30)$$

where $v \in D_A$ and the α_i are constants.

Now by (2.22) of Lemma 2.5 we have $[vz]_\infty = 0$ for all $z \in D_B$. Also from the fact that $\phi_{in} = 0$ for $n > m$, we see that $[\phi_i z]_\infty = 0$ $(1 \leq i \leq 4)$ for all $z \in D_B$. Hence from (3.30) we obtain from all $z \in D_B$

$$[yz]_\infty = 0.$$

Then we can deduce the following modified form of Theorem 3.4:

<u>Theorem 3.5:</u> If the deficiency indices of the operator A are (2,2) then a self-adjoint extension $\bar{A}$ is defined by

$$A \subset \bar{A} \subset B \qquad (3.31)$$

and the boundary conditions that $y \in D_{\bar{A}}$ if and only if

$$[yw_i]_{-1} = 0 \qquad (i = 1,2) \qquad (3.32)$$

where the vectors w_i $(i = 1,2)$, determined by the operator $\bar{A}$, satisfy

$$[w_i w_j]_{-1} = 0 \qquad (i, j = 1,2) \qquad (3.33)$$

The boundary conditions in this case can be written in a form independent of the operator P, analogously to the case of the regular boundary problem discussed in section 3, Theorem 3.1 (Billigheimer (6)). We have the following corollary:

Corollary 3.2: If the deficiency indices of the operator A are (2,2), then a self-adjoint extension $\bar{A}$ of the operator A is defined by

$$a \subset \bar{A} \subset B \qquad (3.34)$$

and the boundary conditions that $y \in D_{\bar{A}}$ if and only if

$$y_{-2} + fy_0 = 0, \; y_{-1} + g(c_{-1}y_0 + d_{-1}y_1) = 0 \qquad (3.35)$$

where f,g are constants.

Proof: Condition (3.32) is equivalent to the boundary conditions

$$\alpha_{i1}y_1 + \alpha_{i0}y_0 + \alpha_{i,-1}y_{-1} + \alpha_{i,-2}y_{-2} = 0 \quad (i = 1,2) \qquad (3.36)$$

where at least one of the coefficients $\alpha_{in}(n = -2,-1,0,1)$ for each fixed i in (3.36) is not equal to 0 and

$$\alpha_{i,-2} = d_{-2}\bar{w}_{io}, \alpha_{i,-1} = d_{-1}\bar{w}_{i1} + c_{-1}\bar{w}_{i0}, \alpha_{i0} = -d_{-2}\bar{w}_{i,-2} - c_{-1}\bar{w}_{i.-1}$$

$$\alpha_{i1} = -d_{-1}\bar{w}_{i,-1} \qquad (i = 1,2) \qquad (3.37)$$

Condition (3.33) implies

$$(\bar{\alpha}_{i1}\alpha_{j,-2} - \bar{\alpha}_{i,-2}\alpha_{j1}) \frac{c_{-1}}{d_{-1}} - (\bar{\alpha}_{i0}\alpha_{j,-2} - \bar{\alpha}_{i,-2}\alpha_{j0}) - (\bar{\alpha}_{i1}\alpha_{j,-1} - \bar{\alpha}_{i,-1}\alpha_{j1})$$
$$= 0 \qquad (i = 1,2) \qquad (3.38)$$

For real coefficients α_{in} $(i = 1,2)$ $(n = -2,-1,0,1)$, corresponding to real boundary conditions (3.36) we have in particular the real boundary conditions in the form (3.35). In this case $\bar{A}$ is a real self-adjoint extension of A, so that if $y \in D_{\bar{A}}$ then also $\bar{y} \in D_{\bar{A}}$ and $A\bar{y} = \overline{Ay}$.

In Lemma 3.1 we gave a necessary condition for A to have deficiency indices (2,2). We now give sufficient conditions. We prove the following theorem:

Theorem 3.6: The deficiency indices of the operator A are (2.2) if any one of the following conditions holds:

(i) The coefficients b_n, c_n, d_n in (2.3) are uniformly bounded.

(ii) The coefficients c_n, d_n in (2.3) are uniformly bounded.

(iii) For $y, z \in D_{B(0)}$ where $B_{(0)}$ is the maximal operator generated by the formal difference operator P defined in (2.3) in H with domain subject to the boundary conditions

$$y_{-2} = y_{-1} = 0 \tag{3.37}$$

we have

$$[yz]_{\infty} = 0. \tag{3.38}$$

Proof. We consider a symmetric extension $\bar{A} = B_{(0)}$ of A, $A \subset \bar{A} \subset B$, with domain $D_{\bar{A}}$ such that $y = \{y_n\}_{-2}^{\infty} \in D_B$ also satisfies $y \in D_{\bar{A}}$ if and only if y satisfies the boundary conditions (3.37). We show that $\bar{A} = B_{(0)}$ is a self-adjoint operator and use Corollary 3.1 to deduce that the deficiency indices of A are (2.2).

(i) If b_n, c_n, d_n are uniformly bounded, then the operator $B_{(0)}$, the maximal operator generated by P with domain subject to (3.37) in H, is a bounded linear operator. Hence $D_{B(0)}$ is all of $H_{(0)}$ the subspace of H subject to (3.37). Hence $B_{(0)}$ is self- adjoint if it is symmetric. From (2.6) we see that, with c_n, d_n bounded, for $y, z \in H_{(0)}$ we have $[y_2]_{\infty} = 0$. Obviously also $[yz]_{-1} = 0$. Hence, by the Lagrange identity (2.5) or the conditions of Theorem 3.1, $B_{(0)} = \bar{A}$ is self-adjoint.

It is easy to check directly that in this case the deficiency indices of $B_{(0)}$ are (0,0). For the equations $B^*_{(0)}y = \lambda y$ with complex

λ and $B^*_{(0)} = B_{(0)}$ can have no non-trivial summable-square solutions with $|\lambda|$ arbitrarily large and $B_{(0)}$ bounded.

(ii) If merely c_n, d_n are uniformly bounded, then $D_{B(0)}$ is the subset of $H_{(0)}$ consisting of sequences $y = \{y_n\}_{-2}^{\infty}$ satisfying (3.37) and such that

$$\sum_{-2}^{\infty} |y_n|^2 < \infty, \quad \sum_{-2}^{\infty} |b_n\, y_n|^2 < \infty \qquad (3.39)$$

$B_{(0)}$ is the maximal operator generated by P in $H_{(0)}$ and as in the proof of (i) c_n, d_n bounded mean that conditions (3.6), (3.7) of theorem 3.1 are satisfied and hence $B_{(0)} = \bar{A}$ is self-adjoint.

(iii) If we merely assume that for $y, z \in D_B$ and hence also for $y, z \in D_{B(0)}$ we have $[yz]_\infty = 0$, we also see that by (3.37) $[yz]_{-1} = 0$ and by Theorem 31 $B_{(0)}$ is self-adjoint.

We can again show directly that in this case the deficiency indices of $B_{(0)}$ are (0,0). For the equations $B^*_{(0)}\, y = \lambda y$ with complex λ and $B^*_{(0)} = B_{(0)}$ can have no non-trivial summable square solutions satisfying (3.37). Otherwise we deduce for a solution y that $y \in D_{B(0)}$ and hence from our assumptions that $[yy]_\infty = 0$ and also $[yy]_{-1} = 0$. Then from the Lagrange identity (2.5) we obtain

$$(B_{(0)}y,y) - (y,\, B_{(0)}y) = (\lambda-\bar{\lambda})(y,y) = 0$$

and hence that $y = 0$ which contradicts the assumptions.

In a sequel to this paper (Billigheimer (7)) we consider the resolvents of the self-adjoint extensions of the symmetric operator A. We show that in the case of the quasi-regular operator P where the deficiency indices of the symmetric operator A are (4,4) the resolvent operator is completely continuous and the spectrum discrete. The corresponding classical Green's functions give the solutions of boundary problems for inhomogeneous recurrence relations of the form $(Py)_n = \lambda y_n + f_n$ $(n \geq 0)$ where $f \in \ell^2$, and the associated resolutions of the identity correspond to classical eigenfunction expansions and related results.

The results can be generalized to the case of the formally self-adjoint difference operator P of order $2p$ $(p > 0)$ acting on sequences $y = \{y_n\}_{-p}^{\infty}$ of complex numbers y_n of the form $(Py)_n = a_{nn+p}y_{n+p} + a_{nn+p-1}y_{n+p-1} + \ldots + a_{nn}y_n + \ldots + a_{nn-p}y_{n-p}$ $(n \geq 0)$ where the coefficients a_{nm} $(n \geq 0,\ n-p \leq m \leq n+p)$ are real and satisfy $a_{nm} = a_{mn}$ and $a_{nn-p}, a_{nn+p} > 0$ (cf. Billigheimer (8)).

References

1. Akhiezer, N.I. and I.M. Glazman. Theory of linear operators in Hilbert space, Moscow-Leningrad (1950); Eng. transl. Ungar, New York I (1961), II (1963).
2. Atkinson, F.V. Discrete and continuous boundary problems, Academic Press, New York (1964).
3. Berezanskii, Ju. M. Expansions in eigenfunctions of self-adjoint operators, Kiev (1965), Eng. transl. Amer. Math. Soc. Transl. (1968).
4. Billigheimer, C.E. Regular boundary problems for a five-term recurrence relation, Pac. J. Math. 35, 1((1970), 23-51.
5. Billigheimer, C.E. Singular boundary problems for a five-term recurrence relation, Amer. J. Math. 91, 4 (1969), 1012-1048.
6. Billigheimer, C.E. Regular self-adjoint fourth order real linear difference operators, (to appear).

7. Billigheimer, C.E. Boundary problems for self-adjoint linear relations, Proc. Canad. Math. Congress, 1971, 264-292.

8. Billigheimer, C.E. Singular self-adjoint linear difference operators, Les 265 communications individuelles, congres international des mathematiciens, Nice (1970), 222.

9. Birkhoff, G.D. Boundary-value problems and expansion problems of ordinary linear differential equations, Trans. Amer. Math. Soc. 9 (1908), 373-395.

10. Coddington, E.A. The spectral representation of ordinary self-adjoint differential operators, Annals of Math. 60 (1954), 191-211.

11. Coddington, E.A. and N. Levinson. Theory of ordinary differential equations, McGraw-Hall, New York (1955).

12. Dunford, N. and J.T. Schwartz. Linear operators: Part I, General theory, Part II, Spectral theory, Interscience Publishers, New York (1964).

13. Everitt, W.N. Fourth order singular differential equations, Math. Annalen 149 (1963), 320-340.

14. Everitt, W.N. Singular differential equations I: The even order case, Math. Annalen 156 (1964), 9-24.

15. Glazman, I.M. On the theory of singular differential operators, Usp. Math. Nauk (N.S.) 5,6 (40) (1955), 102-135. eng. transl. Amer. Math. Soc. Transl. Ser. 1, Vol. 4 (1962), 331-372.

16. Hartman, P. The number of L_2 solutions of $x'' + g(t)x = 0$, Amer. J. Math. 73 (1951), 635-645.

17. Hartman, P. and A. Wintner. Oscillatory and non-oscillatory linear differential equations, Amer. J. Math. 71 (1949), 627-648.

18. Hellinger, E. Zur Stieltjesschen Kettenbruchtheorie, Math. Ann 86 (1922), 18-29.

19. Kodaira, K. On ordinary differential equations of any even order and the corresponding eigenfunction expansion, Amer. J. of Math 72 (1950), 502-544.

20. Krein, M.G. On a one-dimensional singular boundary value problem of even order in the interval (0,∞), Doklady Adad. Nauk. SSSR (N.S) 74 (1950), 9-12.

21. Levinson, N. A simplified proof of the expansion theorem for singular second order differential equations, Duke Math. J. 18 (1951), 57-71, 719-722.

22. Levinson, N. The expansion theorem for singular self-adjoint linear differential operators, Ann of Math. (2) 59 (1954), 300-315.

23. Levitan, B.M. Expansion in eigenfunctions for second order differential equations, GTTL (1950).

24. Naimark, M.A. Linear differential operators, Moscow (1952) Eng. transl. Ungar, New York I (1967), II (1968).

25. Stone, M.H. Linear transformations in Hilbert space, Amer. Math. Soc. Colloquium Publications, Vol. XV, New York (1932).

26. Titchmarsh, E.C. Eigenfunction expansions associated with second order differential equations, Part I, Oxford University Press (1948).

27. Weyl, H. Über gewöhnliche Differentialgleichungen mit Singularitäten und die zugehörigen Entwicklungen willkürlicher Funktionen, Math. Annalen 68 (1910), 220-269.

28. Wintner, A. On the location of continuous spectra, Amer. J. Math. 20 (1948), 22-30.

29. Yosida, K. On Titchmarsh-Kodaira's formula concerning Weyl-Stone's eigenfunction expansion, Nagoya Math. J. 1 (1950), 49-58.

McMaster University
Hamilton
Ontario
Canada

Canadian Mathematical Society
Conference Proceedings
Volume 8 (1987)

SYMMETRIC STURM – LIOUVILLE OPERATORS WITH EIGENVALUE DEPENDING BOUNDARY CONDITIONS

AAD DIJKSMA, HEINZ LANGER, HENK DE SNOO [1]

Dedicated with great admiration to M.G. Krein on the occasion of his 80th birthday

INTRODUCTION. Let L be the second order symmetric Sturm-Liouville differential operator $L = -D^2+q$, where D stands for differentiation, on the interval $[0,b)$, $0 < b \le \infty$, in the Hilbert space $L^2(0,b)$. We assume that q is real valued, $q \in L^1_{loc}[0,b)$ and L is regular at $x = 0$ and limit point at $x = b$, and consider the problem

$$(0.1)\qquad Ly - \ell y = f\ ,\ \alpha(\ell)y(0) + \beta(\ell)y'(0) = 0,$$

where α,β are functions locally holomorphic in ℓ on some set $\mathfrak{D}_{\alpha,\beta} \subset \mathbb{C}\backslash\mathbb{R}$, satisfying some additional conditions (see Section 1). In general, for each $f \in L^2(0,b)$ and for sufficiently many $\ell \in \mathfrak{D}_{\alpha,\beta}$ (see Theorem 1.2), the problem has a unique solution $y \in L^2(0,b)$. If we denote this solution by $y = R(\ell)f$ then $R(\ell)$ is a bounded linear operator on $L^2(0,b)$.

In Section 1 we show that for essentially all ℓ in a bounded simply connected domain $\mathfrak{D}$ with smooth boundary, contained in $\mathfrak{D}_{\alpha,\beta}$, $R(\ell)$ is the compressed resolvent of some selfadjoint operator A in a Krein space $\mathfrak{K}$ containing $L^2(0,b)$: $R(\ell) = P(A-\ell)^{-1}|_{L^2(0,b)}$, $\ell \in \mathfrak{D}\cap\rho(A)$, where P is the orthogonal projection of $\mathfrak{K}$ onto $L^2(0,b)$. This operator A is an extension of the minimal symmetric operator L_{min} in $L^2(0,b)$ associated with the differential expression L and the boundary conditions $y(0) = y'(0) = 0$. That is, $R(\ell)$ is in fact a generalized resolvent of L_{min}, corresponding to a selfadjoint extension A of L_{min} in a Krein space $\mathfrak{K}$. If $\mathfrak{K}$, A are chosen to be "minimal" and if α/β belongs to some class N_κ (see [4], [25], [27] or Section 1 below) then the Krein space becomes a Pontryagin or even a Hilbert space, corresponding to

---

[1] 1980 *Mathematics Subject Classification* (1985 *Revision*) 34B25, 46D05, 47E05.
Work supported in part by the Netherlands Organization of Pure Research (ZWO).

$\varkappa > 0$ and $\varkappa = 0$, respectively. The proof of these facts is given in Section 2; it relies on our earlier results on operator colligations in Krein spaces and their characteristic functions. We mention that in particular cases these results can be considered to be well known: With the extending space $\mathfrak{K}$ being a Hilbert space they go basically back to [18], [19], [21], [37], [38], and for $\mathfrak{K}$ being a Pontryagin space such results were obtained in [5], [34], [35]. In the latter papers where polynomials α,β have been considered as well as in other papers with more special α,β (e.g., linear functions in ℓ) the Pontryagin or Hilbert space $\mathfrak{K}$ has been described in a more explicit way.

The operator $R(\ell)$ can also be calculated explicitly, and it turns out that, as in the classical case of eigenvalue independent boundary conditions, the singularities of $R(\ell)$ in $\mathfrak{D}_{\alpha,\beta}$ or in $\mathbb{C}$ coincide with the singularities of some analytic 2x2 matrix valued function Ω and, in some instances, even with some complex valued function ω. These functions, sometimes called Weyl coefficients, belong to some class $N_{\varkappa}^{2x2}$, $N_{\varkappa}$ respectively, if the extending space $\mathfrak{K}$ is a Pontryagin space of index $\varkappa$. On the other hand, it is well known (see [26]) that as $\omega \in N_{\varkappa}$ it admits a representation by means of a unitary operator $U(\omega)$ or, if some asymptotic condition at ∞ is satisfied, a selfadjoint operator $A(\omega)$ in some $\pi_{\varkappa}$-space $\Pi(\omega)$, and, similarly, for $\Omega \in N_{\varkappa}^{2x2}$ (see [4]). In Sections 3 and 4 we show that under the conditions $\alpha/\beta \in N_{\varkappa}$ and with α,β polynomials, respectively, the operators $A(\Omega)$ and $A(\omega)$ are isomorphically equivalent to the selfadjoint extension A of L_{min}. We mention that if, in particular, $\varkappa = 0$, the functions $\Omega \in N_{0}^{2x2}$, $\omega \in N_{0}$ admit the well known representations

$$\Omega(z) = A + Bz + \int_{\mathbb{R}} ((t-z)^{-1} - t(1+t^2)^{-1})d\Sigma(t), \quad \int_{\mathbb{R}} (1+t^2)^{-1}d\Sigma(t) < \infty,$$

$$\omega(z) = a + bz + \int_{\mathbb{R}} ((t-z)^{-1} - t(1+t^2)^{-1})d\sigma(t), \quad \int_{\mathbb{R}} (1+t^2)^{-1}d\sigma(t) < \infty,$$

with e.g. A,B hermitian 2x2 matrices, $B \geq 0$ and Σ a nondecreasing function on $\mathbb{R}$, (which in our problem even specialize to, for instance $B = 0$ and $b = 0$), and the operators $A(\Omega)$, $A(\omega)$ are isomorphic to the multiplication operators by the independent variable in the space L^2_{Σ}, L^2_{σ}, respectively. In these cases Σ and σ are spectral functions of the problem, defined according to M.G. Krein, by the fact that the Parseval equality with respect to a Fourier transformation holds. The isomorphism of A and those multiplication operators yields, in particular, that $\sigma(A)$, the spectrum of A, is completely characterized by Σ or σ (cf. [2]).

If α,β are such that the extending space $\mathfrak{K}$ is a Pontryagin space of index $\varkappa$, the operator A has an invariant nonpositive subspace $\mathfrak{L}_0$ of dimension $\varkappa$; in particular, it has at least $\varkappa$ eigenvalues (counted according to their multiplicity) in the closed upper half plane $\mathbb{C}^+ \cup \mathbb{R}$. Evidently, the nonreal or isolated real eigenvalues are the poles of ω; the embedded eigenvalues with nonpositive eigenvectors are, according to [27] and the above isomorphism, the generalized poles of negative type of ω. Using this we shall characterize in Section 5 the projection onto $L^2(0,b)$ of the eigenvectors and associated vectors corresponding to each eigenvalue of A, restricted to $\mathfrak{L}_0$, as a chain of solutions of certain initial value problems related to (0.1). For an isolated eigenvalue the definition of the chain coincides with the one given in [31]; here, however, these chains appear as the projections of Jordan chains of the linear operator A in the larger space $\mathfrak{K}$.

The fact that we concentrate on a particular case is no actual restriction. The methods we use also work for leading coefficients of L and weight functions not identically equal to 1 and also for L being limit circle on the left or limit circle or regular on the righthand side of the interval and also for general symmetric differential operators with eigenvalue depending boundary conditions. Here, however, we restrict ourselves to this special case in order to avoid technical complications. In particular, there is in this paper no need for relations, i.e., multivalued operators. Indeed, all operators are or turn out to be single valued. The theory of relations seems to be useful for the case of canonical systems where analogous results can be proved; these will be published elsewhere.

1. GENERALIZED RESOLVENTS. On the interval $[0,b)$, $0 < b \le \infty$, we consider the differential expression $L = -D^2+q$, where $q \in L^1_{loc}[0,b)$ is real valued. We will assume that L is limit point at $x = b$, thus the homogeneous equation

$$Ly - \ell y = 0 \ , \quad \ell \in \mathbb{C}, \tag{1.1}$$

has for $\ell \in \mathbb{C}\backslash\mathbb{R}$, up to scalar multiples, a unique solution $\psi(.;\ell)$ in the Hilbert space $L^2(0,b)$. Introducing the two solutions $u_1(,.\ell)$ and $u_2(.,\ell)$ of (1.1) which satisfy the initial conditions

$$u_1(0,\ell) = 1 \quad , \quad u_1'(0,\ell) = 0,$$

$$u_2(0,\ell) = 0 \quad , \quad u_2'(0,\ell) = -1,$$

we may write

$$\psi(.,\ell) = u_2(.,\ell) + m(\ell)u_1(.,\ell), \qquad \ell \in \mathbb{C}\backslash\mathbb{R},$$

and this relation defines a locally holomorphic function m on $\mathbb{C}\backslash\mathbb{R}$, the Weyl coefficient of the problem (1.1), mapping the upper half plane $\mathbb{C}^+$ into itself; in fact, from the definition of ψ and the initial conditions satisfied by the solutions u_1 and u_2, it follows that

$$\frac{m(\ell)-\overline{m(\lambda)}}{\ell-\bar{\lambda}} = \int_0^\infty \psi(x,\ell)\overline{\psi(x,\lambda)}dx \ , \qquad \ell,\lambda \in \mathbb{C}\backslash\mathbb{R},\ \ell \neq \lambda,$$

see for instance [15].

For $f \in L^2(0,b)$ we consider the inhomogeneous differential equation

(1.2) $\qquad Ly - \ell y = f \quad , \quad \ell \in \mathbb{C}\backslash\mathbb{R},$

and impose at $x = 0$ a boundary condition on its solutions y which depends on the parameter ℓ:

(1.3) $\qquad \alpha(\ell)y(0) + \beta(\ell)y'(0) = 0.$

Here the functions α and β are defined and locally holomorphic on some open set $\mathfrak{D}_{\alpha,\beta} \subset \mathbb{C}\backslash\mathbb{R}$ which is symmetric with respect to $\mathbb{R}$ and satisfy:

(1.4) $\qquad \alpha(\ell)\overline{\beta(\bar{\ell})} - \beta(\ell)\overline{\alpha(\bar{\ell})} = 0 \ , \ |\alpha(\ell)| + |\beta(\ell)| \neq 0,\ \ell \in \mathfrak{D}_{\alpha,\beta}.$

If $\mathfrak{D}$ is some subset of $\mathbb{C}$ we denote by $\mathfrak{D}^* = \{\ \bar{\ell}\ |\ \ell \in \mathfrak{D}\ \}$, thus $\mathfrak{D}_{\alpha,\beta} = \mathfrak{D}^*_{\alpha,\beta}$. We assume that

(1.5) $\qquad \mathfrak{D}_0 = \{\ \ell \in \mathfrak{D}_{\alpha,\beta} |\ \alpha(\ell)m(\ell) \neq \beta(\ell)\ \} \neq \emptyset.$

Then for each $\ell \in \mathfrak{D}_0$ and $f \in L^2(0,b)$, the initial value problem (1.2)-(1.3) has a unique solution $y \in L^2(0,b)$ which we denote by $y = R(\ell)f$, $f \in L^2(0,b)$, see Theorem 1.2 below. This operator $R(\ell)$ on $L^2(0,b)$ is bounded and, on account of (1.4), satisfies

$$R(\bar{\ell})f = R(\ell)^*f \quad , \quad \ell \in \mathfrak{D}_0\cap(\mathbb{C}\backslash\mathbb{R}) \quad , \quad f \in L^2(0,b),$$

(note that $\mathfrak{D}_0 = \mathfrak{D}_0^*$). If $f \in L^2(0,b)$ vanishes near b, a calculation shows

(1.6) $$(R(\ell)f)(x) = \begin{pmatrix} u_1(x;\ell) & u_2(x;\ell) \end{pmatrix} \Omega(\ell) \begin{pmatrix} \int_0^b u_1(t;\ell)f(t)dt \\ \int_0^b u_2(t;\ell)f(t)dt \end{pmatrix}$$
$$- u_2(x;\ell)\int_x^b u_1(t;\ell)f(t)dt - u_1(x;\ell)\int_0^x u_2(t;\ell)f(t)dt,$$

where the 2x2 matrix function Ω is given by

$$\Omega(\ell) = (\beta(\ell)-\alpha(\ell)m(\ell))^{-1} \begin{pmatrix} m(\ell)\beta(\ell) & \beta(\ell) \\ \beta(\ell) & \alpha(\ell) \end{pmatrix} .$$

The function Ω is sometimes called the Weyl coefficient corresponding to the boundary condition (1.3) and the Fourier transform

$$f \mapsto \begin{bmatrix} \int_0^b u_1(t;.)f(t)dt \\ \int_0^b u_2(t;.)f(t)dt \end{bmatrix},$$

with $f \in L^2(0,b)$ vanishing near b. With the exception of Ω all terms on the right hand side of (1.6) are entire functions which are real on the real axis if f is real. We shall study Ω in Section 3. Here we note that

$$\Omega = \begin{bmatrix} m & 1 \\ 1 & 0 \end{bmatrix} \quad \text{and} \quad \Omega = \begin{bmatrix} 0 & 0 \\ 0 & -1/m \end{bmatrix},$$

if $\alpha \equiv 0$, $\beta \equiv 1$, and if $\alpha \equiv 1$, $\beta \equiv 0$, respectively. In the sequel we shall exclude these cases, as they can be considered to be well understood.

Recall that an nxn matrix function Q belongs to the class $N_\kappa^{n\times n}$ if it is locally meromorphic on $\mathbb{C}\backslash\mathbb{R}$, $Q(\bar{\ell}) = Q(\ell)^*$ and is such that the kernel

$$N_Q(\ell,\lambda) = \frac{Q(\ell)-Q(\lambda)^*}{\ell-\bar{\lambda}}, \quad \ell,\lambda \in \mathfrak{D}_Q, \ \ell \neq \bar{\lambda},$$

has κ negative squares. Here $\mathfrak{D}_Q$ denotes the set of complex numbers, where Q is holomorphic and κ is a nonnegative integer. For the basic properties of functions in $N_\kappa^{n\times n}$ we refer to [4], [25], [27]. In particular, the class $N_0^{n\times n}$ consists of all nxn matrix functions Q which are locally holomorphic on $\mathbb{C}\backslash\mathbb{R}$, $Q(\bar{\ell}) = Q(\ell)^*$ and $\mathrm{Im}\ell\cdot\mathrm{Im}Q(\ell) \geq 0$, $\ell \in \mathbb{C}\backslash\mathbb{R}$. We write N_κ for $N_\kappa^{1\times 1}$. The Weyl coefficient m introduced above belongs to the class N_0. We need the following simple result.

LEMMA 1.1. *Let L, α and β be as above with $\mathfrak{D}_{\alpha,\beta} = \mathbb{C}\backslash\mathbb{R}$ and denote by* m *the Weyl coefficient. If $\alpha/\beta \in N_\kappa$, then* (1.5) *holds, and the* 2x2 *matrix function Ω in* (1.6) *belongs to the class $N_\kappa^{2\times 2}$.*

PROOF. It is easily verified that $Q \in N_\kappa^{n\times n}$ and $\det Q \not\equiv 0$ imply that $-Q^{-1} \in N_\kappa^{n\times n}$. Hence if $\alpha/\beta \in N_\kappa$ then $-\beta/\alpha \in N_\kappa$ and therefore $m - \beta/\alpha \not\equiv 0$. Otherwise the kernel N_m would have a finite number of positive squares, hence it would be a rational function, which is not the case. Thus $\mathfrak{D}_0 \neq \emptyset$. Furthermore,

$$-\Omega(\ell)^{-1} = \begin{bmatrix} \alpha(\ell)/\beta(\ell) & -1 \\ -1 & m(\ell) \end{bmatrix}$$

and it is easy to see that $\alpha/\beta \in N_\kappa$ and $m \in N_0$ imply that $-\Omega(\ell)^{-1} \in N_\kappa^{2\times 2}$. Consequently, $\Omega \in N_\kappa^{2\times 2}$. □

Let L_{min} and L_{max} be the minimal and maximal operators associated with L in the Hilbert space $L^2(0,b)$. Then L_{min} is symmetric and $L_{max} = L^*_{min}$. A selfadjoint operator A in a Krein space $\mathfrak{K}$ with nonempty resolvent set $\rho(A)$ is called an extension of L_{min} if $\mathfrak{K}$ contains $L^2(0,b)$ as a subspace and $L_{min} \subset A$. The operator $P(A-\ell)^{-1}|_{L^2(0,b)}$, $\ell \in \rho(A)$, where P is the orthogonal projection of $\mathfrak{K}$ onto $L^2(0,b)$ is called the compressed resolvent of A or a generalized resolvent of L_{min}. The following theorem gives a global representation of the operator $R(\ell)$: on each domain with smooth boundary, contained in $\mathfrak{D}_{\alpha,\beta}$ it is a generalized resolvent of L_{min}. The theorem is a consequence of the results in [10].

THEOREM 1.2. *Let* L, α *and* β *be given as above, denote by* m *the Weyl coefficient of* L *and assume* (1.5). *Then:*

(i) *For each* $\ell \in \mathfrak{D}_0$ *and* $f \in L^2(0,b)$ *the initial value problem* (1.2) - (1.3) *has a unique solution* $y = R(\ell)f \in L^2(0,b)$.

(ii) *If* $\mathfrak{D}$ *is a bounded, simply connected open set with smooth boundary* $\partial\mathfrak{D}$ *such that* $\mathfrak{D} \cup \partial\mathfrak{D} \subset \mathfrak{D}_{\alpha,\beta} \cap \mathbb{C}^+$, *and* $\tilde{\mathfrak{D}} = \mathfrak{D} \cup \mathfrak{D}^*$, *then there exists a selfadjoint extension* A *of* L_{min} *in a Krein space* $\mathfrak{K}$ *such that* $\tilde{\mathfrak{D}} \setminus \rho(A)$ *is empty or consists of isolated points only and*

$$R(\ell)f = P(A-\ell)^{-1}f \quad , \quad \ell \in \rho(A) \cap \tilde{\mathfrak{D}} \quad , \quad f \in L^2(0,b). \tag{1.7}$$

The pair $\mathfrak{K}$,A *can be chosen to be minimal*

$$\mathfrak{K} = \text{c.l.s.}\ \{ \bigcup (A-\ell)^{-1}L^2(0,b) \mid \ell \in \rho(A) \cap \tilde{\mathfrak{D}} \}$$

in which case it is uniquely determined up to weak isomorphisms.

(iii) *If, in particular,* $\mathfrak{D}_{\alpha,\beta} = \mathbb{C}\setminus\mathbb{R}$ *and* $\alpha/\beta \in N_x$, *then* (1.5) *is automatically satisfied, the minimal space* $\mathfrak{K}$ *is a Pontryagin space of index* x *and* (1.7) *holds with* $\tilde{\mathfrak{D}}$ *replaced by* $\mathbb{C}\setminus\mathbb{R}$.

In the proof of part *(ii)*, which we postpone until Section 2, we first construct a compressed resolvent $P(A-\ell)^{-1}|_{L^2(0,b)}$ and show that it yields the unique solution of the initial value problem. Now the proof of part *(i)* follows easily from that of *(ii)*: one takes $\mathfrak{D}$ to be a sufficiently small disc around $\ell \in \mathfrak{D}_0$ and verifies that $\ell \in \rho(A) \cap \mathfrak{D}$. To prove *(iii)* assume $\mathfrak{D}_{\alpha,\beta} = \mathbb{C}\setminus\mathbb{R}$ and $\alpha/\beta \in N_x$. Then Lemma 1.1 implies that $\mathfrak{D}_0 \neq \emptyset$, and Theorem 2.3 of [10] implies that, if the representation of $R(\ell)$ in *(ii)* is minimal, then $\mathfrak{K}$ is a π_x-space. Since a weak isomorphism between two π_x-spaces can be extended to an isomorphism between these spaces, the pair $\mathfrak{K}$,A is now uniquely determined up to an isomorphism. Moreover $\rho(A) \cap \mathbb{C}\setminus\mathbb{R}$ coincides with $\mathbb{C}\setminus\mathbb{R}$ except for finitely many points which remain fixed under the application of an isomorphism. Hence *(iii)* follows from *(ii)* by considering a chain of domains $\mathfrak{D}_1 \subset \mathfrak{D}_2 \subset \ldots \subset \mathfrak{D}_n \subset \ldots$

satisfying the hypotheses of part *(ii)* and such that $\bigcup \mathfrak{D}_n = \mathbb{C}^+$. □

REMARK. A converse to part *(ii)* of Theorem 1.2 also holds: Let a selfadjoint extension A of L_{min} in some Krein space with $\rho(A) \not\equiv \emptyset$ be given and let $\mathfrak{C}$ be a component of $\rho(A) \cap \mathbb{C}^+$. Then there exist functions α,β meromorphic on $\mathfrak{C}$ such that $y = P(A-\ell)^{-1}f$ is the unique $L^2(0,b)$ solution of (1.2) -(1.3) for $\ell \in \mathfrak{C} \cap \mathfrak{D}_{\alpha,\beta}$.

In Sections 3 and 4 we shall make use of part *(iii)* of Theorem 1.2. In particular, in Section 4 we consider the case where α and β are polynomials with real coefficients. Boundary value problems with boundary conditions that contain the eigenvalue parameter of the type (1.2) - (1.3) have been studied in many papers see for instance [1], [5], [6], [11], [12], [13], [21], [22], [32], [36], [39].

A special case that can be found in the recent literature, cf. [13], [16], is given by

$$\alpha(\ell) = \alpha_1 - \ell\alpha_1', \quad \beta(\ell) = -\alpha_2 + \ell\alpha_2',$$

where $\alpha_1, \alpha_1', \alpha_2$ and α_2' are real constants, not simultaneously zero. Note that

$$\frac{\alpha(\ell)\overline{\beta(\lambda)} - \overline{\alpha(\lambda)}\beta(\ell)}{\ell - \bar{\lambda}} = \alpha_1'\alpha_2 - \alpha_1\alpha_2', \qquad \ell,\lambda \in \mathbb{C}.$$

This implies with $\alpha = \alpha_1'\alpha_2 - \alpha_1\alpha_2'$ that $\alpha = 0$ corresponds to canonical selfadjoint extensions, $\alpha > 0$ corresponds to true Hilbert space extensions and $\alpha < 0$ corresponds to true π_1-space extensions. In this case one can give a concrete model for the selfadjoint extension in Theorem 1.2 *(iii)*. For $\alpha \not\equiv 0$ we define $\mathfrak{K} = L^2(0,b) \oplus \mathbb{C}$ with inner product

$$\left[\begin{pmatrix} f \\ \tilde{f} \end{pmatrix}, \begin{pmatrix} g \\ \tilde{g} \end{pmatrix}\right] = [f,g]_{L^2(0,b)} + \frac{1}{\alpha}\bar{\tilde{g}}\,\tilde{f}.$$

Then $\mathfrak{K}$ is a Hilbert space for $\alpha > 0$ and a π_1-space for $\alpha < 0$. It is straightforward to check that

$$A = \left\{ \left\{ \begin{pmatrix} y \\ \alpha_1' y(0) - \alpha_2' y'(0) \end{pmatrix}, \begin{pmatrix} L_{max} y \\ \alpha_1 y(0) - \alpha_2 y'(0) \end{pmatrix} \right\} \;\middle|\; y \in \mathfrak{D}(L_{max}) \right\}$$

is a selfadjoint extension in $\mathfrak{K}$, of L_{min}, for which (1.7) holds true.

2. UNITARY COLLIGATIONS AND THEIR CHARACTERISTIC FUNCTIONS.

Let $\mathfrak{K},\mathfrak{F},\mathfrak{G}$ be Krein spaces and let $U : \binom{\mathfrak{K}}{\mathfrak{F}} \longrightarrow \binom{\mathfrak{K}}{\mathfrak{G}}$ be a unitary operator. Here $\binom{\mathfrak{K}}{\mathfrak{F}}$, for example, denotes the Krein space which is the orthogonal direct sum of $\mathfrak{K}$ and $\mathfrak{F}$. The quadruple $\Delta = (\mathfrak{K},\mathfrak{F},\mathfrak{G};U)$ is called a unitary colligation. Let U have the matrix representation

(2.1) $$U = \begin{pmatrix} T & F \\ G & H \end{pmatrix} : \begin{pmatrix} \mathfrak{K} \\ \mathfrak{J} \end{pmatrix} \longrightarrow \begin{pmatrix} \mathfrak{K} \\ \mathfrak{G} \end{pmatrix},$$

where $T \in L(\mathfrak{K},\mathfrak{K})$, $F \in L(\mathfrak{J},\mathfrak{K})$, $G \in L(\mathfrak{K},\mathfrak{G})$ and $H \in L(\mathfrak{J},\mathfrak{G})$. Here for example $L(\mathfrak{J},\mathfrak{G})$ denotes the set of all bounded linear operators from $\mathfrak{J}$ to $\mathfrak{G}$. Then we sometimes write $\Delta = (\mathfrak{K},\mathfrak{J},\mathfrak{G};T,F,G,H)$ and we call Δ closely connected if $\mathfrak{K}$ is the closed linear span of all elements of the form T^nFf or $T^{*m}G^*g$ with $f \in \mathfrak{J}$, $g \in \mathfrak{G}, m,n = 0,1,2...$. The mapping $\Theta(z) = H + zG(I-zT)^{-1}F$ is called the characteristic function of Δ and is denoted by Θ_Δ, see [8], [10] for further details. Clearly, $\Theta_\Delta \in S(\mathfrak{J},\mathfrak{G})$, the class of all $L(\mathfrak{J},\mathfrak{G})$ valued functions defined and holomorphic on a neighbourhood of 0 in the unit disc $\mathbb{D} = \{z \in \mathbb{C} \mid |z| < 1\}$. For $\Theta \in S(\mathfrak{J},\mathfrak{G})$, we denote by $\mathfrak{D}_\Theta$ ($\hat{\mathfrak{D}}_\Theta$) the domain of holomorphy (meromorphy, respectively) of Θ in $\mathbb{D}$. Thus $\mathfrak{D}_\Theta \subset \hat{\mathfrak{D}}_\Theta \subset \mathbb{D}$ and $\hat{\mathfrak{D}}_\Theta \backslash \mathfrak{D}_\Theta$ is the set of all points in which Θ has a pole of finite multiplicity, see for instance [8], where also other references are given. Any compact subset of $\hat{\mathfrak{D}}_\Theta$ contains at most finitely many points from $\hat{\mathfrak{D}}_\Theta \backslash \mathfrak{D}_\Theta$. The following theorem shows that each $\Theta \in S(\mathfrak{J},\mathfrak{G})$ coincides with the characteristic function of a unitary colligation Δ. It may happen that the domain of holomorphy of Θ extends beyond $\mathbb{D}$, but then it need not have the property that it is symmetric with respect to $\partial\mathbb{D}$, i.e., $\Theta(z)^{-1} = \Theta(\bar{z}^{-1})^*$, whereas Θ_Δ has this property, at least for sufficiently small $|z|$. For this reason we have restricted the domain of holomorphy of $\Theta \in S(\mathfrak{J},\mathfrak{G})$ to $\mathbb{D}$. The theorem presented here is a slight generalization of Theorem 2.2. of [10], which basically stems from T.Ya. Azizov. Here we allow the set $\bar{\mathfrak{D}}$ to contain finitely many points where Θ has a pole, i.e., $\bar{\mathfrak{D}}$ maybe contained in $\hat{\mathfrak{D}}_\Theta$ rather than in $\mathfrak{D}_\Theta$ as in Theorem 2.2 of [10].

THEOREM 2.1. *Let $\Theta \in S(\mathfrak{J},\mathfrak{G})$ and $\mathfrak{D}$ be a simply connected open set with a smooth boundary $\partial\mathfrak{D}$ such that $0 \in \mathfrak{D}$ and $\mathfrak{D} \cup \partial\mathfrak{D} \subset \hat{\mathfrak{D}}_\Theta$. Then there exists a unitary colligation $\Delta = (\mathfrak{K},\mathfrak{J},\mathfrak{G};U)$ such that*

$$\hat{\mathfrak{D}}_{\Theta_\Delta} \supset \mathfrak{D} \text{ and } \Theta = \Theta_\Delta \text{ on } \mathfrak{D} \cap \mathfrak{D}_\Theta.$$

The colligation can be constructed so that it is closely connected in which case it is uniquely determined up to weak isomorphisms. If $\mathfrak{J} = \mathfrak{G}$ and $\mathfrak{J}$ is finite dimensional, then $\mathfrak{D}\backslash\rho(U)$ is empty or consists only of isolated points.

The proof of the theorem may be given by considering the mapping Θ_1 which is obtained from Θ by cancelling its poles in $\mathbb{D}$ by means of a finite Blaschke-Potapov product $B(z)$: $\Theta_1(z) = B(z)\Theta(z)$, as in Section 7 of [8],and applying Theorem 2.2 of [10] and finite dimensional perturbation theory. □

REMARK. Let Θ, $\mathfrak{D}$ and Δ be as in Theorem 2.1 and let U have the matrix representation (2.1). If Δ is closely connected, then the map $z \longmapsto z^{-1}$ establishes a one-to-one correspondence between the (finitely many) poles of Θ in $\mathfrak{D}$ and the spectrum of T in $\mathfrak{D}$. The latter consists of normal eigenvalues only. Moreover, the pole multiplicity of Θ at a pole $z = \alpha$ is equal to the algebraic multiplicity of the eigenvalue α^{-1} of T and the lengths of the chains at these points are the same, see [8].

PROOF of part *(ii)* of Theorem 1.2. Let L_{min} and L_{max} be as in Section 1 and fix $\mu \in \mathbb{C}^+ \cap \mathfrak{D}_o$, where $\mathfrak{D}_o$ is defined by (1.5). Identifying operators with their graphs in $L^2(0,b) \times L^2(0,b)$ we use Von Neumanns identity

$$L_{max} = L_{min} \dotplus M_{\mu} \dotplus M_{\bar{\mu}}, \qquad \text{direct sum},$$

where M_{ℓ} for $\ell \in \mathbb{C}\backslash\mathbb{R}$ is the one dimensional space

$$M_{\ell} = \{ \{f,Lf\} \in L_{max} \mid Lf = \ell f \}$$
$$= \{ \{\psi,\ell\psi\} \mid \psi = \alpha\psi(.;\ell) \text{ for some } \alpha \in \mathbb{C} \}.$$

Now if we put $z(\ell) = (\ell-\mu)/(\ell-\bar{\mu})$ and if we define the linear operator $\Theta(z(\ell)) : \mathfrak{D}(M_{\bar{\mu}}) \longrightarrow \mathfrak{D}(M_{\mu})$ by

$$(2.2) \qquad (\Theta(z(\ell))\psi(.,\bar{\mu}))(x) = \frac{\alpha(\ell)m(\bar{\mu})-\beta(\ell)}{\alpha(\ell)m(\mu)-\beta(\ell)}\,\psi(x;\mu)$$

(compare [10], Section 6), then $\Theta \in S(\mathfrak{D}(M_{\bar{\mu}}),\mathfrak{D}(M_{\mu}))$. Let $\mathfrak{D}$ be a set satisfying the conditions of part *(ii)* of the theorem. Then we may apply Theorem 2.1 with $\mathfrak{F} = \mathfrak{D}(M_{\bar{\mu}})$, $\mathfrak{G} = \mathfrak{D}(M_{\mu})$ and $\mathfrak{D}$ replaced by $z(\mathfrak{D})$ and conclude that there exists a closely connected unitary colligation $\Delta = (\mathfrak{K}_o,\mathfrak{D}(M_{\bar{\mu}}),\mathfrak{D}(M_{\mu});T,F,G,H)$ such that $\Theta(z) = \Theta_{\Delta}(z)$ on $\mathfrak{D} \cap \mathfrak{D}_{\Theta}$. Since the Cayley transform $V = C_{\mu}(L_{min})$ is an isometry from $\mathfrak{D}(V) = \mathfrak{R}(L_{min}-\mu)$ onto $\mathfrak{R}(V) = \mathfrak{R}(L_{min}-\bar{\mu})$ and since, for $\ell \in \mathbb{C}\backslash\mathbb{R}$, $L^2(0,b) = \mathfrak{R}(L_{min}-\ell) \oplus \mathfrak{D}(M_{\bar{\ell}})$, we obtain that

$$\tilde{U} = \begin{pmatrix} T & F & 0 \\ G & H & 0 \\ 0 & 0 & V \end{pmatrix} : \begin{pmatrix} \mathfrak{K}_o \\ \mathfrak{D}(M_{\bar{\mu}}) \\ \mathfrak{D}(V) \end{pmatrix} \longrightarrow \begin{pmatrix} \mathfrak{K}_o \\ \mathfrak{D}(M_{\mu}) \\ \mathfrak{R}(V) \end{pmatrix}$$

is a unitary mapping. The spaces $\mathfrak{D}(M_{\bar{\mu}})$ and $\mathfrak{D}(M_{\mu})$ are one dimensional and identifying them with each other one can verify that Theorem 2.1 implies that $z(\mathfrak{D})\backslash\rho(\tilde{U})$ is empty or consists only of isolated points. The closely connectedness of Δ implies that the inverse Cayley transform $A = F_{\mu}(\tilde{U})$ is a selfadjoint operator in $\mathfrak{K} = \mathfrak{K}_o \oplus L^2(0,b)$, compare Proposition 4.3 in [6]. Clearly, A extends L_{min} and the spectral mapping theorem gives that $\mathfrak{D} \subset \rho(A)$ except for at most isolated points. For $\ell \in \tilde{\mathfrak{D}}$ we define the Štraus subspace

$T(\ell)$ by

$$T(\ell) = \{ \{Pf,Pg\} \mid \{f,g\} \in A,\ g-\ell f \in L^2(0,b) \}.$$

A direct calculation, see [7], shows that for $\ell \in \mathfrak{D}$ we have

$$\begin{aligned} T(\ell) &= L_{min} \dotplus \{ \{\psi-\Theta(z(\ell))\psi, \bar{\mu}\psi-\mu\Theta(z(\ell))\psi\} \mid \psi \in \mathfrak{D}(M_{\bar{\mu}}) \} \\ &= \{ \{y,L_{max}y\} \mid y \in \mathfrak{D}(L_{max}),\ \alpha(\ell)y(0) + \beta(\ell)y'(0) = 0 \}. \end{aligned}$$

It is easy to verify that $T(\bar{\ell}) = T(\ell)^*$, $\ell \in \tilde{\mathfrak{D}}$, and hence using (1.4) we find that for $\ell \in \mathfrak{D}^*$

$$T(\ell) = \{ \{y,L_{max}y\} \mid y \in \mathfrak{D}(L_{max}),\ \alpha(\ell)y(0) + \beta(\ell)y'(0) = 0 \}.$$

Thus this equality holds for all $\ell \in \tilde{\mathfrak{D}}$. On the other hand we have

$$(T(\ell)-\ell)^{-1} = P(A-\ell)^{-1}|_{L^2(0,b)}, \quad \ell \in \tilde{\mathfrak{D}} \cap \rho(A),$$

and this relation shows that $(T(\ell)-\ell)^{-1}$ exists as a bounded operator on $L^2(0,b)$. Putting $R(\ell) = (T(\ell)-\ell)^{-1}$, $\ell \in \tilde{\mathfrak{D}}$, we see that $y = R(\ell)f$, $\ell \in \tilde{\mathfrak{D}}$ and $f \in L^2(0,b)$, is the unique solution of (1.2) - (1.3) in $L^2(0,b)$ and that $R(\ell)f = P(A-\ell)^{-1}f$. This completes the proof of the main part of Theorem 2.1 *(ii)*; the statement concerning the weak uniqueness is left to the reader. □

REMARKS 1. The above proof shows that the function Θ defined by (2.2) determines the extension A of L_{min} up to a weak isomorphism. It is also true that each such A determines on each component of $\rho(A)$ a function Θ and hence the coefficients α,β appearing in the boundary condition (1.3), compare the Remark after Theorem 1.2. For further details we refer to [10].

2. In case $\alpha/\beta \in N_x$, Θ determines the minimal extension A of L_{min} in the π_x-space $\mathfrak{K}$ up to an isomorphism. With $\mu \in \mathbb{C}^+ \cap \mathfrak{D}_0$, $V = C_\mu(L_{min})$, and $U = C_\mu(A)|_{\mathfrak{K}\ominus\mathfrak{D}(V)}$ it is the characteristic function of the closely connected unitary colligation $\Delta = (\mathfrak{K} \ominus L^2(0,b), \mathfrak{D}(V)^\perp, \mathfrak{R}(V)^\perp; U)$.

The characteristic function of an isometry V in a Hilbert space $\mathfrak{H}$ is defined as the characteristic function of the unitary colligation

$$\begin{pmatrix} \tilde{V} & I_{\mathfrak{R}(V)^\perp} \\ P_{\mathfrak{D}(V)^\perp} & 0 \end{pmatrix} : \begin{pmatrix} \mathfrak{H} \\ \mathfrak{R}(V)^\perp \end{pmatrix} \longrightarrow \begin{pmatrix} \mathfrak{H} \\ \mathfrak{D}(V)^\perp \end{pmatrix},$$

where $\tilde{V}$ is the trivial linear extension of V to all of $\mathfrak{H}$:

$$\tilde{V}x = \begin{cases} Vx, & x \in \mathfrak{D}(V), \\ 0, & x \in \mathfrak{D}(V)^\perp. \end{cases}$$

It is denoted by X; hence

$$X(z) = z P_{\mathfrak{D}(V)^{\perp}}(I-z\tilde{V})^{-1}|_{\mathfrak{R}(V)^{\perp}}, \qquad z \in \mathbb{D}.$$

It coincides with the characteristic function of the simple part of V. In [10, Section 3] necessary and sufficient conditions are given which $X \in S(\mathfrak{F},\mathfrak{G})$, where $\mathfrak{F},\mathfrak{G}$ are Hilbert spaces, must satisfy so that it is the characteristic function of a simple isometry V in a Hilbert space $\mathfrak{H}$ containing $\mathfrak{F}$ and $\mathfrak{G}$ as subspaces. Under these conditions X uniquely determines the simple isometry V. Now take $\mathfrak{H} = L^2(0,b)$ and $V = C_{\mu}(L_{min})$. Then V, as is wellknown, is simple and its characteristic function is given by

$$(2.3) \qquad (X(z(\ell))\psi(\cdot,\mu))(x) = \frac{m(\ell)-m(\mu)}{m(\ell)-m(\bar{\mu})}\psi(x,\bar{\mu}), \qquad \ell \in \mathbb{C}^+,$$

(Compare [10], Section 6). In the proof of the theorem in the next section we shall make use of Remark 2 above and formula (2.3).

3. THE OPERATOR A(Ω). Let $Q \in N_{\kappa}^{n\times n}$. In [4] (see also [9], [24]) it is shown that Q admits a representation

$$(3.1) \qquad Q(\ell) = S + i(\mathrm{Im}\mu)\Gamma^*(I+z(\ell)U)(I-z(\ell)U)^{-1}\Gamma$$

where U is a unitary operator in a π_{κ}-space Π, $\Gamma:\mathbb{C}^n \longrightarrow \Pi$ a linear mapping, S a constant nxn matrix with $S = S^*$, $\mu \in \mathfrak{D}_Q \cap \mathbb{C}^+$ and $z(\ell) = (\ell-\mu)/(\ell-\bar{\mu})$. The pair Π,U can be chosen to be minimal:

$$(3.2) \qquad \Pi = \text{c.l.s. } \{ \bigcup (U-\ell)^{-1}\Gamma\mathbb{C}^n \mid \ell \in \rho(U) \}$$

in which case it is uniquely determined up to isomorphisms. Here c.l.s. stands for closed linear span. Clearly, we have $S = \mathrm{Re}\, Q(\mu)$. If Q is defined by the righthand side with given π_{κ}-space Π, and a unitary operator U in Π, $\Gamma:\mathbb{C}^n \longrightarrow \Pi$ and nxn matrix $S = S^*$, then $Q \in N_{\kappa'}^{n\times n}$ for some $\kappa' \le \kappa$ and $\kappa' = \kappa$ if the pair Π,U is minimal. Let A be the inverse Cayley transform of U at μ, $A = F_{\mu}(U)$. Then A is selfadjoint and (3.1) becomes

$$(3.3) \qquad Q(\ell) = S + \Gamma^*(\ell-\mathrm{Re}\mu+(\ell-\bar{\mu})(\ell-\mu)(A-\ell)^{-1})\Gamma$$

and (3.2) is equivalent to

$$\Pi = \text{c.l.s.}\{ \bigcup (A-\ell)^{-1}\Gamma\mathbb{C}^n \cup \Gamma\mathbb{C}^n \mid \ell \in \rho(U) \}.$$

We now briefly describe the construction of the space Π and the operators U,A and Γ, see [25]. Let $Q \in N_{\kappa}^{n\times n}$ be given. Consider the linear space $\mathfrak{L}$ of finite sums

$$\Sigma_{\ell}\, \hat{\varepsilon}_{\ell} f_{\ell}$$

where $\ell \in \mathfrak{D}_Q$, $f_{\ell} \in \mathbb{C}^n$ and $\hat{\varepsilon}_{\ell}$ is a symbol associated with each $\ell \in \mathfrak{D}_Q$ and provide $\mathfrak{L}$ with the (possibly degenerate, indefinite) inner product

$$\left[\Sigma_\ell \hat{\varepsilon}_\ell f_\ell, \Sigma_\lambda \hat{\varepsilon}_\lambda g_\lambda \right] = \Sigma_{\ell,\lambda} \left(\frac{Q(\ell)-Q(\lambda)^*}{\ell-\bar{\lambda}} f_\ell, g_\lambda \right)$$

where (,) denotes the usual inner product on $\mathbb{C}^n$. Consider the "operator" A_o in $\mathfrak{L}$, whose graph in $\mathfrak{L} \times \mathfrak{L}$ is given by

$$A_o = \{ \{ \Sigma_\ell \hat{\varepsilon}_\ell f_\ell, \Sigma_\ell \hat{\varepsilon}_\ell \ell f_\ell \} \mid \Sigma_\ell f_\ell = 0 \}$$

and its "Cayley transform" $U_o = C_\mu(A)$ with $\mu \in \mathbb{C}^+ \cap \mathfrak{D}_Q$:

$$U_o = \{ \{ \Sigma_\ell \hat{\varepsilon}_\ell (\ell-\mu) f_\ell, \Sigma_\ell \hat{\varepsilon}_\ell (\ell-\bar{\mu}) f_\ell \} \mid \Sigma_\ell f_\ell = 0 \}$$

Then A_o is a "symmetric" relation (possibly multivalued operator) in $\mathfrak{L}$, U_o is "isometric" in $\mathfrak{L}$ with "dense" domain $\mathfrak{D}(U_o) = \{ \Sigma_\ell \hat{\varepsilon}_\ell f_\ell \mid f_\mu = 0 \}$ and "dense" range $\mathfrak{R}(U_o)$, and

$$U_o \hat{\varepsilon}_\ell f = \frac{\ell-\bar{\mu}}{\ell-\mu} \hat{\varepsilon}_\ell f - \frac{\mu-\bar{\mu}}{\ell-\mu} \hat{\varepsilon}_\mu f, \quad \ell \in \tilde{\mathfrak{D}}_Q, \ \ell \neq \mu, \ f \in \mathbb{C}^n,$$

so that in particular, $U_o \hat{\varepsilon}_{\bar{\mu}} f = \hat{\varepsilon}_\mu f$. Define $\Gamma_o : \mathbb{C}^n \longrightarrow \mathfrak{L}$ by $\Gamma_o f = \hat{\varepsilon}_\mu f$, $f \in \mathbb{C}^n$. Then a straightforward calculation shows that

$$(3.4) \qquad Q(\ell)f = \operatorname{Re} Q(\mu)f + i(\operatorname{Im}\mu)\Gamma_o^*(I+z(\ell)U_o)(I-z(\ell)U_o)^{-1}\Gamma_o f.$$

Finally let $\mathfrak{L}_o = \mathfrak{L} \cap \mathfrak{L}^\perp$. Then completion of the quotient space $\mathfrak{L}/\mathfrak{L}_o$ with respect to the inner product [,] yields the π_χ-space Π and the operators A_o, U_o and Γ_o give rise to a selfadjoint operator A in Π, a unitary operator U with $U = C_\mu(A)$ and a mapping $\Gamma : \mathbb{C}^n \longrightarrow \Pi$. The representation (3.1) now follows from (3.4). In this section we shall also need the "isometry" V_o in $\mathfrak{L}$ which is the restriction of U_o to $\langle \hat{\varepsilon}_{\bar{\mu}} \rangle^\perp$; it maps $\langle \hat{\varepsilon}_{\bar{\mu}} \rangle^\perp$ onto $\langle \hat{\varepsilon}_\mu \rangle^{\perp o}$. Here $\langle \hat{\varepsilon}_\ell \rangle = \{ \hat{\varepsilon}_\ell f \mid f \in \mathbb{C}^n \}$. In Π it gives rise to an isometry $V \subset U$ with defect spaces $\mathfrak{D}(V)^\perp = U^{-1}\Gamma\mathbb{C}^n$ and $\mathfrak{R}(V)^\perp = \Gamma\mathbb{C}^n$.

Let $m \in N_o$ and $\alpha/\beta \in N_\chi$ and assume (1.5). Then, as in Lemma 1.1, $\Omega \in N_\chi^{2\times 2}$. In what follows we have to deal with the cases $Q = m$, α/β or Ω and therefore we shall use the notations $\mathfrak{L}(Q)$, $\Pi(Q)$, $[\ , \ ]_Q$, $U_o(Q)$, $U(Q)$ etc. to denote the dependence on Q of the spaces and operators we have just considered. Note that $\Pi(m)$ is a Hilbert space as $m \in N_o$ and that the defect spaces of $V(m)$ in $\Pi(m)$, $\mathfrak{D}(V(m))^\perp$ and $\mathfrak{R}(V(m)^\perp$, are one dimensional.

PROPOSITION 3.1. *The isometric operator $V(m)$ is simple and its characteristic function X_m is given by*

$$X_m(z(\ell)) = \frac{m(\ell)-m(\mu)}{m(\ell)-m(\bar{\mu})}, \quad \ell \in \mathbb{C}^+.$$

PROOF. Since the pair $U(m)$, $\Pi(m)$ is minimal and $V(m) \subset U(m)$, $V(m)$ is simple. To determine X_m it suffices to calculate $zP_o(I-z\tilde{V}_o)^{-1}\hat{\varepsilon}_\mu x$, where $x \in \mathbb{C}$,

$$\tilde{V}_o x = \begin{cases} V_o(m)x, & x \in \mathfrak{D}(V_o(m)), \\ 0 & , \ x \in \mathfrak{D}(V_o(m))^\perp = \langle \hat{\varepsilon}_{\bar{\mu}} \rangle, \end{cases}$$

P_o is the projection onto $\langle \hat{\varepsilon}_{\bar{\mu}} \rangle$ in $\mathfrak{L}(m)$ and $z(\ell) = (\ell-\mu)/(\ell-\bar{\mu})$. Then a straightforward calculation shows that for $\ell \neq \mu$:

$$P_o\hat{\varepsilon}_\ell x = \hat{\varepsilon}_{\bar{\mu}} c(\ell)x \quad , \quad c(\ell) = \frac{\mu-\bar{\mu}}{\ell-\mu}(m(\mu)-m(\bar{\mu}))^{-1}(m(\ell)-m(\mu))$$

and

$$\tilde{V}_o\hat{\varepsilon}_\ell\, x = U_o(m)(I-P_o)\hat{\varepsilon}_\ell x = \hat{\varepsilon}_\ell\, z^{-1}\, x + \hat{\varepsilon}_\mu(1-z^{-1}-c(\ell))x \ .$$

That implies that

$$zP_o(I-z\tilde{V})^{-1}\hat{\varepsilon}_\mu x = zP_o\hat{\varepsilon}_\ell(zc(\ell)+1-z)^{-1}x = \hat{\varepsilon}_{\bar{\mu}} zc(\ell)(zc(\ell)+1-z)^{-1}x$$

$$= \hat{\varepsilon}_{\bar{\mu}}(m(\ell)-m(\mu))(m(\ell)-m(\bar{\mu}))^{-1}x,$$

from which the proposition follows. □

PROPOSITION 3.2. *The $\pi_\varkappa$-space $\Pi(\Omega)$ of index $\varkappa$ can be identified with the orthogonal sum of the $\pi_\varkappa$-space $\Pi(\alpha/\beta)$ of index $\varkappa$ and the Hilbert space $\Pi(m)$:*

$$\Pi(\Omega) = \begin{bmatrix} \Pi(\alpha/\beta) \\ \Pi(m) \end{bmatrix} .$$

With this identification $U(\Omega)$ is a unitary extension of $V(m)$. If $U(\Omega)$ has the matrix representation

$$U(\Omega) = \begin{bmatrix} T & F & 0 \\ G & H & 0 \\ 0 & 0 & V(m) \end{bmatrix} : \begin{bmatrix} \Pi(\alpha/\beta) \\ \mathfrak{D}(V(m))^\perp \\ \mathfrak{D}(V(m)) \end{bmatrix} \longrightarrow \begin{bmatrix} \Pi(\alpha/\beta) \\ \mathfrak{R}(V(m))^\perp \\ \mathfrak{R}(V(m)) \end{bmatrix} ,$$

then $\Delta = (\Pi(\alpha/\beta), \mathfrak{D}(V(m))^\perp, \mathfrak{R}(V(m))^\perp; T,F,G,H)$ is a closely connected unitary colligation with characteristic function

$$\Theta_\Delta(z(\ell)) = \frac{\beta(\ell)-m(\bar{\mu})\alpha(\ell)}{\beta(\ell)-m(\mu)\alpha(\ell)} , \qquad \ell \in \mathbb{C}^+,$$

PROOF. We consider the space $\mathfrak{L}(\Omega)$, spanned by elements of the form $\hat{\varepsilon}_\ell \begin{bmatrix} x \\ y \end{bmatrix}$ with $\ell \in \mathfrak{D}_\Omega$, $x,y \in \mathbb{C}$. Each such element can be written as

$$\hat{\varepsilon}_\ell \begin{bmatrix} x \\ y \end{bmatrix} = \hat{\varepsilon}_\ell\, \Omega(\ell)^{-1}\Omega(\ell)\begin{bmatrix} x \\ y \end{bmatrix} = \delta_\ell \frac{m(\ell)\beta(\ell)x+\beta(\ell)y}{n(\ell)} + \varepsilon_\ell \frac{\beta(\ell)x+\alpha(\ell)y}{n(\ell)} ,$$

where $\delta_\ell = \hat{\varepsilon}_\ell \begin{bmatrix} -\alpha(\ell)/\beta(\ell) \\ 1 \end{bmatrix}$, $\varepsilon_\ell = \hat{\varepsilon}_\ell \begin{bmatrix} 1 \\ -m(\ell) \end{bmatrix}$ and $n(\ell) = \beta(\ell)-m(\ell)\alpha(\ell)$. We note that for $x,y \in \mathbb{C}$

$$[\delta_\ell x, \delta_\lambda y]_\Omega = \left(\frac{\alpha(\ell)/\beta(\ell)-(\alpha(\lambda)/\beta(\lambda))^*}{\ell-\bar{\lambda}} x, y \right) ,$$

$$[\delta_\ell x, \varepsilon_\lambda y]_\Omega = 0 ,$$

$$[\varepsilon_\ell x, \varepsilon_\lambda y]_\Omega = \left(\frac{m(\ell)-m(\bar\lambda)}{\ell-\bar\lambda} x, y \right).$$

This implies that $\mathfrak{L}(\alpha/\beta)$ ($\mathfrak{L}(m)$) can be identified with the linear subspaces spanned by the δ_ℓ's (ε_ℓ's, respectively) and that $\mathfrak{L}(\Omega)$ is the orthogonal sum of $\mathfrak{L}(\alpha/\beta)$ and $\mathfrak{L}(m)$:

$$\mathfrak{L}(\Omega) = \begin{bmatrix} \mathfrak{L}(\alpha/\beta) \\ \mathfrak{L}(m) \end{bmatrix}.$$

From

$$\begin{aligned} U_0(\Omega) &= \{ \{ \Sigma_\ell \hat\varepsilon_\ell(\ell-\mu)f_\ell, \Sigma_\ell \hat\varepsilon_\ell(\ell-\bar\mu)f_\ell \} \mid \Sigma_\ell f_\ell = 0 \} \\ &\supset \{ \{ \Sigma_\ell \varepsilon_\ell(\ell-\mu)x_\ell, \Sigma_\ell \varepsilon_\ell(\ell-\bar\mu)x_\ell \} \mid \Sigma_\ell \begin{pmatrix} 1 \\ -m(\ell) \end{pmatrix} x_\ell = 0 \} \\ &= \{ \{ \Sigma_\ell \varepsilon_\ell x_\ell, \Sigma_\ell \varepsilon_\ell \frac{\ell-\bar\mu}{\ell-\mu} x_\ell \} \mid x_\mu = 0, [\Sigma_\ell \varepsilon_\ell x_\ell, \varepsilon_{\bar\mu} x]_m = 0, x \in \mathbb{C} \} \\ &= V_0(m), \end{aligned}$$

we see that $U_0(\Omega)$ is an extension of $V_0(m)$. From these results the first part of the proposition easily follows. To prove the second part we recall that

$$(3.5) \qquad U_0(\Omega)\hat\varepsilon_\ell f = \hat\varepsilon_\ell \frac{\ell-\bar\mu}{\ell-\mu} f - \hat\varepsilon_\mu \frac{\mu-\bar\mu}{\ell-\mu} f, \quad \ell \neq \mu,\ f \in \mathbb{C}^2.$$

From this we find that for $\ell \neq \mu$ and $x, y \in \mathbb{C}$

$$(3.6) \qquad U_0(\Omega) \begin{pmatrix} \delta_\ell x \\ \varepsilon_\ell y \end{pmatrix} = U_0(\Omega)\, \hat\varepsilon_\ell \begin{pmatrix} y-(\alpha(\ell)/\beta(\ell))x \\ x-m(\ell)y \end{pmatrix} = \begin{pmatrix} \delta_\ell \frac{\ell-\bar\mu}{\ell-\mu} x - \delta_\mu \frac{\mu-\bar\mu}{\ell-\mu} u \\ \varepsilon_\ell \frac{\ell-\bar\mu}{\ell-\mu} y - \varepsilon_\mu \frac{\mu-\bar\mu}{\ell-\mu} v \end{pmatrix}$$

where,

$$u = \frac{\beta(\mu)(\beta(\ell)-m(\mu)\alpha(\ell))}{\beta(\ell)n(\mu)} x + \frac{\beta(\mu)(m(\mu)-m(\ell))}{n(\mu)} y ,$$

$$v = \frac{\beta(\ell)\alpha(\mu)-\alpha(\ell)\beta(\mu)}{\beta(\ell)n(\mu)} x + \frac{\beta(\mu)-m(\ell)\alpha(\mu)}{n(\mu)} y .$$

The inclusion $V_0(m) \subset U_0(\Omega)$ yields a matrix representation for $U_0(\Omega)$:

$$U_0(\Omega) = \begin{pmatrix} T_0 & F_0 & 0 \\ G_0 & H_0 & 0 \\ 0 & 0 & V_0(m) \end{pmatrix} : \begin{pmatrix} \mathfrak{L}(\alpha/\beta) \\ \langle \varepsilon_{\bar\mu} \rangle \\ \langle \varepsilon_{\bar\mu} \rangle^\perp \end{pmatrix} \to \begin{pmatrix} \mathfrak{L}(\alpha/\beta) \\ \langle \varepsilon_\mu \rangle \\ \langle \varepsilon_\mu \rangle^\perp \end{pmatrix}$$

where, as follows from formula (3.6)

$$\begin{cases} T_o\delta_\ell x = \delta_\ell \dfrac{\ell-\bar\mu}{\ell-\mu}\, x - \delta_\mu \dfrac{\mu-\bar\mu}{\ell-\mu}\, \dfrac{\beta(\mu)(\beta(\ell)-m(\mu)\alpha(\ell))}{\beta(\ell)n(\mu)}\, x\,, \\[2ex] G_o\delta_\ell x = \varepsilon_\mu \left(-\dfrac{\mu-\bar\mu}{\ell-\mu}\right) \dfrac{\beta(\ell)\alpha(\mu)-\alpha(\ell)\beta(\mu)}{\beta(\ell)n(\mu)}\, x\,, \\[2ex] F_o\varepsilon_{\bar\mu} y = \delta_\mu \dfrac{\beta(\mu)(m(\mu)-m(\bar\mu))}{n(\mu)}\, y\,, \\[2ex] H_o\varepsilon_{\bar\mu} y = \varepsilon_\mu \dfrac{\beta(\mu)-m(\bar\mu)\alpha(\mu)}{n(\mu)}\, y. \end{cases}$$

A straightforward calculation shows that

$$\Theta_o(z(\ell))\varepsilon_{\bar\mu} x = (H_o+z(\ell)G_o(I-z(\ell)T_o)^{-1}F_o)\varepsilon_{\bar\mu} x = \varepsilon_\mu \frac{\beta(\ell)-m(\bar\mu)\alpha(\ell)}{\beta(\ell)-m(\mu)\alpha(\ell)}\, x$$

and from this the formula for the characteristic function of the unitary colligation Δ in the proposition follows. To prove that Δ is closely connected it suffices to show that $\Pi(\alpha/\beta)$ does not contain a nonzero subspace invariant under $U(\Omega)$ (see [8], Proposition 3.2) and this follows from the fact that

$$\text{(3.7)} \qquad \Pi(\Omega) = \text{c.l.s.}\{\, \bigcup U(\Omega)^n(\mathfrak{D}(V(m))^\perp + \mathfrak{R}(V(m))^\perp) \mid n \in \mathbb{Z} \,\}.$$

In order to see (3.7) we note that, by definition, $\Gamma_o(\Omega)\mathbb{C}^2 = \langle \hat\varepsilon_\mu \rangle = \{\, \delta_\mu x + \varepsilon_\mu y \mid x,y \in \mathbb{C} \,\}$ and, that on account of (3.5), $U_o(\Omega)\varepsilon_{\bar\mu} u = \delta_\mu v + \varepsilon_\mu w$ for some nonzero v and $w \in \mathbb{C}$, provided $u \in \mathbb{C}$ is nonzero. This implies that

$$\Gamma_o(\Omega)\mathbb{C}^2 = \{\, U_o(\Omega)\varepsilon_{\bar\mu} x + \varepsilon_\mu y \mid x,y \in \mathbb{C} \,\}\,,$$

cf. [14, Theorem 3]. Now (3.7) follows from the minimality of $\Pi(\Omega), U(\Omega)$, see (3.2), and the fact that $\langle \varepsilon_{\bar\mu} \rangle$ and $\langle \varepsilon_\mu \rangle$ give rise to the spaces $\mathfrak{D}(V(m))^\perp$ and $\mathfrak{R}(V(m))^\perp$, respectively. This completes the proof. □

THEOREM 3.3. *Let* m,α,β *and* Ω *be given as in Section* 1. *Assume* $\mathfrak{D}_{\alpha,\beta} = \mathbb{C}\backslash\mathbb{R}$ *and* $\alpha/\beta \in N_\kappa$, *and fix some* $\mu \in \mathfrak{D}_o$. *Then*

(i) The operator L_{min} *in* $L^2(0,b)$ *is isomorphic to the inverse Cayley transform* $F_\mu(V(m))$ *in* $\Pi(m)$, *and*

(ii) The minimal selfadjoint extension A *of* L_{min} *in the* π_κ*-space* $\mathfrak{K}$ *in the representation* (1.7) *is isomorphic to* $A(\Omega) = F_\mu(U(\Omega))$ *in* $\Pi(\Omega)$ *in the representation* (3.3) *for* $Q = \Omega$.

PROOF. It follows from (2.3) and Proposition 3.1 that $C_\mu(L_{min})$ and $V(m)$ have the same characteristic function. As both operators are simple, *(i)* follows. Similarly *(ii)* follows from Remark 2 after the proof of Theorem 1.2 *(ii)* and Proposition 3.2. □

4. THE OPERATOR A(ω). In this section we shall suppose that α and β in the boundary condition (1.3) are polynomials satisfying (1.4) now for all $\ell \in \mathbb{C}$. (We could even allow α, β to be a certain pair of entire functions, see for instance [28]). Then the kernel

$$N_{\alpha,\beta}(\ell,\lambda) = \frac{\alpha(\ell)\overline{\beta(\lambda)}-\overline{\alpha(\lambda)}\beta(\ell)}{\ell-\bar{\lambda}}$$

has a finite number of positive and negative squares, and hence the minimal selfadjoint operator A in Theorem 1.2 *(iii)* is a finite dimensional extension of L_{min} : $\dim \mathfrak{K} \ominus L^2(0,b)$ equals the sum of the positive and negative squares of $N_{\alpha,\beta}$. According to the second condition in (1.4), α and β have no common divisor and therefore there exist polynomials γ, δ such that

$$\alpha(\ell)\gamma(\ell) + \beta(\ell)\delta(\ell) = 1, \quad \ell \in \mathbb{C}. \tag{4.1}$$

We choose γ and δ so that $\deg\delta \le \deg\alpha$ and $\deg\gamma \le \deg\beta$. Furthermore, if

$$\alpha(\ell) = c \prod_{j=0}^{\deg\alpha} (\ell-\alpha_j), \quad \beta(\ell) = d \prod_{k=0}^{\deg\beta} (\ell-\beta_k), \qquad c,d \in \mathbb{C},$$

then the relation $\alpha(\bar{\ell})/\beta(\bar{\ell}) = \overline{\alpha(\ell)}/\overline{\beta(\ell)}$ implies that the α_j and β_k are real or appear in conjugate pairs, and that c/d is real. Hence we can assume that α and β satisfy $\alpha(\bar{\ell}) = \overline{\alpha(\ell)}$, $\beta(\bar{\ell}) = \overline{\beta(\ell)}$, $\ell \in \mathbb{C}$. Then the Euclidean algorithm shows that γ and δ have the same property. In the sequel we shall always suppose therefore that α,β,γ and δ have real coefficients.

We introduce the solutions $\varphi(.;\ell)$ and $\theta(.;\ell)$ of (1.1) satisfying the initial conditions

$$\varphi(0;\ell) = \beta(\ell), \quad \varphi'(0;\ell) = -\alpha(\ell),$$
$$\theta(0;\ell) = -\gamma(\ell), \quad \theta'(0;\ell) = -\delta(\ell).$$

That is, $\varphi(.;\ell)$ satisfies the boundary conditions (1.3). Clearly, $\varphi = \beta u_1+\alpha u_2$, $\theta = -\gamma u_1+\delta u_2$, $\varphi\theta'-\theta\varphi' \equiv -1$ on $[0,b) \times \mathbb{C}$, and the solution $\psi \in L^2(0,b)$ of (1.1) can be written as

$$\psi(x;\ell) = n(\ell)(\theta(x;\ell) + \omega(\ell)\varphi(x;\ell))$$

with

$$\omega(\ell) = (\gamma(\ell)+m(\ell)\delta(\ell))(\beta(\ell)-m(\ell)\alpha(\ell))^{-1}, \quad \ell \in \mathbb{C}\backslash\mathbb{R}. \tag{4.2}$$

Hence $\theta(.;\ell) + \omega(\ell)\varphi(.;\ell) \in L^2(0,b)$, $\ell \in \mathbb{C}\backslash\mathbb{R}$. We find for $f \in L^2(0,b)$, vanishing near b, that $R(\ell)f$ in (1.6) can be rewritten as

$$(R(\ell)f)(x) = \omega(\ell)\varphi(x;\ell) \int_0^b \varphi(t;\ell)f(t)dt + \tag{4.3}$$
$$+ \varphi(x;\ell) \int_x^b \theta(t,\ell)f(t)dt + \theta(x,\ell) \int_0^x \varphi(t;\ell)f(t)dt.$$

Hence ω is the Weyl coefficient corresponding to the boundary condition (1.3) and the Fourier transform

$$f \longmapsto \int_0^b \varphi(t;.)f(t)dt.$$

At this point let us briefly review the classical situation, i.e. for $\zeta \in \mathbb{R}$ we consider solutions $\varphi(.,\ell;\zeta)$ and $\theta(.,\ell;\zeta)$ by

$$\varphi(0,\ell;\zeta) = \cos\zeta, \qquad \varphi'(0,\ell;\zeta) = \sin\zeta,$$
$$\theta(0,\ell;\zeta) = \sin\zeta, \qquad \theta'(0,\ell;\zeta) = -\cos\zeta.$$

Then $\theta(.,\ell;\zeta) + \omega(\ell;\zeta)\varphi(.,\ell;\zeta)$ is (up to scalar multiples) the only $L^2(0,b)$ solution, where

$$\omega(\ell;\zeta) = \frac{m(\ell)\cos\zeta-\sin\zeta}{m(\ell)\sin\zeta+\cos\zeta}$$

is the Weyl coefficient corresponding to the boundary condition

$$y(0)\sin\zeta - y'(0)\cos\zeta = 0$$

and the Fourier transform

$$f \mapsto \int_0^b f(t)\varphi(t,.;\zeta)dt.$$

It is clear that for each $\zeta\in\mathbb{R}$, $\omega(.;\zeta) \in N_0$. We have the representation

$$\omega(\ell;\zeta) = \int_{\mathbb{R}} \frac{d\rho(t;\zeta)}{t-\ell} - \tan\zeta, \qquad \zeta \neq \tfrac{1}{2}\pi \pmod{\pi},$$

with a non-decreasing function $\rho(.;\zeta)$, that satisfies

$$|\rho(t;\zeta)| \le C_1(\zeta) + C_2(\zeta)|t|^{\frac{1}{2}}, \qquad C_1(\zeta),\ C_2(\zeta) > 0,$$

and a similar, but different result for $\zeta = \frac{1}{2}\pi(\mathrm{mod}\,\pi)$, see for instance [15, Theorem 10.2.4 and Exercise 10.2.14]. A direct consequence of the above is

$$\omega(\ell;\eta) = \frac{\omega(\ell;\zeta)\cos(\zeta-\eta) + \sin(\zeta-\eta)}{\cos(\zeta-\eta) - \omega(\ell;\zeta)\sin(\zeta-\eta)}, \qquad \zeta,\eta \in \mathbb{R},$$

see [3, p.257, problem 8]; in particular $\omega(\ell;\zeta+\frac{1}{2}\pi) = -(\omega(\ell;\zeta))^{-1}$. We obtain

$$\lim_{y\to\infty} m(iy) = 0, \qquad \lim_{y\to\infty} |ym(iy)| = \infty,$$

by using the results in [20], cf. similar results in [21,p.76].

LEMMA 4.1. *Let α,β,γ and δ be polynomials with real coefficients such that $\alpha/\beta \in N_x$, $\alpha\gamma+\beta\delta=1$, $\deg\delta \le \deg\alpha$ and $\deg\gamma \le \deg\beta$. Let $m \in N_0$ be arbitrary and assume* (1.5). *Then ω defined by* (4.2) *belongs to $N_{x'}$ for some $x'\le x$.*

PROOF. We put $N = \{ \bigcup N_x \mid x\ge 0 \}$ In order to prove that $\omega \in N$ we first note that if $Q \in N_\varkappa$ and $Q' \in N_{\varkappa'}$ then $Q + Q' \in N_\tau$ where $\tau \le \varkappa+\varkappa'$ and $-Q^{-1} \in N_\varkappa$, if $Q \not\equiv 0$. Now, ω can be written as

$$\omega = -\delta/\alpha - \alpha^{-2}(m-\beta/\alpha)^{-1}. \tag{4.4}$$

Clearly, $-\delta/\alpha \in N$ and $m - \beta/\alpha \in N_{\varkappa''}$ for some $\varkappa'' \le \varkappa$. It follows for instance from the integral representations of functions in the class N (see [26]) that also $\alpha^2(m-\beta/\alpha) \in N$. Hence $\omega \in N$, i.e. $\omega \in N_{\varkappa'}$. In order to show that $\varkappa' \le \varkappa$ we first assume that m is a meromorphic function and that no pole of m is a zero of α. Then we claim that ω and $m-\beta/\alpha$ belong to the same class $N_{\varkappa''}$. This follows from the fact that at a pole ℓ_0 of ω we have $\alpha(\ell_0) \ne 0$ (otherwise, as $\alpha(\ell_0)m(\ell_0)-\beta(\ell_0)=0$, $\beta(\ell_0)=0$). Hence by (4.4) the main parts of the Laurent series of ω and $-(m-\beta/\alpha)^{-1}$ at ℓ_0 coincide up to some factor $\alpha(\ell_0)^{-2}$ which is positive when ℓ_0 is real, and this means that each such pole gives rise to the same number of negative squares to the kernel N_ω and $N_{m-\beta/\alpha}$. Hence $\omega \in N_{\varkappa''}$. If the function m is not meromorphic we can approximate it by a sequence of meromorphic functions $m_n \in N_0$, which are such that this convergence holds uniformly on some open set $\mathfrak{D}$ having a positive distance from the real axis and being symmetric with respect to the real axis (Consider, e.g., the integral representation of m and approximate the "spectral measure" of m by a stepfunction). If $\omega_n = (\gamma+m_n\delta)(\beta-m_n\alpha)^{-1}$ then $\omega_n \longrightarrow \omega$ uniformly on $\mathfrak{D}$. Hence the number of negative squares of ω is less than or equal to the number of negative squares of ω_n. Therefore $\varkappa' \le \varkappa$. □

REMARK. If in Lemma 4.1 the function m has the properties

$$\lim_{y\to\infty} m(iy) = 0 , \quad \lim_{y\to\infty} y|m(iy)| = \infty,$$

then $\mathfrak{D}_0 \ne \emptyset$ and we have (recall $\alpha \not\equiv 0, \beta \not\equiv 0$):

(i) if $\deg\alpha\beta=0$: $\lim_{y\to\infty} \omega(iy) = \frac{\gamma}{\beta}$,

(ii)$_1$ if $\deg\alpha=1$, $\deg\beta=0$: $\lim_{y\to\infty} y\omega(iy) = -\lim_{y\to\infty} \frac{y}{\alpha(iy)\beta(iy)}$,

(ii)$_2$ if $\deg\alpha=0$, $\deg\beta=1$: $\lim_{y\to\infty} y\omega(iy) = \lim_{y\to\infty} \frac{y}{\alpha(iy)\beta(iy)}$,

(iii) if $1\le\deg\alpha\le\deg\beta$: $\lim_{y\to\infty} y^n\omega(iy) = \lim_{y\to\infty} \frac{y^n\gamma(iy)}{\beta(iy)}$, $n=1,..,\deg\beta-\deg\gamma$,

(iv) if $1\le\deg\beta<\deg\alpha$: $\lim_{y\to\infty} y^n\omega(iy) = -\lim_{y\to\infty} \frac{y^n\delta(iy)}{\alpha(iy)}$, $n=1,..,\deg\alpha-\deg\delta$,

where in the cases *(iii)* and *(iv)* we have chosen γ and δ such that $\deg\gamma < \deg\beta$ and $\deg\delta < \deg\alpha$. For the cases *(ii)* - *(iv)* we observe that $\sup \{ y|\omega(iy)| \mid y > 0 \} < \alpha$ and hence the "spectral measure" of ω is finite near ∞. This generalizes an observation made by Fulton [13]. Moreover, in all cases we have that $\lim_{y\to\infty} y^{-1}\omega(iy) = 0$ and this implies that the selfadjoint relation $A(\omega)$ in the minimal representation (3.3) of $\omega \in N_{x'}$ is actually an operator (see [25]). In particular, since the Weyl coefficient m of problem (1.1) possesses the above mentioned properties our $A(\omega)$, where ω is defined by (4.2), is an operator for which we have the representation

$$(4.5) \qquad \omega(\ell) = S - \Gamma(\omega)^*(\ell-\mathrm{Re}\mu+(\ell-\bar{\mu})(\ell-\mu)(A(\omega)-\ell)^{-1})\Gamma(\omega).$$

THEOREM 4.2. *Let* m *be the Weyl coefficient of* L, α *and* β *in* (1.3) *be polynomials satisfying* (1.4) *and let* ω *be given by* (4.2) *with polynomials* γ,δ *satisfying* (4.1). *Then* $\omega \in N_x$ *and the minimal selfadjoint extension* A *of* L_{min} *in the* π_x*-space* $\mathfrak{K}$ *of index* x *for which the representation* (1.7) *holds is isomorphic to the selfadjoint operator* $A(\omega)$ *in* $\Pi(\omega)$ *in the minimal representation* (3.2) *for* $Q = \omega$.

PROOF. Theorem 1.2 *(iii)* and (4.3) imply that for $f,g \in L^2(0,b)$ vanishing near b

$$(4.6) \qquad [(A-\ell)^{-1}f,g] = \Phi(\bar{g},\ell)\omega(\ell)\Phi(f,\ell) + \ldots$$

where $\Phi(f,\ell) = \int_0^b \varphi(t,\ell)f(t)dt$. Here and in the following +... means that the equality holds up to an entire function of ℓ. Let $\Delta = [p,q]$ be a compact interval such that p and q are not critical points of A or $A(\omega)$ and the spectral function of ω is continuous at p and q, i.e. $p,q \notin \sigma_p(A(\omega))$, see [4]. Then $p,q \notin \sigma_p(A)$, otherwise there would exist elements x,y such that for example $(\ell-p)[(A-\ell)^{-1}x,y]$ tends to a number $\neq 0$ as $\ell \to a$, whereas $(\ell-p)$ times the expression on the righthand side of (4.5) tends to zero as $\ell \to p$. In the following we denote by $E(\Delta)$ ($E_\omega(\Delta)$) the spectral projection of A ($A(\omega)$, respectively) corresponding to Δ, see [23], [29], and write $A^\Delta(\omega)$ for the bounded selfadjoint operator

$$A(\omega)|_{E_\omega(\Delta)\Pi(\omega)}.$$

From Lemma 4.1 we have $\omega \in N_{x'}$ for some $x' \le x$. We decompose ω and write $\omega = \omega_0+\omega_1$, with

$$(4.7) \qquad \omega_0(\ell) = \Gamma_0^*(A^\Delta(\omega)-\ell)^{-1}\Gamma_0 \quad , \quad \Gamma_0 = (A(\omega)-\mu)E(\Delta)\Gamma(\omega)$$

and, $\omega_1(\ell) = \omega(\ell)-\omega_0(\ell)$, where $\omega(\ell)$ is given by (4.5) so that

$$\omega_j \in N_{x_j}, \quad j = 0,1,$$

with $x_0+x_1=x'$, $\sigma(\omega_0) \subset \Delta$, $\sigma(\omega_1) \subset \overline{\mathbb{C}\setminus\Delta}$, ω_0 is holomorphic at ∞ and $\omega_0(\infty) = 0$, compare [4, Proposition 3.1]. We note that the representation (4.7) of $\omega_0(\ell)$ is minimal. If C_Δ denotes the contour $p+i\eta,p-i\eta,q-i\eta,q+i\eta,p+i\eta$ along straight lines in $\mathbb{C}$, where $\eta > 0$ is chosen so small that C_Δ surrounds no nonreal singularity of ω, we find for ζ outside of C_Δ (cf. [29]):

$$-\frac{1}{2\pi i}\int'_{C_\Delta}[(A-\ell)^{-1}f,g](\ell-\zeta)^{-1}d\ell = [(A-\zeta)^{-1}E(\Delta)f,E(\Delta)g]$$

and

$$-\frac{1}{2\pi i}\int'_{C_\Delta}\Phi(\bar{g},\ell)\ \omega(\ell)\Phi(f,\ell)(\ell-\zeta)^{-1}d\ell = -\frac{1}{2\pi i}\int'_{C_\Delta}\frac{\Phi(\bar{g},\ell)-\Phi(\bar{g},\zeta)}{\ell-\zeta}\ \omega(\ell)\Phi(f,\ell)d\ell\ +$$

$$-\frac{1}{2\pi i}\int'_{C_\Delta}\Phi(\bar{g},\zeta)\omega(\ell)\ \frac{\Phi(f,\ell)-\Phi(f,\zeta)}{\ell-\zeta}\ d\ell$$

$$-\frac{1}{2\pi i}\int'_{C_\Delta}\Phi(\bar{g},\zeta)\ \frac{\omega(\ell)}{\ell-\zeta}\ \Phi(f;\zeta)d\ell$$

$$= \Phi(\bar{g},\zeta)\omega_o(\zeta)\Phi(f,\zeta)\ +\ldots$$

and hence (4.6) implies

$$(4.8)\qquad [(A-\ell))^{-1}E(\Delta)f,E(\Delta)g] = \Phi(\bar{g},\ell)\omega_o(\ell)\Phi(f,\ell)\ +\ldots$$

Combining (4.7) and (4.8) we find that

$$(4.9)\qquad [(A-\ell)^{-1}E(\Delta)f,E(\Delta)g] = \Phi(\bar{g},\ell)\Gamma_o^*(A^\Delta(\omega)-\ell)^{-1}\Gamma_o\Phi(f,\ell)$$

$$= \Phi(\bar{g},\ell)\ [(A^\Delta(\omega)-\ell)^{-1}u_o,u_o]\Phi(f,\ell)$$

where $\Gamma_o\Phi(f,\ell) = \Phi(f,\ell)u_o$, u_o being some generating element in $E_\omega(\Delta)\Pi(\omega)$. Multiplying both sides of (4.9) by ℓ^{n+m} and integrating them along C_Δ we obtain

$$(4.10)\qquad [A^mE(\Delta)f,A^nE(\Delta)g] = [(A^\Delta(\omega))^m\Phi(f,A^\Delta(\omega))u_o,(A^\Delta(\omega))^n\Phi(g,A^\Delta(\omega))u_o].$$

The minimality of the space $\mathfrak{K}$ implies that

$$(4.11)\qquad E(\Delta)\mathfrak{K} = \text{c.}\ell\text{.s.}\{\ A^mE(\Delta)f\ \mid\ f\in L^2(0,b)\ \text{vanishing near } b,\ m = 0,1,.\ \}.$$

Also we have that

$$(4.12)\qquad E_\omega(\Delta)\Pi(\omega) = \text{c.}\ell\text{.s.}\{\ (A^\Delta(\omega))^m\Phi(f,A^\Delta(\omega))u_o\ \mid\ f\in L^2(0,b)\ \text{vanishing near } b,\ m = 0,1,.\ \}.$$

In order to see this, we observe that for a δ_o-sequence of smooth elements $f_n \in L^2(0,b)$ which vanish outside some fixed neighbourhood of 0 we have

$$\Phi(f_n,\ell)\longrightarrow\beta(\ell),\quad \Phi(f_n',\ell)\longrightarrow\alpha(\ell),\qquad \text{as } n\longrightarrow\infty,$$

uniformly in ℓ on compact subsets of $\mathbb{C}$. Hence in the norm topology of operators on $E(\Delta)\Pi(\omega)$

$$\Phi(f_n,A^\Delta(\omega))\longrightarrow\beta(A^\Delta(\omega))\ ,\ \Phi(f_n',\ell)\longrightarrow\alpha(A^\Delta(\omega)),$$

and hence the space on the righthand side of (4.12) contains the elements

$(A^{\Delta}(\omega))^{m}\alpha(A^{\Delta}(\omega))u_{o}$ and $(A^{\Delta}(\omega))^{m}\beta(A^{\Delta}(\omega))u_{o}$, therefore also the element $\gamma(A^{\Delta}(\omega))\alpha(A^{\Delta}(\omega))u_{o} + \delta(A^{\Delta}(\omega))\beta(A^{\Delta}(\omega))u_{o} = u_{o}$. Now (4.12) follows from the fact that u_{o} is a generating element of $E(\Delta)\Pi(\omega)$. We define the mapping V_{Δ} by

$$V_{\Delta}A^{m}E(\Delta)f = (A^{\Delta}(\omega))^{m}\Phi(f,A^{\Delta}(\omega))u_{o}.$$

According to (4.10), (4.11) and (4.12) it is well defined and can be extended by continuity to an isomorphism between the spaces $E(\Delta)\mathfrak{K}$ and $E_{\omega}(\Delta)\Pi(\omega)$ and it follows that

$$A|_{E(\Delta)\mathfrak{K}} \quad \text{and} \quad A^{\Delta}(\omega)$$

are isomorphic. As Δ can be made arbitrarily large we see that

$$A|_{E(\mathbb{R})\mathfrak{K}} \quad \text{and} \quad A(\omega)|_{E_{\omega}(\mathbb{R})\Pi(\omega)}$$

are isomorphic. It follows from (4.5) and (4.6) that the nonreal spectra of A and $A(\omega)$ coincide and similar but simpler arguments as above show that also the corresponding algebraic eigenspaces of A and $A(\omega)$ are isomorphic. Hence A in $\mathfrak{K}$ and $A(\omega)$ in $\Pi(\omega)$ are isomorphic. In particular, the indices of the π_{κ}-spaces $\mathfrak{K}$ and $\Pi(\omega)$ are equal, i.e., $\omega \in N_{\kappa}$. □

5. **THE EIGENVALUES OF A AND CORRESPONDING CHAINS.** The isomorphism between A (the selfadjoint extension of L_{min} corresponding to the boundary condition (1.3)), $A(\Omega)$ and $A(\omega)$ implies in particular, that the spectra of these operators (and also the sets of eigenvalues, continuous spectra) coincide. Here we shall formulate some consequences with respect to $A(\omega)$, if α and β are polynomials such that $\alpha/\beta \in N_{\kappa}$ as in Section 4. We will assume this last condition in the rest of the section. As the extending space $\mathfrak{K}$ is a Pontryagin space of index κ, the operator A has an invariant nonpositive subspace $\mathfrak{L}_{o}$ of dimension κ, see [17]. Now $\lambda \in \mathbb{C}^{+}$ ($\lambda \in \mathbb{R}\cup\{\infty\}$) is a pole (a generalized pole of negative type) of ω of multiplicity k if and only if λ is an eigenvalue of A, restricted to $\mathfrak{L}_{o}$, of algebraic multiplicity k, see [27]. Evidently the nonreal eigenvalues of A, restricted to $\mathfrak{L}_{o}$, are just poles of ω, however its real eigenvalues need not be isolated. We introduce the function $\omega_{1} = -\omega^{-1}$, and we note that $\lambda \in \mathbb{C}^{+}$($\lambda \in \mathbb{R}\cup\{\infty\}$) is a zero (a generalized zero of negative type) of ω_{1} of multiplicity k if and only if λ is a pole (a generalized pole of negative type) of ω of algebraic multiplicity k, see [27, Proposition 3.3]. We will use the characterization of generalized zeros of N_{κ} functions as given in [30, Theorem 3.2]: if $\lambda \in \mathbb{R}$ is a generalized zero of negative type of ω_{1} of multiplicity k, then

$$\hat{\lim_{\ell\to\lambda}} \, (\ell-\lambda)^{1-2k}\omega_{1}(\ell)$$

exists and is finite and nonpositive, while

$$\hat{\lim_{\ell\to\lambda}}\ (\ell-\lambda)^{-1-2k}\ \omega_1(\ell)$$

exists and is positive ($\leq\infty$). Here the symbol ^ indicates that the limit is nontangential from the upper half plane. In the following theorem we shall describe the projections of the corresponding chains of eigenvalues $\lambda\in\mathbb{R}$ of A, restricted to $\mathfrak{L}_0$ in terms of the function $\varphi(.,\lambda)$ and its derivatives with respect to λ. Therefore we note that the function $\varphi(.,\ell)$, $\ell\in\mathbb{C}$, is a solution of the differential equation $(L-\ell)y = 0$, which is uniquely determined by the initial conditions

$$\varphi(0,\ell) = \beta(\ell), \qquad \varphi'(0,\ell) = -\alpha(\ell).$$

Hence $\varphi(.,\ell)$ satisfies

$$\alpha(\ell)\varphi(0,\ell) + \beta(\ell)\varphi'(0,\ell) = 0, \qquad \ell\in\mathbb{C}.$$

It is clear that

$$(L-\ell)\ \frac{1}{j!}\ \left(\frac{\partial}{\partial\ell}\right)^j\varphi(.,\ell) = \frac{1}{(j-1)!}\ \left(\frac{\partial}{\partial\ell}\right)^{j-1}\varphi(.,\ell),$$

and

$$\sum_{m=0}^{j}\left[\ \frac{1}{m!}\left(\left(\frac{\partial}{\partial\ell}\right)^m\alpha(\ell)\right)\frac{1}{(j-m)!}\left(\frac{\partial}{\partial\ell}\right)^{j-m}\varphi(0,\ell)\right.$$
$$\left.+\ \frac{1}{m!}\left(\left(\frac{\partial}{\partial\ell}\right)^m\beta(\ell)\right)\frac{1}{(j-m)!}\left(\frac{\partial}{\partial\ell}\right)^{j-m}\varphi'(0,\ell)\ \right] = 0,$$

where $\left(\frac{\partial}{\partial\ell}\right)^{-1}$ is interpreted as the zero operator, and $\ell\in\mathbb{C}$.

THEOREM 5.1. *Let* $\lambda\in\mathbb{R}$ *be a generalized zero of negative type of* ω_1 *of order* k.

(i) *If* $\hat{\lim_{\ell\to\lambda}}\ (\ell-\lambda)^{-1-2k}\omega_1(\ell) = \infty$, *then*

$$\left(\frac{\partial}{\partial\ell}\right)^j\varphi(.,\ell)\big|_{\ell=\lambda}, \qquad j = 0,\ldots,k-1,$$

belong to $L^2(0,b)$ *and there exist unique* $\tilde{y}_0,\tilde{y}_1,\ldots,\tilde{y}_{k-1}\in\mathfrak{D}(A)$, *such that* (*with* $\tilde{y}_{-1} = 0$) *for* $i,j = 0,1,\ldots,k-1$,

$$\begin{cases}(A-\lambda)\tilde{y}_j = \tilde{y}_{j-1},\\[2mm] P\tilde{y}_j = \dfrac{1}{j!}\ \left(\dfrac{\partial}{\partial\ell}\right)^j\varphi(.,\ell)\big|_{\ell=\lambda},\\[2mm] [\tilde{y}_i,\tilde{y}_j] = 0, \qquad\qquad i+j\neq 2k-2,\end{cases}$$

and

$$[\tilde{y}_{k-1},\tilde{y}_{k-1}] = \hat{\lim_{\ell\to\lambda}}\ (\ell-\lambda)^{1-2k}\omega_1(\ell) \leq 0.$$

(ii) If $\hat{\lim}_{\ell\to\lambda} (\ell-\lambda)^{-1-2k}\omega_1(\ell) < \infty$, *then*

$$\left(\frac{\partial}{\partial\ell}\right)^j \varphi(.,\ell)\Big|_{\ell=\lambda}, \qquad j = 0,\dots,k,$$

belong to $L^2(0,b)$ *and there exist unique* $\tilde{y}_0,\tilde{y}_1,\dots,\tilde{y}_k \in \mathfrak{D}(A)$, *such that (with* $\tilde{y}_{-1} = 0$) *for* $i,j,\dots,k$,

$$\begin{cases} (A-\lambda)\tilde{y}_j = \tilde{y}_{j-1}, \\ P\tilde{y}_j = \frac{1}{j!}\left(\frac{\partial}{\partial\ell}\right)^j \varphi(.,\ell)\Big|_{\ell=\lambda} \\ [\tilde{y}_i,\tilde{y}_j] = 0, \qquad i + j \neq 2k, \end{cases}$$

and

$$[\tilde{y}_k,\tilde{y}_k] = \hat{\lim}_{\ell\to\lambda} (\ell-\lambda)^{-1-2k}\omega_1(\ell) > 0.$$

PROOF. We will prove *(i)*. We shall first show that $\left(\frac{\partial}{\partial\ell}\right)^j \varphi(.,\ell)\big|_{\ell=\lambda}$ belong to $L^2(0,b)$, $j = 0,\dots,k - 1$. The rest of *(i)* is then an application of Proposition 5.3, which will be stated at the end of this section. We define the function ψ by $\psi(.,\ell) = \omega_1(\ell)\theta(.,\ell) - \varphi(.,\ell)$ for $\ell \in \mathbb{C}\backslash\mathbb{R}$ at which ω_1 is defined. From the definition of the functions $\theta(.,\ell)$ and $\varphi(.,\ell)$ we obtain the following identity:

$$(5.1) \qquad (\ell-\ell')\int_0^b \psi(x,\ell)\psi(x,\ell')dx =$$
$$\alpha(\ell')\beta(\ell)-\alpha(\ell)\beta(\ell') + \omega_1(\ell)(\alpha(\ell')\gamma(\ell)+\beta(\ell')\delta(\ell))$$
$$- \omega_1(\ell')(\alpha(\ell)\gamma(\ell')+\beta(\ell)\delta(\ell')) + \omega_1(\ell)\omega_1(\ell')(\gamma(\ell')\delta(\ell)-\gamma(\ell)\delta(\ell')).$$

Since $\hat{\lim}_{\ell\to\lambda} (\ell-\lambda)^{1-2k}\omega_1(\ell)$ is finite, we obtain

$$\int_0^b |\varphi(x,\lambda)|^2dx = \hat{\lim}_{\ell\to\lambda} \int_0^b |\psi(x,\ell)|^2dx$$
$$= \hat{\lim}_{\ell\to\lambda} \frac{\overline{\alpha(\ell)}\beta(\ell) - \alpha(\ell)\overline{\beta(\ell)}}{\ell-\bar{\ell}} < \infty,$$

and thus $\varphi(.,\lambda) \in L^2(0,b)$. Differentiating (5.1) j times with respect to ℓ and ℓ' we obtain likewise that

$$\left(\frac{\partial}{\partial\ell}\right)^j \varphi(.,\ell)\Big|_{\ell=\lambda} \in L^2(0,b), \quad j = 0,1,\dots,k - 1.$$

This proves the first part of *(i)*. For the rest of the proof of *(i)* we note that the formula (4.3) can be written as

$$R(\ell)\omega_1(\ell)f = -\varphi(.,\ell)\Phi(f,\ell) + \omega_1(\ell)E(.,\ell).$$

Here we have $\Phi(f,\ell) = \int_0^b f(x)\varphi(x,\ell)dx$ and for each $x \in [0,b)$ the function $\ell \mapsto E(x,\ell)$ is entire. Since $\widehat{\lim_{\ell\to\lambda}} (\ell-\lambda)^{1-2k}\omega_1(\ell)$ exists and is finite we have

$$R(\ell)\omega_1(\ell)f = -\sum_{j=0}^{k-1} \frac{1}{j!}\Big(\frac{\partial}{\partial\ell}\Big)^j(\varphi(.,\ell)\Phi(f,\ell))\Big|_{\ell=\lambda} (\ell-\lambda)^j + o(\ell-\lambda)^{k-1},$$

where because of the previous part of the theorem the o-term is meant in the $L^2(0,b)$ metric. By Proposition 5.3 with $f(\ell) = \omega_1(\ell)f$ there exist $\tilde{x}_0,\dots,\tilde{x}_{k-1} \in \mathfrak{D}(A)$, such that for $i,j = 0,1,\dots,k-1$,

$$\begin{cases} x_j = P\tilde{x}_j = -\dfrac{1}{j!}\Big(\dfrac{\partial}{\partial\ell}\Big)^j(\varphi(.,\ell)\Phi(f,\ell))\Big|_{\ell=\lambda}, \\ (A-\lambda)\tilde{x}_j = \tilde{x}_{j-1}, \\ [\tilde{x}_i,\tilde{x}_j] = 0, \qquad\qquad i+j \neq 2k-2, \end{cases}$$

and

$$[\tilde{x}_{k-1},\tilde{x}_{k-1}] = -[x_0,f]\ \widehat{\lim_{\ell\to\lambda}}\ (\ell-\lambda)^{1-2k}\omega_1(\ell).$$

Since we can find f such that $\Phi(f,\lambda) \neq 0$, statement *(i)* follows by simply taking suitable linear combinations of the elements $\tilde{x}_0,\dots,\tilde{x}_{k-1}$. In particular we note as a consequence of Proposition 5.3

$$\begin{aligned}[\tilde{y}_{k-1},\tilde{y}_{k-1}] &= (|\Phi(f,\lambda)|^2)^{-1}\,[\tilde{x}_{k-1},\tilde{x}_{k-1}] \\ &= -(|\Phi(f,\lambda)|^2)^{-1}[x_0,f]\ \widehat{\lim_{\ell\to\lambda}}\ (\ell-\lambda)^{1-2k}\omega_1(\ell) \\ &= \widehat{\lim_{\ell\to\lambda}}\ (\ell-\lambda)^{1-2k}\omega_1(\ell).\end{aligned}$$

This completes the proof of *(i)*. The proof of *(ii)* is similar, and uses Proposition 5.3 with k replaced by k + 1. This completes the proof. □

Before stating Proposition 5.3 we will first prove a preliminary lemma.

LEMMA 5.2. *Let* (x_n) *be a sequence in a* π_χ*-space* Π. *Then:*

(i) (x_n) *converges weakly in* Π *if and only if* $([x_n,u])$ *is a Cauchy sequence in* $\mathbb{C}$ *for all* u *in a total subset of* Π *and* $([x_n,x_n])$ *is bounded in* $\mathbb{C}$.

(ii) (x_n) *converges strongly in* Π *if and only if* $([x_n,u])$ *is a Cauchy sequence in* $\mathbb{C}$ *for all* u *in a total subset of* Π *and* $([x_n,x_m])$ *converges in* $\mathbb{C}$ *as* $n,m \to \infty$.

In case (ii), evidently, if $x_0 = \lim x_n$, *then* $[x_0,x_0] = \lim\,[x_n,x_n]$, *whereas in case (i)* $[x_0,x_0] \le \underline{\lim}\,[x_n,x_n]$.

PROOF. Part *(ii)* here is *(ii)* in Theorem 2.4 of [17]. The proof of *(i)* follows easily if one observes that the conditions imply the strong convergence of (x_n^-), where x_n^- is the projection of x_n onto a negative finite dimensional subspace contained in the span of the total subset. □

PROPOSITION 5.3. *Let* $\mathfrak{K}$ *be a* π_χ*-space,* $\tilde{A}$ *a selfadjoint operator in* $\mathfrak{K}$ *and* $\mathfrak{H}$ *a Hilbert space contained in* $\mathfrak{K}$ *such that* $\mathfrak{K} = \text{c.l.s.}\{ \bigcup (\tilde{A}-\ell)^{-1}\mathfrak{H} \mid \ell \in \rho(A) \}$*. Let* $R(z) = P(\tilde{A}-z)^{-1}|_{\mathfrak{H}}$*, where* P *is the orthogonal projection of* $\mathfrak{K}$ *onto* $\mathfrak{H}$*. Let* $\lambda \in \mathbb{R}$ *and let* f *be a mapping defined in a sector above* λ*, with values in* $\mathfrak{H}$ *such that, as* $z \xrightarrow{\wedge} \lambda$*,* $(z-\lambda)^{1-2k}f(z)$ *is bounded and*

$$R(z)f(z) = x_0 + (z-\lambda)x_1 + \ldots + (z-\lambda)^{k-1}x_{k-1} + o((z-\lambda)^{k-1})$$

for some $k \geq 1$ *and* $x_0, x_1, \ldots, x_{k-1} \in \mathfrak{H}$*. Then there exist unique* $\tilde{x}_0, \tilde{x}_1, \ldots, \tilde{x}_{k-1} \in \mathfrak{D}(\tilde{A})$ *such that (with* $\tilde{x}_{-1} = 0$*) for* $i,j = 0,1,\ldots,k-1$

$$\begin{cases} (\tilde{A}-\lambda)\tilde{x}_j = \tilde{x}_{j-1}, & \\ P\tilde{x}_j = x_j, & \\ [\tilde{x}_i,\tilde{x}_j] = 0, & i+j \neq 2k-2. \end{cases}$$

In addition

$$[\tilde{x}_{k-1},\tilde{x}_{k-1}] = -\widehat{\lim_{z\to\lambda}}\, [x_0,(z-\lambda)^{1-2k}f(z)],$$

whenever the limit exists.

PROOF. Because of the condition $\mathfrak{K} = \text{c.l.s.}\{ \bigcup (\tilde{A}-\ell)^{-1}\mathfrak{H} \mid \ell \in \rho(A) \}$ the operator $\tilde{A}$ cannot have eigenvectors in $\mathfrak{K} \ominus \mathfrak{H}$. Hence if the elements $\tilde{x}_0, \ldots, \tilde{x}_{k-1} \in \mathfrak{D}(\tilde{A})$ exist, they are uniquely determined. Without loss of generality we may take $\lambda = 0$. We first show:

(i) $(\tilde{A}-z)^{-1}f(z)$ converges weakly, and if $k > 1$ even strongly, to, say, $\tilde{x}_0 \in \mathfrak{K}$, as $z \xrightarrow{\wedge} 0$,

(ii) $\tilde{A}\,(\tilde{A}-z)^{-1}f(z)$ converges strongly to 0 as $z \xrightarrow{\wedge} 0$,

(iii) If, additionally $k > 1$, then $[\tilde{x}_0,\tilde{x}_0] = 0$.

Note that *(i)* and *(ii)* imply that $\tilde{x}_0 \in \mathfrak{D}(\tilde{A})$, $\tilde{A}\tilde{x}_0=0$ and $P\tilde{x}_0=x_0$. We apply Lemma 5.2. The set $\{ (\tilde{A}-\zeta)^{-1}y \mid \zeta \in \rho(A)\backslash\mathbb{R},\ y \in \mathfrak{H} \}$ is total in $\mathfrak{K}$ and

$$\begin{aligned}[(\tilde{A}-z)^{-1}f(z),(\tilde{A}-\zeta)^{-1}y] &= \frac{1}{z-\bar{\zeta}}\,([R(z)f(z),y] - [R(\bar{\zeta})f(z),y]) \\ &\longrightarrow -\,\bar{\zeta}^{-1}\,[x_0,y]\end{aligned}$$

for all $\zeta \in \rho(A)\backslash\mathbb{R}$ and $y \in \mathfrak{H}$. Furthermore,

$$[(\tilde{A}-z)^{-1}f(z),(\tilde{A}-w)^{-1}f(w)] = \frac{[R(z)f(z),f(w)] - [f(z),R(w)f(w)]}{z - \bar{w}},$$

and, as $|z-\bar{w}|^{-1} \leq C|z|^{-1}$ or $C|w|^{-1}$ (we choose z,w in the same halfplane), the righthand side is bounded as $z,w \hat{\longrightarrow} 0$. If additionally $k > 1$, then the righthand side tends to zero, hence $(\tilde{A}-z)^{-1}f(z)$ converges strongly to $\tilde{x}_0$ and $[\tilde{x}_0,\tilde{x}_0] = 0$. This proves the statements *(i)* and *(iii)*. To prove *(ii)*, we write $\tilde{A}(\tilde{A}-z)^{-1}f(z) = f(z) + z(\tilde{A}-z)^{-1}f(z)$ and observe that the righthand side tends to 0 (strongly), as $z \hat{\longrightarrow} 0$, since $f(z) \longrightarrow 0$ and $(\tilde{A}-z)^{-1}f(z)$ is bounded.

When $k > 1$ we continue with induction. Assume $0 \leq i < k-1$ and that there exist $\tilde{x}_0,\tilde{x}_1,\ldots,\tilde{x}_i \in \mathfrak{K}$ such that

$$z^{-n}\left((\tilde{A}-z)^{-1}f(z) - \tilde{x}_0 - \ldots - z^{n-1}\tilde{x}_{n-1} \right) \longrightarrow \tilde{x}_n, \text{ strongly},$$

$$z^{-n}\tilde{A}\left((\tilde{A}-z)^{-1}f(z) - \tilde{x}_0 - \ldots - z^{n-1}\tilde{x}_{n-1} \right) \longrightarrow \tilde{x}_{n-1}, \text{ strongly},$$

as $z \hat{\longrightarrow} 0$, $n = 0,1,\ldots,i$ and

$$[\tilde{x}_j,\tilde{x}_n] = 0, \quad 0 \leq n,j \leq i,$$

so that, for $n = 0,1,\ldots,i$, $\tilde{x}_n \in \mathfrak{D}(\tilde{A})$, $\tilde{A}\tilde{x}_n = \tilde{x}_{n-1}$ and $P\tilde{x}_n = x_n$. We shall show that for some $\tilde{x}_{i+1} \in \mathfrak{K}$:

(iv) $\quad z^{-i-1}\left((\tilde{A}-z)^{-1}f(z) - \tilde{x}_0 - \ldots - z^i\tilde{x}_i \right) \longrightarrow \tilde{x}_{i+1}$, as $z \hat{\longrightarrow} 0$, strongly if $i+1 < k-1$ and weakly if $i+1 = k-1$,

(v) $\quad z^{-i-1}\tilde{A}\left((\tilde{A}-z)^{-1}f(z) - \tilde{x}_0 - \ldots - z^i\tilde{x}_i \right) \longrightarrow \tilde{x}_i$, as $z \hat{\longrightarrow} 0$, strongly if $i+1 \leq k-1$,

(vi) $\quad [\tilde{x}_j,\tilde{x}_{i+1}] = 0$, $j = 0,1,\ldots,i+1$, if $i+1 < k-1$ and $[\tilde{x}_j,\tilde{x}_{i+1}] = 0$, $j = 0,1,\ldots,i$, if $i+1 = k-1$.

It follows then that $\tilde{x}_{i+1} \in \mathfrak{D}(\tilde{A})$, $\tilde{A}\tilde{x}_{i+1} = \tilde{x}_i$ and $P\tilde{x}_{i+1} = x_{i+1}$. Again we apply Lemma 5.2. Using the simple relation

$$(z-\bar{\zeta})(\tilde{A}-\bar{\zeta})^{-1} \sum_{n=0}^{i} \tilde{x}_n z^n = \sum_{n=0}^{i} \tilde{x}_n \left(z^n - \frac{z^{i+1}}{\bar{\zeta}^{i-n+1}} \right),$$

we obtain for $\zeta \in \rho(A)\backslash\mathbb{R}$ and $y \in \mathfrak{H}$:

$$\left[z^{-i-1}\left((\tilde{A}-z)^{-1}f(z) - \sum_{n=0}^{i} \tilde{x}_n z^n \right),(\tilde{A}-\zeta)^{-1}y \right] =$$

$$= z^{-i-1}(z-\bar{\zeta})^{-1}\left([R(z)f(z) - \sum_{n=0}^{i} x_n z^n,y] - [R(\bar{\zeta})f(z),y] + \sum_{n=0}^{i} [x_n \frac{z^{i+1}}{\bar{\zeta}^{i-n+1}},y] \right)$$

and the righthand side converges as $z \hat{\longrightarrow} 0$. Furthermore,

$$z^{-i-1}\bar{w}^{-i-1}\Big[\,(\tilde{A}-z)^{-1}f(z) - \sum_{n=0}^{i}\tilde{x}_n z^n \,,\ (\tilde{A}-w)^{-1}f(w) - \sum_{n=0}^{i}\tilde{x}_n w^n\,\Big] =$$

$$= z^{-i-1}\bar{w}^{-i-1}(z-\bar{w})^{-1}\Big\{\Big[\,R(z)f(z) - \sum_{n=0}^{i} x_n z^n, f(w)\,\Big] +$$

$$+ \sum_{n=0}^{i}\Big[\,x_n \frac{z^{i+1}}{\bar{w}^{i-n+1}}\,,\ f(w)\,\Big] - \Big[\,f(z),\ R(w)f(w) - \sum_{n=0}^{i} x_n w^n\,\Big]$$

$$- \Big[\,f(z),\ \sum_{n=0}^{i} x_n \frac{w^{i+1}}{\bar{z}^{i-n+1}}\,\Big]\Big\}\,,$$

where we have used that $[\tilde{x}_n,\tilde{x}_j] = 0$, $0 \le n,j \le i$. According to the conditions in the theorem the righthand side is bounded if $i+1 < k$, as then e.g. $|z^{-2i-2}(z-\bar{w})^{-1}f(z)| \le C$ (here also, z,w are in the same halfplane). If even $i+1 < k-1$, then it tends to zero, in which case the strong limit exists. This implies *(iv)* and $[\tilde{x}_{i+1},\tilde{x}_{i+1}] = 0$ if $i+1 < k-1$. Concerning *(v)* we note that

$$z^{-i-1}\tilde{A}\,\Big(\,(\tilde{A}-z)^{-1}f(z) - \sum_{n=0}^{i}\tilde{x}_n z_n\,\Big) =$$

$$= z^{-i-1}f(z) + z^{-i}\,\Big(\,(\tilde{A}-z)^{-1}f(z) - \sum_{n=0}^{i-1}\tilde{x}_n z^n\,\Big)$$

$$\longrightarrow \tilde{x}_i$$

strongly, according to the first induction hypothesis. To prove *(vi)* we only have to show that $[\tilde{x}_j,\tilde{x}_{i+1}] = 0$ for $0 \le j < i+1$ as it has been shown above that $[\tilde{x}_{i+1},\tilde{x}_{i+1}] = 0$ if $i+1 < k-1$. On account of *(iv)*, the last induction hypothesis and the boundedness of $z^{1-2k}f(z)$, $z \xrightarrow{\wedge} 0$, we obtain these equalities as follows:

$$\begin{aligned}
[\tilde{x}_j,\tilde{x}_{i+1}] &= \widehat{\lim_{z\to 0}}\Big[\,\tilde{x}_j\,,\ z^{-i-1}\Big(\,(\tilde{A}-z)^{-1}f(z) - \sum_{n=0}^{i}\tilde{x}_n z^n\,\Big)\,\Big] \\
&= \widehat{\lim_{z\to 0}}\,[\,\tilde{x}_j\,,\ z^{-i-1}(\tilde{A}-z)^{-1}f(z)\,] \\
&= -\,\widehat{\lim_{z\to 0}}\sum_{n=0}^{j}\,[\,\tilde{x}_{j-n}\,,\ z^{-i-n-2}f(z)\,] \\
&= 0\,, \quad j = 0,1,..,i+1,
\end{aligned}$$

where we have used the relation

$$(\tilde{A}-z)^{-1}\,\tilde{x}_j = -z^{-1}\sum_{n=0}^{j} z^{-n}\,\tilde{x}_{j-n}.$$

Finally we note using *(iv)* that

$$[\,\tilde{x}_{k-1},\tilde{x}_{k-1}] = \widehat{\lim_{z\to 0}}\Big[\,\tilde{x}_{k-1}\,,\ \frac{(\tilde{A}-z)^{-1}f(z)-\tilde{x}_0-\ldots-z^{k-2}\tilde{x}_{k-2}}{z^{k-1}}\,\Big]$$

$$= \widehat{\lim_{z\to 0}} [\tilde{x}_{k-1}, z^{-k+1}(\tilde{A}-z)^{-1}f(z)]$$

$$= -\widehat{\lim_{z\to 0}} \sum_{n=0}^{k-1} [\tilde{x}_{k-1-n}, z^{-n-k}f(z)]$$

$$= -\widehat{\lim_{z\to 0}} [\tilde{x}_0, z^{1-2k}f(z)]$$

$$= -\widehat{\lim_{z\to 0}} [x_0, z^{1-2k}f(z)] ,$$

whenever this limit exists. This completes the proof. □

BIBLIOGRAPHY

1. A.I. Benedek and R. Panzone, "On Sturm-Liouville problems with the square-root of the eigenvalue parameter contained in the boundary conditions", Notas de Algebra y Analysis, Vol. 10, Universidad Nacional del Sur, Bahia Blanca, 1981.

2. J. Chaudhuri and W.N. Everitt, "On the spectrum of ordinary second-order differential operators", Proc. Roy. Soc. Edinburgh, 68A (1969), 95-119.

3. E.A. Coddington and N. Levinson, *Theory of Ordinary Differential Equations*, McGraw-Hill, New York, 1955.

4. K. Daho and H. Langer, "Matrix functions of the class $N_\varkappa$" , Math.Nachr., 120 (1985),275-294.

5. A. Dijksma, "Eigenfunction expansions for a class of J-selfadjoint ordinary differential operators with boundary conditions containing the eigenvalue parameter", Proc. Roy. Soc. Edinburgh, 86A (1980), 1-27.

6. A. Dijksma, H. Langer and H.S.V. de Snoo, "Selfadjoint $\pi_\varkappa$-extensions of symmetric subspaces: an abstract approach to boundary problems with spectral parameter in the boundary conditions", Integral equations and operator theory, 7 (1984), 459-515.

7. A. Dijksma, H. Langer and H.S.V. de Snoo, "Unitary colligations in $\Pi_\varkappa$-spaces, characteristic functions and Štraus extensions", Pacific J. Math., 125 (1986), 347-362.

8. A. Dijksma, H. Langer and H.S.V. de Snoo, "Characteristic functions of unitary operator colligations in $\Pi_\varkappa$-spaces", Operator Theory: Advances and Applications, 19 (1986), 125-194.

9. A. Dijksma, H. Langer and H.S.V. de Snoo, "Representations of holomorphic functions by means of resolvents of unitary or selfadjoint operators in Krein spaces", Operator Theory: Advances and Applications, to appear.

10. A. Dijksma, H. Langer and H.S.V. de Snoo, "Unitary colligations in Krein spaces and their role in the extension theory of isometries and symmetric linear relations in Hilbert spaces", to appear.

11. B. Friedman, *Principles and Techniques of Applied Mathematics*, Wiley, New York, 1956.

12. C.T. Fulton, "Two-point boundary value problems with eigenvalue parameter contained in the boundary conditions", Proc. Roy. Soc. Edinburgh 77A, (1977), 293-308.

13. C.T. Fulton, "Singular eigenvalue problems with eigenvalue parameter contained in the boundary conditions", Proc. Roy. Soc. Edinburgh 87A, (1980), 1-34.

14. R.C. Gilbert, "Spectral representation of selfadjoint extensions of a symmetric operator", Rocky Mountain J. Math., 2 (1972), 75-96.

15. E. Hille, *Lectures on Ordinary Differential Equations*, Addison-Wesley, Reading, 1969.

16. D. Hinton and K. Shaw, "Spectrum of a Hamiltonian system with spectral parameter in a boundary condition", preprint.

17. I.S. Iohvidov, M.G. Krein and H. Langer, *Introduction to the Spectral Theory of Operators in Spaces with an Indefinite Metric*, Akademie-Verlag, Berlin, 1982 (Reihe: Mathematical Research, vol. 9).

18. I.S. Kac, "On the multiplicity of the spectrum of a second-order differential operator", Dokl. Akad. Nauk. SSSR, 145 (1962), 510-514 (English translation: Sov. Math., 3 (1962), 1035-1039).

19.I.S. Kac, "Spectral multiplicity of a second-order differential operator and expansion in eigenfunctions", Izv. Akad. Nauk. SSSR Ser. Mat., 27 (1963), 1081-1112 (Russian).

20. I.S. Kac and M.G. Krein, "R-functions-analytic functions mapping the upper halfplane into itself", Supplement I to the Russian edition of F.V. Atkinson, *Discrete and continuous boundary problems*, Mir, Moscow, 1968 (English translation: Amer. Math. Soc. Transl., (2) 103 (1974), 1-18).

21. I.S. Kac and M.G. Krein, "On the spectral functions of a string", Supplement II to the Russian edition of F.V. Atkinson, *Discrete and continuous boundary problems*, Mir, Moscow, 1968 (English translation: Amer. Math. Soc. Transl., (2) 103 (1974), 19-102).

22. A.M. Krall and L. Littlejohn, "A singular sixth order differential equation with orthogonal polynomial eigenfunctions", in Ordinary and Partial Differential Equations, Lecture Notes in Mathematics 964, Springer-Verlag, Berlin-Heidelberg-New York, 1982, 435-444.

23. M.G. Krein and H. Langer, "On the spectral function of a selfadjoint operator in a space with indefinite metric", Dokl. Akad. Nauk SSSR, 152 (1963) 39-42 (English translation: Soviet Math. 4 (1963), 1236-1239).

24. M.G. Krein and H. Langer, "Über die verallgemeinerten Resolventen und die charakteristische Funktion eines isometrischen Operators im Raume $\Pi_\varkappa$", Hilbert Space Operators and Operator Algebras (Proc. Int. Conf., Tihany, 1970) Colloquia Math. Soc. János Bolyai, no 5, North-Holland, Amsterdam, 1972 353-399.

25. M.G. Krein and H. Langer, "Über die Q-Funktion eines π-hermiteschen Operators im Raume $\Pi_\varkappa$", Acta Sci. Math., 34 (1973), 191-230.

26. M.G. Krein and H. Langer, "Über einige Fortsetzungsprobleme, die eng mit der Theorie hermitescher Operatoren im Raume $\Pi_\varkappa$ zusammenhängen, Teil I: Einige Funktionenklassen und ihre Darstellungen", Math. Nachr., 77 (1977), 187-236.

27. M.G. Krein and H. Langer, "Some propositions on analytic matrix functions related to the theory of operators in the space $\Pi_\varkappa$", Acta Sci. Math. (Szeged), 43 (1981), 181-205.

28. M.G. Krein and H. Langer, "On some continuation problems which are closely related to the theory of operators in spaces $\Pi_\varkappa$. IV: Continuous analogues of orthogonal polynomials on the unit circle with respect to an indefinite weight and related continuation problems for some classes of functions", J. Operator Theory, 13 (1985), 299-417.

29. H. Langer, "Spectral functions of definitizable operators in Krein spaces", Functional Analysis, Proceedings Dubrovnik 1981, Lecture Notes in Mathematics 948, Springer-Verlag, Berlin-Heidelberg-New York, 1982, 1-46.

30. H. Langer, "A characterization of generalized zeros of negative type of functions of the class $N_\varkappa$", Operator Theory: Advances and Applications, 17 (1986), 201-212.

31. R. Mennicken and M. Möller, "Boundary eigenvalue problems" Notas de Algebra y Analysis, Vol 14, Universidad Nacional del Sur, Bahia Blanca, 1986.

32. H.D. Niessen, "Singuläre S-hermitesche Rand-Eigenwertprobleme", manuscripta math., 3 (1970), 35-68.

33. H. Röh, "Self-adjoint subspace extensions satisfying λ-linear boundary conditions", Proc. Roy. Soc. Edinburgh, 90A (1981), 107-124.

34. E.M. Russakovskii, "Operator treatment of boundary problems with spectral parameters entering via polynomials in the boundary conditions", Functional Anal. Appl., 9 (1975) 358-359.

35. E.M. Russakovskii, "Operator treatment of a boundary-value problem with the spectral parameter appearing rationally in the boundary conditions", Theory of functions, functional analysis and its applications 30, Kharkov, 1978, 120-128 (Russian).

36. A. Schneider, "On spectral theory for the linear selfadjoint equation $Fy = \lambda Gy$", Ordinary and Partial Differential Equations, Proceedings, Dundee Scotland 1980, Lecture Notes in Mathematics 846, Springer-Verlag, Berlin-Heidelberg-New York, 1981, 306-332.

37. A.V. Štraus, "On spectral functions of differential operators", Izvest. Akad. Nauk SSSR Ser. Mat., 19 (1955), 201-220 (Russian).

38. A.V. Štraus, "Extensions and generalized resolvents of a symmetric operator which is not densely defined", Izv. Akad. Nauk SSSR Ser. Mat., 34 (1970), 175-202 (English translation: Math. USSR Izv., 4 (1970), 179-208).

39. J. Walter, "Regular eigenvalue problems with eigenvalue parameter in the boundary conditions", Math. Z., 133 (1973), 301-312.

A. DIJKSMA AND H.S.V. DE SNOO
SUBFACULTEIT WISKUNDE EN INFORMATICA
RIJKSUNIVERSITEIT GRONINGEN
POSTBUS 800
9700 AV GRONINGEN
NEDERLAND

H. LANGER
SEKTION MATHEMATIK
TECHNISCHE UNIVERSITÄT
DDR-8027 DRESDEN
MOMMSENSTRASSE 13
DDR

Canadian Mathematical Society
Conference Proceedings
Volume 8 (1987)

MINORS OF THE WRONSKIAN FOR SOLUTIONS OF THE DIFFERENTIAL EQUATION $L_n y + p(x)y = 0$

Uri Elias

1. The differential equation to be studied here is of the form

$$(1) \qquad L_n y + p(x)y = 0 , \qquad 0 \leqslant x < \infty ;$$

here $p(x)$ is continuous and of one sign, and L_n is the disconjugate differential operator

$$L_n y = \rho_n(\rho_{n-1} \cdots (\rho_1(\rho_0 y)')' \cdots)'$$

with weight functions $\rho_i > 0$, $\rho_i \in C^{n-i}[0,\infty)$, $i=0,\ldots,n$. We denote $L_0 y = \rho_0 y$, $L_i y = \rho_i(L_{i-1})'$, $i=1,\ldots,n$ and call $L_0 y,\ldots,L_n y$ the <u>quasiderivatives</u> of y. This type of equation has recently been studied in several works.

In this work we discuss the role of the minors of the Wronskian in the oscillation theory of Equation (1). In order to explain our problems and results, we start with some known facts about Equation (1) and define a convenient basis for its solution space.

It is known that for any $\{i_1,\ldots,i_k\},\{j_1,\ldots,j_{n-k}\} \subset \{0,\ldots,n\}$, the boundary value conditions

$$L_i y(a) = 0, \quad i = i_1,\ldots,i_k,$$
$$L_j y(b) = 0, \quad j = j_1,\ldots,j_{n-k}, \qquad a < b,$$

1980 Mathematics Subject Classification (1985 Revision). 34C10.

have no nontrivial solutions if $p(x)$ is positive (negative) and $n-k$ is even (odd). Hence, we shall call an integer k, $0 \leqslant k \leqslant n$, of <u>admissible parity</u> if

$$(2) \qquad (-1)^{n-k}p(x) < 0 .$$

Let $a < b$ be two fixed points in $[0,\infty)$. For every admissible k, we define, as in [1], $u_{k-1}(x,b)$, $u_k(x,b)$ to be the unique solutions of (1) which satisfy the systems of boundary value conditions

$$(3)_{k-1} \qquad \begin{aligned} &L_t u(a) = 0, \quad t = 0,\dots,k-2,k, \\ &L_{k-1}u(a) = 1, \\ &L_t u(b) = 0, \quad t = 0,\dots,n-k-2 , \end{aligned}$$

and

$$(3)_{k} \qquad \begin{aligned} &L_t u(a) = 0, \quad t = 0,\dots,k-2,k-1, \\ &L_{k}u(a) = 1, \\ &L_t u(b) = 0, \quad t = 0,\dots,n-k-2 , \end{aligned}$$

respectively. Here and thoughout the paper the following convention will be used: <u>Whenever an admissible</u> k <u>and solutions</u> u_{k-1}, u_k <u>are mentioned, omit</u> u_{-1} (<u>if</u> k=0) <u>or</u> u_n (<u>if</u> k=n), <u>which are not defined at all</u>. It was proved in [1] that the so-obtained set $\{u_0(x,b),\dots,u_{n-1}(x,b)\}$ exists, it is unique and it is a basis of solutions of (1). Moreover, this basis proved to be an intrinsic choice [1,2,6,7]. The Wronskian which is used in the context of Equation (1) will be

$$W(y_0,\dots,y_{n-1}) = \det(L_i y_j)$$

and its minors will be denoted by

$$W\binom{u_{i_1},\dots,u_{i_r}}{j_1,\dots,j_r}(x) = \det\big(L_{j_s}u_{i_t}(x)\big)_{s,t=1,\dots,r} .$$

Minors of the Wronskian play an important role in [6]. The paper reveals relations between disconjugacy types and minors of the Wronskian, though its main aim is the factorization of (1). Here is a typical result

[6, Theorem 13]: Let $\ell < k$ be admissible integers. If $W(u_{\ell-1}(x,b), u_\ell(x,b),\dots,u_{k-1}(x,b))$ vanishes in (a,b), then (1) is not $(k,n-k)$-disconjugate in (a,b). Theorem 14 of [6] provides a partial converse of this statement.

The aim of this work is to derive various properties which are related to minors of the Wronskian of (1). Our first result claims that if y is a solution of Equation (1), then $W(y,u_{\ell-1},\dots,u_k)$ is a solution of a very similar differential equation, dependent only on $u_{\ell-1},\dots,u_k$. For simplicity we shall assume that p^{-1}, $r_0,\dots,r_n$ of Equation (1) belong to $C^\infty[0,\infty)$.

THEOREM 1. Let $u_0(x,b),\dots,u_{n-1}(x,b)$ be the basis of solution space of (1) as defined by $(3)_0$-$(3)_{n-1}$, and $\ell \leqslant k$ be two integers of admissible parity.

(a) For every solution y of (1), the function

$$z = W\binom{y,u_{\ell-1},\dots,u_k}{0,\ 1\ ,\dots,k-\ell+2} \Big/ W\binom{u_{\ell-1},\dots,u_k}{1\ ,\dots,k-\ell+2} \tag{4}$$

is a solution of the differential equation

$$M_n z + p(x)z = 0, \quad a < x < b \tag{5}$$

where $M_0 z = \sigma_0 z$, $M_i z = \sigma_i(M_{i-1}z)'$, $\sigma_i > 0$ on (a,b), $i=0,\dots,n$, $\sigma_0(x),\dots,\sigma_n(x)$ depend only on $u_{\ell-1},\dots,u_k$ and satisfy $\sigma_0\cdot\ldots\cdot\sigma_n = \rho_0\cdot\ldots\cdot\rho_n$.

(b) A basis for the solution space of (5) is given by

$$z_j = W\binom{y_j,u_{\ell-1},\dots,u_k}{0\ ,\ 1\ ,\dots,k-\ell+2} \Big/ W\binom{u_{\ell-1},\dots,u_k}{1\ ,\dots,k-\ell+2},\ j=0,\dots,n-1 \tag{6}$$

where

$$y_j = \begin{cases} \text{a solution of } L_n y+py = pu_j, & j \in \{\ell-1,\dots,k\} \\ u_j\ , & j \notin \{\ell-1,\dots,k\}. \end{cases} \tag{7}$$

$\{u_{\ell-1},\dots,u_k\}$ will be called pivotal block for equation (5).

(c) Conversely, given Equation (5) and the basis $z_0,\dots,z_{n-1}$, then a similar process, in which $\{z_0,\dots,z_{\ell-1},z_{k+1},\dots,z_{n-1}\}$ is used as a pivotal block, yields Equation (1) and $u_0,\dots,u_{n-1}$.

Here we outline some details of the proof. We define the differential operators

$$\bar{L}_0 y = \bar{r}_0 y,\ \bar{L}_i y = \bar{r}_i(\bar{L}_{i-1}y)',\ i=1,\dots,n+k-\ell+2,$$

with

$$\bar{r}_i = \begin{cases} r_i, & i = 0,\dots,n-1, \\ r_0 r_n |r|^{-1}, & i = n, \\ r_{i-n}, & i = n+1,\dots,n+k-\ell+2 . \end{cases}$$

Then it is possible to show that the required $M_i z$ are given by

$$M_i z = \bar{W}\binom{y,u_{\ell-1},\dots,\quad u_k}{i,\ i+1,\dots,i+k-\ell+2} \Big/ \bar{W}\binom{u_{\ell-1},\dots,\quad u_k}{i+1\ ,\dots,i+k-\ell+2},\ i=0,\dots,n-1,$$

and

$$M_n z = |p|\ \bar{W}\binom{y,u_{\ell-1},\dots,\quad u_k}{n,\ n+1,\dots,n+k-\ell+2} \Big/ \bar{W}\binom{u_{\ell-1},\dots,\quad u_k}{n+1\ ,\dots,n+k-\ell+2},$$

where $\bar{W}$ denotes a minor which consists of quasiderivatives $\bar{L}_i$. The weight functions s_i are given by $s_0 \equiv 1$,

$$\bar{s}_i = r_{i+k-1+2}\ *$$

$$\bar{W}\binom{u_{\ell-1},\dots,\quad u_k}{i\ ,\dots,i+k-\ell+1}^2 \Big/ \bar{W}\binom{u_{\ell-1},\dots,\quad u_k}{i-1\ ,\dots,i+k-\ell}\ \bar{W}\binom{u_{\ell-1},\dots,\quad u_k}{i+1\ ,\dots,i+k-\ell+2},$$

for $i=1,\dots,n-1$ and

$$s_n = |p|\bar{r}_{n+k-\ell+2}\ *$$

$$\bar{W}\binom{u_{\ell-1},\dots,\quad u_k}{n\ ,\dots,n+k-\ell+1}^2 \Big/ \bar{W}\binom{u_{\ell-1},\dots,\quad u_k}{n-1\ ,\dots,n+k-\ell}\ \bar{W}\binom{u_{\ell-1},\dots,\quad u_k}{n+1\ ,\dots,n+k-\ell+2}.$$

Several papers have dealt with the so called "dominance problem" for Equation (1). In general terms the problem is to split the solution space into disjoint subsets such that the solutions in one set are, in a certain sense, larger than those in another set. For precise definitions and various results see [1-2,5,7]. While it is not difficult to compare non-oscillatory solutions of (1) [9], it is more intriguing to compare the relative magnitudes of oscillatory solutions. One possible approach is to compare the nonvanishing minors of the Wronskian. Let

$$u_i(x) = \lim u_i(x,b), \quad i = 0,\dots,n-1,$$

where b tends to infinity through a suitably chosen sequence. For the so-obtained basis $\{u_0(x),\ldots,u_{n-1}(x)\}$ the following was proved [2, Theorem 10]: *If* k *is of admissible parity and* $i,j < k$ *then*

$$W(u_i,u_j,u_{k+1},\ldots,u_{n-1}) \,/\, W(u_{k-1},u_k u_{k+1},\ldots,u_{n-1})$$

is bounded as $x \to \infty$. Our next result also reflects the magnitudes of these solutions:

THEOREM 2. *If* k *is of admissible parity and* $i,j < k$, *then*

$$W(u_i,u_j) \,/\, W(u_{k-1},u_k)$$

is bounded as $x \to \infty$.

To prove this result we choose the pivotal block $\{u_0,\ldots,u_k\}$ and construct the corresponding Equation (5). We use the fact that its solutions $z_0,\ldots,z_{n-1}$ satisfy

$$\frac{W(u_i,u_j)}{W(u_{k-1},u_k)} \equiv \frac{W(z_i,z_j,z_{k+1},\ldots,z_{n-1})}{W(z_{k-1},z_k,z_{k+1},\ldots,z_{n-1})}.$$

The details of the proof are more complicated than those of [2, Theorem 10] since the z_i's are defined by minors and not by boundary value conditions and since a is a singular point for Equation (5).

In the study of the solutions of (1) we found it useful to consider the number of the sign changes among the quasiderivatives $L_0y(x),\ldots,L_ny(x)$ of a solution y at a point x. Namely, let $S(c_0,\ldots,c_n)$ denote the number of sign changes in the sequence $c_0,\ldots,c_n$ on nonzero terms. For a solution $y \not\equiv 0$ of (1) we define

$$S(y,x^+) = \lim_{t\to x^+} S\big(L_0y(t),-L_1y(t),\ldots,(-1)^nL_ny(t)\big),$$
$$S(y,x^-) = \lim_{t\to x^-} S\big(L_0y(t),\ L_1y(t),\ldots,\ L_ny(t)\big).$$

The quantity $S(y,x^+)$ has been used recently in several works which deal with Equation (1). One of its useful properties is the following:

Let $x_1,\ldots,x_r$ *be the zeros of the quasiderivatives* $L_0y,\ldots,L_{n-1}y$ *in* (a,b), $n(x_i)$ *the multiplicity of the zero at* x_i *and* $\langle q\rangle$ *the greatest even integer not exceeding* q. *Then*

$$S(y,x^+) + \sum_{a<x_i<b} \langle n(x_i)\rangle + S(y,b^-) \leqslant n.$$

The last inequality has obvious applications in restricting the possible zero-distributions of solutions of (1). Let us mention another fact:

$S(y,x^+)$ *is an integer valued, non-decreasing function of* x *on* $[0,\infty)$.

The ultimate value of $S(y,x^+)$, namely $\lim_{x\to\infty} S(y,x^+)$, can be used to study the behaviour of $y(x)$ near infinity. For various applications of $S(y,x^+)$ see, for example, [1,2,5-7,10].

Here we discuss some properties of the basis $\{u_0,\ldots,u_{n-1}\}$ of Equation (1) which are inherited by Equation (5). The following result is known:

THEOREM [2]. *Let* $u_0(x,b),\ldots,u_{n-1}(x,b)$ *be the basis of solutions of* (1) *as defined by* $(3)_0$-$(3)_{n-1}$. *If* $r \leqslant q$ *are of admissible parity, then*

(8) $\quad r \leqslant S(c_{r-1}u_{r-1}+\ldots+c_q u_q, x^+) \leqslant q$ *on* $[a,b)$,

(9) $\quad W(u_{r-1},u_r,\ldots,u_q) \neq 0$ on (a,b)

(*recall the convention about* u_{-1}, u_n !).

THEOREM 3. *Let* $l \leqslant k$, $r \leqslant q$ *be four integers of admissible parity such that* $\{l-1,\ldots,k\}$, $\{r-1,\ldots,q\}$ *are disjoint. The solutions*

$$z_j = W\binom{u_j,u_{l-1},\ldots,\ u_k}{0\ ,\ 1\ ,\ldots,k-l+2} \Big/ W\binom{u_{l-1},\ldots,\ u_k}{1\ ,\ldots,k-l+2},\quad j=r-1,\ldots,q,$$

of (5) *satisfy*

(10) $\quad r \leqslant S_M(c_{r-1}z_{r-1}+\ldots+c_q z_q, x^+) \leqslant q, \quad a \leqslant x < b,$

(11) $\quad W_M\binom{z_{r-1},\ldots,\ z_q}{i\ ,\ldots,i+q-r+1} \neq 0, \qquad a < x < b.$

In particular, for $r = q$,

$$S_M(cz_{q-1} + dz_q, x^+) \equiv S_L(cu_{q-1}+du_q, x^+) \equiv q, \quad a \leqslant x < b.$$

The subscripts M,L remind us that the quasiderivatives in question are $M_i z$, $L_i u$ respectively. The proof uses the theory of variation diminishing transformations [8].

Theorems 1 and 2 can be generalized in various directions. If, for example, $\{u_i(x,b)\}$ are replaced by

$$u_i(x) = \lim_{b\to\infty} u_i(x,b), \quad i = 0,\dots,n-1,$$

we get analogous theorems on $[0,\infty)$. Also the pivotal block $\{u_{\ell-1},u_\ell,\dots,u_k\}$ can be replaced by the more general

$$\{u_{\ell_1-1},\dots,u_{k_1},u_{\ell_2-1},\dots,u_{k_2},\dots,u_{\ell_r-1},\dots,u_{k_r}\},$$

where $0 \leqslant \ell_1 \leqslant k_1 < \ell_2 \leqslant k_2 < \dots < \ell_r \leqslant k_r \leqslant n$ are integers of admissible parity.

We believe that the relations between Equations (1) and (5) will be useful tools in the study of the oscillation of (1). The proofs of the theorems appear in [3] and [4].

BIBLIOGRAPHY

1. U. Elias, "A classification of solutions of a differential equation according to their asymptotic behaviour", Proc. Roy. Soc. Edinburgh, 83A (1979), 25-38.

2. U. Elias, "A classification of solutions of a differential equation according to their behaviour at infinity, II", Proc. Roy. Soc. Edinburgh, 100A (1985), 53-66.

3. U. Elias, "Minors of the Wronskian for solutions of the differential equation $L_ny+p(x)y = 0$" (preprint).

4. U. Elias, "Minors of the Wronskian for solutions of the differential equation $L_ny+p(x)y = 0$, II: Dominance of solutions" (preprint).

5. G.J. Etgen, G.D. Jones and W.E. Taylor, "Structure of the solution space of certain linear equations", J. Differential Equations, 59 (1985), 229-242.

6. G.J. Etgen, G.D. Jones and W.E. Taylor, "On the factorization of ordinary linear differential operators", Trans. Amer. Math. Soc., 297 (1986), 717-728.

7. G.D. Jones, "Growth properties of solutions of a linear differential equation" (preprint).

8. S. Karlin, Total positivity, Stanford University Press, 1968.

9. W.J. Kim, "On the fundamental system of solutions of $y^{(n)}+py = 0$", Funkc. Ekvac. 25 (1982), 1-17.

10. M. Svec, "Behaviour of a fourth order self adjoint linear differential equation", Annal. Polon. Math. 42 (1983), 333-344.

Department of Mathematics
Technion — Israel Institute of Technology
32000 Haifa, Israel

Canadian Mathematical Society
Conference Proceedings
Volume 8 (1987)

An Oscillation Result for Second Order Differential Systems*

L.H. Erbe
Department of Mathematics
University of Alberta
Edmonton, Alberta T6G 2G1

* Dedicated to the memory of G.J. Butler

Abstract: We generalize a scalar oscillation criterion of Hartman to the second order differential system $Y'' + Q(t)Y = 0$, where Q, Y are $n \times n$ real matrix-valued functions defined on $[a,+\infty)$ with $Q = Q(t)$ symmetric.

Introduction: In this paper we are concerned with the second order differential system

$$Y'' + Q(t)Y = 0, \quad t \in [a,+\infty) \tag{1.1}$$

where $Y(t)$, $Q(t)$ are $n \times n$ real continuous matrix functions and $Q(t)$ is symmetric. We shall be interested in analogues of scalar oscillation criteria for (1.1). For the corresponding scalar equation

$$y'' + q(t)y = 0, \quad t \in [a,+\infty), \tag{1.2}$$

with $q(t)$ a real-valued continuous function, the literature on the oscillation theory and distribution of zeros is very extensive. We refer

---

Research supported by NSERC Grant A-7673.

to some of the earlier surveys of Willett [23], Wong [26], Rab [19] and the books of Swanson [20] and Hartman [9]. For the scalar equation (1.2) the well-known Fite-Wintner Theorem ([8], [25]) states that (1.2) is oscillatory (i.e., every solution of (1.2) has infinitely many zeros in each half line $[b,+\infty)$, $b > a$), in case

$$\int_a^\infty q(s)ds = +\infty . \tag{1.3}$$

Wintner in [25] showed that

$$\lim_{T\to\infty} \frac{1}{T}\int_a^T \int_a^t q(s)dsdt = +\infty \tag{1.4}$$

implies oscillation of (1.2) and Hartman (see [9]) showed that

$$-\infty < \liminf_{T\to\infty} \frac{1}{T}\int_a^T \int_a^t q(s)dsdt < \limsup_{T\to\infty} \frac{1}{T}\int_a^T \int_a^t q(s)dsdt \tag{1.5}$$

is also sufficient for oscillation of (1.2). These results were further extended by Olech, Opial and Wazewski [18] and by Willett [23], [24] using more general averaging techniques. Kwong and Zettl in [15] have also recently given additional criteria involving the notion of L^2 convergence introduced by Coppel [6]. It is therefore of interest to determine to what extent the criteria mentioned above have analogues for the matrix equation (1.1). The vector equation associated with (1.1) is

$$y'' + Q(t)y = 0 , \quad t \in [a,+\infty) \tag{1.6}$$

where $y = \text{col}(y_1,\dots,y_n)$ is an n-vector. Equation (1.6) is said to be <u>disconjugate</u> on an interval $J \subset [a,+\infty)$ if every nontrivial solution of (1.6) vanishes at most once in J and (1.6) is said to be <u>oscillatory</u> if for each $t_0 > a$ there exists $t_1 > t_0$ such that (1.6) is not disconjugate on $[t_0,t_1]$. A solution $Y(t)$ of the matrix equation (1.1) is <u>nontrivial</u> if $\det Y(t) \neq 0$ for at least one $t \in [a,+\infty)$ and a nontrivial solution $Y(t)$ is said to be <u>prepared</u> in case

$$(1.7) \qquad Y^*(t)Y'(t) - Y^{*\prime}(t)y(t) \equiv 0 \ , \quad t \in [a,+\infty)$$

(Here the transpose of a matrix A is denoted by A^*). Equation (1.1) is said to be <u>oscillatory</u> on $[a,+\infty)$ in case the determinant of every nontrivial prepared solution vanishes on $[b,+\infty)$ for each $b > a$. (This is equivalent to oscillation of (1.6) since any solution of (1.6) is of the form $y(t) \equiv Y(t)\alpha$ for some constant vector α and some nontrivial prepared solution $Y(t)$ of (1.1).)

Many recent results [1], [7], [10], [11], [21], [22] have been obtained which show that (1.1) is oscillatory if a corresponding scalar equation obtained by applying a positive linear functional is oscillatory. Other recent oscillation criteria for (1.1) have involved conditions on the eigenvalues of $Q(t)$ (or of its integral). It was conjectured in [11] that (1.1) is oscillatory in case

$$(1.8) \qquad \lim_{t\to\infty} \lambda_1(Q_1(t;a)) = +\infty$$

where $\lambda_1(A) \geq \lambda_2(A) \geq \cdots \geq \lambda_n(A)$ denotes the usual ordering of the eigenvalues of the symmetric matrix A and where we use the notation

$$(1.9) \qquad Q_1(t;a) \equiv \int_a^t Q(s)ds \ .$$

This conjecture was settled with additional assumptions on the rate of growth of the trace of $Q_1(t)$ by Mingarelli [16], [17], Kwong et al [14], Butler and Erbe [2], [3]. The conjecture was finally settled in the case

$n = 2$ by Kaper and Kwong [12] and for arbitrary n by Byers, Harris and Kwong [5]. We refer also to Kaper and Kwong [13] for an alternate proof of a strengthened form of the conjecture.

In [4], additional criteria for oscillation of (1.1) were given based on both Riccati techniques and variational principles. These extended (at least partially) the scalar criteria (1.5),(1.6). In this paper we further extend the above results of [4] and give the complete systems analogue of the scalar criterion (1.5) of Hartman [9].

2. Prior to stating some results of [4] which will be needed, we introduce the (extended real-valued) function L defined on the class of $n\times n$ continuous real symmetric matrices defined on $[a,\infty)$ by

$$(2.1)\qquad L(Q) \equiv \liminf_{T\to\infty} \frac{1}{T}\int_a^T \int_a^t \operatorname{tr}(Q(s))\,ds\,dt$$

We recall (cf [7]) that equation (1.1) is nonoscillatory on $[a,+\infty)$ iff the corresponding Riccati integral equation

$$(2.2)\qquad Z(t) = Z(t_0) + \int_{t_0}^t Z^2(s)\,ds + \int_{t_0}^t Q(s)\,ds$$

has a continuous symmetric solution on $[t_0,\infty)$ for some $t_0 \geq a$. Thus, if $Y = Y(t)$ is a nontrivial prepared solution of (1.1) with $\det Y(t) \neq 0$ for $t \geq t_0$ then $Z = -Y'Y^{-1}$ is a symmetric solution of (2.2) on $[t_0,\infty)$. The following lemma (Lemma 5.1 of [4]) is the systems analogue of a result of Hartman [9]. The proof is given in [4].

<u>Lemma 2.1 ([4])</u>: Assume equation (1.1) is nonoscillatory on $[a,+\infty)$. Then a necessary and sufficient condition that

$$\lim_{T\to\infty} \int_t^T Z^2(s)ds \tag{2.3}$$

exists for any solution $Z = -Y'Y^{-1}$ of (2.2) (where $Y = Y(t)$ is a prepared solution of (1.1)) is that $L(Q) > -\infty$.

We shall also need the following result of [4]:

Theorem 2.2 ([4]): Assume $L(Q) > -\infty$. Then equation (1.1) is oscillatory in case:

$$\limsup_{T\to\infty} \frac{1}{T} \int_a^T \lambda_1(Q_1(t;a))dt = +\infty$$

We introduce the following notation: For each $t_0 \geq a$ we define the functions $\eta(t_0)$, $\mu(t_0)$ by:

$$\eta(t_0) \equiv \limsup_{T\to\infty} \frac{1}{T} \int_{t_0}^T \lambda_1\big(Q_1(t;t_0)\big)dt$$

$$\mu(t_0) \equiv \liminf_{T\to\infty} \frac{1}{T} \int_{t_0}^T \lambda_1\big(Q_1(t;t_0)\big)dt$$

For the scalar equation (1.2) we note that if $a \leq t_0 < t_1 < +\infty$, then $\eta(t_0) = \eta(t_1) + \int_{t_0}^{t_1} q(t)dt$, $\mu(t_0) = \mu(t_1) + \int_{t_0}^{t_1} q(t)dt$ and hence $0 \leq \eta(t_0) - \mu(t_0) = \text{const}$ for all $t_0 \geq a$. For system (1.1), on the other hand, we have

$$\left.\begin{aligned} \eta(t_0) &\leq \eta(t_1) + \lambda_1\big(Q_1(t_1;t_0)\big) \\ \mu(t_0) &\geq \mu(t_1) + \lambda_n\big(Q_1(t_1;t_0)\big) \end{aligned}\right\} ,\quad a \leq t_0 < t_1 < +\infty$$

so that $\eta(t_0) - \mu(t_0) \leq \eta(t_1) - \mu(t_1) + \lambda_1\big(Q_1(t_1;t_0)\big) - \lambda_n\big(Q_1(t_1;t_0)\big)$.

We may now state and prove the following generalization of the scalar criterion (1.5).

Theorem 2.4: Assume $L(Q) > -\infty$. Then equation (1.1) is oscillatory in case there exists $\delta > 0$ such that for all large t_0 we have

$$\eta(t_0) - \mu(t_0) \geq \delta.$$

Proof: The proof is by contradiction. Assume that $Z = -Y'Y^{-1}$ is a solution of (2.2) on $[t_0,\infty)$, $t_0 \geq a$. Since $\lambda_1(A) \geq \frac{1}{n}\operatorname{tr} A$ for any symmetric matrix A, it follows that $\mu(t_0) > -\infty$. If $\eta(t_0) = +\infty$, then the result follows from Theorem 2.2. Therefore the only case to be considered is when both $\eta(t_0)$ and $\mu(t_0)$ are finite. Assume then that this is the case and define

$$A(t;t_0) \equiv Z(t_0) + \int_{t_0}^{t} Z^2(s)ds \tag{2.4}$$

By Lemma 2.1, $\lim_{t\to\infty} A(t;t_0)$ exists and therefore $\lim_{t\to\infty} \operatorname{tr} A(t;t_0)$ exists. Next choose $\varepsilon > 0$ with $\varepsilon < \frac{\delta}{3}$. Since $\lim_{t\to\infty} \int_{t_0}^{t} \operatorname{tr} Z^2(s)ds$ exists, we may assume that t_0 is sufficiently large so that

$$\int_{s}^{t} \operatorname{tr} Z^2(s)ds < \varepsilon \quad \text{for all} \quad t_0 \leq s < t < +\infty . \tag{2.5}$$

Further, since $0 \leq \lambda_n(Z^2(t)) \leq \lambda_1(Z^2(t)) \leq \operatorname{tr} Z^2(t)$, we may also assume that t_0 is chosen so that

$$\lambda_1(Z^2(t_0)) < \varepsilon^2 \tag{2.6}$$

and hence since $|\lambda_n(Z(t))|^2 = |\lambda_1(-Z(t))|^2 \leq \lambda_1(Z^2(t))$ and $|\lambda_1(Z(t))|^2 \leq \lambda_1(Z^2(t))$ we also have that $|\lambda_n(Z(t_0))| < \varepsilon$ and $|\lambda_1(Z(t_0))| < \varepsilon$. From equation (2.2) we have

$$\lambda_1(Z(t)) \leq \lambda_1(A(t;t_0)) + \lambda_1(Q_1(t;t_0)) \tag{2.7}$$

and

$$\lambda_1(Z(t)) \geq \lambda_n(A(t;t_0)) + \lambda_1(Q_1(t;t_0)) \tag{2.8}$$

(since $\lambda_n(A) + \lambda_1(B) \leq \lambda_1(A+B) \leq \lambda_1(A) + \lambda_1(B)$ for symmetric matrices A,B). From (2.6) and (2.7) we have that $\lambda_1(A(t;t_0)) \leq \lambda_1(Z(t_0)$ $+ \lambda_1(\int_{t_0}^{t} Z^2(s)ds) < \varepsilon + \varepsilon = 2\varepsilon$ for all $t \geq t_0$. Further, we see that $\lambda_n(A(t;t_0)) \geq \lambda_n(Z(t_0)) + \lambda_n(\int_{t_0}^{t} Z^2(s)ds) > -\varepsilon$ for all $t \geq t_0$. Choose $T_n \to \infty$ and $\tau_n \to \infty$ so that

$$\lim_{T_n \to \infty} \frac{1}{T_n} \int_{t_0}^{T_n} \lambda_1(Q_1(t;t_0))dt = \eta(t_0) \tag{2.9}$$

and

$$\lim_{\tau_n \to \infty} \frac{1}{\tau_n} \int_{t_0}^{\tau_n} \lambda_1(Q_1(t;t_0))dt = \mu(t_0) \; . \tag{2.10}$$

From (2.2), (2.7) and (2.10) we have

$$\begin{aligned} \frac{1}{\tau_n} \int_{t_0}^{\tau_n} \lambda_1(Z(t))dt &\leq \frac{1}{\tau_n} \int_{t_0}^{\tau_n} \lambda_1(A(t;t_0)dt + \frac{1}{\tau_n} \int_{t_0}^{\tau_n} \lambda_1(Q_1(t;t_0))dt \\ &< 2\varepsilon + \frac{1}{\tau_n} \int_{t_0}^{\tau_n} \lambda_1(Q_1(t;t_0))dt \end{aligned} \tag{2.11}$$

so that

(2.12) $$\limsup_{\tau_n \to \infty} \frac{1}{\tau_n} \int_{t_0}^{\tau_n} \lambda_1(Z(t)) \leq 2\varepsilon + \mu(t_0) .$$

Similarly, from (2.8) and (2.9) we have

(2.13) $$\liminf_{T_n \to \infty} \frac{1}{T_n} \int_{t_0}^{T_n} \lambda_1(Z(t)) \geq -\varepsilon + \eta(t_0) .$$

Now by the Cauchy-Schwartz inequality, we have for all $T > t_0$

(2.14) $$\begin{aligned} \left|\frac{1}{T} \int_{t_0}^{T} \lambda_1(Z(s))ds\right|^2 &\leq \frac{1}{T} \int_{t_0}^{T} (\lambda_1(Z(s))^2 ds \\ &\leq \frac{1}{T} \left(\int_{t_0}^{\infty} \lambda_1(Z^2(s))ds\right) \to 0 \quad \text{as} \quad T \to \infty . \end{aligned}$$

Hence, (2.12),(2.13) and (2.14) imply

(2.15) $$-\varepsilon + \eta(t_0) \leq 0 \leq 2\varepsilon + \mu(t_0)$$

contradicting the choice of ε. This proves the Theorem.

References

[1] W. Allegretto and L. Erbe, Oscillation criteria for matrix differential inequalities, Can. Math. Bull. 16(1973), 5-10.

[2] G.J. Butler and L.H. Erbe, Oscillation results for second order differential systems, SIAM J. Math. Anal., 17(1986), 19-29.

[3] G.J. Butler and L.H. Erbe, Oscillation results for self-adjoint differential systems, J. Math. Anal. Appl., 115(1986), 470-481.

[4] G.J. Butler, L.H. Erbe and A.B. Mingarelli, Riccati techniques and variational principles in oscillation theory for linear systems, to appear, Trans. Amer. Math. Soc.

[5] R. Byers, B.J. Harris and M.K. Kwong, Weighted means and oscillation conditions for second order matrix differential equations, J. Diff. Eqn. 61(1986), 164-177.

[6] W.A. Coppel, "Disconjugacy", Lecture Notes in Mathematics No. 220, Springer-Verlag, New York, 1971.

[7] G.J. Etgen and R.T. Lewis, Positive functionals and oscillation criteria for second order differential systems, Proc. Edinburgh Math. Soc. 22(1979), 277-290.

[8] W.B. Fite, Concerning the zeros of the solutions of certain differential equations, Trans. Amer. Math. Soc. 19(1918), 341-352.

[9] P. Hartman, "Ordinary Differential Equations", Wiley, New York, 1964.

[10] P. Hartman, Oscillation criteria for self-adjoint second order differential systems and 'Principal sectional curvatures', J. Diff. Eqns. 34(1979), 326-338.

[11] D.B. Hinton and R.T. Lewis, Oscillation theory of generalized second order differential equations, Rocky Mtn. J. Math. 10(1980), 751-766.

[12] Man Kam Kwong and H.G. Kaper, Oscillation of two-dimensional linear second-order differential systems, J. Diff. Eqns. 56(1985), 195-205.

[13] Man Kam Kwong and H.G. Kaper, Oscillation theory for linear second-order differential systems, this proceedings,

[14] Man Kam Kwong, H.G.Kaper, K. Akiyama and A.B. Mingarelli, Oscillation of linear second order differential systems, Proc. Amer. Math. Soc. 91(1984), 85-91.

[15] Man Kam Kwong and A. Zettle, Asymptotically constant functions and second order linear oscillation, J. Math. Anal. Appl. 93(1983), 475-494.

[16] A.B. Mingarelli, On a conjecture for oscillation of second order differential systems, Proc. Amer. Math. Soc. 82(1981), 593-598.

[17] A.B. Mingarelli, An oscillation criterion for second order self-adjoint differential systems, C.R.Math. Rep. Acad. Sci. Canada 2(1980), 287-290.

[18] C. Olech, Z. Opial and T. Wazewski, Sur le probleme d'oscillation des ineégrales de l'équation y" + q(t)y = 0, bull. Acad. Polonaise Sci. cl. III, 5(1957), 621-626.

[19] M. Rab, Kriterien fur die Oszillation der Losungen der Differentialgleichung (p(x)y')' + q(x)y = 0, Cas. Pest. Mat. 84(1959), 335-370.

[20] C.A. Swanson, "Comparison and oscillation theory of linear differential equations", Academic Press, New York, 1968.

[21] E.C. Tomastik, Oscillation of systems of second order differential equations, J. Diff. Eqns. 9(1971), 436-442.

[22] T. Walters, A characterization of positive linear functionals and oscillation criteria for matrix differential equations. Proc. Amer. Math. Soc. 78(1980), 198-202.

[23] D. Willett, Classification of second order linear differential equations with respect to oscillation, Adv. in Math. 3(1969), 594-623.

[24] D. Willett, On the oscillatory behaviour of the solutions of second order linear differential equations. Ann. Polon. Math. 21(1969), 175-194.

[25] A. Wintner, A criterion of oscillatory stability, Quart. Appl. Math. 7(1949), 115-117.

[26] J.S.W. Wong, Oscillation and nonoscillation of solutions of second order linear differential equations with integrable coefficients, Trans. Amer. Math. Soc. 144(1969), 197-215.

Canadian Mathematical Society
Conference Proceedings
Volume 8 (1987)

Testing of Error Estimation Procedures for Boundary Value Problems with a Singularity of the First Kind.

Martin L. Gräff, Ewa B. Weinmüller

Abstract. The application of the Iterated Defect Correction Method for the numerical treatment of singular boundary value problems is considered. The quality of the standard procedures for the local error estimation and the effectivness of the mesh-placement algorithm based on the equidistribution of the local discretization error along the interval of integration is tested. The global error on the adapted mesh is compared with the global discretization error on the equidistant mesh, containing the same number of points.

1. Introduction.

Here we deal with singular boundary value problems of the form

$$(1.1)\qquad \begin{aligned} &y''(t) - \frac{A_1(t)}{t}y'(t) - \frac{A_0(t)}{t^2}y(t) = f(t, y(t)), \qquad 0 < t \le 1, \\ &B_0 Y(0) + B_1 Y(1) = \beta, \qquad Y(t) = (y(t), y'(t))^T, \end{aligned}$$

where y and f are vector-valued functions of dimension n, A_0 and A_1 are $n{\times}n$ matrices, B_0, B_1 are $2n{\times}2n$ constant matrices and β is a $2n$-vector. We assume problem (1.1) to be well-posed and to have a solution $y \in C^1$.

For the approximate solution of (1.1) various numerical methods were proposed: shooting based on Adams predictor-corrector methods or Taylor series methods, collocation and finite differences. In many applications rather rigorous tolerance requirements are to be satisfied, but on the other hand we attempt to avoid unreasonably increasing costs. The method of Iterated Defect Correction (IDeC) seems to be suitable for obtaining a very accurate solution under adequate expense. In order to investigate the possible application of IDeC for singular problems, we tested how standard procedures for the estimation of the local discretization error work, if the singularity of the differential operator is present. The main goal was to decide, whether the quality of the estimates delivered by the error estimation procedure is sufficient to serve the mesh placement algorithm.

Model problems showing somewhat troublesome behavior were used to investigate the error estimation routines and the mesh selection strategy. We could increase the degree of the "troublesomeness" by an appropriate choice of the parameter. Applications from the nonlinear theory of thinwalled shells, where the singularity arises at both interval ends, were also considered.

The outline of the paper follows: In §2 we present the numerical method and the mesh placement algorithm. Results of testing are stated and interpreted in §3.

2. Numerical method.

Both algorithms, Iterated Deferred Correction (IDC), see [1], [5], and Iterated Defect Correction (IDeC), see [2], [7], have been proposed to accelerate the convergence of finite difference schemes. Let us consider (1.1) in the form

$$(2.1)\qquad \begin{aligned} &y''(t) = g(t, y(t), y'(t)), \qquad 0 < t \le 1, \\ &y(0) = 0, \quad y(1) = \beta, \end{aligned}$$

where

$$g(t, y(t), y'(t)) = f(t, y(t)) + \frac{1}{t}A_1(t)y'(t) + \frac{1}{t^2}A_0(t)y(t).$$

We denote by Δ the quasi-uniform partition of $[0,1]$

$$\Delta := \{t_i, i = 0(1)N; t_0 = 0, t_N = 1\},$$

such that $t_i = t_{i-1} + h_i$, $i = 1(1)N$, and $\max_i h_i / \min_i h_i \leq H$. Then the finite difference method based on the three-point discretization yields the finite difference scheme

$$N_\Delta[u_\Delta] := \begin{cases} u_0 = 0, & \\ D_i^2(u_\Delta) - g\left(t_i, u_i, D_i^1(u_\Delta)\right) = 0, & i = 1(1)N-1, \\ u_N - \beta = 0, & \end{cases} \tag{2.2}$$

where $u_\Delta := (u_0, u_1, \ldots, u_N)$ and

$$\begin{aligned} D_i^2(u_\Delta) &= 2\frac{h_{i+1}u_{i-1} - (h_i + h_{i+1})u_i + h_i u_{i+1}}{h_i h_{i+1}(h_i + h_{i+1})}, \\ D_i^1(u_\Delta) &= \frac{-h_{i+1}^2 u_{i-1} + (h_{i+1}^2 - h_i^2)u_i + h_i^2 u_{i+1}}{h_i h_{i+1}(h_i + h_{i+1})}. \end{aligned} \tag{2.3}$$

The finite difference scheme (2.2) provides a nonlinear algebraic system for the unknown vector u_Δ and its solution is computed by Newton's method.

We see from the form of g that its partial derivatives with respect to y and y' are, in general, unbounded for $t \to 0$. In [6] the stability of the scheme (2.2) for $f \equiv f(t)$ was studied. Furthermore, bounds for the global discretization error have been derived. It turns out that the order of convergence depends on the smoothness of y and the spectral properties of $A_0(0)$ and $A_1(0)$. If y and f are sufficiently smooth, then we typically have $h^2|\ln h|^p$, $p \geq 0$.

An algorithm of IDC or IDeC type consists of the folowing two main steps.

1. Step. The base solution u_Δ^0 of (2.2) is generated. We start with an equidistant grid and then we adapt it in such a way that the condition

$$\left\| \tilde{l}_\Delta(u_\Delta^0)(t_i) \right\| \approx \text{const} \leq TOL^0, \tag{2.4}$$

holds for all gridpoints t_i. $\tilde{l}_\Delta(u_\Delta^0)(t_i)$ is an estimate for the local discretization error $l_\Delta(y_\Delta)(t_i)$ defined by (2.6), where $y_\Delta := (y(t_0), y(t_1), \ldots, y(t_N))$. This means that the mesh placement strategy is based on the requirement of the *equidistribution of the local discretization error* along the interval of integration.

2. Step. If we knew the local discretization error $l_\Delta(y_\Delta)(t_i)$, $i = 0(1)N$, and if we solved the following system of equations

$$N_\Delta[u_\Delta] := \begin{cases} u_0 = l_\Delta(y_\Delta)(0), & \\ D_i^2(u_\Delta) - g\left(t_i, u_i, D_i^1(u_\Delta)\right) = l_\Delta(y_\Delta)(t_i), & i = 1(1)N-1, \\ u_N - \beta = l_\Delta(y_\Delta)(1), & \end{cases}$$

instead of (2.2), we would obtain the values $u_i = y(t_i)$ of the exact solution of (2.1). This observation suggests the following iteration

$$N_\Delta[u_\Delta^{k+1}] = \tilde{l}_\Delta[u_\Delta^k], \qquad k = 0, 1, 2, \ldots \tag{2.5}$$

which is carried out on the grid associated with the base solution u_Δ^0. In the classical case, if problem (2.1) is sufficiently smooth and the scheme is defined upon an equidistant grid, one can expect that the quality of the approximate solution u_Δ^k increases by h^2 in each iteration. The global error of u_Δ^0 is $O(h^2)$, the error of u_Δ^1 is $O(h^4)$, the error of u_Δ^2 is $O(h^6)$, etc., provided that the quality of the local discretization error increases in an asymptotically correct way,

$$\tilde{l}_\Delta[u_\Delta^k] = l_\Delta[y_\Delta] + O(h^{2k+4}), \qquad k = 0, 1, 2, \ldots.$$

Here, we restrict our attention to the discussion of the first step. In particular we are interested in testing the quality of the error estimates and the efficency of the mesh selection strategy in case the singularity is present.

2.1 Error estimation for the IDC-procedure.

The idea for the error estimation is based on the explicit knowledge of the series representation of the local discretization error. Let

$$\Delta(h) = \{t_i, i = 0(1)N; t_i = i \cdot h, h = 1/N\}$$

be an equidistant partition of $[0,1]$ in N subintervals of length h. We substitute y_Δ into the scheme (2.2) and obtain

$$\begin{aligned} &l_\Delta(y_\Delta)(0) = 0, \\ (2.6) \qquad &l_\Delta(y_\Delta)(t_i) = \frac{y(t_{i-1}) - 2y(t_i) + y(t_{i+1})}{h^2} - g\Big(t_i, y(t_i), \frac{y(t_{i+1}) - y(t_{i-1})}{2h}\Big), \qquad i = 1(1)N-1, \\ &l_\Delta(y_\Delta)(1) = 0. \end{aligned}$$

If y is sufficiently smooth, we apply Taylor's theorem to (2.6) and deduce

$$(2.7) \qquad l_\Delta[y_\Delta] = \begin{cases} l_\Delta(y_\Delta)(t_i) = h^2\left(\alpha_1 y^{(4)}(t_i) - \dfrac{A_1(t_i)}{t_i}\beta_1 y^{(3)}(t_i)\right) + O(h^4), & i = 1(1)N-1, \\ l_\Delta(y_\Delta)(0) = l_\Delta(y_\Delta)(1) = 0. \end{cases}$$

The error estimate is

$$(2.8) \qquad \tilde{l}_\Delta[u_\Delta^0] = \begin{cases} \tilde{l}_\Delta(u_\Delta^0)(t_i) = h^2\left(\alpha_1 D_i^4(u_\Delta^0) - \dfrac{A_1(t_i)}{t_i}\beta_1 D_i^3(u_\Delta^0)\right), & i = 1(1)N-1, \\ \tilde{l}_\Delta(u_\Delta^0)(0) = \tilde{l}_\Delta(u_\Delta^0)(1) = 0, \end{cases}$$

where D_i^4 and D_i^3 are finite differences approximating $y^{(4)}(t_i)$ and $y^{(3)}(t_i)$, respectively. We choose the finite differences so that

$$D_i^4(y_\Delta) - y^{(4)}(t_i) = O(h^2), \qquad D_i^3(y_\Delta) - y^{(3)}(t_i) = O(h^2),$$

which implies

$$(2.9) \qquad \tilde{l}_\Delta(y_\Delta)(t_i) = l_\Delta(y_\Delta)(t_i) + O(h^4) = l_\Delta(y_\Delta)(t_i)\left(1 + O(h^2)\right).$$

The disadvantage of this procedure is the necessity to know the asymptotic expansion of the local discretization error. If this representation has a complicated structure or if f depends nonlinearly on y' a practicable and efficient application of an IDC-code is almost impossible.

2.2 Error estimation for the IDeC-procedure.

In order to estimate the local discretization error at t_i, we consider the neighboring points $t_i \in [t_j, t_{j+m}]$, $j \le i \le j+m$ and construct a polynomial P_i of degree m such that $P_i(t_l) = u_l$ for all $j \le l \le j+m$. Then

$$(2.10) \qquad \begin{aligned} y''(t) &= g(t, y(t), y'(t)) + P_i''(t) - g(t, P_i(t), P_i'(t)), \\ y(0) &= P_i(0), \qquad y(1) = P_i(1), \end{aligned}$$

is a "local neighboring problem" associated with (2.1), solved *exactly* by $P_i(t)$. Consequently, the local discretization error of (2.10) is known and we can use it as an estimate for the unknown local discretization error (2.6)

$$(2.11) \qquad \tilde{l}_\Delta[u_\Delta^0] = \begin{cases} \tilde{l}_\Delta(u_\Delta^0)(t_i) = D_i^2(u_\Delta^0) - g(t_i, u_i^0, D_i^1(u_\Delta^0)) \\ \qquad\qquad - P_i''(t_i) + g(t_i, P_i(t_i), P_i'(t_i)), & i = 1(1)N-1, \\ \tilde{l}_\Delta(u_\Delta^0)(0) = \tilde{l}_\Delta(u_\Delta^0)(1) = 0. \end{cases}$$

Again, smoothness of y and an appropriate choice of m provide (2.9).

2.3 Mesh selection.

Our grid adjustment strategy is a simple version of an automatic mesh selection presented in [4]. We first solve (2.2) on an equidistant grid with $h = 1/20$ and estimate the local discretization error due to (2.11), obtaining $l_i = \tilde{l}_\Delta(u_\Delta)(t_i)$, $i = 0(1)N$. If, as in classical case, the local discretization error may be written as

$$l_\Delta(y_\Delta)(t_i) = h^2 T_1(t_i) + O(h^4), \tag{2.12}$$

where $T_1(t_i)$ is uniformly bounded in $[0,1]$, then we predict the number of points which have to be inserted in the interval $[t_{i-1}, t_i]$ by

$$k_i = \left\lfloor \sqrt{\frac{|l_i|}{TOL^0}} \right\rfloor, \tag{2.13a}$$

or equivalently, the new stepsize in this interval is

$$h_i^{new} = \frac{h}{k_i + 1}. \tag{2.13b}$$

To clarify this statement we note that by (2.12), if the error estimate is asymptotically correct,

$$|l_\Delta^{new}(y_\Delta)(t_i)| \approx \left| \left(\frac{h}{k_i + 1}\right)^2 T_1(t_i) \right| \approx \left| \frac{l_i}{(k_i + 1)^2} \right| \leq TOL^0,$$

if

$$k_i \geq \sqrt{\frac{|l_i|}{TOL^0}} - 1 \quad \Rightarrow \quad k_i = \left\lfloor \sqrt{\frac{|l_i|}{TOL^0}} \right\rfloor.$$

On the new grid we solve (2.2) again and generate the error estimate for u_Δ^0 in order to compare it with TOL^0.

3. Computational results.

Before presenting the numerical results we make a comment on the structure of the local discretization error of singular problems. For this purpose we consider the following scalar equation

$$y''(t) - \frac{\alpha - 1}{t} y'(t) = 0, \quad y(0) = 0, \quad y(1) = 1, \quad \alpha > 0,$$

whose solution is $y(t) = t^\alpha$. From (2.7) we conclude

$$l_\Delta(y_\Delta)(t_i) = \underbrace{\frac{h^2}{12} y^{(4)}(t_i) - \frac{h^2}{6} \frac{\alpha - 1}{t_i} y^{(3)}(t_i)}_{h^2 T_1(t_i)} + \cdots, \tag{3.1}$$

and we can see that the asymptotic behavior of l_Δ depends on the higher derivatives of the solution y. Depending on α we can distinguish between the following cases:

if $\alpha = 1, 2$ then $l_\Delta(y_\Delta)(t_i) = 0$ and the method is exact;

if $\alpha \neq 1, 2$ then $l_\Delta(y_\Delta)(t_i) \sim \text{const} h^2 t_i^{\alpha - 4}$.

Let $\alpha \geq 4$, i.e. $y \in C^4[0,1]$ and $y^{(3)} \sim t^\beta$, $\beta \geq 1$; then $T_1(t_i)$ is uniformly bounded for all $0 \leq t_i \leq 1$ and as in the classical case

$$l_\Delta(y_\Delta) = O(h^2).$$

If $2 < \alpha < 4$ and $y \in C^2[0,1]$, then for the first i^* points, $t_i = i \cdot h$, $i = 1(1)i^*$, near the singularity

$$l_\Delta(y_\Delta)(t_i) \sim \text{const} h^{\alpha - 2}.$$

Nevertheless, it follows from [6] for the global discretization error

$$\|e_\Delta\| := \max_{0\le i\le N} |u_i - y(t_i)| \le \begin{cases} \text{const}h^\alpha, & 0<\alpha<2,\\ \text{const}h^2, & \alpha\ge 2,\end{cases}$$

or equivalently, the global convergence of the method for each $\alpha > 0$.

We recapitlate: fo r singular problems with a smooth solution the local discretization error does not satisfy $l_\Delta(y_\Delta) = O(h^2)$, in general. If $y^{(3)}(t) \sim t^\beta$, $\beta \ge 1$, then

$$l_\Delta(y_\Delta)(t_i) = O(h^2), \qquad \forall t_i \in \Delta(h), \tag{3.2a}$$

and if $y^{(3)}(t) \sim$ constant, then

$$l_\Delta(y_\Delta)(t_i) = O\left(\frac{h^2}{t_i}\right), \qquad i = 1(1)N. \tag{3.2b}$$

We can interpret (3.2b) in two different ways: either $l_\Delta(y_\Delta) = O(h)$, $\forall t_i \in \Delta(h)$, or $l_\Delta(y_\Delta) = O(h^2)$ for all $t_i \in [t^*, 1]$, where t^* is a fix point, $t^* > 0$.

The aim of numerical experiments was to clarify the following questions.

1. We have observed that possibly

$$l_\Delta(y_\Delta)(t_i) = O\left(\frac{h^2}{t_i^\alpha}\right), \qquad i = 1(1)N, \quad \alpha > 0, \tag{3.3}$$

and we ask how (3.3) influences the efficiency of the mesh adaption, based on the assumption (2.12).

2. Is the quality of the error estimates on the coarse grid sufficient for a reliable mesh placement algorithm?

3. Can the asymptotic behavior of the error estimates (2.9) be observed for decreasing h?

Example 1.

In order to study the effects of (3.3) we consider the already mentioned scalar boundary value problem

$$y''(t) - \frac{\alpha-1}{t}y'(t) = 0, \quad y(0) = 0, \quad y(1) = 1, \quad \alpha = \tfrac{1}{2}, 3, 10.$$

For $\alpha = \frac{1}{2}$, $l_\Delta(y_\Delta) = O(h^{-\frac{3}{2}})$ and thus the idea to equidistribute the local discretization error cannot be realized in a reasonable way. Furthermore, we observe a bad quality of the error estimates due to the large global error, $e_\Delta(y_\Delta) = O(\sqrt{h})$ in Problem 1.1. If $\alpha = 3$ then $l_\Delta(y_\Delta) = O(h)$ near the singularity and the mesh-selection algorithm based on the assumption (2.12) generates a grid which is not fine enough there. Consequently, the local error is bounded by 10^{-2} approximately, which exceeds the prescribed tolerance TOL^0 by a factor of 10 in Problem 1.2. For $\alpha = 10$, Problem 1.3, (3.2a) holds and on the adapted mesh the equidistribution of the local error has been achieved; $TOL^0 = 10^{-3}$.

Example 2.

The scalar boundary value problem

$$y''(t) + \frac{1}{t}y'(t) - \frac{1}{t^2}y(t) = \alpha^2 y(t) + ct^{k-2}e^{-\alpha t}(k^2 - 1 - \alpha t(1+2k)),$$
$$y(0) = 0, \qquad y(1) = ce^{-\alpha}$$

is solved by $y(t) = t^k e^{-\alpha t}$, and we choose c in such a way that $\max_{0\le t\le 1}|y(t)| = 1$. Here, we record the results for the cases

$$\underbrace{k=4,\ \alpha=8}_{\text{Problem 2.1}}; \qquad \underbrace{k=64,\ \alpha=128}_{\text{Problem 2.2}}; \qquad \underbrace{k=4,\ \alpha=32}_{\text{Problem 2.3}}.$$

For all problems $l_\Delta(y_\Delta) = O(h^2)$, and $e_\Delta(y_\Delta) = O(h^2)$ and for $h \to 0$ we observe the asymptotic behavior of the error estimates, according to (2.9). The difficulty in Problem 2.2 consists in strongly varying large discretization error in the middle of the interval, in Problem 2.3 near the singularity (higher derivatives of the solution y are large). For all problems the mesh is refined in a reasonable way – the place of the troublesome behavior of the local error is properly indicated – but the quality of the error estimates on the starting grid is not sufficient to bound the local error on the new grid by TOL^0. For Problem 2.1, $TOL^0 = 10^{-3}$ and for Problems 2.2 and 2.3, $TOL^0 = 5\cdot 10^{-3}$.

Notation in tables and plots

The global error, the local discretization error and the errors of the estimation procedures on an equidistant grid with stepsize h are e_h, l_h, $d_h^{IDC} = \tilde{l}_h^{IDC} - l_h$ and $d_h^{IDeC} = \tilde{l}_h^{IDeC} - l_h$, respectively; t_e and t_l denote the gridpoints, where $\|e_h\|$ and $\|l_h\|$, respectively, occur.

References

[1] L. Fox, The Numerical Solution of Two-Point Boundary Value Problems in Ordinary Differential Equations, Oxford Univ. Press, New York, 1957. MR 21#972.

[2] R. Frank, J. Hertling, Ch. Überhuber, An extension of the applicability of iterated deferred corrections, Math. Comp., 31(1977), pp. 907–915.

[3] M. Gräff, E. Weinmüller, Schätzungen des lokalen Diskretisierungsfehlers bei singulären Randwertproblemen, Report 66/86, Institute for Applied and Numerical Mathematics, Technical University, Vienna, 1986.

[4] M. Lentini, V. Pereyra, An adaptive finite difference solver for nonlinear two-point boundary problems with mild boundary layers, SIAM J. Numer. Anal., 14(1977), pp. 91-111.

[5] V. Pereyra, Iterated deferred corrections for nonlinear boundary value problems, Numer. Math., 11(1968), pp. 111-125. MR 37#1091.

[6] E. Weinmüller, A difference method for a singular boundary value problem of second order, Math. Comp., 42(1984), pp. 441–464.

[7] P. E. Zadunaisky, A method for the estimation of errors propagated in the numerical solution of ordinary differential equations, Proc. Internat. Astronom. Union Symposium No.25 (Thessaloniki,1964), Academic Press, New York, 1966.

M. L. Gräff
E. B. Weinmüller
Institut für Angewandte und Numerische Mathematik
Technische Universität Wien
Wiedner Hauptstrasse 6–10
A-1040 Wien
Austria

Example 1.1

h	$\|e_h\|$	t_e	$\|e_h\|/h^{0.5}$	$\|l_h\|$	t_l	$\|l_h\|/h^{-1.5}$
1/10	$7.6925 \cdot 10^{-2}$	0.1000	$2.4326 \cdot 10^{-1}$	$7.3439 \cdot 10^{0}$	0.1000	$2.3223 \cdot 10^{-1}$
1/20	$6.0103 \cdot 10^{-2}$	0.0500	$2.6879 \cdot 10^{-1}$	$2.0772 \cdot 10^{1}$	0.0500	$2.3223 \cdot 10^{-1}$
1/40	$4.5199 \cdot 10^{-2}$	0.0250	$2.8586 \cdot 10^{-1}$	$5.8751 \cdot 10^{1}$	0.0250	$2.3223 \cdot 10^{-1}$
1/80	$3.3259 \cdot 10^{-2}$	0.0125	$2.9748 \cdot 10^{-1}$	$1.6617 \cdot 10^{2}$	0.0125	$2.3223 \cdot 10^{-1}$
1/160	$2.4149 \cdot 10^{-2}$	0.0063	$3.0546 \cdot 10^{-1}$	$4.7001 \cdot 10^{2}$	0.0063	$2.3223 \cdot 10^{-1}$
1/320	$1.7431 \cdot 10^{-2}$	0.0063	$3.1182 \cdot 10^{-1}$	$1.3294 \cdot 10^{3}$	0.0031	$2.3223 \cdot 10^{-1}$
1/640	$1.2570 \cdot 10^{-2}$	0.0031	$3.1800 \cdot 10^{-1}$	$3.7601 \cdot 10^{3}$	0.0016	$2.3223 \cdot 10^{-1}$

$t = 0.025$				
h	$e_h(t)$	$e_h(t)/h^{0.5}$	$l_h(t)$	$l_h(t)/h^2$
1/40	$-4.5199 \cdot 10^{-2}$	$-2.8586 \cdot 10^{-1}$	$-5.8751 \cdot 10^{1}$	$-9.4001 \cdot 10^{4}$
1/80	$-3.2443 \cdot 10^{-2}$	$-2.9018 \cdot 10^{-1}$	$-3.4847 \cdot 10^{0}$	$-2.2302 \cdot 10^{4}$
1/160	$-2.2956 \cdot 10^{-2}$	$-2.9037 \cdot 10^{-1}$	$-7.6935 \cdot 10^{-1}$	$-1.9695 \cdot 10^{4}$
1/320	$-1.6174 \cdot 10^{-2}$	$-2.8933 \cdot 10^{-1}$	$-1.8700 \cdot 10^{-1}$	$-1.9148 \cdot 10^{4}$
1/640	$-1.1390 \cdot 10^{-2}$	$-2.8815 \cdot 10^{-1}$	$-4.6428 \cdot 10^{-2}$	$-1.9017 \cdot 10^{4}$

h	$l_h(t)$	$d_h^{\mathrm{IDeC}}(t)$	$d_h^{\mathrm{IDeC}}(t)/h^{2.5}$	$d_h^{\mathrm{IDeM}}(t)$	$d_h^{\mathrm{IDeM}}(t)/h^{2.5}$
1/40	$-5.8751 \cdot 10^{1}$	$6.1221 \cdot 10^{1}$	$6.1952 \cdot 10^{5}$	$5.8590 \cdot 10^{1}$	$5.9289 \cdot 10^{5}$
1/80	$-3.4847 \cdot 10^{0}$	$-1.6711 \cdot 10^{0}$	$-9.5658 \cdot 10^{4}$	$-5.1268 \cdot 10^{-1}$	$-2.9348 \cdot 10^{4}$
1/160	$-7.6935 \cdot 10^{-1}$	$-7.8697 \cdot 10^{-2}$	$-2.5483 \cdot 10^{4}$	$6.8330 \cdot 10^{-2}$	$2.2126 \cdot 10^{4}$
1/320	$-1.8700 \cdot 10^{-1}$	$-6.8244 \cdot 10^{-3}$	$-1.2501 \cdot 10^{4}$	$-7.0083 \cdot 10^{-4}$	$-1.2838 \cdot 10^{3}$
1/640	$-4.6428 \cdot 10^{-2}$	$-8.2417 \cdot 10^{-4}$	$-8.5401 \cdot 10^{3}$	$-4.6856 \cdot 10^{-4}$	$-4.8553 \cdot 10^{3}$

$t = 0.8$				
h	$e_h(t)$	$e_h(t)/h^{0.5}$	$l_h(t)$	$l_h(t)/h^2$
1/10	$-1.2544 \cdot 10^{-2}$	$-3.9668 \cdot 10^{-2}$	$-1.0330 \cdot 10^{-3}$	$-1.0330 \cdot 10^{-1}$
1/20	$-8.6096 \cdot 10^{-3}$	$-3.8503 \cdot 10^{-2}$	$-2.5648 \cdot 10^{-4}$	$-1.0259 \cdot 10^{-1}$
1/40	$-5.9552 \cdot 10^{-3}$	$-3.7664 \cdot 10^{-2}$	$-6.4011 \cdot 10^{-5}$	$-1.0242 \cdot 10^{-1}$
1/80	$-4.1448 \cdot 10^{-3}$	$-3.7072 \cdot 10^{-2}$	$-1.5996 \cdot 10^{-5}$	$-1.0237 \cdot 10^{-1}$
1/160	$-2.8981 \cdot 10^{-3}$	$-3.6658 \cdot 10^{-2}$	$-3.9982 \cdot 10^{-6}$	$-1.0235 \cdot 10^{-1}$
1/320	$-2.0330 \cdot 10^{-3}$	$-3.6368 \cdot 10^{-2}$	$-9.9897 \cdot 10^{-7}$	$-1.0229 \cdot 10^{-1}$
1/640	$-1.4296 \cdot 10^{-3}$	$-3.6165 \cdot 10^{-2}$	$-2.4867 \cdot 10^{-7}$	$-1.0186 \cdot 10^{-1}$

h	$l_h(t)$	$d_h^{\mathrm{IDeC}}(t)$	$d_h^{\mathrm{IDeC}}(t)/h^{2.5}$	$d_h^{\mathrm{IDeM}}(t)$	$d_h^{\mathrm{IDeM}}(t)/h^{2.5}$
1/10	$-1.0330 \cdot 10^{-3}$	$-1.4298 \cdot 10^{-4}$	$-4.5214 \cdot 10^{-2}$	$-2.8678 \cdot 10^{-5}$	$-9.0689 \cdot 10^{-3}$
1/20	$-2.5648 \cdot 10^{-4}$	$-2.2110 \cdot 10^{-5}$	$-3.9551 \cdot 10^{-2}$	$-1.9999 \cdot 10^{-5}$	$-3.5775 \cdot 10^{-2}$
1/40	$-6.4011 \cdot 10^{-5}$	$-3.6832 \cdot 10^{-6}$	$-3.7271 \cdot 10^{-2}$	$-3.5573 \cdot 10^{-6}$	$-3.5998 \cdot 10^{-2}$
1/80	$-1.5996 \cdot 10^{-5}$	$-6.3258 \cdot 10^{-7}$	$-3.6211 \cdot 10^{-2}$	$-6.2483 \cdot 10^{-7}$	$-3.5767 \cdot 10^{-2}$
1/160	$-3.9982 \cdot 10^{-6}$	$-1.1059 \cdot 10^{-7}$	$-3.5811 \cdot 10^{-2}$	$-1.0993 \cdot 10^{-7}$	$-3.5598 \cdot 10^{-2}$
1/320	$-9.9897 \cdot 10^{-7}$	$-2.0275 \cdot 10^{-8}$	$-3.7140 \cdot 10^{-2}$	$-1.9875 \cdot 10^{-8}$	$-3.6407 \cdot 10^{-2}$
1/640	$-2.4867 \cdot 10^{-7}$	$-8.7751 \cdot 10^{-9}$	$-9.0929 \cdot 10^{-2}$	$-4.6096 \cdot 10^{-9}$	$-4.7765 \cdot 10^{-2}$

Example 1.1

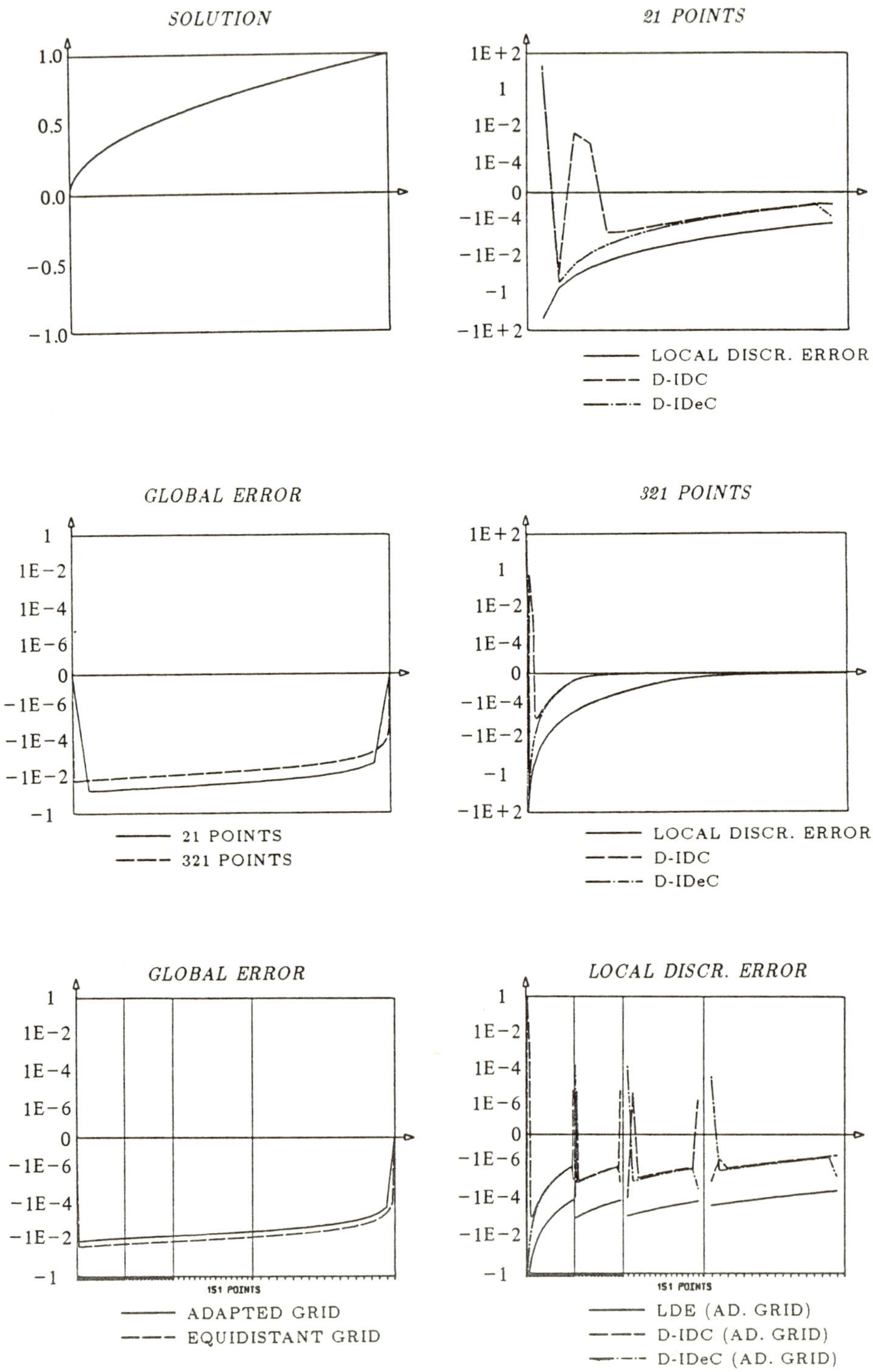
SOLUTION
1.0
0.5
0.0
-0.5
-1.0
21 POINTS
1E+2
1
1E-2
1E-4
0
-1E-4
-1E-2
-1
-1E+2
LOCAL DISCR. ERROR
D-IDC
D-IDeC
GLOBAL ERROR
1
1E-2
1E-4
1E-6
0
-1E-6
-1E-4
-1E-2
-1
21 POINTS
321 POINTS
321 POINTS
1E+2
1
1E-2
1E-4
0
-1E-4
-1E-2
-1
-1E+2
LOCAL DISCR. ERROR
D-IDC
D-IDeC
GLOBAL ERROR
1
1E-2
1E-4
1E-6
0
-1E-6
-1E-4
-1E-2
-1
151 POINTS
ADAPTED GRID
EQUIDISTANT GRID
LOCAL DISCR. ERROR
1
1E-2
1E-4
1E-6
0
-1E-6
-1E-4
-1E-2
-1
151 POINTS
LDE (AD. GRID)
D-IDC (AD. GRID)
D-IDeC (AD. GRID)

Example 1.2

h	$\|e_h\|$	t_e	$\|e_h\|/h^2$	$\|l_h\|$	t_l	$\|l_h\|/h^1$
1/10	$3.8788 \cdot 10^{-3}$	0.6000	$3.8788 \cdot 10^{-1}$	$2.0000 \cdot 10^{-1}$	0.1000	$2.0000 \cdot 10^{0}$
1/20	$9.6241 \cdot 10^{-4}$	0.6000	$3.8496 \cdot 10^{-1}$	$1.0000 \cdot 10^{-1}$	0.0500	$2.0000 \cdot 10^{0}$
1/40	$2.4071 \cdot 10^{-4}$	0.5750	$3.8513 \cdot 10^{-1}$	$5.0000 \cdot 10^{-2}$	0.0250	$2.0000 \cdot 10^{0}$
1/80	$6.0149 \cdot 10^{-5}$	0.5750	$3.8495 \cdot 10^{-1}$	$2.5000 \cdot 10^{-2}$	0.0125	$2.0000 \cdot 10^{0}$
1/160	$1.5035 \cdot 10^{-5}$	0.5750	$3.8491 \cdot 10^{-1}$	$1.2500 \cdot 10^{-2}$	0.0063	$2.0000 \cdot 10^{0}$
1/320	$3.7588 \cdot 10^{-6}$	0.5781	$3.8490 \cdot 10^{-1}$	$6.2500 \cdot 10^{-3}$	0.0031	$2.0000 \cdot 10^{0}$
1/640	$9.3965 \cdot 10^{-7}$	0.5766	$3.8488 \cdot 10^{-1}$	$3.1250 \cdot 10^{-3}$	0.0016	$2.0000 \cdot 10^{0}$

$t = 0.025$				
h	$e_h(t)$	$e_h(t)/h^2$	$l_h(t)$	$l_h(t)/h^2$
1/40	$-1.5625 \cdot 10^{-5}$	$-2.5000 \cdot 10^{-2}$	$-5.0000 \cdot 10^{-2}$	$-8.0000 \cdot 10^{1}$
1/80	$-3.9044 \cdot 10^{-6}$	$-2.4988 \cdot 10^{-2}$	$-1.2500 \cdot 10^{-2}$	$-8.0000 \cdot 10^{1}$
1/160	$-9.7599 \cdot 10^{-7}$	$-2.4985 \cdot 10^{-2}$	$-3.1250 \cdot 10^{-3}$	$-8.0000 \cdot 10^{1}$
1/320	$-2.4399 \cdot 10^{-7}$	$-2.4985 \cdot 10^{-2}$	$-7.8125 \cdot 10^{-4}$	$-8.0000 \cdot 10^{1}$
1/640	$-6.0997 \cdot 10^{-8}$	$-2.4984 \cdot 10^{-2}$	$-1.9531 \cdot 10^{-4}$	$-8.0000 \cdot 10^{1}$

h	$l_h(t)$	$d_h^{IDC}(t)$	$d_h^{IDC}(t)/h^4$	$d_h^{IDeC}(t)$	$d_h^{IDeC}(t)/h^4$
1/40	$-5.0000 \cdot 10^{-2}$	$-3.1270 \cdot 10^{-5}$	$-8.0050 \cdot 10^{1}$	$-1.6569 \cdot 10^{-2}$	$-4.2418 \cdot 10^{4}$
1/80	$-1.2500 \cdot 10^{-2}$	$-1.9534 \cdot 10^{-6}$	$-8.0012 \cdot 10^{1}$	$-1.8078 \cdot 10^{-3}$	$-7.4047 \cdot 10^{4}$
1/160	$-3.1250 \cdot 10^{-3}$	$-1.2207 \cdot 10^{-7}$	$-8.0000 \cdot 10^{1}$	$-9.5612 \cdot 10^{-5}$	$-6.2660 \cdot 10^{4}$
1/320	$-7.8125 \cdot 10^{-4}$	$-7.6263 \cdot 10^{-9}$	$-7.9967 \cdot 10^{1}$	$-6.1047 \cdot 10^{-6}$	$-6.4013 \cdot 10^{4}$
1/640	$-1.9531 \cdot 10^{-4}$	$-3.9595 \cdot 10^{-10}$	$-6.6429 \cdot 10^{1}$	$-3.8192 \cdot 10^{-7}$	$-6.4076 \cdot 10^{4}$

$t = 0.8$				
h	$e_h(t)$	$e_h(t)/h^2$	$l_h(t)$	$l_h(t)/h^2$
1/10	$-2.9091 \cdot 10^{-3}$	$-2.9091 \cdot 10^{-1}$	$-2.5000 \cdot 10^{-2}$	$-2.5000 \cdot 10^{0}$
1/20	$-7.2180 \cdot 10^{-4}$	$-2.8872 \cdot 10^{-1}$	$-6.2500 \cdot 10^{-3}$	$-2.5000 \cdot 10^{0}$
1/40	$-1.8011 \cdot 10^{-4}$	$-2.8818 \cdot 10^{-1}$	$-1.5625 \cdot 10^{-3}$	$-2.5000 \cdot 10^{0}$
1/80	$-4.5007 \cdot 10^{-5}$	$-2.8804 \cdot 10^{-1}$	$-3.9062 \cdot 10^{-4}$	$-2.5000 \cdot 10^{0}$
1/160	$-1.1250 \cdot 10^{-5}$	$-2.8801 \cdot 10^{-1}$	$-9.7656 \cdot 10^{-5}$	$-2.5000 \cdot 10^{0}$
1/320	$-2.8125 \cdot 10^{-6}$	$-2.8800 \cdot 10^{-1}$	$-2.4414 \cdot 10^{-5}$	$-2.5000 \cdot 10^{0}$
1/640	$-7.0308 \cdot 10^{-7}$	$-2.8798 \cdot 10^{-1}$	$-6.1020 \cdot 10^{-6}$	$-2.4994 \cdot 10^{0}$

h	$l_h(t)$	$d_h^{IDC}(t)$	$d_h^{IDC}(t)/h^4$	$d_h^{IDeC}(t)$	$d_h^{IDeC}(t)/h^4$
1/10	$-2.5000 \cdot 10^{-2}$	$-2.5253 \cdot 10^{-4}$	$-2.5253 \cdot 10^{0}$	$-4.4905 \cdot 10^{-4}$	$-4.4905 \cdot 10^{0}$
1/20	$-6.2500 \cdot 10^{-3}$	$-1.5664 \cdot 10^{-5}$	$-2.5063 \cdot 10^{0}$	$-2.7901 \cdot 10^{-5}$	$-4.4642 \cdot 10^{0}$
1/40	$-1.5625 \cdot 10^{-3}$	$-9.7720 \cdot 10^{-7}$	$-2.5016 \cdot 10^{0}$	$-1.7406 \cdot 10^{-6}$	$-4.4559 \cdot 10^{0}$
1/80	$-3.9062 \cdot 10^{-4}$	$-6.1125 \cdot 10^{-8}$	$-2.5037 \cdot 10^{0}$	$-1.0878 \cdot 10^{-7}$	$-4.4557 \cdot 10^{0}$
1/160	$-9.7656 \cdot 10^{-5}$	$-4.0896 \cdot 10^{-9}$	$-2.6802 \cdot 10^{0}$	$-6.8983 \cdot 10^{-9}$	$-4.5208 \cdot 10^{0}$
1/320	$-2.4414 \cdot 10^{-5}$	$-4.3426 \cdot 10^{-10}$	$-4.5535 \cdot 10^{0}$	$-5.8661 \cdot 10^{-10}$	$-6.1510 \cdot 10^{0}$
1/640	$-6.1020 \cdot 10^{-6}$	$-1.1005 \cdot 10^{-8}$	$-1.8463 \cdot 10^{3}$	$-1.5515 \cdot 10^{-9}$	$-2.6031 \cdot 10^{2}$

Example 1.2

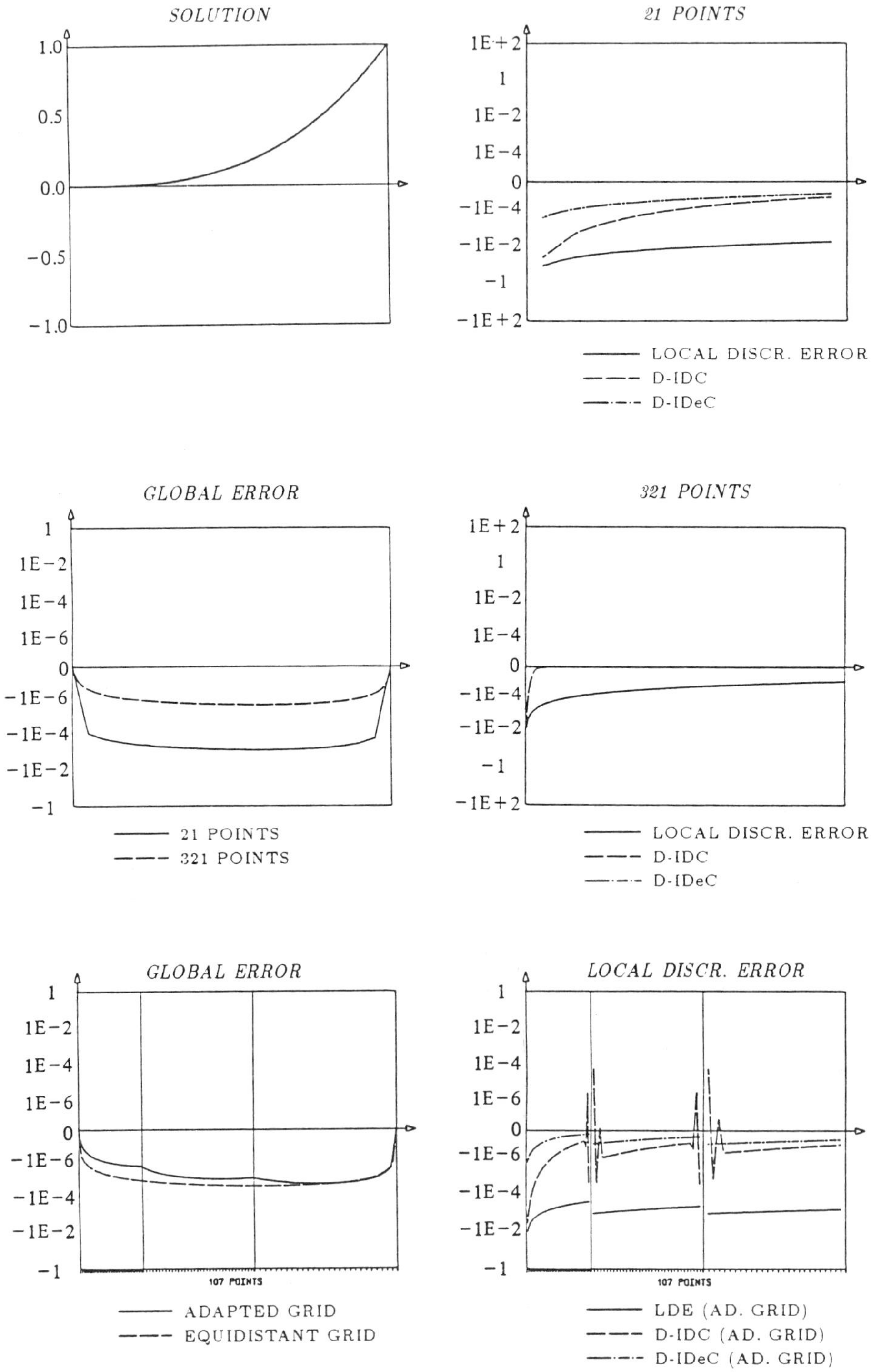
SOLUTION
1.0
0.5
0.0
−0.5
−1.0
21 POINTS
1E+2
1
1E−2
1E−4
0
−1E−4
−1E−2
−1
−1E+2
LOCAL DISCR. ERROR
D-IDC
D-IDeC
GLOBAL ERROR
1
1E−2
1E−4
1E−6
0
−1E−6
−1E−4
−1E−2
−1
21 POINTS
321 POINTS
321 POINTS
1E+2
1
1E−2
1E−4
0
−1E−4
−1E−2
−1
−1E+2
LOCAL DISCR. ERROR
D-IDC
D-IDeC
GLOBAL ERROR
1
1E−2
1E−4
1E−6
0
−1E−6
−1E−4
−1E−2
−1
107 POINTS
ADAPTED GRID
EQUIDISTANT GRID
LOCAL DISCR. ERROR
1
1E−2
1E−4
1E−6
0
−1E−6
−1E−4
−1E−2
−1
107 POINTS
LDE (AD. GRID)
D-IDC (AD. GRID)
D-IDeC (AD. GRID)

Example 1.3

h	$\|e_h\|$	t_e	$\|e_h\|/h^2$	$\|l_h\|$	t_l	$\|l_h\|/h^2$
1/10	$3.8334\cdot 10^{-2}$	0.9000	$3.8334\cdot 10^{0}$	$3.6296\cdot 10^{0}$	0.9000	$3.6296\cdot 10^{2}$
1/20	$8.6871\cdot 10^{-3}$	0.9000	$3.4748\cdot 10^{0}$	$1.2223\cdot 10^{0}$	0.9500	$4.8893\cdot 10^{2}$
1/40	$2.1240\cdot 10^{-3}$	0.9000	$3.3983\cdot 10^{0}$	$3.5502\cdot 10^{-1}$	0.9750	$5.6803\cdot 10^{2}$
1/80	$5.2811\cdot 10^{-4}$	0.9000	$3.3799\cdot 10^{0}$	$9.5671\cdot 10^{-2}$	0.9875	$6.1230\cdot 10^{2}$
1/160	$1.3206\cdot 10^{-4}$	0.8938	$3.3806\cdot 10^{0}$	$2.4832\cdot 10^{-2}$	0.9938	$6.3570\cdot 10^{2}$
1/320	$3.3003\cdot 10^{-5}$	0.8938	$3.3795\cdot 10^{0}$	$6.3256\cdot 10^{-3}$	0.9969	$6.4774\cdot 10^{2}$
1/640	$8.2500\cdot 10^{-6}$	0.8938	$3.3792\cdot 10^{0}$	$1.5963\cdot 10^{-3}$	0.9984	$6.5383\cdot 10^{2}$

$t = 0.8$				
h	$e_h(t)$	$e_h(t)/h^2$	$l_h(t)$	$l_h(t)/h^2$
1/10	$-2.6914\cdot 10^{-2}$	$-2.6914\cdot 10^{0}$	$-1.8065\cdot 10^{0}$	$-1.8065\cdot 10^{2}$
1/20	$-6.3432\cdot 10^{-3}$	$-2.5373\cdot 10^{0}$	$-4.3728\cdot 10^{-1}$	$-1.7491\cdot 10^{2}$
1/40	$-1.5642\cdot 10^{-3}$	$-2.5027\cdot 10^{0}$	$-1.0843\cdot 10^{-1}$	$-1.7349\cdot 10^{2}$
1/80	$-3.8972\cdot 10^{-4}$	$-2.4942\cdot 10^{0}$	$-2.7052\cdot 10^{-2}$	$-1.7313\cdot 10^{2}$
1/160	$-9.7348\cdot 10^{-5}$	$-2.4921\cdot 10^{0}$	$-6.7596\cdot 10^{-3}$	$-1.7304\cdot 10^{2}$
1/320	$-2.4332\cdot 10^{-5}$	$-2.4916\cdot 10^{0}$	$-1.6897\cdot 10^{-3}$	$-1.7302\cdot 10^{2}$
1/640	$-6.0826\cdot 10^{-6}$	$-2.4914\cdot 10^{0}$	$-4.2240\cdot 10^{-4}$	$-1.7302\cdot 10^{2}$

h	$l_h(t)$	$d_h^{\mathrm{IDC}}(t)$	$d_h^{\mathrm{IDC}}(t)/h^4$	$d_h^{\mathrm{IDeC}}(t)$	$d_h^{\mathrm{IDeC}}(t)/h^4$
1/10	$-1.8065\cdot 10^{0}$	$-5.3901\cdot 10^{-1}$	$-5.3901\cdot 10^{3}$	$-7.9882\cdot 10^{-2}$	$-7.9882\cdot 10^{2}$
1/20	$-4.3728\cdot 10^{-1}$	$-2.6876\cdot 10^{-2}$	$-4.3001\cdot 10^{3}$	$1.0970\cdot 10^{-2}$	$1.7552\cdot 10^{3}$
1/40	$-1.0843\cdot 10^{-1}$	$-1.5949\cdot 10^{-3}$	$-4.0830\cdot 10^{3}$	$6.1650\cdot 10^{-4}$	$1.5782\cdot 10^{3}$
1/80	$-2.7052\cdot 10^{-2}$	$-9.8424\cdot 10^{-5}$	$-4.0314\cdot 10^{3}$	$3.7596\cdot 10^{-5}$	$1.5399\cdot 10^{3}$
1/160	$-6.7596\cdot 10^{-3}$	$-6.1323\cdot 10^{-6}$	$-4.0189\cdot 10^{3}$	$2.3352\cdot 10^{-6}$	$1.5304\cdot 10^{3}$
1/320	$-1.6897\cdot 10^{-3}$	$-3.8490\cdot 10^{-7}$	$-4.0360\cdot 10^{3}$	$1.4547\cdot 10^{-7}$	$1.5254\cdot 10^{3}$
1/640	$-4.2240\cdot 10^{-4}$	$-2.9402\cdot 10^{-8}$	$-4.9328\cdot 10^{3}$	$7.0035\cdot 10^{-9}$	$1.1750\cdot 10^{3}$

Example 1.3

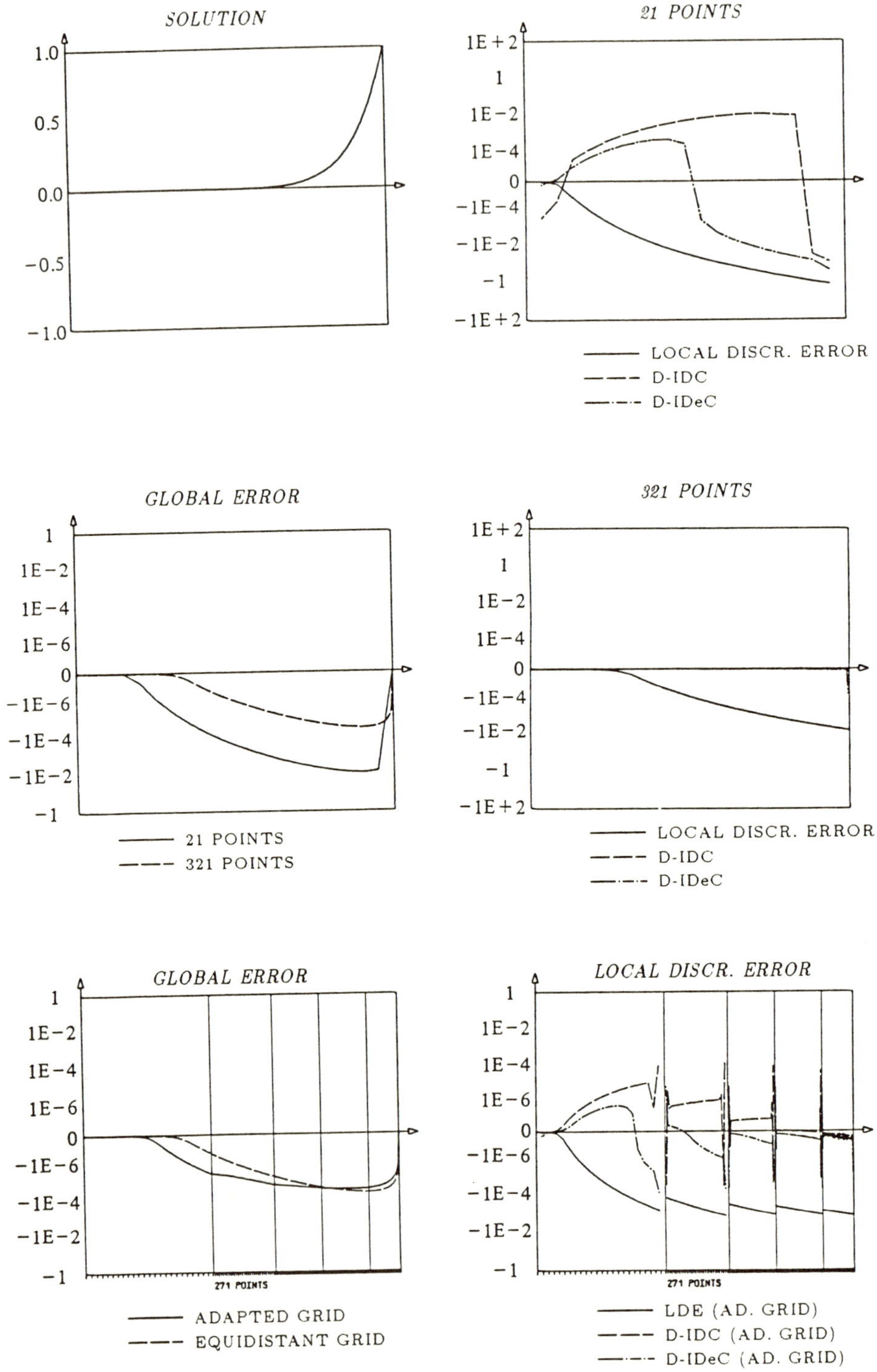
SOLUTION
1.0
0.5
0.0
−0.5
−1.0
21 POINTS
1E+2
1
1E−2
1E−4
0
−1E−4
−1E−2
−1
−1E+2
LOCAL DISCR. ERROR
D-IDC
D-IDeC
GLOBAL ERROR
1
1E−2
1E−4
1E−6
0
−1E−6
−1E−4
−1E−2
−1
21 POINTS
321 POINTS
321 POINTS
1E+2
1
1E−2
1E−4
0
−1E−4
−1E−2
−1
−1E+2
LOCAL DISCR. ERROR
D-IDC
D-IDeC
GLOBAL ERROR
1
1E−2
1E−4
1E−6
0
−1E−6
−1E−4
−1E−2
−1
271 POINTS
ADAPTED GRID
EQUIDISTANT GRID
LOCAL DISCR. ERROR
1
1E−2
1E−4
1E−6
0
−1E−6
−1E−4
−1E−2
−1
271 POINTS
LDE (AD. GRID)
D-IDC (AD. GRID)
D-IDeC (AD. GRID)

Example 2.1

h	$\|e_h\|$	t_e	$\|e_h\|/h^2$	$\|l_h\|$	t_l	$\|l_h\|/h^2$
1/10	$1.2245\cdot 10^{-2}$	0.2000	$1.2245\cdot 10^{0}$	$2.6037\cdot 10^{0}$	0.2000	$2.6037\cdot 10^{2}$
1/20	$3.5952\cdot 10^{-3}$	0.1500	$1.4381\cdot 10^{0}$	$2.2364\cdot 10^{0}$	0.0500	$8.9457\cdot 10^{2}$
1/40	$8.8408\cdot 10^{-4}$	0.1500	$1.4145\cdot 10^{0}$	$1.5241\cdot 10^{0}$	0.0250	$2.4386\cdot 10^{3}$
1/80	$2.2029\cdot 10^{-4}$	0.1500	$1.4099\cdot 10^{0}$	$5.6888\cdot 10^{-1}$	0.0125	$3.6408\cdot 10^{3}$
1/160	$5.5031\cdot 10^{-5}$	0.1500	$1.4088\cdot 10^{0}$	$1.7129\cdot 10^{-1}$	0.0063	$4.3850\cdot 10^{3}$
1/320	$1.3761\cdot 10^{-5}$	0.1531	$1.4091\cdot 10^{0}$	$4.6859\cdot 10^{-2}$	0.0031	$4.7983\cdot 10^{3}$
1/640	$3.4403\cdot 10^{-6}$	0.1516	$1.4091\cdot 10^{0}$	$1.2246\cdot 10^{-2}$	0.0016	$5.0160\cdot 10^{3}$

$t = 0.025$				
h	$e_h(t)$	$e_h(t)/h^2$	$l_h(t)$	$l_h(t)/h^2$
1/40	$3.8490\cdot 10^{-4}$	$6.1585\cdot 10^{-1}$	$1.5241\cdot 10^{0}$	$2.4386\cdot 10^{3}$
1/80	$9.3238\cdot 10^{-5}$	$5.9672\cdot 10^{-1}$	$3.8931\cdot 10^{-1}$	$2.4916\cdot 10^{3}$
1/160	$2.3096\cdot 10^{-5}$	$5.9127\cdot 10^{-1}$	$9.7841\cdot 10^{-2}$	$2.5047\cdot 10^{3}$
1/320	$5.7603\cdot 10^{-6}$	$5.8986\cdot 10^{-1}$	$2.4492\cdot 10^{-2}$	$2.5080\cdot 10^{3}$
1/640	$1.4392\cdot 10^{-6}$	$5.8950\cdot 10^{-1}$	$6.1250\cdot 10^{-3}$	$2.5088\cdot 10^{3}$

h	$l_h(t)$	$d_h^{\mathrm{IDeC}}(t)$	$d_h^{\mathrm{IDeC}}(t)/h^4$	$d_h^{\mathrm{IDeC}}(t)$	$d_h^{\mathrm{IDeC}}(t)/h^4$
1/40	$1.5241\cdot 10^{0}$	$-5.8873\cdot 10^{-2}$	$-1.5071\cdot 10^{5}$	$-3.0690\cdot 10^{-2}$	$-7.8567\cdot 10^{4}$
1/80	$3.8931\cdot 10^{-1}$	$-4.4013\cdot 10^{-3}$	$-1.8028\cdot 10^{5}$	$-1.1963\cdot 10^{-2}$	$-4.9002\cdot 10^{5}$
1/160	$9.7841\cdot 10^{-2}$	$-2.6765\cdot 10^{-4}$	$-1.7541\cdot 10^{5}$	$-4.4918\cdot 10^{-5}$	$-2.9437\cdot 10^{4}$
1/320	$2.4492\cdot 10^{-2}$	$-1.6697\cdot 10^{-5}$	$-1.7508\cdot 10^{5}$	$-1.6433\cdot 10^{-6}$	$-1.7231\cdot 10^{4}$
1/640	$6.1250\cdot 10^{-3}$	$-1.0432\cdot 10^{-6}$	$-1.7502\cdot 10^{5}$	$-1.0403\cdot 10^{-7}$	$-1.7454\cdot 10^{4}$

$t = 0.8$				
h	$e_h(t)$	$e_h(t)/h^2$	$l_h(t)$	$l_h(t)/h^2$
1/10	$-7.4307\cdot 10^{-4}$	$-7.4307\cdot 10^{-2}$	$-1.3334\cdot 10^{-1}$	$-1.3334\cdot 10^{1}$
1/20	$-1.8562\cdot 10^{-4}$	$-7.4248\cdot 10^{-2}$	$-3.3715\cdot 10^{-2}$	$-1.3486\cdot 10^{1}$
1/40	$-4.6415\cdot 10^{-5}$	$-7.4265\cdot 10^{-2}$	$-8.4524\cdot 10^{-3}$	$-1.3524\cdot 10^{1}$
1/80	$-1.1605\cdot 10^{-5}$	$-7.4271\cdot 10^{-2}$	$-2.1146\cdot 10^{-3}$	$-1.3533\cdot 10^{1}$
1/160	$-2.9013\cdot 10^{-6}$	$-7.4273\cdot 10^{-2}$	$-5.2873\cdot 10^{-4}$	$-1.3536\cdot 10^{1}$
1/320	$-7.2532\cdot 10^{-7}$	$-7.4273\cdot 10^{-2}$	$-1.3219\cdot 10^{-4}$	$-1.3536\cdot 10^{1}$
1/640	$-1.8132\cdot 10^{-7}$	$-7.4270\cdot 10^{-2}$	$-3.3050\cdot 10^{-5}$	$-1.3537\cdot 10^{1}$

h	$l_h(t)$	$d_h^{\mathrm{IDeC}}(t)$	$d_h^{\mathrm{IDeC}}(t)/h^4$	$d_h^{\mathrm{IDeC}}(t)$	$d_h^{\mathrm{IDeC}}(t)/h^4$
1/10	$-1.3334\cdot 10^{-1}$	$7.2782\cdot 10^{-3}$	$7.2782\cdot 10^{1}$	$-8.0803\cdot 10^{-3}$	$-8.0803\cdot 10^{1}$
1/20	$-3.3715\cdot 10^{-2}$	$4.5700\cdot 10^{-4}$	$7.3120\cdot 10^{1}$	$1.8163\cdot 10^{-4}$	$2.9061\cdot 10^{1}$
1/40	$-8.4524\cdot 10^{-3}$	$2.8590\cdot 10^{-5}$	$7.3190\cdot 10^{1}$	$1.1432\cdot 10^{-5}$	$2.9267\cdot 10^{1}$
1/80	$-2.1146\cdot 10^{-3}$	$1.7872\cdot 10^{-6}$	$7.3205\cdot 10^{1}$	$7.1972\cdot 10^{-7}$	$2.9480\cdot 10^{1}$
1/160	$-5.2873\cdot 10^{-4}$	$1.1213\cdot 10^{-7}$	$7.3486\cdot 10^{1}$	$4.5049\cdot 10^{-8}$	$2.9523\cdot 10^{1}$
1/320	$-1.3219\cdot 10^{-4}$	$8.0668\cdot 10^{-9}$	$8.4586\cdot 10^{1}$	$3.0761\cdot 10^{-9}$	$3.2255\cdot 10^{1}$
1/640	$-3.3050\cdot 10^{-5}$	$2.6190\cdot 10^{-9}$	$4.3940\cdot 10^{2}$	$2.1435\cdot 10^{-9}$	$3.5963\cdot 10^{2}$

Example 2.1

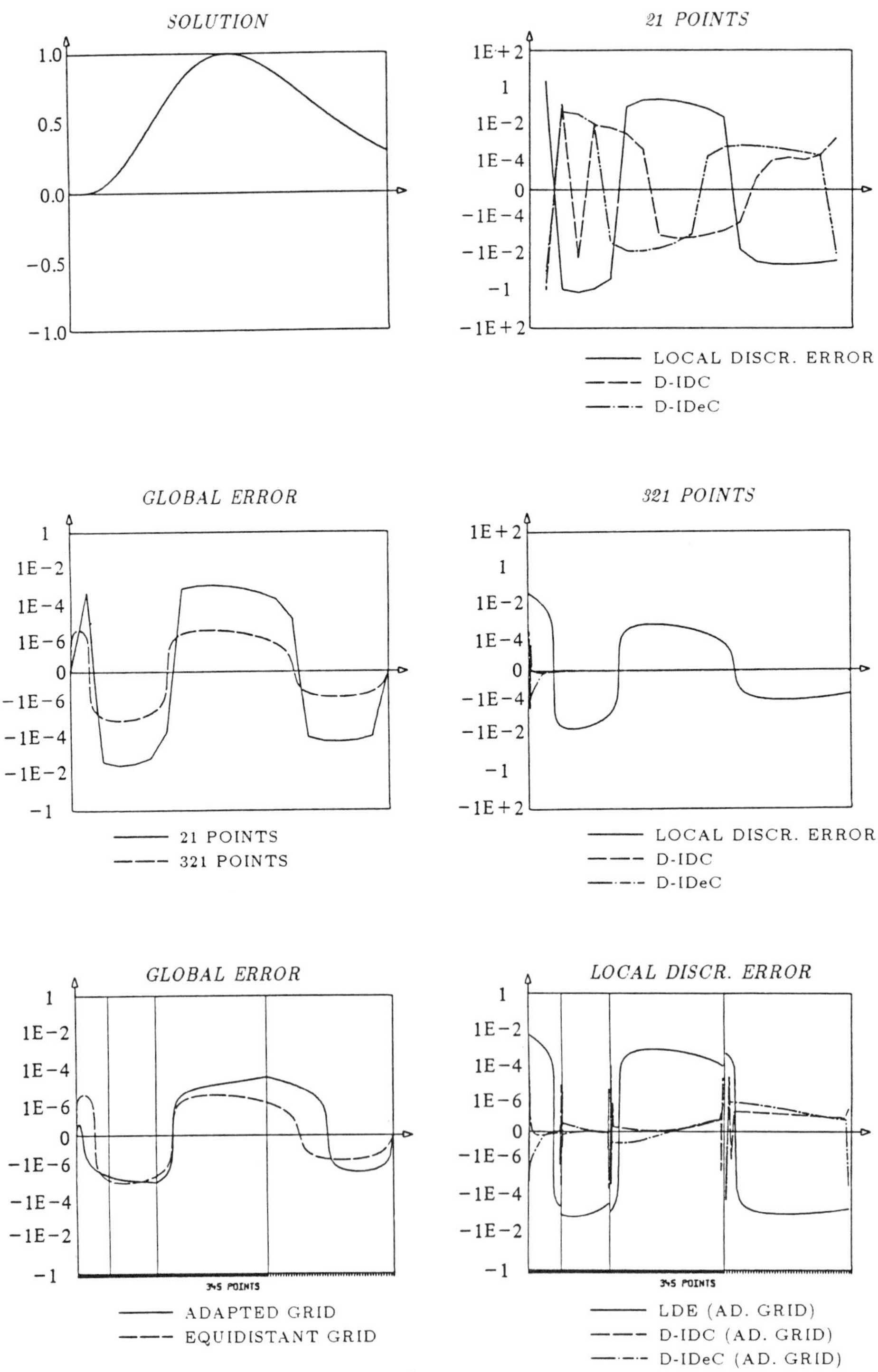
SOLUTION
1.0
0.5
0.0
−0.5
−1.0
21 POINTS
1E+2
1
1E−2
1E−4
0
−1E−4
−1E−2
−1
−1E+2
LOCAL DISCR. ERROR
D-IDC
D-IDeC
GLOBAL ERROR
1
1E−2
1E−4
1E−6
0
−1E−6
−1E−4
−1E−2
−1
21 POINTS
321 POINTS
321 POINTS
1E+2
1
1E−2
1E−4
0
−1E−4
−1E−2
−1
−1E+2
LOCAL DISCR. ERROR
D-IDC
D-IDeC
GLOBAL ERROR
1
1E−2
1E−4
1E−6
0
−1E−6
−1E−4
−1E−2
−1
345 POINTS
ADAPTED GRID
EQUIDISTANT GRID
LOCAL DISCR. ERROR
1
1E−2
1E−4
1E−6
0
−1E−6
−1E−4
−1E−2
−1
345 POINTS
LDE (AD. GRID)
D-IDC (AD. GRID)
D-IDeC (AD. GRID)

Example 2.2

h	$\|e_h\|$	t_e	$\|e_h\|/h^2$	$\|l_h\|$	t_l	$\|l_h\|/h^2$
1/10	$6.6992 \cdot 10^{-3}$	0.5000	$6.6992 \cdot 10^{-1}$	$1.1195 \cdot 10^{2}$	0.5000	$1.1195 \cdot 10^{4}$
1/20	$2.1315 \cdot 10^{-3}$	0.5000	$8.5262 \cdot 10^{-1}$	$3.6740 \cdot 10^{1}$	0.5000	$1.4696 \cdot 10^{4}$
1/40	$5.6547 \cdot 10^{-4}$	0.5000	$9.0475 \cdot 10^{-1}$	$9.8842 \cdot 10^{0}$	0.5000	$1.5815 \cdot 10^{4}$
1/80	$1.4456 \cdot 10^{-4}$	0.4875	$9.2521 \cdot 10^{-1}$	$2.5563 \cdot 10^{0}$	0.4875	$1.6361 \cdot 10^{4}$
1/160	$3.7049 \cdot 10^{-5}$	0.4938	$9.4844 \cdot 10^{-1}$	$6.5383 \cdot 10^{-1}$	0.4938	$1.6738 \cdot 10^{4}$
1/320	$9.2709 \cdot 10^{-6}$	0.4938	$9.4934 \cdot 10^{-1}$	$1.6366 \cdot 10^{-1}$	0.4938	$1.6759 \cdot 10^{4}$
1/640	$2.3183 \cdot 10^{-6}$	0.4938	$9.4957 \cdot 10^{-1}$	$4.0951 \cdot 10^{-2}$	0.4922	$1.6774 \cdot 10^{4}$

$t = 0.3$				
h	$e_h(t)$	$e_h(t)/h^2$	$l_h(t)$	$l_h(t)/h^2$
1/10	$1.2062 \cdot 10^{-3}$	$1.2062 \cdot 10^{-1}$	$2.0663 \cdot 10^{1}$	$2.0663 \cdot 10^{3}$
1/20	$3.0889 \cdot 10^{-4}$	$1.2356 \cdot 10^{-1}$	$5.1625 \cdot 10^{0}$	$2.0650 \cdot 10^{3}$
1/40	$7.7077 \cdot 10^{-5}$	$1.2332 \cdot 10^{-1}$	$1.2364 \cdot 10^{0}$	$1.9782 \cdot 10^{3}$
1/80	$1.9262 \cdot 10^{-5}$	$1.2328 \cdot 10^{-1}$	$3.0462 \cdot 10^{-1}$	$1.9496 \cdot 10^{3}$
1/160	$4.8153 \cdot 10^{-6}$	$1.2327 \cdot 10^{-1}$	$7.5856 \cdot 10^{-2}$	$1.9419 \cdot 10^{3}$
1/320	$1.2038 \cdot 10^{-6}$	$1.2327 \cdot 10^{-1}$	$1.8945 \cdot 10^{-2}$	$1.9400 \cdot 10^{3}$
1/640	$3.0095 \cdot 10^{-7}$	$1.2327 \cdot 10^{-1}$	$4.7351 \cdot 10^{-3}$	$1.9395 \cdot 10^{3}$

h	$l_h(t)$	$d_h^{\mathrm{IDC}}(t)$	$d_h^{\mathrm{IDC}}(t)/h^4$	$d_h^{\mathrm{IDeC}}(t)$	$d_h^{\mathrm{IDeC}}(t)/h^4$
1/10	$2.0663 \cdot 10^{1}$	$-1.8002 \cdot 10^{1}$	$-1.8002 \cdot 10^{5}$	$-1.5852 \cdot 10^{1}$	$-1.5852 \cdot 10^{5}$
1/20	$5.1625 \cdot 10^{0}$	$-5.2055 \cdot 10^{-2}$	$-8.3289 \cdot 10^{3}$	$7.2316 \cdot 10^{-1}$	$1.1570 \cdot 10^{5}$
1/40	$1.2364 \cdot 10^{0}$	$5.7445 \cdot 10^{-2}$	$1.4706 \cdot 10^{5}$	$4.7316 \cdot 10^{-2}$	$1.2113 \cdot 10^{5}$
1/80	$3.0462 \cdot 10^{-1}$	$4.8022 \cdot 10^{-3}$	$1.9670 \cdot 10^{5}$	$2.0638 \cdot 10^{-3}$	$8.4532 \cdot 10^{4}$
1/160	$7.5856 \cdot 10^{-2}$	$3.2021 \cdot 10^{-4}$	$2.0985 \cdot 10^{5}$	$1.1059 \cdot 10^{-4}$	$7.2475 \cdot 10^{4}$
1/320	$1.8945 \cdot 10^{-2}$	$2.0331 \cdot 10^{-5}$	$2.1319 \cdot 10^{5}$	$6.6077 \cdot 10^{-6}$	$6.9287 \cdot 10^{4}$
1/640	$4.7351 \cdot 10^{-3}$	$1.2779 \cdot 10^{-6}$	$2.1440 \cdot 10^{5}$	$4.0801 \cdot 10^{-7}$	$6.8452 \cdot 10^{4}$

$t = 0.8$				
h	$e_h(t)$	$e_h(t)/h^2$	$l_h(t)$	$l_h(t)/h^2$
1/10	$6.4731 \cdot 10^{-5}$	$6.4731 \cdot 10^{-3}$	$1.0368 \cdot 10^{0}$	$1.0368 \cdot 10^{2}$
1/20	$1.3256 \cdot 10^{-5}$	$5.3024 \cdot 10^{-3}$	$2.0423 \cdot 10^{-1}$	$8.1693 \cdot 10^{1}$
1/40	$3.1112 \cdot 10^{-6}$	$4.9779 \cdot 10^{-3}$	$4.7660 \cdot 10^{-2}$	$7.6255 \cdot 10^{1}$
1/80	$7.6478 \cdot 10^{-7}$	$4.8946 \cdot 10^{-3}$	$1.1704 \cdot 10^{-2}$	$7.4908 \cdot 10^{1}$
1/160	$1.9038 \cdot 10^{-7}$	$4.8737 \cdot 10^{-3}$	$2.9130 \cdot 10^{-3}$	$7.4573 \cdot 10^{1}$
1/320	$4.7543 \cdot 10^{-8}$	$4.8684 \cdot 10^{-3}$	$7.2743 \cdot 10^{-4}$	$7.4489 \cdot 10^{1}$
1/640	$1.1883 \cdot 10^{-8}$	$4.8672 \cdot 10^{-3}$	$1.8181 \cdot 10^{-4}$	$7.4468 \cdot 10^{1}$

h	$l_h(t)$	$d_h^{\mathrm{IDC}}(t)$	$d_h^{\mathrm{IDC}}(t)/h^4$	$d_h^{\mathrm{IDeC}}(t)$	$d_h^{\mathrm{IDeC}}(t)/h^4$
1/10	$1.0368 \cdot 10^{0}$	$7.6125 \cdot 10^{-1}$	$7.6125 \cdot 10^{3}$	$2.0965 \cdot 10^{0}$	$2.0965 \cdot 10^{4}$
1/20	$2.0423 \cdot 10^{-1}$	$7.1957 \cdot 10^{-2}$	$1.1513 \cdot 10^{4}$	$2.8583 \cdot 10^{-2}$	$4.5733 \cdot 10^{3}$
1/40	$4.7660 \cdot 10^{-2}$	$4.5315 \cdot 10^{-3}$	$1.1601 \cdot 10^{4}$	$1.5425 \cdot 10^{-3}$	$3.9487 \cdot 10^{3}$
1/80	$1.1704 \cdot 10^{-2}$	$2.8205 \cdot 10^{-4}$	$1.1553 \cdot 10^{4}$	$9.8127 \cdot 10^{-5}$	$4.0193 \cdot 10^{3}$
1/160	$2.9130 \cdot 10^{-3}$	$1.7603 \cdot 10^{-5}$	$1.1536 \cdot 10^{4}$	$6.1825 \cdot 10^{-6}$	$4.0518 \cdot 10^{3}$
1/320	$7.2743 \cdot 10^{-4}$	$1.1012 \cdot 10^{-6}$	$1.1547 \cdot 10^{4}$	$3.8727 \cdot 10^{-7}$	$4.0608 \cdot 10^{3}$
1/640	$1.8181 \cdot 10^{-4}$	$6.6751 \cdot 10^{-8}$	$1.1199 \cdot 10^{4}$	$2.4255 \cdot 10^{-8}$	$4.0693 \cdot 10^{3}$

Example 2.2

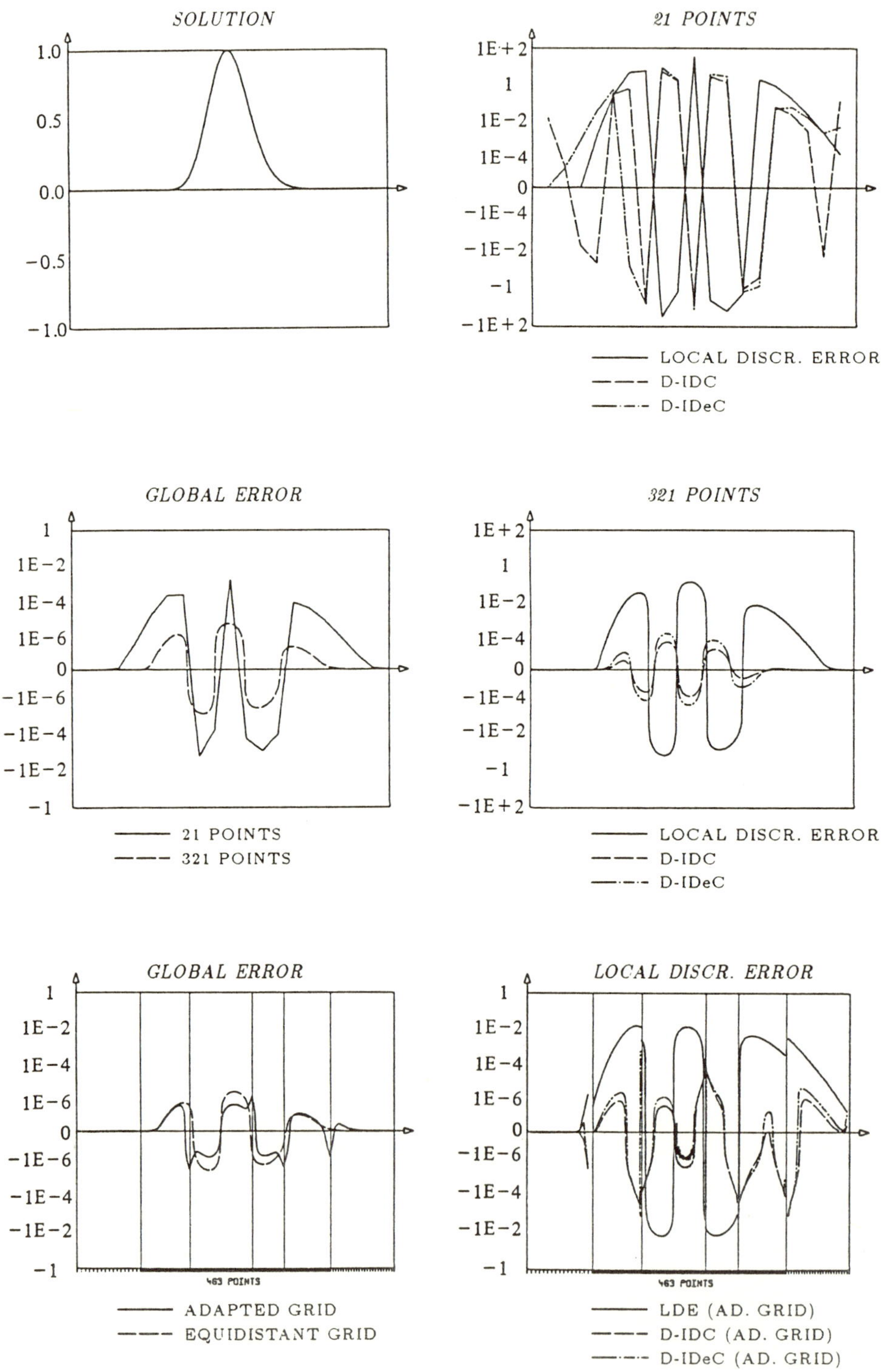
SOLUTION
1.0
0.5
0.0
-0.5
-1.0
21 POINTS
1E+2
1
1E-2
1E-4
0
-1E-4
-1E-2
-1
-1E+2
LOCAL DISCR. ERROR
D-IDC
D-IDeC
GLOBAL ERROR
1
1E-2
1E-4
1E-6
0
-1E-6
-1E-4
-1E-2
-1
21 POINTS
321 POINTS
321 POINTS
1E+2
1
1E-2
1E-4
0
-1E-4
-1E-2
-1
-1E+2
LOCAL DISCR. ERROR
D-IDC
D-IDeC
GLOBAL ERROR
1
1E-2
1E-4
1E-6
0
-1E-6
-1E-4
-1E-2
-1
463 POINTS
ADAPTED GRID
EQUIDISTANT GRID
LOCAL DISCR. ERROR
1
1E-2
1E-4
1E-6
0
-1E-6
-1E-4
-1E-2
-1
463 POINTS
LDE (AD. GRID)
D-IDC (AD. GRID)
D-IDeC (AD. GRID)

Example 2.3

h	$\|e_h\|$	t_e	$\|e_h\|/h^2$	$\|l_h\|$	t_l	$\|l_h\|/h^2$
1/10	$1.0426\cdot10^{-1}$	0.1000	$1.0426\cdot10^{1}$	$1.4023\cdot10^{2}$	0.1000	$1.4023\cdot10^{4}$
1/20	$6.1128\cdot10^{-2}$	0.0500	$2.4451\cdot10^{1}$	$1.4838\cdot10^{2}$	0.0500	$5.9350\cdot10^{4}$
1/40	$1.2247\cdot10^{-2}$	0.0500	$1.9595\cdot10^{1}$	$4.1660\cdot10^{1}$	0.0500	$6.6656\cdot10^{4}$
1/80	$3.5956\cdot10^{-3}$	0.0375	$2.3012\cdot10^{1}$	$3.5783\cdot10^{1}$	0.0125	$2.2901\cdot10^{5}$
1/160	$8.8417\cdot10^{-4}$	0.0375	$2.2635\cdot10^{1}$	$2.4386\cdot10^{1}$	0.0063	$6.2427\cdot10^{5}$
1/320	$2.2032\cdot10^{-4}$	0.0375	$2.2560\cdot10^{1}$	$9.1021\cdot10^{0}$	0.0031	$9.3206\cdot10^{5}$
1/640	$5.5037\cdot10^{-5}$	0.0375	$2.2543\cdot10^{1}$	$2.7406\cdot10^{0}$	0.0016	$1.1226\cdot10^{6}$

$t = 0.1$				
h	$e_h(t)$	$e_h(t)/h^2$	$l_h(t)$	$l_h(t)/h^2$
1/10	$1.0426\cdot10^{-1}$	$1.0426\cdot10^{1}$	$1.4023\cdot10^{2}$	$1.4023\cdot10^{4}$
1/20	$2.0711\cdot10^{-2}$	$8.2845\cdot10^{0}$	$5.4622\cdot10^{1}$	$2.1849\cdot10^{4}$
1/40	$5.3031\cdot10^{-3}$	$8.4850\cdot10^{0}$	$1.4982\cdot10^{1}$	$2.3971\cdot10^{4}$
1/80	$1.3193\cdot10^{-3}$	$8.4433\cdot10^{0}$	$3.8273\cdot10^{0}$	$2.4495\cdot10^{4}$
1/160	$3.2866\cdot10^{-4}$	$8.4138\cdot10^{0}$	$9.6192\cdot10^{-1}$	$2.4625\cdot10^{4}$
1/320	$8.2079\cdot10^{-5}$	$8.4049\cdot10^{0}$	$2.4080\cdot10^{-1}$	$2.4658\cdot10^{4}$
1/640	$2.0514\cdot10^{-5}$	$8.4026\cdot10^{0}$	$6.0219\cdot10^{-2}$	$2.4666\cdot10^{4}$

h	$l_h(t)$	$d_h^{IDC}(t)$	$d_h^{IDC}(t)/h^4$	$d_h^{IDeC}(t)$	$d_h^{IDeC}(t)/h^4$
1/10	$1.4023\cdot10^{2}$	$-1.1644\cdot10^{2}$	$-1.1644\cdot10^{6}$	$-1.2049\cdot10^{2}$	$-1.2049\cdot10^{6}$
1/20	$5.4622\cdot10^{1}$	$-1.7210\cdot10^{1}$	$-2.7536\cdot10^{6}$	$-2.5409\cdot10^{1}$	$-4.0654\cdot10^{6}$
1/40	$1.4982\cdot10^{1}$	$-1.0704\cdot10^{0}$	$-2.7401\cdot10^{6}$	$-1.2733\cdot10^{-1}$	$-3.2597\cdot10^{5}$
1/80	$3.8273\cdot10^{0}$	$-6.7779\cdot10^{-2}$	$-2.7762\cdot10^{6}$	$-1.4207\cdot10^{-2}$	$-5.8190\cdot10^{5}$
1/160	$9.6192\cdot10^{-1}$	$-4.2542\cdot10^{-3}$	$-2.7881\cdot10^{6}$	$-9.4441\cdot10^{-4}$	$-6.1893\cdot10^{5}$
1/320	$2.4080\cdot10^{-1}$	$-2.6619\cdot10^{-4}$	$-2.7912\cdot10^{6}$	$-6.0019\cdot10^{-5}$	$-6.2935\cdot10^{5}$
1/640	$6.0219\cdot10^{-2}$	$-1.6625\cdot10^{-5}$	$-2.7892\cdot10^{6}$	$-3.7491\cdot10^{-6}$	$-6.2900\cdot10^{5}$

$t = 0.5$				
h	$e_h(t)$	$e_h(t)/h^2$	$l_h(t)$	$l_h(t)/h^2$
1/10	$3.3053\cdot10^{-4}$	$3.3053\cdot10^{-2}$	$3.1720\cdot10^{-1}$	$3.1720\cdot10^{1}$
1/20	$8.3881\cdot10^{-5}$	$3.3552\cdot10^{-2}$	$7.3976\cdot10^{-2}$	$2.9590\cdot10^{1}$
1/40	$2.1047\cdot10^{-5}$	$3.3676\cdot10^{-2}$	$1.8158\cdot10^{-2}$	$2.9054\cdot10^{1}$
1/80	$5.2665\cdot10^{-6}$	$3.3706\cdot10^{-2}$	$4.5186\cdot10^{-3}$	$2.8919\cdot10^{1}$
1/160	$1.3169\cdot10^{-6}$	$3.3713\cdot10^{-2}$	$1.1284\cdot10^{-3}$	$2.8886\cdot10^{1}$
1/320	$3.2925\cdot10^{-7}$	$3.3715\cdot10^{-2}$	$2.8201\cdot10^{-4}$	$2.8877\cdot10^{1}$
1/640	$8.2314\cdot10^{-8}$	$3.3716\cdot10^{-2}$	$7.0496\cdot10^{-5}$	$2.8875\cdot10^{1}$

h	$l_h(t)$	$d_h^{IDC}(t)$	$d_h^{IDC}(t)/h^4$	$d_h^{IDeC}(t)$	$d_h^{IDeC}(t)/h^4$
1/10	$3.1720\cdot10^{-1}$	$6.6270\cdot10^{-2}$	$6.6270\cdot10^{2}$	$2.5674\cdot10^{-2}$	$2.5674\cdot10^{2}$
1/20	$7.3976\cdot10^{-2}$	$5.5266\cdot10^{-3}$	$8.8425\cdot10^{2}$	$6.4514\cdot10^{-4}$	$1.0322\cdot10^{2}$
1/40	$1.8158\cdot10^{-2}$	$3.6634\cdot10^{-4}$	$9.3783\cdot10^{2}$	$4.3676\cdot10^{-5}$	$1.1181\cdot10^{2}$
1/80	$4.5186\cdot10^{-3}$	$2.3218\cdot10^{-5}$	$9.5103\cdot10^{2}$	$2.8823\cdot10^{-6}$	$1.1806\cdot10^{2}$
1/160	$1.1284\cdot10^{-3}$	$1.4560\cdot10^{-6}$	$9.5418\cdot10^{2}$	$1.8295\cdot10^{-7}$	$1.1990\cdot10^{2}$
1/320	$2.8201\cdot10^{-4}$	$9.0730\cdot10^{-8}$	$9.5137\cdot10^{2}$	$1.1467\cdot10^{-8}$	$1.2024\cdot10^{2}$
1/640	$7.0496\cdot10^{-5}$	$5.8794\cdot10^{-9}$	$9.8640\cdot10^{2}$	$6.8992\cdot10^{-10}$	$1.1575\cdot10^{2}$

Example 2.3

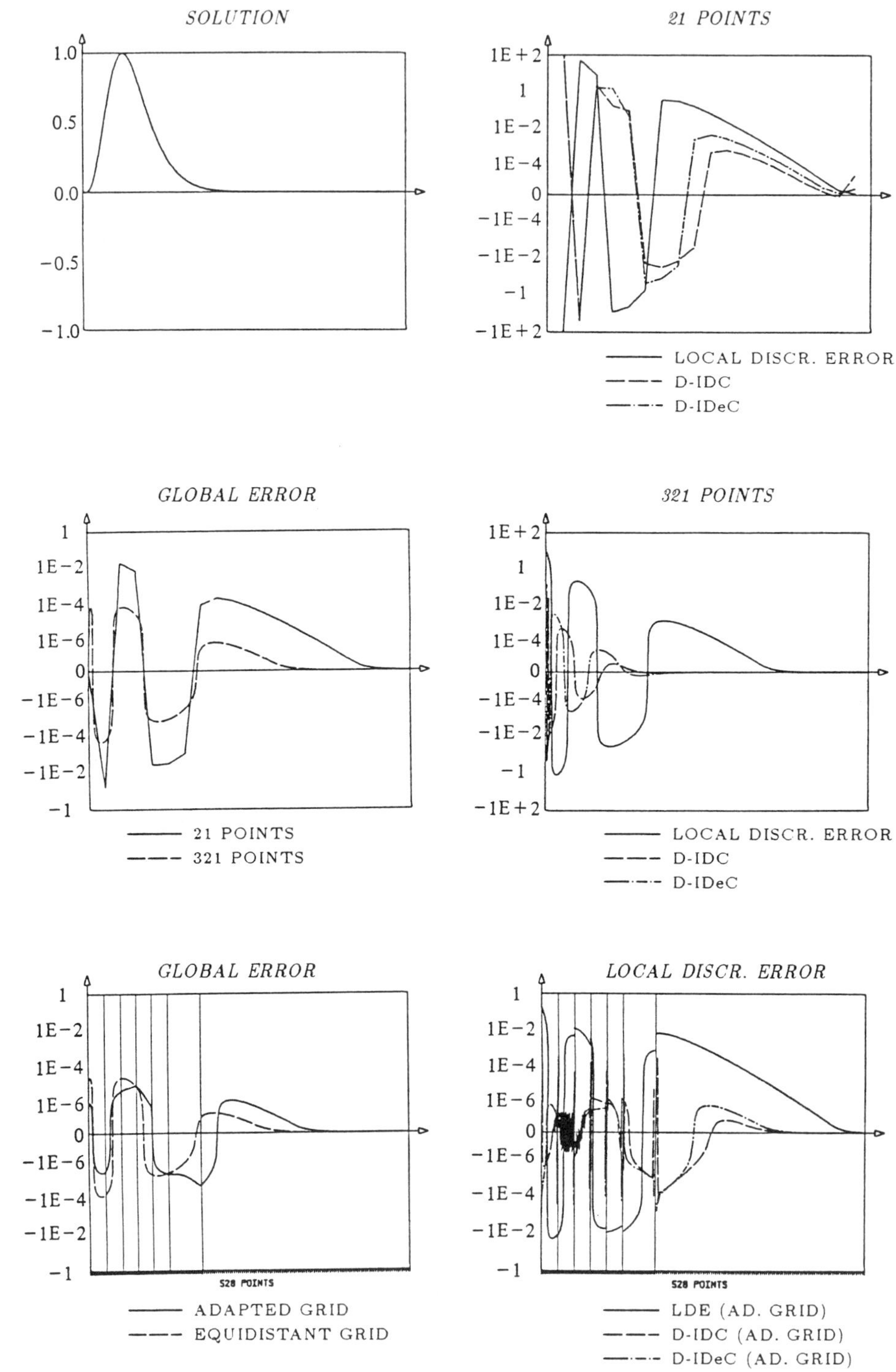
SOLUTION
1.0
0.5
0.0
−0.5
−1.0
21 POINTS
1E+2
1
1E−2
1E−4
0
−1E−4
−1E−2
−1
−1E+2
LOCAL DISCR. ERROR
D-IDC
D-IDeC
GLOBAL ERROR
1
1E−2
1E−4
1E−6
0
−1E−6
−1E−4
−1E−2
−1
21 POINTS
321 POINTS
321 POINTS
1E+2
1
1E−2
1E−4
0
−1E−4
−1E−2
−1
−1E+2
LOCAL DISCR. ERROR
D-IDC
D-IDeC
GLOBAL ERROR
1
1E−2
1E−4
1E−6
0
−1E−6
−1E−4
−1E−2
−1
528 POINTS
ADAPTED GRID
EQUIDISTANT GRID
LOCAL DISCR. ERROR
1
1E−2
1E−4
1E−6
0
−1E−6
−1E−4
−1E−2
−1
528 POINTS
LDE (AD. GRID)
D-IDC (AD. GRID)
D-IDeC (AD. GRID)

Canadian Mathematical Society
Conference Proceedings
Volume 8 (1987)

OSCILLATION AND ASYMPTOTIC BEHAVIOR OF SECOND ORDER NEUTRAL DIFFERENTIAL EQUATIONS WITH DEVIATING ARGUMENTS

by

M.K. Grammatikopoulos[1,2], E.A. Grove[1] and G. Ladas[1]

ABSTRACT. Consider the neutral differential equation

$$\frac{d^2}{dt^2}[y(t) + py(t-\tau)] + qy(t-\sigma) = 0, \quad t \geq t_0 \tag{1}$$

where $q \neq 0$, p, τ, and σ are real numbers.

The asymptotic behavior of nonoscillatory solutions of Eq. (1) is studied. Sufficient conditions are also given to insure that all solutions of Eq. (1) are oscillatory.

1. INTRODUCTION

A neutral differential equation (NDE) is a differential equation in which the highest derivative of the unknown function appears with the argument t (the present state of the system) as well as one or more retarded and/or advanced arguments (the past and/or the future state of the system). In general, the study of neutral differential equations presents complications not seen in nonneutral retarded or advanced differential equations (see, for example, [8] and [9]).

Besides its theoretical interest, the study of the asymptotic and oscillatory behavior of solutions of neutral differential equations has some importance in applications. Neutral differential equations appear, for example, in problems concerning networks containing lossless transmission lines. Such networks arise in high speed computers where lossless transmission lines are used to interconnect switching circuits (see [1]). Second order NDE appear in the study of vibrating masses attached to an elastic bar, and also (as the Euler equation) in some variational problems (see [4, pp. 4-7]).

AMS 1980 Subject Classification: Primary - 34K15; Secondary - 34C10.

[1]Department of Mathematics, University of Rhode Island, Kingston, Rhode Island 02881 - 0816, U.S.A.

[2]Department of Mathematics, University of Ioannina, Ioannina 45332, GREECE.

Recently, Ladas and Sficas [5], and Grammatikopoulos, Grove, and Ladas [2], [3] studied the asymptotic nature of the delayed NDE

$$\frac{d}{dt}[y(t) + py(t-\tau)] + qy(t-\sigma) = 0, \quad t \geqq t_0,$$

and found sufficient conditions for the oscillation of all solutions in the cases where p, τ, σ are real numbers, and q is a nonzero real number. In this paper we study the asymptotic nature of the NDE

$$\frac{d^2}{dt^2}[y(t) + py(t-\tau)] + qy(t-\sigma) = 0, \quad t \geqq t_0, \tag{1}$$

and find sufficient conditions for the oscillation of all solutions, again for the cases p, τ, σ real numbers and q a nonzero real number.

2. SOME BASIC LEMMAS

In this section we group together some lemmas which will be useful in our study of the NDE

$$\frac{d^2}{dt^2}[y(t) + py(t-\tau)] + qy(t-\sigma) = 0, \quad t \geqq t_0, \tag{1}$$

where $p, \tau, \sigma \in \mathbb{R}$ and $q \in \mathbb{R} - \{0\}$.

The first lemma has been extracted from results due to Ladas and Stavroulakis ([6] and [7]).

LEMMA 1. Suppose r and μ are positive numbers such that

$$r^{1/2}\frac{\mu}{2} > \frac{1}{e}.$$

Then the following statements are true:

(i) The inequality

$$x''(t) - rx(t-\mu) \leqq 0$$

has no eventually negative bounded solutions.

(ii) The inequality

$$x''(t) - rx(t+\mu) \leqq 0$$

has no eventually negative unbounded solutions.

(iii) The inequality

$$x''(t) - rx(t-\mu) \geqq 0$$

has no eventually positive bounded solutions.

(iv) The inequality

$$x''(t) - rx(t+\mu) \geqq 0$$

has no eventually positive unbounded solutions.

The following duality lemma will be useful as a source for discovering new results, as well as enabling us to reduce the required number of proofs in this paper.

LEMMA 2. Suppose p and q are nonzero real numbers. Then y(t) is a solution of Eq. (1) if and only if y(t) is a solution of the "dual" equation

$$\frac{d^2}{dt^2}\left[y(t) + \frac{1}{p}\, y(t+\tau)\right] + \frac{q}{p}\, y(t+\tau-\sigma) = 0.$$

PROOF. Clearly, y(t) satisfies Eq. (1) if and only if

$$\frac{d^2}{dt^2}\left[\frac{1}{p}y(t) + y(t-\tau)\right] + \frac{q}{p}\, y(t-\sigma) = 0$$

if and only if

$$\frac{d^2}{dt^2}\left[y(t) + \frac{1}{p}\, y(t+\tau)\right] + \frac{q}{p}\, y(t+\tau-\sigma) = 0.$$

Symbolically, we can express the duality lemma as follows:

$$(p,\ q,\ \tau,\ \sigma) \longleftrightarrow \left(\frac{1}{p},\ \frac{q}{p},\ -\tau,\ \sigma-\tau\right).$$

The next lemma (the proof of which follows directly by substitution) shows how given any solution of Eq. (1), we can construct other solutions of Eq. (1) which are sufficiently differentiable.

LEMMA 3. Let y(t) be a solution of Eq. (1). Set

$$z(t) = y(t) + y(t - \tau).$$

Then z(t) is a twice continuously differentiable solution of the neutral differential equation

$$z''(t) + pz''(t-\tau) + qz(t-\sigma) = 0. \tag{2}$$

The next result establishes some useful facts about the function z(t) when y(t) is an eventually positive solution of Eq. (1).

LEMMA 4. Let y(t) be an eventually positive solution of Eq. (1). Set

$$z(t) = y(t) + py(t - \tau).$$

Then:

(a) $qz''(t) < 0$.

(b) z(t) and z'(t) are monotonic. Moreover, either

$$q \lim_{t\to\infty} z(t) = q \lim_{t\to\infty} z'(t) = -\infty \tag{3}$$

or

$$\lim_{t\to\infty} z(t) = \lim_{t\to\infty} z'(t) = 0, \quad qz(t) < 0, \quad \text{and} \quad qz'(t) > 0. \tag{4}$$

In particular, $qz(t) < 0$.

(c) If $p \leqq -1$, $q < 0$, and $\tau \leqq 0$, then (4) holds.

(d) If $p < -1$, $q < 0$, and $0 \leqq \tau$, then (3) holds.

(e) If $-1 < p < 0$, $0 < q$, and $\tau \leqq 0$, then (3) holds.

(f) If $-1 \leqq p < 0$, $0 < q$, and $0 \leqq \tau$, then (4) holds.

PROOF. (a) Note that

$$z''(t) = -qy(t-\sigma) \tag{5}$$

and so since $q \neq 0$,

$$qz''(t) = -q^2y(t-\sigma) < 0.$$

(b) It follows from (a) that $z'(t)$ and $z(t)$ are strictly monotonic, and either

$$\lim_{t\to\infty} qz'(t) = -\infty$$

or

$$\lim_{t\to\infty} z'(t) = \ell \in \mathbb{R}.$$

If $\lim_{t\to\infty} qz'(t) = -\infty$, then (3) is satisfied, and so it remains to show that $\lim_{t\to\infty} z'(t) = \ell \in \mathbb{R}$ implies (4). By integrating (5) from t_1 to t, and then letting $t \to \infty$, we see that

$$\ell - z'(t_1) = -q \int_{t_1}^{\infty} y(s-\sigma)ds.$$

Thus $y \in L^1[t_0,\infty)$, and so $z \in L^1[t_0,\infty)$. Since z is also monotonic, it follows that

$$\lim_{t\to\infty} z(t) = 0$$

and so as z' is monotonic, we see that

$$\lim_{t\to\infty} z'(t) = 0.$$

From the fact that $qz''(t) < 0$, we see that $qz'(t)$ decreases to zero which implies $qz'(t) > 0$.

Finally, since $\lim_{t\to\infty} qz(t) = 0$ and $qz'(t) > 0$, we see that $qz(t) < 0$, and (4) is satisfied.

(c) Otherwise (3) holds, and in particular,

$$\lim_{t\to\infty} z(t) = \infty.$$

Hence $0 < -py(t-\tau) < y(t)$. Thus $\tau < 0$, and so by iteration

$$0 < (-p)^n y(t-n\tau) < y(t)$$

which implies (since $1 \leqq -p$) that $y(t)$ is bounded. This contradicts $\lim_{t\to\infty} z(t) = \infty$.

(d) Otherwise (4) holds, and in particular, $\lim_{t\to\infty} z(t) = 0$ and $z(t) > 0$ which implies that $0 < -py(t-\tau) < y(t)$. So $0 < \tau$, and it follows by iteration

$$0 < (-p)^n y(t) < y(t+n\tau)$$

and so $\lim_{t\to\infty} y(t) = \infty$. But this contradicts $\lim_{t\to\infty} z(t) = 0$.

By Lemma 2, (e) follows from (d) and (f) follows from (c).

3. ASYMPTOTIC BEHAVIOR OF NONOSCILLATORY SOLUTIONS

In this section we study the asymptotic behavior of the nonoscillatory solutions of the NDE

$$\frac{d^2}{dt^2}[y(t) + py(t-\tau)] + qy(t-\sigma) = 0, \quad t \geqq t_0, \tag{1}$$

where $p, \tau, \sigma \in \mathbb{R}$ and $q \in \mathbb{R} - \{0\}$.

THEOREM 1. Suppose $p < -1$, $q < 0$, and $0 \leqq \tau$. Then every nonoscillatory solution of Eq. (1) tends to $-\infty$ or ∞ as $t \to \infty$.

PROOF. Since the negative of a solution of Eq. (1) is also a solution of Eq. (1), it suffices to assume that $y(t)$ is an eventually positive solution of Eq. (1). Set

$$z(t) = y(t) + py(t-\tau).$$

Then by Lemma 4(d), $\lim_{t\to\infty} z(t) = \infty$. So as $y(t) > z(t)$, the result follows.

The next result is an immediate consequence of Theorem 1 and Lemma 2.

THEOREM 2. Suppose $-1 < p < 0$, $0 < q$, and $\tau \leqq 0$. Then every nonoscillatory solution of Eq. (1) tends to $-\infty$ or ∞ as $t \to \infty$.

THEOREM 3. Suppose $0 < q$ and $0 \leqq \tau$. Then:

(i) if $p = -1$, every nonoscillatory solution of Eq. (1) is bounded;

(ii) if $-1 < p < 0$, every nonoscillatory solution of Eq. (1) tends to zero as $t \to \infty$.

PROOF. It suffices to establish the theorem for an eventually positive solution $y(t)$ of Eq. (1). Set

$$z(t) = y(t) + py(t-\tau).$$

Then by Lemma 4(f), $y(t) < -py(t-\tau)$, and so by iteration,

$$y(t+n\tau) < (-p)^n y(t)$$

from which the theorem is an immediate consequence.

Theorem 3 and Lemma 2 yield the following result.

THEOREM 4. Suppose $q < 0$ and $\tau \leqq 0$. Then:

(i) if $p = -1$, every nonoscillatory solution of Eq. (1) is bounded;

(ii) if $p < -1$, every nonoscillatory solution of Eq. (1) tends to zero as $t \to \infty$.

4. SUFFICIENT CONDITIONS FOR ALL SOLUTIONS TO OSCILLATE

In this section we present conditions under which all bounded solutions of Eq. (1) oscillate, under which all unbounded solutions of Eq. (1) oscillate, and conditions under which all solutions of Eq. (1) oscillate.

From the characteristic equation

$$F(\lambda) = \lambda^2 + p\lambda^2 e^{-\lambda\tau} + qe^{-\lambda\sigma}$$

of Eq. (1), it is easily seen that if $0 \leqq p$ and $q < 0$, then there are always nonoscillatory solutions of Eq. (1) which tend to ∞ if $0 \leqq \sigma$, and there are always nonoscillatory solutions which tend to zero if $\sigma \leqq 0$. On the other hand if $p \geqq 0$ and $q > 0$, then we have the following result.

THEOREM 5. Suppose $p \geqq 0$ and $q > 0$. Then every solution of Eq. (1) oscillates.

PROOF. Assume for the sake of contradiction that $y(t)$ is an eventually positive solution of Eq. (1). Set

$$z(t) = y(t) + py(t-\tau).$$

Then $z(t) > 0$ while Lemma 4(b) implies that $z(t) < 0$. The proof is complete.

The following result is an immediate consequence of Theorems 1 - 4.

THEOREM 6.

(a) Suppose $p \leqq -1$, $q < 0$, and $\tau \leqq 0$. Then every unbounded solution of Eq. (1) oscillates.

(b) Suppose $p < -1$, $q < 0$, and $0 \leqq \tau$. Then every bounded solution of Eq. (1) oscillates.

(c) Suppose $-1 < p < 0$, $0 < q$, and $\tau \leqq 0$. Then every bounded solution of Eq. (1) oscillates.

(d) Suppose $-1 \leqq p < 0$, $0 < q$, and $0 \leqq \tau$. Then every unbounded solution of Eq. (1) oscillates.

The following theorem gives us additional information about the oscillatory behavior of bounded solutions of Eq. (1).

THEOREM 7. Suppose $p < 0$. Then each of the following four conditions is sufficient to imply that every bounded solution of Eq. (1) oscillates:

(i) (C_1) $0 < q$ and $\left(\frac{q}{-p}\right)^{1/2} \frac{\sigma - \tau}{2} > \frac{1}{e}$;

(ii) (C_2) $0 < q$ and $\left(\frac{q}{p^2}\right)^{1/2} \frac{\sigma - 2\tau}{2} > \frac{1}{e}$;

(iii) (C_3) $q < 0$ and $(-q)^{1/2} \frac{\sigma}{2} > \frac{1}{e}$;

(iv) (C_4) $q < 0$ and $(pq)^{1/2} \frac{\sigma + \tau}{2} > \frac{1}{e}$.

PROOF. We first assume Condition (C_1), and we also assume for the sake of contradiction that $y(t)$ is a bounded eventually positive solution of Eq. (1). Set

$$z(t) = y(t) + py(t-\tau).$$

Then by Lemma 4(a) and Eq. (2) we obtain

$$pz''(t-\tau) + qz(t-\sigma) > 0$$

or

$$z''(t) - \left(\frac{q}{-p}\right) z[t - (\sigma-\tau)] < 0. \tag{6}$$

Now by Lemma 1(i), it is impossible for Ineq. (6) to have an eventually negative bounded solution. But $z(t)$ is bounded, and $z(t)$ is negative by Lemma 4. This contradiction proves the theorem.

We next assume condition (C_4), and again for the sake of contradiction, we assume $y(t)$ is a bounded eventually positive solution of Eq. (1). Set

$$z(t) = y(t) + py(t-\tau).$$

Then $z(t)$ is a bounded twice continuously differentiable solution of Eq. (1), and by Lemma 4 we have

$$z(t) > 0, \quad z'(t) < 0, \text{ and } \quad z''(t) = -qy(t-\sigma) > 0.$$

Since $q < 0$, $z(t) > 0$, and $z(t)$ is a solution of Eq. (1), we find

$$z''(t) + pz''(t-\tau) > 0.$$

As $0 < z(t) = y(t) + py(t-\tau) < y(t)$ and $z''(t) = -qy(t-\sigma)$, we see that

$$z''(t-\tau) = -qy[t - (\sigma+\tau)] > -qz[t - (\sigma+\tau)]$$

and so

$$pz''(t-\tau) < -pqz[t - (\sigma+\tau)].$$

Thus

$$0 < z''(t) + pz''(t-\tau) < z''(t) - (pq)z[t - (\sigma+\tau)].$$

But by (C_4) and Lemma 1(iii), this is impossible and the proof is complete in this case also.

Finally note that Conditions (C_1) and (C_3) are dual, and Conditions (C_2) and (C_4) are dual. This establishes the proof in all four cases.

By combining Theorem 7(i) and Theorem 6(d), we obtain the following result.

THEOREM 8. Suppose $-1 \leqq p < 0$, $0 < q$, $0 \leqq \tau$, and condition (C_1) is satisfied. Then every solution of Eq. (1) oscillates.

The following theorem will be a consequence of Ineq. (6) when $\tau - \sigma > 0$.

THEOREM 9. Suppose $-1 < p < 0$, $0 < q$, $\tau \leqq 0$, and

$$(C_5) \quad \left(-\frac{q}{p}\right)^{1/2} \frac{\tau - \sigma}{2} > \frac{1}{e}.$$

Then every solution of Eq. (1) oscillates.

PROOF. Assume for the sake of contradiction that $y(t)$ is an eventually positive solution of Eq. (1). Then as in the proof of Theorem 7, we find

$$z''(t) - \left(\frac{q}{-p}\right) z[t + (\tau - \sigma)] < 0. \tag{7}$$

Now by Lemma 4(e), $z(t)$ is negative and unbounded. But by (C_5) and Lemma 1(ii), it is impossible for Ineq. (7) to have an eventually negative unbounded solution. This contradiction proves the theorem.

By Lemma 2, the dual of the condition (C_1) is

$$(C_3) \quad (-q)^{1/2} \frac{\sigma}{2} > \frac{1}{e}$$

and the dual of the condition (C_5) is

$$(C_6) \quad (-q)^{1/2} \frac{-\sigma}{2} > \frac{1}{e}.$$

The following two theorems follow from the duality principle applied to Theorems 8 and 9.

THEOREM 10. Suppose $p \leqq -1$, $q < 0$, $\tau \leqq 0$, and Condition (C_3) is satisfied. Then every solution of Eq. (1) oscillates.

THEOREM 11. Suppose $p < -1$, $q < 0$, $0 \leqq \tau$, and Condition (C_6) is satisfied. Then every solution of Eq. (1) oscillates.

Theorem 12 follows from the duality principle applied to Theorem 7.

THEOREM 12. Suppose $p < 0$, $q < 0$, and Condition (C_3) is satisfied. Then every bounded solution of Eq. (1) oscillates.

We end this note with an unsolved problem and a conjecture.

PROBLEM. Suppose $p < -1$, $0 < \tau$, $\sigma \leqq 0$, and $0 < q$. Find a sufficient condition to insure that every unbounded solution of Eq. (1) oscillates.

CONJECTURE. Suppose $0 \leqq p$, $q < 0$, $\tau > 0$ and $\sigma < 0$. Then we conjecture that every unbounded solution of Eq. (1) oscillates.

REFERENCES

1. R. K. BRAYTON AND R. A. WILLOUGHBY, On the numerical integration of a symmetric system of difference-differential equations of neutral type, J. Math. Anal. Appl. 18(1967), 182-189.

2. M. K. GRAMMATIKOPOULOS, E. A. GROVE AND G. LADAS, Oscillations of first order neutral delay differential equations, J. Math. Anal. Appl. (to appear).

3. M. K. GRAMMATIKOPOULOS, E. A. GROVE AND G. LADAS, Oscillation and asymptotic behavior of neutral differential equations with deviating arguments, Applicable Anal. (to appear).

4. J. HALE, "Theory of functional differential equations", Springer-Verlag, New York, 1977.

5. G. LADAS AND Y. G. SFICAS, Oscillations of neutral delay differential equations, Canad. Math. Bull. (to appear).

6. G. LADAS AND I. P. STAVROULAKIS, On delay differential inequalities of higher order, Canad. Math. Bull. 25(1982), 348-354.

7. G. LADAS AND I. P. STAVROULAKIS, Oscillations of differential equations of mixed type, J. Math. Phys. Sciences 18(1984), 245-262.

8. M. SLEMROD AND E. F. INFANTE, Asymptotic stability criteria for linear systems of difference-differential equations of neutral type and their discrete analogues, J. Math. Anal. Appl. 38(1972), 399-415.

9. W. SNOW, Existence, uniqueness, and stability for nonlinear differential-difference equations in the neutral case, N.Y.U. Courant Inst. Math. Sci. Rep. IMM NYU 328, (February 1965).

Canadian Mathematical Society
Conference Proceedings
Volume 8 (1987)

STURM COMPARISON THEOREMS FOR NEUTRAL DIFFERENTIAL EQUATIONS

by

E. A. Grove[1], M. R. S. Kulenović[1,2] and G. Ladas[1]

ABSTRACT. Consider the neutral delay differential equations

$$\frac{d}{dt}[x(t) - P_1(t)x(t-\tau_1)] + Q_1(t)x(t-\sigma_1) = 0 \tag{1}$$

and

$$\frac{d}{dt}[y(t) - P_2(t)y(t-\tau_2)] + Q_2(t)y(t-\sigma_2) = 0 \tag{2}$$

where $P_1,P_2, Q_1,Q_2 \in C[[t_0,\infty), \mathbb{R}^+]$ and $\tau_1,\tau_2, \sigma_1,\sigma_2 \in \mathbb{R}^+$. We established a Sturm-type comparison result according to which if Eq. (1) is oscillatory, so is Eq. (2)

1. INTRODUCTION

A neutral delay differential equation (NDDE) is a differential equation in which the highest order derivative of the unknown function appears both with and without delays.

The aim in this paper is to obtain Sturm-type comparison results for the NDDEs

$$\frac{d}{dt}[x(t) - P_1(t)x(t-\tau_1)] + Q_1(t)x(t-\sigma_1) = 0 \tag{1}$$

and

$$\frac{d}{dt}[y(t) - P_2(t)y(t-\tau_2)] + Q_2(t)y(t-\sigma_2) = 0 \tag{2}$$

where

$$P_1,P_2,Q_1,Q_2 \in C[[t_0,\infty), \mathbb{R}^+] \quad \text{and} \quad \tau_1,\tau_2,\sigma_1,\sigma_2 \in \mathbb{R}^+. \tag{3}$$

AMS 1980 Subject Classification: Primary - 34K15; Secondary - 34C10.

[1]Department of Mathematics, University of Rhode Island, Kingston, R. I. 02881, U.S.A.

[2]Department of Mathematics, University of Sarajevo, Sarajevo 71000, Yugoslavia.

In general, the study of NDDEs presents complications and basic results which are true for delay differential equations may be false for neutral equations. For example, it has been shown by Snow [13], see also Slemrod and Infante [12], that even though the characteristic roots of a NDDE may all have negative real parts, it is still possible for some solutions to be unbounded.

The study of NDDEs has some interest in applications. For example, NDDEs appear in networks containing lossless transmission lines. Such networks arise in high speed computers where the lossless transmission lines are used to interconnect switching circuits. See Brayton and Willoughby [1] and Hale [6, pp. 4-7].

Existence and uniqueness of solutions for equations of the form (1) are easily established by the method of steps. Let $m = \max\{\tau_1, \sigma_1\}$ and let $\phi \in C[[t_0-m, t_0], \mathbb{R}]$. By a solution of Eq. (1) with initial function ϕ at t_0, we mean a function $x \in C[[t_0-m,\infty), \mathbb{R}]$ such that $x(t) = \phi(t)$ for $t_0-m \leq t \leq t_0$, $x(t) - P_1(t)x(t-\tau_1)$ is continuously differentiable for $t \geq t_0$ and such that Eq. (1) is satisfied for $t \geq t_0$. Using the method of steps, it follows that for every continuous initial function ϕ, there exists a unique solution of Eq. (1) valid for $t \geq t_0$.

A solution of Eq. (1) is called <u>oscillatory</u> if it has arbitrarily large zeros and <u>nonoscillatory</u> if it is eventually positive or eventually negative. Equation (1) is called <u>oscillatory</u> if for every $t_1 \geq t_0$ and $\phi \in C[[t_1-m, t_1], \mathbb{R}]$ the solution of Eq. (1) with initial function ϕ at t_1 is oscillatory. Otherwise Eq. (1) is called <u>nonoscillatory</u>.

Recently, there has been a lot of activity concerning the oscillatory behavior of NDDEs. See, for example, [2]-[5], [7], [8] and [11]. However, this is the first study concerning Sturm-type comparison results for NDDEs. As in the case of the classical comparison theorem of Sturm [14], we will give conditions under which if Eq. (1) oscillates, so does Eq. (2). For non-neutral equations, that is when $P_1(t) = P_2(t) \equiv 0$, related comparison results follow from results in Myskis [9, Ch. 6], and Philos [10].

2. A STURM-TYPE COMPARISON THEOREM

Consider Eqs. (1) and (2) and assume that (3) holds,

$$P_1 \in C^1[[t_0,\infty), \mathbb{R}^+], \ P_1 \text{ is bounded and } \dot{P}_1(t) \geq 0 \text{ for } t \geq t_0, \tag{4}$$

$$0 \leq P_2(t) \leq 1 \tag{5}$$

and that there exists $K > 0$ such that

$$0 < K \leq Q_2(t) \tag{6}$$

for all $t \geq t_0$.

The following result is a Sturm-type comparison theorem for NDDEs.

THEOREM 1. In addition to Conditions (3) - (6) assume that

$$P_1(t) \leq P_2(t - \sigma_2) \frac{Q_2(t)}{Q_2(t-\tau_2)}, \tag{7}$$

$$Q_1(t) \leq Q_2(t), \tag{8}$$

$$\tau_1 \leq \tau_2 \quad \text{and} \quad \sigma_1 \leq \sigma_2, \tag{9}$$

and either

$$\tau_1 P_1(t) > 0 \quad \text{for} \quad t \geq t_0 \tag{10}$$

or

$$Q_1(t) \not\equiv 0 \text{ on any interval of length } \sigma_1. \tag{11}$$

Then if Eq. (1) is oscillatory, so is Eq. (2).

The proof of Theorem 1 makes use of the following two lemmas.

LEMMA 1. Assume that Conditions (5) and (6) are satisfied. Let $y(t)$ be an eventually positive solution of Eq. (2). Set

$$z(t) = y(t) - P_2(t)y(t - \tau_2).$$

Then eventually,

$$z(t) > 0, \quad \dot{z}(t) < 0 \quad \text{and} \quad \lim_{t\to\infty} z(t) = 0 . \tag{12}$$

PROOF. We have,

$$\dot{z}(t) = -Q_2(t)y(t-\sigma_2) \leq -Ky(t-\sigma_2). \tag{13}$$

Therefore, either

$$\lim_{t\to\infty} z(t) = -\infty \tag{14}$$

or

$$\lim_{t\to\infty} z(t) = L \in \mathbb{R}. \tag{15}$$

We shall first show (14) is impossible. So for the sake of contradiction, suppose (14) holds. Then $y(t)$ must be unbounded; in addition,

$$y(t) - P_2(t)y(t-\tau_2) < 0.$$

But then by (5), we see that

$$y(t) < P_2(t)y(t-\tau_2) \leq y(t-\tau_2)$$

which contradicts the fact that $y(t)$ is unbounded. Hence (15) holds. By integrating (13) from t_1 to t (where t_1 is chosen sufficiently large to insure that $y(t-\sigma_2) > 0$ for $t \geq t_1$) and then taking the limit as $t \to \infty$, we find that

$$L - z(t_1) \leq -K \int_{t_1}^{\infty} y(s-\sigma_2)\,ds$$

and so

$$y \in L^1[t_1,\infty).$$

Thus $z \in L^1[t_1,\infty)$ and so we see $L = 0$. The proof is complete.

LEMMA 2. Consider the differential inequality

$$P_1(t)z(t-\tau_1) + \int_t^{\infty} Q_1(s)z(s-\sigma_1)\,ds \leq z(t), \quad t \geq T \tag{16}$$

where

$$P_1, Q_1 \in C[[T,\infty), \mathbb{R}^+], \ \tau_1 \ \text{and} \ \sigma_1 \in \mathbb{R}^+$$

and either (10) or (11) holds for $t \geq T$.
Let $m = \max\{\tau_1, \sigma_1\}$ and assume that (16) has a continuous positive solution $z\colon [T-m,\infty) \to (0,\infty)$ such that

$$z(T) < z(t) \quad \text{for} \quad T-m \leq t < T. \tag{17}$$

Then there exists a continuous positive solution $x\colon [T-m,\infty) \to (0,\infty)$ of the equation

$$P_1(t)x(t-\tau_1) + \int_t^{\infty} Q_1(s)x(s-\sigma_1)\,ds = x(t), \quad t \geq T \tag{18}$$

or equivalently, of the NDDE (1) for $t \geq T$.

PROOF. Define the set of functions

$$W = \{w \in C[[T-m,\infty), \mathbb{R}^+]\colon \ 0 \leq w(t) \leq z(t), \ t \geq T-m\}.$$

Next define the mapping S on W as follows:

$$(Sw)(t) = \begin{cases} P_1(t)w(t-\tau_1) + \int_t^{\infty} Q_1(s)w(s-\sigma_1)\,ds & \text{for} \quad t \geq T \\ (Sw)(T) + z(t) - z(T) & \text{for} \quad T-m \leq t \leq T. \end{cases}$$

It is clear that $Sw\colon [T-m,\infty) \to \mathbb{R}^+$ is continuous. It is also clear that $w_1, w_2 \in W$ with $w_1 \leq w_2$ implies $Sw_1 \leq Sw_2$. Note also that by (16), $Sz \leq z$. Hence $w \in W$ implies $Sw \leq Sz \leq z$ and so we see $S\colon W \to W$. We now inductively define the following sequence in W:

$$z_0 = z \quad \text{and} \quad z_n = Sz_{n-1} \quad \text{for} \quad n = 1,2,\ldots .$$

It is clear by induction that

$$0 \le z_n(t) \le z_{n-1}(t) \le z(t), \qquad t \ge T\text{-}m.$$

Set

$$x(t) = \lim_{n\to\infty} z_n(t), \qquad t \ge T\text{-}m.$$

Then it follows by Lebesgue's dominated convergence theorem that $x(t)$ satisfies (18). Next we will show that x is continuous. Now if $T\text{-}m \le t \le T$, then

$$\begin{aligned} z_n(t) - z_m(t) &= [z_n(T) + z(t) - z(T)] - [z_m(T) + z(t) - z(T)] \\ &= z_n(T) - z_m(T) \end{aligned}$$

and so we see that $\{z_n(t)\}$ converges uniformly on $[T\text{-}m,T]$. Hence x is continuous on $[T\text{-}m,T]$. So clearly, x is continuous. Finally, it remains to show that $x(t) > 0$ for $t \ge T\text{-}m$. Now if $T\text{-}m \le t < T$, then by (17) and the definition of z_n

$$0 < z(t) - z(T) \le z_n(t)$$

and so

$$0 < z(t) - z(T) \le x(t).$$

Let

$$t^* = \inf\{t \ge T\colon x(t) = 0\}.$$

We shall show $t^* = \infty$. If not, then $t^* \in [T,\infty)$. So $x(t) > 0$ for $T\text{-}m \le t < t^*$ and $x(t^*) = 0$. But

$$x(t^*) = P_1(t^*)x(t^* - \tau_1) + \int_{t^*}^{\infty} Q_1(s)x(s - \sigma_1)ds > 0$$

because of either (10) or (11). This contradiction completes the proof of the Lemma.

PROOF OF THEOREM 1. Assume, for the sake of contradiction, that Eq. (2) has an eventually positive solution $y(t)$. Set $z(t) = y(t) - P_2(t)y(t\text{-}\tau_2)$. Then by Lemma 1, (12) holds. It follows by direct substitution that $z(t)$ satisfies the NDDE

$$\dot z(t) - P_2(t\text{-}\sigma_2)\,\frac{Q_2(t)}{Q_2(t\text{-}\tau_2)}\,\dot z(t\text{-}\tau_2) + Q_2(t)z(t\text{-}\sigma_2) = 0. \tag{19}$$

By using (7), (8) and (12), Eq. (19) yields the inequality

$$\dot z(t) - P_1(t)\dot z(t\text{-}\tau_2) + Q_1(t)z(t\text{-}\sigma_2) \le 0. \tag{20}$$

By integrating (20) from t to t_1 and then letting $t_1 \to \infty$, we find

$$-z(t) + P_1(t)z(t-\tau_2) + \int_t^\infty \dot{P}_1(s)z(s-\tau_2)ds + \int_t^\infty Q_1(s)z(s-\sigma_2)ds \le 0$$

and so by (9) and (4)

$$P_1(t)z(t-\tau_1) + \int_t^\infty Q_1(s)z(s-\sigma_1)ds \le z(t), \quad t \ge t_0 + m.$$

The hypotheses of Lemma 2 are satisfied and so Eq. (1) has a positive solution. This contradicts the hypothesis that every solution of Eq. (1) oscillates. The proof is complete.

REMARK. In the special case where the coefficients of Eqs. (1) and (2) are constants, the hypotheses of Theorem 1 reduce to

$$0 \le P_1 \le P_2 \le 1 \tag{21}$$

$$0 < Q_1 \le Q_2 \tag{22}$$

and

$$0 \le \tau_1 \le \tau_2 \quad \text{and} \quad 0 \le \sigma_1 \le \sigma_2. \tag{23}$$

Now assume that (21)-(23) are satisfied and that Eq. (1) is oscillatory. Then, there exists a positive constant m such that

$$\lambda - \lambda P_1 e^{-\lambda\tau_1} + Q_1 e^{-\lambda\sigma_1} \ge m \quad \text{for all} \quad \lambda \in \mathbb{R} .$$

Also

$$\lambda - \lambda P_2 e^{-\lambda\tau_2} + Q_2 e^{-\lambda\sigma_2} \ge \lambda - \lambda P_1 e^{-\lambda\tau_1} + Q_1 e^{-\lambda\sigma_2} \quad \text{for} \quad \lambda \le 0$$

and so

$$\lambda - \lambda P_2 e^{-\lambda\tau_2} + Q_2 e^{-\lambda\sigma_2} > 0 \quad \text{for} \quad \lambda \ge 0.$$

The above inequalities imply that the characteristic equation of Eq. (2) has no real roots which proves (independently of Theorem 1) that Eq. (2) oscillates [see [5], [7] and [11]].

REFERENCES

1. R. K. BRAYTON and R. A. WILLOUGHBY, On the numerical integration of a symmetric system of difference-differential equations of neutral type, J. Math. Anal. Appl. 18(1967), 182-189.

2. M. K. GRAMMATIKOPOULOS, E. A. GROVE and G. LADAS, Oscillations of first order neutral delay differential equations, J. Math. Anal. Appl. (to appear).

3. M. K. GRAMMATIKOPOULOS, E. A. GROVE and G. LADAS, Oscillation and asymptotic behavior of neutral differential equations with deviating arguments, Applicable Anal. (to appear).

4. M. K. GRAMMATIKOPOULOS, G. LADAS and Y. G. SFICAS, Oscillation and asymptotic behavior of neutral equations with variable coefficients, (to appear).

5. E. A. GROVE, G. LADAS and A. MEIMARIDOU, A necessary and sufficient condition for the oscillation of neutral equations, J. Math. Anal. Appl. (to appear).

6. J. HALE, "Theory of functional differential equations", Springer-Verlag, New York, 1977.

7. M. R. S. KULENOVIĆ, G. LADAS and A. MEIMARIDOU, Necessary and sufficient conditions for oscillations of neutral differential equations, J. Austr. Math. Soc. Series B (to appear).

8. G. LADAS and Y. G. SFICAS, Oscillations of neutral delay differential equations, Canadian Math. Bull. (to appear).

9. A. D. MYSKIS, Linear Differential Equations with Retarded Argument (Russian), Izdatel'stvo "Nauka", 2nd Edition, Moscow, 1972.

10. C. G. PHILOS, On the existence of nonoscillatory solutions tending to zero at ∞ for differential equations with positive delays, Archiv. Math. 36(1980), 168-178.

11. Y. G. SFICAS and I. P. STAVROULAKIS, Necessary and sufficient conditions for oscillations of neutral differential equations, J. Math. Anal. Appl. (to appear).

12. M. SLEMROD and E. F. INFANTE, Asymptotic stability criteria for linear systems of difference-differential equations of neutral type and their discrete analogues, J. Math. Anal. Appl. 38(1972), 399-415.

13. W. SNOW, Existence, uniqueness and stability for nonlinear differential-difference equations in the neutral case, N.Y.U. Courant Inst. Math. Sci. Rep. IMM NYU 328, (February 1965).

14. C. STURM, Sur les equations differentielles linearies du second order, J. Math. Pures Appl. 1(1836), 106-186.

Canadian Mathematical Society
Conference Proceedings
Volume 8 (1987)

Spectrum of a Hamiltonian System with Spectral Parameter in a Boundary Condition

D.B. Hinton and J.K. Shaw

Abstract

A system $J\vec{y}' = (\lambda A+B)\vec{y}$ is considered on an interval $[a,b)$, and is limit point at $x=b$. At $x=a$ is imposed the parameter dependent condition $\alpha_1 y(a)-\alpha_2\hat{y}(a)=\lambda(\alpha_1' y(a)-\alpha_2'\hat{y}(a))$, where $\vec{y} = \begin{pmatrix} y \\ \hat{y} \end{pmatrix}$, the two $n\times 1$ components of $\vec{y}$. The spectrum of this boundary problem is completely characterized in terms of the Titchmarsh-Weyl coefficient for the problem. Using the main theorem, a result on invariance with respect to λ-dependent boundary conditions is proved.

1. Introduction. Various problems in physics and engineering are concerned with eigenvalues of systems in which the spectral parameter λ occurs not only in a differential equation model but also in at least one boundary condition. Atkinson ([4],p.22) cites the examples of a transmission line beginning or ending with a circuit, and the transverse motion of an elastic string with a mass attached at one end. The behavior of an oscillating cable with tip mass is also associated with the feedback control of large flexible space structures ([5]). Numerous examples of parameter-dependent boundary conditions were given by C.T. Fulton in [9], for regular boundary problems, and in [10] for the singular case.

In all the references given so far, the basic differential equation is $-(p(x)y')'+q(x)y=\lambda k(x)y$. Boundary conditions in [9] take the form $\cos\alpha y(a)+\sin\alpha p(a)y'(a)=0$ and $\beta_1 y(b)-\beta_2 p(b)y'(b)=\lambda(\beta_1' y(b)-\beta_2' p(b)y'(b))$. In [10] the problem is set on $(-\infty,\infty)$, is limit circle at $-\infty$ but can be either limit point or limit circle at $+\infty$; a standard boundary condition is imposed at $x=\infty$ (none, if limit point) and $\alpha_1 B_1^{(-)}(y)-\alpha_2 B_2^{(-)}(y)=\lambda(\alpha_1' B_1^{(-)}(y)-\alpha_2' B_2^{(-)}(y))$ is used at $x=-\infty$. The expressions $B_k^{(-)}$ are the usual boundary forms associated with a limit circle endpoint (see [8]). Inasmuch

0731-1036/87 $1.00 + $.25 per page

as a limit circle endpoint is structurally equivalent to a regular endpoint ([8,p.61]), Fulton's boundary condition at $x=-\infty$ is equivalent to a parameter-dependent condition placed at a finite, regular point.

Further discussion of parameter-dependent boundary conditions is given in [13], where discrete spectrum criteria and associated eigenfunction expansion are developed. Walter's fundamental paper [20] developed the operator-theoretic formulation of the problem. Additional references are [12], [11] and [19]; an extensive list of references is given in [10].

Parameter-dependent boundary conditions for fourth-order problems have been considered. The equation $y^{(4)}-(p(x)y')'=\lambda y$, $0<x<1$, coupled with end conditions $y(0)=y'(0)=0$, $y''(1)=0$, $y'''(1)=k_1y'(1)-\lambda k_2y(1)$, describes the transverse motion of a rotating beam with tip mass, such as a helicopter blade ([2],[18]) or a bob pendulum suspended from a wire ([1]).

The presence of higher order parameter-dependent problems suggests a unified treatment of these, together with Sturm-Liouville equations, within the framework of Hamiltonian systems. In this paper we will consider a 2nx1 system

(1.1) $J\vec{y}'=[\lambda A(x)+B(x)]\vec{y}$, $-\infty<a\leqslant x<b\leqslant\infty$, λ complex,

where $J=\begin{pmatrix}0 & -I\\ I & 0\end{pmatrix}$, I is the nxn identity, $A\geqslant 0$ and B are 2nx2n and symmetric and $\vec{y}=\begin{pmatrix}y\\ \hat{y}\end{pmatrix}$ is the 2nx1 vector with nx1 components y and $\hat{y}$. More specific hypothesis are given in §2. We assume that (1.1) is of limit point type at x=b, and at x=a we impose the λ-dependent boundary condition

(1.2) $\alpha_1y(a)-\alpha_2\hat{y}(a)=\lambda(\alpha_1'y(a)-\alpha_2'\hat{y}(a))-u$

where $\alpha_1,\alpha_2,\alpha_1'$ and α_2' are suitably restricted nxn matrices; see our hypotheses in §2. In the light of our earlier remarks on limit circle endpoints, much of our work encompasses Fulton's in [10], particularly with regard to construction of the resolvent operator and the Titchmarsh-Weyl coefficient.

Whereas Fulton was mainly interested in eigenfunction expansions, we will be taking a different direction, characterizing instead the various parts of the spectrum of (1.1)-(1.2) in terms of the associated Titchmarsh-Weyl coefficient, $M(\lambda)$. The connection between the spectrum and $M(\lambda)$ for standard (not λ-dependent) boundary problems has a well-known beginning in the paper [6] of Chaudhuri and Everitt, which considers a selfadjoint operator $Ty=-(p(x)y')'+q(x)y, 0\leq x<\infty$. Membership in the domain $D(T)$ requires $\cos\alpha y(0)+p(0)\sin\alpha y'(0)=0$, together with the usual smoothness and integrability assumptions, and the equation is taken to be limit point at $x=\infty$. Letting $\Theta(x,\lambda)$ and $\varphi(x,\lambda)$ denote the solutions of $-(py')'+qy=\lambda y$ satisfying $\Theta(0,\lambda)=p(0)\varphi'(0,\lambda)=\cos\alpha$, $-\varphi(0,\lambda)=p(0)\Theta(0,\lambda)=\sin\alpha$ for all λ, the Titchmarsh-Weyl $m(\lambda)$ coefficient is the analytic function $m(\lambda)=-\lim_{x\to\infty}(\Theta(x,\lambda)/\varphi(x,\lambda))$, $\mathrm{Im}(\lambda)\neq 0$. Letting $\sigma(T)$ denote the spectrum of T, $\rho(T)$ the resolvent set, $P(T)$ the point spectrum, $C(T)$ the continuous spectrum and $PC(T)$ the point-continuous spectrum, Chandhuri and Everitt obtained the following classification:

(i) $\lambda_0\in\rho(T)$ iff $m(\lambda)$ is analytic at λ_0;

(ii) $\lambda_0\in P(T)$ iff $m(\lambda)$ has a simple pole at λ_0;

(1.3) (iii) $\lambda_0\in C(T)$ iff $m(\lambda)$ is not analytic at λ_0 and $\lim_{\nu\to 0}\nu m(\lambda_0+i\nu)=0$;

(iv) $\lambda_0\in PC(T)$ iff $\lim_{\nu\to 0}\nu m(\lambda_0+i\nu)=s\neq 0$ and $m(\lambda)-s(\lambda-\lambda_0)^{-1}$ is not analytic at λ_0.

The present authors extended this classification to Hamiltonian systems (1.1) in [16], and further to two-singular endpoint problems for $J\vec{y}'=(\lambda A+B)\vec{y}$ in [17].

The purpose of this paper is to establish a classification analogous to (1.3) for the parameter-dependent Hamiltonian system (1.1)-(1.2). To extend (1.3)(i) we derive the resolvent operator in such a way as to include the corresponding resolvent in [10]. The generalization of (1.3)(ii) is of special interest because of the complexity of the inner product, made necessary by the parameter-dependent condition (1.2), and its influence on computing the "reduced resolvent" which acts on the orthogonal complement of the eigenspace corresponding to the pole λ_0. We state the general classification theorem for (1.1)-(1.2) in §2, after introducing some necessary terminology and notation, and give the proof in §3. Quite a few elements of the proof are modifications of corresponding ones for the nonparameter-dependent case in [16], and we will omit these stages whenever possible.

2. Hypotheses and Statements of Results. As in [16] we assume A and B in (1.1) are locally integrable and that A(x) is either invertible for all x or that $A(x) = \begin{pmatrix} A_1(x) & 0 \\ 0 & 0 \end{pmatrix}$, where $A_1(x)$ is $r\times r$ and positive definite for all x. Furthermore, we assume Atkinson's "definiteness hypothesis" ([4,p.253]) $\int_\alpha^\beta \vec{y}^*(x)A(x)\vec{y}(x)dx>0$, $a\leq\alpha<\beta<b$, for all nontrivial solutions $\vec{y}$ of (1.1).

The limit point hypothesis on (1.1) is that for all nonreal λ there are exactly n linearly independent solutions $\vec{y}$ of (1.1) of "integrable square"; i.e., $\int_a^b \vec{y}^*(x)A(x)\vec{y}(x)dx<\infty$. It follows from [14] and [15] that this condition implies

$$\lim_{x\to b} \vec{y}^*(x)J\vec{z}(x)=0 \tag{2.1}$$

for all solutions $\vec{y}$ of (1.1) and $\vec{z}$ of (1.1) (with a possibly different λ) which are of integrable square.

Denote by L_A^2 the set of measurable vectors $\vec{f}(x)$ with $\int_a^b \vec{f}^*A\vec{f}dx<\infty$. Let $E_r = \begin{pmatrix} I_r & 0 \\ 0 & 0 \end{pmatrix}$, with I_r $r\times r$, or $E_r=I$ if $r=2n$, so that $E_rL_A^2=\{E_r\vec{f} \mid \vec{f}\in L_A^2[a,b)\}$ is a Hilbert space. Let an operator T with domain D(T) be defined by saying $\vec{y}\in D(T)$ if and only if (i) $\vec{y}\in L_A^2$, (ii) $\vec{y}$ is locally absolutely continuous on $[a,b)$, (iii) $\begin{pmatrix} A_1^{-1} & 0 \\ 0 & 0 \end{pmatrix}[J\vec{y}'-B\vec{y}]\in E_rL_A^2$ and (iv) $\begin{pmatrix} 0 & 0 \\ 0 & I_{2n-r} \end{pmatrix}[J\vec{y}'-B\vec{y}]=0$, and set $T(\vec{y})=\begin{pmatrix} A_1^{-1} & 0 \\ 0 & 0 \end{pmatrix}[J\vec{y}'-B\vec{y}]$: $D(T)\to E_rL_A^2$. Thus (1.1) is formulated in an operator-theoretic setting by writing $T(\vec{y})=\lambda E_r\vec{y}$.

The $n\times n$ matrices α_i and α_i' in (1.2) will be assumed to satisfy

$$\begin{aligned} &\alpha=\alpha_2\alpha_1'^*-\alpha_1\alpha_2'^*=\alpha^*=\alpha_1'\alpha_2^*-\alpha_2'\alpha_1^*>0, \\ &\alpha_1\alpha_2^*-\alpha_2\alpha_1^*=0=\alpha_1'\alpha_2'^*-\alpha_2'\alpha_1'^*, \\ &\alpha_2^*-\lambda\alpha_2'^* \text{ is invertible for all nonreal } \lambda. \end{aligned} \tag{2.2}$$

Let $R(\vec{y})=\alpha_1 y(a)-\alpha_2\hat{y}(a)$ and $R'(\vec{y})=\alpha_1' y(a)-\alpha_2'\hat{y}(a)$, so that (1.2) becomes $R(\vec{y})-\lambda R'(\vec{y})=-u$. Following Fulton [10] and Walter [20] we place the parameter dependent problem (1.1)-(1.2) in a Hilbert space setting by defining an operator $\tilde{T}$, with domain $D(\tilde{T})$, in

$\tilde{H}=L_A^2\times\mathbb{C}^n$ ($\mathbb{C}$=complex plane) as follows. Denote the members $\tilde{f}\in\tilde{H}$ by $\tilde{f}(x)=\begin{pmatrix}\vec{f}(x)\\ f_3\end{pmatrix}$, so that

$\vec{f}(x)=\begin{pmatrix}f(x)\\ \hat{f}(x)\end{pmatrix}$ is a measurable function and f_3 is an $n\times 1$ vector. Then let

$D(\tilde{T})=\{\tilde{f}\in\tilde{H}\mid \vec{f}\in D(T),\ f_3=R'(\vec{f})\}$, and then define $\tilde{T}$: $D(\tilde{T})\subset L_A^2\times\mathbb{C}^n\to E_r L_A^2\times\mathbb{C}^n$ by

$\tilde{T}(\tilde{y})=\begin{pmatrix}T(\vec{y})\\ R(\vec{y})\end{pmatrix}$. Then (1.1)-(1.2) is equivalent to the operator equation $\tilde{T}(\tilde{y})=\lambda P_r\tilde{y}$, where

$P_r=\begin{pmatrix}E_r & 0\\ 0 & I\end{pmatrix}$ and $F_3=u$.

The inner product associated with $\tilde{H}$ is given by $\langle\tilde{f},\tilde{g}\rangle=\int_a^b\vec{f}^*(x)A(x)\vec{g}(x)dx+f_3^*\alpha^{-1}g_3$; compare [13] and [10].

We may now proceed with introducing the Titchmarsh-Weyl coefficient for (1.1)-(1.2). Let $Y(x,\lambda)$ be the fundamental matrix solution of (1.1), partitioned into $n\times n$ blocks by

$Y(x,\lambda)=\begin{pmatrix}\Theta(x,\lambda) & \varphi(x,\lambda)\\ \hat{\Theta}(x,\lambda) & \hat{\varphi}(\alpha,\lambda)\end{pmatrix}$, with initial value $Y(a,\lambda)=\begin{pmatrix}\alpha_2'^*\alpha^{-1} & \alpha_2^*-\lambda\alpha_2'^*\\ \alpha_1'^*\alpha^{-1} & \alpha_1^*-\lambda\alpha_1'^*\end{pmatrix}$ for all λ.

Substituting $\vec{\varphi}$ into the identity (see [16])

(2.3) $$(\mu-\bar{\lambda})\int_\alpha^\beta\vec{y}^*A\vec{z}\,dx=[\vec{y}^*J\vec{z}]_\alpha^\beta,$$

valid for solutions $\vec{y}=\vec{y}_\lambda$ and $\vec{z}=\vec{z}_\mu$ for (1.1), with λ replaced by μ for $\vec{z}_\mu$, and using (2.2), we obtain $\vec{\varphi}^*(x,\lambda)J\vec{\varphi}(x,\lambda)=2i\operatorname{Im}(\lambda)[\alpha+\int_a^x\vec{\varphi}^*A\vec{\varphi}dx]$. By the definiteness hypothesis and the assumption $\alpha>0$, we have $[\vec{\varphi}^*J\vec{\varphi}(x)/2i\operatorname{Im}(\lambda)]>0$ and it follows that the $n\times n$ component $\varphi(x,\lambda)$ is invertible for $\operatorname{Im}(\lambda)>0$. Therefore we may define the function $M_x(\lambda)=-\varphi^{-1}(x,\lambda)\Theta(x,\lambda)$ for $\operatorname{Im}(\lambda)>0$, $a\leqslant x<b$. Using the identity $Y^*(x,\bar{\lambda})JY(x,\lambda)=J$ (from (1.1)) we can prove $M_x^*(\bar{\lambda})=M_x(\lambda)$ for $\operatorname{Im}(\lambda)\neq 0$. Clearly $M_x(\lambda)$ is therefore meromorphic in the complex plane with real poles only.

Let $Y_0=\begin{pmatrix}\Theta_0 & \varphi_0\\ \hat{\Theta}_0 & \hat{\varphi}_0\end{pmatrix}$ be the fundamental matrix solution of (1.1) with initial values $Y_0(a,\lambda)=I_{2n}$ for all λ. The Titchmarsh-Weyl function $M_0(\lambda)$ for the non-parameter-dependent problem with end condition $y(a)=0$ is given by $M_0(\lambda)=-\lim_{x\to b}\varphi_0^{-1}(x,\lambda)\Theta_0(x,\lambda)$ for $\operatorname{Im}(\lambda)\neq 0$. It is well known that $M_0(\lambda)$ is nonsingular and satisfies $\operatorname{Im}(\lambda)\operatorname{Im}M_0(\lambda)\geqslant 0$, $\operatorname{Im}(\lambda)\neq 0$([14]); the approximation $M_{0,x}(\lambda)=-\varphi_0^{-1}(x,\lambda)\Theta_0(x,\lambda)$ corresponds to $M_x(\lambda)$ above. In fact we can link them explicitly through the formula $Y(x)=Y_0(x)\begin{pmatrix}\alpha_2'^*\alpha^{-1} & \alpha_2^*-\lambda\alpha_2'^*\\ \alpha_1'^*\alpha^{-1} & \alpha_1^*-\lambda\alpha_1'^*\end{pmatrix}$,

$$\text{(2.4)}\qquad -M_x(\lambda)=[(\alpha_1^*-\lambda\alpha_1'^*)-M_{0,x}(\lambda)(\alpha_2^*-\lambda\alpha_2'^*)]^{-1}(\alpha_1'^*-M_{0,x}(\lambda)\alpha_2'^*)\alpha^{-1}$$

to be exact. We are going to let $x\to b$ in (2.4) and assert that the right side converges. It is sufficient to show that the inverted term does so. Writing $Z=(\alpha_1^*-\lambda\alpha_1'^*)(\alpha_2^*-\lambda\alpha_2'^*)^{-1}$ we have by (2.2) $\operatorname{Im}Z=-\operatorname{Im}(\lambda)(\alpha_2-\bar{\lambda}\alpha_2')^{-1}\alpha(\alpha_2^*-\lambda\alpha_2'^*)\leqslant 0$, for $\operatorname{Im}(\lambda)\geqslant 0$, while the first term on the right of (2.4) equals $(\alpha_2^*-\lambda\alpha_2'^*)^{-1}[Z-M_{0,x}(\lambda)]^{-1}$. Since both $\operatorname{Im} M_{0,x}(\lambda)>0$ and $\operatorname{Im}M_0(\lambda)>0$ for $\operatorname{Im}(\lambda)>0$, then $Z-M_{0,x}(\lambda)$ and its limit as $x\to b$ remain invertible. Thus it is permissable to pass to the limit in (2.9). The relation $M^*(\bar{\lambda})=M(\lambda)$ certainly carries over, so as a result we have a Titchmarsh-Weyl coefficient $M(\lambda)$, analytic for $\operatorname{Im}(\lambda)\neq 0$, for the parameter-dependent problem.

As in the non-parameter-dependent case, we use $M(\lambda)$ to construct solutions of (1.1) which lie in L_A^2. Let $\vec{X}_\beta(x,\lambda)=\vec{\Theta}(x,\lambda)+\vec{\varphi}(x,\lambda)M_\beta(\lambda)$, $\operatorname{Im}(\lambda)\neq 0$; by definition of M_β, $\vec{X}_\beta(\beta,\lambda)=0$. Applying (2.3), we have then

$(\mu-\bar{\lambda})\int_a^\beta \vec{X}_\beta^*(t,\lambda)A(t)\vec{X}_\beta(t,\mu)dt=-\vec{X}_\beta^*(a,\lambda)J\vec{X}_\beta(a,\mu)$. Letting $\vec{X}(x,\lambda)=\lim_{\beta\to b}\vec{X}_\beta(x,\lambda)$ and using (2.2) results in the fundamental identity (compare [10, equation (5.7)])

$$(2.5)\qquad (\mu-\bar{\lambda})\int_a^b \vec{X}^*(t,\lambda)A(t)\vec{X}(t,\mu)dt+(\mu-\bar{\lambda})M^*(\lambda)\alpha M(\mu)=M^*(\lambda)-M(\mu),$$

which becomes on setting $\mu=\lambda\,(\mathrm{Im}(\lambda)\neq 0)$

$$(2.6)\qquad \int_a^b \vec{X}^*(t,\lambda)A(t)\vec{X}(t,\lambda)dt+M^*(\lambda)\alpha M(\lambda)=-\frac{\mathrm{Im}M(\lambda)}{\mathrm{Im}(\lambda)}.$$

In particular, $\vec{X}(x,\lambda)\in L_A^2$ and $\mathrm{Im}M(\lambda)\cdot\mathrm{Im}(\lambda)<0$.

Since $\vec{X}(x,\lambda)=Y(x,\lambda)\begin{pmatrix} I \\ M(\lambda)\end{pmatrix}$ and $M(\lambda)$ is nonsingular, then $\vec{X}(x,\lambda)$ has rank n. Its size is 2nxn, and so the columns of $\vec{X}$ form a basis for the n-dimensional space of L_A^2 solutions of (1.1); recall that (1.1) is limit point at x=b.

Based on (2.2) and the relevant definitions, we assert that

$$(2.7)\qquad R(\vec{\Theta})=-I,\ R(\vec{\varphi})=\lambda\alpha,\ R(\vec{X})=-I+\lambda\alpha M,$$
$$R'(\vec{\Theta})=0,\ R'(\vec{\varphi})=\alpha,\ R'(\vec{X})=\alpha M.$$

In particular the vector $\tilde{X}=\begin{pmatrix}\vec{X}\\ \alpha M\end{pmatrix}\in D(\tilde{T})$, and (2.6) may be expressed as $\langle\tilde{X},\tilde{X}\rangle=-\mathrm{Im}M(\lambda)/\mathrm{Im}(\lambda)$.

Following [16] and [10] we proceed to construct to resolvent operator of $\tilde{T}$. For $\mathrm{Im}(\lambda)\neq 0$ let

$$(2.8)\qquad G(x,t,\lambda)=\begin{cases}\vec{X}(x,\lambda)\vec{\varphi}^*(t,\bar{\lambda}), & t<x\\ \vec{\varphi}(x,\lambda)\vec{X}^*(t,\bar{\lambda}), & t>x\end{cases},\quad \tilde{G}(x,t,\lambda)=\begin{pmatrix}G^*(x,t,\lambda)\\ \alpha\vec{X}^*(x,\lambda)\end{pmatrix}$$

To solve the inhomogeneous equation $(\lambda\tilde{P}_r-\tilde{T})\tilde{y}=\tilde{F}$, given $\tilde{F}\in E_rL_A^2\times\mathbb{C}^n$ we put

$\vec{y}(x)=\int_a^b G(x,t,\lambda)A(t)\vec{F}(t)dt+\vec{X}(x,\lambda)F_3=\vec{X}(x,\lambda)\int_a^x\vec{\varphi}^*(t,\bar{\lambda})A(t)\vec{F}(t)dt$

$+\vec{\varphi}(x,\lambda)\int_x^b\vec{X}^*(t,\bar{\lambda})A(t)\vec{F}(t)dt+\vec{X}^*(x,\lambda)F_3=\langle\tilde{G}(x,\cdot,\lambda),\tilde{F}\rangle$ and $y_3=R'(\vec{y})$. By direct differentiation and substitution into (1.1) it is easily seen that formally $(T-\lambda E_r)\vec{y}=-\vec{F}(x)$; the identity $Y(x,\lambda)JY^*(x,\bar{\lambda})=J$ is useful for this computation. It follows from (2.7) and (2.2) that $(R-\lambda R')(\vec{y})=-F_3$, and so we will have $\tilde{y}\in D(\tilde{T})$ if we show that $\tilde{y}\in\tilde{H}$. For this purpose we note first that $\tilde{T}$ is symmetric, for let $\tilde{y}$ and $\tilde{z}$ be in $D(\tilde{T})$ and note that $\langle\tilde{T}(\tilde{y}),\tilde{z}\rangle-\langle\tilde{y},\tilde{T}(\tilde{z})\rangle=\vec{y}^*J\vec{z}(a)+R(\vec{y})^*\alpha^{-1}R'(\vec{z})-R'(\vec{y})^*\alpha^{-1}R(\vec{z})$. The right side of this equation vanishes on account of (2.2). Now the standard technique for Hilbert space operators shows $\|\tilde{y}\|\leq|\mathrm{Im}(\lambda)|^{-1}\|\tilde{F}\|$. Therefore $\tilde{y}\in D(\tilde{T})$ and $(\lambda P_r-\tilde{T})\tilde{y}=\tilde{F}$.

Following the usual definition, we will say that a complex number λ lies in the <u>resolvent set</u> $\rho(\tilde{T})$ of $\tilde{T}$ if $\lambda P_r-\tilde{T}$ has a bounded inverse defined on all of $E_rL_A^2\times\mathbb{C}^n$, and we write $\tilde{R}_\lambda=(\lambda P_r-\tilde{T})^{-1}$ for the resolvent operator. The above calculation shows that $\lambda\in\tilde{P}(T)$ for all nonreal λ and that $\tilde{R}(\tilde{F})=\tilde{y}$, where $\vec{y}(x)=\langle\tilde{G}(x,\cdot,\lambda),\tilde{F}\rangle, y_3=R'(\vec{y})$. The <u>spectrum</u> $\sigma(\tilde{T})$ is defined to be the complement of $\rho(\tilde{T})$. The <u>point spectrum</u> $P(\tilde{T})$ is the set of isolated points of $\sigma(\tilde{T})$, $E(\tilde{T})=\sigma(\tilde{T})-P(\tilde{T})$ is the <u>essential spectrum</u>, $PC(\tilde{T})$ denotes the <u>point-continuous spectrum</u> (those $\lambda\in E(\tilde{T})$ such that $(\lambda P_r-\tilde{T})\tilde{y}=0$ has a nontrivial solution in $D(\tilde{T})$) and $C(\tilde{T})=E(\tilde{T})-PC(\tilde{T})$ is the <u>continuous spectrum</u>. Our main result may be stated as follows.

<u>Theorem 1</u>. The following classification holds:

(i) $\lambda_0\in\rho(\tilde{T})\Leftrightarrow M(\lambda)$ is analytic at λ_0;

(ii) $\lambda_0\in P(\tilde{T})\Leftrightarrow M(\lambda)$ has a simple pole at λ_0;

(iii) $\lambda_0\in C(\tilde{T})\Leftrightarrow M(\lambda)$ is not analytic at λ_0 and $\lim_{\nu\to 0}\nu M(\lambda_0+i\nu)=0$;

(iv) $\lambda_0\in PC(\tilde{T})\Leftrightarrow\lim_{\nu\to 0}\nu M(\lambda_0+i\nu)=S\neq 0$ and $M(\lambda)-iS(\lambda-\lambda_0)^{-1}$ is not analytic at λ_0.

Before giving the proof, we require a few more preliminaries. Since $\operatorname{Im}M(\lambda)\operatorname{Im}(\lambda)<0$, $M(\lambda)$ belongs to the matrix Pick-Nevanlinna class ([3]). The derivations of the Herglotz representation and Titchmarsh-Kodaira formula (compare [16]) for scalar Pick-Nevanlinna functions carry over to the matrix case ([7]) and we state them here without proof.

Theorem A. There exists an nxn Hermitian, right-continuous and nondecreasing spectral function $\tau(\lambda)$ such that

(i) $M(\lambda)=-\int_{-\infty}^{\infty}\left(\frac{1}{\mu-\lambda}-\frac{\mu}{1+\mu^2}\right)d\tau(\mu)+K_1\lambda+K_2$, $\operatorname{Im}(\lambda)\neq 0$, for Hermitian constants K_1 and K_2,

(ii) $-\lim_{\nu\to 0}\nu M(\lambda_0+i\nu)=\tau(\lambda_0+)-\tau(\lambda_0-)$,

(iii) $-(1/\pi)\lim_{\epsilon\to 0}\int_{\alpha}^{\beta}\operatorname{Im}M(\mu+i\epsilon)d\mu=\tau(\beta)-\tau(\alpha)$,

at continuity points α and β of τ.

The following is proved exactly as in [16].

Theorem B. The resolvent set $\rho(\tilde{T})$ is open and, for $\tilde{F},\tilde{G}\in E_r L_A^2\times\mathbb{C}^n$, the function $\langle\tilde{R}_{\bar{\lambda}}(\tilde{F}),\tilde{G}\rangle$ is analytic on $\rho(\tilde{T})$.

3. Proof of Theorem 1. We establish the four parts in the order (i), (ii), (iv), (iii).

(i) Since $M(\lambda)$ is analytic at λ_0 and $\lambda_0\in\rho(\tilde{T})$ for nonreal λ_0, we suppose that λ_0 is real. First suppose $M(\lambda)$ is analytic at λ_0. Let $\tilde{g}_\lambda$ be the operator defined below (2.8) by $\tilde{y}=\tilde{g}_\lambda(\tilde{F})$ if $\vec{y}(x)=\langle(G(x,\cdot,\lambda),\tilde{F}\rangle$ and $\vec{y}_3=R'(y)$. We know the resolvent operator $\tilde{R}_\lambda=\tilde{g}_\lambda$ for nonreal λ. We will now prove that $\tilde{g}_{\lambda_0}$ is defined and $\tilde{R}_{\lambda_0}=\tilde{g}_{\lambda_0}$ for real $\lambda_0\in\rho(\tilde{T})$. Certainly $\vec{X}(x,\lambda_0)$ is defined since $M(\lambda)$ is analytic at λ_0. Moreover, the right side of (2.6) approaches $M'(\lambda_0)$ as $\lambda=\lambda_0+i\nu\to\lambda_0$. Consequently, $\vec{X}(x,\lambda_0)\in L_A^2$ and the operator $\tilde{g}_{\lambda_0}$ is at least defined for $\tilde{F}\in E_r L_A^2 \times \mathbb{C}^n$. Letting $\tilde{y}=\tilde{\mathcal{G}}_{\lambda_0}(\tilde{F})$, direct differentiation shows that all properties placing $\tilde{y}\in D(\tilde{T})$ hold, save possibly $\tilde{y}\in L_A^2\times\mathbb{C}^n$, and formally $(\lambda P_r-\tilde{T})\tilde{y}=\tilde{F}$. Once we show $\tilde{y}\in L_A^2\times\mathbb{C}^n$ it will follow that $\tilde{\mathcal{G}}_{\lambda_0}$ is an onto, right inverse for $\lambda P_r-\tilde{T}$. Upon showing that $\tilde{\mathcal{G}}_{\lambda_0}$ is bounded, it will follow that $\tilde{R}_{\lambda_0}=\tilde{\mathcal{G}}_{\lambda_0}$; i.e., $\lambda_0\in\rho(\tilde{T})$. Now $\tilde{y}=\begin{pmatrix}\vec{y}\\ y_3\end{pmatrix}$ where $\vec{y}(x)=\int_a^b G(x,t,\lambda_0)A(t)\vec{F}(t)dt+\vec{X}(x,\lambda_0)F_3$ and $y_3=R'(\vec{y})=\alpha\int_a^b \vec{X}^*(t,\lambda_0)A(t)\vec{F}(t)dt+\alpha M(\lambda_0)F_3$. Also, $\langle\tilde{y},\tilde{y}\rangle=\int_a^b \vec{y}^*A\vec{y}+y_3^*\alpha^{-1}y_3=\|\vec{y}\|_A^2+\|y_3\|_\alpha^2$ which is finite if $\vec{y}\in L_A^2$. Since $\vec{X}(x,\lambda_0)\in L_A^2$, all we need to know is $\vec{z}(t)=\int_a^b G(x,t,\lambda_0)A(t)\vec{F}(t)dt\in L_A^2$, and the proof of this is given explicitly in [16]; thus $\tilde{y}\in L_A^2\times\mathbb{C}^n$ and $\tilde{y}\in D(\tilde{T})$. To prove that $\tilde{\mathcal{G}}_{\lambda_0}$ is bounded we note that $\|\vec{y}\|_A^2\leqslant 2\|\vec{z}\|_A^2+2\|\vec{X}(\cdot,\lambda_0)F_3\|_A^2$ ([4,p.239]) and that $\|\vec{z}\|_A^2\leqslant\gamma\|\vec{F}\|_A^2$ (γ=constant) from the corresponding proof in [16]. Thus $\|\vec{z}\|_A^2\leqslant\gamma\|\vec{F}\|^2$. Next, $\|\vec{X}(\cdot,\lambda_0)F_3\|_A^2=F_3^*\int_a^b\vec{X}^*(t,\lambda_0)A(t)\vec{X}(t,\lambda_0)dtF_3\leqslant K_1F_3^*\alpha^{-1}F_3=K_1\|F_3\|_\alpha^2\leqslant K_1\|\tilde{F}\|^2$, if K_1 is a sufficiently large constant. Finally, we need to prove that $\|y_3\|_\alpha^2$ is bounded by a multiple of $\|\tilde{F}\|^2$. As for the part $\alpha M(\lambda_0)F_3$ of y_3, $F_3^*M(\lambda_0)^*\alpha\alpha^{-1}\alpha M(\lambda_0)F_3^*\leqslant K_2F_3^*\alpha^{-1}F_3=K_2\|F_3\|_\alpha^2\leqslant K_2\|\tilde{F}\|^2$, as above, for some suitably large constant K_2, so there remains only the term $\alpha\int_a^b\vec{X}^*A\vec{F}$. This is bounded by a constant times the norm of $(\int_a^b \vec{X}^* A\vec{X})(\int_a^b \vec{F}^* A\vec{F})$, and this is not less that some multiple of $\|\vec{F}\|_A^2 \leqslant \|\tilde{F}\|^2$.

Hence $\tilde{g}_{\lambda_0}$ is bounded, $\tilde{R}_{\lambda_0} = \tilde{g}_{\lambda_0}$ and $\lambda_0 \in \rho(\tilde{T})$.

Next suppose that $\lambda_0 \in \rho(\tilde{T})$; we have to show $M(\lambda)$ continues analytically to a real interval containing λ_0. The proof we give is similar to the corresponding one in [16]. Replacing λ by $\bar{\lambda}$ in (2.6) leads to

(3.1) $$M(\lambda)[(\lambda-\mu)\alpha M(\mu)+I] = M(\mu)-(\lambda-\mu)\int_a^b \vec{X}^*(t,\bar{\lambda})A(t)\vec{X}(t,\mu)dt.$$

Consider the factor $(\lambda-\mu)\alpha M(\mu)+I$. Putting $\mu=\lambda_0+i\nu$ we claim $\nu M(\lambda_0+i\nu)\to 0$ as $\nu\to 0$. The limit exists by Theorem A(ii), and if it had value $C\neq 0$ then multiplying (2.6) with $\lambda=\lambda_0+i\nu$ by ν^2 and letting $\nu\to 0$ would yield $\int_a^b (\vec{\varphi}C)^* A(\vec{\varphi}C)+(\alpha C)^*\alpha^{-1}(\alpha C) = -\operatorname{Im}C$, which is to say that there is at least one eigenvector $\tilde{y}$ of $\tilde{T}$ where $\vec{y}=\vec{\varphi}C$ and $y_3=\alpha C$. This is impossible since $\lambda_0 \in \rho(\tilde{T})$. Then $\nu M(\lambda_0+i\nu)\to 0$ so given $\epsilon>0$ we can pick ν so small that $\|\nu\alpha M(\lambda_0+i\nu)\|<\epsilon$, where $\|\cdot\|$ is the Euclidean operator norm. Hence for $\mu=\lambda_0+i\nu$ the factor $(\lambda-\mu)\alpha M(\mu)+I$ is invertible when $\lambda=\lambda_0$. Therefore it is invertible in a λ-neighborhood of λ_0. Its inverse is of course analytic in the same neighborbood of λ_0, so we can solve for $M(\lambda)$ in (3.1) and our proof will be complete if we show that the right side of (3.1) is analytic at λ_0. To see this we note that $\vec{X}(t,\bar{\lambda})=\vec{X}(t,\mu)-(\mu-\bar{\lambda})g(t,\bar{\lambda},\vec{X}(\cdot,\mu))$, as in [16], where $g(x,\lambda,\vec{F})=\int_a^b G(x,t,\lambda)A(t)\vec{F}(t)dt$ for a $2n\times n$ matrix $\vec{F}$. Then the integral on the right side of (3.1) may be expressed as $\int_a^b \vec{X}^*(t,\mu)A(t)\vec{X}(t,\mu)dt-(\mu-\lambda)\int_a^b g^*(t,\bar{\lambda},\vec{X}(\cdot,\mu))A(t)\vec{X}(t,\mu)dt$. The second integral here is an expression of the form $\langle \tilde{R}_{\bar{\lambda}}(\tilde{F}),\tilde{F}\rangle$, where $\tilde{F}$ is $2n\times n$ and $F_3=0$; i.e., we think of the resolvent as applying to the columns of $\tilde{F}$. By Theorem A, this quantity is analytic at $\lambda_0 \in \tilde{\rho}(T)$. Therefore (3.1) provides the analytic continuation of $M(\lambda)$ to λ_0, and this completes the proof of (i).

(ii) If $\lambda_0 \in \tilde{P}(T)$ then λ_0 is an isolated singularity of $M(\lambda)$ by definition of $\tilde{P}(T)$ and part (i) above. Since $\operatorname{Im}M(\lambda)\cdot\operatorname{Im}(\lambda)<0$, $\operatorname{Im}M(\lambda)=0$ in certain intervals (λ_1,λ_0) and (λ_0,λ_2) by continuity. By Theorem A(iii), $\tau(\mu)$ is constant on (λ_1,λ_0) and (λ_0,λ_2). Part (i) of Theorem A shows therefore that $M(\lambda) = [(\tau(\lambda_0-)-\tau(\lambda_0+)]/(\lambda-\lambda_0]+H(\lambda)$, where $H(\lambda)$ is analytic at λ_0. Consequently, $M(\lambda)$ has a simple pole at λ_0.

Conversely, suppose $M(\lambda)=\sigma_{-1}(\lambda-\lambda_0)^{-1}+\sigma_0+\sigma_1(\lambda-\lambda_0)+\dots$. Putting $\lambda=\lambda_0+i\nu$ in (2.6), multiplying both sides by ν^2 and letting $\nu\to 0$ results in $\int_a^b(\vec{\varphi}(t,\lambda_0)\sigma_{-1})^*A(t)\vec{\varphi}(t,\lambda_0)\sigma_{-1}dt+\sigma_{-1}^*\alpha\sigma_{-1}=\sigma_{-1}$; i.e., $\binom{\vec{\varphi}(\cdot,\lambda_0)}{\alpha}\sigma_{-1}$ comprises a basis of eigenvectors of $\tilde{T}$ for $\lambda=\lambda_0$, so $\lambda_0 \in \tilde{P}(T)$.

(iv) Suppose $\lim_{\nu\to 0}\nu M(\lambda_0+i\nu)=S\neq 0$ and that $M(\lambda)-iS(\lambda-\lambda_0)^{-1}$ is not analytic at λ_0. Then $\lambda_0 \notin \tilde{\rho}(T)$ as otherwise $\nu M(\lambda_0+i\nu)\to 0$; $\lambda_0 \notin \tilde{P}(T)$ as otherwise $\sigma_{-1}=iS$ and $M(\lambda)-\sigma_{-1}(\lambda-\lambda_0)^{-1}$ would be analytic at λ_0. Thus $\lambda_0 \in \tilde{E}(T)$ and there remains only to produce an eigenfunction corresponding to λ_0, and this is accomplished exactly as in part (ii) above. That is, we multiply (2.6) by ν^2, where $\lambda=\lambda_0+i\nu$, and making $\nu\to 0$ gives $\int_a^b(\vec{\varphi}(t,\lambda_0)S)^*A(t)\vec{\varphi}(t,\lambda_0)Sdt+S^*\alpha S<\infty$, implying that the columns of $\begin{pmatrix}\vec{\varphi}(t,\lambda_0)\\ \alpha\end{pmatrix}S$ are eigenfunctions and $\lambda_0 \in \tilde{PC}(T)$.

Let $\lambda_0 \in \tilde{PC}(T)$. By Theorem A(i), $\lim_{\nu\to 0}\nu M(\lambda_0+i\nu)$ exists, and we suppose by way of contradiction that it has value 0. Let $\vec{z}(t)$ be an eigenfunction corresponding to λ_0 and apply (2.3) with $\mu=\lambda_0$, $\lambda=\lambda_0+i\nu$ and with $\vec{y}$ replaced by the 2nxn matrix $\vec{X}(t,\lambda_0+i\nu)$. Using the limit point hypothesis (2.1) we obtain $(i\nu)\int_a^b \vec{X}^*(t,\lambda)A(t)\vec{z}(t)dt = -\vec{X}^*(a,\lambda)J\vec{z}(a)=-[I,M^*(\lambda)]Y^*(a,\lambda)J\vec{z}(a)$. That $\vec{z}$ satisfies the boundary condition $(R-\lambda_0R')\vec{z}(a)=0$ is equivalent to $\vec{z}(a)=\begin{pmatrix}\alpha_2^*-\lambda_0\alpha_2'^*\\ \alpha_1^*-\lambda_0\alpha_1'^*\end{pmatrix}\vec{c}$, for some nx1 vector $\vec{c}\neq 0$, since $Y(a,\lambda_0)$ is a fundamental matrix. Multiplying out the indicated quantities, $-[I,M^*(\lambda)]Y^*(a,\lambda)J\vec{z}(a)=-[I+i\nu M^*(\lambda)\alpha]\vec{c}$. Letting $\nu\to 0$, this tends to $-\vec{c}$. The kth entry of the quantity it equals is $(i\nu)\int_a^b \vec{X}^{(k)*}(t,\lambda)A(t)\vec{z}(t)dt$, where $\vec{X}^{(k)}$ is the kth column, but this quantity (by the Cauchy-Schwarz inequality) does not exceed $\nu\|\vec{X}^{(k)}(\cdot,\lambda)\|\|\vec{z}\|$. But $\nu\|\vec{X}^{(t)}(\cdot,\lambda)\|\to 0$ as $\nu\to 0$ by our assumption, so we have a contradiction; i.e., we have proved $\nu M(\lambda_0+i\nu)\to S\neq 0$. If $M(\lambda)-iS(\lambda-\lambda_0)^{-1}$ were analytic at λ_0 then λ_0 would be a pole of $M(\lambda)$, contrary to $\lambda_0\in \widetilde{PC(T)}$ by part (iv).

(iii) Let $\lambda_0\in\widetilde{C(T)}$. Since $\widetilde{C(T)}\subset\widetilde{\sigma(T)}$ then $M(\lambda)$ is not analytic at λ_0 by part (i). Also, $M(\lambda)-iS(\lambda-\lambda_0)^{-1}$ cannot be analytic for any S since otherwise the singularity would not be isolated. Thus $\nu M(\lambda_0+i\nu)\to S\neq 0$ is ruled out by (iv); i.e, $\nu M(\lambda_0+i\nu)\to 0$.

If $M(\lambda)$ is not analytic at λ_0 then $\lambda_0\notin\rho(T)$. If $\nu M(\lambda_0+i\nu)\to 0$ then $M(\lambda)$ cannot have a pole since the limit is the residue times $-i$. Thus $\lambda_0\in\widetilde{PC(T)}\cup\widetilde{C(T)}$, and the limit condition excludes $\widetilde{PC(T)}$ by part (iv).

This completes the proof of Theorem 1. We close with two remarks which extend the theorem, by computing the reduced resolvent at a pole, and apply it to an invariance principle.

Remark 1. Let $\lambda_0 \in P(\tilde{T})$. Even though the resolvent operator $\tilde{R}_\lambda$ does not exist at $\lambda=\lambda_0$, we can still solve the inhomogeneous problem $(\lambda_0\tilde{P}_r-\tilde{T})\tilde{y}=\tilde{F}$, where $\tilde{F}\in E_r L_A^2 \times \mathbb{C}^n$ lies in the orthogonal complement of the columns of $\begin{pmatrix}\vec{\varphi}(t,\lambda_0)\\ \alpha\end{pmatrix}\sigma_{-1}$, whose linear span comprises the eigenvectors. Let $\mathrm{Im}(\lambda)\neq 0$ and to the standard resolvent formula

$$\vec{y}(x)=[\vec{\Theta}(x)+\vec{\varphi}(x)M(\lambda)]\int_a^x \vec{\varphi}^*(t,\lambda)A(t)\vec{F}(t)dt+\vec{\varphi}(x,\lambda)\int_x^b [\vec{\Theta}(t,\lambda)+\vec{\varphi}(t,\lambda)M(\lambda)]^*A(t)\vec{F}(t)dt$$

$$+[\vec{\Theta}(x,\lambda)+\vec{\varphi}(x,\lambda)M(\lambda)]F_3$$ add the term

$$0=[\vec{\varphi}(x,\lambda_0)/(\lambda-\lambda_0)]\{\int_a^b(\vec{\varphi}(t,\lambda_0)\sigma_{-1})^*A(t)\vec{F}(t)dt+\sigma_{-1}F_3\},$$ where $M(\lambda)=\sigma_{-1}(\lambda-\lambda_0)^{-1}+\sigma_0+\sigma_1(\lambda-\lambda_0)+\ldots$. Now rearrange terms and argue as in [16] that as $\lambda\to\lambda_0$ the above expression for $\vec{y}(x)$ approaches

$$\vec{y}(x)=[\vec{\Theta}(x,\lambda_0]+\vec{\varphi}(x,\lambda_0)\sigma_0]^*\int_a^x \vec{\varphi}^*(t,\lambda_0)A(t)\vec{F}(t)dt$$

$$+\vec{\varphi}(x,\lambda_0)\int_a^x(\frac{\partial}{\partial\lambda}\vec{\varphi}(t,\lambda_0)\sigma_{-1})^*A(t)\vec{F}(t)dt+\vec{\varphi}(x,\lambda_0)\int_x^b[\vec{\Theta}(t,\lambda_0)$$

$$+\frac{\partial}{\partial\lambda}\vec{\varphi}(t,\lambda_0)\sigma_{-1}+\vec{\varphi}(t,\lambda_0)\sigma_0]^*A(t)\vec{F}(t)dt+[\vec{\Theta}(x,\lambda_0)+\vec{\varphi}(x,\lambda_0)\sigma_0]F_3.$$ This solves the inhomogeneous problem for $\lambda=\lambda_0$, and is the parameter-dependent analogue of the corresponding formula in [16] (and [6] for the scalar case).

Remark 2. Letting $x\to b$ in (2.4) results in a linking formula between $M(\lambda)$ for problem (1.1)-(1.2) and $M_0(\lambda)$ for the non-parameter-dependent problem (1.1) with boundary condition $y(a)=0$, which we refer to here as the "standard problem". Let T_0 be the corresponding operator with spectrum $\sigma(T_0)$, point spectrum $P(T_0)$, etc., for the standard problem. The analogue of Theorem 1 in [16] asserts the same conclusions as Theorem 1 for $M_0(\lambda)$ and

$\rho(t_0)$, $P(T_0)$, $C(T_0)$, $PC(T_0)$. A point $\lambda_0 \in \rho(T_0) \cup P(T_0)$ if and only if λ_0 is either a regular point or pole of $M_0(\lambda)$, and (2.4) (with x=b) along with Theorem 1 shows that $\lambda_0 \in \rho(\tilde{T}) \cup P(\tilde{T})$. The converse follows from the inverse linking formulas. Therefore the sets $\rho(T_0) \cup P(T_0)$ and $E(T_0)$ agree respectively with $\rho(\tilde{T}) \cup P(\tilde{T})$ and $E(\tilde{T})$.

References

1. H.J. Ahn, Vibrations of a pendulum consisting of a bob suspended from a wire, Q. Appl. Math. 39(1981), 109-117.

2. H.J. Ahn, On random transverse vibrations of a rotating beam with tip mass, Q. J. Mech. Appl. Math. 36(1983), 97-109.

3. N.I. Akhiezer, "The Classical Moment Problem", Oliver and Boyd, Edinburgh, 1965.

4. F.V. Atkinson, "Discrete and Continuous Boundary Problems," Academic Press, New York 1964.

5. J.A. Burns and E.M. Cliff, An approximation technique for the control and identification of hybrid systems, in "Dynamics and Control of Large Flexible Space Structures", L. Meirovitch, editor, Virginia Tech, Blacksburg, Virginia, 1981, pp. 269-284.

6. J. Chaudhuri and W.N. Everitt, On the spectrum of ordinary second order differential equations, Proc. Roy. Soc. Edin. 68A(1968), 95-119.

7. S. Clark and D.B. Hinton, unpublished manuscript.

8. C. Fulton, Parameterizations of Titchmarsh's m(λ)-function in the limit circle case, Trans. A.M.S. 229(1977), 51-63.

9. C. Fulton, Two point boundary value problems with eigenvalue parameter contained in the boundary conditions, Proc. Roy. Soc. Edin. 77A(1977), 293-308.

10. C. Fulton, Singular eigenvalue problems with eigenvalue parameter contained in the boundary conditions, Proc. Roy. Soc. Edin. 87A (1980), 1-34.

11. C. Fulton and S. Pruess, Numerical methods for a singular eigenvalue problem with eigenparameter in the boundary conditions, J. Math. Anal. Appl. 71 (1979), 431-462.

12. D.B. Hinton, An expansion theorem for an eigenvalue problem with eigenvalue parameter in the boundary condition, Q.J. Math. Oxford (2)30(1979), 33-42.

13. D.B. Hinton, Eigenfunction expansions for a singular eigenvalue problem with eigenparameter in the boundary condition, SIAM J. Math. Anal. 12(1981), 572-584.

14. D.B. Hinton and J.K. Shaw, On Titchmarsh-Weyl M(λ)-functions for linear Hamiltonian

systems, J. Diff. Eq.s 40(1981), 316-342.

15. D.B. Hinton and J. K. Shaw, On boundary value problems for Hamiltonian systems with two singular points, SIAM J. Math. Anal. 15(1984), 272-286.

16. D.B. Hinton and J.K. Shaw, On the spectrum of a singular Hamiltonian system, Quaes. Math. 5(1982), 29-81.

17. D.B. Hinton and J.K. Shaw, On the spectrum of a singular Hamiltonian system, II, Quaes. Math., to appear.

18. L.H. Jones, The transverse vibration of a rotating beam with tip mass, Q. Appl. Math. 23 (1975), 193-203.

19. A. Schneider, A note of eigenvalue problems with eigenvalue parameter in the boundary conditions, Math Z. 136 (1974), 163-167.

20. J. Walter, Regular eigenvalue problems with eigenvalue parameter in the boundary condition, Math. Z. 133 (1973), 301-312.

D.B. Hinton
Mathematics Department
University of Tennessee
Knoxville, TN 37996-1300

J.K. Shaw
Mathematics Department
Virginia Tech
Blacksburg, VA 24061-4097

Canadian Mathematical Society
Conference Proceedings
Volume 8 (1987)

OSCILLATION THEORY FOR LINEAR SECOND-ORDER DIFFERENTIAL SYSTEMS

Hans G. Kaper and Man Kam Kwong[1]

ABSTRACT. This article is concerned with the oscillatory behavior at infinity of the solution $y : [a, \infty) \to \mathbf{R}^n$ of a system of second-order differential equations, $y''(t) + Q(t)y(t) = 0$, $t \in [a, \infty)$; Q is a continuous function on $[a, \infty)$, whose values are real symmetric matrices of order n. It is shown that the solution is oscillatory at infinity if the largest eigenvalue of the matrix $\int_a^t Q(t)\, dt$ is sufficiently large on a sufficiently large set of t-values. This result confirms and strengthens a conjecture of Hinton and Lewis [*Rocky Mountain J. Math.* **10** (1980), 751-766].

1. INTRODUCTION. We are concerned with the differential equation

$$y''(t) + Q(t)y(t) = 0, \qquad t \in [a, \infty), \tag{1.1}$$

for vector-valued functions $y : [a, \infty) \to \mathbf{R}^n$. Here, Q is a continuous function on $[a, \infty)$, whose values are real symmetric matrices of order n. Throughout this article $Q_1(t)$ denotes the matrix of the once integrated elements of Q,

$$Q_1(t) = \int_a^t Q(s)\, ds. \tag{1.2}$$

Two points $\alpha, \beta \in [a, \infty)$ are said to be *conjugate relative to* (1.1) if there exists a nontrivial function y which satisfies (1.1) and vanishes at α and β. Equation (1.1) is said to be *oscillatory at infinity* if, for any point $\alpha \in [a, \infty)$, there exists a point $\beta \in (\alpha, \infty)$ such that α and β are conjugate relative to (1.1).

The oscillation theory for (1.1) has received considerable attention in the past; see, for example, Reid [**1**, Chapter V]. It has been conjectured by Hinton and Lewis [**2**] that (1.1) is oscillatory at infinity whenever

$$\lim_{t \to \infty} \lambda_1\{Q_1(t)\} = \infty. \tag{1.3}$$

Here, $\lambda_1\{\,\cdot\,\}$ denotes the largest eigenvalue of the matrix argument. For arbitrary n, the

1980 *Mathematics Subject Classification.* Primary 34A30, 34C10.

Key words and phrases. Matrix differential equation, oscillation theory, matrix Riccati equation, Riccati inequality.

[1]Work supported by the Applied Mathematical Sciences subprogram of the Office of Energy Research, U.S. Department of Energy, under contract W-31-109-Eng-38.

conjecture was first established under various growth conditions on the trace of $Q_1(t)$ by Mingarelli [**3, 4**] and Kwong et al. [**5**]. (Generalizations of these results were reported in two recent articles by Butler and Erbe [**6, 7**].) Subsequently, Atkinson, Kaper, and Kwong [**8**] showed that (1.1) is oscillatory at infinity if (at least) $n-1$ eigenvalues of $Q_1(t)$ tend to infinity as $t \to \infty$. Eventually, Hinton and Lewis' conjecture was proven for $n = 2$ by Kaper and Kwong in [**9**] and for arbitrary n by Byers, Harris, and Kwong in [**10**].

In this article we give another proof of the conjecture for arbitrary n. Our results strengthen the conjecture, in the sense that they establish oscillatory behavior at infinity for the system (1.1) under conditions that are much weaker than (1.3). Although our results are similar to those of [**10**], the proofs are quite different. (For the record: The proof given here predates the proof published in [**10**], but was written up later.)

The oscillatory properties of (1.1) are usually studied by means of the prepared or conjoined solutions of the associated matrix differential equation, i.e., those solutions Y of the equation

$$(1.4) \qquad Y''(t) + Q(t)Y(t) = 0, \qquad t \in [a, \infty),$$

for which $Y'(t)Y^{-1}(t)$ is real symmetric. An alternative approach, which we have used in our earlier studies [**5, 8, 9**] and which we shall use again in the present study, is based on the solution of a non-linear Riccati equation for the matrix-valued function R,

$$(1.5) \qquad R(t) = R(a) + Q_1(t) + \int_a^t R^2(s)\, ds, \qquad t \in [a, \infty).$$

The solutions of (1.4) and (1.5) are related via the change of variables

$$(1.6) \qquad R(t) = -\, Y'(t)Y^{-1}(t).$$

This identity shows that $R(t)$ blows up when $Y(t)$ becomes singular.

Equation (1.1) is non-oscillatory at infinity if and only if there exists a neighborhood of infinity where (1.4) has a non-singular conjoined solution; cf., for example, Etgen and Pawlowski [**11**]. Hence, a necessary and sufficient condition for (1.1) to be non-oscillatory at infinity is that (1.5) has a continuous solution in some neighborhood of infinity. Since a in (1.1) is arbitrary, there is no loss of generality if we fix our attention on the neighborhood $[a, \infty)$ and establish criteria for the existence or non-existence of continuous solutions of (1.5) on $[a, \infty)$.

We shall prove that (1.1) is oscillatory at infinity whenever $\lambda_1\{Q_1(t)\}$ is sufficiently large on a sufficiently large set of t-values. A more precise statement is given in Section 3. This result implies the validity of the conjecture that (1.3) is sufficient for oscillatory behavior at infinity.

In the following analysis, M_n denotes the set of real symmetric matrices of order n and I_n (or I, if no confusion is possible) the unit matrix of order n. We use the standard partial ordering in M_n, viz., $A \geq B$ if $A - B$ is non-negative.

2. PRELIMINARY RESULTS. Our main result will follow from an ordering result for the quadratic term in the matrix Riccati equation

$$R(t) = F(t) + \int_a^t R^2(s)\,\mathrm{d}s, \qquad t \geq a, \tag{2.1}$$

where F is a given M_n-valued function, which is continuous for $t \geq a$.

Let b be fixed, $a < b \leq \infty$. The largest eigenvalue $\lambda_1\{F(\cdot)\}$ of $F(\cdot)$ on (a, b) defines a set-valued function S,

$$S(a, b) = \{t \in (a, b) : \lambda_1\{F(t)\} \geq 1\}. \tag{2.2}$$

The thrust of the following lemmas is that the size of $S(a, b)$ is critical for the existence or non-existence of a continuous solution R of (2.1) on $[a, b)$.

LEMMA 1. *There exists a finite number T_n such that*

$$\int_a^b R^2(t)\,\mathrm{d}t \geq \frac{1}{10} I, \tag{2.3}$$

whenever mes $S(a, b) \geq T_n$.

Before we prove this lemma, we state and prove several of its consequences. First we show that the constant 1/10 in (2.3) is immaterial if the set S is sufficiently large.

LEMMA 2. *If* mes $S(a, b) \geq 10T_n$, *then*

$$\int_a^b R^2(t)\,\mathrm{d}t \geq I. \tag{2.4}$$

Proof. The interval (a, b) contains at least ten disjoint subintervals (a_i, b_i) where $\lambda_1\{F(t)\} \geq 1$ on a set whose measure is at least T_n. Lemma 1 implies that $\int_{a_i}^{b_i} R^2(t)\,\mathrm{d}t \geq \frac{1}{10} I$ for $i = 1, \cdots, 10$. ■

Let S_α, $\alpha > 0$, be the following generalization of S:

$$S_\alpha(a, b) = \{t \in (a, b) : \lambda_1\{F(t)\} \geq \alpha\}. \tag{2.5}$$

LEMMA 3. *If* mes $S_\alpha(a, b) \geq (10/\alpha)T_n$ *for some* $\alpha > 0$, *then*

$$\int_a^b R^2(t)\,\mathrm{d}t \geq \alpha I. \tag{2.6}$$

Proof. If R satisfies (2.1) on $[a, b)$, then $\tilde{R}$, defined by the expression $\tilde{R}(t) = (1/\alpha)R(t/\alpha)$, satisfies the matrix Riccati equation

$$\tilde{R}(t) = \tilde{F}(t) + \int_{\alpha a}^{t} \tilde{R}^2(s)\, ds$$

on $[\alpha a, \alpha b)$, where $\tilde{F}(t) = (1/\alpha)F(t/\alpha)$. Let the set $\tilde{S}(\alpha a, \alpha b)$ be defined by

$$\tilde{S}(\alpha a, \alpha b) = \{t \in (\alpha a, \alpha b) : \lambda_1\{\tilde{F}(t)\} \geq 1\}.$$

If $\lambda_1\{F(t)\} \geq \alpha$ for some $t > a$, then $\lambda_1\{\tilde{F}(t')\} \geq 1$ for $t' = \alpha t > \alpha a$, so mes $\tilde{S}(\alpha a, \alpha b) = \alpha$ mes $S_\alpha(a, b)$. The assertion now follows from Lemma 2. ■

The following lemma relates the size of $S(a, b)$ to the existence or non-existence of a continuous solution R of (2.1) on $[a, b)$.

LEMMA 4. *If* mes $S(a, b) > 20T_n$, *then (2.1) does not have a continuous solution on* $[a, b)$.

Proof. The proof is by contradiction. Assume that (2.1) has a continuous solution on $[a, b)$. Let $a_1 = a$ and choose b_1 such that mes $S(a_1, b_1) = 10T_n$. Then

$$\int_{a_1}^{b_1} R^2(t)\, dt \geq I,$$

according to Lemma 2. Next, let $a_2 = b_1$ and choose b_2 such that mes $S(a_2, b_2) = (10/2)T_n$. Then

$$R(t) = F_2(t) + \int_{a_2}^{t} R^2(s)\, ds, \qquad t \geq a_2,$$

where

$$F_2(t) = F(t) + \int_{a_1}^{b_1} R^2(t)\, dt.$$

Clearly, $\lambda_1\{F_2(t)\} \geq 2$ for all $t \in S(a_2, b_2)$, so Lemma 3 implies that

$$\int_{a_2}^{b_2} R^2(t)\, dt \geq 2I.$$

Next, let $a_3 = b_2$ and choose b_3 such that mes $S(a_3, b_3) = (10/4)T_n$. Then

$$R(t) = F_3(t) + \int_{a_3}^{t} R^2(s)\, ds, \qquad t \geq a_3,$$

where

$$F_3(t) = F(t) + \int_{a_1}^{b_1} R^2(t)\,dt + \int_{a_2}^{b_2} R^2(t)\,dt.$$

Now, $\lambda_1\{F_3(t)\} \geq 4$ for all $t \in S(a_3, b_3)$, so, again applying Lemma 3, we find that

$$\int_{a_3}^{b_3} R^2(t)\,dt \geq 4I.$$

This procedure can be continued indefinitely, because mes $S(a, b) > 20T_n$. Clearly, $b_k \to \bar{b}$ for some $\bar{b} < b$, where mes $S(a, \bar{b}) = 20T_n$. Hence, $\int_a^t R^2(s)\,ds$ exceeds any multiple of I as $t \to b$, so $R(t)$ blows up as $t \to b$. This conclusion contradicts the assumption that R is a continuous solution of (2.1) on $[a, b)$. ■

The following lemma relates the size of $S_\alpha(a, b)$ to the existence or non-existence of a continuous solution R of (2.1) on $[a, b)$.

LEMMA 5. *If* mes $S_\alpha(a, b) \geq (20/\alpha)T_n$, *then (2.1) does not have a continuous solution on* $[a, b)$.

Proof. The assertion follows from Lemma 4 after scaling, as in the proof of Lemma 3. ■

We need one more consequence of Lemma 1. It is concerned with the existence or non-existence of a solution of a scaled version of (2.1), viz.,

$$R(t) = \delta\Big[F(t) + \int_a^t R^2(s)\,ds\Big], \qquad t \geq a. \tag{2.7}$$

Here δ is any positive constant.

LEMMA 6. *If* mes $S(a, b) > (20/\delta^2)T_n$, *then (2.7) does not have a continuous solution on* $[a, b)$.

Proof. If R satisfies (2.7) on $[a, b)$, then $\tilde{R}$, defined by the expression $\tilde{R}(t) = (1/\delta)R(t/\delta^2)$, satisfies the matrix Riccati equation

$$\tilde{R}(t) = \tilde{F}(t) + \int_{\delta^2 a}^{t} R^2(s)\,ds$$

on $[\delta^2 a, \delta^2 b)$, where $\tilde{F}(t) = F(t/\delta^2)$. Let the set $\tilde{S}(\delta^2 a, \delta^2 b)$ be defined by

$$\tilde{S}(\delta^2 a, \delta^2 b) = \{t \in (\delta^2 a, \delta^2 b) : \lambda_1\{\tilde{F}(t)\} \geq 1\}.$$

Suppose that $\lambda_1\{F(t)\} \geq 1$ for some $t > a$. Then $\lambda_1\{\tilde{F}(t')\} \geq 1$ for $t' = \delta^2 t > \delta^2 a$, so mes $\tilde{S}(\delta^2 a, \delta^2 b) = \delta^2$ mes $S(a, b)$. The assertion then follows from Lemma 4. ■

It remains to prove Lemma 1. The proof will be by induction on n. Notice that Lemmas 2 through 6 follow from Lemma 1 as long as n, the dimension of the system, is kept fixed. Hence, if we assume the validity of Lemma 1 for some n, we assume implicitly the validity of Lemmas 2 through 6 for that same n.

Proof of Lemma 1. Consider the scalar case $n = 1$. Then $\lambda_1\{F(t)\} = F(t)$, so $R(t) \geq 1$ whenever $t \in S(a, b)$. Hence,

$$\int_a^b R^2(t)\, dt \geq \int_{S(a,\, b)} R^2(t)\, dt \geq \text{mes}\, S(a, b). \tag{2.8}$$

The assertion of the lemma follows with $T_1 = 1/10$.

Suppose that the assertion of the lemma is true in the n-dimensional case. Then we shall prove that it is also true in the $(n + 1)$-dimensional case, with $T_{n+1} = 2000T_n + 40n^2(n + 1)$. The proof is by contradiction.

Consider the matrix Riccati equation (2.1) in M_{n+1}. Suppose that there exists a b, $a < b \leq \infty$, such that (2.1) has a continuous solution R on $[a, b)$ and mes $S(a, b) \geq T_{n+1}$, but R does not satisfy the ordering relation (2.3). Without loss of generality we may assume that $\int_a^b R^2(t)\, dt$ is diagonal,

$$\int_a^b R^2(t)\, dt = \text{diag}\, (\sigma_1, \cdots, \sigma_{n+1}), \tag{2.9}$$

with $\sigma_1 \geq \cdots \geq \sigma_{n+1}$. The fact that R does not satisfy (2.3) implies that $\sigma_{n+1} < 1/10$. We use the following notation:

$$R(t) = \Big[\, r_{ij}(t) \,\Big]_{i,\, j = 1, \cdots, n+1}, \qquad P(t) = \int_a^t R^2(s)\, ds = \Big[\, \rho_{ij}(t) \,\Big]_{i,\, j = 1, \cdots, n+1}. \tag{2.10}$$

Because $\sigma_{n+1} = \sum_{j=1}^{n+1} \int_a^b r^2_{n+1,j}(t)\, dt < \frac{1}{10}$, it follows that

$$\int_a^b r^2_{n+1,j}(t)\, dt < \frac{1}{10} \qquad \text{for } j = 1, \cdots, n + 1. \tag{2.11}$$

These inequalities put an upper bound on the measure of the set of points $t \in (a, b)$ where $r_{n+1,j}(t)$ exceeds a given value. In particular, we have

$$\text{mes}\, \{t \in (a, b) : |r_{n+1,j}(t)| \geq \frac{1}{20n}\} < 40n^2 \qquad \text{for } j = 1, \cdots, n + 1. \tag{2.12}$$

Thus, if we eliminate from $S(a, b)$ those points t where $|r_{n+1,j}(t)| \geq 1/(20n)$ for some j and call the remaining set $S^{\#}(a, b)$,

(2.13) $$S^{\#}(a, b) = \{t \in (a, b) : \lambda_1\{F(t)\} \geq 1 \text{ and } |r_{n+1,j}(t)| < \frac{1}{20n},\ j = 1, \cdots, n+1\},$$

then

(2.14) $$\text{mes } S^{\#}(a, b) > T_{n+1} - 40n^2(n+1) = 2000T_n.$$

Next, we introduce the following partitions of $R(t)$ and $P(t)$:

(2.15) $$R(t) = \begin{bmatrix} R^{\#}(t) & r^{\#}(t) \\ r^{\#}(t)^T & \alpha(t) \end{bmatrix}, \quad P(t) = \begin{bmatrix} P^{\#}(t) & p^{\#}(t) \\ p^{\#}(t)^T & \beta(t) \end{bmatrix},$$

where $R^{\#}(t)$ and $P^{\#}(t)$ are $n \times n$ matrices, $r^{\#}(t)$ and $p^{\#}(t)$ n-dimensional column vectors, $r^{\#}(t)^T$ and $p^{\#}(t)^T$ their transposes, and $\alpha(t)$ and $\beta(t)$ scalars. Thus,

(2.16) $$P^{\#}(t) = \int_a^t R^{\#2}(s)\, ds + \int_a^t r^{\#}(s)r^{\#}(s)^T\, ds.$$

A straightforward computation shows that $F(t)$, which is equal to the difference $R(t) - P(t)$, is the algebraic sum of four matrices,

(2.17) $$F(t) = A(t) - U(t) - V(t) - W(t),$$

where

$$A(t) = \begin{bmatrix} R^{\#}(t) - \frac{1}{10}\int_a^t R^{\#2}(s)\, ds + \frac{9}{10}I_n & 0 \\ 0 & \frac{19}{20} \end{bmatrix}, \quad U(t) = \begin{bmatrix} \frac{1}{10}\int_a^t r^{\#}(s)r^{\#}(s)^T\, ds & 0 \\ 0 & 0 \end{bmatrix},$$

$$V(t) = \begin{bmatrix} \frac{9}{10}P^{\#}(t) & p^{\#}(t) \\ p^{\#}(t)^T & \frac{8}{10} + \beta(t) \end{bmatrix}, \quad W(t) = \begin{bmatrix} \frac{9}{10}I_n & -r^{\#}(t) \\ -r^{\#}(t)^T & \frac{3}{20} - \alpha(t) \end{bmatrix}.$$

Clearly, $U(t) \geq 0$ for all $t \in (a, b)$. We shall show that $V(t) \geq 0$ and $W(t) \geq 0$ for all $t \in S^{\#}(a, b)$. We need the following inequalities, which hold on $S^{\#}(a, b)$:

(2.18) $$|r_j^{\#}(t)| = |r_{n+1,j}(t)| < \frac{1}{20n}, \quad j = 1, \cdots, n,$$

(2.19) $$|\alpha(t)| = |r_{n+1,n+1}(t)| < \frac{1}{20n} \leq \frac{1}{20},$$

(2.20) $$0 \leq \beta(t) = \rho_{n+1,n+1}(t) \leq \rho_{n+1,n+1}(b) = \sigma_{n+1} < \frac{1}{10}.$$

Consider the matrix $V(t)$ for $t \in S^{\#}(a, b)$. For any $(n+1)$-dimensional vector $x = (x_1, \cdots, x_{n+1})^T = (x^{\#}, x_{n+1})^T$ we have

$$\langle x, V(t)x\rangle = \frac{9}{10}\langle x^{\#}, P^{\#}(t)x^{\#}\rangle + 2\langle x^{\#}, p^{\#}(t)x_{n+1}\rangle + (\frac{8}{10} + \beta(t))x_{n+1}^2$$

$$\geq \frac{1}{9}<x^{\#}, P^{\#}(t)x^{\#}> + 2<x^{\#}, p^{\#}(t)x_{n+1}> + 9\beta(t)x_{n+1}^2 + 8(\frac{1}{10} - \beta(t))x_{n+1}^2$$

$$= <y, P(t)y> + 8(\frac{1}{10} - \beta(t))x_{n+1}^2, \tag{2.21}$$

where $y = (\frac{1}{3}x^{\#}, 3x_{n+1})^T$. The first term in the last member of (2.21) is certainly non-negative, because $P(t) \geq 0$ for all $t \in (a, b)$. The second term is non-negative, because of (2.20). Hence, $V(t) \geq 0$ for all $t \in S^{\#}(a, b)$, as claimed.

Next, consider the matrix $W(t)$ for $t \in S^{\#}(a, b)$. We have

$$<x, W(t)x> = \frac{9}{10}\|x^{\#}\|^2 - 2<x^{\#}, r^{\#}(t)x_{n+1}> + (\frac{3}{20} - \alpha(t))x_{n+1}^2$$

$$\geq \frac{1}{10}\|x\|^2 - 2\sum_{j=1}^{n} |r_j^{\#}(t)|\ \|x\|^2 + (\frac{1}{20} - \alpha(t))x_{n+1}^2. \tag{2.22}$$

The difference between the first and second term of the lower bound is (strictly) positive, because of (2.18). The third term is non-negative, because of (2.19). Hence, $W(t) \geq 0$ for all $t \in S^{\#}(a, b)$, as claimed.

We conclude that $A(t) = F(t) + Z(t)$, where $Z(t) \geq 0$ on $S^{\#}(a, b)$. Since $\lambda_1\{F(t)\} \geq 1$ on the set $S(a, b)$, which includes $S^{\#}(a, b)$, it follows that $\lambda_1\{A(t)\} \geq 1$ on $S^{\#}(a, b)$. Considering the definition of the matrix $A(t)$, we see that any eigenvalue of $A(t)$ that is greater than 1 must also be an eigenvalue of the $n \times n$ submatrix in the top left corner. Hence,

$$\lambda_1\{R^{\#}(t) - \frac{1}{10}\int_a^t R^{\#2}(s)\, ds + \frac{9}{10}I_n\} \geq 1, \qquad t \in S^{\#}(a, b). \tag{2.23}$$

Now, let the M_n-valued function $F^{\#}$ be defined in terms of $R^{\#}$ by the expression

$$F^{\#}(t) = 10\big[R^{\#}(t) - \frac{1}{10}\int_a^t R^{\#2}(s)\, ds\big], \qquad t \geq a. \tag{2.24}$$

Then it follows from (2.23) that

$$\lambda_1\{F^{\#}(t)\} \geq 1, \qquad t \in S^{\#}(a, b). \tag{2.25}$$

But here we have arrived at a contradiction. If we interpret the definition (2.24) of $F^{\#}$ as a matrix Riccati equation for $R^{\#}$ in M_n,

$$R^{\#}(t) = \frac{1}{10}\big[F^{\#}(t) + \int_a^t R^{\#2}(s)\, ds\big], \qquad t \geq a, \tag{2.26}$$

then the existence of a continuous solution R of (2.1) on $[a, b)$ implies the existence of a continuous solution $R^{\#}$ of (2.26) on $[a, b)$. On the other hand, because mes $S^{\#}(a, b) > 2000T_n$, Lemma 6 (whose validity in the n-dimensional case follows from that of Lemma 1) implies that (2.26) cannot have a continuous solution on $[a, b)$. We therefore conclude that the solution of (2.1) must satisfy the ordering relation (2.3), so the assertion of the lemma is also true in the $n + 1$-dimensional case. ■

3. MAIN RESULTS. We now consider the matrix differential equation (1.1) and the associated Riccati equation (1.5). Let

(3.1) $$S_\alpha(a, \infty) = \{t \in (a, \infty) : \lambda_1\{Q_1(t)\} \geq \alpha\}$$

for any positive α.

THEOREM. *There exists a finite number T, whose value depends on the dimension n, such that (1.1) is oscillatory at infinity whenever*

(3.2) $$\liminf_{\alpha \to \infty} \{\alpha \text{ mes } S_\alpha(a, \infty)\} \geq T.$$

Proof. We shall prove the assertion of the theorem with $T = 80T_n$, where T_n is the number introduced in Lemma 1.

Suppose that (3.2) is satisfied. Then mes $S_\alpha(a, \infty) \geq (40/\alpha)T_n$ for α sufficiently large. Furthermore, when α is sufficiently large and $\lambda_1\{Q_1(t)\} \geq \alpha$ for some t, then also $\lambda_1\{R(a) + Q_1(t)\} \geq \alpha/2$. Hence,

$$\text{mes } \{t \in (a, \infty) : \lambda_1\{R(a) + Q_1(t)\} \geq \alpha/2\} \geq (40/\alpha)T_n.$$

But now Lemma 5 implies that the solution of the matrix Riccati equation (1.5) blows up at some finite t, so the matrix differential equation (1.4) does not have a non-singular conjoined solution on $[a, \infty)$ and (1.1) is oscillatory at infinity. ■

The theorem expresses the fact that (1.1) is oscillatory at infinity whenever $\lambda_1\{Q_1(t)\}$, the largest eigenvalue of the once integrated coefficient matrix Q, is sufficiently large on a sufficiently large set of t-values. In particular, the condition is satisfied if this eigenvalue tends to infinity as $t \to \infty$, as was conjectured by Hinton and Lewis. A more general statement is given in the following corollary.

COROLLARY 1. *If there is a measurable subset J of $[a, \infty)$ of infinite measure, such that*

(3.3) $$\lim_{t \to \infty,\, t \in J} \lambda_1\{Q_1(t)\} = \infty,$$

then (1.1) is oscillatory at infinity.

The next corollary shows that oscillatory behavior at infinity may result even if $\lambda_1\{Q_1(t)\}$ is negative everywhere except on a set of finite measure.

COROLLARY 2. *If there is a measurable subset J of $[a, \infty)$ of finite measure and a $\gamma \in (0, 1)$, such that*

(3.4) $$\int_J [\lambda_1\{Q_1(t)\}]_+^\gamma \, dt = \infty,$$

where $[\,\cdot\,]_+ = \max\,\{0, \cdot\,\}$, *then (1.1) is oscillatory at infinity.*

Proof. Let $J_n = \{t \in J : n \le \lambda_1\{Q_1(t)\} < n+1\}$ for $n = 0, 1, \cdots$. Then

$$\int_J [\lambda_1\{Q_1(t)\}]_+^\gamma \, dt = \sum_{n=0}^{\infty} \int_{J_n} (\lambda_1\{Q_1(t)\})^\gamma \, dt < \sum_{n=0}^{\infty} (n+1)^\gamma \operatorname{mes} J_n$$

$$= \sum_{n=0}^{\infty} (n+1)^\gamma \{\operatorname{mes} J \cap S_n(a, \infty) - \operatorname{mes} J \cap S_{n+1}(a, \infty)\}$$

(3.5) $$\le \operatorname{mes} J + \sum_{n=1}^{\infty} \{(n+1)^\gamma - n^\gamma\} \operatorname{mes} S_n(a, \infty).$$

Since the integral in the first term of (3.5) diverges and mes J is finite, the infinite series in the last term of (3.5) must diverge. However, the series $\sum_{n=0}^{\infty} n^{-1}\{(n+1)^\gamma - n^\gamma\}$ converges for all $\gamma \in (0, 1)$, so it must be the case that $n \operatorname{mes} S_n(a, \infty)$ grows beyond any finite bound as $n \to \infty$. The condition (3.2) is therefore certainly satisfied, whatever the value of T. ■

BIBLIOGRAPHY

1. W. T. Reid, Sturmian theory for ordinary differential equations, Springer-Verlag, New York, 1980.

2. D. Hinton and R. T. Lewis, "Oscillation theory for generalized second-order differential equations," Rocky Mountain J. Math., 10 (1980), 751-766.

3. A. B. Mingarelli, "An oscillation criterion for second-order self-adjoint differential systems," C. R. Math. Rep. Acad. Sci. Canada, 2 (1980), 287-290.

4. A. B. Mingarelli, "On a conjecture for oscillation of second-order ordinary differential systems,pq Proc. Amer. Math. Soc., 82 (1981), 593-598.

5. Man Kam Kwong, H. G. Kaper, K. Akiyama, and A. B. Mingarelli, "Oscillation of linear second-order differential systems" Proc. Amer. Math. Soc., 91 (1984), 85-91.

6. G. J. Butler and L. H. Erbe, "Oscillation results for second order differential systems," SIAM J. Math. Anal., 17 (1986), 19-29.

7. G. J. Butler and L. H. Erbe, "Oscillation results for self adjoint differential systems," J. Math. Anal. and Applic., 115 (1986), 470-481.

8. F. V. Atkinson, H. G. Kaper, and Man Kam Kwong, "An oscillation criterion for linear second-order differential systems," Proc. Amer. Math. Soc., 94 (1985), 91-96.

9. Man Kam Kwong and H. G. Kaper, "Oscillation of two-dimensional linear second-order differential systems," J. Diff. Eq., 56 (1985), 195-205.

10. R. Byers, B. J. Harris, and Man Kam Kwong, "Weighted means and oscillation conditions for second order matrix differential equations," J. Diff. Eq., 61 (1986), 164-177.

11. G. J. Etgen and J. F. Pawlowski, "Oscillation criteria for second-order self-adjoint differential systems," Pacific J. Math., 66 (1976), 99-110.

MATHEMATICS AND COMPUTER SCIENCE DEPARTMENT
ARGONNE NATIONAL LABORATORY
ARGONNE, ILLINOIS 60439-4844 (Current address of H. G. Kaper)

DEPARTMENT OF MATHEMATICAL SCIENCES
NORTHERN ILLINOIS UNIVERSITY
DEKALB, ILLINOIS 60115 (Current address of M. K. Kwong)

Canadian Mathematical Society
Conference Proceedings
Volume 8 (1987)

On the variation- diminishing property of Schrödinger operators

Martin Klaus

1. Introduction

Consider the Schrödinger operator

$$Ly = -y'' + q(x)y \qquad 0 \leqslant x < \infty. \tag{1.1}$$

where q is a real measurable function. Further assumptions on q will be stated below. We will only consider operators which are in the limit point case at infinity and in the limit circle case at 0 where we have a boundary condition $y(0) = 0$ so that L becomes a self-adjoint operator on $L^2[0,\infty]$. For any real function $y \in L^2$ we denote by $V[y]$ the number of sign changes of y on $[0,\infty)$ which is defined as the minimum of the number of sign changes of all functions which are almost everywhere equal to y. Let $\varphi(\lambda)$ be a real bounded measurable function defined on the spectrum $\sigma(L)$ of L. Hence $\varphi(L)$ is a bounded self- adjoint operator. Here we address the following question:

Which functions φ are variation- diminishing, i.e. $V[\varphi(L)y] \leqslant V[y]$ for all $y \in L^2$? We first learned about this problem from C. Prather who also showed us a paper by I. I. Hirschman [5] on this subject. Hirschman proved the following remarkable result:

Theorem 1.1: Suppose that q is bounded, continuous and obeys

$$\int_0^\infty (1 + x^2)\ |q(x)|\ dx < \infty. \tag{1.2}$$

0731-1036/87 $1.00 + $.25 per page

Then $\varphi(L)$ is variation- diminishing if and only if either

$$\text{a)}\quad \varphi(\lambda) = de^{-c\lambda} \prod_K (1 + \frac{\lambda}{b_K})^{-1} \tag{1.3}$$

where $c \geqslant 0$, $\Sigma > -b_1 \geqslant -b_2 \geqslant \ldots$, $\sum_K b_K^{-1} < \infty$, $d \neq 0$

or (if d=0)

$$\text{b)}\quad \varphi(\lambda) = 0 \qquad \lambda \in \sigma(L) \setminus \{\Sigma\}. \tag{1.4}$$

Here $\Sigma \leqslant 0$ is the lowest point of the spectrum of L. Note that L has continuous spectrum $[0,\infty)$ and at most finitely many negative eigenvalues.

Remark. Hirschman proved the theorem for general homogeneous boundary conditions at 0. The continuity and boundedness assumptions are not needed.

The interesting thing about this theorem is that q enters only marginally through the value of Σ. Therefore one would expect this theorem to be true for a much larger class of potentials. In Hirschman's proof the condition (1.2) is needed, firstly, to ensure the finiteness of the discrete spectrum and, secondly, to control certain contour integrals which arise from the use of an eigenfunction expansion. But we observe that there is still room for extending the theorem even if one wants to keep the discrete spectrum finite. A weaker condition which preserves this property is

$$\int_0^\infty (1+x)|q(x)|dx < \infty. \tag{1.5}$$

In Section 4 (Theorem 4.1) we extend Theorem 1.1 using this condition. As a by-product of our analysis we obtain precise information about the spectral density as the spectral parameter λ

goes to 0. Since this result may be of independent interest we give its proof in Section 5. In Section 3 we describe the main ideas behind Hirschman's proof. Next we need to introduce some notation and to recall a few facts about solutions of the spectral problem.

2. Notation and properties of solutions

Let $y(x,\lambda)$ denote the solution of

$$Ly = \lambda y \qquad y(0,\lambda) = 0,\ y'(0,\lambda) = 1. \tag{2.1}$$

We set $\lambda = s^2$, $s=\sigma + i\tau$. The solution $y(x,\lambda)$ is entire in λ. For $\tau > 0$ we have that

$$\lim_{x\to\infty} e^{isx}\, y(x,s^2) = M(s) \tag{2.2}$$

where $M(s)$ is analytic for $\tau > 0$ and has continuous boundary values as $\tau \downarrow 0$ (by (4.3) below). We put $-2i\sigma M(\sigma) = A(\sigma) \exp(i\phi(\sigma))$, with A,ϕ real. For $s = \sigma$ we have

$$y(x,\sigma^2) = \frac{A(\sigma)}{\sigma} \sin(\sigma x-\phi(\sigma)) + o(1) \qquad x\to\infty. \tag{2.3}$$

Moreover, the equation $Ly=\lambda y$ has a unique solution $y_1(x,s)$ (Jost solution) satisfying

$$\lim_{x\to\infty} y_1(x,s)e^{-isx} = 1. \tag{2.4}$$

$y_1(x,s)$ is analytic for $\tau > 0$ and continuous for $\tau \geqslant 0$. In terms of y_1 we can write

$$y(x,\sigma^2) = \bar{M}(\sigma)\, y_1(x,\sigma) + M(\sigma)\bar{y}_1(x,\sigma) \qquad \sigma\neq 0. \tag{2.5}$$

Furthermore, $y(x,\sigma) = y(x,-\sigma)$, $\bar{y}_1(x,\sigma) = y_1(x,-\sigma)$, $\bar{M}(\sigma) = M(-\sigma)$, $A(\sigma) = A(-\sigma)$ and $\phi(-\sigma) = -\phi(\sigma)$. The zeros of $M(s)$ in the upper half plane must all lie on the imaginary axis and if $i\tau_K$ is such a zero then $-\tau_K^2=\lambda_K$ is an eigenvalue of L. The point $\lambda = 0$ is never an eigenvalue under condition (1.5).

3. Outline of Hirschman's proof

We only consider the "only if" direction of the proof. For the "if" direction we refer the reader to [5].

Step 1: Write $\varphi(L)$ in terms of eigenfunctions of L as

$$(\varphi(L)f)(x) = \int_0^\infty y(x,\lambda)\hat{f}(\lambda)\varphi(\lambda)d\rho(\lambda)$$

$$+ \sum_K y(x,\lambda_K)\hat{f}(\lambda_K)\varphi(\lambda_K)\gamma_K \tag{3.1}$$

where

$$\gamma_K = 1/\int_0^\infty y(x,\lambda_K)^2 dx, \tag{3.2}$$

$$d\rho/d\lambda = (4\pi)^{-1}\lambda^{-1/2}|M(\sqrt{\lambda})|^{-2}, \tag{3.3}$$

$$\hat{f}(\lambda) = \int_0^\infty y(x,\lambda)f(x)dx, \tag{3.4}$$

$$f(x) = \int_{-\infty}^{\infty} y(x,\lambda)f(\lambda)d\rho(\lambda). \tag{3.5}$$

By using (2.5) and $\gamma_K = M_K$ where $-i\ M_K$ is the residue of $1/M(s)$ at $s = i\tau_K$ (one does not need a contour integration as in [5] to calculate this, see e.g [2,p.64]) one finds that

$$(\varphi(L)f)(x) = \int_0^\infty K(x,t)f(t)dt$$

where

$$K(x,t) = \frac{1}{2\pi}\int_{-\infty}^{\infty} y_1(x,\sigma)y(t,\sigma^2)\varphi(\sigma^2)M(\sigma)^{-1}d\sigma +$$

$$\sum_K y_1(x,i\tau_K)y(t,\lambda_K)\varphi(\lambda_K)M_K. \tag{3.6}$$

In the next step we will assume that φ has been replaced by $e^{-\lambda}\varphi = \psi$ in order to prevent convergence problems. Since $e^{-\lambda}$ is variation-diminishing this causes no loss of generality.

Step 2: Suppose that $K(x,t)$ is a variation- diminishing kernel. One then shows by means of the asymptotic formulas (2.3) and (2.4) that

$$\lim_{r\to\infty} K(x+r,\ t+r) = L(x - t) \tag{3.7}$$

exists for each x and t and that L is a variation- diminishing kernel defined for all $-\infty < x,\ t < \infty$ mapping L^1 into the bounded functions. It turns out that

$$L(x) = \frac{1}{\pi} \int_0^\infty \psi(\sigma^2)\cos x\sigma \, d\sigma \ (\psi=\varphi e^{-\lambda}). \tag{3.8}$$

Now by a result of Schoenberg [11] (see also [5],[6],[7]) it follows that φ is of the form (1.3) with $0<b_1\leqslant b_2\leqslant\ldots$, provided that $\lambda \geqslant 0$. If L has no eigenvalues we are done. If there are negative eigenvalues one needs

Step 3: By looking at certain contour integrals one shows that for $\lambda=\lambda_K$ $\varphi(\lambda)$ must necessarily be given by (1.3) and that the b_K's are all greater than $-\sum$. The exact details of these arguments are somewhat involved so we must refer the reader to [5]. One also sees that if $d=0$, then the only nontrivial φ that remains is one that sends all functions into a multiple of the eigenfunction belonging to $\sum$.

4. Extension to potentials obeying (1.5)

Theorem 4.1: Suppose that q obeys (1.5). Then statements a) and b) of Theorem 1.1 hold.

In proving this theorem we want to follow Hirschman's proof as closely as possible. This means that we need detailed information about the behavior of the function M(s) as s approaches zero from the upper half plane. Specifically, Hirschman's proof relies heavily on the following result due to Levinson [9]:

$$2sM(s) = \begin{cases} a+o(1) & a\neq 0 \qquad (4.1a) \\ \text{or else} & s\to 0 \\ bs+o(s) & b\neq 0 \qquad (4.1b) \end{cases}$$

and

$$2sM(s) = i+o(1) \qquad s\to\infty \qquad (4.2)$$

uniformly for $\tau \geq 0$. These asymptotics are used to justify certain deformations of the line of integration in step 3 above. Note that the function $M(\sigma)^{-1}$ appears in (3.6). If we can verify the relations (4.1a), (4.1b) and (4.2) under condition (1.5), then Hirschman's proof carries over. Therefore our initial way of attacking the problem consisted in pushing through Levinson's result by merely assuming condition (1.5). It turns out that only relation (4.1b) poses any problems. As a result, we were able to find an expression for a and b in terms of the potential q and the solution of Ly=0. This, in turn, yields the asymptotics of $d\rho/d\lambda$ as $\lambda \downarrow 0$. We have subsequently learned that in the book of

Agranovich and Marchenko [1] the relation (4.1b) had already been established earlier, although in a rather indirect way. This follows from Lemmas 2.4.1, 5.6.2, 5.6.3, and the conclusions on p. 138 of [1]. However, there are no expressions for a and, more importantly, none for b in [1]. Due to the results in [1] the proof of Theorem 4.1 became much shorter than we initially had anticipated.

Proof of Theorem 4.1. It is only necessary to prove the "only if" direction because the proof of the "if" direction is the same as that in [5]. From [9] we have the integral representation

$$F(s)\equiv-2isM(s)=1+\int_0^\infty e^{isx}q(x)y(x,\lambda)dx. \tag{4.3}$$

Assuming only $\int_0^\infty x|q(x)|dx<\infty$ the proof of (4.2) is already contained in [9]. Moreover, under the same assumption, Levinson derived the bound

$$|y(x,\lambda)| \leqslant Kxe^{|\tau|x} \qquad 0\leqslant x<\infty \tag{4.4}$$

which is valid for all λ. Hence (4.1a) is an immediate consequence of the dominated convergence theorem. Explicitly,

$$-ia = 1+\int_0^\infty q(x)y(x,0)dx. \tag{4.5}$$

As discussed above, (4.1b) follows from results in [1]. This establishes (4.1a), (4.1b), (4.2) under condition (1.5) and hence, as mentioned before, Hirschman's proof goes through.

Remark. The case (4.1b) occurs precisely when $\lambda = 0$ is a 'virtual level' (i.e. eq. (2.1) has a bounded solution) or, equivalently, when g=1 is a 'critical coupling constant' (or 'coupling constant threshold') for $L(g) = - d^2/dx^2+gq(x)$ (i.e. L(g) has a negative eigenvalue which converges to 0 as $g \downarrow 1$). The equivalence of these two notions and their equivalence with (4.1b) can be proved by means of the Birman-Schwinger kernel (see [8] for related results).

5. Small-λ behavior of the spectral density

Theorem 5.1: Suppose that q satisfies (1.5). Then

$$\frac{d\rho(\lambda)}{d\lambda} \sim \begin{cases} \dfrac{\sqrt{\lambda}}{\pi|a|^2} & (5.1a) \\ & \lambda \downarrow 0 \\ \dfrac{1}{\pi b^2\sqrt{\lambda}} & (5.1b) \end{cases}$$

where a is given by (4.5) and $b = -\int_0^\infty xy(x,o)q(x)dx$.

Proof. (5.1a) follows immediately from (3.3) and (4.1a). To prove (5.1b) we write

$$F(\sigma)=\int_0^\infty(e^{i\sigma x}-1)y(x,0)q(x)dx$$

$$+ \int_0^\infty e^{i\sigma x}[y(x,\sigma^2)-y(x,0)]q(x)dx = I_1+I_2. \qquad (5.2)$$

Levinson estimated both integrals and used (1.2) for both of them. The first step is to show that $y(x,0)$ is bounded. By simply integrating the differential equation $-y'' + q(x)\, y = 0$ we find that

$$-y'(x) + 1 + \int_0^x q(x)\, y(x,0)\, dx = 0. \qquad (5.3)$$

Since $a = 0$ $y'(x,0)\to 0$ as $x\to\infty$. It is well-known that $y(x,0)$ must either grow linearly or else approach a constant as $x\to\infty$ [4]. So in this case y approaches a constant. Thus

$$I_1=i\sigma\int_0^\infty xy(x,0)q(x)dx + i\sigma\int_0^\infty \frac{e^{i\sigma x}-1-i\sigma x}{i\sigma x}xy(x,0)q(x)dx. \tag{5.4}$$

The last integral is o(1) by the dominated convergence theorem. The first term on the r.h.s. of (5.4) gives the value of b as

$$b = -\int_0^\infty x\ y(x,0)\ q(x)\ dx. \tag{5.5}$$

It is shown in [9] that $b\neq 0$. The problem now is to prove that I_2 is $o(\sigma)$. We start from the integral equation

$$y(x,\lambda) = \frac{\sin\sigma x}{\sigma} + \frac{1}{\sigma}\int_0^x \sin\sigma(x-t)\ q(t)y(t,\lambda)dt \qquad (\lambda=\sigma^2) \tag{5.6}$$

which for $\sigma = 0$ reduces to

$$y(x,0) = x + \int_0^x (x-t)\ q(t)\ y(t,0)\ dt \tag{5.7}$$

We now take the difference between $y=y(x,\lambda)$ and $y_0=y(x,0)$ and rewrite it using $a = 0$ as

$$\begin{aligned} y-y_0 &= \left(x - \frac{\sin\sigma x}{\sigma}\right)\int_x^\infty q y_0 dt \\ &+ \int_0^x q y_0\left(t-\frac{\sin\sigma t}{\sigma}\right)dt - \frac{\cos\sigma x-1}{\sigma}\int_0^x \sin\sigma t\ q y_0 dt \\ &+ \frac{\sin\sigma x}{\sigma}\int_0^x(\cos\sigma t-1)qy dt \\ &+ \frac{\sin\sigma x}{\sigma}\int_0^x q(y-y_0)dt - \frac{\cos\sigma x}{\sigma}\int_0^x \sin\sigma t\ q(y-y_0)dt. \end{aligned} \tag{5.8}$$

Estimating the r.h.s. we obtain (K is a suitable constant)

$$\begin{aligned} |y-y_0| &\leqslant K\sigma x\int_x^\infty t|q||y_0|dt + K\sigma x\int_0^x |q||y_0|t dt + K\sigma x\int_0^x t|q||y_0|dt \\ &+ K\sigma x\int_0^x t|q||y|dt + Kx\int_0^x |q||y-y_0|dt + K\int_0^x t|q||y-y_0|dt. \end{aligned} \tag{5.9}$$

Since y_0 is bounded we get

$$|y-y_0| \leqslant K\sigma x + K\sigma x\int_0^x t|q||y|dt + Kx\int_0^x |q||y-y_0|dt. \tag{5.10}$$

Now we use the fact that y satisfies the integral equation

$$y(x,\lambda) = \frac{A(\sigma)}{\sigma}\sin(\sigma x-\phi(\sigma)) + \frac{1}{\sigma}\int_x^\infty \sin\sigma(t-x)\ qy(t,\lambda)dt. \tag{5.11}$$

from which we get the bound

$$|y(x,\lambda)| \leqslant K\ \frac{A(\sigma)}{\sigma}. \tag{5.12}$$

By using this bound to estimate the second term on the r.h.s. of (5.10) we find that

$$|y-y_0| \leqslant K\sigma x + KA(\sigma)x + Kx\int_0^x |q||y-y_0|dt. \tag{5.13}$$

So by Gronwall's inequality

$$\int_0^x |q||y-y_0|dt \leqslant K(\sigma+A(\sigma))$$

and hence

$$|y-y_0| \leqslant Kx(\sigma+A(\sigma)). \tag{5.14}$$

Now we return to the integral I_2. We break it up into two parts, going from 0 to R, the other from R to ∞. The former is bounded by $C(R)\sigma^2$ since we are on a bounded interval and can use variation of parameters with $\sigma^2 y$ as the nonhomogeneous term (as in [9]). The latter integral is bounded by

$$\int_R^\infty e^{i\sigma t}|y-y_0||q|dt \leqslant K(\sigma+A(\sigma))\int_R^\infty t|q|dt$$
$$= K(\sigma+A(\sigma))\epsilon(R). \tag{5.15}$$

So we arrive at the inequality

$$|I_2| \leqslant C(R)\sigma^2 + K(\sigma+A(\sigma))\epsilon(R). \tag{5.16}$$

Of course, $C(R)\to\infty$ as $R\to\infty$, in general. Since $F = I_1 + I_2$ and $I_1 = -ib\sigma + \delta(\sigma)$ where $\delta(\sigma)=o(\sigma)$ we conclude that

$$|F(\sigma)| = A(\sigma) \leqslant |b|\sigma + |\delta(\sigma)| + C(R)\sigma^2$$
$$+ K(\sigma+A(\sigma))\epsilon(R). \tag{5.17}$$

Thus for R large enough

$$A(\sigma) \leqslant (1-K\epsilon(R))^{-1} (|b|\sigma + |\delta(\sigma)| + C(R)\sigma^2 + K\sigma\epsilon(R)). \quad (5.18)$$

On substituting this estimate in (5.16) and letting $R\to\infty$ and $\sigma\to 0$ we find that $I_2/\sigma\to 0$ as $\sigma\to 0$. Therefore

$$\lim_{\sigma\to 0} \frac{F(\sigma)}{\sigma} = -i\ b \quad (5.19)$$

and hence (5.1b) follows from (3.3). Theorem 5.1 is proved.

Remarks. 1. The analysis above can be carried out with σ replaced by s (Ims>0) by using an appropriate integral equation and then it follows that (5.19) holds uniformly in s as $s\to 0$ from the upper half plane. A quicker way of arriving at this conclusion is by using a Phragmen-Lindelöf type theorem [3,p.3] since (5.19) holds also as $\sigma \uparrow 0$.

2. The behavior of the Jost function (which is just F(s)) as s tends to 0 in a sector $S = \{s: \epsilon<\arg s<\pi-\epsilon,\ \epsilon>0\}$ has also been studied by Newton [10] (assuming (1.2)).

References

[1] Z.S. Agranovich and V.A. Marchenko, The Inverse Problem of Scattering Theory, Gordon and Breach 1963

[2] V. de Alfaro and T. Regge, Potential Scattering, North-Holland 1965

[3] R.P. Boas, Entire Functions, Academic Press 1954

[4] E.Hille, Non-oscillation theorems, TAMS, 64 (1948) 234-252

[5] I.I. Hirschman, Variation-Diminishing Transformations and Sturm-Liouville Systems, Comment. Math,. Helv. 36 (1961)

214- 233

[6] I.I. Hirschman and D.V. Widder, The Convolution Transform, Princeton University Press 1955

[7] S. Karlin, Total Positivity, Stanford University Press 1968

[8] M.Klaus and B. Simon, Coupling Constant Thresholds in Nonrelativistic Quantum Mechanics, I. Short-Range Two-Body Case, Ann. Phys. 130 (1980) 251 - 281

[9] N. Levinson, Det Kgl. Danske Videnskabernes Selskab, Matematisk-Fysiske Meddelelser, 25 (1949), no. 9

[10] R. Newton, Analytic properties of radial wave-functions, J. Math. Phys. 1 (1960) 319 - 347

[11] I.J. Schoenberg, On Polya frequency functions, II, Acta Sci. Math. Szeged, 12 (1950) 97 -106.

Department of Mathematics

Virginia Polytechnic Institute and State University

Blacksburg, VA 24061

Canadian Mathematical Society
Conference Proceedings
Volume 8 (1987)

Semigroup Smoothing by Highly Singular Operators

Mark A. Kon, * Columbia University and Boston University

In this note we consider Sobolev smoothing properties of Schrödinger operators $H = -\Delta + V$ on R^n with highly singular potentials V. By highly singular potentials we mean potentials which have local singularities causing them to be unbounded in L^2 relative to the Laplacian. Relatively bounded potentials include functions V in $L^{\frac{n}{2}+r}$, with $r > 0$. In the class of homogeneous potentials $V = r^{-\alpha}$, $r = |x|$, V is relatively bounded for $\alpha < 2$, but not for $\alpha > 2$. For $\alpha = 2$ the question is delicate; V is relatively bounded in some cases (see [RS]).

Study of very singular potentials has included distributional potentials [HS,H,D]. In addition, the study of operators in divergence form with measurable coefficients (in particular work related to Kato's conjecture) has implicitly involved second order operators with distributional coefficients.

In another paper, we will show that for potentials in $L^{\frac{n}{2}+r}$, $r > 0$, the smoothing behavior of the semigroup in the scale of Sobolev spaces is deterministically tied to the smoothness of the potential (Theorem 1); this affirmatively answers a conjecture of Simon [Si]. Precisely, it is shown that under the above assumption (close to the assumption of relative boundedness), the semigroup sends $L^p(\mathbf{R}^n)$ into $L^p_{s+2} \equiv (1-\Delta)^{-\frac{s}{2}} L^p$ if and only if $V \in L^p_s$ $(1 < p < \infty)$. Here we show (by examples) that without the control that relative boundedness gives to the Laplacian over the potential, essentially anything can happen. That is, there exist positive super-singular Schrödinger operators whose semigroups smooth to all orders (i.e., map L^2 into $L^2_\infty \equiv \bigcap_{s>0} L^2_s$), and such operators whose semigroups fail to add more than two derivatives to functions. An important component of the degree of singularity which the semigroup imparts to functions in this case depends on the singular support of the potential. In particular, if the singular support is "evenly spread" on a variety in $\mathbf{R}^n$, then the singularity of the potential (i.e., its L^p class), together with the dimension of the variety seem to largely determine the smoothness of the semigroup range. It would be nice to quantify this statement precisely, but this is not done here.

* Research partially supported under N.S.F. grant DMS-8509458

The smoothing action on functions of the semigroup generated by an operator is a good indicator of the smoothness of the operator's (generalized) eigenfunctions, and how much the operator's resolvent smooths (together with its powers). For example, an operator must have (generalized) eigenfunctions which are at least as smooth as the range of its semigroup. The resolvent and the semigroup of H are related by

$$R_\zeta = \int_0^\infty e^{-tH}\,dt,$$

and the inverse relationship is

$$e^{-tA} = \frac{1}{2\pi i}\int_\Gamma (\zeta - A)^{-1} e^{-t\zeta}\,d\zeta,$$

where Γ is an appropriate contour enclosing the spectrum of H (see [G1]).

We remark that super-singular potentials arise from ideal electric multipoles. For example, an ideal dipole in $\mathbf{R}^3$ has a potential of the form $\Phi \equiv \frac{\mathbf{p}\cdot\mathbf{x}}{|x|^3}$, where $\mathbf{p}$ is the dipole moment. A quadrupole has a potential which is of order $\frac{1}{x^3}$ near 0, and in general an ideal 2^n-pole in $\mathbf{R}^3$ has a potential

$$V \equiv \sum_{i_1,\ldots,i_n} M_{i_1,\ldots,i_n} x_{i_1} x_{i_2} \ldots x_{i_n} |x|^{-(2n+1)}.$$

In any dimension, ideal multipoles of sufficiently high order generate highly singular potentials.

In order to precisely gauge L^p class versus L^p smoothing here, we define classes of "sharp" L^p spaces as follows:

$$L^{p*}_s \equiv \{V \in L^p_s : V \notin L^q_s \text{ for } q > p\}. \tag{1}$$

$$L^p_{s*} \equiv \{V \in L^p_s : V \notin L^p_t \text{ for } t > s\}. \tag{2}$$

Finally, $L^{p*}_{s*} \equiv L^{p*}_s \cap L^p_{s*}$. Our main result states that outside the class of relatively bounded potentials (where smoothing is completely a function of the Sobolev class of the potential), any type of Sobolev smoothing is possible in L^2 for potentials in L^{p*}. We restrict our examples to positive potentials, since there is no question in these cases of the definition of the operator and its semigroup.

We will require a Theorem [Ko]:

THEOREM 1: *Assume* $V \in L^{\frac{n}{2}+\epsilon}$, $\epsilon > 0$, *and* $1 < p < \infty$. *Then* $e^{-tH} : L^p \to L^p_{s+2,\mathrm{loc}}$ *if and only if* $V \in L^p_{s,\mathrm{loc}}$, *for* $s \geq 0$.

DEFINITION: For $1 \leq p \leq \infty$, we define the *L^p-semigroup Sobolev range* to be the smallest interval $[s_1, s_2]$ such that the Schrödinger semigroup corresponding to a potential $V \in L^{p*}$ maps all positive functions in L^p into $\bigcup_{s_1 < s < s_2} L^{p*}_{s*}$.

Note that mapping all functions into $\bigcup_{s_1 \leq s \leq s_2} L^{p*}_{s*}$ would be impossible; certainly it would not hold for the 0 function.

THEOREM 2: *For $1 \leq p < \frac{n}{2}$ the L^p-semigroup Sobolev range is $[2, \infty]$. For $p > \frac{n}{2}$, this range collapses to the single point $\{2\}$.*

Proof of Theorem 2

The second statement of Theorem 2 follows from Theorem 1.

Our proof of the first part will require the Feynman-Kac formula. Let $b_x(t)$ denote standard Brownian motion in $\mathbf{R}^n$ starting at time 0 at $x \in \mathbf{R}^n$. Let $-\Delta+V$ be a Schrödinger operator on $\mathbf{R}^n$. If the semigroup is well-defined, the formula gives

$$(e^{tH} f)(x) = \mathcal{E}_x \left(e^{-\int_0^t V(b(s))ds} f(b(t)) \right), \tag{3}$$

where $\mathcal{E}_x$ denotes mathematical expectation with respect to Brownian motion starting at x.

We now define our class of operators. For expediency we will deal with potentials which are bounded below, though our essential results will carry over to potentials with negative parts in the class K_n (see below). Let $V(x)$ be a locally L^1 function which is bounded below. The operator $-\Delta + V$ as a quadratic form is well-defined on the intersection of the form domains of $-\Delta$ and V, since both operators are positive. Thus, since the sum is a closed semibounded form, it corresponds uniquely to a self-adjoint operator in L^2 (see [RS, VIII.6]). Furthermore, it is easy to show that the operator restricted to $L^p \cap C^2$, $1 \leq p < \infty$, is closeable, and we denote the L^p closure by H as well. The Trotter product formula and the positivity of V show easily that the semigroup is L^p-contractive, since the Laplacian itself generates a contractive semigroup in L^p.

The Kato classes K_n are defined by $V \in K_n$ if and only if

(i) $\lim_{\epsilon \downarrow 0} \left[\sup_x \int_{|x-y| \leq \epsilon} |x-y|^{-(n-2)} |V(y)| d^n y \right] = 0 \qquad (n \geq 3)$

(ii) $\lim_{\epsilon \downarrow 0} \left[\sup_x \int_{|x-y| \leq \epsilon} \ln(|x-y|^{-1}) |V(y)| d^2 y \right] = 0 \qquad (n = 2)$

(iii) $\sup_x \int_{|x-y| \leq \epsilon} |V(y)| dy < \infty \qquad (n = 1).$

These spaces are natural in many senses for Schrödinger theory. For example, we have the following two theorems from [AS]:

THEOREM 3: *$V \in K_n$ if and only if*

$$\lim_{t \downarrow 0} \sup_x \mathcal{E}_x \left(\int_0^t |V(b_x(s))| ds \right) = 0. \tag{4}$$

The proof depends on a precise study of the integral kernel of the operator on V defined by the expectation in (4), and in particular the kernel's singular behavior on the diagonal.

We also need

THEOREM 4: *If $V \in K_n$ and $f \in L^p$, $1 \le p \le \infty$, then $e^{-tH}f$ is a continuous function.*

Theorem 3 implies

COROLLARY: *If $V \in K_n$, then for $f \ge 0$, f not identically 0, $e^{-tH}f$ is everywhere positive.*

Indeed, assume $V \in K_n$. Then $(e^{-tH}f)(x) = 0$ is equivalent to $e^{-\int_0^t V(b(s))ds} f(b(t)) = 0$ (almost surely is always assumed here for statements involving Brownian motion), which is equivalent to the statement that $\int_0^t V(b(s))ds$ is infinite whenever $f(b(t)) \ne 0$. Since $f(b(t)) \ne 0$ has positive probability for all $t > 0$, Theorem 3 implies that for all x, $e^{-tH}f(x) > 0$ for t sufficiently small. However, the semigroup e^{-tH} is analytic in the sector $-\frac{\pi}{2} \le \arg t \le \frac{\pi}{2}$ [RS, Corollary to Theorem X.52]. Thus, by a standard analytic continuation argument, $e^{-tH}f > 0$ almost everywhere for $t > 0$.

We now prove a Lemma. Note that for V bounded below, it is well known that $e^{-tH}: L^p \to L^\infty$.

LEMMA 1: *Suppose $V \in L^{p*}$ is bounded below. If $e^{-tH}f > 0$ $(t > 0,\ x \in \mathbf{R}^n)$ for some $f \in L^p$, then $e^{-tH}f \in L^{p*}_{2*}$.*

Proof: In this case, $He^{-tH}f = e^{-\frac{tH}{2}}He^{-\frac{tH}{2}}f \in L^p$. Thus,

$$\Delta e^{-tH}f = Ve^{-tH}f - e^{-\frac{tH}{2}}He^{-\frac{tH}{2}}f \in L^p, \tag{5}$$

so that $e^{-tH}f \in L^p_2$. Thus, if we again sandwich H between powers of the semigroup, we conclude that $He^{-tH}f \in L^p_2$. Since $V \in L^{p*}$, (5) together with the positivity of $e^{-tH}f$ gives that $\Delta e^{-tH}f \in L^{p*} \subset L^{p*}_{0*}$, the last inclusion following from Sobolev imbedding. Thus, $e^{-tH}f \subset L^{p*}_{2*}$.■

To prove Theorem 2, we fix p, $1 \le p < \frac{n}{2}$. We will consider potentials in L^p which are not in the class K_n. Let B denote the unit ball in $\mathbf{R}^n$, and χ_G denote the characteristic function of the set G. Initially, we will consider potentials $V = r^{-\alpha}$, $r \equiv |x|$, with α such that $V \in L^p$, $p < \frac{n}{2}$. For the potential to be highly singular we also must have $\alpha > 2$, so that $2 < \alpha < \frac{n}{p}$.

We define for $G \subset \mathbf{R}^n$

$$L^p_s(G) = \{f \in L^p(G) : \text{there exists } \tilde{f} \in L^p_s(\mathbf{R}^n) \text{ such that } \tilde{f}|_G = f\}. \tag{6}$$

Let B_a denote the ball of radius a centered at 0. First, observe that if $H = -\Delta + r^{-\alpha}$, and $f \in L^p(\mathbf{R}^n)$, then $e^{-tH} f \in L^p_s(\mathbf{R} \backslash B_a)$ for $a > 0$, $s \geq 0$. Indeed the above bootstrapping argument, using (5) and restricting to $\mathbf{R}^n \backslash B_a$, shows initially that $e^{-tH} f \in L^p_2(\mathbf{R}^n \backslash B_a)$, and on the nth iteration that $e^{-tH} f \in L^p_{2n}(\mathbf{R}^n \backslash B_a)$. Therefore, to show $e^{-tH} f \in L^p_\infty$, it suffices to show $e^{-tH} f \in C^\infty$ in some neighborhood of 0. For this it suffices to show that $e^{-tH} f$ decays near 0 faster than any power of r (by using the difference quotient definition of derivatives at 0). To this end, first we may assume $\|f\|_\infty \leq C$ for some $C > 0$, since the range of the semigroup is in any case in L^∞.

For Brownian motion b_x, define the event

$$E_x = \{|b_x(s) - x| \leq |x| \text{ for } 0 \leq s \leq |x|^\alpha\}, \tag{7}$$

and E^c as the complement. Choose $\epsilon > 0$ such that $\alpha - \epsilon > 2$. By the Feynman-Kac formula and positivity of V, for x small,

$$\begin{aligned} |e^{-tH} f(x)| &= \left| \mathcal{E}\left(e^{-\int_0^t V(b_x(s))ds} f(b_x(t)) \right) \right| \\ &\leq C\mathcal{E}\left(e^{-\int_0^{|x|^{\alpha-\epsilon}} V(b_x(s))ds} \middle| E_x \right) \mathcal{P}(E_x) + C\mathcal{E}\left(e^{-\int_0^{|x|^{\alpha-\epsilon}} V(b_x(s))ds} \middle| E_x^c \right) \mathcal{P}(E_x^c) \\ &\leq C\left[e^{-|x|^{\alpha-\epsilon}(2|x|)^{-\alpha}} + \mathcal{P}(E_x^c) \right], \end{aligned} \tag{8}$$

where $\mathcal{P}$ denotes probability. The probability

$$\mathcal{P}(E_x^c) = \mathcal{P}\left(|b_x(s) - x| \geq |x| \text{ for some } s,\ 0 \leq s \leq |x|^\alpha\right) \tag{9}$$

can be estimated by observing that once a Brownian path leaves a half-space H, the probability of its being outside H at any later time is $\leq \frac{1}{2}$ (reflection principle). Clearly the same holds (in fact with strict inequality) when H is replaced by a ball. Thus

$$\mathcal{P}(E_x^c) \leq 2\mathcal{P}(|b_x(s) - x| \geq |x| \text{ for } s = |x|^\alpha) = 2\mathcal{P}\left(\sqrt{|x|^\alpha}|Z_n| \geq |x|\right), \tag{10}$$

where Z_n denotes a standard n-dimensional Gaussian random variable. Therefore

$$\mathcal{P}(E_x^c) \leq C_1 e^{-\frac{1}{2}|x|^{2-\alpha}}, \tag{11}$$

so

$$e^{-tH} f \leq C_2 \left[e^{-2^{-\alpha}|x|^{-\epsilon}} + e^{-\frac{1}{2}|x|^{2-\alpha}} \right], \tag{12}$$

and the faster than polynomial decay near 0 in (12) implies that $e^{-tH} f$ is in C^∞ near 0, and hence everywhere. Thus we have

LEMMA 2: *If $V(x) = r^{-\alpha}$, $\alpha > 2$, then e^{-tH} is infinitely smoothing in the scale of Sobolev spaces.*

Clearly, the same holds for any positive potential V which is smooth in $\mathbf{R}^n\setminus\{0\}$, and behaves like $|x|^{-\alpha}$ for x sufficiently small. Note that if $V_1 \geq V_2$, and $H_i = -\Delta + V_i$, then $e^{-tH_1}f \leq e^{-tH_2}f$, for $f \geq 0$ (monotonicity property). Note that $V(x) = r^{-\alpha}$ is locally in L^p for $p < \frac{n}{\alpha}$; and

$$V(x) = |x|^{-\alpha}\big|\ln|x|\big|^{-2} \in L^{\frac{n}{\alpha}*}_{\mathrm{loc}}. \tag{13}$$

We can use Lemma 2 and monotonicity to construct potentials (of the form (13) near 0) in L^{p*}, for all $p < \frac{n}{2}$, whose semigroup is infinitely smoothing.

However, there also exist positive potentials in L^{p*} whose semigroups only manage to smooth into L^p_2. To construct these, we require

LEMMA 3: *There exist $V \in L^{p*}$, $1 \leq p < \frac{n}{2}$, such that $e^{-tH}f \in L^{p*}_{2*}(\mathbf{R}^n)$, for $f \geq 0$, f not identically 0.*

The construction for Lemma 3 is as follows. Let x_1 be the first coordinate of x, and $V_1(x_1) \in L^{p*}[-\frac{1}{2}, \frac{1}{2}]$. Let

$$V(x) = \begin{cases} V_1(x_1) & \text{if } |x| \leq 1 \\ 0 & \text{elsewhere} \end{cases}. \tag{14}$$

Then $V \in L^{p*}(\mathbf{R}^n)$. Furthermore, letting $\bar{y} \equiv (y_2, y_3, ..., y_n)$, we have (for $n \geq 3$; the argument is similar otherwise)

$$\begin{aligned} \lim_{\epsilon\to 0}\int_{|x-y|\leq\epsilon} |x-y|^{-(n-2)}V(y)dy &\leq \lim_{\epsilon\to 0}\int_{\substack{|y_1|\leq\epsilon,\\ |\bar{y}|\leq 1}} |y|^{-(n-2)}V(y+x)dy \\ &\leq \lim_{\epsilon\to 0}\int_{\substack{|y_1|\leq\epsilon,\\ |\bar{y}|\leq 1}} |y|^{-(n-2)}V_1(y_1+x_1)dy \\ &\leq \lim_{\epsilon\to 0}\int_{|y_1|\leq\epsilon} V_1(y_1+x_1)dy = 0. \end{aligned} \tag{15}$$

In the last inequality we have used the fact that

$$\int_{|\bar{y}|\leq 1} |y|^{-(n-2)}d\bar{y} = O(1) \qquad (y_1 \to 0), \tag{16}$$

and that $V_1 \in K_1(\mathbf{R}^1)$. Equation (15) shows that $V \in K_n$. Thus, $e^{-tH} > 0$ for $f \geq 0$, $f \neq 0$, and Lemma 1 shows that for such f, $e^{-tH}f \in L^{p*}_{s*}$. Lemmas 2 and 3 together serve to complete the proof of Theorem 2.

Remark: The results in this paper can be extended to perturbations $A + V$ of elliptic second order symmetric differential expressions $A = \sum_{i,j}\left(\frac{\partial}{\partial x_i}\right)a_{ij}(x)\left(\frac{\partial}{\partial x_j}\right)$. For these we require results on kernels of uniformly and strictly elliptic operators [G1,G2]. See [CFZ] for some potential theoretic results for such operators.

References

[AS] M. Aizenmann and B. Simon, Brownian motion and Harnack's inequality for Schrödinger operators, *Comm. Pure Appl. Math.* 35 (1982), 209-273.

[CFZ] M. Cranston, E. Fabes, Z. Zhao, Potential theory for the Schrödinger equation, *Bull. Am. Math. Soc. 15 (1986)* , 213-216.

[D] B. Davies, Eigenfunction expansions for singular Schrödinger operators, Arch. Rational Mech. Anal. 63 (1977), 261-291.

[G1] David Gurarie, Kernels of elliptic operators: bounds and summability, *J. Differential Equations* 55 (1984), 1-29.

[G2] David Gurarie, Finite propagation speed, maximal functions, and kernels of strictly elliptic operators, preprint.

[H] Rhonda J. Hughes, Singular perturbations in the interaction representation, II, *J. Functional Analysis* 49 (1982), 293-314.

[HS] Rhonda J. Hughes and Irving E. Segal, Singular perturbations in the interaction representation, *J. Functional Analysis* 31 (1980), 71-98.

[Ko] Mark A. Kon, Exact conditions for Sobolev smoothing of Schrödinger semigroups, in preparation.

[RS] M. Reed and B. Simon, *Methods of Mathematical Physics,* Vol. 1 and 2, Academic Press, New York, 1975.

[Si] Simon, B., Schrödinger semigroups in the scale of Sobolev spaces, *Pacific J. Math.* 122 (1986), 475-480.

Canadian Mathematical Society
Conference Proceedings
Volume 8 (1987)

Some Negative L^p Results for Eigenfunction Expansions Associated with Scattering Theory

Mark A. Kon, * Columbia University and Boston University
Louise A. Raphael, ** Howard University

In [Ag1, Ag2], Agmon studies a class of eigenfunctions and related expansions for Schrödinger operators with short range potentials, first studied by Ikebe [Ik]. Results are described which relate the continuum spectra and eigenfunctions for such operators to those of the Laplacian. In particular it is shown that the continuous portion of a Schrödinger operator with short range potential is unitarily equivalent to the Laplacian.

In this note, we study the relationship between the eigenfunction expansions of the two operators with regard to pointwise and L^p behavior on $\mathbf{R}^n$. In particular, we will study two questions:

(1) Equiconvergence: Under what conditions do the two eigenfunction expansions (using, say, spherical means) converge at the same points? In the same L^p spaces?

(2) Equisummability: What if summation methods are introduced into the expansions? Do the results change?

We will answer the second question in more detail than the first. We obtain necessary conditions (derived from the second question) for question (1).

We now give a very brief description of the expansion theorem associated with such operators. Let $V(x)$ be in $L^r_{\text{loc}}(\mathbf{R}^n)$, $r > \frac{n}{2}$, with $V \in [\text{SR}]$, where $[\text{SR}] = \{V(x) : V(x) = O(|x|^{-1-\epsilon})$ as $x \to \infty$ for some $\epsilon > 0\}$ denotes the class of short range potentials. Let $\mathbf{k} \in \mathbf{R}^n$ be represented by $\mathbf{k} = k\omega$, where ω is a unit vector, and $k = |\mathbf{k}|$. One first seeks eigenfunctions $e(x,\omega;k)$, $x \in \mathbf{R}^n$, which

(i) satisfy

$$-\Delta u + Vu = k^2 u$$

(ii) are perturbations of eigenfunctions of the Laplacian:

* Research partially supported under N.S.F. grant DMS-8509458
** Research supported by Army Research Office grant number DAAG-29-84-G0004

$$e(x,\omega;k) = e^{ik\omega\cdot x} + v, \tag{1}$$

where v is small in some sense as $x \to \infty$. Such eigenfunctions can be constructed by at least two means ([Ag2]); and there are two sets of complete independent eigenfunctions, denoted by $e(x,\omega;\pm k)$, where $k \in \mathbf{R}^+$ (the two sets of course depend linearly on each other). The resulting expansions are "distorted" Fourier transforms for $f \in L^2(\mathbf{R}^n)$:

$$\tilde{f}^{\pm}(\xi) = (2\pi)^{-\frac{n}{2}} \int_{\mathbf{R}^n} f(x)e_{\pm}(x,\xi)^* dx, \tag{2a}$$

where $*$ denotes conjugation, and $e_{\pm}(x,\xi) = e(x, \frac{\xi}{|\xi|}; \pm|\xi|)$. Henceforth in this paper $\epsilon > 0$ will be assumed; this will be assumed, in particular, in all of the $\epsilon \to 0$ limits. Letting $\chi_n(\xi)$ denote the characteristic function of the unit ball in $\mathbf{R}^n$, the transforms are inverted by

$$f(x) = \lim_{\epsilon\to 0}(2\pi)^{-\frac{n}{2}} \int_{\mathbf{R}^n} \chi_n(\epsilon\xi)\tilde{f}^{\pm}(\xi)e_{\pm}(x,\xi)d\xi + \sum \chi_1(\epsilon\lambda_j)c_j\psi_j(x), \tag{2b}$$

with $\{\psi_j\}$ an orthonormal set of discrete eigenfunctions, and $c_j = \int f(\xi)\psi_j^*(\xi)d\xi$. Under our assumptions it can be shown ([Ag2]) that the spectrum of $H \equiv -\Delta + V$ is bounded below in L^2. The limit above is taken in L^2, and the presence of χ_n in the inverting integral in (2b) is equivalent to the understanding that the integral (without χ_n in it) is taken as the L^2 limit of integrals over balls of increasing diameter.

It is well-known that multidimensional L^2 eigenfunction expansions in general fail to converge in other L^p spaces or almost everywhere [Fe]. In fact, even summation using the various Riesz methods fails in some L^p spaces [KST].

The presence of χ in the definition of the inverse operator in (2b) suggests that more robust convergence in (2b) might be obtained if we replace χ by other functions with "better" properties. Indeed, this is known to be true for one-dimensional cases [Ti,KR], and, more importantly, in higher dimensions [GK,G1,G2]. For these reasons, we will be more interested in defining the inverse transform in (2) more generally, replacing χ by a function $\phi : \mathbf{R} \to \mathbf{R}$, such that $\phi(0) = 1$, $\phi \geq 0$, and ϕ is continuous near 0 and bounded. Since the information about behavior of (2b) as $\epsilon \to 0$ is not always complete, question (2) arises here. It can be formulated more precisely as:

QUESTION: For a given ϕ, how do the expansions of a function f in the Laplacian eigenfunctions $e_0(x,\xi) \equiv e^{-ix\cdot\xi}$ and the perturbed ones $e_{\pm}(x,\xi)$ compare?

One of the relationships ([Ag2]) between the perturbed and unperturbed problems is that they are unitarily equivalent when restricted to their continuous spectra. This fact has to do with Hilbert space isomorphism of the two operators. On the other hand, the above question addresses the actual relationship of the two operators on the base space $\mathbf{R}^n$. The question makes sense for any ϕ for which the integrals in (2b) converge pointwise (this will certainly occur if $\phi(|\xi|) \in L^2(\mathbf{R}^n))$, but we will also consider those for which (2b) converges only in L^2 for a given ϵ. For the latter to occur, we need only $\phi \in L^\infty$, which follows in any case from our hypotheses. These weaker hypotheses on ϕ will be useful later in the proofs of negative results when $\phi = \chi_n$; but they can always be strengthened if pointwise convergence in (2b) is needed.

Without loss of generality we will assume that the discrete spectrum of $-\Delta + V$ is empty; the arguments proceed similarly if the discrete part of the spectrum is taken into account.

Define

$$f^0_{\phi,\epsilon}(x) = (2\pi)^{-\frac{n}{2}} \int \phi(\epsilon|\xi|)e^{-i\xi\cdot x}d\xi, \tag{3}$$

and $f^\pm_{\phi,\epsilon}(x)$ similarly, with $e^{-i\xi\cdot x}$ replaced by $e_\pm(x,\xi)$.

Initially we assume that ϕ extends to an analytic function in a region Ω_θ which consists of a bounded neighborhood of 0 with smooth boundary, together with a wedge $\{z \in \mathbf{C} : |\arg z| \leq \theta\}$ for some $\theta > 0$. In addition we assume that $\phi(z) = O(|z|^{-\delta})$ in Ω_θ, for some $\delta > 0$, and that ϕ is of bounded variation in $\mathbf{R}^+$. Thus, let $V \in [\mathrm{SR}] \cap L^r_{\mathrm{loc}}(\mathbf{R}^n)$, with $r > \frac{n}{2}$. Let $H = -\Delta + V$ denote the unique self-adjoint realization of the indicated operator. The following theorem follows from Theorem 2 in [GK]:

THEOREM 1: *Let H be as above. Let $1 \leq q \leq \infty$, $f \in L^2$, and*

$$\frac{1}{q} > \frac{1}{2} + \frac{1}{r} - \frac{2}{n}. \tag{4}$$

Then the Fourier transform of f and its scattering expansion (in the eigenfunctions $e_\pm(\xi, x)$) are ϕ-equisummable in L^q, i.e.,

$$f^\pm_{\phi,\epsilon}(x) - f^0_{\phi,\epsilon} \underset{\epsilon\to 0}{\longrightarrow} 0. \tag{5}$$

in L^q. Precisely, if (4) holds, then

$$\|f^\pm_{\phi\epsilon} - f^0_{\phi\epsilon}(x)\|_q = O(\epsilon^{1-\frac{n}{2}(\frac{1}{2}+\frac{1}{r}-\frac{1}{q})}).$$

Furthermore, the scale in (4) is the best possible in that for any r for which the opposite inequality holds, there exists $V \in L^r$ such that (5) fails in L^q.

The proof uses operator calculus expressions for $f^\pm_{\phi,\epsilon}(x)$ and $f^0_{\phi,\epsilon}(x)$:

$$\begin{aligned} f^\pm_{\phi,\epsilon}(x) &= \phi(\epsilon\sqrt{H})f(x) \\ f^0_{\phi,\epsilon}(x) &= \phi(\epsilon\sqrt{-\Delta})f(x), \end{aligned} \tag{6}$$

with

$$\phi(H) = \frac{1}{2\pi i}\int_\Gamma \phi(\zeta)(\zeta - H)^{-1}d\zeta, \tag{7}$$

where Γ is a positively oriented countour in Ω_θ containing the spectrum of H. The result relies on bounds for the kernel of $(\zeta - H)^{-1}$ (see [GK]).

We will elaborate on the meaning of Theorem 1. Recall that the integral (2b) (with χ replaced by ϕ) defining $f^{\pm}_{\phi,\epsilon}(x)$ is generally defined in the L^2 sense, though for ϕ decaying rapidly enough (say, $|\phi(z)| \le C|z|^{-\frac{n}{2}-\delta}$, $\delta > 0$, for $|z|$ large) it is also defined pointwise. Note that the scale of q allowed by (4) is of the form $q \in [2, q_{\max}]$, where $q_{\max}$ is the supremum of the values allowed by (4). If $\frac{1}{2} + \frac{1}{r} - \frac{2}{n} < 0$, then $q_{\max} = \infty$; in that case, by (5), the expansion in $e_{\pm}(x,\xi)$ and the Fourier transform behave identically pointwise as $\epsilon \to 0$.

It can in fact be shown that the scale (4) of L^p spaces is the best possible: for $\frac{1}{q} < \frac{1}{2} + \frac{1}{r} - \frac{2}{n}$, there exists a potential of the form $V = \chi_n |x|^{-\alpha} \in L^{\frac{n}{2}+\epsilon}$ for some $\alpha > 0$ (χ_n is again the characteristic function of the unit ball) and $f \in L^2$, such that (5) fails in L^q (see [GK]).

The main result of this paper has to do with an implication of the above failure for ordinary equiconvergence, i.e., (5) for $\phi = \chi_n$.

LEMMA 1: *If $f \in L^2(\mathbf{R}^n)$, and*

$$f^{\pm}_{\phi,\epsilon}(x) - f^{0}_{\phi,\epsilon}(x) \underset{\epsilon\to 0}{\not\to} 0 \tag{8}$$

for ϕ as above, then the same holds if ϕ is replaced by χ_n, the characteristic function of the unit ball in $\mathbf{R}^n$.

Proof: Suppose

$$f^{\pm}_{\chi_n,\epsilon}(x) - f^{0}_{\chi_n,\epsilon}(x) \underset{\epsilon\to 0}{\longrightarrow} 0$$

in L^q. Let $P_H(\lambda)$ be the family of spectral projections corresponding to H, and χ_1 be the unit ball of $\mathbf{R}^1$. We have

$$f^{\pm}_{\chi_n,\epsilon} = \chi_1(\epsilon\sqrt{H})f(x) = P_H\left(\frac{1}{\epsilon^2}\right)f,$$

so that

$$f^{\pm}_{\chi_n,\epsilon}(x) - f^{0}_{\chi_n,\epsilon}(x) = P_H\left(\frac{1}{\epsilon^2}\right)f - P_{-\Delta}\left(\frac{1}{\epsilon^2}\right)f \underset{\epsilon\to 0}{\longrightarrow} 0 \tag{9}$$

in L^q. (Recall that without loss we have assumed that H has no discrete spectrum, so that $H \ge 0$.) By the spectral theorem and integration by parts, if a is below the spectra of $-\Delta$ and H,

$$\begin{aligned} f^{\pm}_{\phi,\epsilon} - f^{0}_{\phi,\epsilon} &= \phi(\epsilon\sqrt{H}) - \phi(\epsilon\sqrt{-\Delta}) \\ &= \int_{0-}^{\infty} \phi(\epsilon\sqrt{\lambda})d(P_H(\lambda) - P_{-\Delta}(\lambda))f \\ &= \int_{0-}^{\infty} (P_H(\lambda) - P_{-\Delta}(\lambda))\frac{d}{d\lambda}\phi(\epsilon\sqrt{\lambda})d\lambda \\ &= \int_{0-}^{\infty} [P_H\left(\frac{\lambda}{\epsilon^2}\right) - P_{-\Delta}\left(\frac{\lambda}{\epsilon^2}\right)f]\frac{d}{d\lambda}\phi(\sqrt{\lambda})d\lambda. \end{aligned} \tag{10}$$

Since $\frac{d}{d\lambda}\phi(\sqrt{\lambda}) \in L^1$ (recall ϕ is of bounded variation on $\mathbf{R}^+$), the dominated convergence theorem implies that (10) converges to 0 in L^p. This completes the proof. ∎

We remark that there are only slight changes in the above proof when H has a discrete spectral component. Namely, the integrals in (10) will have discrete components as well.

The second paragraph of Theorem 1 together with Lemma 1 gives:

THEOREM 2: *Let $r, q \in [1, \infty]$. If*

$$\frac{1}{q} < \frac{1}{2} + \frac{1}{r} - \frac{2}{n},$$

then there exists $V \in [\mathrm{SR}] \bigcap L^r(\mathbf{R}^n)$ such that the Fourier integral and the $H = -\Delta + V$ eigenfunction expansion are not *equiconvergent in L^q. Precisely,*

$$\int_{|\xi|\leq R} \tilde{f}^{\pm}(\xi) e_{\pm}(x,\xi) d\xi - \int_{|\xi|\leq R} \tilde{f}^0(\xi) e^{i\xi\cdot x} d\xi \underset{R\to\infty}{\not\to} 0 \tag{11}$$

in L^q.

COROLLARY: *For any $n > 4$ there exists $V \in L^\infty(\mathbf{R}^n) \cap [\mathrm{SR}]$ such that the Fourier integral and the scattering expansion (with respect to $-\Delta + V$) are not pointwise equiconvergent. That is, there exist V in this class for which*

(i) The difference (11) of the expansions in general fails to vanish uniformly almost everywhere as $R \to \infty$.

(ii) There exist $f \in L^2$ for which the difference (11) diverges at 0 as $R \to \infty$.

Proof: (*i*) Set $q = \infty$ in the Theorem.

(*ii*) This follows from examination of the proof after Theorem 2 in [GK] that (4) is optimal.

Remark: In general, of course, the scale of dimensions in the Corollary depends on the L^r class of $V \in L^r(\mathbf{R}^n) \cap [\mathrm{SR}]$. In particular, if r is close to $\frac{n}{2}$, then pointwise equiconvergence generally fails in all dimensions. (i.e., there exist $V \in L^r_{\mathrm{loc}} \cap [\mathrm{SR}]$ for which it fails). These results do not indicate one way or another whether or not the result of the Corollary holds in dimensions 1 through 4.

References

[Ag1] S. Agmon, Spectral theory of Schrödinger on Euclidean and non-Euclidean spaces, to appear, *Comm. Pure Appl. Math.*

[Ag2] S. Agmon, Spectral properties of Schrödinger operators and scattering theory. *Ann. Scuola Norm. Sup. Pisa* 2 (1975), 151-218.

[Fe] C. Fefferman, The new multiplier theorem for the ball, *Ann. of Math.* 94 (1971), 330-336.

[GK] David Gurarie and Mark A. Kon, Resolvents and regularity properties of elliptic operators, in *Operator Theory: Advances and Applications*, C. Apostol, Ed., Birkhäuser Verlag, 1983.

[G1] Gurarie, David, Kernels of elliptic operators: bounds and summability, *J. Differential Equations* 55 (1984), 1-29.

[G2] Gurarie, David, Finite propagation speed, maximal functions, and kernels of strictly elliptic operators, preprint.

[Ik] T. Ikebe, Eigenfunction expansions associated with the Schrödinger operators and their applications in scattering theory, *Arch. Rational Mech. Anal.* 5 (1960), 1-34.

[KST] C.E. Kenig, R.J. Stanton, and P.A. Tomas, Divergence of eigenfunction expansions, *J. Functional Analysis* 46 (1982), 28-44.

[KR] Mark Kon and Louise Raphael, New multiplier methods for summing classical eigenfunction expansions, *J. Differential Equations* 3 (1983), 391-406.

[Ti] A.N. Tikhonov, Stable methods for the summation of Fourier series, *Soviet Math. Dokl.* 5 (1964), 641-646.

Canadian Mathematical Society
Conference Proceedings
Volume 8 (1987)

DISTRIBUTIONAL SOLUTIONS OF ORDINARY DIFFERENTIAL EQUATIONS

by

Allan M. Krall[1]

R. P. Kanwal[1]

L. L. Littlejohn[2]

ABSTRACT. Distributional solutions involving the Dirac delta function and its derivatives are discovered for orthogonal polynomial weight differential equations, the confluent hypergeometric equation, the hypergeometric equation, Bessel's equations and other second order linear differential equations with analytic coefficients. Results are sometimes surprising with hitherto unexpected solutions to some very classic problems.

I. INTRODUCTION. Several years ago a number of people became interested in a new set of orthogonal polynomials: the Bessel polynomials, [3], [6]. As their properties were discovered, one remained elusive: that of finding a suitable weight function and interval with respect to which they would be orthogonal. What was discovered instead was that they were orthogonal in a complex sense with weight $e^{-2/z}$ around any path encircling the origin. This seemed to imply from a theorem of Stieltjes [9] that the real weight had support only at 0. This did not make sense.

In 1975 when the first author reexamined the problem, it occurred to him that perhaps a distributional weight function, such as the Dirac delta function, might do, since its support is only at 0. Subsequent investigation included the delta functions derivatives as well, until finally

$$w(x) = -\sum_{n=0}^{\infty} \frac{2^{n+1}\delta^{(n)}(x)}{n!(n+1)!}$$

1. Department of Mathematics, McAllister Building
The Pennsylvania State University, University Park, PA 16802.

2. Department of Mathematics, Utah State University,
Logan, UT 84322.

was found, and, surprisingly it actually worked [8].

When the technique used to derive the formula above was applied to other polynomials, similar weights were also found for these, and in these cases connections with the classical weight functions were found. Later the third author discovered that all these weights could be found as distributional solutions to certain differential equations, satisfied by symmetry factors for the differential equations of the orthogonal polynomials.

These differential equations are all singular at the ends of the orthogonality intervals. The singularities generate discontinuities and, even point masses in their distributional solutions. These are needed to make the polynomials orthogonal.

Finally the second and third authors applied the distributional techniques to the confluent hypergeometric equation and came up with some very interesting solutions [7]. The second author has also considered Bessel's equation in this light and again found some interesting new solutions [4].

Our purpose here is to summarize these different looking results, to apply the techniques to the hypergeometric equation, and to encourage others to assist with the exploration. We hope some who read this will help with an explanation of what is actually going on behind the scenes. There are many questions still to be answered. Considerable assistance is needed.

II. FUNDAMENTALS. We remind the reader of the various spaces of test functions $D \subset S \subset P \subset E$, consisting of infinitely differentiable functions with compact support, which decay rapidly, which grow slowly, or which have no growth constraints at all [2] [8]. The spaces of continuous linear functionals which act upon them are $D' \supset S' \supset P' \supset E'$. They have no growth constraints, grow slowly, decay rapidly, or have compact support. Whenever possible we shall work in these spaces of distributions. Occasionally, in order to use the Fourier transform, we shall step outside, but this will be brief. We not that if we find a distribution in D' with a certain, unique property, there can be no additional distributions in the other spaces of functionals with the same property.

The spaces P and P' were invented in [8], and are not so well known as the others [2]. We refer the reader to [8] for an enlarged discussion:

There are two basic theorems need to solve singular ordinary differential equations.

II.1 THEOREM. _If_ $t' = 0$ _in_ D', _then_ $t = c$, _a constant_.

II.2 THEOREM. _If_ $xt = 0$ _in_ D', _then_ $t = c\delta(x)$, _where_ $\delta(x)$ _is the Dirac delta function_.

These are well known, their proofs may be found in [11].

Note that $t = c$ exists in D' and S'. It may not exist in P' or E' however. Only when the interval of definition for D, S, P and E is finite will it also be in the latter two spaces.

III. WEIGHTS FOR ORTHOGONAL POLYNOMIALS. A collection of (rather general) orthogonal polynomials may be defined as follows. Let $\{\mu_n\}_{n=0}^{\infty}$ be a collection of real numbers with the property that $\Delta_0 = \mu_0 \neq 0$,

$$\Delta_n = \begin{vmatrix} \mu_0 & \cdots & \mu_n \\ \vdots & & \vdots \\ \mu_n & \cdots & \mu_{2n} \end{vmatrix} \neq 0, \qquad n = 1, \ldots .$$

Then set $p_0 = 1$,

$$p_n(x) = [1/\Delta_{n-1}] \begin{vmatrix} \mu_0 & & \mu_n \\ \vdots & & \vdots \\ \mu_{n-1} & \cdots & \mu_{2n-1} \\ 1 & \cdots & x^n \end{vmatrix}, \qquad n = 1, \ldots .$$

The numbers μ_n are called moments. The polynomials $p_n(x)$ have highest coefficient 1, and are therefore monic orthogonal polynomials.

The collection $\{p_n(x)\}_{n=0}^{\infty}$ may be made orthogonal by any linear functional w which has the property $\langle w, p_n p_m \rangle = 0$, $n \neq m$, $\langle w, p_n^2 \rangle \neq 0$. Such functionals are by no means unique. There is one functional, w_0, generated by an ordinary function not identically zero on $[0,\infty)$ which satisfies $\langle w_0, x^n \rangle = 0$ for all $n = 0,1,\ldots$. Without additional constraints, uniqueness is out of the question.

There is one satisfactory functional which can be written down immediately. It is

$$w = \sum_{n=0}^{\infty} (-1)^n \mu_n \delta^{(n)}(x)/n!$$

It is, in fact, a functional Taylor's series, since if a test function ϕ is real analytic, then

$$\begin{aligned} \langle w, \phi \rangle &= \langle w, \sum_{n=0}^{\infty} \phi^{(n)}(0) x^n/n! \rangle \\ &= \sum_{n=0}^{\infty} \langle w, x^n \rangle \langle (-1)^n \delta^{(n)}, \phi \rangle /n! \\ &= \langle \sum_{n=0}^{\infty} (-1)^n \mu_n \delta^{(n)}/n!, \phi \rangle \end{aligned}$$

where $\mu_n = \langle w, x^n \rangle$, $u=0,1,\ldots$ are the moments associated with w. It is an easy matter to show

$$\langle w, p_n p_m \rangle = 0 \quad , \quad n \neq m,$$
$$\langle w, p_0^2 \rangle = \mu_0,$$
$$\langle w, p_n^2 \rangle = \Delta_n/\Delta_{n-1}, \; n = 1, 2, \ldots \; .$$

See [5], [8].

Continuity of $w = \sum_{n=0}^{\infty} (-1)^u \mu_n \delta^{(n)}/n!$ is more difficult to establish. If

$$|\mu_n| < cM^n n! \;, \quad n = 0, 1, \ldots \;,$$

then w is continuous on the space $Z_{M\varepsilon}$ consisting of test functions ψ satisfying the constraint

$$|x+iy|^q \; |\psi(x+iy)| < C_q e^{a|y|}$$

where $a \leq M+\varepsilon$ and $q \geq 0$ is arbitrary [8]. Further work needs to be done here.

If the polynomials satisfy a differential equation of second order, it must be of the form

$$py'' + qy' = \lambda_n y$$

where

$$p = ax^2 + bx + c \quad , \quad q = dx + e,$$

if p and q are to be independent of n.

Let y and z represent two different polynomials satisfying

$$py'' + qy' = \lambda y,$$

$$pz'' + qz = \mu z.$$

Then if $W = yz' - y'z$, the Wronskian,

$$pW' + qW = (\mu-\lambda) yz.$$

If w satisfies

$$-(pw)' + qw = 0,$$

then, eliminating the terms involving q,

$$(pwW)' = (\mu-\lambda) yzw.$$

Integrating from $-\infty$ to ∞,

$$(\mu-\lambda)\int_{-\infty}^{\infty} yzwdx = 0,$$

assuming that $pwW|_{-\infty}^{\infty} = 0$.

Therefore the problem of finding a weight function w can also be solved if the equation

$$-(pw)' + qw = 0$$

has a solution which vanishes rapidly enough at $\pm\infty$.

EXAMPLE 1. THE LEGENDRE POLYNOMIALS. The Legendre differential equation is

$$(1-x^2)y'' - 2xy' + n(n+1)y = 0$$

Hence w should satisfy

$$(1-x^2)w' = 0.$$

To solve this distributionally, we note that dividing by $(1-x)$ and $(1+x)$ results in

$$w' = A\,\delta(x+1) + B\,\delta(x-1),$$

and, thus, if $H(x)$ is the Heaviside function,

$$w = A\,H(x+1) + B\,H(x-1) + C.$$

The only way w can vanish at $\pm\infty$ is to have $C = 0$ and $B = -A$. Setting $A = 1$,

$$w = H(x+1) - H(x-1),$$

or

$$\begin{aligned} w &= 1 \quad , \quad -1 \le x \le 1, \\ &= 0 \quad , \quad \text{otherwise}. \end{aligned}$$

The moments μ_n associated with the Legendre polynomials are $\mu_{2n} = 2/(2n+1)$, $\mu_{2n+1} = 0$. The series for w presented earlier is therefore

$$w = \sum_{n=0}^{\infty} 2\,\delta^{(2n)}(x)/(2n+1)!\ .$$

Since its inverse Fourier transform is

$$F^{-1}w = -\frac{1}{\pi}\sum_{n=0}^{\infty}\frac{(it)^{2n}}{(2n+1)!}$$
$$= (\sin t)/\pi t,$$

we again recover the first form of w by an application of the Fourier transform.

EXAMPLE 2. THE LAGUERRE POLYNOMIALS. The generalized Laguerre polynomials satisfy

$$xy'' + (\alpha+1-x)y' + ny = 0.$$

Hence w should satisfy

$$xw' + (x-\alpha)w = 0.$$

This may be rewritten as

$$x^{\alpha+1}e^{-x}(x^{-\alpha}e^{x}w)' = 0.$$

As long as $\alpha > -1$, this has the solution

$$w = x^{\alpha}e^{-x}[AH(x) + B].$$

In order to have w vanish at $-\infty$, $B = 0$. Setting $A = 1$, we find

$$w = x^{\alpha}e^{-x}H(x),$$

or

$$\begin{aligned} w &= x^{\alpha}e^{-x} \quad , \quad 0 \le x < \infty \\ &= 0 \quad , \quad -\infty < x < 0. \end{aligned}$$

When $\alpha < -1$ and is not an integer, a Cauchy regularization is required. In this case when $-j-1 < \alpha < -j$,

$$\langle w,\phi\rangle = \frac{(-1)^j}{\Gamma(\alpha+j+1)}\int_0^{\infty} x^{\alpha+j}[e^{-x}\phi(x)]^{(j)}dx.$$

The moments associated with the Laguerre polynomials are $\mu_n = \Gamma(\alpha+n+1)/\Gamma(\alpha+1)$. The functional Taylor's series is

$$w = \sum_{n=0}^{\infty}\frac{(-1)^n\Gamma(\alpha+n+1)}{\Gamma(\alpha+1)\ n!}\ \delta^{(n)}(x).$$

Connection to the formulas above is again made through the inverse and forward

Fourier transform.

EXAMPLE 3. THE HERMITE POLYNOMIIALS. The Hermite equation is

$$y'' - 2x\, y' + 2n\, y = 0.$$

Hence w should satisfy

$$w' + 2xw = 0,$$

or

$$e^{-x^2}(e^{x^2}w)' = 0.$$

The solution to this equation is $w = e^{-x^2}$, $-\infty < x < \infty$. The series

$$w = \sum_{n=0}^{\infty} \frac{\sqrt{\pi}\ \delta^{(2n)}(x)}{4^n\ n!}$$

also satisfies the equation above and can be connected to e^{-x^2} via the Fourier transform.

EXAMPLE 4. THE BESSEL POLYNOMIALS. The starter of it all, the ordinary Bessel polynomials satisfy

$$x^2y'' + (2x+2)y' + n(n+1)y = 0.$$

Hence w should satisfy

$$x^2w' - 2w = 0,$$

or

$$x^2e^{-2/x}(e^{2/x}w)' = 0.$$

Since $e^{-2/x}$ has an essential singularity at 0, a distributional solution to this equation has so far remained enigmatic. The classical solution, $w = e^{-2/x}$, does not vanish at $\pm\infty$, and is therefore not a real weight function. Since the moments here are

$$\mu_n = (-2)^{n=1}/(n+1)!\quad ,\quad n=0,1,\ldots,$$

the series

$$w = -\sum_{n=0}^{\infty} \frac{2^{n+1}\delta^{(n)}(x)}{(n+1)!\ n!}$$

is a weight function and distributionally satisfies w's differential equation. A real weight function, however, remains elusive.

There are extensions of these problems involving orthogonal polynomials satisfying fourth, sixth and eighth order differential equations. Weights in the fourth order case must satisfy two differential equations of orders one and three simultaneously. In the sixth order case, weights satisfy three differential equations of orders one, three and five. In the eight order case, the weight satisfies four differential equations of orders one, three, five and seven.

In all cases the technqiue is essentially the same. Find solutions in D' if possible, then see if there is a particular solution in P'.

IV. THE CONFLUENT HYPERGEOMETRIC EQUATION. Ordinary solutions of the confluent hypergeometric equation

$$xy'' + (m-x)y' - py = 0$$

are well known. If $(p)_0 = 1$,

$$(p)_r = p(p+1)\dots(p+r-1) \quad , \quad r = 1,\dots,$$

then

$$y_1 = {}_1F_1(p;m;x) = \sum_{r=0}^{\infty} \frac{(p)_r\ x^r}{(m)_r\ r!}$$

$m \neq 0,-1,\dots,$ and

$$y_2 = x^{1-m}{}_1F_1(1-m+p;2-m;x) = \sum_{r=0}^{\infty} \frac{(1-m+p)_r x^{1+r-m}}{(2-m)_r\ r!}$$

$m \neq 2,3,\dots$ are solutions. The functions ${}_1F_1$ have a number of nice properties. For instance,

$${}_1F_1(p;p;x) = e^x.$$

We refer the reader to [10] for details.

We search here for solutions of the type

$$y = \sum_{n=0}^{\infty} a_n\delta^{(n)}(x).$$

Whenever possible we shall then convert the series into an ordinary function.

We note that if ϕ is a test function in D,

$$\langle \delta^{(n)}, x\phi'' \rangle = \langle -n\delta^{(n+1)}, \phi \rangle ,$$
$$\langle \delta^{(n)}, x\phi' \rangle = \langle n\delta^{(n)}, \phi \rangle ,$$
$$\langle \delta^{(n)}, \phi' \rangle = \langle -\delta^{(n+1)}, \phi \rangle ,$$

Therefore

$$\begin{aligned}
&\langle xy'' + (m-x)y' - py, \phi \rangle \\
&= \langle y, x\phi'' + x\phi' + (2-m)\phi' + (1-p)\phi \rangle \\
&= \sum_{n=0}^{\infty} a_n \langle \delta^{(n)}, x\phi'' + x\phi' + (2-m)\phi' + (1-p)\phi \rangle \\
&= \sum_{n=0}^{\infty} a_n \langle -n\delta^{(n+1)} + n\delta^{(n)} + (2-m)\delta^{(n+1)} + (1-p)\delta^{(n)}, \phi \rangle \\
&= \langle \sum_{r=1}^{\infty} [a_{r-1}(m-r-1) + a_r(r-p+1)]\delta^{(r)} + (1-p)a_0\delta, \phi \rangle .
\end{aligned}$$

If y is a solution, then

$$(1-p)a_0 = 0,$$
$$a_{r-1}(m-r-1) - a_r(p-r-1) = 0, \quad r=1,2,\ldots .$$

In order to find $\{a_r\}_{r=1}^{\infty}$, we must consider four cases.

Case 1. If p is not a positive integer, then $p-r-1 \neq 0$, and

$$a_r = \frac{a_{r-1}(m-r-1)}{(p-r-1)} .$$

Since $a_0 = 0$, we find $a_r = 0$, $r = 1,\ldots,$ and $y = 0$.

Case 2. If p is a positive integer, then $p-r-1 = 0$ when $r = p-1$. If m is not a positive integer, then $a_0 = 0$, $a_1 = 0$, $\ldots$, $a_{p-2} = 0$. But a_{p-1} is arbitrary. Setting $a_{p-1} = A$, we find

$$a_{p+n} = \frac{A(p+1-m)(p+2-m)\ldots(p+n+1-m)}{(n+1)!} .$$

Therefore, setting $A = 1$,

$$\begin{aligned}
y &= \sum_{n=-1}^{\infty} \frac{(p+1-m)_{n+1}}{(n+1)!} \delta^{(p+n)}(x), \\
&= \sum_{r=0}^{\infty} \frac{(p+1-m)_r}{r!} \delta^{(p+r-1)}(x).
\end{aligned}$$

Case 3. If p is a positive integer, m is a positive integer, and

$m \leqq p$, then $a_0 = 0$, $a_1 = 0$, ..., $a_{p-2} = 0$. Again $a_{p-1} = A$ is arbitrary, and, if $A = 1$,

$$y = \sum_{r=0}^{\infty} \frac{(p+1-m)_r}{r!} \delta^{(p+r-1)}(x).$$

Case 4. If p is a positive integer, m is a positive integer, and $m > p$, then $a_0 = 0$, $a_1 = 0$, ..., $a_{p-2} = 0$. $a_{p-1} = A$ is arbitrary,

$$a_p = \frac{A(p+1-m)}{1},$$

$$\cdots$$

$$a_{p+n} = \frac{A(p-m+1)\ldots(p+n+1-m)}{(n+1)!}$$

$n = 1,\ldots,m-2$. But then $a_{m-1} = 0$, and $a_n = 0$, $n = m,m+1,\ldots$. Setting $A = 1$, we find

$$y = \sum_{n=-1}^{m-p-2} \frac{(p+1-m)_{n+1}}{(n+1)!} \delta^{(p+n)}(x).$$

In summary,

THEOREM. *Let* $Z+$ *denote the positive integers*.

1. *If* $p \notin Z+$, *then* $y = 0$.
2. *If* $p \in Z+$, $m \notin Z+$, *then* $y = \sum_{r=0}^{\infty} \frac{(p+1-m)_r}{r!} \delta^{(p+r-1)}(x)$.
3. *If* $p \in Z+$, $m \in Z+$, $m \leq p$, *then* $y = \sum_{r=0}^{\infty} \frac{(p+1-m)_r}{r!} \delta^{(p+r-1)}(x)$.
4. *If* $p \in Z+$, $m \in Z+$, $m > p$, *then* $y = \sum_{r=0}^{m-p-1} \frac{(p+1-m)_r}{r!} \delta^{(p+r-1)}(x)$.

In all cases $\frac{(p+1-m)_r}{r!} = \binom{p-m+r}{r}$, the binomial coefficient.

Some examples are in order.

1. If $p = 1$, $m = 2$, $y = \delta(x)$.
2. If $p = 3$, $m = 10$,

$$y = \delta^{(2)} - 6\delta^{(3)} + 15\delta^{(4)} - 20\delta^{(5)} + 15\delta^{(6)} - 6\delta^{(7)} + \delta^{(8)}.$$

3. If $p = 7$, $m = 4$,

$$y = \sum_{r=0}^{\infty} \frac{(4)_r}{r!} \delta^{(r+6)}.$$

4. If $p = 1$, $m = 3$,

$$y = \delta(x) - \delta'(x).$$

More can be said in cases 2 and 3. We recall that in the space of distributions S', the Fourier transform of $\delta^{(j)}$ is

$$\mathcal{F}\delta^{(j)} = (it)^j/\sqrt{2\pi}$$

Therefore we find

$$\begin{aligned}\mathcal{F}y &= \sum_{r=0}^{\infty} \frac{(p+1-m)_r}{\sqrt{2\pi}\; r!} (it)^{p+r-1} \\ &= \frac{(it)^{p-1}}{\sqrt{2\pi}} (1-it)^{m-1-p}.\end{aligned}$$

Taking the inverse transform,

THEOREM. _If_ $p \in Z+$, $m \notin Z+$, _then_

(1) _If_ $p-m > 0$ _and_

$$w = \frac{(-x)^{p-m} e^x H(-x)}{\Gamma(p-m-1)}$$

where

$$\begin{aligned}H(-x) &= 1, \quad -\infty < x \le 0 \\ &= 0, \quad 0 < x < \infty,\end{aligned}$$

then y _is the_ p-1 st _distributional derivative_

$$y = \frac{d^{p-1}}{dx^{p-1}} w(x).$$

(2) _If_ $0 > p-m > -1$, _then_ y _is a Cauchy regularization of the_ p-1-st _derivative of_ $w(x)$.

(3) _If_ $-1 > p-m$, _then, again_, y _is a (more complicated) Cauchy regularization of the_ p-1-st _derivative of_ $w(x)$.

The formulas are not very enlightening. We refer the reader to [7] for details.

Again some examples are in order.

1. If $p = 1$, $m = 3/2$,

$$\langle y, \phi \rangle = \int_{-\infty}^{0} (-x)^{-1/2} e^x \phi(x)\,dx.$$

2. If $p = 1$, $m = 10/3$,

$$\langle y,\phi\rangle = \int_{-\infty}^{0} (-x)^{-7/3}e^{x}[\phi(x) - \phi(0) - x\phi'(0)]dx$$
$$- (9/4)\Gamma(2/3)\langle\delta,\phi\rangle - 3\Gamma(2/3)\langle\delta',\phi\rangle.$$

THEOREM. *If* $p \in Z+$, $m \in Z+$, $m \leqq p$, *then*

$$y = \sum_{k=0}^{p-m} \frac{\binom{p-1}{k}}{(p-m-k)!} x^{p-m-k}e^{x}H(-x)$$
$$- \sum_{k=0}^{m-2} \binom{p-k-2}{m-k-2} \delta^{(k)}(x).$$

Again some examples:

1. If $p = m = 1$, $y = e^{x}H(-x)$.
2. If $p = m = 2$, $y = e^{x}H(-x)-\delta(x)$.
3. If $p = 7$, $m = 4$,

$$y = \sum_{k=0}^{3} \frac{\binom{6}{k}x^{3-k}}{(3-k)!} e^{x}H(-x) - \sum_{k=0}^{2}\binom{5-k}{2-k}\delta^{(k)}(x)$$
$$= \left[\frac{1}{6} x^3 + 3x^2 + 15x + 20\right]e^{x}H(-x)$$
$$-[10\delta(x) + 4\delta'(x) + \delta''(x)].$$

V. THE HYPERGEOMETRIC EQUATION. Ordinary solutions of the hypergeometeric equation

$$x(1-x)y'' + [\gamma-(\alpha+\beta+1)x]y' - \alpha\beta y = 0$$

are easily calculated. If γ is not a negative integer or 0,

$$y = {}_2F_1(\alpha,\beta;\gamma;x) = \sum_{n=0}^{\infty} \frac{(\alpha)_n(\beta)_n}{(\gamma)_n\, n!} x^n$$

is one solution. If γ is not a positive integer,

$$y = x^{1-\gamma}\ {}_2F_1(1-\gamma+\alpha, 1-\gamma+\beta; 2-\gamma; x),$$
$$= \sum_{n=0}^{\infty} \frac{(1-\gamma+\alpha)_n(1-\gamma+\beta)_n}{(2-\gamma)_n\, n!} x^{1-\gamma+n}$$

is another. Like ${}_1F_1$, ${}_2F_1$ has a number of interesting properties. For example, if $\alpha = 1$, $\beta = \gamma$, then ${}_2F_1(1,\gamma;\gamma;x) = (1-x)^{-1}$. Again [10] is an excellent reference.

We are interested in solutions of the type

$$y = \sum_{n=0}^{\infty} a_n \delta^{(n)}(x),$$

however. If the series is inserted in the differential equation, then

$$a_0(1-\alpha)(1-\beta) = 0,$$
$$a_n(n+1-\alpha)(n+1-\beta) + a_{n-1}(n+1-\gamma) = 0, \quad n=1,2,\dots .$$

We must examine various cases.

Case 1. If neither α or β is a positive integer, then all the coefficients are 0 and $y = 0$.

Case 2. If α is a positive integer, but β is not, then $a_0 = 0, \dots, a_{\alpha-2} = 0$, but $a_{\alpha-1}$ is arbitrary. With $a_{\alpha-1} = 0$, the recurrence relation shows

$$a_{\alpha+m-1} = \frac{(-1)^m(\alpha-\gamma+1)_m}{m!\ (\alpha-\beta+1)_m}$$

and

$$y = \sum_{m=0}^{\infty} \frac{(-1)^m(\alpha-\gamma+1)_m}{m!\ (\alpha-\beta+1)_m} \delta^{(\alpha+m-1)}(x).$$

If γ is a positive integer, $\alpha < \gamma$, the series terminates, and

$$y = \sum_{m=0}^{\gamma-\alpha-1} \frac{(-1)^m(\alpha-\gamma+1)_m}{m!\ (\alpha-\beta+1)_m} \delta^{(\alpha+m-1)}(x).$$

Case 3. If both α and β are positive integers, we may assume that $\alpha \geqq \beta$. If γ is not an integer, there is no change from case 2,

$$y = \sum_{m=0}^{\infty} \frac{(-1)^m(\alpha-\gamma+1)_m}{m!\ (\alpha-\beta+1)_m} \delta^{(\alpha+m-1)}(x).$$

If γ is an integer, $\alpha \geqq \beta > \gamma$, there is no change from case 2,

$$y = \sum_{m=0}^{\infty} \frac{(-1)^m(\alpha-\gamma+1)_m}{m!\ (\alpha-\beta+1)_m} \delta^{(\alpha+m-1)}(x).$$

If γ is an integer, $\alpha \geqq \gamma > \beta$, $a_0, \dots, a_{\beta-2}$ are 0, $a_{\beta-1}$ is arbitrary, and determines $a_\beta, \dots, a_{\gamma-2}$. $a_{\gamma-1}, \dots, a_{\alpha-2}$ are 0. $a_{\alpha-1}$ is arbitrary and determines $a_\alpha, a_{\alpha+1}, \dots$. There are two solutions

$$y_1 = \sum_{m=0}^{\gamma-\beta-1} \frac{(-1)^m(\beta-\gamma+1)_m}{m!\ (\beta-\alpha+1)_m} \delta^{(\beta+m-1)}(x),$$

and,

$$y_2 = \sum_{m=0}^{\infty} \frac{(-1)^m(\alpha-\gamma+1)_m}{m!\ (\alpha-\beta+1)_m} \delta^{(\alpha+m-1)}(x).$$

If γ is an integer, $\gamma > \alpha \geqq \beta$, then

$$y = \sum_{m=0}^{\gamma-\alpha-1} \frac{(-1)^m(\alpha-\gamma+1)_m}{m!\ (\alpha-\beta+1)_m} \delta^{(\alpha+m-1)}(x).$$

THEOREM. *Let* $Z+$ *denote the positive integers.*

1. *If* $\alpha, \beta \notin Z+$, *then* $y = 0$.
2. *If* $\alpha \in Z+$, $\beta \notin Z+$, *then*

$$y = \sum_{m=0}^{\infty} \frac{(-1)^m(\alpha-\gamma+1)_m}{m!\ (\alpha-\beta+1)_m} \delta^{(\alpha+m-1)}(x).$$

If $\alpha, \gamma \in Z+$, $\gamma > \alpha$, $\beta \notin Z+$, *then*

$$y = \sum_{m=0}^{\gamma-\alpha-1} \frac{(-1)^m(\alpha-\gamma+1)_m}{m!\ (\alpha-\beta+1)_m} \delta^{(\alpha+m-1)}(x).$$

3. *If* $\alpha, \beta \in Z+$, $\alpha \geqq \beta$, $\gamma \notin Z+$, *then*

$$y = \sum_{m=0}^{\infty} \frac{(-1)^m(\alpha-\gamma+1)_m}{m!\ (\alpha-\beta+1)_m} \delta^{(\alpha+m-1)}(x).$$

If $\alpha, \beta, \gamma \in Z+$, $\alpha \geqq \beta > \gamma$, *then*

$$y = \sum_{m=0}^{\infty} \frac{(-1)^m(\alpha-\gamma+1)_m}{m!\ (\alpha-\beta+1)_m} \delta^{(\alpha+m-1)}(x).$$

If $\alpha, \beta, \gamma \in Z+$, $\alpha \geqq \gamma \geq \beta$, *then*

$$y_1 = \sum_{m=0}^{\gamma-\beta-1} \frac{(-1)^m(\beta-\gamma+1)_m}{m!\ (\beta-\alpha+1)_m} \delta^{(\beta+m-1)}(x),$$

$\gamma > \beta$, *and*

$$y_2 = \sum_{m=0}^{\infty} \frac{(-1)^m(\alpha-\gamma+1)_m}{m!\ (\alpha-\beta+1)_m} \delta^{(\alpha+m-1)}(x)$$

<u>are solutions</u>.

<u>If</u> $\alpha,\beta,\gamma \in Z+$, $\gamma > \alpha \geqq \beta$, <u>then</u>

$$y = \sum_{m=0}^{\gamma-\alpha-1} \frac{(-1)^m(\alpha-\gamma+1)_m}{m!\ (\alpha-\beta+1)_m} \delta^{(\alpha+m-1)}(x).$$

Some examples are interesting

1. If $\alpha = \beta = \gamma = 1$,

$$y = \sum_{m=0}^{\infty} \frac{(-1)^m}{m!} \delta^{(m)}(x)$$

satisfies

$$x(1-x)y'' + (1-3x)y' - y' = 0.$$

2. If $\alpha = \beta = \gamma = 2$,

$$y = \sum_{m=0}^{\infty} \frac{(-1)^m}{m!} \delta^{(m+1)}(x)$$

satisifes

$$x(1-x)y'' + (2-5x)y' - 4y = 0.$$

3. If $\alpha = \beta = 2$, $\gamma = 1$,

$$y = \sum_{m=0}^{\infty} \frac{(-1)^m(m+1)}{m!} \delta^{(m+1)}(x)$$

satisfies

$$x(1-x)y'' + (1+5x)y' - 4y = 0.$$

4. If $\alpha = 2$, $\beta = 1$, $\gamma = 2$, then $y_1 = \delta(x)$ satisfies

$$x(1-x)y'' + (2-4x)y' - 2y = 0.$$

More can be said in cases 2 and 3 when $\alpha \geqq \gamma > \beta$ or $\beta \geqq \gamma > \alpha$. If we take the Fourier transform of

$$y = \sum_{m=0}^{\infty} \frac{(-1)^m(\alpha-\gamma+1)_m}{m!\ (\alpha-\beta+1)_m} \delta^{(\alpha+m-1)}(x),$$

we find

$$\mathfrak{F}y = \sum_{m=0}^{\infty} \frac{{}^{m}\ \alpha\ \gamma\ \ {}_{m}\ \ {}^{\alpha+m-1}}{m!\ (\alpha+\beta+1)_m \sqrt{2\pi}}$$

$$= \frac{(it)^{\alpha-1}}{\sqrt{2\pi}} \sum_{m=0}^{\infty} \frac{(\alpha-\gamma+1)_m (-it)^m}{m!\ (\alpha-\beta+1)_m}$$

$$= \frac{(it)^{\alpha-1}}{\sqrt{2\pi}}\ {}_1F_1(\alpha-\gamma+1;\alpha-\beta+1;-it).$$

Since from the Bateman tables [1],

$$\mathfrak{F}({}_1F_1(\mu;\mu+\nu;-it)/\sqrt{2\pi}) =$$

$$\frac{(1-x)^{\nu-1}(x)^{\mu-1}[H(x)-H(x-1)]}{B(\mu,\nu)}, \quad \mu > 0, \quad \nu > 0,$$

we find

THEOREM. _If_ $\alpha \in Z+$, $\alpha \geqq \gamma > \beta$, _then_

$$y = \frac{d^{\alpha-1}}{dx^{\alpha-1}} \left[\frac{(1-x)^{\gamma-\beta-1}(x)^{\alpha-\gamma}[H(x)-H(x-1)]}{B(\alpha-\gamma+1,\gamma-\beta)} \right].$$

_If_ $\alpha \in Z+$, $\alpha \geq \beta > \gamma$, _then_

$$y = \frac{d^{\alpha-1}}{dx^{\alpha-1}} \left[\frac{(1-x)^{\beta-\gamma-1}(x)^{\alpha-\beta}[H(x)-H(x-1)]}{B(\alpha-\beta+1,\beta-\gamma)} \right].$$

As a final example, let $\alpha = 2$, $\beta = 1$, $\gamma = 2$, so $\alpha \geqq \gamma > \beta$. Then

$$y = \frac{d}{dx}\,[H(x)-H(x-1)]/B(1,1),$$
$$= [\delta(x)-\delta(x-1)]/B(1,1)$$

satisfies

$$x(1-x)y'' - (2+4x)y' - 2y = 0.$$

The Fourier transform can also be used to simplify the examples.

1. If $\alpha = \beta = \gamma = 1$,

$$y = \sum_{m=0}^{\infty} \frac{(-1)^m}{m!}\,\delta^{(m)}(x) = \delta(x-1).$$

2. If $\alpha = \beta = \gamma = 2$,

$$y = \sum_{m=0}^{\infty} \frac{(-1)^m}{m!} \delta^{(m+1)}(x) = \delta'(x-1).$$

3. If $\alpha = \beta = 2$, $\gamma = 1$

$$y = \sum_{m=0}^{\infty} \frac{(-1)^m (m+1)}{m!} \delta^{(m+1)}(x) = \delta'(x-1) - \delta''(x-1).$$

VI BESSEL'S EQUATION. The ordinary solutions of the Bessel equation

$$x^2 y'' + xy' + (x^2 - \nu^2) y = 0$$

are $J_\nu(x)$ and $J_{-\nu}(x)$ when ν is not an integer, $J_\nu(x)$ and $Y_\nu(x)$ when ν is an integer. If, however

$$y = \sum_{n=0}^{\infty} a_n \delta^{(n)}(x)$$

is inserted, the recurrence relation

$$a_{n+2} = -a_n \frac{(n+1-\nu)(n+1+\nu)}{(n+1)(n+2)}$$

is recovered, and, two solutions

$$y_1 = a_0 \sum_{n=0}^{\infty} (-1)^n \frac{\prod_{j=1}^{n} (2j-1-\nu) \prod_{j=1}^{n} (2j-1+\nu)}{(2n)!} \delta^{(2n)}(x),$$

$$y_2 = a_1 \sum_{n=0}^{\infty} \frac{(-1)^n \prod_{j=1}^{n} (2j-\nu) \prod_{j=1}^{n} (2j+\nu)}{(2n+1)!} \delta^{(2n+1)}(x)$$

are found. If $\nu = 2N+1 > 0$ is an odd integer, then $a_{2N+2} = 0$, and the first series terminates. If $\nu = 2N > 0$ is an even integer, then $a_{2N+1} = 0$ and the second series terminates. For example, when

$\nu = 1$, $y_1 = a_0 \delta(x)$,

$\nu = 2$, $y_2 = a_0 \delta'(x)$,

$\nu = 3$, $y_1 = a_0(\delta(x) + 4\delta''(x))$,

$\nu = 4$, $y_2 = a_1(\delta'(x) + 2\delta'''(x))$,

$\nu = 5$, $y_1 = a_0(\delta(x) + 12\delta''(x) + 16\delta^{(iv)}(x))$,

$\nu = 6$, $y_2 = a_1(\delta'(x) + \frac{16}{3}\delta'''(x) + \frac{16}{3}\delta^{(v)}(x))$.

If $\nu = 0$, the Laplace transform [1] of y_1 is

$$\mathfrak{L}y_1 = a_0 \sum_{n=0}^{\infty} \frac{(-1)^n (2n)!}{2^{2n}(n!)^2} s^{2n} = a_0 (1+s^2)^{-1/2}.$$

Since $(1+s^2)^{-1/2}$ is the Laplace transform of $J_0(x)$, $y_1 = a_0 J_0(x)$.

Likewise if $\nu = 0$, the Laplace transform of y_2 is

$$\mathfrak{L}y_2 = a_1 \sum_{n=0}^{\infty} \frac{(-1)^n 2^{2n}(n!)^2}{(2n+1)!} s^{2n+1} = a_1 \frac{\sinh^{-1} s}{(1+s^2)^{1/2}} .$$

Since $(\sinh^{-1} s)/(1+s^2)^{1/2}$ is the Laplace transform of $-(\pi/2)Y_0(x)$, the Bessel function of the second kind, $y_2 = -a_1(\pi/2)Y_0(x)$.

For arbitrary $\nu > -1$

$$\mathfrak{L}(J_\nu) = ((1+s^2)^{1/2} - s)^\nu / (1+s^2)^{1/2}.$$

If $a_0 = 1$, $a_1 = -\nu$ and $y = y_1 + y_2$, then

$$\mathfrak{L}y = ((1+s^2)^{1/2} - s)^\nu / (1+s^2)^{1/2}$$

as well and

$$J_\nu(x) = \sum_{n=0}^{\infty} \frac{(-1)^n \prod_{j=1}^{n} (2j-1-\nu) \prod_{j=1}^{n} (2j-1+\nu) \delta^{(2n)}(x)}{(2n)!}$$
$$- \sum_{n=0}^{\infty} \frac{(-1)^n \nu \prod_{j=1}^{n} (2j-\nu) \prod_{j=1}^{n} (2j+\nu) \delta^{(2n+1)}(x)}{(2n+1)!}$$

$J_{-\nu}(x)$ is represented by the same formula with the sign on the second part reversal.

When $\nu > 0$ is an integer, Y_ν does not possess a Laplace transform. Instead the finite delta series are found. These are singular distributions.

VII FORMAL SOLUTIONS. The obvious generalization of the previous sections is to assume that

$$\left\{\sum_{j=0}^{\infty} P_j x^j\right\} y'' + \left\{\sum_{j=0}^{\infty} Q_j x^j\right\} y' + \left\{\sum_{j=0}^{\infty} R_j x^j\right\} y = 0$$

and

$$y = \sum_{n=0}^{\infty} a_n \delta^{(n)}(x).$$

After considerable algebra, a recurrence relation is found:

$$\sum_{k=r}^{\infty} [P_{k-r+2}(k+2)(k+1) - Q_{k-r+1}(k+1) + R_{k-r}](-1)^k k! a_k = 0,$$

$r = 0,1,\ldots$.

If $P_\gamma = 0$, $\gamma = L+1,\ldots, Q_\gamma = 0$, $\gamma = M+1,\ldots$, $R_\gamma = 0$, $\gamma = N+1,\ldots$, and

$$T = \sup\{L-2, M-1, N\},$$

then the recurrence relation becomes

$$\sum_{k=r}^{T+r} [P_{k-r+2}(k+2)(k+1) - Q_{k-r+1}(k+1) + R_{k-r}](-1)^k k! a_k = 0,$$

and

$$a_{T+r} = - \frac{\sum_{k=r}^{T+r-1} [P_{k-r+2}(k+2)(k+1) - Q_{k-r+1}(k+1) + R_{k-r}](-1)^k k! a_k}{[P_{T+2}(T+2+r)(T+1+r) - Q_{T+1}(T+r+1) + R_T](-1)^{T+r}(T+r)!}$$

$r = 0,1,\ldots$.

The only difficulties in determining the coefficients occur when the denominator vanishes. Different circumstances may arise, as we have seen, leading to different sorts of series. The technique we have used can surely be extended to other problems. We refer the reader to [4] for some interesting additional examples.

REFERENCES

1. H. Bateman, TABLES OF INTEGRAL TRANSFORMS, vol. 1, McGraw Hill, New York, 1954.

2. H. Bremermann, DISTRIBUTIONS, COMPLEX VARIABLES, AND FOURIER TRANSFORMS, Addison-Wesley, Reading, Mass. 1965.

3. E. Grosswald, BESSEL POLYNOMIALS, Lect. Notes Math. 698, Springer-Verlag, Heidelberg, 1978.

4. R.P. Kanwal, GENERALIZED FUNCTIONS: THEORY AND TECHNIQUE, Academic Press, New York, 1983.

5. A.M. Krall, "Orthogonal polynomials through moment generating functionals," SIAM J. Math. Anal. 9(1978), 600-603.

6. H.L. Krall and O. Frink, "A new class of orthogonal polynomials: The Bessel polynomials," Trans. Amer. Math. Soc. 65(1949), 100-115.

7. L.L. Littlejohn and R.P. Kanwal, "Distributional solutions of the hypergeometric equations", J. Math. Analys. Applics., to appear.

8. R.D. Morton and A.M. Krall, "Distributional weight functions for orthogonal polynomials", SIAM J. Math. Anal. 9(1978), 604-626.

9. O. Perron, DIE LEHRE VON DEN KETTENBRÜCHEN, B. G. Teubner, Leibzig and Berlin, 1929.

10. E.D. Rainville, SPECIAL FUNCTIONS, Macmillan, New York, 1960.

11. I. Stakgold, BOUNDARY VALUE PROBLEMS OF MATHEMATICAL PHYSICS, Macmillan, New York, 1968.

Canadian Mathematical Society
Conference Proceedings
Volume 8 (1987)

STURM-LIOUVILLE OPERATORS AND ORTHOGONAL POLYNOMIALS

A. M. KRALL
and
L. L. LITTLEJOHN

ABSTRACT. In 1929, S. Bochner classified all orthogonal polynomial solutions of the Sturm-Liouville equation $-(py')' + qy = \lambda\omega y$ and implicitly asked the question: can we classify all orthogonal polynomial solutions to higher order differential equations? These differential equations are of considerable interest because they serve as models for the current research done in singular boundary value problems. In particular, when the order r of the equation is greater than 2, a λ-dependent boundary condition is needed. In this paper, we wish to discuss recent results of Littlejohn and Littlejohn and Krall. Besides this, we prove a new result: under rather general conditions, if $\{\phi_m(x)\}$ is a sequence of orthogonal polynomial solutions to

$$L_{2n}(y) = : \sum_{i=1}^{r} a_i(x)y^{(i)}(x) = \lambda y(x) ,$$

then $L_{2n}(y)$ can be made symmetric. Thus the well-developed theory of self adjoint extensions of symmetric operators may be applied.

§1 INTRODUCTION

Recently, there has been increasing interest in finding differential equations having a sequence of orthogonal polynomial solutions. (e.g. see [9], [10], [11]). The differential equations themselves serve as models to illustrate the general limit-m theory of singular differential equations. Moreover, the known differential equations can be made formally self adjoint; consequently, boundary conditions must be prescribed at one or more

AMS SUBJECT CLASSIFICATION: Primary 33A65, 34B20, 34B25, 34B30 of the endpoints of the interval of orthogonality in order to pose a self adjoint boundary problem. Thus, from the viewpoint of singular boundary value problems, these differential equations are important. In fact, when the order of the equation is greater than two, it is generally true that at least one of these boundary conditions is a λ-dependent boundary condition. The subject of λ-dependent boundary value problems is of much current interest; for example see [3], [4], [5] and [13].

Salomon Bochner, in 1929 [3], essentially asked the following question: can we classify all differential equations (up to a linear change of variable) of the form:

$$(1.1) \qquad \sum_{i=1}^{r} a_i(x)y^{(i)}(x) = \lambda y(x)$$

that have an orthogonal polynomial sequence (OPS) as eigenfunctions. He realized that it is necessary that $a_i(x)$ be a polynomial of degree $\leq i$. Using this fact and studying the location of the roots of $a_2(x)$, Bochner solved this problem when $r=2$. He showed that there are only four sets of orthogonal polynomials satisfying a second order equation of the form (1.1): the Jacobi, Laguerre, Hermite and another sequence that was not explicitly found until 1949 - the Bessel polynomials. In 1940, H.L. Krall solved this classification problem when $r=4$, finding three new OPS's [8]. In order to obtain this result, Krall used his 1938 "classification" theorem [7]:

<u>THEOREM 1</u>

Suppose $\phi_m(x)$ is a polynomial of degree m, $m=0,1,\ldots$. Then $\{\phi_m(x)\}$ is an OPS and $\phi_m(x)$ satisfies

$$(1.1) \qquad \sum_{i=1}^{r} a_i(x)y^{(i)}(x) = \lambda_m y(x) \quad , \quad m=0,1,\ldots$$

where

$$a_i(x) = \sum_{j=0}^{i} \ell_{ij}x^j \quad , \quad i=1,2,\ldots r,$$

and

$$\lambda_m = m\ell_{11} + m(m-1)\ell_{22} + \ldots + m(m-1)\ldots(m-r+1)\ell_{rr} ,$$

if and only if the moments $\{\mu_m\}$ satisfy:

i)
$$\begin{vmatrix} \mu_0 & \mu_1 & \cdots & \mu_m \\ \mu_1 & \mu_2 & \cdots & \mu_{m+1} \\ \vdots & & & \\ \mu_m & \mu_{m+1} & \cdots & \mu_{2m} \end{vmatrix} \neq 0 \quad , \quad m=0,1,\ldots \quad \underline{\text{AND}}$$

(ii)

$$(1.2) \quad S_k(m) = \sum_{i=2k+1}^{r} \sum_{u=0}^{i} \begin{bmatrix} i-k-1 \\ k \end{bmatrix} P(m-2k-1,i-2k-1)\ell_{i,i-u}\,\mu_{m-u} = 0 ,$$

$2k+1 \leq r$, $m = 2k+1, 2k+2,\ldots$, where $P(n,k) = n(n-1)\ldots(n-k+1)$. Furthermore, r is necessarily even. That is to say, if r is the smallest order of a differential equation of the form (1.1) having $\{\phi_m(x)\}$ as solutions, then r is even. □

From this last statement, we shall assume henceforth that $r = 2n$. Thus, there are n recurrence relations in (1.2); we shall call these n equations the MOMENT EQUATIONS FOR $\{\phi_m(x)\}$.

There are now known ten differential equations of the form (1.1) having orthogonal polynomial solutions. Littlejohn found two sixth order differential equations [9], [11] and he and A. M. Krall have very recently discovered an eighth order example [12].

In 1984, Littlejohn discovered a method [10] for constructing an orthogonalizing weight distribution for an OPS that satisfies an equation of type (1.1). More specifically, he showed that, under certain conditions, a weight may be found by simultaneously solving a system of distributional differential equations. Recently, this method was improved by Krall and Littlejohn [12]. We shall briefly discuss these improvements in the next section and then show how this theory leads to some applications in §3. Lastly, in §4, we develop a new result that says, under rather general conditions, if an OPS satisfies an equation of the form (1.1), then this equation may be made formally self adjoint. Consequently, the differential equation is a Sturm-Liouville equation and therefore the theory of self adjoint extensions of symmetric operators may be applied.

§2 BRIEF SUMMARY OF PREVIOUS RESULTS

DEFINITION 1

Let

$$L(y) = \sum_{k=0}^{r} a_k(x) y^{(k)}(x) .$$

If $f(x)L(y)$ is symmetric (formally self adjoint), then $f(x)$ is called a SYMMETRY FACTOR for $L(y)$.

It is well-known that if $a_k(x)$ is real valued, $k = 0,1,\ldots r$, then r is necessarily even. In [12], the following result was established.

THEOREM 2

Let

$$L(y) = \sum_{k=0}^{2n} a_k(x) y^{(k)}(x) ,$$

where $a_k \in C^k(I)$ is real valued, $k = 0,1,\ldots 2n$ and $a_{2n}(x) \neq 0$ for all $x \in I$, I being some interval of the real line R. Then $f(x)$ is a symmetry factor for $L(y)$ if and only if $f(x)$ simultaneously satisfies the following system of n homogeneous differential equations:

$$(2.1) \qquad \sum_{i=2k+1}^{2n} (-1)^i \begin{bmatrix} i-k-1 \\ k \end{bmatrix} \Big[a_i(x) f(x) \Big]^{(i-2k-1)} = 0 , \quad k = 0,1,\ldots n-1. \quad \square$$

These n equations are called the SYMMETRY EQUATIONS for $L(y)$. We remark that these equations have orders $1,3,\ldots(2n-1)$. In fact, the first order equation is $na_{2n}(x)y'(x) + (na'_{2n}(x)-a_{2n-1}(x))y(x) = 0$ and consequently

$$f(x) = \frac{\exp\left[\frac{1}{n} \int^x \frac{a_{2n-1}(t)dt}{a_{2n}(t)} \right]}{a_{2n}(x)} .$$

DEFINITION 2

A distribution Λ weakly satisfies a system $A(\Lambda) = 0$ of equations if $\langle A(\Lambda), \phi \rangle = 0$ for all polynomials ϕ. Here, the notation $\langle \Lambda, \phi \rangle$ denotes the action of the distribution Λ on the test function ϕ.

THEOREM 3

Suppose $\{\phi_m(x)\}$ is an OPS that satisfies the differential equation (1.1) of order $r = 2n$. Suppose Λ is a nontrivial weak solution to the system (2.1) and suppose $\Lambda(x) \to 0$ as $|x| \to \infty$. Then Λ is an orthogonalizing weight distribution for $\{\phi_m(x)\}$. □

A proof of this theorem may be found in [12]. If the equations (2.1) are solved weakly on polynomials, we shall refer to these equations as the WEIGHT EQUATIONS for $\{\phi_m(x)\}$.

EXAMPLE

The Legendre type polynomials satisfy the fourth order equation

$$(x^2-1)^2y^{(4)} + 8x(x^2-1)y^{(3)} + (4\alpha+12)(x^2-1)y'' + 8\alpha xy' = \lambda_m y ,$$

where

$$\lambda_m = m^4 + 2m^3 + (4\alpha-1)m^2 + (4\alpha-2)m.$$

In this case, the weight equations are:

$$\text{(i)} \quad (x^2-1)^2\Lambda' = 0$$

$$\text{(ii)} \quad (x^2-1)^2\Lambda^{(3)} + 4x(x^2-1)\Lambda'' + (4\alpha x^2-4\alpha-8)\Lambda' = 0.$$

These are easily solved to yield, up to a constant multiple,

$$\Lambda(x) = \alpha/2 + \tfrac{1}{2}\delta(x+1) + \tfrac{1}{2}\delta(x-1) , \quad -1 < x < 1 .$$

§4 APPLICATIONS

(A) APPLICATION OF A RESULT OF A.M. KRALL AND R.D. MORTON

In [6], Krall and Morton show that, formally at least, the "distribution"

$$w(x) = \sum_{m=0}^{\infty} \frac{(-1)^m \mu_m \delta^{(m)}(x)}{m!} \tag{4.1}$$

generates orthogonality of an OPS with moment sequence $\{\mu_m\}$. Using the Fourier transform, they were able to obtain classical representations of (4.1) in most cases. In [12], Littlejohn and Krall proved the following result:

THEOREM 4

Suppose $\{\phi_m(x)\}$ is an OPS with corresponding moment sequence $\{\mu_m\}$. Suppose $\{\phi_m(x)\}$ satisfies (1.1). Then $\omega(x)$ given in (4.1) formally satisfies the weight equations (2.1). □

(B) APPLICATION OF A RESULT OF ATKINSON AND EVERITT

A problem of related interest to the classification problem that we are considering is the following: find all differential equations of the form

$$A_2(x,m)y''(x) + A_1(x,m)y'(x) + A_0(x,m)y(x) = 0 \tag{4.2}$$

that have an OPS for solutions. In [1], Atkinson and Everitt prove the following important result:

THEOREM 5

Suppose $\{\phi_m(x)\}$ is an OPS with respect to $d\mu(x) = d\mu_1(x) + d\mu_2(x)$ where $d\mu_1(x) = f(x)dx$ and $\mu_2(x)$ is a nondecreasing step function with a finite number of jumps. Suppose $f(x)$ is non-negative and continuous on some interval (a,b), where $-\infty \leq a < b \leq \infty$, and suppose $f(x)$ satisfies a first order equation

$$\sigma_1 f' + \sigma_0 f = 0 ,$$

where σ_0 and σ_1 are polynomials. Then $\phi_m(x)$, $m = 0,1,\ldots,$ satisfies a differential equation of the form (4.2) with polynomial coefficients. □

From this, we get the following theorem [12]:

THEOREM 6

Suppose μ and f satisfy the conditions of Theorem 5. Suppose $\{\phi_m(x)\}$ is an OPS with respect to $\Lambda(x)dx = d\mu(x)$ and $\phi_m(x)$, $m = 0,1,\ldots,$ satisfies an equation of the form (1.1) of order $r = 2n$. Assume $\Lambda(x)$ satisfies the weight equations (2.1). Then $\{\phi_m(x)\}$ satisfies a second order equation of the form (4.2). □

The proof of this theorem is trivial since the first order weight equation is

$$na_{2n}(x)y'(x) + (na_{2n}'(x)-a_{2n-1}(x))y(x) = 0 .$$

(C) APPLICATION: HOW TO FIND NEW ORTHOGONAL POLYNOMIAL SEQUENCES SATISFYING DIFFERENTIAL EQUATIONS OF THE FORM (1.1)

So far, this paper has shown how to find an orthogonalizing weight distribution given the differential equation. This process may also be reversed. Indeed, to illustrate this statement, let

$$w(x) = x^2e^{-x} + (1/R)\delta(x) \quad , \quad x > 0 . \tag{4.3}$$

Since the polynomials orthogonal with respect to $e^{-x} + (1/R)\delta(x)$ satisfy a fourth order equation and the polynomials orthogonal with respect to $xe^{-x} + (1/R)\delta(x)$ satisfy a sixth order equation (see [10] and [11]), we shall assume the polynomials orthogonal with respect to (4.3) satisfy an eighth order equation $L_8(y) = \lambda_m y$. In addition, we shall assume

(i) $x^2e^{-x}L_8(y)$ is symmetric, and

(ii) $w(x)$ satisfies the four weight equations (2.1).

We leave out the details; they are much too tedious for this article. The reader is encouraged to consult [11] for similar calculations. We find that the differential equation is $L_8(y) =:$

$$x^4y^{(8)} + (-4x^4+24x^3)y^{(7)} + (6x^4-84x^3+168x^2)y^{(6)} + (-4x^4+108x^3-504x^2+336x)y^{(5)}$$
$$+ (x^4-60x^3+540x^2-840x)y^{(4)} + (12x^3-240x^2+720x)y^{(3)} + (36x^2-(48R+240)x)y''$$
$$+ ((24+48R)x-144R)y' = (m^4+6m^3+11m^2+(6+48R)m)y \ .$$

The polynomials are easily found to be

$$\phi_m(x) = \frac{(m+2)!}{2} \sum_{k=0}^{m} \frac{(-1)^k[12R(k+3)+(m+1)(m+2)(m+3)k]x^k}{k!(k+3)!(m-k)!} \ .$$

Applying the method of Frobenius, we see that $x=0$ is in the limit-6 case and $x=\infty$ is in the limit-4 case. Consequently, two boundary conditions are necessary at $x=0$ for a self adjoint boundary value problem. If we let $[y,\bar{z}]'$ denote the right side of Lagrange's identity, then the appropriate boundary conditions may be prescribed using the concomitant $[y,\bar{z}]$. For "good" functions, we see that

$$[y,1](0) = -144y'(0)$$
$$[y,x](0) = 144y(0)$$
$$L_8(y)(0) = 144Ry'(0) = \lambda y(0) \ .$$

Hence, one boundary condition at $x=0$ is the λ-dependent condition

$$144R[y,1](0) = \lambda[y,x](0) \ .$$

The other boundary condition is $\lim_{x\to 0} [y,x^2] = 0$.

Define $H = L^2(0,\infty;x^2e^{-x}) \otimes \mathbb{C}$. Vectors $Y,Z \in H$ are given by

$$Y = (y(x),\ y_0)^T$$
$$Z = (z(x),\ z_0)^T$$

and an inner product on H is defined by:

$$(Y,Z) = \int_0^\infty y(x)\bar{z}(x)x^2e^{-x}dx + \frac{1}{R}\, y_0\bar{z}_0 \ .$$

DEFINITION 3

We denote by D_A those elements $Y = (y(x),\ y_0)^T \in H$ satisfying

(i) $y \in L^2(0,\infty;x^2e^{-x})$

(ii) y , y' , y'' , $y^{(3)}$, $y^{(4)}$, $y^{(5)}$, $y^{(6)}$, $y^{(7)}$ exist and $y^{(7)}$ is absolutely continuous on compact subsets of $(0,\infty)$

(iii) $L_8(y)$ exists a.e. and is in $L^2(0,\infty;x^2e^{-x})$

(iv) $y_0 = [y,x](0)$

(v) $\lim_{x\to 0} [y,x^2] = 0$

Define the operator A on D_A by $A(Y) = \begin{bmatrix} L_8(y) \\ 144R[y,1](0) \end{bmatrix}$

THEOREM 7

A in H is self adjoint. □ (see [13])

§5 A NEW APPLICATION: SYMMETRY OF THE DIFFERENTIAL EQUATIONS

THEOREM 8

Suppose $\{\phi_m(x)\}$ is an OPS with respect to $\Lambda(x)dx = d\mu(x) = f(x)dx + d\mu_1(x)$ on an interval $I \subset R$ where $f(x) \geq 0$, $x \in I$, is continuous and $\mu_1(x)$ is a nondecreasing step function with a finite number of jumps.

(a) If I is compact and $\{\phi_m(x)\}$ satisfies the differential equation

$$L_{2n}(y) = \sum_{i=1}^{2n} a_i(x)y^{(i)}(x) = \lambda_m y(x)$$

and if $f \in C^{2n-1}(I)$, then $f(x)L_{2n}(y)$ is symmetric.

(b) Suppose $\Lambda(x)$ simultaneously satisfies the equations

$$\sum_{i=2k+1}^{2n} (-1)^i \begin{bmatrix} i-k-1 \\ k \end{bmatrix} (a_i\Lambda)^{(i-2k-1)} = 0 ,$$

$k = 0,1,\ldots n-1$, weakly on polynomials where $a_i(x)$ is a polynomial of degree $\leq i$, $i = 1,2,\ldots 2n$, and suppose $\Lambda(x) \to 0$ as $|x| \to \infty$. Then $\{\phi_m(x)\}$ satisfies

$$L_{2n}(y) = \sum_{i=1}^{2n} a_i(x)_y{}^{(i)}(x) = \lambda_m y(x)$$

and $f(x)L_{2n}(y)$ is symmetric.

PROOF

(a) Suppose $\{\phi_m(x)\}$ is orthogonal on $I = [a,b]$. For any $p,m = 0,1,\ldots$, we have

$$0 = (\lambda_p-\lambda_m)\int_a^b \phi_p\phi_m\Lambda dx$$

$$= \int_a^b (\lambda_p\phi_p\phi_m - \lambda_m\phi_m\phi_p)\Lambda dx$$

$$= \int_a^b [L_{2n}(\phi_p)\phi_m - \phi_p L_{2n}(\phi_m)]\Lambda dx$$

That is to say, we have:

$$(5.1) \qquad 0 = \int_a^b \left[L_{2n}(\phi_p)\phi_m - \phi_p L_{2n}(\phi_m)\right] f dx + \int_a^b \left[L_{2n}(\phi_p)\phi_m - \phi_p L_{2n}(\phi_m)\right] d\mu_1(x)$$

On the other hand, Green's formula asserts

$$(5.2) \qquad \int_a^b \left[fL_{2n}(\phi_p)\phi_m - (fL_{2n})^+(\phi_m)\phi_p\right] dx = \left[\tilde{k}(\phi_m)Pk(\phi_p)\right]_a^b, \text{ where}$$

$$k(\phi) = \begin{bmatrix} \phi \\ \phi' \\ \vdots \\ \phi^{(2n-1)} \end{bmatrix},$$

$\tilde{k}(\phi)$ denotes the transpose of $k(\phi)$, P is the $2n \times 2n$ concomitant matrix and $(fL_{2n})^+$ is the Lagrange adjoint of fL_{2n}. Subtracting (5.1) from (5.2) yields

$$(5.3) \qquad \int_a^b \left[fL_{2n}(\phi_m) - (fL_{2n})^+(\phi_m)\right]\phi_p dx = \left[\tilde{k}(\phi_m)Pk(\phi_p)\right]_a^b + \int_a^b \left[L_{2n}(\phi_p)\phi_m - L_{2n}(\phi_m)\phi_p\right] d\mu_1(x) .$$

Observe that (5.3) holds with ϕ_p replaced by any polynomial ϕ. To show $fL_{2n} = (fL_{2n})^+$, it suffices to prove that $fL_{2n}(\phi_m) = (fL_{2n})^+(\phi_m)$, $m = 0,1,\ldots$.

Fix $m \geq 0$ and suppose there exists $x_0 \in (a,b)$ such that

$$fL_{2n}(\phi_m(x_0)) - (fL_{2n})^+(\phi_m(x_0)) > 0 .$$

Since $fL_{2n}(\phi_m) - (fL_{2n})^+(\phi_m)$ is continuous, there exists $\epsilon > 0$ and an interval $I' = (a', b') \subset I$ containing x_0 such that

$$fL_{2n}(\phi_m(x)) - (fL_{2n})^+(\phi_m(x)) > \epsilon \text{ for all } x \in I' .$$

Without loss of generality, assume $\mu_1(x)$ has no jumps in the closure of I' (if x_0 is a jump of μ_1, modify I' slightly: i.e. let $I' = (x_0+\delta, b')$ for some sufficiently small $\delta > 0$).

Let $\varphi(x) \geq 0$ be any C^∞ function with compact support $[a', b']$ such that $\varphi(x) = 1$ for $x \in I'' = [a'', b''] \subset [a', b']$

Since polynomials are dense in $C[a,b]$, there exists a sequence of polynomials $\{p_k(x)\}$ such that $p_k(x) \to \varphi^{(2n-1)}(x)$ uniformly on $[a,b]$.

Define

$$p_{k,-1}(x) = \int_a^x p_k(x)dx \qquad k = 0,1,\ldots$$

Since

$$\begin{aligned} |p_{k,-1}(x)-\varphi^{(2n-2)}(x)| &= \left| \int_a^x p_k(x)dx - \int_a^x \varphi^{(2n-1)}(x)dx \right| \\ &\leq \int_a^x |p_k(x)-\varphi^{(2n-1)}(x)|dx \\ &\leq \|p_k-\varphi^{(2n-1)}\|_\infty (b-a) \to 0 \text{ as } k \to \infty , \end{aligned}$$

we see that $p_{k,-1}(x) \to \varphi^{(2n-2)}(x)$ uniformly on $[a,b]$.

By induction, the polynomials

$$p_{k,-j}(x) =: \int_a^x p_{k,-j+1}(x)dx \qquad (p_{k,0} = p_k) , \; j = 1,2,\ldots$$

have the property that $p_{k,-j}(x) \to \varphi^{(2n-j-1)}(x)$ uniformly on $[a,b]$ as $k \to \infty$.

Define $\varphi_k(x) = p_{k,-2n+1}(x)$; by construction, $\varphi_k^{(j)}(x) \to \varphi^{(j)}(x)$ uniformly on $[a,b]$, $j = 0,1,\ldots(2n-1)$.

Hence,

$$\begin{aligned} 0 < \epsilon(b'' - a'') &< \int_{a''}^{b''} \left[fL_{2n}(\phi_m) - (fL_{2n})^+(\phi_m)\right]dx \\ &= \int_{a''}^{b''} \left[fL_{2n}(\phi_m) - (fL_{2n})^+(\phi_m)\right]\varphi(x)dx \\ &\leq \int_{a'}^{b'} \left[fL_{2n}(\phi_m) - (fL_{2n})^+(\phi_m)\right]\varphi(x)dx \\ &= \int_a^b \left[fL_{2n}(\phi_m) - (fL_{2n})^+(\phi_m)\right]\varphi(x)dx \end{aligned}$$

$$= \lim_{k\to\infty} \int_a^b \left[fL_{2n}(\phi_m) - (fL_{2n})^+(\phi_m)\right]\psi_k(x)dx$$

$$= \lim_{k\to\infty} \left\{ \left[\tilde{k}(\phi_m)Pk(\psi_k)\right]_a^b + \int_a^b \left[L_{2n}(\psi_k)\phi_m - L_{2n}(\phi_m)\psi_k\right]d\mu_1(x)\right\}$$

$$= \left[\tilde{k}(\phi_m)Pk(\psi)\right]_a^b + \int_a^b \left[L_{2n}(\psi)\phi_m - L_{2n}(\phi_m)\psi\right]d\mu_1(x)$$

$= 0$, a contradiction.

Hence $(fL_{2n})(\phi_m) = (fL_{2n})^+(\phi_m)$, $m = 0,1,2,\ldots$ and, consequently, fL_{2n} is symmetric.

(b) Suppose $\Lambda(x)$ satisfies $\displaystyle\sum_{i=2k+1}^{2n} (-1)^i \begin{bmatrix} i-k-1 \\ k \end{bmatrix} \left[a_i(x)\Lambda(x)\right]^{(i-2k-1)} = 0$

weakly on polynomials ϕ .

Since $\Lambda(x) \to 0$ when $|x| \to \infty$,

$$0 = \left\langle \sum_{i=2k+1}^{2n} (-1)^i \begin{bmatrix} i-k-1 \\ k \end{bmatrix} (a_i\Lambda)^{(i-2k-1)} , \phi \right\rangle$$

$$= \left\langle \Lambda, \sum_{i=2k+1}^{2n} \begin{bmatrix} i-k-1 \\ k \end{bmatrix} a_i\phi^{(i-2k-1)} \right\rangle, \text{ for all polynomials } \phi$$

Let $A_{k,\phi} = \displaystyle\sum_{i=2k+1}^{2n} \begin{bmatrix} i-k-1 \\ k \end{bmatrix} a_i\phi^{(i-2k-1)}$ so $\langle \Lambda, A_{k,\phi} \rangle = 0$ for all

polynomials ϕ, $k = 0,1,\ldots(n-1)$.

It is not too difficult to establish:

$$0 = \sum_{k=0}^{n-1} (-1)^k \left\langle \Lambda, A_{k,\phi_p^{(k+1)}\phi_m^{(k)} - \phi_m^{(k+1)}\phi_p^{(k)}} \right\rangle$$

$$= \langle \Lambda, \ \phi_m L_{2n}(\phi_p) - \phi_p L_{2n}(\phi_m) \rangle$$

$$= \langle \Lambda, \ \phi_m L_{2n}(\phi_p) - \phi_p a_{2n}\phi_m^{(2n)} - \phi_p a_{2n-1}\phi_m^{(2n-1)} - \dots - \phi_p a_1 \phi_m' \rangle$$

But $\langle \Lambda, \ \phi_p\left[a_{2n}\phi_m^{(2n)}\right] \rangle = \langle \Lambda, \ \phi_p\left[a_{2n-1}\phi_m^{(2n-1)}\right] \rangle = \dots = \langle \Lambda, \ \phi_p(a_1\phi_m') \rangle$

$= 0$, since ϕ_p is orthogonal to all polynomials of degree $< p$.

Hence,

$\langle \Lambda, \ \phi_m L_{2n}(\phi_p) \rangle = 0$ for any $0 \le m < p$.

i.e. $L_{2n}(\phi_p)$ is orthogonal to $\phi_0, \phi_1, \dots \phi_{p-1}$ so $L_{2n}(\phi_p)$ must be a multiple of ϕ_p, i.e. $L_{2n}(\phi_p) = \lambda_p \phi_p$.

If we write

$$a_i(x) = \sum_{j=0}^{i} \ell_{ij} x^j ,$$ comparison of coefficients yields

$$\lambda_p = p\ell_{11} + p(p-1)\ell_{22} + \dots + p(p-1)\dots(p-2n+1)\ell_{2n,2n} .$$

Finally, the solution to the system

$$\sum_{i=2k+1}^{2n} (-1)^i \begin{bmatrix} i-k-1 \\ k \end{bmatrix} (a_i \Lambda)^{(i-2k-1)} = 0 \tag{5.4}$$

is classical if we restrict x to an interval where there are no jumps of $\mu_1(x)$. If we further restrict this interval (if necessary) so that $a_{2n}(x) \neq 0$, we see that $f(x)$ is a classical solution to (5.4). Hence, by Theorem 2, $f(x)L_{2n}(y)$ is symmetric. □

BIBLIOGRAPHY

1. F.V. Atkinson and W.N. Everitt, "Orthogonal Polynomials which satisfy second order differential equations," in E.B. Christoffel, ed. P.L. Butzer and F. Feher, Birkhauser, Basel, 1981, 173-181.

2. S. Bochner, "Uber Sturm-Liouvillesche Polynomsysteme," Math. Z., 29, 1929, 730-736.

3. H.S.V. de Snoo, H. Langer, A. Dijksma, "Boundary problems with boundary conditions that contain the eigenvalue parameter," Colloquium talk at Utah State University, July, 1986.

4. C.T. Fulton, "Parametrizations of Titchmarsh's $m(\lambda)$-functions in the limit circle case," Trans. Amer. Math. Soc., 229, 1977, 51-63.

5. ________ and A.M. Krall, "Self-adjoint 4th order boundary value problems in the limit-4 case," Ord. Diff. Eq. and Oper. Conference, Dundee, 1982, W.N. Everitt and R.T. Lewis eds., Lecture Notes in Math., No. 1032, Springer-Verlag, New York.

6. A.M. Krall and R.D. Morton, "Distributional weight functions for orthogonal polynomials," SIAM J. Math. Anal.,9, 1978, 604-626.

7. H.L. Krall, "Certain differential equations for Tchebycheff polynomials," Duke Math. J., 4, 1938, 705-718.

8. ________, "On orthogonal polynomials satisfying a certain fourth order equation," The Pennsylvania State College Studies, No. 6, The Pennsylvania State College, State College, Pa., 1940.

9. L.L. Littlejohn, "The Krall polynomials: a new class of orthogonal polynomials," Quaestiones Math., 5, 1982, 255-265.

10. ________, "On the classification of differential equations having orthogonal polynomial solutions," Ann. Mat. Pura Appl., 4, 1984, 35-53.

11. ________, "An application of a new theorem in orthogonal polynomials and differential equations," Quaestiones Math., 10, 1986, 49-61.

12. ________, and A.M. Krall, "On the classification of differential equations having orthogonal polynomial solutions," submitted.

13. ________, "Orthogonal polynomials and singular Sturm-Liouville systems II," submitted.

ALLAN M. KRALL
DEPARTMENT OF MATHEMATICS
PENN STATE UNIVERSITY
UNIVERSITY PARK, PA 16802

LANCE L. LITTLEJOHN
DEPARTMENT OF MATHEMATICS
UTAH STATE UNIVERSITY
LOGAN, UT 84322-4125

Canadian Mathematical Society
Conference Proceedings
Volume 8 (1987)

LINEARIZED OSCILLATION THEORY FOR SECOND ORDER DIFFERENTIAL EQUATIONS

M. R. S. Kulenovic[1,2] and G. Ladas[1]

ABSTRACT. Consider the delay differential equation

$$\ddot{y}(t) + \alpha\dot{y}(t) + \beta f(y(t-r)) = 0, \quad t \geq 0 \tag{1}$$

where f is a continuous function such that

$$uf(u) > 0 \quad \text{for} \quad u \neq 0 \quad \text{and} \quad \lim_{|u|\to\infty} \frac{f(u)}{u} = 1.$$

We obtained linearized oscillation results, that is, we found conditions for the oscillation of all solutions of Eq. (1) in terms of the characteristic equation

$$\lambda^2 + \alpha\lambda + \beta e^{-\lambda r} = 0$$

of the linearized equation

$$\ddot{z}(t) + \alpha\dot{z}(t) + \beta z(t - r) = 0.$$

1. INTRODUCTION

Recently, Kulenović and Ladas [4] obtained linearized oscillation results for the delay differential equation (DDE)

$$\ddot{y}(t) + \alpha\dot{y}(t) + \beta f(y(t-r)) = 0, \ t \geq 0 \tag{1}$$

where

$$\alpha,\beta,r \in R^+ \tag{2}$$

and f is a continuous function such that

$$uf(u) > 0 \quad \text{for} \quad u \neq 0 \tag{3}$$

and

$$\lim_{u\to 0} \frac{f(u)}{u} = 1. \tag{4}$$

AMS 1980 Subject Classification: Primary - 34K15; Secondary - 34C10

[1]Department of Mathematics, University of Rhode Island, Kingston, R.I. 02881, U.S.A.

[2]Department of Mathematics, University of Sarajevo, 71000 Sarajevo, Yugoslavia.

More precisely, we found conditions for the oscillation of Eq. (1) in terms of the characteristic equation

$$\lambda^2 + \alpha\lambda + \beta e^{-\lambda r} = 0 \tag{5}$$

of the linearized equation

$$\ddot{z}(t) + \alpha\dot{z}(t) + \beta z(t-r) = 0. \tag{6}$$

In this paper we are concerned with linearized oscillation results for Eq. (1) when

$$\alpha < 0, \quad \beta > 0 \quad \text{and} \quad r \geqq 0. \tag{7}$$

Changing the sign of α causes a substantial change in the asymptotic behavior of the nonoscillatory solutions of Eq. (1) and to obtain linearized oscillation results, Condition (4) will be replaced by

$$\lim_{|u|\to\infty} \frac{f(u)}{u} = 1. \tag{8}$$

For first order equations, linearized oscillation results were recently obtained by Kulenović, Ladas and Meimaridou [3] and Kulenović, Ladas and Sficas [5].

The oscillation theory of Eq. (1) was extensively developed by several authors, when the function f satisfies (3) and some additional, strongly nonlinear conditions such that

$$\int_{\pm 1}^{\pm\infty} du/f(u) < \infty \quad \text{or} \quad \int_{\pm 0}^{\pm 1} du/f(u) < \infty. \tag{9}$$

The main characteristic of these results is the independence of the oscillatory behavior of solutions of Eq. (1) from the delay r. See [2] and [6]. The main feature of our results is that the oscillation of Eq. (1) depends strongly on the delay r. Another important feature of our results is that they hold under the Condition (8), which is, in a sense, complementary to the Conditions (9). As far as we know this is the first oscillation result for Eq. (1) subject to Condition (8).

The following fixed point theorem of Knaster and Tarski [7] will be used in this paper.

LEMMA 0. (Knaster-Tarski fixed point theorem) Let E be a partially ordered set with ordering $\leqq$. Assume that inf E $\in$ E and that every non-empty subset of E has a supremum (which belongs to E). Let S: E $\to$ E be an increasing mapping (that is, $x \leqq y$ implies $Sx \leqq Sy$). Then S has a fixed point in E.

As is customary, a solution of Eq. (1) is called oscillatory if it has arbitrarily large zeros and nonoscillatory if it is eventually positive or eventually negative.

2. MAIN RESULT

In order to prove our main result we will first establish two lemmas.

LEMMA 1. Consider the DDE (1) and assume that Conditions (3) and (7) are satisfied. Let $y(t)$ be an eventually positive solution of Eq. (1). Then, eventually

$$\dot{y}(t) > 0 \tag{10}$$

and

$$\lim_{t\to\infty} y(t) = \infty . \tag{11}$$

PROOF. In view of (3) and (7)

$$\ddot{y}(t) + \alpha\dot{y}(t) < 0$$

which implies that the functions

$$u(t) \equiv \dot{y}(t) + \alpha y(t) \tag{12}$$

and

$$v(t) \equiv \dot{y}(t)e^{\alpha t} \tag{13}$$

are both strictly decreasing. In particular, it follows from (13) that $\dot{y}(t)$ is eventually of constant sign. We claim that (10) holds. Otherwise,

$$\dot{y}(t) < 0$$

and

$$\ddot{y}(t) < -\alpha\dot{y}(t) \leqq 0$$

which leads to the contradiction that $y(t) < 0$.
Next, we claim that (11) holds. Otherwise, $\lim_{t\to\infty} y(t)$ exists and is a positive number ℓ. But then Eq. (1) implies that

$$\lim_{t\to\infty} \dot{u}(t) = -\beta f(\ell) < 0$$

and so

$$\lim_{t\to\infty} u(t) = -\infty. \tag{14}$$

On the other hand, as we see from (12)

$$u(t) > \alpha y(t) \geqq \alpha\frac{\ell}{2} \in R \tag{15}$$

which contradicts (14) and completes the proof.

LEMMA 2. Assume that (7) holds and that the characteristic equation (5) has no positive roots. Then there exist positive numbers μ and ε such that

$$\lambda^2 + \alpha\lambda + \beta(1-\varepsilon)e^{-\lambda r} \geqq \mu \quad \text{for} \quad \lambda \geqq 0. \tag{16}$$

PROOF. Set

$$F(\lambda) = \lambda^2 + \alpha\lambda + \beta e^{-\lambda r}.$$

Since $F(\infty) = \infty$, $F(0) = \beta > 0$ and $F(\lambda)$ has no positive roots, it follows that

$$m \equiv \min_{\lambda \geqq 0} F(\lambda) > 0$$

and so

$$F(\lambda) \geqq m \quad \text{for} \quad \lambda \geqq 0.$$

Also, there exists $\lambda_0 > 0$ such that

$$\lambda^2 + \alpha\lambda + \tfrac{1}{2}\beta e^{-\lambda r} \geqq \frac{m}{2} \quad \text{for} \quad \lambda \geqq \lambda_0. \tag{17}$$

Choose

$$\varepsilon = \min\{\ \frac{1}{2}, \frac{m}{2\beta}\ \}.$$

Then for $\lambda \geqq \lambda_0$, using (17), we have

$$\lambda^2 + \alpha\lambda + \beta(1-\varepsilon)e^{-\lambda r} \geqq \lambda^2 + \alpha\lambda + \tfrac{1}{2}\beta e^{-\lambda r} \geqq \frac{m}{2} \tag{18}$$

while for $0 \leqq \lambda < \lambda_0$ we get

$$\begin{aligned} \lambda^2 + \alpha\lambda + \beta(1-\varepsilon)e^{-\lambda r} &\geqq \lambda^2 + \alpha\lambda + \beta e^{-\lambda r} - \tfrac{1}{2}m e^{-\lambda r} \\ &\geqq m - \tfrac{1}{2}m = \frac{m}{2}. \end{aligned} \tag{19}$$

From (18) and (19) we see that (16) holds with

$$\mu = \tfrac{1}{2}m.$$

Now we are ready to establish the following linearized oscillation result.

THEOREM 1. Consider Eq. (1) subject to Conditions (3), (7) and (8). Assume that every unbounded solution of the linearized equation (6) oscillates. Then every solution of Eq. (1) also oscillates.

PROOF. Assume, for the sake of contradiction, that Eq. (1) has an eventually positive solution. Then, by Lemma 1, (10) and (11) are satisfied. The hypothesis that every unbounded solution of Eq. (6) oscillates implies that Eq. (5) has no positive roots. Therefore, by Lemma 2, there exist positive numbers μ and ε with $\varepsilon \leqq 1/2$ such that (16) holds. Choose $T \geqq t_0$ so large that

$$y(t-r) > 0 \quad \text{and} \quad \frac{f(y(t-r))}{y(t-r)} > 1-\varepsilon \quad \text{for} \quad t \geqq T, \tag{20}$$

which is possible because of (8) and (11). Then, from Eq. (1) and (20) we see that

$$\ddot{y}(t) + \alpha\dot{y}(t) + \beta(1-\varepsilon)y(t-r) \leqq 0, \quad t \geqq T, \tag{21}$$

that is

$$(e^{\alpha t}\dot{y}(t))^{\cdot} + \beta(1-\varepsilon)e^{\alpha t}y(t-r) \leqq 0, \; t \geqq T. \tag{22}$$

Integrating (22) from $t \geqq T$ to ∞ and taking into account the fact that

$$\lim_{t\to\infty} e^{\alpha t}\dot{y}(t) = c\varepsilon[0,\infty),$$

(indeed, $e^{\alpha t}\dot{y}(t) > 0$ and $(e^{\alpha t}\dot{y}(t))^{\cdot} < 0$) we conclude that

$$e^{\alpha t}\dot{y}(t) \geqq \beta(1-\varepsilon)\int_t^\infty e^{\alpha s}y(s-r)ds, \quad t \geqq T$$

and so

$$\dot{y}(t) \geqq \beta(1-\varepsilon)e^{-\alpha t}\int_t^\infty e^{\alpha s}y(s-r)ds, \; t \geqq T.$$

Integrating this inequality from $T+r$ to $t > T+r$, we get

$$y(t) \geqq y(T+r) + \int_{T+r}^t [\beta(1-\varepsilon)e^{-\alpha s}\int_s^\infty e^{\alpha\xi}y(\xi-r)d\xi]ds, \quad t \geqq T+r. \tag{23}$$

Now, we are ready to apply Lemma 0.

Let X be the set of all non-negative and nondecreasing functions x defined on $[T,\infty)$ such that

$$x(t) = y(t) \quad \text{for} \quad T \leqq t \leqq T+r \tag{24}$$

and

$$x(t) \leqq y(t) \quad \text{for} \quad t > T+r. \tag{25}$$

If x_1 and x_2 belong to X we will say that $x_1 \leqq x_2$ if and only if $x_1(t) \leqq x_2(t)$ for $t \geqq T$. Clearly, X is a partially ordered set. Define the mapping T on X as follows

$$(Tx)(t) = \begin{cases} y(t) & , \quad T \leqq t \leqq T+r \\ y(T+r) + \int_{T+r}^t [\beta(1-\varepsilon)e^{-\alpha s}\int_s^\infty e^{\alpha\xi}x(\xi-r)d\xi]ds, & t > T+r. \end{cases}$$

In view of (23) and (25)

$$(Tx)(t) \leqq (Ty)(t) \leqq y(t) \quad \text{for} \quad t > T+r,$$

while in view of the definition of T

$$(Tx)(t) = y(t) \quad \text{for} \quad T \leqq t \leqq T+r.$$

Also Tx is a non-negative and nondecreasing function for $t \geqq T$. So $T: X \to X$. Clearly, all other conditions of Lemma 0 are satisfied and consequently T has a fixed point $x \in X$. Thus

$$(Tx)(t) = x(t) \quad \text{for} \quad t \geqq T$$

or

$$x(t) = \begin{cases} y(t) & , \quad T \leqq t \leqq T+r \\ y(T+r) + \int_{T+r}^{t} [\beta(1-\varepsilon)e^{-\alpha s}\int_{s}^{\infty} e^{\alpha\xi}x(\xi-r)d\xi]ds, & t > T+r. \end{cases}$$

Obviously, $x(t) > 0$ and $\dot{x}(t) > 0$ for $t \geqq T$. Differentiating for $t \geqq T+r$ we see that x satisfies the differential equation

$$(e^{\alpha t}\dot{x}(t))^{\cdot} + \beta(1-\varepsilon)e^{\alpha t}x(t-r) = 0, \quad t \geqq T+r$$

and so

$$\ddot{x}(t) + \alpha\dot{x}(t) + \beta(1-\varepsilon)x(t-r) = 0, \quad t > T+r.$$

This, in turn, implies that the characteristic equation

$$\lambda^2 + \alpha\lambda + \beta(1-\varepsilon)e^{-\lambda r} = 0$$

has a positive root which contradicts (16). The proof is complete.

3. CONCLUDING REMARKS

Using an argument similar to that used in the proof of the above result we obtain the following partial converse to Theorem 1.

THEOREM 2. Consider Eq. (1) subject to Conditions (3) and (7). Assume that

$$f(u) \leqq u \tag{26}$$

and

$$f \text{ is nondecreasing} \tag{27}$$

hold for positive, sufficiently large u and that the characteristic equation (5) of the majorant equation (6) has a positive root. Then, Eq. (1) has an unbounded nonoscillatory solution.

PROOF. Let μ be a positive root of Eq. (5). Then, Eq. (6) has the unbounded nonoscillatory solution $y(t) = e^{\mu t}$. Hence, in view of (26), the linear differential inequality

$$\ddot{y}(t) + \alpha\dot{y}(t) + \beta f(y(t-r)) \leqq 0$$

has y(t) as a solution for large t.
Now, an argument similar to that used in the proof of Theorem 1 implies that the corresponding linear equation (6) has a nonoscillatory solution z(t), which, by Lemma 1, is unbounded. The proof is complete.

REMARK 1. A more general result on the existence of nonoscillatory solutions of Eq. (1) was established in [1].

REMARK 2. In Theorems 1 and 2 we restricted our attention to unbounded solution only. In view of Lemma 1 this is not a real restriction because under Conditions (3) and (7), every nonoscillatory solution of Eq. (1) is unbounded. Thus, Theorems 1 and 2 can be formulated as global oscillation results, which hold for all solutions of Eq. (1). Combining Theorems 1 and 2 we obtain the following necessary and sufficient condition for the oscillation of Eq. (1).

THEOREM 3. Assume that Conditions (3), (7) and (8) hold and that Conditions (26) and (27) are satisfied for sufficiently large u. Then every solution of Eq. (1) oscillates if and only if the characteristic equation (5) of the linearized equation (6) has no real roots.

NOTE. Because of Condition (7), the characteristic equation (5) cannot have negative roots.

Finally, combining Theorem 3 with the result in [4] mentioned in the Introduction, we get the following global oscillation result, which holds for every real α.

THEOREM 4. Assume that Conditions (3), (4), (8), (26) and (27) are satisfied. Let $\alpha \in R$, $\beta > 0$ and $r \geq 0$. Then every solution of Eq. (1) oscillates if and only if the characteristic equation (5) of the linearized equation (6) has no real roots.

In the special case, when $r = 0$ this necessary and sufficient condition can be expressed in terms of the coefficients α and β.

COROLLARY. Consider Eq. (1) with $r = 0$, subject to Conditions (3), (4), (8), (26) and (27). If $\beta > 0$, then every solution of Eq. (1) oscillates if and only if $\alpha^2 - 4\beta < 0$.

REFERENCES

1. M. R. KULENOVIĆ and M. K. GRAMMATIKOPOULOS, Oscillation and asymptotic behavior of nonlinear differential inequalities and equations with deviating arguments, Ukrain. Mat. Z. 36(3), 1984, 309-316 (Russian).
2. M. R. KULENOVIĆ and M. K. GRAMMATIKOPOULOS, Stability of asymptotic behavior of solutions of nonlinear differential inequalities with respect to argument delay, Ukrain. Mat. Z. 36(4), 1984, 437-443 (Russian).
3. M. R. KULENOVIĆ, G. LADAS and A. MEIMARIDOU, On oscillation of nonlinear delay equations, Quart. Appl. Math. (to appear).
4. M. R. KULENOVIĆ, and G. LADAS, Oscillations of the sunflower equation (to appear).
5. M. R. KULENOVIĆ, G. LADAS and Y. G. SFICAS, Comparison results for oscillations of delay equations (to appear).
6. W. E. MAHFOUD, Comparison theorems for delay differential equations, Pacific J. Math. 83(1979), 187-197.
7. A. TARSKI, A lattice theoretical fixed point theorem and its applications, Pacific J. Math. 5(1955), 285-309.

Canadian Mathematical Society
Conference Proceedings
Volume 8 (1987)

Sign-variations of the solutions in a boundary value problem

Pu Fuquan , Chen Shuilian , Shao Mingfeng
(Tsinghua University, Beijing, China)

Abstract

The variation of signs of the solution of nonhomogeneous linear differential equation of second order with boundary value conditions is discussed. We follow the method used by R. Bellman [2] , [3] .

1. Introduction

We consider the problem

$$u'' + q(x)u = f(x) \quad , \tag{1}$$

$$\begin{cases} \alpha u(0) + \alpha' u'(0) = 0 \ , \\ \beta u(1) + \beta' u'(1) = 0 \ . \end{cases} \tag{2}$$

The number of sign-variations of the solutions on the interval [0, 1] is discussed. We assume that $q(x)$, $f(x)$ are continuous on [0, 1] , α , α', β, β' are constants which guarantee the existence of the problem (1) and (2) ([1] P.278). The main result is that under the condition $q(x) \leq \pi^2$ the maximum number of sign-variations of the solutions of (1) and (2) on [0,1] is not greater than that of $f(x)$ on [0, 1] plus two.

R. Bellman in [2], [3] discussed the nonnegativity of Green function by the method of calculus of variations. We use similar method to investigate the problem when $f(x)$ has variation of signs on [0 , 1], the estimation of sign-variations of the solutions. The physical meaning of the problem lies in that under certain conditions on the forcing term $f(x)$ the oscillations of the free vibration on an elastic support can be diminished.

2. Lemmas

We give some lemmas first, then the main theorem.

Lemma 1. Assume

(i) $w(x) \in C[0,1]$, $w'(x) \in L^2[0, 1]$ (3)

(ii) $w(0)=w(1)=0$, (4)

(iii) $q(x) \leq \pi^2$, $q(x) \not\equiv \pi^2$, (5)

then the functional

$$Q(w) = \int_0^1 (w'^2 - q(x)w^2)dx > 0 \tag{6}$$

Proof: We are to show that the smallest eigenvalue λ_0 of the sturm-Lionville problem

$$w'' + \lambda q(x) w = 0 \tag{7}$$

$$w(0) = w(1) = 0$$

is greater than one. For, otherwise

$$\lambda_0 q(x) \leq \pi^2, \tag{8}$$

and by comparison theorem, every solution of

$$u'' + \pi^2 u = 0 \tag{9}$$

must have at least one zero in (0, 1) because the eigenfunction w_0 of

$$w_0'' + \lambda_0 q(x) w_0 = 0 \tag{10}$$

is zero at x=0 and x=1. But $u(x)=\sin \pi x$ is a solution of (9) and a positive function in (0,1). This is a contradiction and therefore $\lambda_0 > 1$.

Now for w satisfying $\int_0^1 q(x) w^2 dx \leq 0$, $Q(w)$ is always positive definite. So suppose $w_1(x)$ is a function satisfying (i) and (ii) and $\int_0^1 q(x)w_1^2 dx = c > 0$, for some constant . Since

$$\int_0^1 w'^2 dx = w w' \Big|_0^1 - \int_0^1 w w'' dt$$
$$= \lambda \int_0^1 q(x) w^2 dx$$
$$= \lambda$$

so λ_0 is the minimum of $\int_0^1 w'^2 dx$.

$$(1/c)\int_0^1 w_1'^2 dx \geq \lambda_0 > 1,$$

i.e. $$\int_0^1 w'^2 dx > c, \tag{11}$$

i.e. $$Q(w_1) = \int_0^1 w_1'^2 dx - \int_0^1 q(x) w_1^2 dx > 0. \tag{12}$$

The lemma is thus proved.

Lemma 2. The solution u(x) of the problem (1), (2) is the minimum function of the functional

$$J(u) = \int_0^1 ((u'^2-q(x))u^2+2f(x)u) dx \qquad (13)$$

Subjecting to the conditions (2) and (5).

Proof: Suppose there exists an arbitrary continuous function v(x) on [0,1]and $v\ (x) \in L^2\ [0,1]$, $v(0)=v(1)=0$, then

$$J(u+v)=J(u)+2\int_0^1(u'v' -q(x)uv+f(x)v)dx + \int_0^1 (v'^2-q(x)v^2)dx$$
$$=J(u)+ \int_0^1 (v'^2-q(x)v^2)dx$$

On account of u(x) being the solution of

$$\int_0^1 (u'v' -q(x)uv+f(x)v)dx=u'v \Big|_0^1 - \int_0^1 v(u'' +q(x)u-f(x))dx=0$$

So, by Lemma 1, we have $J(u+v) > J(u)$

Lemma 2 is thus proved.

Here we introduce a notation NUS F(x) [a,b] , it signifies the number of variation of signs of the function F(x) on [a,b] . A point of sign variation x_i means $F(x_i)=0$ and $F(x_i-\varepsilon)F(x_i+\varepsilon) < 0$, $\varepsilon > 0$ sufficiently small.

Lemma 3. Assume $\varphi(x) \in C[0,1]$ all zeros are points of variation of sign and $\varphi(0)\varphi(1)=0$, then we have

(i) NUS $\varphi(x)$ [0,1]= 2k , iff $\varphi(0)\varphi(1) > 0$,

(ii) NUS $\varphi(x)$ [0,1]=2k+1 , iff $\varphi(0)\varphi(1) < 0$,

where k is any integer including zero.

Proof: Let $0<x_1<x_2\cdots<x_n<1$ be the points of variation of sign of $\varphi(x)$ (Fig. 1)

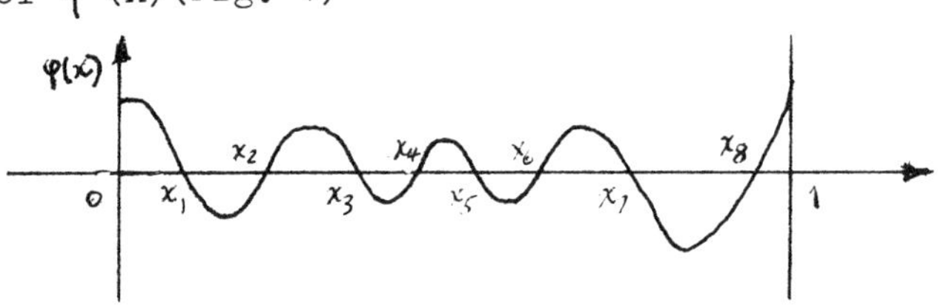

Fig. 1

Let us define y_i such that $x_i < y_i < x_{i+1}$, i=1,2,... , n-1.

We have

$$\varphi(0)\varphi(y_1) < 0,\ \varphi(y_1)\varphi(y_2) < 0, \ldots\ \varphi(y_{n-1})\varphi(1) < 0,$$

$$\operatorname{sgn}\left[\varphi(0)\varphi^2(y_1)\ldots\ \varphi^2(y_{n-1})\varphi(1)\right] = \operatorname{sgn}(-1)^n.$$

Here NUS $\varphi(x)[0,1] = n$, so

$$\varphi(0)\varphi(1) > 0 \quad \text{iff} \quad \operatorname{sgn}(-1)^n > 0,$$

$n = 2k$;

$$\varphi(0)\varphi(1) < 0 \quad \text{iff} \quad \operatorname{sgn}(-1)^n < 0,$$

$n = 2k+1$.

3. Main theorem

Theorem 1. Assume that the function $f(x)$ of equation (1) satisfies

(i) $f(\alpha)f(\beta) \neq 0$,
(ii) N $Sf(x)[\alpha,\beta] = 0$,

where $[\alpha,\beta] \subset [0,1]$, then the minimum function $u(x)$ of the functional (13) satisfies

$$\text{NUS}u(x)[\alpha,\beta] \leq 2.$$

In fact,

(a) $\text{NUS}u(x)[\alpha,\beta] = 0$ when $f(\alpha)u(\alpha) < 0$, $f(\beta)u(\beta) < 0$;

(b) $\text{NUS}u(x)[\alpha,\beta] = 1$ when $f(\alpha)u(\alpha) \gtrless 0$, $f(\beta)u(\beta) \lessgtr 0$.

(c) $\text{NUS}u(x)[\alpha,\beta] = 0$ or 2 when $f(\alpha)u(\alpha) > 0$, $f(\beta)u(\beta) > 0$.

Proof: Case (a). For simplicity, let $f(x) > 0$,

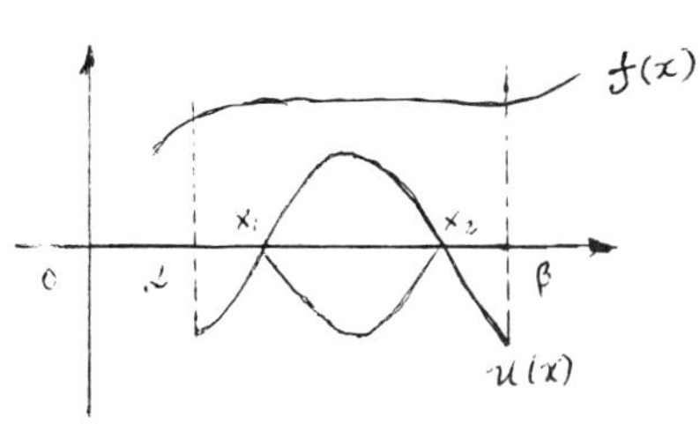

Fig 2 if u(x) has points of sign-variation in (a), there is a contradiction

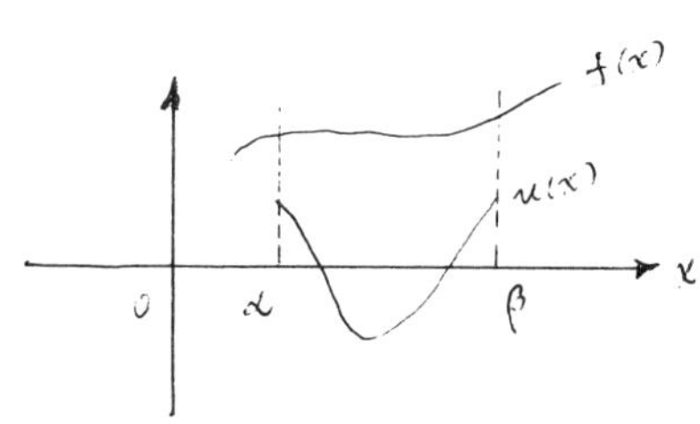

Fig. 3 Case (c) NUS $u(x)[\alpha,\beta] = 2$

We have then $u(\alpha) < 0$, $u(\beta) < 0$. If there exists a point of sign-variation of $u(x)$, by lemma 3, there exist at least two

(Fig. 2), We construct a new function $v(x)$ such that

$$v(x) = \begin{cases} -u(x) & x \in [x_1, x_2], \\ u(x) & x \bar{\in} [x_1, x_2] \end{cases}$$

so we have

$$J(v)=\int_0^1 (v'^2 -q(x)v +2f(x)v(x))dx$$
$$< \int_0^1 (u'^2 -q(x)u +2f(x)u(x))dx=J(u)$$

This is a contradiction since $u(x)$ is the minimum function of $J(u)$. Hence $NUSu(x)[\alpha,\beta] =0$.

Case (b) and (c) can be proved similarly.

Corollary 1. Under the conditions of theorem 1 and if $u(\alpha)= u(\beta)=0$, then $u(x)=0$, $u(x)f(x)<0$ $\forall x \in (\alpha,\beta)$.

Suppose $NUSf(x)[0,1]<\infty$, x_i being the points of sign-variation of $f(x)$ and all zeros of $f(x)$ are points of sign-variation. We define z_i, y_i as follows:

If x_i is not a point of sign-variation of $u(x)$, z_i and y_i satisfy

$$x_{i-1} < z_i < x_i < y_i < x_{i+1} \text{ , and } y_i < z_{i+1}$$

and there is no sign-variation of $u(x)$ in the intervals (z_i,x_i). Owing to the continuity of $u(x)$, there exist y_i and z_i ;

If x_i is a point of sign-variation, $y_i = x_i$.

Corollary 2.

(a) If $f(y_{i-1})u(y_i) > 0$, $f(y_i)u(y_i)> 0$ or $f(y_{i-1})u(y_{i-1})< 0$, $f(y_i)u(y_i)< 0$, then $NUSu(x)[y_{i-1}, y_i]= 1$;

(b) If $f(y_{i-1})u(y_{i-1}) > 0$, $f(y_1)u(y_i) < 0$, then $NUSu(x)[y_{i-1}, y_i]=2k$, $k=0,1$;

(c) If $f(y_{i-1})u(y_{i-1}) < 0$, $f(y_1)u(y_i) > 0$, then $NUSu(x)[y_{i-1}, y_i]=0$.

Proof: We are going to prove (b) only, the cases (a) and (b) can be proved similarly. Without loss of generality, let $f(y_{i-1})> 0$, then $u(y_{i-1}) > 0$, $u(y_i) > 0$ since $f(y_i)< 0$, and $NUSu(x)(y_{i-1},y_i)$ $2k$, $k=0,1$, (14)

Since if $NUSu(x)(y_{i-1},y_i)$ 2, this implies $NUSu(x)(y_{i-1},z_i)> 2$. Observing that $f(x)$ does not change sign in (y_{i-1},z_i), by

Theorem 1, $NUSu(x)(y_{i-1},z_i) > 0$ is impossible. So that we have (14).

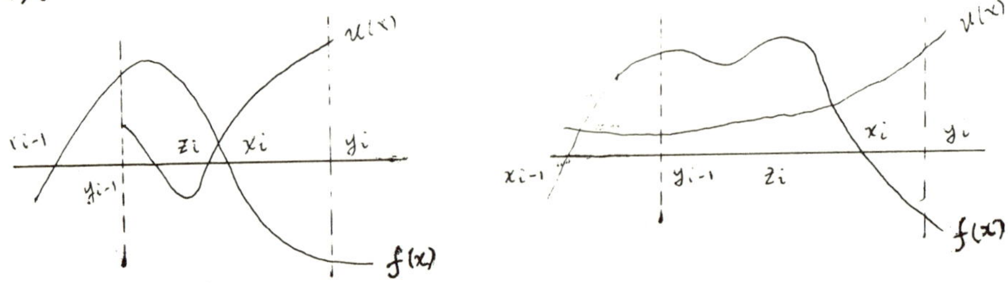

(a) $NUSu(x)[y_{i-1},y_i] = 2$ (b) $NUSu(x)[y_{i-1},y_i]=0$

Fig. 4

Corollary 2′ If $u(x_i)=0$, then $f(y_i)u(y_i)=0$, we have

(a) $f(y_{i-1})u(y_{i-1})< 0$, then $NUSu(x)[y_{i-1},y_i] = 0$

(b) $f(y_{i-1})u(y_{i-1})> 0$, then $NUSu(x)[y_{i-1},y_i]=$ or 1.

By Corollary 2 and 2', we readily have

Corollary 3.

(a) If $f(y_{i-1})u(y_{i-1}) > 0$, $f(y_i)u(y_i)< 0$, $f(y_{i+1})u(y_{i+1})> 0$, or $f(y_{i-1})u(y_{i-1})< 0$, $f(y_i)u(y_i)> 0$, $f(y_{i+1})u(y_{i+1})< 0$, then Max $(NUSu(x)[y_{i-1}, y_{i+1}]) = 2$.

(b) when $u(x_i)=0$, we have $NUSu(x)[y_{i-1},y_{i+1}] = 1$,

Corollary 3' Assume y_i y_j y_l being not adjacent and not points of sign-variation of $f(x)$ and $u(x)$. If $f(y_i)u(y_i) > 0$, $f(y_j)u(y_j)< 0$, $f(y_1)u(y_1) < 0$ or $f(y_i)u(y_i)< 0$, $f(y_j)u(y_j)> 0$, $f(y_l)u(y_l) > 0$, and $f(x)u(x) > 0$, $x \in [y_i,y_{j-1}]$, $f(x)u(x)< 0$, $x \in [y_j,y_{l-1}]$, then we have

$$NUSu(x)[y_{j-1},y_j] + NUSu(x)[y_{l-1},y_l] = 2k,\ k=0,1.$$

Corollary 4.

(a) If $f(0)u(0)> 0$, $f(y_n)u(y_n)> 0$ or $f(0)u(0)< 0$, $f(y_n)u(y_n)< 0$, then $NUSu(x)[0,y_n] \leq NUSf(x)[0,y_n] = n$;

(b) If $f(0)u(0) > 0$, $f(y_n)u(y_n)< 0$, then $NUSu(x)[0,y_n] \leq NUSf(x)[0,y_n] + 1 = n+1$;

(c) If $f(0)u(0)< 0$, $f(y_n)u(y_n)> 0$, then $NUSu(x)[0,y_n]$ $NUS f(x)[0,y_n] -1=n-1$.

Proof: (a), we consider the case $f(0)u(0)> 0$, $f(y_n)u(y_n)>0$. Suppose there is a point y_k in the interval $(0,y_n)$ which is the first point we have $f(y_k)u(y_k) < 0$. By the assumption $f(y_n)u(y_n) > 0$, there must exist a point y_1, $y_k < y_1 \leq y_n$, such that $f(y_l)u(y_l) > 0$, and

$f(y_i)u(y_i) > 0$, when $0 \leq y_i \leq y_k$;

$f(y_i)u(y_i)$ 0, when $y_k \leq y_i < y_l$;

$f(y_i)u(y_i)$ 0, when $y_l \leq y_i \leq y_n$.

By Corollary 3', $NUSu(x)[y_{k-1}, y_k] + NUSu(x)[y_{l-1}, y_l] = 2$, by Corollary 2, we have

$NUSu(x)[0, y_{k-1}] = NUSf(x)[0, y_{k-1}] = k-1$;

$NUSu(x)[y_k, y_{l-1}] = NUSf(x)[y_k, y_{l-1}] = 1-1-k$;

$NUSu(x)[y_1, y_n] = NUSf(x)[y_1, y_n] = n-1$.

Therefore, $NUSu(x)[0, y_n] = NUSu(x)[0, y_{k-1}] + NUSu(x)[y_k, y_{l-1}] + NUSu(x)[y_l, y_n] + NUSu(x)[y_{k-1}, y_k] + NUSu(x)[y_{l-1}, y_l]$

$= (k-1)+(1-1-k)+(n-1)+2 = n$

The case $f(0)u(0) < 0$, $f(Y_n)u(y_n) > 0$ can be done by the same method.

The proof of (b) and (c) can be proved similarly.

Let us now consider the sign-variation of $u(x)$ on the interval $[y_n\ 1]$. By the definition of y_n, $NUSf(x)[y_n, 1] = 0$, so we have

(a) $NUSu(x)[y_n, 1] = 0$, when $f(y_n)u(y_n) < 0$, $f(1)u(1) < 0$;

(b) $NUSu(x)[y_n, 1] = 1$, when $f(y_n)u(y_n) > 0$, $f(1)u(1) < 0$;

(c) $NUSu(x)[y_n, 1] = 2k$, $k=0,1$, when $f(y_n)u(y_n) > 0$, $f(1)u(1) > 0$.

Combining the results of corollary 4, we have

(1) $NUSu(x)[0.1] \leq NUSf(x)[0,1] + 2 = n+2$, when $f(0)u(0) > 0$, $f(1)u(1) > 0$;

(2) $NUSu(x)[0,1] \leq NUSf(x)[0,1] + 1 = n+1$, when $f(0)u(0) \gtrless 0$, $f(1)u(1) \lessgtr 0$;

(3) $NUSu(x)[0,1] \leq NUSf(x)[0,1] = n$, when $f(0)u(0) < 0$, $f(1)u(1) < 0$.

We obtain the following main theorem.

Theorem 2. (main theorem) If the problem (1) and (2) satisfies

(i) $q(x) \lessgtr \pi^2$, $q(x) \not\equiv \pi^2$,

(ii) $NUSf(x)[0,1] = n$,

(iii) All zeros of $f(x)$ are points of sign-variation,

then $NUSu(x)[0,1] \leq n+2$.

The generalized problem

$$\frac{d}{dx}\left(p(x)\frac{du}{dx}\right) + q(x)u = f(x),$$

$$\begin{cases} \alpha u(a) + \alpha' u'(a) = 0, \\ \beta u(b) + \beta' u'(b) = 0 \end{cases}$$

can be discussed similarly and the conclusioniis the same.

References

1. G. Birkhoff and G. C. Rota, Ordinary Differential Equations, John Wiley, 1969 (2nd edition)
2. R. Bellman, On the nonnegativity of Green's functions, Bull. Bull. Un. Mat. Ital. 12, 1957 P.P. 411-413
3. R. Bellman, On variation, Diminishing Properties of Green's functions, Bull. Un. Mat. Ital. 16, 1961 P.P.164-166
4. R. Bellman, Introduction to the Mathematical Theory of Control Processes Vol 1. Academic Press, New York, 1967

Canadian Mathematical Society
Conference Proceedings
Volume 8 (1987)

A NOTE ON A LIMIT-CIRCLE CRITERION OF KUPCOV

D. Race

Department of Mathematics
University of Surrey
Guildford, Surrey. GU2 5XH
England

Synopsis

A well-known, very general limit-circle criterion due to Kupcov is shown to extend to the differential expression $w^{-1}(-(pu')' + qu)$ on $[a,b)$ where (i) p and q may be complex-valued on $[a,b)$, (ii) w is real-valued and positive a.e. on $[a,b)$ and (iii) p^{-1}, q, w are locally Lebesgue integrable on $[a,b)$, where $-\infty < a < b \leqslant +\infty$.

1. Introduction

We consider differential expressions of the type

$$\ell(u) = (w(x))^{-1}\{ - (p(x)u'(x))' + q(x)u(x)\} ,x \in [a,b) \qquad (1.1)$$

where $-\infty < a < b \leqslant \infty$, $w > 0$ a.e. and p^{-1}, q and $w \in L_{loc}[a,b)$. The limit-point/limit-circle dichotomy for real-valued coefficients p and q is well-known and in [5, Theorem 1], Kupcov gave a very general criterion for ℓ to be in the limit-circle case (when $w=p = 1$ on $[a,\infty)$). This is known to extend easily to cover $p \neq 1$ and $w \neq 1$ on $[a,b)$ for $b \leqslant \infty$ and to include many of the known limit-circle criteria, by means of suitable choices of the arbitrary functions in Kupcov's result (see [3] for further details and references). In the results cited so far, p is assumed to be positive on $[a,b)$.

In the recent work [1] of Everitt, Knowles and Read, which indeed motivated our present discussion, it is pointed out that the requirement that p be positive, is not needed in the derivation of the above-mentioned results. Hence the Kupcov criterion is extended in [1, Theorem 2] to give one in which there is no restriction on the sign of p or on how often p may change sign. For further discussion of the uses and implications of this result, we refer the reader to [1, Sections 1 and 3].

In recent years, there has been increased interest shown in expressions of the form (1.1), in which p and q are permitted to be complex-valued (see [4, 8] and the references contained therein). The classical Weyl limit-point/limit-circle dichotomy has a natural analogue, discussed in [4, Section 2], but given in its greatest generality in [8, Section 5]. For the present discussion, we need only note that, in the terminology of [8], ℓ is in the limit-circle case if every solution of $\ell(y) = \lambda y$ is square-integrable with respect to the weight function w for some (and hence every) complex number λ. i.e. $\int_a^b w|y|^2 < \infty$ for every solution y. Our aim here is to extend Kupcov's result beyond the version given in [1] to give a limit-circle criterion in this, the most general situation for the expression (1.1).

The inequality from which the criterion is derived, was generalized in Knowles and Race [4, Section 4] to cover complex-valued p and q, and the result we give below was stated and proved for w = 1 on $[a,\infty)$ in the unpublished thesis [7, Theorem 5.1]. For p = w = 1, the extension of some of the best-known limit-circle criteria, to complex-valued q, were given in [7, Theorems 5.3–5.10]. For the sake of brevity we illustrate our result below by just a few simple examples.

2. Main Result

We assume only that p, q and w satisfy the basic assumptions stated at the beginning of the last section. Let y be a solution of $\ell(y) = 0$ on [a,b) and define the quadratic form E on [a,b) by

$$E(x) = |p(x)\, y'(x) + \gamma(x)y(x)\,|^2 + \phi^2(x)\, |y(x)|^2 \tag{2.1}$$

where γ and $\phi > 0$ are arbitrary. locally absolutely continuous functions on [a,b) with γ being permitted to be complex-valued. In [4] an additional arbitrary function Θ multiplied the term $p(x)y'(x)$ in (2.1) and the expression $E'(x)/E(x)$ was then calculated and expressed as a quadratic form involving a Hermitian matrix. Its value was then estimated in terms of the eigenvalues of this matrix. Upon integrating, this gave an upper bound for E(x). Setting $\Theta \equiv 1$ in [4, (4.5)] yields that, for y satisfying $\ell(y) = 0$ and for $x \in [a,b)$,

$$E(x) \leqslant E(a)(\phi(a))^{-1}.\phi(x)\exp\Big[\int_a^x \{\,|\tfrac{\phi}{\phi}' - 2\mathrm{Re}(\tfrac{\gamma}{p})|^2 + \tfrac{1}{\phi^2}|q + \tfrac{\phi^2}{\bar{p}} - \tfrac{\gamma^2}{p} + \gamma'\,|^2\}^{\frac{1}{2}}\Big] \tag{2.2}$$

where Re(.) denotes the real part of a complex number and $\bar{p}$ is the complex conjugate of p. Since $w|y|^2 \leqslant w\,\phi^{-2}E$, we may deduce from (2.2), a condition which ensures that y must belong to the weighted Hilbert space $L^2_w\,[a,b)$. Since y is an arbitrary solution of $\ell(y) = 0$, this forces ℓ to be in the limit-circle case. We thus have the following result.

THEOREM Let p, q and w satisfy the basic conditions above and suppose there exist functions γ and $\phi > 0$ which are locally _absolutely continuous on_ [a.b), _such that_

$$\int_a^b \frac{w}{\phi} \exp \left[\int_a^x \left\{ \left| \frac{\phi'}{\phi} - 2\mathrm{Re}\left(\frac{\gamma}{p}\right) \right|^2 + \frac{1}{\phi^2} \left| q + \frac{\phi^2}{\bar{p}} - \frac{\gamma^2}{p} + \gamma' \right|^2 \right\}^{\frac{1}{2}} \right] < \infty \tag{2.3}$$

Then ℓ _is of limit-circle type at b_.

It is easy to see from [4] that there is no loss of generality in our having set $\Theta = 1$ in [4] to obtain this result.

We demonstrate the extent to which the complex-valued coefficient p may be highly oscillatory, by means of some simple examples:

Examples. In each of the following, ℓ is of limit-circle type at b.

(1) $p(x) = \exp(i\, f(x))$, where f is real, monotonic increasing and unbounded on $[a,\infty)$. $q(x) = g(x) - \exp(i\, f(x))$, where g and $w \in L^1[a,\infty)$.

(Set $\phi = 1$, $\gamma = 0$ above; this includes the case $p = \exp(i\, x^\alpha)$, $q = -\exp(i\, x^\alpha)$ for $\alpha > 0$, when $w \in L^1[a,\infty)$.)

(2) $p(x) = \exp(i\, f(x))$, where f is real, monotonic increasing, locally absolutely continuous and unbounded on $[a,b)$ for $b < \infty$. $q(x) = f'(x) \exp(i\, f(x)) + g(x)$ where g and $w \in L^1[a,b)$.

(Set $\phi = 1$, $\gamma = i\, p$ above; this includes the case $p = \exp(i/(b-x)^\alpha)$, $q = \alpha\, (b-x)^{-\alpha-1} \exp(i/(b-x)^\alpha)$ for $\alpha > 0$ when $w \in L^1[a,b)$).

3. Additional Consequences

The examples above illustrate how the theorem may be applied to complex-valued, highly oscillatory coefficients p. Examples were given in [1] to illustrate how it may be applied to real-valued coefficients p, whose sign changes at infinitely-many points of discontinuity. Of the many possible corollaries that one could formally derive from the theorem, we isolate three here, which illustrate the consequences of permitting the coefficient q to be complex-valued. Throughout this section we set

$$p = w = 1 \quad \text{on} \quad [a,\infty)$$

and define $F(x) \geq 0$ by

$$F^2 = \left|\frac{\phi'}{\phi} - 2\,\mathrm{Re}\,\gamma\right|^2 + \frac{1}{\phi^2}\left|q + \phi^2 - \gamma^2 + \gamma'\right|^2 \qquad (3.1)$$

COROLLARY 1 <u>Suppose</u> Re q <u>is negative, locally absolutely continuous and satisfies</u>

$$\frac{\mathrm{Re}\, q'(t)}{\mathrm{Re}\, q\,(t)} \geq \frac{2+\varepsilon}{t} \quad \underline{\text{on}}\ [a,\infty) \qquad (3.2)$$

<u>for some</u> $\varepsilon > 0$. <u>If also</u>,

$$\int_a^\infty |\mathrm{Im}\, q|\ |\mathrm{Re}\, q|^{-\frac{1}{2}} < \infty \qquad (3.3)$$

<u>then</u> ℓ <u>is of limit-circle type at</u> ∞ .

<u>Proof</u> Set $\phi = (-\mathrm{Re}\, q)^{\frac{1}{2}}$ and $\gamma(t) = (2+\varepsilon)(4t)^{-1}$. Now (3.2) implies Re $q' < 0$ and for some constant $k > 0$,

$$|\mathrm{Re}\, q(t)| \geq kt^{2+\varepsilon}. \qquad (3.4)$$

Thus,

$$F^2(t) = \left[\frac{2+\varepsilon}{2t} - \frac{\mathrm{Re}\, q'}{2\mathrm{Re}\, q}\right]^2 + \frac{1}{|\mathrm{Re}\, q|}\cdot\left|i\mathrm{Im}\, q - \frac{(2+\varepsilon)^2}{16t^2} - \frac{(2+\varepsilon)}{4t^2}\right|^2$$

Hence from (3.2) and (3.4),

$$F(t) \leq \frac{\mathrm{Re}\, q'}{2\mathrm{Re} q} - \frac{2+\varepsilon}{2t} + \frac{|\mathrm{Im}\, q|}{|\mathrm{Re}\, q|^{\frac{1}{2}}} + k_1\, t^{-3}$$

for some constant $k_1 > 0$. Hence using (3.3) we obtain for some constant k_2,

$$\int_a^x \phi^{-1} \exp [\int_a^t F] \leq k_2 \int_a^x \phi^{-1}\exp[\tfrac{1}{2} \log|\mathrm{Re}\ q| - (1 + \varepsilon/2)\log t]$$

$$= k_2 \int_a^x t^{-(1+\varepsilon/2)}$$

Thus (2.3) is satisfied and the result follows □

Corollary 1 was proved by Lidskii for real-valued q in [6] (c.f. [5].)

COROLLARY 2 *Suppose* Re q *is negative and that both* Re q *and* Re q' *are locally absolutely continuous. Suppose also that*

$$|\mathrm{Re}\ q|^{-\frac{1}{2}},\ |\mathrm{Im}\ q|.|\mathrm{Re}\ q|^{-\frac{1}{2}},\left|\frac{\mathrm{Re}\ q''}{|\mathrm{Re}\ q|^{3/2}} + \frac{5}{4}\frac{(\mathrm{Re}^2 q')}{|\mathrm{Re}\ q|^{5/2}}\right| \in L^1[a,\infty)$$

then ℓ *is of limit-circle type at* ∞.

Proof. Set $\phi = (-\mathrm{Re}\ q)^{\frac{1}{2}}$ and $\gamma = \mathrm{Re}\ q'.(4\ \mathrm{Re}\ q)^{-1}$, then $\phi^{-1} \in L^1[a,\infty)$ so for (2.3) it suffices to show $F \in L^1[a,\infty)$. However, γ and ϕ have been chosen so that the first term in (3.1) vanishes, so

$$F = |\mathrm{Re}\ q|^{-\frac{1}{2}}.|i\ \mathrm{Im}\ q - (\mathrm{Re}\ q')^2(4\mathrm{Re}\ q)^{-2} + (\mathrm{Re}\ q'(4\mathrm{Re}\ q)^{-1})'|$$

$$\leq |\mathrm{Im}\ q|.|\mathrm{Re}\ q|^{-\frac{1}{2}} + \tfrac{1}{4}|\mathrm{Re}\ q|^{-\frac{1}{2}}.\left|\frac{\mathrm{Re}\ q''}{\mathrm{Re}\ q} - \frac{5}{4}.\left(\frac{\mathrm{Re}\ q'}{\mathrm{Re}\ q}\right)^2\right|$$

which belongs to $L^1[a,\infty)$ by virtue of our assumptions.

[]

This is an extension to complex q of results known already for real q. (see [1] and the references mentioned there). Finally, we give an extension to complex q of results given by Halvorsen in [2].

COROLLARY 3 *Suppose there exist positive constants* k, n, δ *such that either*

$$\text{(i)}\ |q(t) + k^2 t^{2(n+1)}| \leq (n-\delta)\, kt^n \text{ on } [1,\infty)\ ; \tag{3.5}$$

or $$\text{(ii)}\ |q(t) + k^2 e^{nt}| \leq k\, e^{nt/2} \left(\frac{n}{2} - \frac{1+\delta}{t}\right) \text{ on } [1,\infty) \tag{3.6}$$

then ℓ *is of limit-circle type at* ∞.

Proof. If (i) holds, set $\phi(t) = k\, t^{n+1}$ and $\gamma(t) = (n+1)/2t$ then the first term of (3.1) vanishes, so

$$F(t) = (k\, t^{n+1})^{-1}\, |q+k^2\, t^{2(n+1)} - (n+1)(n+3)(4t^2)^{-1}|$$

$$\leq (n-\delta)\, t^{-1} + (n+1)(n+3)(4kt^{n+3})^{-1}.$$

Hence in (2.3), for a suitable constant K, we have

$$\int_1^\infty \phi^{-1} \exp\left(\int_1^t F\right) \leq K\int_1^\infty t^{-(n+1)} .\ t^{n-\delta} < \infty$$

which proves the result when (i) holds.

If (ii) holds, set $\phi(t) = k\, e^{nt/2}$ and $\gamma(t) = n/4$ then

$$F(t) = (ke^{nt/2})^{-1}|q + k^2\, e^{nt} - n^2/16|$$

$$\leq \left(\frac{n}{2} - \frac{1+\delta}{t}\right) + \frac{n^2}{16k}\, e^{-nt/2}$$

Hence

$$\int_1^\infty \phi^{-1} \exp\left(\int_1^t F\right) \leq K\int_1^\infty e^{-nt/2} .\ e^{nt/2} .\ t^{-(1+\delta)} < \infty$$

as required in our theorem.

□

It is easy to see from the above proof, that (3.5) and (3.6) may be replaced by the (independent) conditions,

$$\text{(i)}' \quad |\mathrm{Re}\; q(t) + k^2\, t^{2(n+1)}| \leqslant (n-\delta)kt^n \text{ and } \int_1^\infty |\mathrm{Im}\; q|.t^{-(n+1)} < \infty\;;$$

$$\text{(ii)}' \quad |\mathrm{Re}\; q(t) + k^2\, e^{nt}| \leqslant k\, e^{nt/2}\left(\frac{n}{2} - \frac{1+\delta}{t}\right) \text{ and } \int_1^\infty |\mathrm{Im}\; q|e^{-nt/2} < \infty\;.$$

Acknowledgements

The author wishes to thank his PhD supervisor, Professor I.W. Knowles for many useful discussions, especially during the preparation of his thesis, [7]. He is also grateful to the authors of reference [1] for permitting him to see their manuscript at an early stage.

References

1. W.N. Everitt, I.W. Knowles, T.T. Read. Limit-point and limit-circle criteria for Sturm-Liouville equations with intermittently negative principal coefficients. Proc. Roy. Soc. Edinburgh Sect. A, 103(1986), 215-228.

2. S.G. Halvorsen. Bounds for solutions of second order linear differential equations with applications to L^2-boundedness. Norske Vid. Selsk. Forh. (Trondheim) 36 (1963) Nr.8, 36-40.

3. I.W. Knowles. On the number of L^2 - solutions of second order linear differential equations. Proc. Roy. Soc. Edinburgh Sect.A 80 (1978), 1-13.

4. I.W. Knowles and D. Race. On the point spectra of complex Sturm-Liouville operators. Proc. Roy. Soc. Edinburgh Sect. A 85 (1980), 263-289.

5. N.P. Kupcov. Conditions for non-selfadjointness of a second order linear differential operator. Dokl. Akad. Nauk. SSSR 138(1961), 767-770. (translated as Soviet Maths 2 (1961), 710-713)

6. V.B. Lidskii. On the number of square integrable solutions of the system of differential equations $-y'' + P(t)y = \lambda y$. Dokl. Akad. Nauk SSSR 95 (1954), 217-220.

7. D. Race. Spectral theory of complex Sturm-Liouville operators. PhD thesis, University of the Witwatersrand, Johannesburg, 1980.

8. D. Race, The theory of J-selfadjoint extensions of J-symmetric operators. J. Diff. Equations 57 (1985), 258-274.

Canadian Mathematical Society
Conference Proceedings
Volume 8 (1987)

TIME-PERIODIC SOLUTIONS OF NONLINEAR WAVE EQUATIONS IN BALLS

M. W. Smiley*

ABSTRACT. The problem of time-periodic solutions of the equation $u_{tt} - u_{xx} + g(u) = f$, $|x| < a$, is an example of a hyperbolic boundary value problem. The present work considers a generalization of this problem, in which the interval $(-a,a)$ is replaced by the ball $B_n = \{x \in \mathbf{R}^n : \|x\| < a\}$. Assuming spherical symmetry of the forcing term f, it is possible to establish results analogous to those known for the 1-dimensional case. The method of proof employs a recently discovered Hilbert space theory for hyperbolic boundary value problems, which requires certain estimates on the spectrum of the associated linear operator. These estimates can be established by using an argument from oscillation theory derived from Sturm's original work.

1. INTRODUCTION

In this paper we shall consider the existence, uniqueness, and continuous dependence of radially symmetric time-periodic solutions of the semilinear wave problem

$$u_{tt} - \Delta u + g(u) = f, \quad (t,x) \in \mathbf{R} \times B_a^n,$$

$$u(t,x) = 0, \qquad (t,x) \in \mathbf{R} \times S_a^{n-1},$$

$$u(t+T,x) = u(t,x), \qquad (t,x) \in \mathbf{R} \times B_a^n.$$

Here we use $\Delta = \sum \partial^2/\partial x_i^2$, $B_a^n = \{x \in \mathbf{R}^n : \|x\| < a\}$ and $S_a^{n-1} = \{x \in R^n : \|x\| = a\}$, with $\|x\| = (\sum x_i^2)^{1/2}$. The forcing term $f : \mathbf{R} \times B_a^n \to \mathbf{R}$ is assumed to be T-periodic in time, radially symmetric on B_a^n, and in the Lebesque space $L^2((0,T) \times B_a^n)$. The nonlinear term g is assumed to be a Lipschitz continuous real-valued function on $\mathbf{R}$, which is strictly monotone. Since boundary conditions are being prescribed on all surfaces of the time-space cylinder $(0,T) \times B_a^n$ this problem falls into the class of hyperbolic boundary value problems.

1980 Mathematics Subject Classification (1985 Revision). 35L70

* Department of Mathematics, Iowa State University, Ames, Iowa, 50011, USA

When $n = 1$ the problem is equivalent to the problem of the periodically forced semilinear vibrating string with an additional constraint of symmetry about the mid-point of the string. With and without this additional constraint, the problem has received considerable attention over the last two decades (cf. [2]). Here we wish to show that some of the known results for $n = 1$ generalize to the case $n > 1$, under the assumption of radial symmetry. In particular we show, when a/T is rational and g is a strictly monotone Lipschitz continuous function, that solutions of the problem exist for every T-periodic forcing function f which is radially symmetric. When a/T is irrational both the associated linear operator and its partial inverse are unbounded; one encounters a problem of small divisors. In this case no general results are known, even when $n = 1$ (cf. [2]). With the assumption of radial symmetry in force the problem can be written in the form

$$(1.1) \qquad u_{tt} - u_{rr} - \frac{(n-1)}{r} u_r + g(u) = f \ , \qquad (t,r) \in \mathbf{R} \times (0,a)$$

$$(1.2) \qquad u(t,a) = 0, \qquad t \in \mathbf{R},$$

$$(1.3) \qquad u(t+T,r) = u(t,r), \qquad (t,r) \in \mathbf{R} \times (0,a) \ .$$

Without the symmetry hypothesis, there are no results known to the author when the spatial domain is a ball. However when the spatial domain is a hyper-rectangle in $\mathbf{R}^n$, similar results on existence, uniqueness and continuous dependence have been obtained in [8].

By a solution of (1.1)-(1.3) we mean a weak solution in the following sense. Let D denote the class of functions $\phi \in C^\infty(\mathbf{R} \times B_a^n)$ which are T-periodic in time for each $x \in B_a^n$, and are radially symmetric having compact support in B_a^n for each $t \in \mathbf{R}$. Let H denote the space of functions $u : \mathbf{R} \times B_a^n \to \mathbf{R}$ which are T-periodic, radially symmetric and belong to $L^2((0,T) \times B_a^n)$. We equip H with the L^2-topology making it a Hilbert space. We say that $u \in H$ is a weak solution of (1.1)-(1.3) provided that

$$\int_0^T \int_0^a [u(\phi_{tt} - \phi_{rr} - \frac{n-1}{r} \phi_r) + (g(u) - f)\phi] r^{n-1} dr dt = 0 \ ,$$

for every $\phi \in D$. Under the assumption that $g \in C(\mathbf{R})$ satisfies

$$(1.4) \qquad |g(u) - g(v)| \leq \beta_1 |u - v| , \qquad u,v \in \mathbf{R},$$

$$(1.5) \qquad [g(u) - g(v)](u - v) \geq \beta_0 |u - v|^2 , \qquad u,v \in \mathbf{R},$$

we shall prove that (weak) solutions of (1.1)-(1.3) do exist. In particular we will obtain the following perturbation result as an immediate corollary.

THEOREM 1.1 Let $h \in C(\mathbf{R})$ and satisfy (1.4)-(1.5). Set $g(u) = \varepsilon h(u)$ where $\varepsilon > 0$. Suppose that the ratio a/T of the radius of the ball to the period of oscillation is a rational number. Then for each dimension $n = 1,2,3,\ldots$ there is a number $\varepsilon_n > 0$ such that (1.1)-(1.3) has a unique solution $u \in H$, for every $f \in H$ and ε in the interval $(0,\varepsilon_n)$. Moreover u depends continuously on the pair ε, f.

In order to state the stronger version of this result we require some additional notation and definitions. We introduce these in the next section.

2. A GENERAL THEORY

The following framework provides a general setting for nonlinear hyperbolic boundary value problems. We follow [7], [8] in presenting some basic facts. We use H to denote a real separable Hilbert space, L and G to denote linear and nonlinear operators in H. We are concerned with the problem

$$(2.1) \qquad Lu + Gu = f, \quad u \in \operatorname{dom}(L) \subset H,$$

where $f \in H$. The domain of G is assumed to be all of H. The linear operator L is assumed to be closed, densely defined and symmetric, with eigenvalues $\{\lambda_i : i \in I\}$ and eigenvectors $\{\phi_i : i \in I\}$. Here I denotes a countable index set (e.g. a set of multi-indices). Further we assume completeness: the set $\{\phi_i : i \in I\}$ forms a complete orthonormal basis for H. Letting $\|\cdot\|$ and $(\cdot,\cdot)$ denote the norm and inner product in H we set $H_0 = \{u \in H : Lu = 0, \ u \in \operatorname{dom}(L)\}$ and $H_1 = H_0^{\perp} = \{u \in H : (u,v) = 0$ for all $v \in H_0\}$. Let $I_0 = \{i \in I : \lambda_i = 0\}$, $I_1 = \{i \in I : \lambda_i \neq 0\}$, and $\Lambda_1 = \{\lambda_i : \lambda_i \neq 0\} \subset \mathbf{R}$. From [7] we have

THEOREM 2.1 If there exists $\delta > 0$ such that $(-\delta,\delta) \cap \Lambda_1 = \emptyset$, the empty set, then:

i) There is a Hilbert subspace $\mathcal{H}_1 \subset H_1$, with dual $\mathcal{H}_1^*$, such that $\mathcal{H}_1 \subset H_1 \subset \mathcal{H}_1^*$ algebraically, topologically and densely.

ii) $L : \mathcal{H}_1 \to \mathcal{H}_1^*$ is an isometric isomorphism.

iii) The operator L has a continuous linear partial inverse $K : H_1 \to \mathcal{H}_1$ with $LK = id_{H_1}$ and $KL = id_{\mathcal{H}_1}$ on $dom(L)$.

REMARKS: 1) The above theorem provides a hyperbolic version of the Lax-Milgram theorem.

2) The norm in $\mathcal{H}_1$ is defined by

$$\|u\|_{\mathcal{H}_1} = \{\sum_{i \in I_1} |\lambda_i| u_i^2\}^{1/2}, \quad u_i = (u,\phi_i) .$$

We can define a Hilbert subspace $\mathcal{H} \subset H$ when Theorem 2.1 applies, by setting $\mathcal{H} = H_0 + \mathcal{H}_1$ and using the norm

$$\|u\|_{\mathcal{H}} = \{\sum_{i \in I} (1 + |\lambda_i|) u_i^2\}^{1/2} .$$

This will be the space to which our solutions will belong.

In the hyperbolic case we expect that the set Λ_1 is doubly infinite, that is containing subsequences which tend to both $+\infty$ and $-\infty$. In this case the sets $\Lambda_1 \cap (-\infty,-\delta)$ and $\Lambda_1 \cap (\delta,+\infty)$ will be non-empty for any $\delta > 0$; and the constants $\kappa_- = \inf\{|\lambda_i| : \lambda_i \in \Lambda_1 \cap (-\infty,-\delta)\}$, $\kappa_+ = \inf\{|\lambda_i| : \lambda_i \in \Lambda_1 \cap (\delta,+\infty)\}$ will satisfy $\delta \leq \kappa_+,\kappa_- < +\infty$. Let $J_0 = \{i \in I : 0 \leq \lambda_i < \kappa_-\}$ and $J_1 = \{i \in I : |\lambda_i| \geq \kappa_-\}$ so that $J_0 \cup J_1 = I$. We need at least the first of the following hypotheses on G: there are constants $\beta_0, \beta_1 > 0$ such that

(2.2) $\quad \|Gu - Gv\| \leq \beta_1 \|u - v\|, \qquad \forall u,v \in H,$

(2.3) $\quad (Gu-Gv, u-v) \geq \beta_0 \|u-v\|^2, \qquad \forall u,v \in H.$

From [8] we record the following.

THEOREM 2.2 Let L, H be as above and suppose G satisfies (2.2). Further assume that there exists $\delta > 0$ such that $\Lambda_1 \cap (-\delta,\delta) = \emptyset$ and $\Lambda_1 \cap (-\infty,-\delta) \neq \emptyset$, with $\emptyset$ denoting the empty set.

i) (nonresonance) If $I_0 = \emptyset$ and $\beta_1 < \delta$ then for all $f \in H$ there is a unique solution $u \in \mathcal{H}$ of (2.1) having the a priori bounds

$$\|u\| \leq (\delta - \beta_1)^{-1} \|f\|, \quad \|u\|_{\mathcal{H}} \leq (\delta^{-2} + \delta^{-1})^{1/2} (\|f\| + \beta_1 \|u\|).$$

In addition u depends continuously on f.

ii) (resonance) Let $\kappa = \kappa_-$ and suppose that G satisfies (2.3). If $I_0 \neq \emptyset$ so that $J_0 \neq \emptyset$, and if $\beta_1^2 < \kappa\beta_0$ then for all $f \in H$ there is a unique solution $u \in \mathcal{H}$ of (2.1) having the a priori bounds

$$\|u\| \leq (\kappa + \beta_1)(\kappa\beta_0 - \beta_1^2)^{-1}\|f\|, \quad \|u\|_{\mathcal{H}} \leq c[\kappa\beta_*(\kappa\beta_0 - \beta_1^2)]^{-1}\|f\|,$$

where $\beta_* = \min(1,\beta_0)$ and c is a positive constant. In addition u depends continuously on f.

REMARKS: 1) The proof of part ii) of the above theorem uses the splitting technique of the method of alternative problems (cf. [3], [4]). The proof is constructive in nature.

2) The condition (2.3) can be reversed to $(Gu-Gv, u-v) \leq -\beta_0\|u-v\|^2$. In this case we set $\kappa = \kappa_+$.

3. AN EIGENVALUE PROGLEM

Knowledge of the spectral properties of the linear operator associated with (1.1)-(1.3) is essential in applying the general theory of the previous section. Therefore we need to consider the eigenvalue problem

(3.1) $$\phi_{tt} - \left(\phi_{rr} + \frac{n-1}{r}\phi_r\right) = \lambda\phi, \qquad (t,r) \in \mathbf{R} \times (0,a),$$

(3.2) $$\phi(t,a) = 0, \qquad t \in \mathbf{R},$$

(3.3) $$\phi(t+T,r) = \phi(t,r), \qquad (t,r) \in \mathbf{R} \times (0,a).$$

Employing the classical method of separation of variables we set $\phi(t,r) = \tau(t)\rho(r)$ and derive that ρ must satisfy

(3.4) $$r^2\rho'' + (n-1)r\rho' + r^2\mu^2\rho = 0, \qquad 0 < r < a,$$

(3.5) $$\rho(a) = 0 \text{ and } \rho \text{ bounded on } [0,a],$$

where $\mu^2 = \lambda + (2\pi k/T)^2$ for any integer $k \geq 0$. The function τ_k is taken to be any one of the corresponding functions $\cos(2k\pi t/T)$, $\sin(2k\pi t/T)$, or 1 if $k = 0$. The change of variables $\psi(r) = r^{(n-2)/2}\rho(r)$ transforms (3.4)-(3.5) into

$$(3.6)\qquad r^2\psi'' + r\psi' + \left[\mu^2 r^2 - \left(\frac{n-2}{2}\right)^2\right]\psi = 0, \qquad 0 < r < a,$$

$$(3.7)\qquad \psi(a) = 0, \quad \psi(r) = O\left(r^{n-2/2}\right) \quad \text{as} \quad r \to 0^+.$$

This is the classical eigenvalue problem for Bessel's equation of order $\nu = (n-2)/2$. Let $J_\nu(x)$ denote the Bessel function of the first kind of order $\nu = (n-2)/2$ (cf. [1], [9]), then $y = J_\nu(x)$ satisfies

$$(3.8)\qquad x^2y'' + xy' + (x^2 - \nu^2)y = 0, \quad x > 0,$$

and $J_\nu(x) = O(x^\nu)$ as $x \to 0^+$. It is well-known that J_ν has an infinite sequence of distinct positive zeros $\{x_{\nu,j}\}_{j=1}^\infty$ tending to infinity. The eigenvalues for problem (3.6)-(3.7) are seen to be $\mu_{n,j}^2 = \left(\alpha_{n,j}/a\right)^2$ where $\alpha_{n,j}$ is the j-th positive zero of J_ν when $\nu = (n-2)/2$. The corresponding eigenfunctions are $\psi_{n,j}(r) = J_\nu\left(\alpha_{n,j}r/a\right)$ Hence problem (3.1)-(3.3) has eigenvalues and eigenfunctions

$$(3.9)\qquad \lambda_{j,k}^n = \left(\frac{\alpha_{n,j}}{a}\right)^2 - \left(\frac{2k\pi}{T}\right)^2,$$

$$(3.10)\qquad \phi_{j,k}^n(t,r) = \begin{Bmatrix} \cos\left(2k\pi t/T\right) \\ \\ \sin\left(2k\pi t/T\right) \end{Bmatrix} r^{(2-n)/2} J_{(n-2)/2}\left(\alpha_{n,j}r/a\right),$$

for $k \geq 0$ and $j \geq 1$. It is clear that for each $n \geq 1$ the sequence $\{\lambda_{j,k}^n\}$ is unbounded both above and below. This is characteristic of hyperbolic boundary value problems.

4. RADIALLY SYMMETRIC TIME-PERIODIC SOLUTIONS

We assume throughout this section that the ratio a/T is a rational number, and write $a/T = p/q$ where p,q are nonnegative integers. We are interested in the nature of the spectrum near the origin. In particular we need to establish a lower bound for the set of numbers $\{|\lambda_{j,k}^n| : \lambda_{j,k}^n \neq 0\}$. The numbers $\lambda_{j,k}^n$ are the eigenvalues given in (3.9) and it is clear that $|\lambda_{j,k}^n| = a^{-2}|\alpha_{n,j} - 2k\pi p/q|\left(\alpha_{n,j} + 2k\pi p/q\right) \geq a^{-2}\alpha_{n,j}|\alpha_{n,j} - 2k\pi p/q|$. Using the notation of the Appendix we write $\alpha_{n,j} = b_{\nu,j} - \varepsilon_{\nu,j}$ with $\nu = (n-2)/2$ and

$b_{\nu,j} = (j + \frac{1}{2}\nu - \frac{1}{4})\pi$. Let $R_{njk} = b_{\nu,j}/\pi - 2kp/q = j + (n-3)/4 - 2kp/q$. From the rational form of R_{njk} we know that either $R_{njk} = 0$ or $|R_{njk}| \geq (4q)^{-1}$. Employing this notation we have

$|\lambda^n_{j,k}| \geq a^{-2}\alpha_{n,j}|\alpha_{n,j} - 2k\pi p/q| = a^{-2}\alpha_{n,j}|\pi R_{njk} - \varepsilon_{\nu,j}|$.

Suppose that $\lambda^n_{j,k} \neq 0$ so that $\pi R_{njk} - \varepsilon_{\nu,j} \neq 0$. If $n = 1$ or 3 then $\varepsilon_{\nu,j} = 0$ and it follows that $|\lambda^n_{j,k}| \geq \pi\alpha_{n,j}/4a^2q \geq \pi\alpha_{n,1}/4a^2q$. If $n \neq 1$ or 3 then $\varepsilon_{\nu,j} \neq 0$, but $\varepsilon_{\nu,j} \to 0$ as $j \to +\infty$. Let j_0 be so large that $|\varepsilon_{\nu,j}| < \pi/8q$ for all $j \geq j_0$ (of course $j_0 = j_0(n)$). If $\lambda^n_{j,k} \neq 0$, $j \geq j_0$, and if $R_{njk} \neq 0$ then

$|\lambda^n_{j,k}| \geq a^{-2}\alpha_{n,j}\,\pi/8q \geq \pi\alpha_{n,1}/8a^2q$. On the other hand if $R_{njk} = 0$ then we may use one of the estimates (A.1) or (A.2) from the Appendix to see that (recall $\nu = (n-2)/2$)

$$(4.1) \qquad |\lambda^n_{j,k}| \geq a^{-2}\alpha_{n,j}|\varepsilon_{\nu,j}| \geq |\nu^2 - \tfrac{1}{4}|\pi x_{\nu,j}/2a^2\delta_{\nu,j}x_{\nu,j+1},$$

where $x_{\nu,j}$ denotes the j-th positive zero of $J_\nu(x)$ and $\delta_{\nu j} = d_{\nu,j} = x_{\nu,j+1} - x_{\nu,j}$ if $\nu > \frac{1}{2}$ (i.e. $n > 3$) or $\delta_{\nu,j} = d_{\nu,j+1}$ if $|\nu| < \frac{1}{2}$ (i.e. $n = 2$). In any case we have $\delta_{\nu,j} \to \pi$ and $x_{\nu,j}/x_{\nu,j+1} \to 1$ as $j \to +\infty$. Hence for all j sufficiently large, whenever $R_{njk} = 0$ and $\lambda^n_{j,k} \neq 0$ we deduce that

$$|\lambda^n_{j,k}| \geq |\nu^2 - \tfrac{1}{4}|/8a^2 = |n^2-4n+3|/32a^2 .$$

We have therefore verified that there is an integer $j_0 > 0$ and a constant $\delta > 0$, independent of j and k, such that either $|\lambda^n_{j,k}| \geq \delta$ or $\lambda^n_{j,k} = 0$ for all $j \geq j_0$, $k \geq 0$ Of the remaining eigenvalues, there can be at most a finite number which are equal to or closely approximate the value zero. It follows that $\lambda = 0$ is not a cluster point of the spectrum $\{\lambda^n_{j,k}\}$.

THEOREM 4.1 Let $g \in C(\mathbf{R})$ and satisfy (1.4)-(1.5) with $\beta_0, \beta_1 > 0$. Suppose that the ratio a/T is a rational number. Then for each dimension $n = 1,2,3,\ldots$ the numbers

$$\delta_n = \inf\{|\lambda^n_{j,k}| : \lambda^n_{j,k} \neq 0 \text{ with } j \geq 1, \ k \geq 0\}, \text{ and}$$

$$\kappa_n = \inf\{|\lambda^n_{j,k}| : \lambda^n_{j,k} < 0 \text{ with } j \geq 1, \ k \geq 0\}$$

are positive. If $\beta_1^2 < \kappa_n \beta_0$ then there is a unique solution $u \in H$ of (1.1)-(1.3) for every $f \in H$. Moreover u depends continuously on f.

REMARKS: 1) The solution actually belongs to the smaller space $\mathcal{H} \subset H$ which was described in section 2. In the 1-dimensional case it has been shown [7] that $\mathcal{H}$ is not one of the usual Sobolev spaces.

2) Theorem 1.1 in the introductory section follows immediately from the above theorem since $(\varepsilon\beta_1)^2 < \kappa(\varepsilon\beta_0)$ is equivalent to $0 < \varepsilon < \varepsilon_n = \kappa_n\beta_0/\beta_1^2$.

3) The above argument does not immediately carry over to the 2-dimensional problem without radial symmetry. In this problem the eigenvalues are $\lambda_{mjk} = (x_{m,j}/a)^2 - (2\pi k/T)^2$ where $x_{m,j}$ is the j-th positive zero of $J_m(x)$. For each m we may choose N_m sufficiently large so that $|\lambda_{mjk}| \geq \delta > 0$ for all $j \geq N_m$ and all k. However estimates for small j are not as precise and do not seem to exclude the possibility of the existence of a subsequence $\{j_m, k_m\}_{m=1}^{\infty}$, with j_m small, such that $\lambda_{m,j_m,k_m} \to 0$ as $m \to \infty$.

4) There may or may not be resonance in the problem (i.e. $\lambda^n_{jk} = 0$ for some j,k). If there is no resonance ($\lambda^n_{jk} \neq 0$ for all j,k) then we only need to require that g satisfies (1.4) where $\beta_1 < \kappa_n$. In the lower dimensions $n = 1,2,3$ we see both resonance and non-resonance. Suppose that $n = 1$ or 3. It is well known that any solution of Bessel's equation of order $\nu = \pm 1/2$ has the form $y = (A \cos x + B \sin x)/\sqrt{x}$; recall that $\nu = (n-2)/2$. Since $J_\nu(x) = O(x^\nu)$ as $x \to 0^+$ we have $J_{-1/2}(x) = x^{-1/2}\cos x$ and $J_{1/2}(x) = x^{-1/2}\sin x$. Hence $\alpha_{1,j} = (2j-1)\pi/2$, $j \geq 1$, and $\alpha_{3,j} = j\pi$, $j \geq 1$. From (3.9) we see that the eigenvalues of (3.1)-(3.3) are

$$\lambda^1_{j,k} = \left(\frac{\pi}{a}\right)^2 \left[\frac{(2j-1)^2}{4} - \left(\frac{2ka}{T}\right)^2\right], \quad \text{and}$$

$$\lambda^3_{j,k} = \left(\frac{\pi}{a}\right)^2 \left[j^2 - \left(\frac{2ka}{T}\right)^2\right] .$$

The case studied most often (cf. [2]) when $n = 1$ is that of $T = 2\pi$ and $2a = \pi$; then $T/2a = 2$. In this case $\lambda^1_{j,k} = (2j-1)^2 - k^2$ and $\lambda^3_{j,k} = 4j^2 - k^2$. Clearly $\lambda = 0$ is an eigenvalue of infinite multiplicity in either of there dimensions. For the sake of comparison to known results we point out that $|\lambda^1_{jk}| \geq 1$ whenever $\lambda^1_{j,k} \neq 0$, and $|\lambda^1_{j,k}| \geq 3$ whenever $\lambda^1_{j,k} < 0$. Thus $\kappa = 3$.

When the dimension $n = 2$ the situation changes. Here the numbers $\alpha_{2,j}$ arise as the zeros of $J_0(x)$. In particular it is known (cf. [1], [9]) that $(j-1/4)\pi < \alpha_{2,j} < (j-\frac{1}{8})\pi$ for all $j \geq 1$. Therefore when $T = 2\pi$, $2a = \pi$ we find that for all $j \geq 1$, $k \geq 0$

$$|\lambda^2_{j,k}| = \left(\frac{2}{\pi}\right)^2 \left|\alpha^2_{2,j} - \left(\frac{k\pi}{2}\right)^2\right| \geq \left(\frac{2}{\pi}\right)^2 \left(\frac{\pi}{8}\right)\alpha_{2,j} \geq \frac{\alpha_{2,1}}{2\pi},$$

where $\alpha_{2,1}$ denotes the 1st positive zero of $J_0(x)$. Hence there is no resonance.

REFERENCES

[1] M. Abramowitz and I. Stegun (eds.), Handbook of Mathematical Functions, Dover, New York, 1965.

[2] H. Brezis, Periodic solutions of nonlinear vibrating strings and duality principles, Bull. Amer. Math. Soc., 8(1983) no. 3, 409-426.

[3] L. Cesari, Functional analysis, nonlinear differential equations, and the alternative method, in Nonlinear Functional Analysis and Differential Equations, Cesari-Kannan-Schuur, eds., Marcel Dekker, New York, 1976, 1-197.

[4] S. Chow and J. Hale, Methods of Bifurcation Theory, Springer-Verlag, New York, 1982.

[5] L. Gatteschi, On the zeros of certain functions with applications to Bessel functions, Nederl. Wetensch. Proc. Ser. A, 55(1952), 224-229.

[6] H. Hethcote, Error bounds for asymptotic approxmations of zeros of transcendental functions, SIAM J. Math. Anal., 1(1970), no. 2, 147-152.

[7] M. Smiley, Hyperbolic Boundary Value Problems: A Lax-Milgram approach and the vibrating string, Bollettino Unione Mat. Ital. (6) 5-A (1986), 23-32.

[8] M. Smiley, Eigenfunction methods and hyperbolic boundary value problems, to appear in Journal of Math. Anal. Appl.

[9] G. Watson, A Treatise on the theory of Bessel functions, Cambridge University Press, 1922.

APPENDIX, A Sturmian Argument

Our goal is to present some lower and upper bound on the differences $\varepsilon_{\nu,j} = b_{\nu,j} - x_{\nu,j}$, where $b_{\nu,j} = (j + \frac{1}{2}\nu - \frac{1}{4})\pi$ and $x_{\nu,j}$ is the j-th positive zero (in ascending order) of $J_\nu(x)$. By $J_\nu(x)$ we mean the Bessel function of the first kind of order ν. Although an asymptotic form of the lower bounds used in section 4 can be obtained from ([1]) McMahon's expansion $x_{\nu,j} = b_{\nu,j} - (4\nu^2-1)/(8b_{\nu,j}) + O(j^{-3})$, as $j \to +\infty$, we wish to present a derivation which uses an elementary variation of the arguments originating with Sturm. Despite the simplicity of the arguments we obtain estimates which are valid for all $j \geq 2$ and $\nu > -\frac{1}{2}$, showing the power and utility of the Sturmian methods. Upper bounds, established by less elementary means, can be found in [5], [6] and references therein.

We begin with the observation that $u(x) = \sqrt{x}\, J_\nu(x)$ satisfies

$$u'' + \left(1 + \frac{1 - 4\nu^2}{4x^2}\right)u = 0, \quad x > 0.$$

If $v(x) = \sin(\sigma x + \theta)$, with σ, θ real, then $[vu' - u'v]' = wuv$, where $w = w(\sigma,\nu,x) = 1 - \sigma^2 + (1 - 4\nu^2)/4x^2$. Let $x_{\nu,j}$ and $x_{\nu,j+1}$ be consecutive zeros of J_ν, $\sigma = \pi/(x_{\nu,j+1} - x_{\nu,j}) = \pi/d_{\nu,j}$ and $\theta = -\sigma x_{\nu,j}$. Then

$$\int_{x_{\nu,j}}^{x_{\nu,j+1}} w(\sigma,\nu,x)u(x)v(x)dx = 0 ,$$

which implies that $w(\sigma,\nu,z_{\nu,j}) = 0$ for some $z_{\nu,j}$ between $x_{\nu,j}$ and $x_{\nu,j+1}$. Since $\sigma = \pi/d_{\nu,j}$, where $d_{\nu,j}$ is the distance between $x_{\nu,j}$ and $x_{\nu,j+1}$, it follows that $d_{\nu,j} = h(z_{\nu,j},\nu)$ where

$h(z,\nu) = 2\pi z/\sqrt{1 + 4(z^2-\nu^2)}$. Clearly $h(z, \pm\frac{1}{2}) = \pi$ and $\lim_{z \to +\infty} h(z,\nu) = \pi$, $\nu \geq -\frac{1}{2}$. Since $\partial h/\partial z = 2\pi[1 + 4(z^2-\nu^2)]^{-3/2}(1 - 4\nu^2)$ it follows that $d_{\nu,j}$ is monotone increasing (decreasing) when $|\nu| < \frac{1}{2}$ $(\nu > \frac{1}{2})$. From the identity $\varepsilon_{\nu,j+1} - \varepsilon_{\nu,j} = \pi - d_{\nu,j}$ it follows that $\varepsilon_{\nu,j}$ is monotone increasing (decreasing) when $|\nu| < \frac{1}{2}$ $(\nu > \frac{1}{2})$; and also that

$$\varepsilon_{\nu,j} = \varepsilon_{\nu,1} + \sum_{i=1}^{j-1} (\pi - d_{\nu,i}) .$$

Since $\lim_{j \to +\infty} \varepsilon_{\nu,j} = 0$ by McMahon's expansion we can eliminate $\varepsilon_{\nu,1}$ above to obtain

$$\varepsilon_{\nu,j} = \sum_{i=j}^{\infty} (d_{\nu,i} - \pi)$$

Using the functions $f(z) = h(z,\nu) - \pi$ and $F(z) = [2z - \sqrt{1 + 4(z^2-\nu^2)}]\pi/2$, and the elementary technique of integral comparison we obtain for $j \geq 2$, $\nu > \frac{1}{2}$

$$\varepsilon_{\nu,j} = \sum_{i=j}^{\infty} (d_{\nu,i}-\pi) < \int_{x_{\nu,j-1}}^{+\infty} f(x)dx = F(x_{\nu,j-1})$$

$$\varepsilon_{\nu,j} = \sum_{i=j}^{\infty} (d_{\nu,i} - \pi) > \frac{1}{d_{\nu,j}} \int_{x_{\nu,j+1}}^{+\infty} f(x)dx = \frac{1}{d_{\nu,j}} F(x_{\nu,j+1})$$

The estimates $|\nu^2 - \frac{1}{4}|\pi/2z \leq |F(z)| \leq |\nu^2 - \frac{1}{4}|\pi/z$, $z > 0$, can now be used to show

$$\frac{|\nu^2 - \frac{1}{4}|\pi}{2d_{\nu,j}\, x_{\nu,j+1}} < \varepsilon_{\nu,j} < \frac{|\nu^2 - \frac{1}{4}|\pi}{x_{\nu,j-1}} \qquad (\nu > \tfrac{1}{2},\ j \geq 2) \tag{A.1}$$

Similar arguments can be used to show

$$\frac{|\nu^2 - \frac{1}{4}|\pi}{2d_{\nu,j+1}\, x_{\nu,j+1}} < |\varepsilon_{\nu,j}| < \frac{|\nu^2 - \frac{1}{4}|\pi}{x_{\nu,j-1}} \qquad (|\nu| < \tfrac{1}{2},\ j \geq 2) \tag{A.2}$$

Canadian Mathematical Society
Conference Proceedings
Volume 8 (1987)

AN OSCILLATION CRITERION FOR SECOND ORDER SUBLINEAR DIFFERENTIAL EQUATIONS

JAMES S. W. WONG

ABSTRACT. An oscillation criterion is given for the second order sublinear differential equation $x''+a(t)|x|^{\gamma}\operatorname{sgn} x = 0$, $0<\gamma<1$, where the coefficient a(t) is not assumed to be non-negative for all large values of t. The result extends a condition obtained by Kamenev for the linear differential equation.

Consider the second order sublinear differential equation

(1) $$x''+a(t)|x|^{\gamma}\operatorname{sgn} x = 0, \qquad 0<\gamma<1,$$

where $a(t)\in C[0,\infty)$. We restrict our attention to solutions of (1) which exist on some ray $[t_0,\infty)$, where $t_0\geq 0$ may depend on the particular solution. Such a solution is said to be oscillatory if it has arbitrarily large zeros. Equation (1) is called oscillatory if all such solutions are oscillatory. For a general discussion on sublinear oscillation problems, we refer the reader to [9]. We are here concerned with sufficient conditions on a(t) for the oscillation of (1) when a(t) is allowed to assume negative values for arbitrarily large values of t. In the linear case, the well-known Wintner's oscillation theorem states that if a(t) satisfies

(2) $$\lim_{T\to\infty}\frac{1}{T}\int_0^T\int_0^t a(s)\,ds\,dt = +\infty,$$

then equation (1) is oscillatory for $\gamma=1$, see [7]. Recently, Bulter [3] proved that Wintner's theorem remains valid for equation (1) where $\gamma>1$. In the sublinear case, i.e. $0<\gamma<1$, condition (2), can be relaxed to

(3) $$\limsup_{T\to\infty}\frac{1}{T}\int_0^T\int_0^t a(s)\,ds\,dt = +\infty,$$

1980 Mathematics Subject Classification : Primary 34C10, 34C15.

an earlier result due to Kamenev [4]. We note that Condition (3) alone is not sufficient for oscillation for $\gamma \geq 1$, in the linear case, see Willett [8].

In a later paper, Kamenev [5] obtained an extension of Wintner's result and proved the following oscillation criterion for the linear equation, i.e. for some $\alpha > 1$,

(5) $$\limsup_{T\to\infty} \frac{1}{T^\alpha} \int_0^T (T-s)^\alpha a(s)\,ds = +\infty.$$

It is therefore natural to ask whether condition (5) would imply oscillation also for the sublinear equation (1). The purpose of this note is to answer this in the affirmative. Indeed, we can prove somewhat more, namely,

THEOREM. Let $\varphi : [0,\infty) \to [0,\infty)$ be a positive nondecreasing concave function, i.e. $\varphi > 0$, $\varphi' \geq 0$, $\varphi'' \leq 0$.

For some $\alpha > 1$, if $a(t)$ satisfies

(6) $$\limsup_{T\to\infty} \frac{1}{T^\alpha} \int_0^T (T-s)^\alpha \varphi^\gamma(s) a(s)\,ds = +\infty,$$

Then equation (1) is oscillatory.

PROOF. Assume the contrary that there exists a nonoscillatory solution $y(t)$ which may be assumed to be positive on $[t_0,\infty)$ for some $t_0 \geq 0$. For $t \geq t_0$, define $z(t) = [y(t)/\varphi(t)]^\gamma$ which is clearly positive. Let $\beta = 1/\gamma > 1$, then $y(t) = \varphi(t) z^\beta(t)$ and we can easily verify

(7) $$\frac{1}{z}(\varphi z^\beta)'' = \frac{\beta}{\beta-1}(\varphi z^{\beta-1})'' + \beta\varphi z^{\beta-3} z'^2 + \frac{1}{1-\beta}\varphi'' z^{\beta-1},$$

an identity which has been used in our earlier paper [6]. Note from (1)

(8) $$\frac{(\varphi z^\beta)''}{z} = \frac{y''}{z} = -a\varphi^\gamma.$$

Since $\varphi'' \leq 0$ and $\beta > 1$, the last two terms in (7) are non-negative, hence we may combine (7) and (8) to obtain the following inequality:

(9) $$\frac{\beta}{\beta-1}(\varphi z^{\beta-1})'' \leq -\varphi^\gamma a.$$

In order to apply condition (6), we first note the following identity for any C^2 function $f(t)$:

$$-\frac{1}{T^{\alpha}}\int_{t_o}^{T}(T-s)^{\alpha}f''(s)\,ds = \frac{1}{T^{\alpha}}\Big\{(T-t_o)^{\alpha}f'(t_o)+\alpha(T-t_o)^{\alpha-1}f(t_o) - \alpha(\alpha-1)\int_{t_o}^{T}(T-s)^{\alpha-2}f(s)\,ds\Big\} \tag{10}$$

The last integral in (10) exists because $\alpha>1$. Letting $f(t)=\varphi(t)z^{\beta-1}(t)$ in (10), and noting that φ, z are positive, we find

$$-\frac{1}{T^{\alpha}}\int_{t_o}^{T}(T-s)^{\alpha}(\varphi z^{\beta-1})''(s)\,ds \leqslant \Big\{(\varphi z^{\beta-1})'(t_o) + \frac{\alpha}{T}(\varphi z^{\beta-1})(t_o)\Big\} \tag{11}$$

Using (9) in (11) and $\beta>1$, we obtain

$$-(\varphi z^{\beta-1})'(t_o) - \frac{\alpha}{T}(z^{\beta-1})(t_o) \leqslant \Big(\frac{1-\beta}{\beta}\Big)\frac{1}{T^{\alpha}}\int_{t_o}^{T}(T-s)^{\alpha}\varphi^{\gamma}(s)a(s)\,ds,$$

which gives the desired contradiction upon applying condition (6).

REMARK 1. As a typical example, we let $\varphi(t)\equiv t^{\sigma}$, $0\leqslant\sigma\leqslant 1$. In particular for $\sigma=0$, condition (6) reduces to condition (5) showing that Kamenev's oscillation cirterion remains valid for the sublinear equation (1).

REMARK 2. In [10], we showed that condition (5) together with a bounded below condition

$$\liminf_{t\to\infty}\int_{o}^{t} a(s)\,ds = -\lambda > -\infty, \qquad \lambda > 0, \tag{12}$$

are sufficient for the oscillation of (1) in the superlinear case when $\gamma>1$.

REMARK 3. We also note that the Theorem remains valid with $\alpha=1$ in condition (6) although the present proof would fail. This is in fact another result of ours [6] extending the well known theorem of Belohorec [1], [2].

REFERENCE

[1] S. Belohorec, "Oscillatory Solutions of Certain Nonlinear Differential Equations of Second Order", Mat. Fyz. Casopis, Sloven, Akad. Vied., 11(1961), 250-255.

[2] S. Belohorec, "Two Remarks on the Properties of Solutions of a Nonlinear Differential Equation", Acta Fac. Rerum. Natur. Univ. Comen. Mathematica, XXVII (1969), 19-26.

[3] G.J. Butler, "Integral Averages and the Oscillation of Second Order Ordinary Differential Equations", SIAM J. Math. Anal. 11(1980), 190-200.

[4] I.V. Kamenev, "Certain Specificially Nonlinear Oscillation Theorems", Mat. Zametki, 10(1971), 129-134. (Russian).

[5] I.V. Kamenev, "Integral Criterion for Oscillations of Linear Differential Equations of Second Order", Mat. Zametki, 23 (1978), 249-251. (Russian).

[6] M.K. Kwong and J.S.W. Wong, "On a Theorem of Belohorec", SIAM J. Math Anal., 14(1983), 474-476.

[7] A. Wintner, "A Criterion of Oscillatory Stability", Quart. Appl. Math. 7(1949), 114-117.

[8] D. Willett, "Classification of Second Order Linear Differential Equation with Respect to Oscillation", Advances in Math, 3(1969), 594-623.

[9] J.S.W. Wong, "Oscillation Theorems for Second Order Nonlinear Differential Equations", Bull. Inst. Math. Acad. Sinica 3 (1975), 283-309.

[10] J.S.W. Wong, "An Oscillation Criterion for Second Order Nonlinear Differential Equations", Proc. Amer. Math. Soc. 98 (1986), (to appear).

CHINA DYEING WORKS, LTD.
819 Swire House
Hong Kong

UNIVERSITY OF HONG KONG
Pokfulam
Hong Kong

Part 2
Bifurcation and Chaos

Canadian Mathematical Society
Conference Proceedings
Volume 8 (1987)

STABILITY OF A DIFFERENTIAL-DELAY EQUATION WITH TWO TIME LAGS

Jacques Bélair[1]

ABSTRACT. The local asymptotic stability of the null solution of a linearized differential-delay equation with two time delays is determined.The ratio of the time lags is fixed, and the coefficients are considered as the main parameters.An example occuring in the modeling of a physiological control system is provided as an application of the analysis.

1. INTRODUCTION. The equation

$$x'(t) = p(x(t)) - d(x(t-\tau)) \quad (1)$$

with $x(\cdot)$ a scalar, real-valued function and τ a fixed, positive number, occurs in models from population biology and physiology [3,6,9,13]. Irregular behavior of solutions has been observed [3,9,13], when $p(\cdot)$ is linear and $d(\cdot)$ is a nonmonotonic function. Equation (1) only includes dependence on one fixed instant in the past. More sophisticated models would postulate the influence of several former levels for the dynamics. In particular, the equation

$$x'(t) = - a_0 x(t) + f(x(t-\tau_1), x(t-\tau_2)) \quad (2)$$

has recently been considered in the modeling of platelet production in humans [1,2]. In this model, there are two time delays whose ratio has a value, in normal conditions, of about two. By scaling time, there is no loss of generality in taking one of the delays to be one.

The first step in the study of the solutions of equation (2) is to determine the linear stability of its stationary solutions. In generic situations, this will determine their local asymptotic stability.The purpose of this paper is the investigation of the asymptotic stability of the null solution of the linear equation

$$x'(t) = - a_0 x(t) + a_1 x(t-1) + a_2 x(t-\tau) \quad (3)$$

where a_i are constants and a_0 is non-negative. This stability is determined by the location in the complex plane of the roots of the characteristic equation

$$z + a_0 = a_1 e^{-z} + a_2 e^{-z\tau} \quad (4)$$

obtained by substituting $x(t) = e^{zt}$ in equation (3). If all these roots have negative real parts, then

1980 Mathematics Subject Classification (1985 Revision). 34K20, 92A09.
[1]Supported by NSERC (Canada), FCAR and MES (Québec)

the zero solution of equation (3) is asymptotically stable.

The location of the roots of equation (4) has been investigated in many special cases. When either one of a_1 or a_2 is zero, equation (3) contains only one delay and the problem of the loss of stability of equation (4) has been completely solved by Hayes [12]. A more general result is available [15] when τ is rational, but is of limited practical use: when τ is the irreducible ratio of integers m and n, polynomials of degrees n and m must be investigated for the location of their roots. The case $a_0 = 0$ has been considered by several authors. Ruiz-Claeyssen [16] assumes $a_1 = a_2 = -1$ and geometrically locates all purely imaginary roots of equation (4). Nussbaum [14] has studied the case $\tau = 2$, with a_1 and a_2 both positive. Braddock and van den Driessche [4,5] have performed a more complete analysis which is much in line with the one presented here.

Two directions of investigation have been pursued in the past. One [11] consists in determining the parameter values where "absolute stability" occurs: this corresponds to asymptotic stability for all values of the delays. The main drawback of this method is to eliminate the dependence of the region of stability on the values of the delays. Another approach is to take the non-normalized time lag as the bifurcation parameter : it is in this context that "stability switches" are observed [6,7]. We will consider another way to evaluate the influence of the delays on the bifurcation structure by considering variations in the values of the coefficients for a fixed value of the (ratio of the) delays. This leads to a direct interpretation of the bifurcations in terms of the parameters present in the model motivating this analysis [2].

We obtain in section 2 the location of the positive real roots of equation (4), for all possible values of the parameters. In the following section , restrictions are placed on the values of a_0 and τ to determine the region of stability in the plane of the coefficients a_1 and a_2 taken as parameters. This is accomplished by using the D-decomposition method [8]: using the continuous dependence of the roots of equation (4) on the coefficients a_1 and a_2 , we identify a connected set D, the set of parameter values corresponding to asymptotic stability. The boundary of D is characterized by the existence of purely imaginary roots of equation (4) at these parameter values.

The problem we address is not new: it has recently been labelled 'venerable' [6]. However, getting precise estimates of the parameter values giving asymptotic stability of the null solution of equation (3) is recognized as a difficult problem [5,6,10,11,14]. We believe that the role of equation (13) in solving this problem is crucial, and that although it has been obtained in a special case [5], its importance has not been properly emphasized. Further complications arise, in the applications, when the coefficents themselves are functions of more easily identifiable parameters. This is illustrated in section 4, on an equation occuring in the modeling of the control of platelet production [2].

2. REAL ROOTS. We consider the values of the parameters a_1 and a_2 for which equation (4) posesses no real, positive root . We assume henceforth that $a_1a_2 \neq 0$, the case where either one of a_1 or a_2 is zero being included in the results of Hayes on the equation with one delay. We also suppose, without loss of generality, that $\tau > 1$.

Let us set $f(z) = a_1e^{-z} + a_2e^{-z\tau}$; the roots of equation (4) are then the points of intersection of the straight line defined by $z + a_0$ and the sum of exponentials f(z). We have the following

THEOREM 1: Let $a_0 > 0$ and $\tau > 1$ be fixed; then the values of a_1 and a_2 such that equation (4) has no real, positive root satisfy (i) $a_2 \le a_0 - a_1$, if $a_1 \le (a_0\tau+1)/(\tau-1)$, and (ii) $a_2 \le g(a_1)$, if $a_1 \ge (a_0\tau+1)/(\tau-1)$, where $a_2=g(a_1)$ is parametrically defined by

(5) $$a_1 = [1+\tau(a_0+u)]e^{u}/[\tau-1] , \text{ and } a_2 = [1+ a_0+ u]e^{u\tau}/[1-\tau] , \qquad u \ge 0.$$

Proof: Since $f(0) = a_1+ a_2$ and $f(\infty) = 0$, a sufficient condition for the existence of a positive root of equation (4) is $a_0 < a_1+ a_2$. If this last inequality is not satisfied and f(z) is monotonic for positive values of z, there cannot exist a positive root to equation (4). Suppose now that $a_0 > a_1 + a_2$ but f(z) posesses an extremum at a nonnegative value of z ; since $f'(z) = -(a_1e^{-z} + \tau a_2e^{-z\tau})$, the latter condition is equivalent to $(a_1/\tau a_2) \ge -1$. If $a_1 > -a_2\tau$, then $f'(0) < 0$, $f(\infty) = 0$ and since f(z) cannot have more than one extremum, there is no positive root of equation (4).

The only remaining possibility is that a positive root appears via the tangential intersection of the graph of f(z) with the line $z + a_0$, as shown in Figure 1. This is equivalent to the existence of a $u > 0$ such that $f'(u) = 1$ and $f(u) = u+ a_0$, or

(6) $$a_1e^{-u} + a_2\tau e^{-u\tau} + 1 = 0 , \text{ and } a_1e^{-u}+ a_2e^{-u\tau} - a_0 - u = 0 .$$

Considering equations (6) as a linear system of two equations in the two unknowns a_1 and a_2 , its determinant is $(1-\tau)e^{-u(1+\tau)}$, and it can be inverted to yield the explicit relations of equation (5).These equations can be viewed as the parametric representation of a curve in the plane (a_1,a_2) with $u \ge 0$ as the parameter. The value of u then corresponds to the point at which the tangency illustrated in Figure 1 occurs. It is an exercise in planar differential geometry to establish

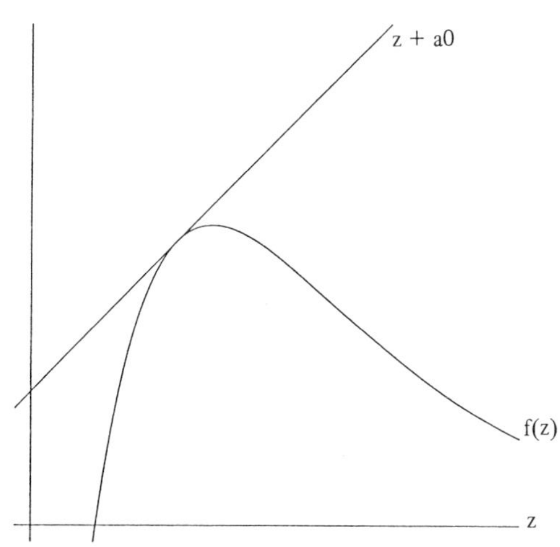

Figure 1 : The tangency condition set by equation (6) [$a_0 = .8$; $\tau = 1.9$; $u = .9$]

that equations (5) define a curve $a_2 = g(a_1)$, such that g is a monotone decreasing, unbounded, concave function defined for $a_1 \geq (a_0\tau+1)/(\tau-1)$, such that $g((a_0\tau+1)/(\tau-1))=(a_0+1)/(1-\tau)$ and $dg/da_1((a_0\tau+1)/(\tau-1)) = -1$.

The region of existence of a positive real root of equation (4) is illustrated in Figure 2, which summarizes the conclusion of Theorem 1.

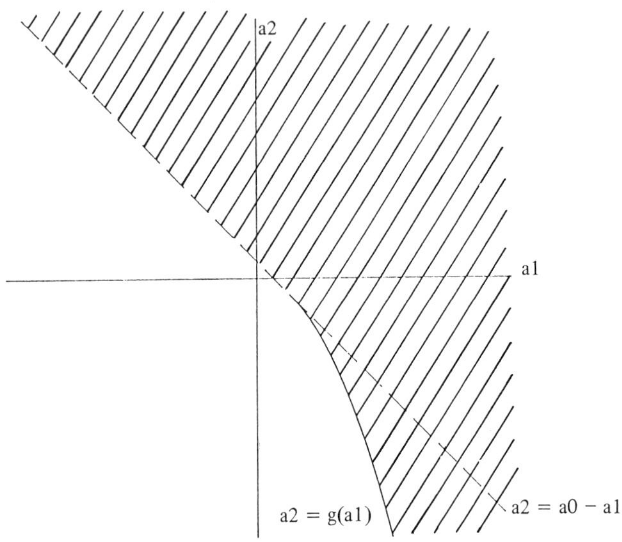

Figure 2. Region (hatched) of existence of a positive root of equation (4) [$a_0 = 1.3$; $\tau = 2.5$]

3. COMPLEX ROOTS. Turning to the analysis of the complex roots of equation (4), we determine the values of the parameters a_1 and a_2 leading to the existence of purely imaginary roots. Letting $z = \mu + i\omega$ in equation (4), separating real and imaginary parts leads to

(7a) $$a_0 + \mu = a_1e^{-\mu} \cos \omega + a_2e^{-\mu\tau}\cos(\tau\omega)$$

(7b) $$- \omega = a_1e^{-\mu} \sin \omega + a_2e^{-\mu\tau}\sin(\tau\omega) \ .$$

Pure imaginary roots of equation (4) will thus exist when

(8a) $$a_0 = a_1\cos \omega + a_2\cos(\tau\omega)$$

(8b) $$- \omega = a_1\sin \omega + a_2\sin(\tau\omega)$$

hold, for some values of τ and a_0. It is clear from equations (7) that complex roots come in complex conjugate pairs; we only consider roots with positive imaginary parts.

As noticed by Braddock and van den Driessche [4,5], bounds on the parameter values necessary for roots of equation (4) to exist can be obtained from equations (7) in a straightforward manner. Use of equation (7b), for example, leads to $-1 = a_1e^{-\mu\tau}\sin\omega/\omega + a_2e^{-\mu\tau}\tau \sin(\tau\omega)/\tau\omega$. For all $\mu \geq 0$ and $\omega > 0$, the right-hand side of this last equation is bounded above by $|a_1|e^{-\mu} + \tau|a_2|e^{-\mu\tau}$ and bounded below by $-|a_1|e^{-\mu} - \tau|a_2|e^{-\mu\tau}$. It is thus impossible to find solutions of equation (7b) when the values of a_1 and a_2 satisfy $|a_1| + |a_2|\tau < 1$.

Necessary conditions for the existence of purely imaginary roots of equation (4) are obtained by squaring both sides of both equations (8) and adding them, yielding

$$(9) \qquad a_0^2 + \omega^2 = a_1^2 + a_2^2 + 2a_1a_2\cos(\omega[\tau-1]) .$$

For fixed values of a_0, a_1, a_2 and τ, equation (9) defines as a function of ω the intersection of a parabola, on the left-hand side, and a vertical translation of a trigonometric function, on the right-hand side. The function $p(\omega)= a_1^2 + a_2^2 + 2a_1a_2\cos(\omega[\tau-1])$ takes values between $(a_1-a_2)^2$ and $(a_1+a_2)^2$. There are two possibilities, depending on the sign of the product a_1a_2. When a_1 and a_2 are of the same sign, $p(\omega)$ has a maximum value of $(a_1+ a_2)^2$ at $\omega = 0$, and there are no roots of equation (9) whenever $a_0^2 > (a_1+ a_2)^2$. There is thus a bifurcation creating such a root when $a_0 = |a_1 + a_2|$. When a_1 and a_2 are of opposite signs, $p(\omega)$ has a minimum value at $\omega = 0$ and solutions of equation (9) are created at parameter values giving rise to the tangency condition illustrated in Figure 3. This geometric condition is described algebraically by simultaneously

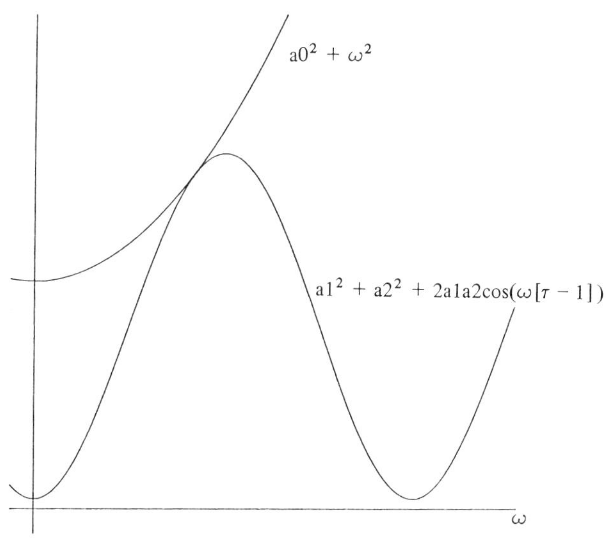

Figure 3. Tangency condition leading to equation (9) [$a_0 = .96$; $a_1 = .5$; $a_2 = -.7$; $\tau = 5$]

solving equation (9) and the equation obtained by taking derivatives on both sides of it. Solving this set of equations is then equivalent to the condition

$$(10) \qquad \cos(\omega[\tau-1]) = [-1 + (1- (\tau-1)^2 (a_1^2 + a_2^2 - a_0^2 - a_1^2a_2^2(\tau-1)^2))^{1/2}]/a_1a_2(\tau-1) .$$

When $a_1 = a_2$, a purely analytical treatment is possible. We write equations (8) as

(11a) $a_0 = 2a_1\cos[\omega(1+\tau)/2]\cos[\omega(1-\tau)/2]$

(11b) $-\omega = 2a_1\sin[\omega(1+\tau)/2]\cos[\omega(1-\tau)/2]$.

Assuming that $\cos[\omega(1-\tau)/2] \neq 0$, which must hold if $\omega \neq 0$, substitution of equation (11a) into equation (11b) leads to

(12) $-\omega = a_0 \tan[\omega(1+\tau)/2]$.

Stability of the null solution of equation (3) requires that all roots of equation (4) have negative real parts; instability is thus produced when the first pair of eigenvalues crosses the imaginary axis. We are thus lead to a restriction in the interval $[\pi/(1+\tau), 2\pi/(1+\tau)]$ for the location of the first root ω of equation (4): this can be seen by considering perturbations off small values of a_1 . If ω^* denotes the unique root of equation (12) in the interval $[\pi/(1+\tau), 2\pi/(1+\tau)]$, substitution back into equation (8b) yields $a_1 = -\omega^* / (\sin(\omega^*) + \sin(\omega^*\tau))$. Assuming, in addition, that $a_0 = 0$, then, from equation (11a), $\omega^* = (2k+1)\pi/(1+\tau)$, for some integer k, and substitution into equation (11b) thus provides the explicit relation $a_1 = -\omega^* / 2\sin(\omega^*)$. The first neutral root of equation (4) thus appears at $a_1^* = \pi/2(1+\tau)\sin[\pi/(1+\tau)]$; the null solution of equation (3), for $a_1 = a_2$ and $a_0 = 0$, will thus be stable exactly when $0 < a_1 < a_1^*$. It is easy to see that these regions of stability are continuous, as a function of a_0 . Indeed, as a_0 approaches 0, ω^* approaches $\pi/(1+\tau)$ and a_1^* tends to $-\pi/2(1+\tau)\cos[\pi(1-\tau)/2(1+\tau)] = \pi/(2(1+\tau)\sin[\pi/(1+\tau)])$.

A clearer picture emerges geometrically by allowing a_1 and a_2 to take arbitrary values, and considering equations (8) as a system of linear equations in two unknowns a_1 and a_2 . Provided that $\sin[\omega(\tau-1)] \neq 0$, we can invert equations (8) to

(13a) $a_1 = [a_0\sin(\omega\tau) + \omega\cos(\omega\tau)] / [\sin\{\omega(\tau-1)\}]$

(13b) $a_2 = [-a_0\sin\omega - \omega\cos\omega] / [\sin\{\omega(\tau-1)\}]$.

For fixed values of a_0 and τ, these equations define a curve Γ in the (a_1,a_2) plane, parametrized by $\omega \in [0,\infty)$. At the values $\omega = j\pi/(\tau-1)$, j a positive integer, both a_1 and a_2 become unbounded. The curve Γ is thus made up of infinitely many branches corresponding to different subsets of its domain of definition, namely intervals $I_j = (j\pi/(\tau-1), (j+1)\pi/(\tau-1))$, for $j \in N$.

Our problem becomes the determination of the connected domain D, in the space of the parameters a_1 and a_2 , containing the origin, $a_1 = a_2 = 0$, and bounded by branches of the curve defined by equations (13). This set D will give the set of parameter values a_1 and a_2 for which equation (4) has no complex root with positive real part. The null steady state of equation (3) is stable when, in addition to this condition, no real root of equation (4) is nonnegative. The complete stability problem is thus solved by adding to the conditions set in Theorem 1 the limits determined by the appropriate branches of equations (13).

Finding which branches are appropriate is a nontrivial task. Numerical calculations show that the domain D is highly dependent on the values of a_0 and τ. A complete solution would involve a two-dimensional analysis in the parameter space (τ, a_0) to locate the branches of the curve defined by equations (13), and thus, a global description would have to be set up in a four-dimensional parameter space. Since this could become intractable, we will concentrate instead on parameter values of a_0 and τ relevant for the model presented in the next section: we assume that τ has a value "close to" 2, and regard a_0 as progressively increased from an initial value of zero.

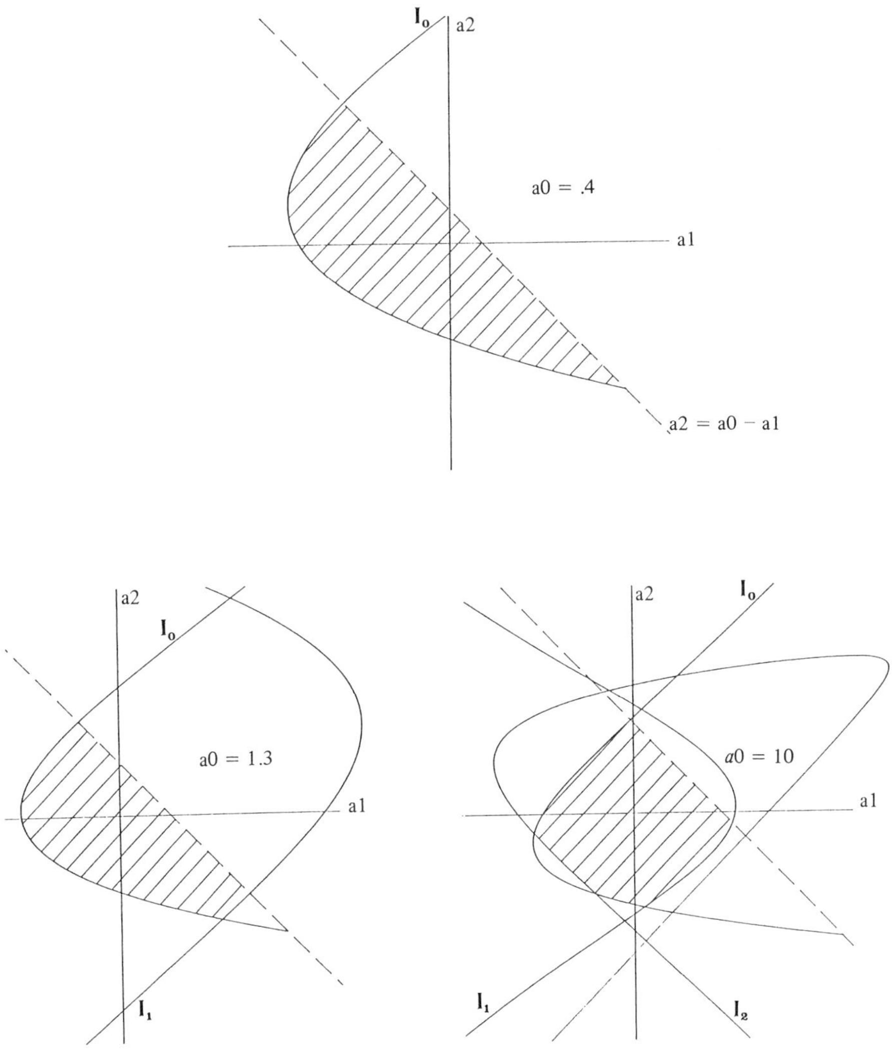

Figure 4. Region (hatched) of stability of equation (4) as determined by the conditions in Theorem 1 and equations (13). The labels indicate the domain I_j of each branch of the curve Γ [$\tau = 1.8$, and a_0 takes values shown].

We present in Figures 4 and 5 numerically computed graphical representations of the bifurcations of the set D, bifurcations due to qualitative changes in the geometry of the curve Γ. A crucial role is played by the delay τ : we can analytically explain the drastic change, in the branch of Γ associated with the domain I_0 ,from Figure 5b to Figure 5c, and prove that a similar change cannot occur when $\tau < 2$. It suffices to write the condition for $a_2(\omega)$ to be zero when ω takes values in the interval $[0,\pi/(\tau-1))$, namely $\tan \omega = -\omega/a_0$. It is geometrically clear that a solution of this last equation will exist when $\tau < 2$, for all values of a_0 ; but there need not be such a solution when $\tau > 2$, if a_0 is sufficiently large. This helps explain the qualitative stability changes taking place at integer values of the delays in multiple-delayed differential equations [14].

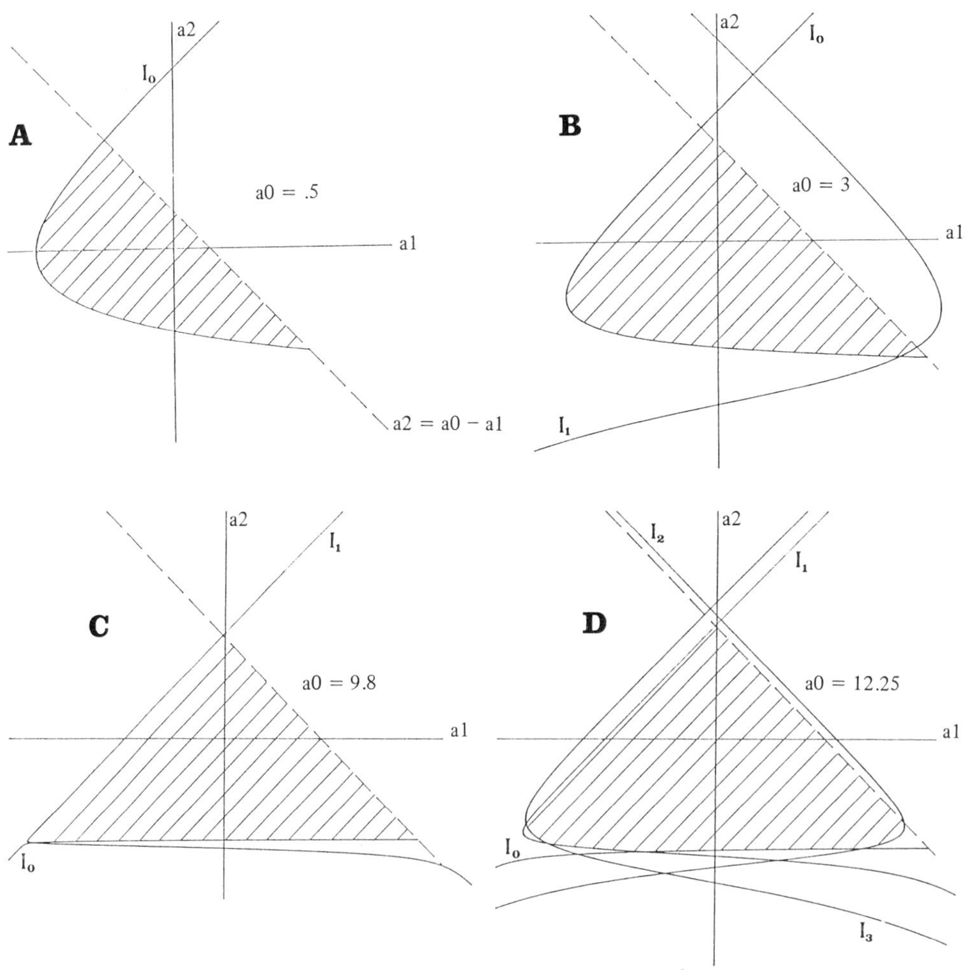

Figure 5. Same as in Figure 4, except $\tau = 2.1$.

Instead of presenting general analytical results on the topological properties of the curve Γ, such as those depicted in Figures 4 and 5, we present in the next section an illustration of our analysis in the investigation of an equation in which the coefficients a_0, a_1 and a_2 are functions of two more fundamental parameters.

4. APPLICATION. We consider the equation

$$dP(t)/dt = -\gamma P(t) + \beta(P(t-T_M)) - \beta(P(t-T_M-T_S))\exp(-\gamma T_S) \tag{14}$$

where $\beta(u) = \beta_0\theta^n u/(\theta^n+u^n)$. This equation is derived in a model for the regulation of the level of platelets $P(t)$ in an individual [2]. The delays correspond to the lifespan T_S of a platelet before its death due to senescence, and the maturation time T_M of a platelet precursor, the megakaryocyte, from the moment it is morphologically identifiable to the time it releases platelets. The parameter γ accounts for a postulated random destruction of the platelets in pathological conditions, and the function $\beta(\cdot)$ represents the influence of the platelet level on the production of maturating precursors in the megakaryocyte compartment. This feedback is of mixed type [13], inducing low levels of maturation when platelet level is either high or very low: this follows from a "saturation" effect at high platelet concentration, and the observation that $\beta(0)$ should be 0. The postulated form for $\beta(\cdot)$ provides the "simplest" continuous function compatible with these conditions, a function with only one maximum. The range of values for T_S (respectively T_M) in a normal individual is accepted to lie between 7 and 11 days (respectively 7 and 10 days). We assume, for definiteness, that $T_S < T_M$.

Equation (14) always possesses the trivial steady state $P_1 = 0$. Linearization about this solution yields

$$T_M x'(t) = -\gamma x(t) + \beta_0 x(t-1) - \beta_0 x(t-\tau)\exp(-\gamma T_S) \tag{15}$$

where $\tau = 1+T_S/T_M$. When $\beta_0 > \gamma/(1-\exp(-\gamma T_S))$, another stationary solution appears, namely $P_2 = \theta[\beta_0(1-\exp(-\gamma T_S))/\gamma - 1]^{1/n}$, and the corresponding linear equation is

$$T_M x'(t) = -\gamma x(t) + \beta_1 x(t-1) - \beta_1 x(t-\tau)\exp(-\gamma T_S) \tag{16}$$

with $\beta_1 = \gamma[\gamma n - \beta_0(n-1)(1-\exp(-\gamma T_S))]/[\beta_0(1-\exp(-\gamma T_S))^2]$. Both equations (15) and (16) are particular cases of equation (4), each one having its coefficients satisfying $a_2 = -a_1\exp(-\gamma T_S)$. The coefficients a_1 and a_2 thus have opposite signs and are constrained to lie on a straight line in the second and fourth quadrants of the parameter plane (a_1,a_2).

Let us consider equation (15) first. When $a_1 > 0$, we assert, on the basis of numerical investigations, that the boundary of the region of stability D on the line $a_2 = -a_1\exp(-\gamma T_S)$ is the "real" boundary, corresponding to the loss of stability through a real root of equation (4) becoming positive. The conditions thus satisfied are, simultaneously, $a_2 = -a_1 + a_0$ and $a_2 = -a_1\exp(-\gamma T_S)$, or

$$a_1 = a_0/(1 - \exp(-\gamma T_S)). \tag{17}$$

For the stability of the zero solution P_1, $a_1 = \beta_0/T_M$ and equation (17) thus becomes $\beta_0 = \gamma / (1-\exp(-\gamma T_S))$. It is thus clear that the loss of stability of the null steady state P_1 occurs precisely when the non-trivial stationary solution appears. In the case of the solution P_2, equation (17) is equivalent to $\beta_0 = -n\gamma/([n-2][1-\exp(-\gamma T_S)])$; the steady state P_2 is thus stable when the coefficient β_1/T_M is positive, precisely when $\beta_0 < \gamma n/([n-1][1-\exp(-\gamma T_S)])$.

When $a_1 < 0$, similar numerical results indicate that the boundary of the stability domain D is the branch of the curve Γ from equation (13) with domain $[0,\pi/(\tau-1))$. The defining equations for this loss of stability are $a_2 = -a_1\exp(-\gamma T_S)$ and $a_2 = (a_0\sin\omega + \omega\cos\omega)/\sin[\omega(\tau-1)]$, the last equation having to be solved in ω. Bounds on the appropriate value $\underline{\omega}$ of ω can be obtained by noticing that $\underline{\omega}$ has to be located between the value of ω giving a zero of a_2 in the interval I_0, and the value of ω leading to the second root of a_1 in the same interval I_0.

Numerical integration of equation (14) for parameter values "beyond" the loss of linear stability of P_2 suggests the presence of higher order bifurcations and irregular behavior of solutions. These more global aspects of the bifurcations present in equation (14) are discussed elsewhere [2].

5. ACKNOWLEDGEMENTS. The author has benefitted from discussions with Michael Mackey.

BIBLIOGRAPHY

1. J. Bélair and M. Mackey, "A model for the regulation of mammalian platelet production," Ann. N.Y. Acad. Sci. To appear (1987).

2. J. Bélair and M. Mackey, "Oscillatory properties of platelet control," In preparation (1987).

3. S. Blythe, R. Nisbet and W. Gurney, "Instability and complex behaviour in population models with long time delays," Theor. Pop. Biology, 22 (1982), 147-176.

4. R. Braddock and P. van den Driessche, "A population model with two time delays," Math. Scient., 5 (1980), 55-66.

5. R. Braddock and P. van den Driessche, "On a two lag differential delay equation," J. Austral. Math. Soc., 24 (1983), 292-317.

6. K. Cooke, "Stability of delay differential equations with applications in biology and medicine," in Mathematics in Biology and Medicine, ed. V. Capasso et al. Lecture Notes in Biomathematics #57, Springer, New York, 1985.

7. K. Cooke and P. van den Driessche, "On zeroes of some transcendental equations," Funkcialaj Ekvacioj, 29 (1986), 77-90.

8. L. E. El'sgol'ts, "Introduction to the theory of differential equations with deviating arguments" Holden-Day, San Francisco, 1966.

9. L. Glass and M. Mackey, "Pathological conditions resulting from instabilities in physiological control systems," Ann. N.Y. Acad. Sci., 316 (1979), 214-235.

10. J. Hale, "Nonlinear oscillations in equations with delays", in Nonlinear Oscillations in Biology, ed F. Hoppensteadt. Lectures in Applied Mathematics #17, AMS, Providence, 1979.

11. J. Hale, E. Infante and P. Tsen, "Stability in linear delay equations," J. Math. Anal. Appl., 105 (1985), 533-555.

12. N. Hayes, "Roots of the transcendental equation associated with a certain differential-difference equation," J. London Math. Soc., 25 (1950), 226-232.

13. U. an der Heiden and M. Mackey, "The dynamics of production and destruction: analytic insight into complex behavior," J. Math. Biology, 16 (1982), 75-101.

14. R. Nussbaum, "Differential delay equations with two time lags," Mem. Amer. Math. Soc.,16 (1) (1975), 1-62.

15. L. Pontrjagin, "On the zeros of some elementary transcendental equations," Amer. Math. Soc. Transl. Ser. 2 (1) (1955), 95-110.

16. J. Ruiz-Claeyssen, "Effects of delays on functional differential equations," J. Differential Equations, 20 (1976), 404-440.

DEPARTEMENT DE MATHEMATIQUES ET DE STATISTIQUE
AND
CENTRE DE RECHERCHES MATHEMATIQUES
UNIVERSITE DE MONTREAL
MONTREAL, QUEBEC, H3C 3J7 CANADA

Canadian Mathematical Society
Conference Proceedings
Volume 8 (1987)

HARVESTING IN DELAYED-RECRUITMENT POPULATION MODELS

Fred Brauer

ABSTRACT: We formulate a first order differential-difference equation as a model for a population with delayed recruitment under constant-yield harvesting. We consider the existence and stability of equilibria, conditions under which bifurcations to periodic solutions do not occur, the possibility of period-doubling and chaotic behaviour, and the effect of harvesting on stability of equilibria.

1. The formulation of single-species population models under the assumption that birth and death rates depend only on total population size leads to ordinary differential equations. Such models, while relatively manageable mathematically, neglect time lags which may be caused by a variety of biological mechanisms and which may have a substantial effect on the behavior of the system. A full age-structured model, such as the nonlinear version of the McKendrick - von Foerster equation introduced in [6] may be considerably more "realistic", but may be quite intractable mathematically.

In this paper we shall examine a class of differential-difference equations introduced in [2] to describe the size of an adult population whose members have a fixed maturation time, in which the birth and adult death rates at any time depend only on the adult population size at that time and the pre-adult death rate at any time depends only on age and the number of pre-adult members of the same age (not on the total pre-adult population size). We shall see that such models exhibit a much richer variety of possible behaviours than ordinary differential equations without time-lags but are more tractable than the partial differential equations or functional equations given by fully age-structured models.

Populations with delayed recruitment have frequently been modelled by difference equations; see for example [9; Chapter 2]. It is known that the solutions of such models can exhibit a wide variety of qualitative

---

1980 Mathematics Subject Classification (1985 Revision). 34K15, 92A15.

behaviours, including asymptotically stable equilibria, asymptotically stable periodic solutions, period doubling, and chaos [8, 10]. It has been suggested that difference equation models are appropriate for populations with discrete generations, but we shall see that continuous models of the form to be studied can exhibit essentially discrete generations as well as the various behaviours exhibited by discrete models.

Over-harvesting can lead to a catastrophe, whose biological interpretation is the collapse of the population system. Before a catastrophic harvest rate is reached, the effect of harvesting is to stabilize the equilibrium of the population system, even if the unharvested system exhibits chaotic behaviour.

2. We wish to study the behaviour of solutions of the differential-difference equation

$$x'(t) = R[x(t-\tau)] - D[x(t)] \tag{1}$$

under the assumptions

(i) the functions $R(x)$ and $D(x)$ are continuously differentiable for $0 \leq x < \infty$ and $\tau > 0$ is a constant.

(ii) $D(0) = 0$, $D'(x) > 0$ for $0 \leq x < \infty$, $\lim_{x\to\infty} D(x) = +\infty$

(iii) $R(0) = 0$, $0 \leq R(x) \leq \rho$ for $0 \leq x < \infty$

(iv) either (a) $R(x)$ has a unique local maximum ρ achieved for $x = \xi$ with $R'(x) > 0$ for $0 \leq x \leq \xi$, $R'(x) < 0$ for $x > \xi$, and $\lim_{x\to\infty} R(x) = 0$, or (b) $R'(x) > 0$ for $0 \leq x < \infty$ and $\lim_{x\to\infty} R(x) = \rho < \infty$.

In addition, we specify initial data

$$x(t) = x_0(t), \quad -\tau \leq t \leq 0 \tag{2}$$

The model (1), (2) has been proposed by Blythe, Nisbet, and Gurney [2] to describe the size of a population whose members have a fixed maturation time τ in which the adult death rate at any time depends only on the adult population size at that time and the rate of recruitment of new adult members of age τ at time t depends only on the adult population size at the time $(t - \tau)$ of birth of these new adult members. Thus $D[x(t)]$ represents the adult death rate at time t and $R[x(t - \tau)]$ represents the rate of recruitment of new adult members at time t. The cases (a) and (b) of hypothesis (iv) correspond to "scramble" and "contest" competition respectively among juvenile members of the population. These cases, described in [2], appear to occur in laboratory populations.

The assumption that the recruitment rate of new adults at time t is a function of the number of adult members at time $(t - \tau)$ is satisfied if the birth rate at any time depends only on the adult population size at that time and the juvenile death rate at any time depends only on the age of the juveniles and on the number of juveniles of the same age. To see this, we let $y(s)$ be the number of (juvenile) members born at time $(t - \tau)$ of age s $(o \leq s \leq \tau)$. Then by assumption $y(0)$ is a function of the adult population size $x(t - \tau)$ at time $t - \tau$. Also by hypothesis the death rate $y'(s)$ depends only on s and $y(s)$; thus there is a function F such that

$$y'(s) = F[s,y(s)]$$

Then $y(s)$ is a solution of the ordinary differential equation $y' = F(s,y)$, and $y(\tau)$, the recruitment rate of new adults at time t, is a function of the initial value $y(0)$, and this is a function of $x(t - \tau)$. A derivation of the model (1) under the assumptions (i)-(iv) with juvenile mortality depending only on the age of the juveniles and the number of juveniles of the same age has been given in [2] based on the McKendrick model (see [6], [11]), for age-structured populations; the above derivation is more elementary.

To the basic model (1), (2) we shall add a constant harvest rate of H adult members in unit time. Thus we consider the differential- difference equation

$$x'(t) = R[x(t - \tau)] - D[x(t)] - H, \tag{3}$$

together with the initial data (2). Since $x(t)$ represents population size, we are interested only in solutions of (3) with $x(t) \geq 0$ for $0 \leq t < \infty$. If there were no time-lag τ, the model would be an ordinary differential equation and it would be possible for a solution which is not identically zero to reach zero in finite time if $H > 0$. In such a case we would consider the population system to have collapsed and the population to have become extinct. With a time-lag, however, it is possible for $x(t)$ to reach zero without the collapse of the system because even if the adult population is wiped out there may still be juveniles who will reach the adult stage. In order to allow for this, we must modify the equation (3) slightly by specifying that whenever $x(t) = 0$,

$$x'(t) = \max\{R[x(t-\tau)] - D[x(t)] - H, 0\} = \max\{R[x(t-\tau)] - H, 0\}$$

This modification has the effect of keeping the system in operation by not permitting $x'(t) < 0$ when $x(t) = 0$. Of course, if $x(t) = 0$ on a time interval of length τ there is no possibility of regeneration of the adult population and we shall consider the population to have become extinct.

3. In this section we shall derive bounds for solutions of (3) as well as some stability results obtained by equilibrium analysis.

THEOREM 1: (Blythe, Nisbet & Gurney [2]). Define $M > 0$ by $D(M) + H = \rho$ and suppose that the initial data $x_0(t)$ satisfies $0 \le x_0(t) \le M$ for $-\tau \le t \le 0$ and that the hypotheses (i)-(iv) are satisfied. Then the solution $x(t)$ of (3) satisfies $0 \le x(t) \le M$ for $0 \le t < \infty$.

The essential idea behind the proof is that if $x(t) > M$, we must have $x'(t) < 0$ and thus the solution can not enter the region $x(t) > M$. The proof in [2] is given for the case $H = 0$ of unharvested populations but extends to the harvested case without change. In order to establish a lower bound, as conjectured in [2], we must assume no harvesting, and we must also restrict ourselves to scramble competition. However, for contest competition we will be able to establish more precise results by equilibrium analysis.

THEOREM 2: Define $m > 0$ by $D(m) = R(M)$, where M is as in Theorem 1. Suppose that $M > \xi$ with ξ defined by $R(\xi) = \max_{0 \le x < \infty} R(x) = \rho$, and that $R(x) > D(x)$ for $0 < x \le \xi$. Assume that the initial data $x_0(t)$ satisfies $m \le x_0(t) \le M$ for $-\tau \le t < 0$, that $H = 0$, and that the hypotheses (i)-(iii), (iva) are satisfied. Then the solution $x(t)$ of (3) satisfies $m \le x(t) \le M$ for $0 \le t < \infty$.

PROOF: Since $D(x)$ is monotone increasing and $\xi < M$, $D(\xi) < D(M) = \rho = R(\xi)$. For large x, $D(x) > R(x)$ and thus there must be at least one $x_\infty > \xi$ for which $R(x_\infty) = D(x_\infty)$. In fact, since $R(x)$ is monotone decreasing while $D(x)$ is monotone increasing for $\xi \le x < \infty$, this x_∞ must be unique. For $m \le x \le x_\infty$, $R(x) \ge D(x) \ge D(m) = R(m)$, and for $x_\infty \le x \le M$, $D(x) \ge R(x) \ge R(M)$, using the definition of m and the fact that $R(x)$ is monotone decreasing for $x \ge \xi$. Thus we have $R(x) \ge R(M)$ for $M \le x \le M$.

Now suppose that $x(t) \geq m$ for $0 \leq t \leq t_0$ with some $t_0 \geq 0$, that $x(t_0) = m$, and that $x(t) < m$ on some interval $t_0 < t < t_0 + \epsilon < t_0 + \tau$. But then for $t_0 < t < t_0 + \epsilon$, $x'(t) = R(x(t - \tau)) - D(x(t)) >$ $R(M) - D(m) = 0$ because $R(x(t - \tau)) > R(M)$ and $D(x(t)) < D(m)$. This shows that $x(t)$ can not enter the region $x < m$ and proves Theorem 2.

The proof of Theorem 2 depends on the fact that $R(x) \geq R(m)$ for $m \leq x < M$, which is not necessarily true if $M \leq \xi$ or if $H > 0$. The conclusion of Theorem 2 remains true if $R(x)' \geq R(M)$ for $m \leq X \leq M$, with m defined by $D(m) + H = R(M)$ if $H > 0$. However, this definition may give $m \leq 0$ and then Theorem 2 is vacuous. In order to examine the question of lower bounds for solutions if $M \leq \xi$ and/or if $H > 0$, we will need some equilibrium analysis.

An equilibrium of the differential-difference equation (3) is a value x_∞ for which

$$(4) \qquad R(x_\infty) = D(x_\infty) + H,$$

that is, the abscissa of an intersection of the recruitment curve $y = R(x)$ and the removal curve $y = D(x) + H$. Sometimes we may denote an equilibrium by $x_\infty(H)$ to emphasize its dependence on the harvest rate H.

THEOREM 3: Let the hypotheses (i)-(iv) be satisfied and suppose that the differential-difference equation (3) has no equilibrium $x_\infty > 0$. Then every solution of (3) tends to zero as $t \to \infty$ if $H = 0$, and every solution of (3) reaches zero in finite time and remains at zero if $H > 0$.

PROOF: We define M as in Theorem 1 by $D(M) + H = \rho = \sup_{0 \leq x < \infty} R(x)$; then for every t such that $x(t) > M$ we have $x'(t) < 0$. (Of course, by Theorem 1, $x(t) > M$ is possible only if the initial data $x_0(t) > M$ for some t, $-\tau \leq t \leq 0$). In any case, there exists $t_0 \geq 0$ such that $x(t) \leq M$ for $t \geq t_0$. Now define $M_1 < M$ by $D(M_1) + H = R(M)$. Then if $x(t) > M_1$ for some $t \geq t_0 + \tau$ we have $D(x(t)) + H > D(M_1) + H$ and $R(x(t - \tau) \leq R(M)$ because $x(t - \tau) \leq M$ and both $D(x)$ and $R(x)$ are monotone increasing functions for $M_1 \leq x \leq M$. Thus if $x(t) > M_1$ for $t \geq t_0 + \tau$ we have $x'(t) < 0$, and there exists $t_1 \geq t_0$ such that $x(t) \leq M_1$ for $t \geq t_1$. We continue inductively, defining $M_{j+1} < M_j$ by $D(M_{j+1}) + H = R(M_j)$ for $j = 1, 2, \dots,$ and then showing that $x(t) \le M_{j+1}$ for $t \ge t_{j+1}$. If $H = 0$, we obtain a decreasing sequence $M_j \to 0$ as $j \to \infty$ and we conclude that $x(t) \to 0$ as $t \to \infty$. If $H > 0$, we obtain $M_j \le 0$ after a finite number of steps, and we conclude that $x(t) = 0$ for all sufficiently large t, completing the proof of Theorem 3.

Theorem 3 provides a condition, namely $R(x) < D(x) + H$ for $0 < x < \infty$, under which there can not be a positive lower bound for solutions of (3), and no analogue of Theorem 2 can hold. A stability analysis of equilibria can be used in some cases in which Theorem 2 does not apply directly to prove the existence of a positive lower bound for solutions.

The linearization of (1) at an equilibrium x_∞ is

$$u'(t) = -D'(x_\infty)u(t) + R'(x_\infty)u(t - \tau). \tag{5}$$

It is well-known [1] that the equilibrium x_∞ of (3) is locally asymptotically stable if the equilibrium solution $u = 0$ of (5) is asymptotically stable, and that the equilibrium x_∞ of (3) is unstable if the equilibrium solution $u = 0$ of (5) is unstable. The analysis of the asymptotic stability of the equilibrium solution $u = 0$ of (5) has been carried out ([7]; see also [1], p. 336) and this yields the following result frequently cited in the biological literature, usually under some additional restrictions.

<u>THEOREM 4</u>: Define

$$p = -D'(x_\infty)\tau, \quad q = R'(x_\infty)\tau \tag{6}$$

and let $z = z(p)$ be the unique solution in $0 < z < \pi$ of the transcendental equation

$$z = p \tan z \tag{7}$$

if $p \ne 0$, with $z(0) = \frac{\pi}{2}$. Then an equilibrium x_∞ of (3) is asymptotically stable if and only if

$$-p \sec z < q < -p. \tag{8}$$

Solutions of (3) near x_∞ are oscillatory if $q < -e^{p-1}$ and monotone (tending to x_∞ in the case of asymptotic stability and away from x_∞ in the case of instability) if $q > -e^{p-1}$. Since $-p \sec z = -z \csc z$, if $p = 0$ the expression $-p \sec z$ is replaced by $-\frac{\pi}{2} \csc \frac{\pi}{2} = -\frac{\pi}{2}$.

If $R'(x_\infty) > 0$, so that $q > 0$ then $q > -e^{p-1}$, and we can not have oscillations near x_∞. This gives the following useful application.

<u>COROLLARY TO THEOREM 4</u>: Under the hypotheses (i)-(iv) an equilibrium x_∞ of the differential-difference equation (3) is exponentially asymptotically stable if $0 < R'(x_\infty) < D'(x_\infty)$, and exponentially unstable if $R'(x_\infty) > D'(x_\infty) > 0$.

If (3) models a population with contest competition (hypothesis iv(b)), then the above corollary implies that oscillatory solutions are impossible. This fact was noted in [2] for unharvested systems $(H = 0)$, but it holds also for $H > 0$. Unless $R(x) < D(x)$ for all x, $0 < x < \infty$, in which case all solutions tend to zero by Theorem 3, the hypotheses $\lim_{x\to\infty} D(x) = +\infty$, $\lim_{x\to\infty} R(x) < \infty$ imply that $R(x) < D(x)$ for large x, and thus there must be a largest equilibrium $x_\infty > 0$ with $0 < R'(X_\infty) < D'(x_\infty)$, exponentially asymptotically stable by the above corollary. Thus solutions with suitable initial values are bounded away from zero, and this remark serves as a replacement for Theorem 2 in the case of contest competition.

Another situation in which Theorem 4 gives a replacement for Theorem 2 is that of a population with an Allee effect, where there exists a threshold population size α such that $R(x) < D(x)$ for $x < \alpha$, $R(\alpha) = D(\alpha)$, and $R(x) > D(x)$ for $x > \alpha$ and sufficiently close to α. Then $x = \alpha$ is an equilibrium for the unharvested system $(H = 0)$ with $0 < D'(\alpha) < R'(\alpha)$, and by the corollary to Theorem 4 this equilibrium is unstable. Thus populations with initial data less than α will become extinct but populations with initial data greater than α will remain greater than α for all $t \geq 0$. Thus α serves as a lower bound for population sizes in place of a lower bound as given by Theorem 2. If H increases from zero, then, at least for small positive H, there is an unstable equilibrium $\alpha(H)$ depending continuously on H with $\alpha(0) = \alpha$ which serves as a lower bound for populations with sufficiently large initial data, while populations with initial data less than $\alpha(H)$ become extinct in finite time. Since $R'(\alpha) > 0$ it is easy to see that $\alpha(H)$ is an increasing function of H.

Thus harvesting a population with an Allee effect increases the minimum population size necessary for the population to persist, as one would expect, but it also increases the minimum population size possible for populations large enough to persist. This prediction, for which we know of neither verification nor disproof by experiment, runs counter to naive intuition.

4. In this section we examine the dependence of the stability of an equilibrium on the time-lag and the harvest rate. First we shall consider x_∞, H, $R'(x_\infty)$, and $D'(x_\infty)$ as fixed and allow τ to vary. As we have seen already, if $R'(x_\infty) > D'(x_\infty)$, then the equilibrium x_∞ is (exponentially) unstable for $0 \leq \tau < \infty$.

If $|R'(x_\infty)| < D'(x_\infty)$, then $|q| < -p$ for all τ, and $p < q < -p$. Since $p < 0$, $\frac{\pi}{2} < z < \pi$, and thus $\sec z < -1$. Then $q > p > -p \sec z$, and the stability condition (8) is satisfied. Thus if $|R'(x_\infty)| < D'(x_\infty)$, the equilibrium x_∞ is asymptotically stable for $0 \leq \tau < \infty$.

If $R'(x_\infty) < -D'(x_\infty)$, then $q < -p$ for all τ. The other stability condition becomes

$$\sec z < \frac{q}{p} \frac{R'(x_\infty)}{D'(x_\infty)}, \tag{9}$$

and $R'(x_\infty)/D'(x_\infty)$ is a constant (independent of τ) less than -1. As $\tau \to 0$, $\sec z \to -\infty$ and (9) is satisfied, while as $\tau \to \infty$, $p \to -\infty$ and $\sec z \to -1$, so that (9) is not satisfied. Thus, by Theorem 4, the equilibrium x_∞ is asymptotically stable for sufficiently small τ and unstable for large τ. In fact, from $p = z \cot z$ we obtain

$$\frac{dp}{dz} = -\frac{z - \sin z \cos z}{\sin^2 z},$$

and from $\sin z < z$, $\cos z < 1$ we have $\frac{dp}{dz} < 0$. Thus

$$\frac{d}{d\tau} \sec z = \sec z \tan z \frac{dz}{dp} \cdot \frac{dp}{d\tau},$$

and since $z > \frac{\pi}{2}$ and p is a decreasing function of τ, $\frac{d}{dz} \sec z > 0$. Thus there is a unique value τ^* of τ for which the equilibrium x_∞

becomes unstable, and if $R'(x_\infty) < -D'(x_\infty)$ the equilibrium x_∞ is asymptotically stable for $0 \le \tau < \tau^*$ and unstable for $\tau > \tau^*$. Numerical simulations such as those reported in [2] suggest that differential-difference equations of the form (3) tend to exhibit periodic solutions, period doubling, and chaotic behavior as τ is increased in the case $R'(x_\infty) < -D'(x_\infty)$.

Numerical simulations carried out by the author on a personal microcomputer, also indicate that continuous population models of the form (3) may exhibit discrete generations. If $H = 0$, $x(t)$ must remain positive but rapid increases and crashes to very small population sizes can be exhibited. If $H > 0$, $x(t)$ may crash to zero and then regenerate rapidly after remaining zero for a time interval less than τ. While difference equations are often appropriate for delayed-recruitment population models where there is an annual recruitment mechanism, especially for fish and whale populations (see for example [5]), difference equations are sometimes used to model populations which exhibit discrete generations merely because difference equations are better understood than differential-difference equations. Differential-difference equations may be more appropriate models in such cases.

Next, we consider the effect of varying H while holding the recruitment function $R(x)$, the mortality function $D(x)$, and the delay τ fixed. There is then a maximum value H_c for which the recruitment curve $y = R(x)$ and the mortality curve $y = D(x) + H$ intersect at an equilibrium $x_\infty(H_c)$. It is clear that $H_c > 0$ unless $D(x) > R(x)$ for all $x > 0$ and if $H_c > 0$ then $R'(x_\infty(H_c)) = D'(x_\infty(H_c)) > 0$. For $H < H_c$ but close to H_c there are in general two intersections of the curves $y = R(x)$ and $y = D(x) + H$; the larger of these intersections is x_∞, and $R'(x_\infty) < D'(x_\infty)$. By Theorem 4, the equilibrium x_∞ is exponentially asymptotically stable for H sufficiently close to H_c. Thus harvesting tends to stabilize a population system modelled by (3) even if the unharvested system behaves chaotically. More specifically, x_∞ is exponentially asymptotically stable if $x_\infty < \xi$, that is if $H > R(\xi) - D(\xi)$, while

$$H_c = \max_{0 \le x < \infty} [R(x) - D(x)] = \max_{0 \le x \le \xi} [R(x) - D(x)] \ge R(\xi) - D(\xi). \tag{10}$$

By Theorem 3, if $H > H_c$ then every solution of (3) vanishes for all large t; thus H_c represents a harvest rate at which the qualitative behavior of the system (1) undergoes a drastic change. Such an occurrence is called a (mathematical) catastrophe.

5. We have considered only the simplest type of continuous delayed-recruitment population model with a fixed time-lag, and with juvenile mortality rates depending only on age and the number of juveniles of the same age. If the juvenile mortality rate depends on the total juvenile population size the derivation of the model (1) in Section 2 is not valid, and it would be necessary to formulate a two-stage model as in [12]. If the adult mortality rate is a function of age rather than a function of adult population density, then the adult population size is governed by a Volterra integral equation of the form

$$x(t) = \int_0^\infty R[x(t - s - \tau)]P(s)ds = \int_{-\infty}^{t-\tau} R[x(u)]P(t - u - \tau)du$$

if there is no harvesting [4]. Here, $x(u)$ on $-\infty \leq u \leq 0$ is regarded as a known initial function, $R[x(u)]$ is rate of recruitment to the adult stage of members born at time u, and $P(s)$ is the probability that a member who has reached the adult stage at age τ will survive to age $(s + \tau)$. It is not clear how to modify this model to incorporate harvesting.

If the age at which members reach the adult stage is not constant, as for example if maturity is determined by size rather than age, the population size is described by an integro-differential equation of the form

$$x'(t) = \int_0^\infty B(x(t - s))a(s)ds - D(x(t)) - H$$

with $B(x)$ the birth rate and $D(x)$ the mortality rate when the adult population size is x and $a(s)$ is the probability of survival and entrance to the adult stage at age s. Such models were introduced in [3] but have not been studied analytically.

The simplest continuous population models are formulated as ordinary differential equations without regard for separate recruitment and mortality functions. The logistic differential equation is introduced as the easiest way to model resource limitations which produce a differential equation with bounded solutions. The introduction of delays by arbitrarily modifying such an equation would seem to have little biological justification. The point of studying models such as (1) is that they separate recruitment and mortality terms and thus make it possible to model delays with more biological justification. A natural project would be to construct two-species models, including predator-prey, host-parasite, competitive, and mutualistic interactions in which recruitment, natural mortality, and interaction-induced mortality terms are separated. It would then be possible to introduce delays in these models on the basis of realism rather than mathematical curiosity.

BIBLIOGRAPHY

1. R. E. Bellman & K. L. Cooke, Differential-Difference Equations, Academic, New York, 1963.

2. S. P. Blythe, R. M. Nisbet & W. S. C. Gurney, "Instability and complex dynamic behaviour in population models with long time delays," Theor. Pop. Biol. 22 (1982), 147-176.

3. ______________________________, Formulating population models with differential aging, Population Biology, Proceedings, Edmonton 1982, H. Freedman & C. Strobeck, eds, Springer-Verlag, Berlin, 1984, 133-140.

4. F. Brauer, A class of Volterra integral equations arising in delayed-recruitment population models, to appear.

5. C. W. Clark, "A delayed-recruitment model of population dynamics, with an application to baleen whale populations," J. Math. Biol. 3 (1976), 381-391.

6. M. E. Gurtin & R. C. MacCamy, "Non-linear age-dependent population dynamics," Arch. Rational Mech. Anal. 3 (1974), 281-300.

7. N. D. Hayes, "Roots of the transcendental equation associated with a certain differential-difference equation," J. London Math. Soc. 25 (1950), 226-232.

8. T. Y. Li & J. A. Yorke, "Period three implies chaos," Am. Math. Monthly, 82 (1975), 985-992.

9. R. M. May, Theoretical Ecology, Principles and Applications, Saunders, Philadelphia, 1976.

10. R. M. May & G. F. Oster, "Bifurcations and dynamic complexity in simple ecological models," Am. Naturalist 110 (1976), 573-599.

11. R. M. Nisbet & W. S. C. Gurney, Modelling Fluctuating Populations, Wiley, Chichester, 1982.

12. ______________________________, "Stage-structure" models of uniform larval competition, Mathematical Ecology, Proceeding, Trieste 1982, S. A. Levin & T. G. Hallam eds., Springer Verlag, Berlin, 1984, 97-113.

DEPARTMENT OF MATHEMATICS
UNIVERSITY OF WISCONSIN
MADISON, WISCONSIN 53706

Canadian Mathematical Society
Conference Proceedings
Volume 8 (1987)

BIFURCATION PROBLEMS ARISING IN GEOCHEMISTRY[1]

J. Chadam[2] and P. Ortoleva

ABSTRACT. We delineate several phenomena in non-equilibrium geochemistry which, through self-organization, lead to patterned states (e.g. reaction-infiltration instability leading to scalloped reaction fronts, surface tension effects in crystal growth, oscillatory zoning in plagioclase feldspars). The first two phenomena are modelled mathematically as moving free boundary problems. Bifurcation techniques are used to examine the stability of these interfaces with respect to shape changes.

1. INTRODUCTION. In recent years social and economic factors have attracted many theoreticians to study problems which in the past might have been characterized as industrial or engineering concerns (e.g. the diagenesis and evolution of mineral, oil and gas reservoirs, the dynamics of nuclear and chemical waste repositories, in situ coal gasification, etc.). Typically, the essential geochemical phenomena in each of these situations can be modelled mathematically as a system of coupled, highly nonlinear reaction-transport equations (c.f.[1, 2] and the references therein). In general, however, even simplified versions of these equations are too complicated to be treated abstractly, analytically or numerically [3, sections 2, 3, 5]. One approach is to restrict attention to physically important problems for which the effective reaction zone (where the serious complications appear) is much smaller and less interesting than the scale of the phenomenon being studied. Mathematically this amounts to doing matched asymptotic expansions (in a parameter which is small for physical reasons) to obtain a more amenable moving free boundary problem for the reaction interface (c.f. [3, 4, 5] as

1980 Mathematics Subject Classification. 35R35, 35B32 86A60, 73HXX.

[1] Various parts of this work were supported by the DOE(USA), NSF (USA), NATO and NSERC (Canada).

[2] Lecturer and preparer of this report. Others who contributed to various parts of the work described here are: G. Auchmuty, J.-B. Baillon, M. Bertsch, J. Hettmer, D. Hoff, S. Howison, E. Merino, C. Moore, L. Peletier, E. Ripley and A. Sen.

well as references to combustion in [6], electrochemical forming and machining in [7], etc.).

In this note we shall describe two such problems--reaction infiltration [2, 3, 8] and solidification [5, 9, 10] instabilities. In both cases the problem of physical interest is the shape instability of the reaction interface (fingering, dendritic growth respectively). The associated mathematical problems will be treated using the techniques of bifurcation and stability analysis. This will provide an opportunity to indicate how these methods are modified to handle this new class of free boundary (as opposed to the more standard reaction-transport) problems. Similarly, the techniques of Hopf bifurcation can be used to treat the problem of oscillatory zoning in plagioclase feldspars [4] but this will not be presented here. For each of the two examples presented we shall briefly describe the physical phenomenon, write down the associated free boundary problem and summarize the results that have been obtained using bifurcation and stability analysis.

2. REACTION INFILTRATION INSTABILITIES

2.1 THE MODELS

2.1.1 THE PHYSICAL MODEL. Consider an aquifer consisting of an insoluble porous matrix (e.g. quartz sandstone) with some soluble mineral (e.g. calcite) partially filling the pores. If water is forced through this porous medium, the soluble component will be dissolved out upstream and the water will become saturated sufficiently far downstream. Between these extremes

Fig. 1. Focusing of flow to tip of porosity level curve.

there is a dissolution zone across which the soluble mineral content -- and hence the porosity -- changes from its original downstream value to the final, altered value upstream. The question of interest is whether the shape of this dissolution zone is stable. Notice that if a bump (in the porosity level curves) in the reaction zone exists at some time, the flow of the undersaturated waters tends to be focused to the tip of the bump via Darcy's law since inside the bump (on the upstream side) the permeability is greater than in the neighbouring regions (see Fig. 1). Thus dissolution is enhanced at the tip causing it to advance more rapidly. This is the porosity change/flow destabilization mechanism. On the other hand, diffusion from the sides of the tip raises the concentration of the solute in the water which is focusing at the tip and hence will decelerate this advancement. The competition between these two processes can lead to decay of the bump, restabilization to a morphologically more complicated dissolution zone or possibly to complete destabilization. A thorough understanding of this phenomenon and the resulting shapes is central to many important applications [1, 2, 3] (e.g. the locating of roll-front ore deposits for which the restabilized shape outlines the deposits; the possible fingering of the flow of reactive nuclear wastes which have broken through their underground confinement).

2.1.2 THE MATHEMATICAL MOVING FREE BOUNDARY MODEL. In order to obtain a mathematically tractable problem, we take the so-called large solid density limit, $c_{eq}/\rho \to o$, where ρ is the density of the dissolving mineral and c_{eq} is its equilibrium concentration when dissolved in the water. The dissolution zone, typically of width $(c_{eq}/\rho)^{1/2}$, collapses to a dissolution interface. Off this interface only transport processes are operative since there can be no reaction (upstream the mineral is completely dissolved out and downstream the water has reached its equilibrium concentration). If the dissolution interface is taken to be located at $x=R(y,t)$ with R unknown the equations upstream are

$$\left.\begin{aligned} &(2.1)\qquad \Delta\gamma + \nabla\gamma\cdot\nabla p = 0 \\ &(2.2)\qquad \phi = \phi_f \\ &(2.3)\qquad \Delta p = 0 \end{aligned}\right\} \quad \text{in } x < R(y,t),\ 0 \leqslant y \leqslant \pi.$$

while downstream they are

$$\left.\begin{aligned} &(2.4)\qquad \gamma \equiv 1 \\ &(2.5)\qquad \phi = \phi_0 \\ &(2.6)\qquad \Delta p = 0 \end{aligned}\right\} \quad \text{in } x > R(y,t),\ 0 \leqslant y \leqslant \pi$$

where $\gamma(x,y,t)$ is the scaled concentration of the solute with equilibrium value of $\gamma=1$, $p(x,y,t)$ is the scaled pressure (the scaled velocity from Darcy's law being $-\nabla p$) and ϕ is the porosity (the fraction of the volume of rock which is pore space). Notice that the porosity assumes only two values, the final value ϕ_f after all the mineral has been dissolved out and the initial value ϕ_0 with all the mineral present. We have taken ϕ_0 to be constant to show that morphological instabilities will even occur in this spatially homogeneous situation.

Across the dissolution interface (essentially defining it) the porosity changes discontinuously from its unaltered value ϕ_0 to its final value $\phi_f > \phi_0$. Using matched asymptotics, the other boundary conditions on the unknown moving dissolution interface are

$$
\left.
\begin{array}{ll}
(2.7) & \gamma = 1 \\
(2.8) & p- = p+ \\
(2.9) & \left(\frac{\partial p}{\partial x} - \frac{\partial p}{\partial y}\frac{\partial R}{\partial y}\right)_- = \Gamma\left(\frac{\partial p}{\partial x} - \frac{\partial p}{\partial y}\cdot\frac{\partial R}{\partial y}\right)_+ \\
(2.10) & \frac{\partial \gamma}{\partial x} - \frac{\partial \gamma}{\partial y}\cdot\frac{\partial R}{\partial y} = R_t
\end{array}
\right\}
\begin{array}{l}
\text{on} \\
x = R(y,t) \\
0 \leqslant y \leqslant \pi
\end{array}
$$

where $0 \leqslant \Gamma = \phi_0\kappa_0/\phi_f\kappa_f \leqslant 1$ is a measure of the porosity change ($\kappa_0 = \kappa(\phi_0)$ and $\kappa_f = \kappa(\phi_f)$ are permeabilities). Equation (2.10) relates the rate of advancement of the moving dissolution interface to the flux of the concentration and is called a Stefan condition.

To complete the problem we add the asymptotic conditions

$$(2.11) \qquad \gamma \to 0,\ \frac{\partial p}{\partial x} \to -\nu_f \text{ as } x \to -\infty, \qquad \gamma \to 1,\ \frac{\partial p}{\partial x} \to -\nu_0 \text{ as } x \to +\infty,$$

and zero flux conditions for all the variables on the aquifer walls:

$$(2.12) \qquad \gamma_y = 0,\ p_y = 0 \text{ for } y = 0, \pi,\ -\infty < x < \infty$$

Here the imposed gradient (equivalently imposed velocity via Darcy's law) at $x=-\infty$ is, as a result of the scalings, $-\nu_f = -\frac{v_f L}{D_f \pi}$ where $v_f = v(\phi_f) = -\kappa(\phi_f)\nabla p$ is the imposed (unscaled) velocity at $x=-\infty$, $D_f = D(\phi_f)$ is the diffusion coefficient in the porous medium when the porosity is ϕ_f and L is the original transverse dimension. The similar pressure gradient at $x=+\infty$, $-\nu_0$, is to be determined.

2.2 SHAPE INSTABILITIES. In this section we describe our analytical results [3, 8] in the context of the large solid density problem (2.1-12). Here the planar, constant velocity solution can be obtained explicitly and completely, including the concentration and pressure profiles which were not available for the more general c_{eq}/ρ finite, reaction-transport model. The linearized stability of this solution is then described, giving a precise value of the parameter ν_f (in terms of Γ) for which the planar solution loses stability to another, more structured, solution. In the language of bifurcation theory we determine the critical parameter value for which the spectrum of the linearized problem changes sign from negative to positive, thus determining the location of a possible bifurcation point. Finally we sketch the bifurcation analysis to show that the linear instabilities are restabilized by the nonlinearities to a morphologically more complicated solution. In the bifurcation terminology, we obtain a Landau equation for the amplitude of the linearly unstable mode indicating a pitchfork bifurcation diagram.

2.2.1 PLANAR SOLUTION. Denoting the planar state quantities with a super bar, one can easily check [3,8] that the following constant velocity solution satisfies problem (2.1-12):

$$\bar{R}(t) = \bar{V}t \tag{2.13}$$

$$\bar{\gamma}(x,t) = \begin{cases} e^{-\nu_f(x-\bar{V}t)} & x < \bar{V}t \quad (2.14a) \\ 1 & x > \bar{V}t \quad (2.14b) \end{cases}$$

$$\bar{p}(x,t) = \begin{cases} -\nu_f(x-\bar{V}t) & x < \bar{V}t \quad (2.15a) \\ -\nu_0(x-\bar{V}t) & x > \bar{V}t. \quad (2.15b) \end{cases}$$

where $\nu_f = v_f L/D_f \pi$, $\nu_0 = \phi_f \nu_f / \phi_0$ (from (2.12) and (2.9)) and the velocity of the planar interface $\bar{V} = \nu_f$ from (2.12).

2.2.2 LINEAR SHAPE INSTABILITY. In order to examine the stability of the above planar solutions (2.13-15) with respect to bumps we consider perturbations of the type (i.e. a generic term in the Fourier cosine decomposition -- see equation (2.12))

$$R(y,t) = \bar{V}t + \delta\, r_{1m}(t) \cos my \tag{2.16a}$$

$$\gamma(x,y,t) = \bar{\gamma}(x,t) + \delta\, \gamma_{1m}(x)\, r_{1m}(t) \cos my \tag{2.16b}$$

$$p(x,y,t) = \bar{p}(x,t) + \delta\, p_{1m}(x)\, r_{1m}(t) \cos my. \tag{2.16c}$$

Considering δ to be small, the linearized version of equations (2.1-12) can be derived [3] (i.e. retain terms in first power of δ). We do not write them here explicitly because they are rather complicated. A simpler example of linearized equations of this sort is given in the next section. The present equations, however, can be solved explicitly for γ_{1m} and p_{1m} and the Stefan condition gives [3,8] the following condition on the amplitude $r_{1m}(t)$ of the cos my bump:

$$r'_{1m}(t) = \frac{\nu_f}{1+\Gamma}\left[\nu_f - (\nu_f^2 + 4m^2)^{1/2} + (1-\Gamma)|m|\right] r_{1m}(t) \tag{2.17}$$

This differential equation indicates that the amplitude of the bump grows or decays depending on the sign of the coefficient. The connection with the equivalent, more conventional viewpoint follows by expressing $r_{1m}(t) = re^{\sigma(m)t}$ in terms of the spectrum $\sigma(m)$ of the linearized problem and obtaining from (2.17) the following dispersion relation

THEOREM 2.1

$$\sigma(m) = \frac{\nu_f}{1+\Gamma}\left[\nu_f - (\nu_f^2 + 4m^2)^{1/2} + (1-\Gamma)|m|\right]. \tag{2.18}$$

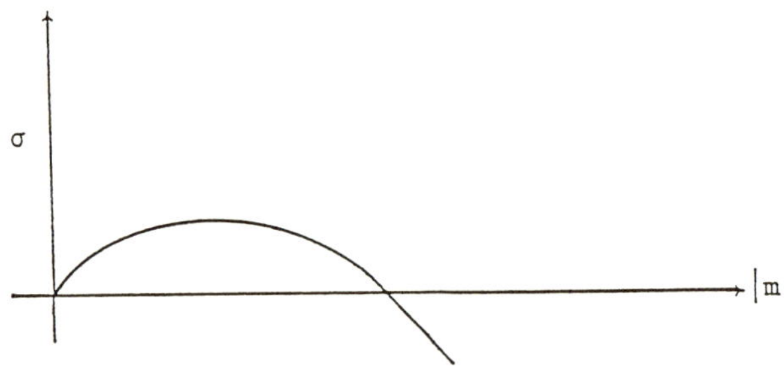

Fig. 2. Graph of dispersion relation (2.18).

The m-dependence of σ is shown in Fig. 2 revealing clearly that the planar solution (2.13-15) is linearly unstable to long wavelength perturbations (because $\Gamma < 1$) and stable to short wavelength perturbations. The critical wave number ($|m_0|$ at which $\sigma(m_0) = 0$) is given by

$$|m_0| = \frac{2(1-\Gamma)}{(3-\Gamma)(1+\Gamma)}\nu_f \tag{2.19}$$

Since our channel width has been normalized to π, the first mode which can be carried is $|m_0| = 1$ giving, from (2.19), the critical parameter value (of $\nu_f = v_f L/D_f \pi$)

$$\nu_c = \nu_c(\Gamma) = \frac{(3-\Gamma)(1+\Gamma)}{2(1-\Gamma)}. \tag{2.20}$$

From this we see that the instability does indeed arise analytically and that, as is physically realistic, larger flow speeds, larger transverse dimensions, larger porosity/permeability changes promote the instability while larger diffusion coefficients inhibit the instability (i.e. diffusion is stabilizing, as mentioned earlier). The limit of $\Gamma \to 1$ (i.e. no porosity change) suggests it is very difficult to produce instabilities. This has been verified by a separate analysis [11]. Thus this instability can occur only if "significant" amounts of the soluble mineral are dissolved.

2.2.3 NON-LINEAR RESTABILIZATION. We begin by scaling the independent variables. Because the instability occurs at finite wavelength none is required for the spatial variables while, as is common for pitchfork bifurcations,

$$t_2 = \varepsilon^2 t, \tag{2.21}$$

(since the growth rate σ in (2.18) is linear in $\nu_f - \nu_c$ which is in turn quadratic in the amplitude ε -- see (2.23)). Additionally, we write

$$\nu_f = \nu_c + \varepsilon\,\nu_1 + \varepsilon^2 \nu_2 + ---. \tag{2.22}$$

We find [8] at $O(\varepsilon^2)$ that, for consistency, $\nu_1 = 0$ (in analogy with (2.17) one obtains $\nu_1 r_{11}(t)\; f(\nu_c) = 0$ with $f(\nu_c) \neq 0$ which requires $\nu_1 = 0$ since $r_{11}(t)$ is the amplitude we are studying -- as usual for pitchfork bifurcations) so that the physical significance of the small parameter ε is

$$\varepsilon \simeq (\nu_f - \nu_c)^{1/2} \tag{2.23}$$

where we have taken, without loss of generality, $\nu_2 = 1$. Thus

$$\nu_f = \nu_c + \varepsilon^2 + \varepsilon^3 \nu_3 + ---. \tag{2.24}$$

The stability calculation then proceeds by expanding all of the dependent variables in terms of ε (suppressing the sub-2 in the new t_2 variable):

$$\begin{aligned} R(y,t) = {} & \bar{V}(\varepsilon)t \\ & + \varepsilon(r_{10}(t) + r_{11}(t)\cos y + r_{12}((t)\cos 2y + ---) \\ & + \varepsilon^2(r_{20}(t) + r_{21}\cos y + r_{22}(t)\cos 2y + ---) \\ & + \varepsilon^3(r_{30}(t) + r_{31}\cos y + r_{23}(t)\cos 2y + ---) \\ & + O(\varepsilon^4), \end{aligned} \tag{2.25a}$$

$$\begin{aligned} \gamma(x,y,t) = {} & \bar{\gamma}(x,t;\varepsilon) \\ & + \varepsilon(\gamma_{10}(x,t) + \gamma_{11}(x,t)\cos y + \gamma_{12}(x,t)\cos 2y + ---) \\ & + \varepsilon^2(\gamma_{20}(x,t) + \gamma_{21}(x,t)\cos y + \gamma_{22}(x,t)\cos 2y + ---) \\ & + \varepsilon^3(\gamma_{30}(x,t) + \gamma_{31}(x,t)\cos y + \gamma_{32}(x,t)\cos 2y + ---) \\ & + O(\varepsilon^4), \end{aligned} \tag{2.25b}$$

and similarly for $p(x,y,t)$. At each order the resulting problems have homogeneous asymptotic conditions and couple to previous orders through very complicated inhomogeneous terms in the differential equations and the conditions at the interface. These can be solved explicitly (rather than obtaining the compatibility conditions via the adjoint problem which is difficult to write down) to obtain a Landau differential equation for the amplitude of the unstable mode, $r_{11}(t)$:

THEOREM 2.2

$$r'_{11}(t) = w\, r_{11}(t) - \Lambda\, r_{11}(t)^3 \tag{2.26}$$

where

$$w = w(\Gamma) = \frac{\nu_c}{1+\Gamma} \frac{[\nu_c^2 + 4)^{1/2} - \nu_c]}{(\nu_c^2 + 4)^{1/2}} \geqslant 0 \tag{2.27}$$

and the Landau constant, $\Lambda = \Lambda(\Gamma)$, which is algebraically very complicated, is given in Fig. 3.

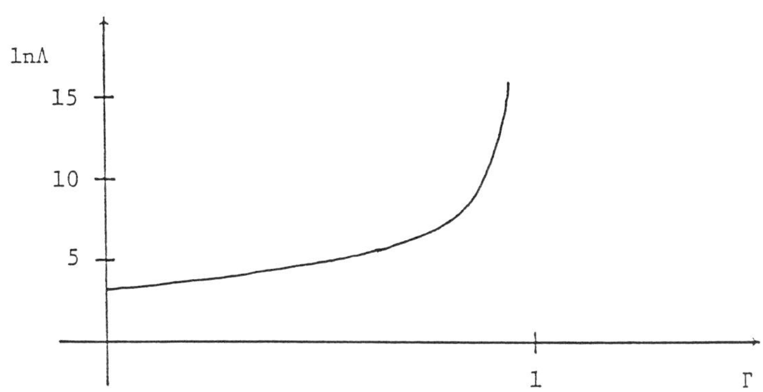

Fig. 3. Graph of the logarithm of the Landau constant versus $\Gamma = \phi_0 \kappa_0 / \phi_f \kappa_f$

The positivity of Λ indicates that in the vicinity of the critical point the linearized instabilities (from $w \geqslant 0$) are restabilized by the non-linearities at the next highest order, and (from (3.13)) that the bifurcation diagram is the symmetric pitchfork. The asymptotic amplitude of the bump can be obtained from (2.26) to be $(w/\Lambda)^{1/2}$.

3. SOLIDIFICATION INSTABILITIES

3.1 THE MODEL. We shall examine the growth of a pure amorphous solid from a diffusing growth material in the context of a model which is a generalization of the classic Mullins and Sekerka model [12, 13] (see also the excellent review by Langer [14] and the references therein). In this case, assuming the solidification process to occur infinitely fast, matched asymptotics gives the following free boundary problem. Let c be the concentration of the solute and the solid-solute interface be $S(\underline{x},t) = 0$; then the model problem adopted is to find these two quantities subject to

(3.1) $c_t = D\Delta c$ in $S(\underline{x},t) > 0$

(3.2) $c = c_{eq}(1+\gamma\kappa(S))$ on $S(\underline{x},t) = 0$,

(3.3) $D\frac{\partial c}{\partial n} = (\rho - c)\, V_n$ on $S(\underline{x},t) = 0$,

(3.4) $c \to c_\infty$ as $|\underline{x}| \to \infty$,

(3.5) $c(\underline{x},0) = c_0(\underline{x})$ in $S(\underline{x},0) = S_0(\underline{x}) > 0$

for given initial functions $c_0(\underline{x})$, $S_0(\underline{x})$. Here ρ is the density of the solid, c_{eq} is the equilibrium solidification concentration on a planar interface and c_∞ is the ambient concentration far from the solid. Also, $\underline{n}$ is the normal to $S(\underline{x},t) = 0$ and $V_n = -S_t/|\nabla S|$ is the normal velocity of the interface. The function $\kappa(S)$ is the mean curvature of the surface $S = 0$ and $\gamma \geq 0$ is a measure of the interfacial energy. Thus we are assuming, as in the Mullins and Sekerka model, that the interaction at the interface is so rapid that it is maintained at equilibrium as given by the Gibbs-Thompson relation (3.3) (effects of kinetic under-cooling are ignored). On the other hand we do not assume that the diffusion is so rapid that it is maintained at a quasi-steady-state and so we replace the Laplace equation in the Mullins and Sekerka model with the diffusion equation.

When $\gamma=0$, problem (3.1-5) is the standard Stefan problem (the classic stable melting version if $c_\infty < c_{eq}$ and the little studied, unstable solidification version if $c_\infty > c_{eq}$). In the solidification case of interest here, diffusion is a destabilizing effect causing the body to shoot dendrites straight out to regions of higher concentration. When $\gamma \neq 0$ equation (3.2) implies that a higher concentration is required to maintain such a highly curved configuration. The interplay of the diffusions shape destabilization and the surface tension's flattening action provides an interesting solidif ication/melting dynamics. Our results on the subject [5,9,10] are presented in the next section.

3.2. SHAPE INSTABILITIES

3.2.1 PLANAR SOLUTION. Using similarity methods it is straightforward to find a solution of the planar version of equations (3.1-5). For nearly planar solution with $S(\underline{x},t) = x - R(y,t)$, equation (3.4) is interpreted as $c \to c_\infty$ as $x \to +\infty$ with appropriate behavior in the y direction to ensure the existence of the Fourier transform of $R(y,t)$ and $c(x,y,t)$ if the transverse direction is infinite or the zero flux conditions (2.12) if the melt is confined between walls of finite extent. One finds (with $D = 1$ here and in the remainder of this section)

(3.6a) $\bar{R}(t) = \alpha\sqrt{t}$,

(3.6b) $\bar{c}(x,t) = -(\rho - c_{eq})\alpha e^{\alpha^2/4}\int_{x/2\sqrt{t}}^{\infty} e^{-y^2}\,dy + c_\infty$,

where α satisfies the equation

(3.6c) $\alpha e^{\alpha^2/4}\int_{\alpha/2}^{\infty} e^{-y^2}\,dy = (c_\infty - c_{eq})/(\rho - c_{eq})$.

This last equation has a unique solution provided that $-\infty < (c_\infty - c_{eq})/(\rho - c_{eq}) < 1$ [5]. Thus in the melting case, $c_\infty < c_{eq}$, a solution exists for all values of the ratio while for solidification we must have the physically reasonable constraint $c_\infty < \rho$. These special ($\bar{R}(0)=0$, $\bar{c}(x,0)=c_\infty$) solutions suggest the following result about the <u>global</u> existence of solutions to the planar problem with <u>arbitrary</u> data $R(0)$ and $c_0(x)$ [9].

THEOREM 3.1. Suppose $c_0 \in C^1[R(0),\infty)$, $c_0(R(0)) = c_{eq}$ and $c_0(x) - c_\infty \in L^1(R(0),\infty)$ (i.e. is of finite mass)

a) The planar versions of equations (3.1-5) have a unique global solution if

(i) $c_{eq} \geq c_0(x) \geq c_\infty$ (the strictly melting version) or

(ii) $\|c_0\|_\infty \leq \rho/8$ if c_0 is anywhere larger than c_{eq} (the solidifying case).

Furthermore as $t \to \infty$, $R(t) \approx \bar{R}(t) = \alpha\sqrt{t}$ as defined in (3.6a,c).

b) If, on the other hand, $c_{eq} \leq c_0(x) \leq c_\infty$ with $c_\infty > \rho$ and (for technical reasons) $c_0'' \leq 0$, then there exists a $T < \infty$ such that

$$\lim_{t \to T-0} \dot{R}(t) = +\infty.$$

This theorem which summarizes our results concerning planar solutions is a fully nonlinear result, part (i) following from maximum principles while part (ii) requires some complicated estimates. Part b) is a finite-time blow-up result indicating that some smallness condition as in part (ii) is required on the initial data, otherwise the reaction front experiences a catastrophic instability at some finite time T when the entire surface moves off with infinite velocity. On the other hand when global existence obtains, part a) shows that all planar solutions tend asymptotically to the one (3.6) with the corresponding value of $(c_\infty - c_{eq})/(\rho - c_{eq})$. Thus the asymptotic stability analysis of any planar solution with respect to shape perturbations reduces to studying the morphological stability of the similarity solution (3.6).

3.2.2 LINEAR SHAPE INSTABILITY. Using (3.6) and writing (with δ small)

(3.7a) $$c_\delta = \bar{c} + \delta c,$$

(3.7b) $$R_\delta = \bar{R} + \delta R,$$

the linearized versions of equations (3.1-5) for the planar case are

(3.8) $$c_t = \Delta c \qquad x > \bar{R}(t),\ t > T$$

(3.9) $$c = -\bar{c}_x R - \frac{\gamma}{2}\,\Delta R \qquad x = \bar{R}(t),\ t>T$$

(3.10) $$c_x = (\rho - c_{eq})R_t - \bar{c}_{xx} R \qquad X = \overline{R}(t),\ t>T$$

(3.11) $$c(\underline{x},T) = \psi(\underline{x}) \qquad x > \bar{R}(T)$$

(3.12) $$c \to 0 \quad \text{as } x \to +\infty$$

Although the original problem (3.1-5) has been greatly simplified in that (3.8-12) has the boundary data given on the known planar surface $x=\bar{R}(t)$, the latter problem is still formidable because the boundary is moving and all coefficients are time-dependent. We should not expect a simple expression for the spectrum of the linearized problem (as in (2.18) for example). More precisely, taking Fourier transforms (series if the melt is confined in $[0,\pi]$) if

(3.13a) $$c = \delta r_m(t) c_m(x) \cos my,$$

(3.13b) $$R = \delta\ r_m(t)\ \cos my,$$

(and similarly for the sin my, $m \neq 0$, modes) we will not obtain a linear constant coefficient differential equation for $r_m(t)$ (as in (2.17)) because of the $x/2\sqrt{t}$ dependence of the planar solution (3.6). Indeed for t large an asymptotic analysis [5] gives that $\hat{r}_m(t) = e^{m^2 t} r_m(t)$ satisfies the integro-differential equation

$$(3.14)\qquad \frac{d\,\hat{r}_m}{dt}(t) - \hat{r}_m(t) = I(t) + (\rho - c_{eq}) \int_T^t \frac{d\,\hat{r}_m}{d\tau}(\tau)\,[P(t,\tau)+\Gamma(t,\tau)]\,d\tau$$

where I(t) is a well-behaved function of the initial data, and the kernels

$$(3.15a)\qquad P(t,\tau) = \frac{\alpha(\rho - c_{eq})\tau^{1/2}}{2\sqrt{\pi}(t-\tau)^{1/2}} - \frac{\alpha^2(\rho - c_{eq})}{8t} + O((t-\tau)^{1/2}),$$

$$(3.16b)\qquad \Gamma(t,\tau) = -\frac{\gamma m^2}{2\sqrt{\pi}\,(t-\tau)^{1/2}} - \frac{\alpha\,\gamma m^2}{8\sqrt{t}} + O((t-\tau)^{1/2})$$

as $t \to \tau + 0$, where α is the solution of (3.6c). This gives rise to [5]

THEOREM 3.2 For $|m|>0$ and $t\to\infty$, the amplitude $r_m(t)$ satisfies the asymptotic estimates:

$$r_m(t) \approx \begin{cases} \text{const. } t^{\alpha^2/4} e^{\alpha|m|t^{1/2}}, & \gamma = 0 \qquad (3.17a) \\ \text{const. } e^{-b(1+b)^{-1} m^2 t}, & b = \dfrac{\gamma|m|}{2(\rho - c_{eq})} > 0 \qquad (3.17b) \end{cases}$$

Thus, for arbitrary perturbations, the melting problem ($\alpha<0$) is asymptotically stable regardless of the presence of surface tension effects, while every mode of the solidification problem is asymptotically unstable without surface tension ($\gamma=0$) and is stabilized by its inclusion. More precise estimates on equation (3.14) are still required, however, for $t\to T$, γ small, to characterize those long-wavelength perturbations which become unstable during solidification.

3.3 GROWING SPHERES. For problem (3.1-5) in a spherical geometry results entirely analogous to those in section 3.2.1 have been obtained [10] even though the surface tension effects contribute on this high symmetry level. On the other hand, the (linearized) shape stability of these growing spheres with respect to <u>arbitrary</u> perturbations is the subject of current study. The following partial results [10] which challenge the classical results [12,13] indicate the importance of obtaining a complete linearized stability analysis.

To begin, there are similarity solutions in then variable $x/2\sqrt{t}$ of the spherical versions of equations (3.1-5) with

$$(3.18a)\qquad \overline{R}(t) = \beta\sqrt{t}$$

where β satisfies

$$\text{(3.18b)} \qquad \frac{\beta^2}{2}\left[1-\beta e^{\beta^2/4}\int_{\beta/2}^{\infty} e^{-y^2}\,dy\right]=(c_\infty - c_{eq})/(\rho - c_{eq})$$

and the precise form of $\overline{c}$ is abailable but unimportant. Furthermore, as in Theorem 3.1, part a), all global spherical solutions are asymptotic to the one of these with the corresponding value of $(c_\infty - c_{eq})/(\rho - c_{eq})$ [12]. The first variational equations of (3.8-12), linearized with respect the spherical solutions (3.18), have special solutions (i.e. with special initial data expressed in terms of Whittaker functions) for which the amplitude of the $Y_{\ell m}(\theta,\phi)$ bump is of the form

$$\text{(3.19)} \qquad r_\ell(t)=d_0 t^{(\ell-1)/2}+d_1 t^{(\ell-2)/2}+ - - - + d_{\ell+1} t^{-1},$$

where the coefficients d_i, $i=0, \ell+1$ are explicitly calculated. The importance of these special solutions is that when one calculates the (unique, it turns out) time $t_{c,\ell}$ when $\dot{r}_\ell$ changes sign from negative to positive (i.e. the time at which the $Y_{\ell m}$-mode loses stability) the associated critical radius of the growing sphere, $R_{c,\ell}=\beta\sqrt{t_{c,\ell}}$, turns out to be quantitatively and qualitatively different from the values obtained classically [12,13]. The classical result, which corresponds to $\beta\to 0$ in our model, gives for the first unstable mode $(\ell=2)$, $\overline{R}_{c,2}=7R^*$ where $R^*=2\gamma c_{eq}/(c_\infty - c_{eq})$ is the so-called critical nucleation radius. Our model for $\beta\to 0$ gives $R_{c,2}=6.4R^*$; that is, our special solutions indicate that the growing sphere loses stability <u>sooner</u> than that predicted classically [12,13]. Moreover, the classical theory gives that $\overline{R}_{c,\ell}=O(\ell^3)$ as $\ell\to\infty$ while our model shows that this is only true for $\beta\to 0$. Indeed for $\beta\to\infty$, $R_{c,\ell} < 2.36R^*$ <u>for all</u> ℓ. Thus it is essential to do a complete linearized stability analysis (i.e. for all data) along the lines of the planar analysis of section 3.2 in order to obtain a more precise answer to this question which is of importance in material science.

BIBLIOGRAPHY

1. J. Chadam, P. Ortoleva, E. Merino and C. Moore, "Geochemical Self-Organization I: Feedback Mechanisms and Modeling Approach", accepted for publication, Amer. J. Sci. (1987).

2. J. Chadam, P. Ortoleva, E. Merino and A. Sen, "Self-Organization in Water-Rock Interaction Systems, II: The Reactive-Infiltrate Instability", accepted for publication, Amer. J. Sci. (1987).

3. J. Chadam, P. Ortoleva and A. Sen, "Reactive infiltration Instabilities", accepted for publication, IMA J. of Appl. Math (1986).

4. C. S. Haase, J. Chadam, D. Feinn and P. Ortoleva, "Oscillatory Zoning in Plagioclase Feldspars", Science 209 (1980), 272-274.

5. J. Chadam and P. Ortoleva, "The stabilizing effect of surface tension on the development of the free boundary in a planar one-dimensional Cauchy-Stefan problem", IMA J. of Appl. Math. 30 (1983).

6. B. J. Matkowsky and G. I. Shivashinsky, "Propogation of a Pulsating Reaction Front in Solid Fuel Combustion", SIAM J. Appl. Math. 35 (1978), 465-478.

7. J. A. McGeough, Principles of Electrochemical Machining, Chapman and Hall, London (1974).

8. J. Chadam, P. Ortoleva and A. Sen, "Weakly nonlinear stability of reaction-percolation interfaces", submitted to SIAM J. of App. Math. (1986).

9. J. Chadam, J.-B. Baillon, M. Bertsch, P. Ortoleva and L. A. Peletier, "Existence, uniqueness and asymptotic behaviour of solutions of the planar, supersaturated solidification, Cauchy-Stefan problem", Lions-Brezis Seminar, Vol VI College de France. Research Notes in Math, 109, 27-47, Pitman, London (1984).

10. J. Chadam, S. Howison and P. Ortoleva, "Spherical Crystals Growing in a Supersaturated Solution", accepted for publication, IMA J. Appl. Math. (1987).

11. J. Chadam, G. Auchmuty, E. Merino, P. Ortoleva and E. Ripley, "The structure and stability of moving redox fronts", SIAM J. Appl. Math. 46 (1986), 588-604.

12. W. W. Mullins and R. F. Sekerka, "Morphological Stability of a Particle Growing by Diffusion or Heat Flow", J. Appl. Phys. 34 (1964), 323-329.

13. W. W. Mullins and R. F. Sekera, "Stability of a Planar Interface During Solidification of a Dilute Alloy", J. Appl. Phys. 35 (1964), 444-451.

14. J. S. Langer, "Instabilities and Pattern Formation in Crystal Growth", Rev. Mod. Phys. 52 (1980), 1-28.

DEPARTMENT OF MATHEMATICS AND STATISTICS
McMASTER UNIVERSITY
HAMILTON, ONTARIO, L8S 4K1, CANADA

DEPARTMENTS OF CHEMISTRY AND GEOLOGY
INDIANA UNIVERSITY
BLOOMINGTON, INDIANA, 47405, U.S.A.

Canadian Mathematical Society
Conference Proceedings
Volume 8 (1987)

HOPF BIFURCATION IN THE PRESENCE OF SYMMETRY, CENTER MANIFOLD AND LIAPUNOV-SCHMIDT REDUCTION

Pascal Chossat

Martin Golubitsky

ABSTRACT: Assume that the linear part of a vector field X is semisimple and has eigenvalues at $\pm\ \omega i$. We show that if the quadratic terms of X vanish when restricted to the center subspace, then to third order the Liapunov-Schmidt reduction for finding periodic solutions of X is identical to a center manifold reduction of X followed by putting the reduction in normal form. Several examples of systems with symmetry that satisfy this hypothesis are discussed.

I. INTRODUCTION

Local bifurcation theory, as developed through the Liapunov-Schmidt decomposition method [9], provides a simple and efficient way for computing zeroes of families of nonlinear maps near singularities. However, when applied to evolution equations (ODE's or PDE's) this method does not, in general, give information on the dynamics close to the bifurcated solutions, except in situations where certain exchange of stability principles are valid, such as at simple eigenvalues [9]. The "natural" setting for the study of dynamics close to a bifurcation is a combination of a center manifold reduction along with a normal form analysis [11]. The relationship between the Liapunov-Schmidt decomposition and the stability of bifurcated solutions has been studied, from a general point of view, in [10]. Our purpose is different. We shall show that if a simple condition is realized (no quadratic terms for the Taylor expansion of the equation on the center manifold), the Liapunov- Schmidt method for Hopf bifurcation leads to an equation which is identical to the normal form equation on the center manifold, up to cubic order (assuming semisimple critical eigenvalues). This can be interesting in problems invariant under a symmetry group, because symmetry can force the

1980 Mathematics Subject Classification (1985 Revision)
58F22, 34C29, 58F14

quadratic terms to be identically zero, as we shall see in Section III. We now recall the Liapunov-Schmidt method and the center manifold reduction process.

Let $f(u,\lambda)$ be a C^k map ($k \geq 1$) from $\mathbb{R}^n \times \mathbb{R}^p$ to $\mathbb{R}^n$ such that $f(0,0) = 0$ and $L = D_u f(0,0)$ is not invertible. Let V be the kernel of L and let P,Q be projections in $\mathbb{R}^n$ such that $\operatorname{Im} P = V$ and $\operatorname{Ker} Q = \text{range } L$. We note that, if 0 is a semisimple eigenvalue of L one can choose $P \equiv Q$. Now set $u = x+y$ where $x \in V$ and $y \in \ker P$; then the equation $f(u,\lambda) = 0$ splits as follows

$$Qf(x+y,\lambda) = 0 \tag{1.1}$$

$$(1-Q)f(x+y,\lambda) = 0 \tag{1.2}$$

Notice that we can write equation (1.2) as follows:

$$(1-Q)Ly = (1-Q)N(x+y,\lambda) \tag{1.3}$$

where

$$N(u,\lambda) = f(u,\lambda) - Lu.$$

It is easy to check that the operator $(1-Q)L$ restricted to a complement of V in $\mathbb{R}^m$ is invertible. Hence the implicit function theorem applies to (1.3) and gives $y = y(x,\lambda)$ of class $\mathbb{C}^k$ in a neighborhood of $(0,0)$. Replacing y by $y(x,\lambda)$ in (1.1) leads to the "bifurcation equation":

$$g(x,\lambda) = QN(x+y(x,\lambda),\lambda) = 0 \tag{1.4}$$

This method has proved to be powerful. For example, the Hopf bifurcation theorem can be proved by looking for the zeroes of a functional equation, within a class of periodic functions. Other methods are possible, but the Liapunov-Schmidt method has the advantage of providing a simple and systematic rule for computing the Taylor expansion of the solution just by matching powers in (1.3) and (1.4).

On the other hand, let X_0 be a smooth (C^k) vector field in $\mathbb{R}^n$, and let 0 be a singular point for X_0, i.e., $X_0(0) = 0$. It is well-known that if no eigenvalues of $A = DX_0(0)$ have zero real part, then the dynamics of the flow of X_0 in a neighborhood of 0 is determined by the eigenvalues of A. In particular, if all of the eigenvalues have negative real part, then 0 is an attractor or sink of X_0. If, however, some eigenvalues lie on the imaginary axis, then the dynamics of the flow near 0 will be determined by higher order terms in the Taylor expansion of X_0. More precisely, let V_0 be the (generalized) eigenspace of A associated to eigenvalues with real part equal to zero , and let P_0 be the projection onto V_0 such that $AP_0 = P_0A$. Then there exists a C^k map on $\mathbb{R}^n$

$$\phi_0 : V_0 \to (1-P_0)\mathbb{R}^n,$$

such that $\phi_0(0) = D\phi_0(0) = 0$ and whose graph is a <u>center manifold</u> M_0, i.e. M_0 is a locally invariant manifold for the flow defined by X_0 [11]. If, in addition, the remaining eigenvalues of A are in the left-half plane, then M_0 is also an attractor for the flow. We can therefore restrict the study of the flow to its action on M_0. In the case of a family X_λ of vector fields ($\lambda \in \mathbb{R}^p$), there exists a map $\phi : V_0 \times \mathbb{R}^p \to (1-P_0)\mathbb{R}^n$ such that $\phi(.,0) = \phi_0$. The Taylor expansion of $\phi(x,\lambda)$ can be computed by a matching rule, slightly more complicated than the rule for $y(x,\lambda)$ in the Liapunov-Schmidt method. Also, the study of the dynamics generally requires the vector field on M_λ to be in "normal form" (the normal form is a k-th order Taylor expansion obtained from X_λ by near identity changes of variables and which is as simple as possible [8]).

II. STATEMENT AND PROOF OF THE MAIN RESULT

Let us consider the equation

$$\frac{du}{dt} = F(u,\lambda), \qquad (2.1)$$

where $F : D \times \mathbb{R} \to E$ of class C^{k+1} ($k \geq 3$) in a neighborhood of $(0,0)$, $F(0,0) = 0$, E is a real Banach space and $D \subset E$ is the domain of the linear operator $A = D_uF(0,0)$. We assume that standard hypotheses are valid for A and F, so that the center manifold theorem applies to (2.1) at $(u,\lambda) = (0,0)$ [11]. We also assume that A has semisimple eigenvalues $\pm i\omega_0$ of finite multiplicity, and that the rest of the spectrum lies in the half-space $\{\zeta \in \mathbb{C} : \operatorname{Re}\zeta \leq \xi < 0\}$. We denote the invariant subspace associated with eigenvalues $\pm i\omega_0$ by V_0 , and let $E = V_0 \oplus W$, where W is such that $A(W \cap D) \subset W$. Finally we let

$$x = P_0 u \quad \text{and} \quad y = (1-P_0)u$$

where P_0 is the projection on V_0 associated to the foregoing decomposition of E. From the center manifold theorem, equation (2.1) is reduced to the equation:

$$\frac{dx}{dt} = G(x,\lambda) \qquad (2.2)$$

where

$$G(x,\lambda) = P_0F(x+y(x,\lambda),\lambda)$$

and $y(x,\lambda)$ is the C^k solution of

$$\frac{dy}{dt} = (1-P_0)F(x+y,\lambda).$$

We now write

$$G(x,\lambda) = Lx + R(x,\lambda)$$

where L is the linear part of G. The remainder term R has the form

$$R(x,\lambda) = \sum_{\substack{p,q \\ p+q\geq 2}} \lambda^p R_{pq}(x)$$

where $R_{p,q}(x)$ is a polynomial homogeneous of degree q in x.

Theorem 2.1: If $R_{02} \equiv 0$, then the normal form of (2.2) and the Taylor expansion of the Liapunov-Schmidt bifurcation equation are identical at order $0(|\lambda||x| + |x|^3)$.

Remark 2.2: The simplest way to check that $R_{02} = 0$ is as follows. Let $\rho:E \to V_0$ be a projection and let $G:V_0\times\mathbb{R} \to V_0$ be ρ composed with the restiction of F to $V_0\times\mathbb{R}$. If the quadratic terms of G are all zero, then R_{02} also vanishes. That is, if the restiction of (2.1) to its center subspace has no quadratic terms, then the Liapunov-Schmidt reduction gives the same equations as reduction to the center manifold coupled with changes of coordinates to normal form.

We split the proof of Theorem 2.1 into two lemmas. In the following we consider the action of the circle group S^1 defined on V_0 by

$$(\phi,x) \longrightarrow e^{\phi L/\omega_0}\, x \quad \text{for all} \quad \phi\varepsilon S^1,\ x\varepsilon V_0.$$

For every vector field X in V, we set

$$\rho(X)[x] = \omega_0/2\pi \int_0^{2\pi/w_0} e^{-\phi L/\omega_0}\, X(e^{\phi L/\omega_0}\, x)d\phi.$$

The operator ρ is a projection onto the space of S^1-equivariant vector fields.

Lemma 2.3: The normal form equation for (2.2) with $R_{02} = 0$ takes the form

$$\frac{dx}{dt} = Lx + \lambda\rho(R_{11})(x) + \rho(R_{03})(x) + 0(|\lambda|^2|x| + |\lambda||x|^3 + |x|^4).$$

Proof: According to the Normal Form process [8], we look for a near identity change of variables with Taylor expansion

$$x \longrightarrow \tilde{x} = x + \lambda h_1(x) + h_3(x,x,x) + \ldots .$$

where h_p is symmetric and p-linear. Then, the new vector field has the following Taylor expansion up to order $|\lambda||x|$ and $|x|^3$:

$$\tilde{R}_{11}(x) = R_{11}(\tilde{x}) - ad_L(h_1)(\tilde{x})$$

$$\tilde{R}_{03}(x) = R_{03}(\tilde{x}) - ad_L(h_3)(\tilde{x})$$

where

$$ad_L(X)(\tilde{x}) = DX(\tilde{x})L\tilde{x} - LX(\tilde{x}).$$

Denoting by $\mathbb{H}_k$ ($\mathbb{G}_k$) the space of vector fields in V_0 whose components are homogeneous polynomials of degree k (S^1-equivariant such vector fields), we know [5] that $\mathbb{H}_k = \mathrm{Im}\, ad_L \oplus \mathbb{G}_k$, the corresponding projection being precisely ρ. Hence $\tilde{R}_{11}=\rho(R_{11})$ and $\tilde{R}_{03} = \rho(R_{03})$. □

We now recall the Liapunov-Schmidt decomposition for equation (2.2). By the change of scale $s = \omega_0 t$, equation (2.2) is rewritten as

$$Jx = R(x,\lambda) - (\omega-\omega_0)\frac{dx}{ds} \qquad (2.3)$$

where ω is the unknown frequency and $J \equiv \omega_0\, d/ds - L$. We look for a function $x(s)$ in $C^1(S^1,V_0)$ where $S^1 = \mathbb{R}/2\pi\mathbb{Z}$. Let $\{\zeta_j\}$ be an orthonormal basis of the eigenspace associated to the eigenvalue $\omega_0 i$. Then, in $C^0(S^1,V_0)$

$$\ker J = \{\sum_j (x_j e^{is}\zeta_j + \bar{x}_j e^{-is}\bar{\zeta}_j): x_j \in \mathbb{C}\}.$$

For

$$x(s) = \sum_j (x_j(s)\zeta_j + \bar{x}_j(s)\bar{\zeta}_j)$$

we define

$$(Px)(s) = \sum_j (\alpha_j e^{is}\zeta_j + \bar{\alpha}_j e^{-is}\bar{\zeta}_j) \qquad (2.4)$$

where

$$\alpha_j = \frac{1}{2\pi}\int_0^{2\pi} x_j(s)e^{-is}ds.$$

This projection on $\ker J$ satisfies the properties required for decomposing (2.3) according to the Liapunov-Schmidt method: set

$$(Px)(s) = \hat{x}(s) \quad \text{and} \quad y = (I-P)x;$$

Thus, $x = \hat{x} + y$ and we obtain

$$\begin{aligned} &(a) \quad y = (I-P)\tilde{J}^{-1}\left[R(\hat{x}+y,\lambda) - (\omega-\omega_0)\frac{dy}{ds}\right] \\ &(b) \quad 0 = PR(\hat{x}+y,\lambda) - (\omega-\omega_0)\frac{d\hat{x}}{ds}. \end{aligned} \qquad (2.5)$$

where $\tilde{J}^{-1}$ is the inverse of J defined on $(I-P)C^0(S^1,V_0)$. Equation (2.5a) is solved for $y = y(\hat{x},\omega,\lambda)$ and then equation (2.5b) gives the bifurcation equation on ker J.

Lemma 2.4: Let $f: V_0 \rightarrow V_0$ and $\hat{x} \in \ker J$. Then

$$P[f(\hat{x}(s)] = \rho(f)(\hat{x}(s)).$$

Proof: Note that we can write

$$\hat{x}(s) = e^{sL/\omega_0}\,\hat{x}_0$$

for some $\hat{x}_0 \in V_0$. We now set $f = \sum_j f_j \zeta_j$. Then

$$P[f(\hat{x}(s)] = \sum_j \frac{1}{2\pi}\left(\int_0^{2\pi} f_j(e^{tL/\omega_0}\,\hat{x}_0)e^{-i(t-s)}dt\right)\zeta_j$$

$$= \frac{1}{2\pi}\int_0^{2\pi}\left(\sum_j f_j(e^{(\tilde{t}+s)L/\omega_0}\,\hat{x}_0)e^{-i\tilde{t}}\,\zeta_j\right)d\tilde{t}$$

where $\tilde{t} = t-s$

$$= \frac{1}{2\pi}\int_0^{2\pi} e^{-\tilde{t}L/\omega_0}\,f(e^{\tilde{t}L/\omega_0}\,\hat{x}(s))d\tilde{t} = \rho(f)[\hat{x}(s)].$$ □

We can now write the expansion of the bifurcation equation to third order in $\hat{x}$ and λ. Since $R_{02} \equiv 0$ and $\frac{d\hat{x}}{ds} = \omega_0^{-1}L\,\hat{x}$, it is easy to check that

$$\omega\,\frac{d\hat{x}}{ds} = L\hat{x} + \lambda\rho(R_{11})(\hat{x}) + \rho(R_{03})(\hat{x}) + \ldots .$$

We therefore obtain the same equation as in Lemma 2.1, which proves Theorem 2.1 upon substituting x for $\hat{x}$. □

Remarks 2.5: (1) We would obtain the same result by applying the Liapunov-Schmidt reduction directly to (2.1) instead of to (2.2). In this case, the projection P on ker J has to be replaced by the projection $P'=PP_0$ of $C^0(S^1,E)$ onto ker J. Then everything else works as above.

(2) We obtain the same result if L has an additional zero eigenvalue or two pairs of imaginary eigenvalues. Such a situation occurs in multiparameter bifurcation of the Couette-Taylor system considered in Example 3.4 below. In the first case, for example, V_0 is the eigenspace associated to the eigenvalues 0 and $\pm\,\omega i$,

$$x(s) = \sum_j (x_j(s)\zeta_j + \overline{x}_j(s)\overline{\zeta}_j) + \sum_k y_k\xi_k$$

where $\{\xi_k\}$ is an orthonormal basis of ker L, and

$$(Px)(s) = \sum_j (\alpha_j e^{is}\zeta_j + \bar{\alpha}_j e^{-is}\bar{\zeta}_j) + \sum_k \beta_k \xi_k$$

with

$$\beta_k = \frac{1}{2\pi}\int_0^{2\pi} y_k(s)ds.$$

The proof of Theorem 2.1 given above then applies to this case.

III. APPLICATIONS OF THEOREM 1

Assume there exits a group Γ acting on E such that (2.1) is invariant under this action, that is, we assume

$$F(\gamma u,\lambda) = \gamma F(u,\lambda), \quad \forall\ \gamma \in \Gamma \text{ and } (u,\lambda) \in D\times\mathbb{R}^p$$

(F is Γ-equivariant). It is well known that the invariance properties carry over to the bifurcation equation [14], which we now write as:

$$g(x,\lambda) = 0, \quad x \in V_0, \ \lambda \in \mathbb{R} .$$

We assume $V_0 = V\oplus V$, where V is Γ-invariant and Γ acts diagonally on $V\oplus V$ (see [14] for a discussion of this point).

<u>Lemma 3.1</u>: Let $P_2(\Gamma,V)$ ($P_2(\Gamma,V_0)$) be the space of Γ-equivariant quadratic maps of $V \to V$ ($V_0 \to V_0$). Then:

$$\dim P_2(\Gamma,V_0) = 6 \dim P_2(\Gamma,V).$$

<u>Proof</u>: Let $f:V\oplus V \to V\oplus V$ be Γ-equivariant. We write $f = (f_1,f_2)$ where $f_j:V\oplus V \to V$ is Γ-equivariant. We need only show that

$$\dim P_2(\Gamma,V\oplus V,V) = 3 \dim P_2(\Gamma,V)$$

where $P_2(\Gamma,V\oplus V,V)$ is the space of quadratic Γ-equivariant maps $V\oplus V \to V$. Now, define $\phi:\{P_2(\Gamma,V)\}^3 \to P_2(\Gamma,V\oplus V,V)$ by

$$\phi(a,b,c)(v+w) = \tilde{a}(v,v) + \tilde{b}(v,w) + \tilde{c}(w,w), \ \forall\ (v,w) \in V\oplus V,$$

where $\tilde{a}$, $\tilde{b}$, $\tilde{c}$ are the symmetric bilinear forms associated with the quadratic mappings a,b,c respectively. We claim that ϕ is a linear isomorphism, as we now prove.

(i) ϕ is 1:1. Evaluating separately at $w=0$ and at $v=0$ shows that

$$\phi(a,b,c) = 0 \Rightarrow a=b=c=0.$$

(ii) ϕ is onto. Let A be in $P_2(\Gamma,V\oplus V,V)$, then

$$A(v+w) = \tilde{A}(v,v) + 2\tilde{A}(v,w) + \tilde{A}(w,w)$$

where $\tilde{A}$ is the symmetric bilinear form associated with A. Hence, $A =$

$\phi(A,2A,A)$. Clearly ϕ itself is Γ-equivariant; hence we have proved the claim.

Corollary 3.2: $P_2(\Gamma, V\oplus V)$ is nonzero $\Leftrightarrow$ $P_2(\Gamma,V)$ is nonzero.

This corollary gives a simple way to characterize the vanishing of the quadratic part due to symmetry. We now turn to some examples.

Example 3.3: Hopf bifurcation with $O(2)$ or D_n symmetry $(n \geq 4)$. Assume that equation (2.1) is invariant under a representation of the group $\Gamma = O(2)$ or D_n such that its restriction to V is irreducible and 2-dimensional. Then there exists a basis of V such that the $O(2)$ action in V is generated by the matrices

$$R_\phi = \begin{bmatrix} \cos\phi & -\sin\phi \\ \sin\phi & \cos\phi \end{bmatrix} \quad \text{and} \quad S = \begin{bmatrix} 1 & 0 \\ 0 & 1 \end{bmatrix}$$

where either $\phi \in SO(2)$ or $\phi = 2\pi/n \in D_n$. This corresponds to the case of critical eigenvalues $\pm i\omega_0$ which are semisimple and double; the assumption of irreducibility of V is, in some sense, generic (see [13]). This problem was studied in [5] and [6], where the stability of bifurcated solutions was analysed by means of normal form. The stability conditions involve only terms up to cubic order in equation (2.2). There is no quadratic term in (2.2) if $\Gamma = O(2)$ or $\Gamma = D_n$ with $n \geq 4$ (this can be easily checked). Therefore, using Corollary 3.2 and Theorem 2.1, we conclude that the stability conditions in [5] and [6] are directly obtained from the Liapunov-Schmidt bifurcation equation. The case $\Gamma = D_3$ should provide an example where the Liapunov-Schmidt method does not give the correct stability assignments for bifurcating branches of periodic solutions.

Example 3.4: The Couette-Taylor problem. This is the problem of the onset of vortex flow in a fluid between two coaxial cylinders which rotate at independent constant speeds. When the cylinders are counterrotating, the following critical situations can occur:

(1) one pair of eigenvalues $\pm i\omega_0$,

(2) one 0 eigenvalue and one pair $\pm i\omega_0$ eigenvalues,

(3) two pairs of eigenvalues $\pm i\omega_1$, $\pm i\omega_2$, $\omega_1 \neq \omega_2$.

Case 1 was studied in [4], case 2 (in a theoretical framework) in [7], and case 3 in [2]. The symmetries are the following: $SO(2)$ (rotation around the axis of cylinders) and O(2) (translations and reflections along the axis of cylinders). The O(2) symmetry comes from the assumption of infinite cylinders and periodicity of the flow in the axial direction (translations are then equivalent to rotations). The eigenvalues are double as a consequence of the reflectional symmetry and hence the dimension of equation (2.2) can be 4,6 or 8. In cases 1 and 3, the analysis in [4] and [2] was done directly on equation (2.2). We remark that equation (2.2) is already in normal form in this problem (for analysis of periodic bifurcations), because the rotational symmetry acts identically with the S^1 action required for the normal form. Here again, one can easily check that, because of symmetry, quadratic terms vanish in (2.2), hence stability conditions are obtained directly from the bifurcation equation (see Remark 2.5 (2) in case 2; an equivalent remark could be given for case 3).

<u>Example 3.5</u>: Onset of convection for a fluid in a rotating self-gravitating spherical shell [1]. When the shell does not rotate, bifurcation occurs at a zero eigenvalue with multiplicity $2\ell+1$, where ℓ is the "degree" of the associated irreducible representation of the group of rotations SO(3) (under which the problem is invariant). If the system is now slowly rotating, the 0 eigenvalue splits in one real eigenvalue and ℓ pairs of complex eigenvalues, leading to a complex diagram of Hopf bifurcations and interactions. In this problem, quadratic terms can be zero due to the SO(3) symmetry if ℓ is odd (see [14]) or, if ℓ is even. because of some physical condition in the problem (see [1]). In such a case, the stability analysis reduces to the bifurcation equations in the $(2\ell+1)$-dimensional invariant subspace of the nonrotating limit case. Numerical applications are given in [3].

<u>ACKNOWLEDGEMENTS</u>

We thank Xianwu Zeng for prodding us to consider precisely how the Liapunov-Schmidt and the center manifold reductions compare when symmetries are present. We are grateful to the referee for suggesting a number of helpful changes in the exposition. The research of M.G. was supported in part by the ACMS Program of DARPA, NASA Grant 2-279 and by NSF Grant DMS-8402604. The research of P.C. was supported in part by the ACMS Program of DARPA.

BIBLIOGRAPHY

[1] P. Chossat, Le probleme de Benard dans une couche spherique. These d'Etat, Universite de Nice, 1982.

[2] P. Chossat, Y. Demay and G. Iooss, Interaction de modes azimutaux dans le probleme de Couette-Taylor. Arch. Rational Mech. Anal. To appear.

[3] P. Chossat and J.P. Giraud, Calcul des coefficients de bifurcation dans une probleme de convection en symetrie spherique. J. de Mecanique Theorique et Appliquee, 2, 5 (1983) 799-828.

[4] P. Chossat and G. Iooss, Primary and secondary bifurcations in the Couette-Taylor problem. Japan J. Appl. Math. 2 (1985) 249-288.

[5] M. Golubitsky and I. Stewart, Hopf bifurcation in the presence of symmetry. Arch. Rational Mech. & Anal. 87, 2 (1985) 107-165.

[6] M. Golubitsky and I. Stewart, Hopf bifurcation with dihedral symmetry and coupled nonlinear oscillators. In: *Multiparameter Bifurcation Theory* (M. Golubitsky and J. Guckenheimer, eds), Contemporary Mathematics 56, American Math. Soc., Providence, 1986, 131-174.

[7] M. Golubitsky and I. Stewart, Symmetry and stability in Taylor-Couette flow. SIAM J. Math. Anal. 17, 2 (1986) 249-288.

[8] J. Guckenheimer and P. Holmes, *Nonlinear Oscillations, Dynamical Systems and Bifurcation of Vector Fields*. Appl. Math. Sci. Ser. 42, Springer-Verlag, 1983.

[9] G. Iooss and D.D. Joseph, "*Elementary Bifurcation and Stability Theory*. Undergraduate Texts Math., Springer-Verlag, 1980.

[10] H. Kielhofer and R. Lauterbach, On the principle of reduced stability. Preprint, 1983.

[11] J. Marsden and M. McCracken, *Hopf Bifurcation and Its Applications*. Appl. Math Sci. Ser. 19, Springer-Verlag, 1976.

[12] G.H. Pimbley, Jr., Eigenfunction Branches of Nonlinear Operators and Their Bifurcation. Lec. Notes in Math 104, Springer-Verlag, 1969.

[13] D. Ruelle, Bifurcation in the presence of a symmetry group. Arch. Rational Mech. Anal. 51 (1973) 136-152.

[14] D.H. Sattinger, *Group Theoretic Methods in Bifurcation Theory*. Lect. Notes in Math. 762, Springer-Verlag, 1979

[15] M. Vainberg and V. A. Trenogin, *Theory of Branching of Solutions of Nonlinear Equations*. Noordhoff, 1974.

Pascal Chossat
Laboratoire de Mathematiques
Universite de Nice
Parc Valrose
F-06034 Nice Cedex
France

Martin Golubitsky
Department of Mathematics
University of Houston
University Park
Houston, TX 77004
USA

Canadian Mathematical Society
Conference Proceedings
Volume 8 (1987)

Invariant Theory and Normal Form of Hamiltonian Vectorfields with Nilpotent Linear Part

Richard Cushman
Jan A. Sanders *

ABSTRACT

A method is presented to compute the normal form for the C^{∞}-jet of a Hamiltonian vectorfield at an equilibrium with nilpotent linear part, using representation theory of $sl(2,\mathbb{R})$. We show that invariant theory can be applied to handle the actual computations.

1. Introduction

In this paper we want to show that normal form theory for Hamiltonian vectorfields with nilpotent linear part is equivalent to classical invariant theory. We use this fact to compute the general normal form for several lower dimensional examples. For higher dimensions, say for six degrees of freedom, the covariant algebra is too complicated to be of any use. Then we apply the obtained results to the case where the linear part has a semisimple-nilpotent decomposition with nonzero semisimple part.

The analysis is somewhat complicated by the fact that the symplectic structure and the invariant theoretic setting require different linear normal forms.

The main goal of this paper is not so much the computation of the normal forms, interesting in its own right as it may be, but to give an illustration of a technique which can also be used in the general vectorfield case, which seems to be well adapted to handle problems with symmetry. This method also gives us the possibility to compute general normal forms automatically, although this will need some heavy machinery (also in mathematical terms) to be carried out in practice.

It should be mentioned that our method, abstract as it is, already gives

1980 **Mathematics Subject Classification** (1985 **Revision**).58F21,14D25.
* Partially supported by the Netherlands Organization for the Advancement of Pure Research (ZWO), while visiting Michigan State University, winter 1986.

0731-1036/87 $1.00 + $.25 per page

the most efficient algorithm known to us to bring a given vectorfield into normal form.

2. Normal Form

Let (V,ω) be a real symplectic vectorspace with symplectic form ω. Let $(\mathbb{R}[[V]],\cdot)$ be the algebra of formal power series on V. Recall that $f\in\mathbb{R}[[V]]$ may be written as $f=f_0+f_1+\cdots+f_n+\cdots$ where $f_i\in P_i(V,\mathbb{R})$, that is f_i is a homogeneous polynomial of degree i. Let $\{f,g\}(v)=\omega(dg(v),df(v))$. Then $(\mathbb{R}[[V]],\cdot,\{,\})$ is a **Poisson algebra,** that is, $(\mathbb{R}[[V]],\{,\})$ is a Lie algebra and

$$ad_f(g\cdot h)=(ad_f g)\cdot h+g\cdot(ad_f h) \quad \textit{for all } f,g,h\in\mathbb{R}[[V]]$$

where $ad_f g=\{f,g\}$. Since $ad_f=-L_{X_f}$ which is a derivation on $\mathbb{R}[[V]]$, ad_f corresponds to a vectorfield $-X_f$ on (V,ω) X_f is called the **Hamiltonian vectorfield** of f. The flow of $-X_f$ is the one parameter group of invertible formal power series on V given by

$$\phi_f^t(v)=\exp t\, ad_f(v)=\sum_{n\geqq 0}\frac{1}{n!}t^n ad_f^n(v)$$

where

$$ad_f(v)=ad_f(v_1,\ldots,v_{2n})=(\{f,v_1\},\ldots,\{f,v_{2n}\})$$

and $(v_1,\ldots,v_{2n})$ are the components of v with respect to a symplectic basis of (V,ω). Here each v_i is thought of as a linear function on V. Since

$$\{g\circ\phi_f^t,h\circ\phi_f^t\}=\{g,h\}\circ\phi_f^t,$$

ϕ_f^t is symplectic, that is, $(\phi_f^t)^*\omega=\omega$.

Given the formal power series Hamiltonian

$$H=H_2+H_3+\cdots+H_n+\cdots$$

on (V,ω), the basic problem of normal form theory is to find a near identity invertible symplectic formal power series ϕ_F which *simplifies* H. For fixed $m\geqq 3$ consider $\phi_{F_m}=\exp ad_{F_m}$, where $F_m\in P_m(V,\mathbb{R})$. Then because

$$(H\circ\phi_{F_m}^t)(v)=(\exp t\; ad_{F_m}(H))(v)$$

we see that the mth order term in $H\circ\phi_{F_m}$ is $H_m+ad_{F_m}H_2=H_m-ad_{H_2}F_m$. Because ad_{H_2} maps $P_m(V,\mathbb{R})$ into itself, if we choose a complement C_m to the image of ad_{H_2} in $P_m(V,\mathbb{R})$, then we may write

$$H_m=\mathcal{H}_m\oplus H_m'$$

where $\mathcal{H}_m\in C_m$ and $H_m'\in im\; ad_{H_2}$. Choosing F_m so that $ad_{H_2}F_m=H_m'$, we see that the mth order term of $H\circ\phi_{F_m}$ is $\mathcal{H}_m$ which lies in C_m. Thus we say that H is in **normal form with respect to** H_2 if for all $m\geqq 3$ H_m lies in C_m.

Suppose that the term H_2 in H is the nilpotent N_2, that is, ad_{N_2} is a nilpotent linear map of $P_2(V,\mathbb{R})$ into itself. Another way of saying this is that the linear Hamiltonian vectorfield X_{N_2} on (V,ω) is a nilpotent real infinitesimally symplectic linear mapping, that is,

$$\omega(X_{N_2}(v),w)+\omega(v,X_{N_2}(w))=0 \quad \textit{for all } v,w\in V$$

The reason for this equivalence is that the Lie algebra $(P_2(V,\mathbb{R}),\{,\})$ is isomorphic to the Lie algebra $(sp(V,\omega),[\ ])$ of infinitesimally symplectic linear maps on (V,ω) (see (Cu1974a)). Because $sp(V,\omega)$ is semisimple, the theorem of Jacobson-Morozov (He1982a) states that there are $T_2,M_2\in(P_2(V,\mathbb{R}),\{,\})$

such that $\{T_2,M_2,N_2\}$ form a Lie algebra of $(P_2(V,\mathbb{R}),\{,\})$ which is isomorphic to $sl_2(\mathbb{R})$, that is,

$$\{T_2,N_2\}=-2N_2,$$

$$\{T_2,M_2\}=2M_2,$$

and

$$\{M_2,N_2\}=T_2.$$

Since $ad:(sl_2(\mathbb{R}),\{,\})\rightarrow(\mathbb{R}[[V]],\{,\}):F\rightarrow ad_F$ is a homomorphism of Lie algebras which preserves degree, $\{ad_{T_2}|P_n, ad_{M_2}|P_n, ad_{N_2}|P_n\}$, where $P_n=P_n(V,\mathbb{R})$, is a representation of $sl_2(\mathbb{R})$ on P_n. Since every finite dimensional representation of $sl_2(\mathbb{R})$ is completely reducible (Hu1972a) and for every irreducible representation ϱ_n of $sl_2(\mathbb{R})$ on V_n we have

$$V_n=ker\ \varrho_n(X)\oplus im\ \varrho_n(Y),$$

where $Y=\begin{pmatrix}0&0\\1&0\end{pmatrix}$ and $X=\begin{pmatrix}0&1\\0&0\end{pmatrix}$, we conclude that

$$P_n=ker\ ad_{M_2}|P_n\oplus im\ ad_{N_2}|P_n.$$

In other words, we may choose $ker\ ad_{M_2}|P_n$ as a complement C_n to the image of ad_{N_2} in P_n. Thus $H=N_2+H_3+\cdots+H_n+\cdots$ is in **normal form with respect to** N_2 if for every $n\geqq 3$ $H_n\in ker\ ad_{M_2}|P_n$, where $\{T_2,N_2,M_2\}$ form a Lie subalgebra of $(P_2(V,\mathbb{R}),\{,\})$ which is isomorphic to $sl_2(\mathbb{R})$(Cu1982a). Note that $(ker\ ad_{M_2},\cdot,\{,\})$ is a Poisson subalgebra of $(\mathbb{R}[[V]],\cdot,\{,\})$ called the **normal form algebra** of H. One of the main problems of normal form theory is to describe $ker\ ad_{M_2}$ in terms of generators and relations.

3. Infinitesimally Symplectic Representations of $sl_2(\mathbb{R})$

Let $\varrho : sl_2(\mathbb{R}) \to sp(V,\omega)$ be a Lie algebra homomorphism. To fix notation let

$$T = \begin{pmatrix} 1 & 0 \\ 0 & -1 \end{pmatrix}, \quad X = \begin{pmatrix} 0 & 1 \\ 0 & 0 \end{pmatrix}, \quad Y = \begin{pmatrix} 0 & 0 \\ 1 & 0 \end{pmatrix}$$

be the standard basis of $sl_2(\mathbb{R})$, that is, $[T,X]=2X, [T,Y]=-2Y$ and $[X,Y]=T$. Moreover let $\varrho(T)=H, \varrho(X)=M$ and $\varrho(Y)=N$. The first step in understanding ϱ is to decompose V into a sum of $sl_2(\mathbb{R})$-indecomposable pieces. Let V_i be symplectic subspaces of (V,ω) such that $V = \sum_i \oplus V_i$ and $\omega(V_i,V_j)=0$ for $i \neq j$. If V_i are invariant under $\varrho(sl_2(\mathbb{R}))$ and in addition have no proper $\varrho(sl_2(\mathbb{R}))$-invariant symplectic subspaces, then V is decomposed into a sum of $sl_2(\mathbb{R})$-**indecomposable** subspaces V_i.

The problem is to decompose (V,ω) into a sum of $sl_2(\mathbb{R})$-indecomposable subspaces. To solve this problem we need the notion of an N-indecomposable subspace. An N-invariant symplectic subspace W of (V,ω) is N**-indecomposable** if it contains no proper N-invariant symplectic subspace. Clearly every $sl_2(\mathbb{R})$-indecomposable subspace is N-indecomposable. The following argument shows how to construct $sl_2(\mathbb{R})$-indecomposable subspaces from N-indecomposable ones.

The theory of normal forms for real infinitesimally symplectic linear mappings (Bu1977a) not only gives a decomposition of (V,ω) into a symplectically orthogonal direct sum of N-indecomposable subspaces, but also gives the following classification of N-indecomposable subspaces W of (V,ω). Let m be the unique nonnegative integer such that $N^m \mid W \neq 0$ while $N^{m+1} \mid W = 0$. The classification splits into two cases:

1. $(m=2r)$: Then there are nonzero vectors $e,f \in W$ such that $\{e_1, \ldots, e_{r+1}, f_1, \ldots, f_{r+1}\}$ with $e_{i+1}=N^r e$, $f_{i+1}=(-1)^i N^{2r-i} f$ is a basis for W_i; moreover $\omega(e_{i+1},e_{j+1})=\omega(f_{i+1},f_{j+1})=0$ and $\omega(e_{i+1},f_{j+1})=\delta_{ij}$ for $i,j=0,1,\ldots,r$.
2. $(m=2r+1)$: Then there is a nonzero vector $e \in W$ and ϵ, with $\epsilon^2=1$ such that $\{e_1,\ldots,e_{r+1}, f_1, \ldots, f_{r+1}\}$ with $e_{i+1}=N^i e$ and $f_{i+1}=(-1)^i \epsilon N^{2r+1-i} e$, $i=0,\ldots,r$ is a basis of W; moreover $\omega(e_{i+1},e_{j+1})=\omega(f_{i+1},f_{j+1})=0$ and $\omega(e_{i+1},f_{j+1})=\delta_{ij}$ for $i,j=0,1,\ldots,r$. Note that $\epsilon=\omega(e,N^r e)$.

In terms of the above basis for W the matrix of $\omega \mid W$ is in $(r+1)\times(r+1)$ block form)

$$\begin{pmatrix} 0 & -I_{r+1} \\ I_{r+1} & 0 \end{pmatrix}$$

while the matrix of N is

$$\begin{pmatrix} P & 0 \\ Q & R \end{pmatrix}$$

where

$$P=\begin{pmatrix} 0 & & & & \\ 1 & 0 & & & \\ & 1 & \cdot & & \\ & & \cdot & 0 & \\ & & & 1 & 0 \end{pmatrix}, \quad Q=\begin{pmatrix} 0 & & & & \\ & 0 & & & \\ & & 0 & & \\ & & & 0 & \\ & & & & \epsilon \end{pmatrix}, \quad R=\begin{pmatrix} 0 & -1 & & & \\ & 0 & -1 & & \\ & & 0 & \cdot & \\ & & & \cdot & -1 \\ & & & & 0 \end{pmatrix}$$

for $m=2r+\epsilon^2, \epsilon=-1,0,1$. The above change of basis for W is natural when using Hamiltonian mechanics because the matrix of the symplectic form ω is in standard shape. However this basis is quite unnatural when doing invariant theory as we shall be doing in §5. Therefore when doing invariant theory we make the following choices:

1. $(m=2r)$: Let $e_{i+1}=\frac{1}{i!}N^i e$ and $f_{i+1}=\frac{1}{i!}N^i f$ for $i=0,\ldots,r$.

2. $(m=2r+1)$: Let $e_{i+1}=\frac{1}{i!}N^i e$ for $i=0,\ldots,2r+1$.

In terms of the invariant theoretic basis (the Gordan normal form!) the matrix of N is

$$N'=\begin{pmatrix} P' & 0 \\ 0 & P' \end{pmatrix} \quad \textit{where } P'=\begin{pmatrix} 0 & & & & \\ 1 & 0 & & & \\ & 2 & \cdot & & \\ & & \cdot & 0 & \\ & & & r & 0 \end{pmatrix}$$

when $m=2r$, and

$$N'=\begin{pmatrix} 0 & & & & \\ 1 & 0 & & & \\ & 2 & \cdot & & \\ & & \cdot & 0 & \\ & & & 2r+1 & 0 \end{pmatrix}$$

when $m=2r+1$.

The matrix for the symplectic form ω in the invariant theoretic basis will not be written down, though it is easy to find.

We now embed N' into a subalgebra of $sp(W,\omega)$, which is isomorphic to $sl_2(\mathbb{R})$, as follows. Using the invariant theoretic basis we define M' and H' as

$$M'=\begin{pmatrix} S & 0 \\ 0 & S \end{pmatrix} \quad \textit{where } S=\begin{pmatrix} 0 & r & & & \\ & 0 & r-1 & & \\ & & \cdot & \cdot & \\ & & & 0 & 1 \\ & & & & 0 \end{pmatrix}$$

and

$$H' = \begin{pmatrix} H'' & 0 \\ 0 & H'' \end{pmatrix} \text{ where } H'' = \begin{pmatrix} r+1 & & & & \\ & r-1 & & & \\ & & \cdot & & \\ & & & -(r-1) & \\ & & & & -(r+1) \end{pmatrix}$$

when $m = 2r$; while

$$M' = \begin{pmatrix} 0 & 2r+1 & & & \\ & 0 & 2r & & \\ & & \cdot & \cdot & \\ & & & 0 & 1 \\ & & & & 0 \end{pmatrix}$$

and

$$H' = \begin{pmatrix} 2r+2 & & & & \\ & 2r & & & \\ & & \cdot & & \\ & & & -(2r) & \\ & & & & -(2r+2) \end{pmatrix}.$$

when $m = 2r+1$.

Let T be the linear symplectic map on (W,ω) such that $N = TN'T^{-1}$. Put $\tilde{M} = TM'T^{-1}$ and $\tilde{H} = TH'T^{-1}$. Then $\{N, \tilde{M}, \tilde{H}\}$ is a subalgebra of $sp(V,\omega)$ which is isomorphic to $sl_2(\mathbb{R})$. Repeating this construction on each N-indecomposable subspace (V,ω) gives a representation $\tilde{\varrho}$ of $sl_2(\mathbb{R})$ with $\tilde{\varrho}(T) = \tilde{H}$, $\tilde{\varrho}(X) = \tilde{M}$, $\tilde{\varrho}(Y) = N$ which has the same N term as the representation ϱ. By a standard result from Lie algebras (see (Fr1969a) there is a linear symplectic map Q of (V,ω) such that

$$QNQ^{-1} = N \text{ , } Q\tilde{M}Q^{-1} \text{ and } Q\tilde{H}Q^{-1} = H.$$

In other words, $\tilde{\varrho}$ is symplectically equivalent to ϱ. Thus we have decomposed ϱ into $sl_2(\mathbb{R})$-indecomposable pieces. Moreover we have shown that there is a one to one correspondence between N-indecomposable subspaces of $(V\omega)$ and $sl_2(\mathbb{R})$-indecomposable subspaces.

Note that an $sl_2(\mathbb{R})$ indecomposable subspace is *not necessarily* a representation space for an irreducible $sl_2(\mathbb{R})$ representation.

4. Covariants and the Cayley correspondence

Here we consider the problem of describing $ker\, L_{\varrho(X)}$, where ϱ is a representation of $sl_2(\mathbb{R})$ on the vectorspace V. Throughout this section we shall suppose that ϱ is the infinitesimalization of a representation R of $SL_2(\mathbb{R})$ on V, that is, $\varrho = dR(e)$, where e is the identity in $SL_2(\mathbb{R})$.

Let $V = V_d = \mathbb{R}^{d+1}$. Consider the space $\mathcal{P}^d_{n,m} = P_m((\mathbb{R}^2)^*, P_n(V_d, \mathbb{R})) = P_n(V_d, \mathbb{R}) \otimes P_m((\mathbb{R}^2)^*, \mathbb{R})$ of binary forms of degree m with coefficients which are homogeneous real valued polynomials of degree n on V_d. In symbols, $Q \in \mathcal{P}^d_{m,n}$ iff

$$Q(v;X,Y) = \sum_{i=0}^{m} \binom{m}{i} p_{i+1}(v) X^{m-i} Y^i$$

where $v \in V_d$ and X,Y are coordinates in $(\mathbb{R}^2)^*$ with respect to the standard dual basis. Let $q_n \otimes s_m \in \mathcal{P}^d_{n,m}$. Then $G = SL_2(\mathbb{R})$ acts on $\mathcal{P}^d_{n,m}$ as follows: for every $g \in G$

$$[g \cdot (q_n \otimes s_m)](v;X,Y) = q_n(R_m(g^{-1})v) \otimes s_m(g^t(X,Y)). \tag{4.1}$$

It is straightforward to check that (4.1) defines a left action. The set $((\mathcal{P}^d_{n,m})^G$ of G-invariant elements of $\mathcal{P}^d_{n,m}$ is the space $\mathcal{R}^d_{n,m}$ of **covariants on V_d of degree m and order n.** Since G is connected, $g \cdot Q = Q$ for every $g \in G$ iff

$$0 = (ad_Z Q)(v;X,Y) = -(ad_{\varrho(Z)} q_n) \otimes s_m + q_n \otimes ad_{Z^t} s_m \tag{4.2}$$

for every $Z \in \ell = sl_2(\mathbb{R})$. Here ad_Z is minus the Lie derivative with respect to the vectorfield Z.

We are now in a position to motivate the definition of the Cayley correspondence. Let $Q(v;X,Y) = \sum_{i=0}^{m} \binom{m}{i} p_{i+1}(v) X^{m-i} Y^i$ be an element of $\mathcal{R}^d_{n,m}$. Then for $T = \begin{pmatrix} 1 & 0 \\ 0 & -1 \end{pmatrix} \in \ell$,

$$0 = -ad_T Q =$$

$$= \sum_{i=0}^{m} \binom{m}{i} [ad_{\varrho(T)} p_{i+1} X^{m-i} Y^i + p_{i+1}(X\frac{d}{dX} - Y\frac{d}{dY})] X^{m-i} Y^i$$

$$= \sum_{i=0}^{m} \binom{m}{i} (ad_{\varrho(T)} - (m-2i)) p_{i+1} X^{m-i} Y^i .$$

Therefore $ad_{\varrho(T)} p_{i+1} = (m-2i) p_{i+1}$, that is, the highest term of Q is an eigenvector of $ad_{\varrho(T)}$ corresponding to the eigenvalue m.

Similarly for $X = \begin{pmatrix} 0 & 1 \\ 0 & 0 \end{pmatrix} \in \ell$,

$$0=-ad_X Q=\sum_{i=0}^{m}\binom{m}{i}[ad_{\varrho(X)}p_{i+1}X^{m-i}Y^i+p_{i+1}X\frac{d}{dY}X^{m-i}Y^i]$$

$$=\sum_{i=0}^{m}\binom{m}{i}ad_{\varrho(X)}p_{i+1}X^{m-i}Y^i+\sum_{i=1}^{m}\binom{m}{i}ip_{i+1}X^{m-i+1}Y^{i-1}$$

$$=\sum_{i=0}^{m}\binom{m}{i}ad_{\varrho(X)}p_{i+1}X^{m-i}Y^i+\sum_{i=0}^{m-1}\binom{m}{i+1}(i+1)p_iX^{m-i}Y^i$$

$$=\sum_{i=0}^{m}\binom{m}{i}ad_{\varrho(X)}p_{i+1}X^{m-i}Y^i+\sum_{i=0}^{m-1}\binom{m}{i}(m-i)p_iX^{m-i}Y^i$$

It follows that $ad_{\varrho(X)}p_{i+i}=-(m-i)p_i$ and $p_{m+1}\in ker\ ad_{\varrho(X)}$. The same reasoning leads to

$$ad_{\varrho(Y)}p_{i+1}=-ip_i$$

From the above discussion we may define the **Cayley correspondence**

$$Cay:\mathcal{R}^d_{n,m}\rightarrow\mathcal{E}^d_n(m)\cap ker(ad_{\varrho(X)}):Q\mapsto p_{m+1}$$

as the map which associates to each covariant Q of degree m and order n on V_d its highest coefficient p_{m+1}. Note that p_{m+1} lies in the intersection of $ker(ad_{\varrho(X)})$ and the eigenspace $\mathcal{E}^d_n(m)$ of $ad_{\varrho(T)}$ on $P_n(V_d,\mathbb{R})$ corresponding to the eigenvalue m.

The following construction shows that the Cayley correspondence is invertible and hence is an isomorphism of vectorspaces. Suppose that $p\in P_n(V_d,\mathbb{R})$ with $ad_{\varrho(X)}p=0$ and $ad_{\varrho(T)}p=mp$. Let

$$Q(v;X,Y)=\sum_{i=0}^{m}\frac{(-1)^{m-i}}{(m-i)!}ad^{m-i}_{\varrho(Y)}p(v)X^{m-i}Y^i \tag{4.3}$$

Q is constructed by taking the eigenvalue m and applying the rule $ad_{\varrho(Y)}p_{i+1}=-ip_i$ repeatedly, starting with $p_{m+1}=p$. Thus Q sits by construction in $ker\ ad_Y$ and we see that its ad_T-eigenvalue is 0. Hence it also is in $ker\ ad_T$. It follows from the basic representation theory of $sl_2(\mathbb{R})$ that Q automatically sits in $ker\ ad_X$.

In fact, what we have been doing is constructing an irreducible $m+1$-dimensional $sl_2(\mathrm{R})$ representation. Thus we may identify irreducible representations and covariants. This leads to some covariants which come with an arbitrary representation. If the representation is irreducible,

$$Q(v;X,Y)=\sum_{i=0}^{d}\binom{d}{i}v_iX^{m-i}Y^i$$

is a covariant. Q is classically known as the **groundform,** since it can be used to construct a $d+1$-dimensional irreducible representation from a given 2-dimensional one. Simply induce the group action on the dual space, such that the form $x_1X + x_2Y$ remains invariant, and then induce it on the v_i's. If the representation is reducible, there are lower degree groundforms which are associated to each irreducible summand. A covariant is called a **joint covariant** of these groundforms. We shall see examples of this in §5.

5. Examples

In this section we compute the normal form algebras for the Hamiltonian vectorfields whose linear part is one of the following:

1.

$$X_{N_2} = \begin{pmatrix} 0 & 0 & 0 & 0 \\ 0 & 0 & 0 & 0 \\ 1 & 0 & 0 & 0 \\ 0 & 1 & 0 & 0 \end{pmatrix}$$

2.

$$X_{N_2} = \begin{pmatrix} 0 & 0 & 0 & 0 \\ 1 & 0 & 0 & 0 \\ 0 & 0 & 0 & -1 \\ 0 & \epsilon & 0 & 0 \end{pmatrix}$$

3.

$$X_{N_2} = \begin{pmatrix} 0 & 0 & 0 & 0 & 0 & 0 \\ 1 & 0 & 0 & 0 & 0 & 0 \\ 0 & 1 & 0 & 0 & 0 & 0 \\ 0 & 0 & 0 & 0 & -1 & 0 \\ 0 & 0 & 0 & 0 & 0 & -1 \\ 0 & 0 & 0 & 0 & 0 & 0 \end{pmatrix}$$

Here the symplectic form ω is in standard shape $\begin{pmatrix} 0 & -I \\ I & 0 \end{pmatrix}$.

6. Example 1

The embedding of X_{N_2} in $sl_2(\mathbb{R})$ is given by

$$X_{T_2} = \begin{pmatrix} -1 & 0 & 0 & 0 \\ 0 & -1 & 0 & 0 \\ 0 & 0 & 1 & 0 \\ 0 & 0 & 0 & 1 \end{pmatrix}, \quad X_{M_2} = \begin{pmatrix} 0 & 0 & 1 & 0 \\ 0 & 0 & 0 & 1 \\ 0 & 0 & 0 & 0 \\ 0 & 0 & 0 & 0 \end{pmatrix} \quad \textit{and } X_{N_2} = \begin{pmatrix} 0 & 0 & 0 & 0 \\ 0 & 0 & 0 & 0 \\ 1 & 0 & 0 & 0 \\ 0 & 1 & 0 & 0 \end{pmatrix}.$$

Applying the change of basis

$$P=\begin{pmatrix}1&0&0&0\\0&0&1&0\\0&1&0&0\\0&0&0&1\end{pmatrix}$$

brings $\{X_{H_2},X_{M_2},X_{N_2}\}$ into invariant theoretic form $\{X'_{T_2},X'_{M_2},X'_{N_2}\}$ where

$$X'_{T_2}=\begin{pmatrix}1&0&0&0\\0&1&0&0\\0&0&-1&0\\0&0&0&-1\end{pmatrix},\quad X'_{M_2}=\begin{pmatrix}0&1&0&0\\0&0&0&0\\0&0&0&1\\0&0&0&0\end{pmatrix},$$

and

$$X'_{N_2}=\begin{pmatrix}0&0&0&0\\1&0&0&0\\0&0&0&0\\0&0&1&0\end{pmatrix}.$$

In terms of the symplectic coordinates $(\xi_1,\xi_2,\eta_1,\eta_2)$ the invariant theoretic variables (x_1,x_2,x_3,x_4) are

$$\begin{pmatrix}x_1\\x_2\\x_3\\x_4\end{pmatrix}=P^{-1}\begin{pmatrix}\xi_1\\\xi_2\\\eta_1\\\eta_2\end{pmatrix}=\begin{pmatrix}\xi_1\\\eta_1\\\xi_2\\\eta_2\end{pmatrix}.$$

Since the $sl_2(\mathbb{R})$, $\{X'_{T_2},X'_{M_2},X'_{N_2}\}$, is the direct sum of two 2-dimensional irreducible representations, we need to determine the joint covariant algebra C of the two linear forms

$$\pi_1=x_1X+x_2Y$$

$$\pi_2=x_3X+x_4Y.$$

But this is classical (Sa1859a): C is freely generated by π_1,π_2 and $\pi_3=\tau_1(\pi_1,\pi_2)=x_1x_4-x_2x_3$, where

$$\tau_1(f,g)(X,Y)=\frac{\partial f}{\partial X}\frac{\partial g}{\partial Y}-\frac{\partial f}{\partial Y}\frac{\partial g}{\partial X}$$

is the first transvectant (Sp1977a). By the Cayley correspondence the leading coefficient of π_1,π_2,π_3 is x_2,x_4, and $x_1x_4-x_2x_3$ respectively. In terms of the symplectic variables (ξ,η), the leading coefficient is

$$\tau_1=\eta_1\ ,\ \tau_2=\eta_2\ ,\ \tau_3=\xi_1\eta_2-\xi_2\eta_1\,.$$

Thus the normal form algebra $ker\ ad_{M_2}$ is freely generated by τ_1,τ_2 and τ_3.

7. Example 2

The embedding of X_{N_2} in $sl_2(\mathbb{R})$ is given by

$$X_{T_2}=\begin{pmatrix}3&0&0&0\\0&1&0&0\\0&0&-1&0\\0&0&0&-3\end{pmatrix},\ X_{M_2}=\begin{pmatrix}0&3&0&0\\0&0&0&4\epsilon\\0&0&0&0\\0&0&-3&0\end{pmatrix},\ X_{N_2}=\begin{pmatrix}0&0&0&0\\1&0&0&0\\0&0&0&-1\\0&\epsilon&0&0\end{pmatrix}.$$

Applying the basis change

$$P=\begin{pmatrix}1&0&0&0\\0&1&0&0\\0&0&0&-\epsilon/6\\0&0&\epsilon/2&0\end{pmatrix}$$

brings the $sl_2(\mathbb{R})$, $\{X_{T_2},X_{M_2},X_{N_2}\}$ into the invariant theoretic form $\{X'_{T_2},X'_{M_2},X'_{N_2}\}$ where

$$X'_{T_2}=\begin{pmatrix}-3&0&0&0\\0&-1&0&0\\0&0&1&0\\0&0&0&3\end{pmatrix},X'_{M_2}=\begin{pmatrix}0&3&0&0\\0&0&2&0\\0&0&0&1\\0&0&0&0\end{pmatrix},X'_{N_2}=\begin{pmatrix}0&0&0&0\\1&0&0&0\\0&2&0&0\\0&0&3&0\end{pmatrix}.$$

In terms of the symplectic coordinates (ξ,η) the invariant theoretic variables x are

$$\begin{pmatrix}x_1\\x_2\\x_3\\x_4\end{pmatrix}=P^{-1}\begin{pmatrix}\xi_1\\\xi_2\\\eta_1\\\eta_2\end{pmatrix}=\begin{pmatrix}\xi_1\\\xi_2\\2\epsilon\eta_2\\-6\epsilon\eta_1\end{pmatrix}$$

Since the $sl_2(\mathbb{R})$, $\{X'_{T_2},X'_{M_2},X'_{N_2}\}$, is irreducible of dimension 4, we need to determine the covariant algebra of the binary cubic form(Sal859a). The generators of C and the Cayley correspondence is summarized in the following table, where $\pi_1=x_1X^3+3x_2X^2Y+3x_3XY^2+x_4Y^3$.

Covariant	Leading term	Element in $ker\ ad_{M_2}$
π_1	x_4	$\tau_1=\eta_1$
$\pi_2=\tau_2(\pi_1,\pi_1)$	$x_2x_4-x_3^2$	$\tau_2=3\epsilon\xi_2\eta_1+2\eta_2^2$
$\pi_3=\tau_1(\pi_1,\pi_2)$	$x_1x_4^2-3x_2x_3x_4+2x_3^3$	$\tau_3=9\xi_1\eta_1^2$ $+9\xi_2\eta_1\eta_2+4\epsilon\eta_2^3$
$\pi_4=\tau_2(\pi_2,\pi_2)$	$x_1^2x_4^2-6x_1x_2x_3x_4$ $+4x_2^3x_4+4x_1x_3^3-3x_2^2x_3^2$	$\tau_4=9\xi_1^2\eta_1^2+18\xi_1\xi_2\eta_1\eta_2$ $-6\epsilon\xi_2^2\eta_1+8\epsilon\xi_1\eta_2^3-3\xi_2^2\eta_2^2$

where $\tau_2(f\ ,f\ )(X,Y)=\dfrac{\partial^2 f}{\partial X^2}\dfrac{\partial^2 f}{\partial Y^2}-(\dfrac{\partial^2 f}{\partial X\partial Y})^2$ is the second transvectant.

As an algebra $ker\ ad_{M_2}$ is generated by $\tau_i, i=1,\ldots,6$. From the relation

$$\pi_3^2=-4\pi_2^3+\pi_1^2\pi_4$$

among the covariants, we obtain the relation

$$\tau_3^2=-\tau_2^3+9\tau_1^2\tau_4$$

among the generators of $ker\ ad_{M_2}$. Hence the normal form is the formal power series Hamiltonian $H(\xi,\eta)=N_2+\overline{H}(\xi,\eta)$ where

$$\overline{H}(\xi,\eta)=F_1(\tau_1,\tau_2,\tau_4)+\tau_3F_2(\tau_1,\tau_2,\tau_4)$$

and F_1,F_2 are formal power series; moreover F_2 has no constant term and F_1 has no constant or linear term in τ_1 (because H begins with quadratic terms).

8. Example 3

The embedding of X_{N_2} into $sl_2(\mathbb{R})$ is given by

$$X_{T_2}=\begin{pmatrix}2&0&0&0&0&0\\0&0&0&0&0&0\\0&0&-2&0&0&0\\0&0&0&-2&0&0\\0&0&0&0&0&0\\0&0&0&0&0&2\end{pmatrix},\quad X_{M_2}=\begin{pmatrix}0&1&0&0&0&0\\0&0&1&0&0&0\\0&0&0&0&0&0\\0&0&0&0&0&0\\0&0&0&-1&0&0\\0&0&0&0&-1&0\end{pmatrix},$$

$$X_{N_2}=\begin{pmatrix}0&0&0&0&0&0\\1&0&0&0&0&0\\0&1&0&0&0&0\\0&0&0&0&-1&0\\0&0&0&0&0&-1\\0&0&0&0&0&0\end{pmatrix}.$$

Applying the basis change

$$P = \begin{pmatrix} 1 & 0 & 0 & 0 & 0 & 0 \\ 0 & 1 & 0 & 0 & 0 & 0 \\ 0 & 0 & \frac{1}{2} & 0 & 0 & 0 \\ 0 & 0 & 0 & 0 & 0 & \frac{1}{2} \\ 0 & 0 & 0 & 0 & -1 & 0 \\ 0 & 0 & 0 & 1 & 0 & 0 \end{pmatrix}$$

brings the $sl_2(\mathbb{R})$, $\{X_{T_2}, X_{M_2}, X_{N_2}\}$, into the invariant theoretic form $\{X'_{T_2}, X'_{M_2}, X'_{N_2}\}$ where

$$X'_{T_2} = \begin{pmatrix} 1 & 0 & 0 & 0 & 0 & 0 \\ 0 & 0 & 0 & 0 & 0 & 0 \\ 0 & 0 & -1 & 0 & 0 & 0 \\ 0 & 0 & 0 & 1 & 0 & 0 \\ 0 & 0 & 0 & 0 & 0 & 0 \\ 0 & 0 & 0 & 0 & 0 & -1 \end{pmatrix}, \quad X'_{M_2} = \begin{pmatrix} 0 & 2 & 0 & 0 & 0 & 0 \\ 0 & 0 & 1 & 0 & 0 & 0 \\ 0 & 0 & 0 & 0 & 0 & 0 \\ 0 & 0 & 0 & 0 & 2 & 0 \\ 0 & 0 & 0 & 0 & 0 & 1 \\ 0 & 0 & 0 & 0 & 0 & 0 \end{pmatrix},$$

$$X'_{N_2} = \begin{pmatrix} 0 & 0 & 0 & 0 & 0 & 0 \\ 1 & 0 & 0 & 0 & 0 & 0 \\ 0 & 2 & 0 & 0 & 0 & 0 \\ 0 & 0 & 0 & 0 & 0 & 0 \\ 0 & 0 & 0 & 1 & 0 & 0 \\ 0 & 0 & 0 & 0 & 2 & 0 \end{pmatrix}.$$

In terms of the symplectic coordinates (ξ,η) the invariant theoretic variables x are

$$\begin{pmatrix} x_1 \\ x_2 \\ x_3 \\ x_4 \\ x_5 \\ x_6 \end{pmatrix} = P^{-1} \begin{pmatrix} \xi_1 \\ \xi_2 \\ \xi_3 \\ \eta_1 \\ \eta_2 \\ \eta_3 \end{pmatrix} = \begin{pmatrix} \xi_1 \\ \xi_2 \\ 2\xi_3 \\ \eta_3 \\ -\eta_2 \\ 2\eta_1 \end{pmatrix}.$$

Since the $sl_2(\mathbb{R})$ $\{X_{T_2}, X_{M_2}, X_{N_2}\}$ is the sum of two irreducibles of dimension 2+1, we need to determine the joint covariant algebra C of two quadratic forms (Sa1859a). The generators of C and the Cayley correspondence is summarized in the following table.

Joint covariants of two quadratic forms	leading term	element in $ker\ ad_{M_2}$
$\pi_1=x_1X^2+2x_2XY+x_3Y^2$	x_3	$\tau_1=\xi_3$
$\pi_2=x_4X^2+2x_5XY+x_6Y^2$	x_6	$\tau_2=\eta_1$
$\pi_3=\tau_1(\pi_1,\pi_1)$	$x_2x_6-x_3x_5$	$\tau_3=\xi_2\eta_1+\xi_3\eta_2$
$\pi_4=\tau_2(\pi_1,\pi_1)$	$x_1x_3-x_2^2$	$\tau_4=2\xi_1\xi_3-\xi_2^2$
$\pi_5=\tau_2(\pi_2,\pi_2)$	$x_4x_6-x_5^2$	$\tau_5=2\eta_1\eta_3-\eta_2^2$
$\pi_6=\tau_2(\pi_1,\pi_2)$	$x_1x_6+x_3x_4-2x_2x_5$	$\tau_6=\xi_1\eta_1+\xi_2\eta_2+\xi_3\eta_3$

where $\tau_2(f,g)=\tau_2(f+g,f+g)-\tau_2(f,f)-\tau_2(g,g)$ is the **polarization** of the second transvectant. Therefore $\tau_i, i=1,\ldots,6$ generate the algebra $ker\ ad_{M_2}$. From the relation

$$\pi_3^2=-\pi_1^2\pi_5+2\pi_1\pi_2\pi_6-\pi_2^2\pi_4$$

among the covariants, we obtain the relation

$$\tau_3^2=2\tau_1\tau_2\tau_6-\tau_1^2\tau_5-\tau_2^2\tau_4$$

among the generators of $ker\ ad_{M_2}$.

Hence the normal form is the formal power series Hamiltonian

$$H(\xi,\eta)=N_2(\xi,\eta)+F_1(\tau_1,\tau_2,\tau_4,\tau_5,\tau_6)+\tau_3F_2(\tau_1,\tau_2,\tau_4,\tau_5,\tau_6)$$

where F_1 and F_2 are formal power series with F_1 having no constant terms or linear terms in τ_1 and τ_2 (since H begins with quadratic terms).

9. Normal form revisited

Suppose $H=H_2+H_3+\cdots+H_n+\cdots\in\mathbb{R}[[V]]$ is a formal power series Hamiltonian on the symplectic vectorspace (V,ω). Let $H_2=S_2+N_2$ be the semisimple-nilpotent decomposition of H_2, that is, $\{S_2,N_2\}=0$, where X_{S_2} is semisimple and X_{N_2} is nilpotent. Then H is in **normal form with respect to** H_2 if for all $n\geq 3$, $H_n\in ker\ ad_{M_2}\cap ker\ ad_{S_2}|P_n(V,\mathbb{R})$ where $\{T_2,M_2,N_2,S_2\}$ is a subalgebra of $(P_2(V,\mathbb{R}),\{,\})$ which is isomorphic to $sl_2(\mathbb{R})\times\mathbb{R}$. Since S_2 commutes with every element of $sl_2(\mathbb{R})$, ad_{S_2} maps the normal form algebra $ker\ ad_{X_{M_2}}$ into itself. Hence $ker\ ad_{S_2}|ker\ ad_{M_2}=ker\ ad_{M_2}\cap ker\ ad_{S_2}$ which is a subalgebra of $\mathbb{R}[[V]]$.

The following refinement of the Jacobson-Morozov lemma shows that N_2 can be embedded into $sl_2(\mathbb{R})\times\mathbb{R}$. Since X_{S_2} is a semisimple element of $sp(V,\omega)$, the set of all $Y_{K_2}\in sp(V,\omega)$ such that $[Y_{K_2},X_{S_2}]=0$, (which is equivalent to the set of all $K_2 \in P_2(V,\mathbb{R})$ such that $\{K_2,S_2\}=0$), forms a reductive Lie subalgebra ℓ of $sp(V,\omega)$. Now X_{N_2} is a nilpotent element of $\mathbf{g}$. Because the Jacobson-Morozov lemma is valid for all real reductive Lie algebras (see (He1982a)), there is a subalgebra $\{X_{T_2},X_{M_2},X_{N_2}\}$ of ℓ which is isomorphic to $sl_2(\mathbb{R})$. This completes the construction of the desired embedding.

We now work out two examples of the general normal form algebra.

Example 1

Take

$$X_{N_2}=\begin{pmatrix}0&0&0&0\\0&0&0&0\\1&0&0&0\\0&1&0&0\end{pmatrix}\quad and\ X_{S_2}=\begin{pmatrix}0&-1&0&0\\1&0&0&0\\0&0&0&-1\\0&0&1&0\end{pmatrix}.$$

The embedding of X_{N_2} into a subalgebra of $sp(4,\mathbb{R})$ which is isomorphic to $sl_2(\mathbb{R})\times\mathbb{R}$ is given by

$$X_{T_2}=\begin{pmatrix}-1&0&0&0\\0&-1&0&0\\0&0&1&0\\0&0&0&1\end{pmatrix},\ X_{M_2}=\begin{pmatrix}0&0&1&0\\0&0&0&1\\0&0&0&0\\0&0&0&0\end{pmatrix},$$

and

$$X_{N_2}=\begin{pmatrix}0&0&0&0\\0&0&0&0\\1&0&0&0\\0&1&0&0\end{pmatrix},\ X_{S_2}=\begin{pmatrix}0&-1&0&0\\1&0&0&0\\0&0&0&-1\\0&0&1&0\end{pmatrix}.$$

Note that the $sl_2(\mathbb{R})$ part is the same as in example 1 of §5. Therefore the algebra $ker\ ad_{M_2}$ is freely generated by

$$\tau_1=\eta_1\ ,\ \tau_2=\eta_2,\ and\ \tau_3=\xi_1\eta_2-\xi_2\eta_1.$$

Since

$$ad_{S_2}=\xi_2\frac{\partial}{\partial\xi_1}-\xi_1\frac{\partial}{\partial\xi_2}+\eta_2\frac{\partial}{\partial\eta_1}-\eta_1\frac{\partial}{\partial\eta_2},$$

we have

$$\dot\tau_1=\tau_2$$

$$\dot\tau_2=-\tau_1$$

$$\dot\tau_3=0.$$

Thus ad_{S_2} is the linear vectorfield

$$\frac{d}{dt}\begin{pmatrix}\tau_1\\ \tau_2\\ \tau_3\end{pmatrix}=\begin{pmatrix}0&1&0\\ -1&0&0\\ 0&0&0\end{pmatrix}\begin{pmatrix}\tau_1\\ \tau_2\\ \tau_3\end{pmatrix}$$

on $ker\ ad_{M_2}$. Hence the normal form algebra $ker\ ad_{S_2}\cap ker\ ad_{M_2}$ is freely generated by

$$\tau_1^2+\tau_2^2 \text{ and } \tau_3$$

which generate all the integrals of ad_{S_2} in $ker\ ad_{M_2}$. Therefore the normal form of H is $F(\tau_1^2+\tau_2^2,\tau_3)$ where F is a formal power series without constant term.

Example 2

Let

$$X_{N_2}=\begin{pmatrix}0&0&0&0&0&0\\ 1&0&0&0&0&0\\ 0&1&0&0&0&0\\ 0&0&0&0&-1&0\\ 0&0&0&0&0&-1\\ 0&0&0&0&0&0\end{pmatrix} \text{ and } X_{S_2}=\epsilon\begin{pmatrix}0&0&0&0&0&1\\ 0&0&0&0&-1&0\\ 0&0&0&1&0&0\\ 0&0&-1&0&0&0\\ 0&1&0&0&0&0\\ -1&0&0&0&0&0\end{pmatrix}.$$

The embedding of X_{N_2} into $sl_2(\mathbb{R})\times\mathbb{R}$ is completed by

$$X_{T_2}=\begin{pmatrix}-2&0&0&0&0&0\\ 0&0&0&0&0&0\\ 0&0&2&0&0&0\\ 0&0&0&2&0&0\\ 0&0&0&0&0&0\\ 0&0&0&0&0&-2\end{pmatrix} \text{ and } X_{M_2}=\begin{pmatrix}0&1&0&0&0&0\\ 0&0&1&0&0&0\\ 0&0&0&0&0&0\\ 0&0&0&0&0&0\\ 0&0&0&-1&0&0\\ 0&0&0&0&-1&0\end{pmatrix}.$$

Note that the $sl_2(\mathbb{R})$ part is the same as in example 3 of §5. Therefore the algebra $ker\ ad_{M_2}$ is generated by

$$\tau_1=\xi_3$$

$$\tau_2=\eta_1$$

$$\tau_3=\xi_2\eta_1+\xi_3\eta_2$$

$$\tau_4=2\xi_1\xi_3-\xi_2^2$$

$$\tau_5=2\eta_1\eta_3-\eta_2^2$$

$$\tau_6=\xi_1\eta_1+\xi_2\eta_2+\xi_3\eta_3$$

subject to the relation

$$\tau_3^2=2\tau_1\tau_2\tau_6-\tau_1^2\tau_5-\tau_2^2\tau_4.$$

Since

$$ad_{S_2}=-\epsilon(\eta_3\frac{\partial}{\partial\xi_1}-\eta_2\frac{\partial}{\partial\xi_2}+\eta_1\frac{\partial}{\partial\xi_3}-\xi_3\frac{\partial}{\partial\eta_1}+\xi_2\frac{\partial}{\partial\eta_2}-\xi_1\frac{\partial}{\partial\eta_3}),$$

we have

$$\dot\tau_3=0$$
$$\dot\tau_1=\tau_2$$
$$\dot\tau_6=\tau_5-\tau_4$$
$$\dot\tau_4+\dot\tau_5=0$$
$$\dot\tau_2=-\tau_1$$
$$\dot\tau_5-\dot\tau_1=-4\tau_6$$

Thus ad_{S_2} is the linear vectorfield

$$\frac{d}{dt}\begin{pmatrix}\tau_3\\ \tau_1\\ 2\tau_6\\ \tau_5+\tau_4\\ \tau_2\\ \tau_5-\tau_4\end{pmatrix}=\begin{pmatrix}0&0&0&0&0&0\\0&0&0&0&1&0\\0&0&0&0&0&2\\0&0&0&0&0&0\\0&-1&0&0&0&0\\0&0&-2&0&0&0\end{pmatrix}\begin{pmatrix}\tau_3\\ \tau_1\\ 2\tau_6\\ \tau_5+\tau_4\\ \tau_2\\ \tau_5-\tau_4\end{pmatrix}$$

which is the 0:1:2 linear harmonic oscillator. Thus the algebra $ker\ ad_{S_2}|ker\ ad_{X_{M_2}}$ has generators given in the following table

generators	formal polynomial
$\sigma_1=\tau_3$	u^2t^2
$\sigma_2=\tau_5+\tau_4$	t^2
$\sigma_3=\tau_1^2+\tau_2^2$	u^4t^2
$\sigma_4=4\tau_6^2+(\tau_5-\tau_4)^2$	t^4
$\sigma_5=\tau_6(\tau_1^2-\tau_2^2)+\tau_1\tau_2(\tau_5-\tau_4)$	u^4t^2

In the formal polynomial the exponent of u is the eigenvalue of ad_{T_2}, while the exponent of t is the degree of the generator as a polynomial in (ξ,η). From the relation

$$4[\tau_6(\tau_1^2-\tau_2^2)+\tau_1\tau_2(\tau_5-\tau_4)]^2+[4\tau_1\tau_2\tau_6+(\tau_5-\tau_4)(\tau_1^2-\tau_2^2)]^2=$$
$$=(\tau_1^2+\tau_2^2)^2(4\tau_6^2+(\tau_5-\tau_4)^2)$$

for the generators which are in 1:2-resonance and using the relation

$$\tau_3^2=2\tau_1\tau_2\tau_6-\tau_1^2\tau^5-\tau_2^2\tau_4,$$

we obtain the relation

$$4\sigma_5^2=\sigma_3^2\sigma_4-(2\sigma_1^2+\sigma_2\sigma_3)^2. \tag{6.1}$$

From the table and (6.1) we find that the two variable table function (the Molien function of the tensor representation) (Cu1986a) for the algebra $ker\ ad_{S_2} | ker\ ad_{M_2}$ is

$$T(u,t)=\frac{1+u^4t^4}{(1-u^2t^2)(1-t^2)(1-u^4t^2)(1-t^4)}.$$

Since

$$\frac{\partial}{\partial u}(uT(u,t))\,|_{u=1}=\frac{1+4t^2+t^4}{(1-t^2)^5}=\sum_{k=0}^{\infty}\binom{k+2}{2}^2 t^{2k}$$

is the Poincaré series for the algebra $ker\ ad_{S_2}$ (see (Fe1986a)), we deduce that as an algebra $ker\ ad_{S_2} | ker\ ad_{M_2}$ is generated by σ_i, $i=1,\ldots,5$ subject to the relation (6.1).

Therefore the normal form of H is

$$F_1(\sigma_1,\sigma_2,\sigma_3,\sigma_4)+\sigma_5F_2(\sigma_1,\sigma_2,\sigma_3,\sigma_4)$$

where F_1, F_2 are formal power series and F_1 has no constant term.

References

Bu1977a. Burgoyne,N. and Cushman,R., "Normal forms for real linear Hamiltonian systems," pp. 483-529 in *The 1976 Ames Research Center (NASA) Conference on Geometric Control Theory*, ed. C.Martin and R.Hermann, Math Sci Press, Brookline, Mass (1977).

Cu1974a. Cushman,R., "The momentum mapping of the harmonic oscillator," pp. 323-342 in *Symposia Mathematica, 14*, Instituto Nazionale di Alta Matematica (1974).

Cu1982a. Cushman,R., "Reduction of the 1:1 nonsemisimple resonance," *Hadronic Journal* **5**, pp. 2109-2124 (1982).

Cu1986a. Cushman,R. and Sanders,J.A., "Nilpotent normal forms and representation theory of $sl(2,\mathbb{R})$," pp. 31-51 in *Multi-parameter Bifurcation Theory*, ed. M. Golubitsky and J. Guckenheimer, American Mathematical Society, Contemporary Mathematics 56, Providence (1986).

Fe1986a. Fekken,A., "On the normal form of a fully resonant Hamiltonian function," Preprint #317, Free University Amsterdam (1986).

Fr1969a. Freudenthal,H. and de Vries,H., *Linear Lie Groups*, Academic Press, New York, London (1969).

He1982a. Helgason,S., *Differential Geometry, Lie Groups and Symmetric Spaces,* Academic Press, New York (1982).

Hu1972a. Humphreys,J., *Introduction to Lie Algebras and Representation Theory,* Springer-Verlag, New York (1972).

Sa1859a. Salmon,G., *Lessons introductory to the Modern Higher Algebra,* (reprinted by Chelsea, New York, 1964), Dublin (1859).

Sp1977a. Springer,T.A., *Invariant Theory,* Lecture Notes in Mathematics 585, Springer-Verlag (1977).

Department of Mathematics and Computer Science
Free University
Postbus 7161
1007 MC Amsterdam

Canadian Mathematical Society
Conference Proceedings
Volume 8 (1987)

ASYMPTOTIC STATISTICAL PROPERTIES OF CHAOTIC POWER SPECTRA

Randall S. Dumont[1]

ABSTRACT. The grassy character of the power spectra (or periodograms) of a chaotic flow is rationalized and reconciled with the Wiener-Khinchin Theorem, in terms of asymptotic power spectrum statistics. Using a strong multiple mixing assumption (assumption of chaos) the power spectrum, $I_T(x,\omega)$, is shown to be asymptotically (as $T\to\infty$) Poisson distributed, with respect to a nonequilibrium sampling of initial conditions, x. $I_T(x,\omega)$ and $I_T(x,\omega')$, as well as $I_T(x,\omega)$ and $I_{T'}(x,\omega)$, are shown to be asymptotically independent. Furthermore, progress is made towards extending these results to Bernoulli flows and dynamical variables with sufficiently rapid decay of correlations.

1. INTRODUCTION AND STATEMENT OF RESULTS. The power spectrum or periodogram of a dynamical variable, f, defined by

$$(1)\quad I_T(x,\omega) = \frac{1}{T} \left| \int_0^T dt\, e^{i\omega t} f(x_t)\right|^2,$$

gives a useful characterization of the time series of f along the bounded trajectory, $\{x_t\}_{t=0}^{T}$. It has been found that quasiperiodic and chaotic motions can be distinguished by the qualitative appearance of their power spectra.[1] Specifically, a quasiperiodic power spectrum is recognizable as a simple line spectrum (consisting of sharp peaks[2] at a few fundamental frequencies and some of their lower order combinations and overtones), while chaotic spectra are notoriously grassy or erratic (having many peaks[3] with no obvious pattern of central frequencies and amplitudes).

Interest in the power spectrum also stems from the Wiener-Khinchin theorem[4], which asserts the weak L^2 (R) convergence of $I_T(x,\omega)$ to the spectral density, if the latter belongs to L^2.

$$(2)\quad \underset{T\to\infty}{\text{l.i.m.}}\ I_T(x,\omega) = S(\omega)$$

---

[1]1980 Mathematics Subject Classification (1985 Revision). 93E14, 58F13.

The spectral density,

(3) $S(\omega) = \int_{-\infty}^{\infty} dt\ e^{i\omega t} < f\ f_t >_\mu$,

is just the Fourier transform of the autocorrelation,

(4) $<f\ f_t >_\mu = \int_M d\mu(x)\ f(x)\ f(x_t)$

The invariant measure, μ, is defined by

$$\int_M d\mu(x)\ g(x) = \lim_{T\to\infty} \frac{1}{T} \int_0^T dt\ g(x_t)$$

where M is the minimal invariant manifold[5] containing x (i.e. the closure of $\{x_t\}_{t=-\infty}^{\infty}$). Since the autocorrelation generally decays rapidly to zero, as $t\to\infty$, when the dynamics are chaotic,[6] the chaotic spectral density is a smooth function. (The average value of f is subtracted out-i.e. $<f>_\mu=0$.) However, we have already noted that the chaotic power spectrum is an increasingly (with T) grassy function. This is consistent with (2) only if $\lim_{T\to\infty} I_T(x,\omega)$ does not exist. Since $I_T(x,\omega)$ is thus a poor estimate of $S(\omega)$, attempts to use (2) for the purpose of estimating the spectral density require that I_T be subject to some form of averaging. The Wiener-Khinchin theorem implies that smoothing in frequency induces the desired convergence behavior. In addition, phase space averaging (i.e. averaging over some x distribution) of (2) converts the weak limit to the more useful pointwise limit. For the phase space averaging to work, $I_T\ (x,\omega)$ must be highly erratic with respect to variations of x as well as of ω.

The results of this paper provide the mathematical foundations for the behavior of the chaotic power spectrum just described. The assumption of chaos adopted in 2, 3 and 4 is a strong multiple mixing condition based upon the typical rapid decay of correlations exhibited by chaotic systems. Defining the arbitrarily long, increasing sequence of times, $\{t_j\}_{j=1}^{J}$, we require

(5) $<f_{t_1} f_{t_2} \ldots f_{t_J}>_\lambda = <f_{t_1} \ldots f_{t_k}>_\lambda <f_{t_{k+1}} \ldots f_{t_J}>_\mu$,

whenever $t_{k+1} > t_k + \tau$, for some $\tau>0$, $k=1, \ldots, J-1$. Note the use of the noninvariant measure λ, with associated density $\ell(x)>0$ (i.e. $\ell(x) = d\lambda(x)/d\mu(x)$), and the inclusion of the case of $k=0$, by setting $t_0=0$ and $<1>_\lambda \equiv <\ell>_\mu = 1$. Equation (5) corresponds to an assumption of strict statistical independence of the random variables, $\ell(x)$ or $f(x)$ and $f(x_{t+\tau})$, $t>0$, with respect to μ.

Given this assumption of chaos, we calculate, in section 2, the probability distribution for the value of the power spectrum, with random sampling of x with respect to the normalized measure λ, absolutely continuous relative to μ:

$$\rho_T(\alpha;\omega) \equiv \int_M d\lambda(x)\ \delta[\alpha - I_T(x,\omega)]$$
$$= \frac{d}{d\alpha}\ \lambda\{I_T(x,\omega)<\alpha\}$$

With the dynamical variable, f, and the density, ℓ, associated with λ, assumed to be bounded as well as to satisfy Eq. (5), we derive

$$(\nu_\delta * \rho_T)\ (\alpha) = \int_0^\infty d\alpha'\nu_\delta(\alpha-\alpha')\rho_T(\alpha';\omega)$$
$$\sim \begin{cases} [S(\omega)]^{-1}e^{-\alpha/S(\omega)}, & \omega\neq 0 \\ [2\pi S(o)\alpha]^{-1/2}e^{-\alpha/[2S(o)]}, & \omega=0 \end{cases}$$

as $T\to\infty$ and $\delta\to 0$, such that $T\delta\to\infty$, where $\nu_\delta(\alpha)$ is an L^1-normalized smoothing distribution with support of width 2δ, centered about $\alpha=0$. (ν_δ is required to satisfy some smoothness requirements which are specified in Appendix B- (B3), (B4) and (B5).) The broad Poisson distribution[7] (we assume that $0<S(\omega)<\infty$) explains the erratic x dependence of $I_T(x,\omega)$ and is consistent with the stabilizing effect of phase space averaging.

In section 3, the asymptotic independence of $I_T(x,\omega)$ and $I_T(x,\omega')$, with respect to the same measure, λ, is established. This result, in combination with the large breadth of the distribution of $I_T(x,\omega)$, explains the grassy character of the chaotic power spectrum. That the grassiness increases with T is evident from the result that the separation of ω and ω', required for independence, is decreasing with T ($|\omega-\omega'|>T^{-\epsilon}$, $0<\epsilon<1$).

In section 4, the asymptotic independence of $I_T(x,\omega)$ and $I_{T'}(x,\omega)$ is established. This result verifies the <u>λ-almost everywhere divergence of</u> $\lim_{T\to\infty} I_T(x,\omega)$.

The results of 2 and 3 have previously been established for discrete time stationary processes satisfying strong mixing assumptions.[8] In addition to extending these results to continuous time and supplementing them with the result of section 4, the present work considers the case of a <u>nonequilibrium sampling</u> of x (i.e. with respect to λ instead of μ). This corresponds to a special nonstationary process not considered by ref. 8. Furthermore, in section 5, progress is made towards extending these results to dynamical systems and variables satisfying more traditional assumptions of chaos.

Note that the $\omega=0$ specialization of the result of 2 is just the central limit theorem (with nonequilibrium sampling). The central limit theorem has previously been shown[9] to hold for Hölder continuous dynamical variables under the dynamics of a geodesic flow on a surface of negative curvature.

In ref. 10 and 21, numerical calculations on a system known empirically to exhibit chaotic behavior are shown to support the analytical results of this paper.

2. THE ASYMPTOTIC PROBABILITY DISTRIBUTION. Of primary concern in this analysis is the joint distribution, $p_T(u,v)$, of the real and imaginary finite Fourier components of $f(x_t)$:

(8) $p_T(u,v) = \int_M d\lambda(x)\ \delta[u-r_T(x)]\delta[v-j_T(x)]$,

where

(9) $r_T(x) + ij_T(x) = T^{-1/2}\int_0^T dt\ e^{-i\omega t} f(x_t)$

defines the real functions, r_T and j_T.

(Note the suppression of the ω dependence.) The probability distribution $\rho_T(\alpha)$, associated with the power spectrum, $I_T(x,\omega)$, is given as

(10) $\rho_T(\alpha) = \int_{R^2} dudv\ \delta[\alpha-(u^2+v^2)]p_T(u,v)$.

The first step is the standard probability theoretic expression of a distribution in terms of moments.[11] This proceeds by introducing the Fourier transform of the probability distribution:

(11) $\tilde{p}_T(\xi,\eta) = \int_{R^2} dudv\ e^{i(\xi u+\eta v)}\ p_T(u,v)$

$= \int_M d\lambda(x)\ e^{ih_T(x;\xi,\eta)}$

where

(12) $h_T(x;\xi,\eta) = \xi r_T(x) + \eta j_T(x)$.

The exponential in (11) is Taylor series expanded and the λ integration is interchanged with the resulting summation. (The validity of this latter operation is verified, for finite T, by the convergence of the ensuing series, which is shown below.) This yields[12]

(13) $\tilde{p}_T(\xi,\eta) = \sum_{n=0}^{\infty} \frac{i^n}{n!} \int_M d\lambda(x)\ h_T^n(x;\xi,\eta)$

Note that the n=0 term of the above sum is trivially unity, so that the nth moment discussed below corresponds to n>0.

Before proceeding further, it is desirable to introduce some simplifying symbolic notation. The top row of Table 1A gives the symbols which replace the operators beneath them, while Table 1B provides similar simplified notation for frequently appearing functions.

Table 1. Simplifying Notation[(a)]

A.					
	$\overset{\uparrow}{j}$	$\overset{\updownarrow}{j}$	$\overset{\underline{\uparrow}}{j}$	$\overset{I}{j}$	$\overset{\overline{\downarrow}}{j}$
	$\int_{t_{j-1}}^{T} dt_{j-1}$	$\int_{t_{j-2}}^{T} dt_j$	$\int_{t_{j-1}+\tau}^{T} dt_j$	$\int_{t_{j-1}}^{t_{j-1}+\tau} dt_j$	$\int_{t_{j-2}}^{t_{j-1}+\tau} dt_j$

B.[(b),(c)]					
	$c_j^{q,p}$	$s_j^{q,p}$	e_j	$S/2 + i\sigma$[(d)]	$<1,2,\ldots,j>_{\lambda(\text{or }\mu)}$
	$\cos^q[p\omega t_j]$	$\sin^q[p\omega t_j]$	$\xi c_j - \eta s_j$	$\int_0^{\tau} dt e^{-i\omega t} <ff_t>_{\mu}$	$<f_{t_1} f_{t_2} \ldots f_{t_j}>_{\lambda(\text{or}\mu)}$

(a) $t_o = 0$

(b) The absence of the subscript j implies that t_j is replaced by T.

(c) The default on p and/or q is 1.

(d) $S = S(\omega)$ is the spectral density of f on M.

A. THE MOMENTS OF h_T. The next step is the expression of the nth moment of h_T in terms of the nth order correlations of f, with respect to λ. This is accomplished by expanding h_T^n in terms of $f(x_t)$, then interchanging the finite λ integration with the finite t integrations. The result is

$$(14) \quad \int_M d\lambda h_T^n = T^{-n/2} \int_0^T dt_1 \ldots \int_0^T dt_j \ldots \int_0^T dt_n e_1 \ldots e_n <1,\ldots n>_{\lambda}$$
$$= n!\, T^{-n/2} \overset{\uparrow}{1}\, \overset{\uparrow}{2} \ldots \overset{\uparrow}{n}\, e_1 \ldots e_n <1,\ldots,n>_{\lambda},$$

where the multiple t integral has been desymmetrized by ordering the t_j's and multiplying by n!.

Because of the multiple mixing conditions satisfied by f and ℓ, (5), and the fact the $<j>_{\mu} = 0$, the correlation in (14) restricts the range of t integration. Focussing only on the restricted range, the integral can be decomposed into a set of elementary contributions, given by the branches of the tree shown in Fig. 1 (with the integrand $e_1 \ldots e_n <1, \ldots, n>_{\lambda}$ appended at the right.

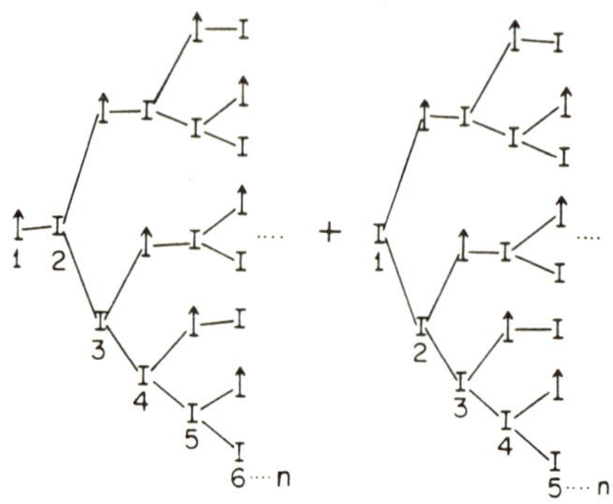

Fig. 1. The decomposition of $\uparrow_1 \uparrow_2 \ldots \uparrow_n$ into elementary contributions.

This binary tree is constructed by starting with $\uparrow_1 = \underline{\uparrow}_1 + I_1$, then following each I_j by $\underline{\uparrow}_{j+1}$ and each $\underline{\uparrow}_j$ by $\underline{\uparrow}_{j+1} + I_{j+1}$. (The latter sum is represented by splitting the branch). Note that only branches ending in I_n have a nonzero contribution and the total number of branches is less than 2^n. Because $|e_j| \leqq |\xi|+|\eta| \leqq [2(\xi^2+\eta^2)]^{1/2}$ and $|f| \leqq b$ [since $f \in L^\infty(M)$], each nonzero branch is bounded by

$$(15) \quad [(\xi^2+\eta^2)d^2/4]^{n/2} \tau^{n-K} \uparrow_1 \uparrow_2 \ldots \uparrow_K 1 = [(\xi^2+\eta^2)d^2/4]^{n/2} \tau^{n-K} T^K/K!$$

where K is the number of $\underline{\uparrow}_j$'s appearing in the branch and $d = 2\sqrt{2}\, b$. (Note that $K \leqq \lfloor n/2 \rfloor$, i.e. the integer part of n/2.) In the following, time is measured in units of τ, which is henceforth set equal to 1. Furthermore, until otherwise specified n will be restricted such that

$$(16) \quad n < T^{1/2}$$

The two cases of odd and even n are distinguished here:

(1) When n is odd, $K \leqq (n-1)/2$ and because of (16), $T^K/K! < T^{(n-1)/2}/[(n-1)/2]!$. Since there are fewer than 2^n contributing branches, this bound combines with that of (15) to give

$$(17) \quad \left|\int_M d\lambda h_T^n\right| < [(\xi^2+\eta^2)d^2)]^{n/2} \frac{n!}{(\frac{n-1}{2})!} T^{-1/2}.$$

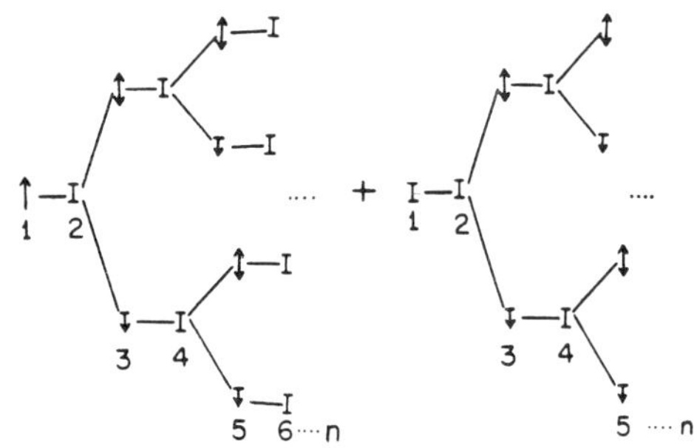

Fig. 2. The further decomposition of the top branch of Fig. 1.

(2) When n is even, we exclude, for the moment, the top branch, so that $K \leqq n/2 - 1$. The overall contribution of these lower branches is bounded by

$$(18)\quad [(\xi^2+\eta^2)d^2]^{n/2}\ \frac{T^{n/2}}{(\frac{n}{2}-1)!}\ T^{-1}.$$

The top branch contribution is given by

$$(19)\quad B_T = \underset{1}{\underline{\uparrow}}\ \underset{2}{I}\ \underset{3}{\underline{\uparrow}}\ \underset{4}{I}\ \dots\ \underset{n-1}{\underline{\uparrow}}\ \underset{n}{I}\ e_1\dots e_n\langle 1,2\rangle_\mu\dots\langle n-1,n\rangle_\mu$$

Using $\underset{j}{\underline{\uparrow}} = \underset{j}{\uparrow} - \underset{j}{\underline{\downarrow}}$ and $\underset{1}{\underline{\uparrow}} = \underset{1}{\uparrow} - \underset{1}{I}$, B_T can be expanded into the binary tree in Fig. 2. The interpretation of this tree is the same as that of Fig. 1, except for an additional factor, $(-1)^{n-K}$, applied to each branch. The exponent n-K is just the number of $\underset{j}{\underline{\downarrow}}$ (including $\underset{1}{I}$) occurrences in the branch. (K is thus the corresponding number of $\underset{j}{\updownarrow}$'s, including $\underset{1}{\uparrow}$.)

Not including the top subbranch, $K \leqq n/2 - 1$, so that just as above the total of all lower contributions to B_T, and hence $\int_M d\lambda h_T^n$, is bounded by (18). The contribution of the top subbranch is given by

$$(20)\quad \tilde{B}_T = \underset{1}{\uparrow}\ \underset{2}{I}\ \underset{3}{\updownarrow}\ \underset{4}{I}\dots\ \underset{n-1}{\updownarrow}\ \underset{n}{I}\ e_1\dots e_n\langle 1,2\rangle_\mu\dots\langle n-1,n\rangle_\mu$$

Since none of the integration ranges depend upon t_j, for even j, all of these t_j integrals can be evaluated explicitly. Thus,

$$(21)\quad \tilde{B}_T = \underset{1}{\uparrow} e_1\ \underset{2}{I}\ e_2\langle 1,2\rangle_\mu\ \underset{3}{\updownarrow} e_3\ \underset{4}{I} e_4\langle 3,4\rangle_\mu\dots\underset{n-1}{\updownarrow}\ e_{n-1}\underset{n}{I} e_n\langle n-1,n\rangle_\mu$$

$$= \underset{1}{\uparrow}B_1\underset{2}{\uparrow}B_2\dots\underset{n/2}{\uparrow}\ B_{n/2}$$

where

(22) $B_{j/2} = e_{j-1} \underset{j}{I} e_j \langle j-1,j\rangle_\mu$

$$= \xi^2[c_{j-1}^2 \frac{S}{2} + c_{j-1}s_{j-1}\sigma] - \xi\eta[2c_{j-1}s_{j-1}\frac{S}{2} - (c_{j-1}^2 - s_{j-1}^2)\sigma$$

$$+\eta^2[s_{j-1}^2 \frac{S}{2} - c_{j-1}s_{j-1}\sigma]$$

Symmetrizing (21) yields

(23) $\tilde{B}_T = (\underset{1}{\dagger} B_1)^{n/2}/(n/2)!$

$$=\{(\xi^2+\eta^2)\frac{ST}{4} + w^{-1}[\xi^2(s'^2\frac{S}{8} + s^2\frac{\sigma}{2}) - \xi\eta(s^2\frac{S}{2} - s'^2\frac{\sigma}{2})$$

$$-\eta^2(s'^2\frac{S}{8} + s^2\frac{\sigma}{2})]\}^{n/2}/(n/2)!$$

Two cases are distinguished: either $w=0$ or $w\neq 0$. In the latter case, for T sufficiently large, w necessarily satisfies $|w| > T^{-1/2}$.

(i) When $w=0$, $w^{-1}s'^2 = 2T$, $w^{-1}s^2 = 0$ and $\sigma = 0$, so that

(24) $\tilde{B}_T = \xi^n(ST/2)^{n/2}/(n/2)!$

In this case, the even nth moment of h_T is given by

(25) $\int_M d\lambda h_T^n = \{[\xi^2 S/2]^{n/2} + T^{-1/2}[(\xi^2+\eta^2)d^2]^{n/2}E/2\}\frac{n!}{(\frac{n}{2})!}$, where $|E|<1$.

(ii) When $|w|>T^{-1/2}$,

(26) $\tilde{B}_T = [(\xi^2+\eta^2)ST/4]^{n/2}(1 + T^{-1/2}E')^{n/2}/(n/2)!$

where $|E'|<D'$, which is independent of T. Since

$$|(1+T^{-1/2}E')^{n/2}-1| < T^{-1/2}(1+D')^{n/2},$$

the nth moment of h_T is given, for even n, by

(27) $\int_M d\lambda h_T^n = \{[(\xi^2+\eta^2)S/4]^{n/2} + 2T^{-1/2}[(\xi^2+\eta^2)D^2]^{n/2}E\}\frac{n!}{(\frac{n}{2})!}$,

where $|E| <1$ and $D^2 = \max\{S\frac{(1+D')}{4}, d^2\}$.

B. THE TAYLOR SERIES for $\tilde{p}_T(\xi,\eta)$. Following the discussion of Appendix A, we <u>exclude</u> from M those x associated with <u>large deviations</u> of h_T and I_T. The following bound is adopted:

(28) $|h_T| < 2D(\beta \ell n T)^{1/2} \ell n[\ell n T]$,

where $\beta>0$ is as of yet unspecified. This restriction on $|h_T|$ affords a very high order bound[13] on the large n contribution to $\tilde{p}_T(\xi,\eta)$ (i.e. the contribution of those terms of (13) such that (16) is violated) and allows the results of section A to be used to obtain the asymptotic form of $\tilde{p}_T(\xi,\eta)$ from (13). Of course, by adopting (28) we tacitly assume (A4) and exclude the contribution from the large argument part of $\tilde{p}_T(\xi,\eta)$. This anticipates the need for the smoothing of $\rho_T(\alpha)$ which is introduced in section C.

Now, substituting (17) and either (25) or (27) into (13), and neglecting the large n terms gives

$$\tilde{p}_T(\xi,\eta) = F_0'(\xi,\eta) + F_1'(\xi,\eta),$$

where
$$F_0'(\xi,\eta) = \begin{cases} \sum_{m=0}^{m_>} \frac{[-\xi^2 S/2]^m}{m!}, & \omega = 0 \\ \sum_{m=0}^{m_>} \frac{[-(\xi^2+\eta^2)S/4]^m}{m!}, & \omega \neq 0 \end{cases}$$

$$|F_1'(\xi,\eta)| < T^{-1/2}\{T^{1/8} \sum_{m=0}^{m_>-1} \frac{[(\xi^2+\eta^2)d^2]^m}{m!} + 2\sum_{m=0}^{m_>} \frac{[(\xi^2+\eta^2)D^2]^m}{m!}\}$$

and $m_> = \lfloor T^{1/2}/2 \rfloor$. Extending the sum in F_0' to $m=\infty$ introduces high order errors in the same fashion as the neglect of the large n terms in the full expression for $\tilde{p}_T$.[13] Thus,

(29) $$\tilde{p}_T(\xi,\eta) = F_0(\xi,\eta) + F_1(\xi,\eta),$$

where
$$F_0(\xi,\eta) = \begin{cases} \exp[-\xi^2 S/2], & \omega = 0 \\ \exp[-(\xi^2+\eta^2)S/4], & \omega \neq 0 \end{cases}$$

and $|F_1(\xi,\eta)| < T^{-3/8}\exp[(\xi^2+\eta^2)d^2] + T^{-1/2}\exp[(\xi^2+\eta^2)D^2]$

C. THE INVERSE FOURIER TRANSFORM AND SMOOTHING. Because of the exponential growth in $(\xi^2+\eta^2)$ of the above bound on $|F_1(\xi,\eta)|$, the inverse Fourier transform of (29) does not provide the asymptotic form of $p_T(u,v)$. For this reason, it is necessary to smooth $\rho_T(\alpha)$ and effectively suppress the large argument contribution of $\tilde{p}_T(\xi,\eta)$.
This is discussed in Appendix B, where the error due to neglecting the large argument contribution is determined to be $O\{(\ell nT)^{-1/28}[\ell n(\ell nT)]^{7/2}\}$, when $\delta = O[(\ell nT)^{-1/28}]$. (Recall that 2δ is the width of the support of the smoothing function, ν_δ, which appears in (7).) In this section we simply retain (A4), i.e. $\zeta = (\xi^2+\eta^2)^{1/2} < (\beta \ell nT)^{1/2}$, and exclude large deviations in I_T:

(30) $I_T < 8D^2[\ell n(\ell nT)]^2$,

as discussed in Appendix A. The latter qualification is required for the analysis of Appendix B. Thus,

$$(31)\quad \rho_{T,\delta}(\alpha) \sim \frac{1}{4\pi^2}\int_{R^2} dudv\ \nu_\delta[\alpha-(u^2+v^2)]\int_{\zeta<K} d\xi d\eta e^{-i(\xi u+\eta v)}\tilde{p}_T(\xi,\eta),$$

where $K = (\beta \ell nT)^{1/2}$. Substituting (29) yields $\rho_{T,\delta}(\alpha) \sim G_0'(\alpha)+ G_1'(\alpha)$, where G_0' and G_1' correspond to F_0 and F_1, respectively. Note that in the $w=0$ case, the η and v integrations can be dropped entirely from (31), because j_T is identically zero and does not contribute to the distribution. Further note that in either case ($w=0$ or $w\neq 0$) extending the (ξ,η) integration to all of R^2 for the F_0 term only introduces an error which is bound in absolute value by (since ν_δ is L^1 normalized)

$$\frac{1}{2\pi}\int_K^\infty d\zeta\zeta e^{-\zeta^2 S/2} = (2\pi S)^{-1} T^{-\beta S/2}.$$

This is of higher order than the errors incurred by the original exclusion of $\zeta>K$, as long as $S\neq 0$. If S were equal to zero, then $\rho_{T,\delta}(\alpha)$ would clearly be clustered close to $\alpha=0$. Furthermore, since δ is chosen in Appendix B to be equal to $K^{-1/14}$, we have

$$(32)\quad \rho_{T,\delta}(\alpha) \sim G_0(\alpha) + G_1(\alpha),$$

where $G_0(\alpha) = \begin{cases} (2\pi S\alpha)^{-1/2} e^{-\alpha/2S}, & w = 0 \\ S^{-1} e^{-\alpha/S}, & w \neq 0 \end{cases}$

and $|G_1(\alpha)| < \frac{1}{2\pi}\int_0^K d\zeta\zeta (T^{-3/8} e^{\zeta^2 d^2} + T^{-1/2} e^{\zeta^2 D^2})$

$$< \frac{1}{2\pi}\left[\frac{T^{-3/8+\beta d^2}}{d^2} + \frac{T^{-1/2+\beta D^2}}{D^2}\right],$$

which is asymptotically decreasing if $\beta < \min\{3/8d^2, 1/2D^2\}$. Thus, $\rho_{T,\delta}(\alpha)$ is asmptotically Poissonian, for $w\neq 0$. For $w=0$, we have the central limit theorem.

3. THE ASYMPTOTIC INDEPENDENCE OF $I_T(x,w)$ and $I_T(x,w')$. The asymptotic independence of $I_T(x,w)$ and $I_T(x,w')$ is established by determining the joint distribution of the two variables, $\rho_T(\alpha,\alpha')$, in the same manner that $\rho_T(\alpha)$ is determined in 2. The analysis of 2 is modified as follows:

(1) $p_T(u,v)$ is replaced by $p_T(u,v,u',v')$, i.e. the joint distribution for the real and imaginary parts of the finite Fourier transform of $f(x_t)$ at the two frequencies, w and w'.

(2) $\tilde{p}_T(\xi,\eta)$ is replaced by $\tilde{p}_T(\xi,\eta,\xi',\eta')$, the Fourier transform of $p_T(u,v,u',v')$.

(3) The notation introduced in Table 1 is modified by the addition of prime superscripted c_j, s_j, S and σ to denote the replacement of w by prime superscripted w, i.e. w'. A + or - superscript on c_j and s_j denotes the

replacement of ω by $\omega + \omega'$ or $\omega-\omega'$, respectively. The same superscripts on S and σ denote their replacement by $(S\pm S')/4$ and $(\sigma\pm\sigma')/2$, respectively. The table is further altered by redefining e_j to be given by $\xi c_j-\eta s_j+\xi' c_j'-\eta' s_j'$.

(4) The analysis of 2A holds identically with the modified notation up to (21). The expression for B_j, given by (22), is altered such that

$$(33)\quad \underset{1}{\uparrow} B_1 = (\xi^2+\eta^2)\frac{ST}{4} + \omega^{-1}[\xi^2(s'^2\frac{S}{8} + s^2\frac{\sigma}{2}) - \xi\eta(s^2\frac{S}{2}-s'^2\frac{\sigma}{2})-\eta^2(s'^2\frac{S}{8}+s^2\frac{\sigma}{2})]$$

$$+(\omega-\omega')^{-1}\{\xi\xi'[s^-S^+-(c^--1)\sigma^-] + \xi'\eta[(c^--1)S + s^-\sigma^-]-\xi\eta'[(c^--1)S^++s^-\sigma^-]$$

$$+\eta\eta'[s^-S^+-(c^--1)\sigma^-]\} + (\omega+\omega')^{-1}\{\xi\xi'[s^+S^+-(c^+-1)\sigma^+] + \xi'\eta[(c^+-1)S^++s^+\sigma^+]$$

$$+\xi\eta'[(c^+-1)S^++s^+\sigma^+]+\eta\eta'[-s^+S^++(c^+-1)\sigma^+]\} + (\xi'^2+\eta'^2)\frac{ST}{4}$$

$$+ \omega'^{-1}[\xi'^2(s'^{2\prime}\frac{S'}{8} +s^{2\prime}\frac{\sigma'}{2}) - \xi'\eta'(s^{2\prime}\frac{S'}{2} - s'^{2\prime}\frac{\sigma'}{2})-\eta'^2(s'^2\frac{S'}{8} + s^{2\prime}\frac{\sigma'}{2})]$$

If none of ω,ω', $\omega+\omega'$ and $\omega-\omega'$ are zero, then (27) holds, for $n<T^{1/2}$, with $\zeta^2=\xi^2+\eta^2$ replaced by $\zeta^2+\zeta'^2 = \xi^2+\eta^2+\xi'^2+\eta'^2$.

(5) The analyses of 2B and Appendix A proceed unchanged, with the above replacement of ζ.

(6) Define $p_{T,\delta}(\alpha,\alpha')$ as the smoothed distribution by convoluting independently in α and α' with ν_δ. By expressing this distribution in terms of the inverse Fourier transform of $\tilde{p}_T(\xi,\eta,\xi',\eta')$, we decompose it into four components described by (a) $\zeta<K$ and $\zeta'<K$, (b) $\zeta<K$ and $\zeta'>K$, (c) $\zeta>K$ and $\zeta'<K$ and (d) $\zeta>K$ and $\zeta'>K$. The first component gives the asymptotic form of $\rho_{T,\delta}(\alpha,\alpha')$, as discussed in 2C:

$$(34)\quad \rho_{T,\delta}(\alpha,\alpha') \sim \exp[-\alpha/S - \alpha'/S']/(SS')$$

The other three components, ρ^b, ρ^c and ρ^d have asymptotically decreasing bounds, as follows. First note that

$$|\rho^b + \rho^c + \rho^d| < |\rho^b + \rho^d| + |\rho^c + \rho^d| + |\rho^d|$$

The first of these three terms can be written as

$$|\int_M d\lambda\ \nu_\delta(\alpha-I_T)\int_o^\infty dy\ y\ \frac{1}{2\pi}\int_o^{2\pi} d\theta\nu_\delta\{\alpha'-[(y\cos\theta + r')^2 + (y\sin\theta + j')^2]\}$$

$$\cdot\int_K^\infty d\zeta\zeta J_o(\zeta y)|$$

Because of (B3), and from the results of Appendix B, this term is bound by $O\{(\ell nT)^{-1/28}[\ell n(\ell nT)]^{7/2}\}$, if $\delta = K^{-1/28} = O[(\ell nT)^{-1/56}]$. (Note that this choice for δ differs from that of Appendix B.) The second term is similarily bounded, while the third is of higher order. This follows because the third term can be written as

$|\int_M d\lambda\, \nu_\delta^{>}.\nu_\delta^{>\prime}|$, where

$\nu_\delta^{>} = \int_0^\infty dy\, y \frac{1}{2\pi}\int_0^{2\pi} d\theta \nu_\delta\{\alpha - [(y\cos\theta + r)^2 + (y\sin\theta + j)^2]\}$

$\cdot \int_K^\infty d\zeta \zeta J_0(\zeta y)$

Since the analysis of Appendix B does not rely upon the λ integration for destructive interference, this third term is clearly bounded by $O\{(\ell nT)^{-3/56}[\ell n(\ell nT)]^7\}$ (when $\delta = K^{-1/28}$).

Thus, (34) shows that the asymptotic distribution factors into two parts depending only on α and α', respectively. This proves the asymptotic independence of $I_T(x,\omega)$ and $I_T(x,\omega')$. Consider point (4) above, which required that $\omega \neq \omega'$. This condition is elaborated for finite time, by $|\omega - \omega'| > T^{-1/2}$. (Note that $|\omega - \omega'| > T^{-\epsilon}$, for any $0<\epsilon<1$, would serve equally well, so long as $\beta < \epsilon/D^2$ - cf 2A point (ii) and (32).) Thus, the independence is seen to be limited to frequencies separated by an interval of width decreasing with T. This is to be expected, as $I_T(x,\omega)$ is clearly a continuous function of ω.

4. THE ASYMPTOTIC INDEPENDENCE OF $I_T(x,\omega)$ and $I_T{}'(x,\omega)$. Because of the independence of $\{f(x_t)\}_{t=0}^{T}$ and $\{f(x_t)\}_{t=T_1}^{T'}$,

where $T' > T_1 = T+\tau$, it is clear that $I_T(x,\omega)$ and $I_{T'-T_1}(x_{T_1},\omega)$ are independent. Thus, if (as is shown below) $|I_T{}'(x,\omega) - I_{T'-T_1}(x_{T_1},\omega)|$ approaches zero asymptotically, with probability approaching unit, then $I_T(x,\omega)$ and $I_T{}'(x,\omega)$ are asymptotically independent.

Now,

$$(35)\quad |I_T{}'(x,\omega) - I_{T'-T_1}(x_{T_1},\omega)| < \frac{T_1}{T'}\ I_{T'-T_1}(x_{T_1},\omega)$$
$$+ \frac{T_1}{T'} I_{T_1}(x,\omega) + 2\left[T_1\frac{(T'-T_1)}{T'}\right]^{1/2} I_{T_1}^{1/2}(x,\omega) I_{T'-T_1}^{1/2}(x_{T_1},\omega)$$

Choosing $T' = T^{\upsilon}$, $\upsilon>1$, and noting (A6), this bound on the difference between the two power spectra is itself bound by $O\{T^{1-\upsilon/2}[\ell n(\ell nT)]^2\}$, with probability $1 - O[(\ell nT)^{-1}]$. Thus, $I_{T'}(x,\omega)$ is asymptotically independent of $I_T(x,\omega)$ if $T' > T^{\upsilon}$, where $\upsilon>2$.

Since $I_T(x,\omega)$ is asymptotically Poisson distributed, the following behavior is evident. If T is allowed to increase indefinitely, then for λ-almost all x, $I_T(x,\omega)$ is arbitrarily large or small, infinitely often.

5. THE CLASS OF DYNAMIAL SYSTEMS AND VARIABLES. This section considers the possibility of extending the result of section 2 to a more general class of dynamical variables and sampling densities. The approach adopted makes use of that above result via an approximation argument.

Consider a class of dynamical systems which admit a sequence of sub-σ-algebras, increasing to B (the full σ-algebra of μ-measureable sets), such that each member has a finite range of dependence. That is, $\exists\{A^\tau\}_\tau$, such that $A^\tau \uparrow B$ and $A^\tau_{t+\tau} \perp A^\tau$, $\forall$ t>0, where the sequence corresponds to an unbounded, increasing sequence of τ's. (A_t denotes evolution of A through time, $A \perp A'$ implies $\mu(A \cap A') = \mu(A)\mu(A')$, $\forall$ A$\in A$ and A'$\in A'$.) Known systems of this kind include the Bernoulli automorphism[14] and the infinite entropy Bernoulli flow.[15] The present author conjectures that the finite entropy Bernoulli flow[14], the model of chaos which we would like to adopt, belongs to this class. If this is the case, then meaningful physical systems as well as paradigm chaotic systems, such as the hard sphere gas in a box, geodesic flow on a surface with negative curvature, and billiards in a concave enclosure, can be shown (as we will see) to have the asymptotic statistical power spectrum properties described above.

Corresponding to the sequence of σ-algebras are sequences of finite range of dependence approximations to f(x) and ℓ(x), $\{f^\tau(x)\}$ and $\{\ell^\tau(x)\}$, respectively. These are given by the conditional expectations with respect to A^τ, i.e.

$$f^\tau(x) = \int_{\pi^\tau(x)} d\mu(x')f(x'),$$

where $\pi^\tau(x)$ is the smallest element of A^τ containing x. The increasing Martingale theorem[16] provides the following convergence property of the sequences:

$$\lim_{\tau\to\infty} f^\tau(x) = f(x) \qquad \mu \text{ a.e.},$$

$\forall$ f $\in L^1(M,\mu)$. Since f^τ and ℓ^τ satisfy (5), the results of 2, 3 and 4 apply if they replace f and ℓ. Thus, these approximations provide the means of extending the results.

A. THE FIRST MOMENT OF THE POWER SPECTRUM DISTRIBUTION AND THE CONVERGENCE OF THE SPECTRAL DENSITY. The asymptotic statistical properties of the power spectrum depend upon the existence of the spectral density and the convergence to it (in T) of the first moment of the power spectrum distribution. These requirements are met by a sufficiently rapid decay of the 3rd order correlation,

$\langle f_{t_1} f_{t_1+t_2} \rangle_\lambda = \langle \ell f_{t_1} f_{t_1+t_2} \rangle_\mu$, in both t_1 and t_2. For example, set

$$\langle \ell f_{t_1} f_{t_1+t_2} \rangle_\mu = \langle f f_{t_2} \rangle_\mu + \Gamma(t_1, t_2)$$

The following bounds are sufficient:

$$|\langle f f_{t_2} \rangle_\mu| < \Lambda \, \|f\|^2 \phi(t_2)$$

and $|\Gamma(t_1,t_2)| < \kappa \|\ell^2\|^{1/2} \|f^2\| \; \gamma(t_1,t_2)$, $t_j>0$, $j=1$ or 2,

where $\|f\|$ is the $L^2(M,\mu)$ norm of f, $\phi(0) = \gamma(0,0) = 1$ and both ϕ and γ are monotonically decreasing functions of their arguments such that $\phi(t_2) < t_2^{-(1+\epsilon)}$ and $\gamma(t_1,t_2) < t_1^{-(1+\epsilon_1)} t_2^{-(1+\epsilon_2)}$.

One further condition must be satisfied before the result of section 2 can be extended. The spectral densities, $S^\tau(\omega)$, associated with the approximations, f^τ, must converge to $S(\omega)$, as $\tau\to\infty$. This requirement can be achieved if the set of approximating dynamical variables, f^τ, together with f, all belong to a class of functions, F, satisfying the following strong mixing condition:

$$\forall g, g' \in F, \quad |\langle g g'_t \rangle_\mu| < \Lambda \|g\| \; \|g'\| \; \phi(|t|)$$

With this assumption, the strong $L^2(M,\mu)$ convergence of f^τ to f (which follows from the μ.a.e. convergence of $f^\tau(x)$ and the compactness of M) implies the convergence of the spectral densities:

$$|S(\omega) - S^\tau(\omega)| = |2\mathrm{Re} \int_{-\infty}^{\infty} dt \; e^{i\omega t} \langle [f - f^\tau] f^\tau_t \rangle_\mu$$
$$+ \int_{-\infty}^{\infty} dt \; e^{i\omega t} \langle [f-f^\tau][f-f^\tau] \rangle_\mu |$$
$$< 2\Lambda(2\|f^\tau\| + \|f-f^\tau\|) \|f-f^\tau\| [1+1/\epsilon] \to 0, \text{ as } \tau\to\infty.$$

B. THE CONVERGENCE OF THE POWER SPECTRUM. If $I^\tau_T(x,\omega)$ is the power spectrum associated with the approximation, f^τ, then

$$(37) \quad |I_T(x,\omega) - I^\tau_T(x,\omega)| = |2\mathrm{Re}[h^\tau_T(x,\omega) \tilde{h}^\tau_T(x,\omega)]$$
$$+ |\tilde{h}^\tau_T(x,\omega)|^2 | < [2|h^\tau_T(x,\omega)| + |\tilde{h}^\tau_T(x,\omega)|] \, |\tilde{h}^\tau_T(x,\omega)| ,$$

where $h^\tau_T(x,\omega) = \dfrac{1}{T^{1/2}} \int_0^T dt e^{i\omega t} f^\tau(x_t)$

and $\tilde{h}^\tau_T(x,\omega) = \dfrac{1}{T^{1/2}} \int_0^T dt \; e^{i\omega t} [f(x_t) - f^\tau(x_t)]$.

We define $g = |\tilde{h}^\tau_T(x,\omega)|$ as a random variable with respect to μ and apply Chebyshev's inequality to obtain $\langle g^2 \rangle_\mu \geq \beta^2 \; \mu\{g>\beta\}$

Setting $\beta = \langle g^2\rangle_\mu^{1/4}$, this becomes

(38) $\langle g^2\rangle_\mu^{1/2} \geqq \mu\{g\rangle \langle g^2\rangle_\mu^{1/4}\}$

Since $\langle g^2\rangle_\mu \sim \tilde{S}^\tau$, the spectral density associated with $f-f^\tau$, as $T\to\infty$, and $\tilde{S}^\tau < 2\Lambda\|f-f^\tau\|^2[1+1/\epsilon]$, the μ-measure of the set on which g is not smaller than a decreasing (to zero) function of τ is itself decreasing to zero, as $\tau\to\infty$. Furthermore, since $I_T^\tau(x,\omega)$ is asymptotically Poisson distributed, the μ-measure associated with increasingly large deviations of $h_T^\tau(x,\omega)$ is decreasing to zero. Thus, for each τ value, M can be partitioned into two sets; $M = M_> \cup M_<$. On $M_<$, $|I_T(x,\omega)-I_T^\tau(x,\omega)|$ is bounded by a function of τ, which decreases to zero, while the μ-measure of $M_>$ also decreases to zero, as $\tau\to\infty$.

C. THE CONVERGENCE OF THE DISTRIBUTION. Finally, consider the smoothed distribution of $I_T(x,\omega)$, with respect to λ,

(39) $$\begin{aligned}\rho_{T,\delta}(\alpha) &= \int_M d\mu(x)\ell(x)\nu_\delta[\alpha-I_T(x,\omega)]\\ &= \int_M d\mu(x)\ell^\tau(x)\nu_\delta[\alpha-I_T^\tau(x,\omega)] + \int_M d\mu(x)\ell^\tau(x)\{\nu_\delta[\alpha-I_T(x,\omega)]\\ &\quad - \nu_\delta[\alpha-I_T^\tau(x,\omega)]\} + \int_M d\mu(x)[\ell(x)-\ell^\tau(x)]\nu_\delta[\alpha-I_T(x,\omega)]\end{aligned}$$

The first of the three terms contributing to $\rho_{T,\delta}$ is the smoothed distribution, $\rho_{T,\delta}^\tau$, shown in 2 to be asymptotically Poissonian. The third of these terms is decreasing, since both $\ell-\ell^\tau$ and ν_δ belong to $L^2(M,\mu)$ (they are bounded and M is compact) and $\ell-\ell^\tau$ converges to zero in L^2. (This requires $\delta\to 0$ sufficiently slowly as $\tau\to\infty$, since $\|\nu_\delta\|\sim\delta^{-1/2}$.) The second of the three terms splits into two contributions, according to the partitioning of M described at the end of section B, above. The contribution from $M_>$ is bounded in absolute value by $2\mu(M_>)$ L B/δ, where L is the bound of $\ell(x)$ which also bounds $\ell^\tau(x)$, since the latter is a conditional expectation. So long as δ is chosen to decrease less rapidly than $\mu(M_>)$, as $\tau\to\infty$ and $T/\tau\to\infty$, this contribution is decreasing and vanishes in the limit. The contribution of $M_<$ is bound in absolute value by

$$\int_{M_<} d\mu(x)\ell^\tau(x)\{A_1+A_2|I_T(x,\omega)-I_T^\tau(x,\omega)|/\delta\}|I_T(x,\omega)-I_T^\tau(x,\omega)|/\delta$$

So long as δ decreases less rapidly than the bound on $|I_T-I_T^\tau|$, this contribution vanishes in the limit. Therefore, if δ is chosen to decrease sufficiently slowly, $\rho_{T,\delta}(\alpha)$ is asymptotically Poissonian.

6. SUMMARY. For dynamical variables, f, and densities, ℓ, satisfying a strong multiple mixing condition, the following results have been established:

(1) The power spectrum of f, $I_T(x,\omega)$, is asymptotically Poissonian, as $T\to\infty$, with respect to random sampling according to the density, $\ell(x)$.

(2) $I_T(x,\omega)$ and $I_T(x,\omega')$ are asymptotically independent, so long as $|\omega-\omega'| > T^{-\epsilon}, 0<\epsilon<1$.

(3) $I_T(x,\omega)$ and $I_{T'}(x,\omega)$ are asymptotically independent so long as $T'>T^{\upsilon}, \upsilon>2$.

Progress has been made towards extending the first result to known models of chaotic dynamics, i.e. Bernoulli flows with dynamical variables satisfying simple strong mixing conditions.

Acknowledgements: I thank Prof. Paul Brumer for his critical reading of this manuscript. The financial support of NSERC and Gulf Canada Limited is appreciated.

APPENDIX A: THE PROBABILITY OF LARGE DEVIATIONS OF h_T AND I_T. From the results of section 2A, it is easily seen that the 1st and 2nd moments of h_T are $O(T^{-1/2})$ and $(\xi^2+\eta^2)S/2 + O(T^{-1})$, respectively. (The latter asymptotic form corresponds to the case of ω bounded away from zero.) Therefore, it is most probable for h_T to be $O(1)$, as $T\to\infty$. However, it is clearly possible for $|h_T|$ to take on values as large as $O(T^{1/2})$. For this reason, it is desirable to give some indication of the likelihood of very large deviations of h_T.

The results obtained in this appendix for h_T translate directly into similar results for I_T, the power spectrum. For example,

$$\int_M d\lambda h_T^2 = \xi^2\int_M d\lambda r_T^2 + 2\xi\eta\int_M d\lambda r_T j_T + \eta^2\int_M d\lambda j_T^2$$
$$= (\xi^2 + \eta^2)S/2 + O(T^{-1})$$

for all $(\xi,\eta)\in R^2$, implies that $\int_M d\lambda r_T^2 = \int_M d\lambda j_T^2 = S/2 + O(T^{-1})$ and

$$\int_M d\lambda r_T j_T = O(T^{-1}).$$ Therefore
$$\int_M d\lambda(x) I_T(x,\omega) = S(\omega) + O(T^{-1})$$

Similarly, setting (ξ,η) to (1,0) and (0,1), converts the large deviations results for $|h_T|$ into analogous results for $|r_T|$ and $|j_T|$, respectively. An analogous result for I_T follows because I_T is large only when $|r_T|$ or $|j_T|$ is large.

In section 2A, an analysis of the nth moment of h_T is given which relies upon the restriction on n, (16). If this restriction were dropped, the following bound on the moment would still hold:

(A1) $|\int_M d\lambda h_T^n| < n![(\xi^2+\eta^2)D^2]^{n/2}$

We let $J(\upsilon) = \int_M d\lambda e^{\upsilon h_T}$ and note that (A1) gives rise to

(A2) $$|J(\upsilon)| < \sum_{n=0}^{\infty} \tilde{\upsilon}^n = (1-\tilde{\upsilon})^{-1},$$

if $|\tilde{\upsilon}|<1$, where $\tilde{\upsilon} = \upsilon\zeta D$ and $\zeta = (\xi^2+\eta^2)^{1/2}$. Proceeding in a fashion completely analogous to the derivation of Bernstein's inequality[17] we consider the random variable, $g(x)=\exp[\upsilon h_T(x)]/J(\upsilon)$ on the probability space M, with respect to the probability measure, λ. Since $g>0$ and its expectation value is 1, the following form of Chebyshev's inequality holds:

$1 = \int_0^\infty dG(\alpha)\alpha \geq \int_\delta^\infty dG(\alpha) = \delta\lambda\{g>\delta\}$

where $G(\alpha)$ is the cumulative probability for g and $\lambda\{g>\delta\}$ is the probability that $g>\delta$. Thus,

$\lambda\{g>\delta\} < \delta^{-1}$ and $\lambda\{h_T > \upsilon^{-1}\ell n[\delta J(\upsilon)]\} < \delta^{-1}$.

Since $-h_T$ could just as well have been used, we have

(A3) $\lambda\{|h_T| > \upsilon^{-1}\ell n[\delta J(\upsilon)]\} < 2\delta^{-1}$

This formula affords an entire class of large deviation probability bounds. For example, if for some $\beta>o$

(A4) $\delta = (\ell nT)/2$, $\upsilon^{-1} = 2D(\beta\ell nT)^{1/2}$ and $\zeta < (\beta\ell nT)^{1/2}$, then

(A5) $J(\upsilon) < 2$ and $\lambda\{|h_T| > 2D(\beta\ell nT)^{1/2}\ell n[\ell nT]\} < 4(\ell nT)^{-1}$

The corresponding result for I_T is[18]

(A6) $\lambda\{I_T>8D^2[\ell n(\ell nT)]^2\} < 8(\ell nT)^{-1}$

Because the λ-measures of those subsets of M associated with large deviations are decreasing as $T\to\infty$, their contribution to the probability distribution will be neglected. Specifically, M is replaced by its subset on which both $|h_T|<2D(\beta\ell nT)^{1/2}\ell n[\ell nT]$ and $I_T<8D^2[\ell n(\ell nT)]^2$. Note that this approximation has <u>no effect</u> on $\rho_T(\alpha)$ for $\alpha<8D^2[\ell n(\ell nT)]^2$.

Appendix B: SMOOTHING AND THE SUPPRESSION OF THE FAR REACHES OF FOURIER SPACE. In this section, smoothing $\rho_T(\alpha)$ is shown to effectively suppress its contribution via $\tilde{p}_T(\xi,\eta)$, from the $\zeta = (\xi^2+\eta^2)^{1/2} > K = (\beta\ell nT)^{1/2}$ region of Fourier space. This requires the exclusion of large deviations of I_T: M is implicitly restricted such the I_T satisfies (30) (see Appendix A). α is also so restricted. Denoting the contribution of concern by $\rho^{>}_{T,\delta}(\alpha)$, we have, formally,

$$\rho^{>}_{T,\delta}(\alpha) = \frac{1}{4\pi^2} \int_M d\lambda \int_{R^2} dudv\, \nu_\delta[\alpha-(u^2+v^2)] \int_{\zeta>K} d\xi d\eta\, e^{-i[\xi(u-r_T)+\eta(v-j_T)]}$$

Dropping the subscripts on r_T and j_T and changing variables, $u \to y\cos\theta+r$ and $v \to y\sin\theta+j$, gives

$$\text{(B1)} \quad \rho^{>}_{T,\delta}(\alpha) = \frac{1}{2\pi} \int_M d\lambda \int_0^\infty dy y \int_0^{2\pi} d\Theta \nu_\delta\{\alpha-[(y\cos\Theta+r)^2+(y\sin\Theta+j)^2]\} \cdot \int_K^\infty d\zeta \zeta J_0(\zeta y),$$

where J_0 is the zeroth order Bessel function of the first kind. In the following analysis, some properties of the Hankel transform are required:

$$\text{(H1)} \quad \int_K^\infty d\zeta\zeta J_0(\zeta y) = [\delta_+(y) - KJ_1(Ky)]/y$$

$$\text{(H2)} \quad \int_0^a dy KJ_1(Ky) = 1-J_0(Ka) \sim 1 - \left(\frac{2}{\pi Ka}\right)^{1/2}\cos[Ka-\pi/4]$$

$$\text{(H3)} \quad \int_0^a dy y KJ_1(Ky) = \frac{\pi a}{2}[J_1(Ka)H_0(Ka) - H_1(Ka)J_0(Ka)] \sim - \left(\frac{2a}{\pi K}\right)^{1/2}\cos[Ka-\pi/4] \text{ and}$$

$$\text{(H4)} \quad \int_0^a dy y^2 KJ_1(Ky) = a^2J_2(Ka) \sim - \left(\frac{2a^3}{\pi K}\right)^{1/2}\cos[Ka-\pi/4]$$

The first formula combines the closure property of the Bessel functions with a formula found in ref. 19. The other formulae are found in ref. 19 and 20. (H_0 and H_1 are Struve functions.) It will be necessary to use the large argument asymptotic forms given here as though they were upper bounds. This is valid for sufficiently large T if the coefficients are increased slightly. (This is not carried out explicitly, as it has no effect on the final result.)

Substituting (H1) into (B1) yields

$$\text{(B2)} \quad p^{>}_{T,\delta}(\alpha) = \int_M d\lambda \{\nu_\delta[\alpha-(r^2+j^2)] - \frac{1}{2\pi}\int_0^{2\pi} d\Theta \int_0^\infty dy \nu_\delta\{\alpha-[(y\cos\Theta+r)^2+(y\sin\Theta+j)^2]\} \cdot KJ_1(Ky)\}$$

We now wish to utilize the smoothness properties of ν_δ, which are given as follows:

$$\text{(B3)} \quad \nu_\delta(\alpha) < B/\delta, \quad \alpha \in R$$

$$\text{(B4)} \quad |[\nu_\delta(\alpha')-\nu_\delta(\alpha)](\alpha'-\alpha) - \tilde{A}_1(\alpha)/\delta| < A_2|\alpha'-\alpha|/\delta^2,$$

$$|\tilde{A}_1(\alpha)| < A_1, \quad \alpha,\alpha' \in R.$$

Of course, as discussed in section 1, there is the additional restriction,

(B5) $\nu_\delta(\alpha) = 0, \ |\alpha|>\delta$

In order to apply these inequalitites it is necessary to partition the range of y integration:

(B6) $[0,\infty) = \bigcup_{k=0}^{N-1} [k^2a, (k+1)^2a) \bigcup [N^2a,\infty),$

where N is chosen such that

(B7) $9D\ell n(\ell nT) < N^2a < 16D\ell n(\ell nT),$

and $a = K^{-5/7}$. Out of all the y intervals, three cases are distinguished:

(i) In the k=0 interval, y<a, so that (B4) gives

(B8) $\nu_\delta\{\alpha-[(y\cos\Theta+r)^2+(y\sin\Theta+j)^2]\} = \nu_\delta[\alpha-(r^2+j^2)]$

$\quad + \delta^{-1}\tilde{A}_1 y[y+2(r\cos\Theta+j\sin\Theta)] + \delta_\nu$

where $|\delta_\nu| < 81A_2\delta^{-2}a^2[D\ell n(\ell nT)]^2 \equiv 81A_2\delta^{-2}a^2J^2$. (Use was made of (30) and the fact that a is small.) Grouping the contribution to $\rho^>_{T,\delta}(\alpha)$ from this interval with the first term in the braces of (B2) and making use of (B8) gives

(B9) $\int_M d\lambda\nu_\delta[\alpha-(r^2+j^2)]\{1-\int_0^a dyKJ_1(Ky)\} - \delta^{-1}\int_M d\lambda\tilde{A}_1\int_0^a dyKJ_1(Ky).$

$\quad \cdot\frac{1}{2\pi}\int_0^{2\pi} d\Theta[y^2+2y(r\cos\Theta+j\sin\Theta)] - \int_M d\lambda\int_0^a dyKJ_1(Ky).\ \frac{1}{2\pi}\int^{2\pi} d\Theta\ \delta_\nu$

Using (H2) and (B3) for the first term, (H4) for the second term and $|J_1(Ky)|<1$ for the third term, these terms are, respectively, b.i.a.v. (bound in absolute value) by

(B10) $(\frac{2}{\pi})^{1/2}B\ \delta^{-1}K^{-1/7}$

(B11) $(\frac{2}{\pi})^{1/2}A_1\delta^{-1}K^{-11/7}$ and

(B12) $81A_2\delta^{-2}K^{-8/7}J^2$

(ii) In the kth (k≠0) interval, (B4) yields (B13) $\nu_\delta\{\alpha-[(y\cos\Theta+r)^2+(y\sin\Theta+j)^2]\} = \nu_\delta\{\alpha-[(k^2a\cos\Theta+r)^2+(k^2a\sin\Theta+j)^2]\}$

$\quad +\delta^{-1}\tilde{A}_1.(y-k^2a)[(y+k^2a)+2(r\cos\Theta+j\sin\Theta)] + \delta_\nu,$

where $|\delta_\nu|<28596A_2(aJk/\delta)^2$. (Use was made of (30) and (B7).) The contribution of the kth interval to $\rho^>_{T,\delta}(\alpha)$ is given by

(B14) $-\int_M d\lambda \frac{1}{2\pi}\int_0^{2\pi} d\Theta \nu_\delta\{\alpha-[(k^2 a\cos\Theta+r)^2+(k^2 a\sin\Theta+j)^2]\}(\frac{2}{\pi K a})^{1/2}\{\frac{\cos[Kk^2a-\pi/4]}{k}$

$- \frac{\cos[K(k+1)^2a-\pi/4]}{(k+1)}\}$

$- \delta^{-1}\int_M d\lambda \frac{1}{2\pi}\int^{2\pi} d\Theta \tilde{A}_1 \int_{k^2a}^{(k+1)^2a} dy[(y^2-k^4a^2)+2(y-k^2a)(r\cos\Theta+j\sin\Theta)]KJ_1(Ky)$

$- \int_M d\lambda \int_{k^2a}^{(k+1)^2a} dy KJ_1(Ky)\frac{1}{2\pi}\int^{2\pi} d\Theta\ \delta_\nu,$

where (H2) has been applied to the first term, which is b.i.a.v. by

(B15) $2(\frac{2}{\pi})^{1/2} B\ \delta^{-1} K^{-1/7}/k$

The second term requires more effort, but can be shown to be b.i.a.v. by (for the details see ref. 21)

(B16) $15(\frac{2}{\pi})^{1/2} A_1 \delta^{-1} K^{-11/7} k^2 + 16(\frac{2}{\pi})^{1/2} A_1 \delta^{-1} K^{-6/7} J$

Finally, the last of the three terms of (B14) is b.i.a.v. by (using the asymptotic form of $J_1(Ky)$)

(B17) $28596 A_2 \delta^{-2} a^2 K^{1/2} J^2 k^2 \int_{k^2a}^{(k+1)^2a} dy/y^{1/2} < 57192 A_2 \delta^{-2} K^{-9/7} J^2 k^2$

To obtain a bound on the total of the contributions from k=1 to N-1, the bounds given by (B15), (B16) and (B17) must be summed over k. Note that from (B7), $N < 4K^{5/14} J^{1/2}$. Furthermore,

$$\sum_{k=1}^{N-1} k^{-1} < 1+\ell n N$$

and $\sum_{k=1}^{N-1} k^2 < N^3/3$

The sums of (B15), (B16) and (B17) are bounded (for T sufficiently large) by

(B18) $\frac{6}{7}(\frac{2}{\pi})^{1/2}\ \frac{B}{D}\delta^{-1} K^{-1/7} J,$

(B19) $240(\frac{2}{\pi})^{1/2} A_1 \delta^{-1} K^{-1/2} J^{3/2} + 64(\frac{2}{\pi})^{1/2} A_1 \delta^{-1} K^{-1/2} J^{3/2}$ and

(B20) $915072 A_2 \delta^{-2} K^{-3/14} J^{7/2}$

(iii) The contribution to $\rho^{>}_{T,\delta}(\alpha)$ from $y>9D\ell n(\ell n T)$ is zero. This results because of (B5) and the fact that, from (30),

$$|\alpha-[(y\cos\Theta+r)^2+(y\sin\Theta+j)^2]| > J^2 > \delta.$$

We conclude by choosing $\delta = K^{-1/14}$ and noting that in this case the total of all contributions to $\rho^{>}_{T,\delta}(\alpha)$ is bounded by $O(K^{-1/14}J^{7/2}) = O((\ell nT)^{-1/28}[\ell n(\ell nT)]^{7/2})$. Thus, smoothing $\rho_T(\alpha)$ with asymptotically increasing resolution ($\delta = O[(\ell nT)^{-1/28}]$) is sufficient to ensure its convergence to the Poisson distribution.

REFERENCES AND FOOTNOTES

1. See D.W. Noid, M.L. Koszykowski and R.A. Marcus, J. Chem. Phys.,67 (1977), 404-408; ref. 10 and the references therein for applications to conservative models for bound molecular motion; also, see J. Guckenheimer and P. Holmes, Nonlinear Oscillations, Dynamical Systems and Bifurcations of Vector Fields, Springer-Verlag, N.Y., 1983; J.P. Eckmann and D. Ruelle, Rev. Mod. Phys. 57 (1985), 617-656 and the references therein for applications to dissipative systems.

2. The peaks of a quasiperiodic spectrum persist with increasing T, their heights increasing, O(T), and widths decreasing, O(1/T).

3. The peaks of a chaotic spectrum generally do not persist with increasing T. They eventually break up into two or more subpeaks. Since this occurs on all large times scales, the chaotic spectrum is increasingly grassy with increasing T. At any given time, however, the peaks are indistinguishable from those of an incompletely resolved quasiperiodic spectrum. For the details of both the regular and chaotic power spectrum see ref. 21 and G.E. Powell and I.C. Percival J. Phys. A12 (1979), 2053-2071.

4. W. Feller, An Introduction to Probability Theory and its Applications, Vol. II, Wiley, N.Y., 1966.

5. V.V. Nemytiski and V.V. Stepanov, Qualitative Theory of Differential Equations, Princeton Univ. Press, Princeton, 1960.

6. See, for example, D. Ruelle, Phys. Rev. Lett. 56 (1986), 405-407.

7. This terminology derives from renewal theory where the standard example, the Poisson process, has a simple exponential p.d.f. (probability distribution function). See for example D.R. Cox and H.D. Miller, The Theory of Stochastic Processes, Methuen, London, 1965.

8. D.R. Brillinger, Time Series Data Analysis and Theory, expanded ed., Holden-Day, San Francisco, 1981; E.J. Hannan, Multiple Time Series, Wiley, N.Y., 1970 [see also, E.J. Hannan, Stochastic Proces. Appl. 9 (1979), 281-289]. Hannan assumes a generalized linear process. This requirement is generally too restrictive for applictions to chaotic

dynamical systems. Brillinger's multiple mixing assumption is given in terms of cumulants. It seems sufficiently general to admit application to Bernoulli automorphisms. The discrete time analogue of the corresponding assumption of the present paper is encompassed by it, in the case of $\lambda = \mu$.

9. M. Ratner, Israel J. of Math. 16 (1973), 181-198.

10. R.S. Dumont and P. Brumer (to be published).

11. P.E. Pfeiffer, Concepts of Probability Theory, pg. 258, Dover, N.Y., 1978.

12. The joint distribution of r_T and j_T is thus expressed in terms of the joint moments by inverse Fourier transformation of (13) and expansion of h_T^n.

13. For example, consider the case of the much weaker bound, $|h_T| < T^{1/4}$, which follows with probability, $1-0[\exp(-T^{1/2}/2)]$ (see Appendix A, (A3) - set $\delta = T^{1/2}/2$, $\upsilon = T^{-1/2}$ and $\varsigma < T^{1/2}/2D$):

$$\left| \sum_{n=N}^{\infty} \frac{i^n}{n!} \int_M d\lambda h_T^n \right| < \sum_{n=N}^{\infty} \frac{T^{n/4}}{n!} < \sum_{n=N}^{\infty} [eT^{-1/4}]^n < 2[eT^{-1/4}]^{T^{1/2}}$$

$= 0([T^{1/4}!]^{-T^{1/4}})$, where $N = \lfloor T^{1/2} \rfloor + 1$ and use is made of $n! > n^n e^{-n}$ and $n! = 0(n^n e^{-n})$, as $n \to \infty$.

14. See D. Ornstein, in Studies in Probability Theory, MAA Studies in Mathematics, Vol. 18, ed. M. Rosenblatt, MAA, 1978 and the references therein.

15. For example, the Bernoulli automorphism possesses an independent generating partition (see ref. 14). Letting $P^{[0]}$ be such a partition, the joining of $P^{[0]}$ with its iterates from time $-\tau/2$ to $\tau/2$, provides the τth member of the sequence of σ-algebras (generated by the partitions). (The joining of partitions is the partition containing every intersection of elements from the constituent partitions.)

16. W. Parry, Topics in Ergodic Theory, pg. 30, Cambridge Univ. Press, Cambridge, 1981.

17. See, I.A. Ibragimov and Yu. V. Linnik, Independent and Stationary Sequences of Random Variables, pg. 169, Walters-Noordhoff, Groninger, 1971.

18. In the case of I_T, since only the $(\xi,\eta) = (1,0)$ and $(0,1)$ instances of (A5) are required, the choices for δ and υ are modified as follows: $\delta = (\ell nT)/2$ and $\upsilon = 1/(2D)$. The result, (A6), obtained with these choices is required in Appendix B.

19. I.H. Sneddon, The Use of Integral Transforms, pg. 523, McGraw-Hill, N.Y., 1972.

20. I.S. Gradsteyn and I.M. Ryzhik, Table of Integrals, Series and Products, corrected and enlarged ed., pgs. 666, 683, 961 and 983, Academic Press, N.Y., 1980.

21. R.S. Dumont, Ph.D. Dissertation, University of Toronto, 1986.

CHEMICAL PHYSICS THEORY GROUP
DEPARTMENT OF CHEMISTRY
UNIVERSITY OF TORONTO
TORONTO, ONTARIO. M5S 1A1
Current Address:
DEPARTMENT OF CHEMISTRY
COLUMBIA UNIVERSITY
NEW YORK, N.Y. 10027
U.S.A.

Canadian Mathematical Society
Conference Proceedings
Volume 8 (1987)

DEGENERATE HOPF BIFURCATIONS IN THE CSTR WITH REACTIONS A→B→C

W. W. Farr and R. Aris

Abstract. In this paper we use the work of Golubitsky and Langford (1981) and Golubitsky and Schaeffer (1985) to investigate the local dynamics of a three dimensional model describing a continuous stirred tank reactor in which consecutive reactions A→B→C occur. This system can have up to 5 steady states and 4 Hopf points. We describe local codimension 2 and 3 degenerate bifurcations involving periodic branch limit points as well as up to three Hopf points. We also discuss a codimension 3 family with one modal parameter which involves transcritical type bifurcations of periodic solution branches. Also reported (but not analyzed) are degenerate dynamic singularites of the nilpotent double zero eigenvalue (Takens-Bogdanov) type.

§1. Introduction

In this paper we use the work of Golubitsky and Langford (1981) and Golubitsky and Schaeffer (1985) on degenerate Hopf bifurcations to analyze the local dynamics of a three dimensional ODE model, which describes the behavior of a continuous stirred tank reactor (CSTR) in which two consecutive, exothermic reactions occur. This model has been around in the chemical engineering literature for some time and exhibits behavior ranging from steady state multiplicity (Balakotaiah and Luss, 1982) to periodic solutions (Halbe and Poore, 1981) to chaos (Jorgensen and Aris, 1983), but large regions of parameter space are still essentially untouched.

Several parameters appear in this model; we will choose one as the distinguished, or bifurcation, parameter and regard the others as free parameters. Our results are often presented in terms of bifurcation diagrams, i.e. plots of a state variable (or other measure of the solution) versus the distinguished parameter, but we want to take a qualitative approach to the dynamics, in the spirit of singularity theory, so our aim is to find conditions leading to qualitative changes in the bifurcation diagrams.

In terms of the steady states of our model, the concept of qualitatively different digrams is well established in terms of the transition sets first defined in Golubitsky and Schaeffer (1979). The dynamic situation, unfortunately, is not so simple and we are forced to admit that our results are far from complete. We set ourselves the lesser task of investigating dynamics locally near Hopf points, by implementing numerically the formulas presented in Golubitsky and Langford (1981). We might point out that we have rederived those formulas without, as they did, scaling the imaginary eigenvalues to $\pm i$ as well as extending their treatment to higher order. The details are in Farr (1986).

Using these methods we will be able to predict a considerable variety of interesting behavior in our model, but it seems appropriate at this time to describe some of the phenomena we are unable investigate. Our methods are local, and apply only near the Hopf point so we have no hope of predicting secondary bifurcations which do not interact with the Hopf points. We also make no serious effort to analyze the F_1 (in the notation of Arnold, 1983), or nilpotent double-zero eigenvalue bifurcations, which occur in our model although we do locate them and make some conjectures.

1980 MATHEMATICS SUBJECT CLASSIFICATION (1985 REVISION) 80A32 (Secondary 58F14)

§2. Theoretical Background

The details of the singularity theory aproach to degenerate Hopf bifurcation are contained in the 1981 paper of Golubitsky and Langford and the the recent (1985) book of Golubitsky and Schaeffer. We are constrained here by space limitations to giving an heuristic description of the main results. Interested readers should consult the book and paper for details. We will provide enough description to make sense of the results, but will not do justice to the underlying theory.

We start by stating the Hopf bifurcation theorem in the context of ordinary differential equations. We consider n-dimensional vector fields which depend smoothly on the state variables as well as on a vector of free parameters. One of these parameters is chosen to be the distinguished bifurcation parameter, denoted by λ. In what follows we will often supress the dependence on the other parameters for convenience.

Hopf theorem

suppose $\dot{u} = f(u,\lambda)$ with $f(0,0)=0$

If

1. the Jacobian of f near (0,0) has simple eigenvalues $\sigma(\lambda) \pm i\omega(\lambda)$ with $\sigma(0) = 0$, $\omega(0) \neq 0$ and no other eigenvalues on the imaginary axis at (0,0)
2. $\sigma'(0) \neq 0$
3. $\mu_2 \neq 0$, where μ_2 is the stability coefficient defined below

Then

There exists a unique branch of periodic solutions emanating from (0,0).

The stability and direction of the branch follow from the amplitude expansion

$\lambda = \mu_2 x^2$

and the exchange of stability principle.

We will call a Hopf point a degenerate Hopf point if (1) is satisfied, but one (or both) of (2), (3) is not satisfied. (The cases of interest for our system where (1) is not satisfied produce the F_1 bifurcation.) The achievement of Golubitsky and Langford was to show how these degeneracies could be analyzed by using the Liapounov-Schmidt reduction and symmetry to reduce the bifurcation problem to a single implicitly defined equation. Singularity theory is then used to analyze the solutions of this equation near a degeneracy resulting in unfoldings that can give qualitative information about the behavior of the original system. Below we will examine the normal forms and unfoldings of several of the important bifurcations as examples to demonstrate this process clearly.

The resulting function describing Hopf bifurcation is shown by G&L to have the $\mathbb{Z}_2$ equivariant form

(1) $a(x^2,\lambda)\, x = 0$, with $a(0,0)=0$

where x=0 for the steady state, and solutions of a=0 for $x \neq 0$ represent periodic solutions of the original system. Singularity theory is then used to investigate these non-trivial solutions of eq. (1), but the form of the equation enables one to use a special formulation that takes advantage of the dependence of a on x^2. This special formulation is described in Golubitsky and Schaeffer (1985) and we will not give the full details but will only draw from the results given therein.

In describing the solutions of eq. (1) it is convenient to define $z = x^2$ and treat $a(x^2,\lambda)$ as a function $a(z,\lambda)$. This substitution also simplifies analysis of the normal forms and their unfoldings. It is very important for mathematical reasons too, as discussed by G&S in their book, but we lack the background at this point to address them. We will use the variable z in defining the transition sets; however it must be kept in mind that z must be greater than or equal to zero to make sense. We will still use the variable x in normal forms and their unfoldings, since the amplitude of the periodic solution is proportional to it.

To return to our analysis of degenerate Hopf bifurcation, the non-degeneracy conditions (2) and (3) of the Hopf theorem were shown by G&L to be equivalent to $a_\lambda \neq 0$ and $a_z \neq 0$, respectively, with

(2) $\sigma'(0) = -a_\lambda$, and $\mu_2 = -a_z/a_\lambda$

but their treatment allows one to qualitatively predict the effects of degeneracies, limited only by one's ability to obtain derivatives of the function a. Even determining the first derivatives is difficult to do analytically, but they can be evaluated numerically without great effort. Before describing some of the degeneracies, we introduce a useful notation for derivatives of a by

(3) $a_{ij} = \partial^{i+j} a/\partial^i z \partial^j \lambda$

in order to motivate the short-hand notation introduced below for degenerate Hopf bifurcations.

For readers not familiar with the book by Golubitsky and Schaeffer (1985), we feel it is necessary to include some of the normal forms and unfoldings they have obtained for bifurcation problems with $\mathbb{Z}_2$ symmetry. Unfortunately, we need quite a few of them but our space is limited, so we will not be able to describe them in detail. The interested reader should consult the book for further details.

The analysis of degenerate Hopf bifurcations via singularity theory proceeds in three steps. In the first step a degenerate Hopf bifurcation is identified with a simple polynomial, called a normal form. This identification is in terms of equalities, called defining conditions, and inequalities, called non-degeneracy conditions, both in terms of derivatives of the function $a(z,\lambda)$ we defined above. The theory then provides the minimum number of parameters, which enter linearly, required in the universal unfoldings of these normal forms.

The second step is the analysis of the unfolded normal form to determine all qualitatively different bifurcation diagrams which occur near the singularity. Golubitsky and Schaeffer (1985) show how to systematically determine all the diagrams in terms of the transition set, which is the union of the five varieties defined by

(4) $\mathbf{B}_0 = \{p\text{: there exists } \lambda \text{ such that } a = a_\lambda = 0 \text{ at } (0,\lambda;p)\}$.

(5) $\mathbf{H}_0 = \{p\text{: there exists } \lambda \text{ such that } a = a_z = 0 \text{ at } (0,\lambda;p)\}$.

(6) $\mathbf{H}_1 = \{p\text{: there exist } z,\lambda \text{ with } z > 0 \text{ such that } a = a_z = a_{zz} = 0 \text{ at } (z,\lambda:p)\}$

(7) $\mathbf{D} = \{p\text{: there exist } \lambda,z_1,z_2 \text{ with } z_1 \neq z_2 \text{ and } z_1,z_2 \geq 0 \text{ such that}$

$a = za_z = 0 \text{ at } (z_1,\lambda;p) \text{ and } (z_2,\lambda;p)\}$.

(8) $\mathbf{B}_1 = \{p\text{: there exist } z, \lambda \text{ with } z>0 \text{ such that } a = a_z = a_\lambda = 0 \text{ at } (z,\lambda;p)\}$

In each case p refers to the vector of free parameters. We will see examples of each of these varieties below. By utilizing these sets the unfolded normal forms become relatively easy to analyze due to their polynomial nature, but they are extremely useful in understanding and predicting the local behavior of the original problem. These transition varieties divide up the parameter space of the unfolded normal forms into regions corresponding to all of the local diagrams, but they do not do quite the same for the periodic branches because other bifurcations not accounted for by the theory may appear.

The third step is to identify the parameters of the real problem with those that appear in the unfoldings. This is straightforward to do locally, but it would be better to calculate the transition set numerically for the real problem. Two of the five varieties are calculated at the Hopf point (i.e. at z=0), and we do so below, but the other three present difficult numerical problems which have not been addressed yet.

The singularities we need to describe in this section can be organized into three families. The first family is concerned with degeneracies in the second hypothesis of the Hopf theorem. We therefore require $a_{10} \neq 0$, but consider degeneracies in the strict crossing or transversality hypothesis. The members of this family can be ordered by codimension as follows. For convenience we call the codimension m member of this family an H_{0m} singularity. The defining conditions are

(9) $a_{01} = a_{02} = .. = a_{0m} = 0$

while the non-degeneracy conditions are

(10) $a_{10} \neq 0,\ a_{0m+1} \neq 0.$

We will need only the first two members of this family; their normal forms and unfoldings are given in Table 1, and the transition varieties and local diagrams are given in Fig. 1. We can describe this family qualitatively by saying that the unfolding of an H_{0m} singularity involves interactions between m+1 Hopf bifurcation points.

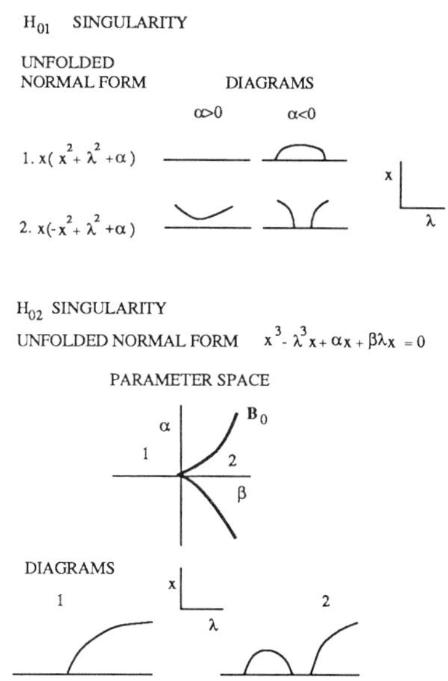

Fig. 1. Unfolded normal forms, transition sets, and diagrams for the H_{01} and H_{02} singularities.

<u>Table 1</u>

Normal forms and unfoldings for H_{01} and H_{02} singularities

A. H_{01} singularity

<u>Normal form</u>	<u>Unfolding</u>
$x(x^2 + \lambda^2) = 0$	$x(x^2 + \lambda^2 + \alpha) = 0$
$x(-x^2 + \lambda^2) = 0$	$x(-x^2 + \lambda^2 + \alpha) = 0$

B. H_{02} singularity

<u>Normal form</u>	<u>Unfolding</u>
$x(x^2 + \lambda^3) = 0$	$x(x^2 + \lambda^3 + \alpha + \beta\lambda) = 0$
$x(x^2 - \lambda^3) = 0$	$x(x^2 - \lambda^3 + \alpha + \beta\lambda) = 0$

The second family is a kind of converse to the first one. We require $a_{01} \neq 0$, but allow degeneracies in the z derivatives of a. A singularity which satisfies the defining conditions

(11) $a_{10} = a_{20} = ... = a_{m0} = 0$

and the non-degeneracy conditions

(12) $a_{01} \neq 0, \ a_{m+10} \neq 0$

we designate as the H_{m0} singularity in this paper. Golubitsky and Langford (1981) proved that the codimension is m. We only need the first two members of this family as well, but we have also found the H_{30} singularity in the model we are using (Farr, (1986)). In simplest terms, this family of bifurcations leads to limit points on the periodic branch and multiple periodic solutions. The normal forms and their unfoldings are given in Table 2 and the transition varieties and local diagrams are shown in Fig. 2.

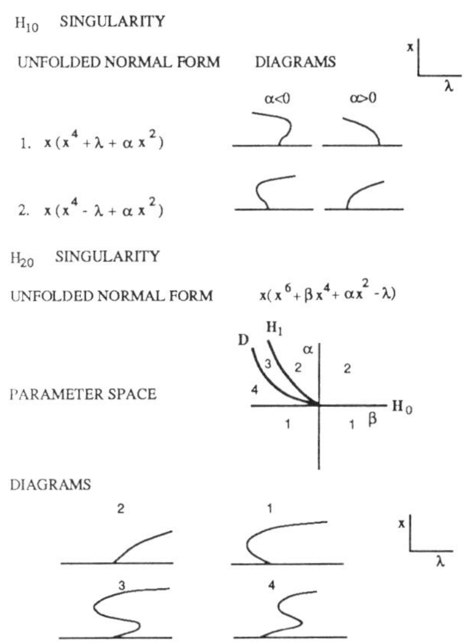

Fig. 2. Unfolded normal forms, transition sets, and local diagrams for the H_{10} and H_{20} singularities.

Table 2

Normal forms and unfoldings for H_{10} and H_{20} singularities

A. H_{10} singularity

Normal form	Unfolding
$x(x^4 + \lambda) = 0$	$x(x^4 + \lambda + \alpha x^2) = 0$
$x(x^4 - \lambda) = 0$	$x(x^4 - \lambda + \alpha x^2) = 0$

B. H_{20} singularity

Normal form	Unfolding
$x(x^6 + \lambda) = 0$	$x(x^6 + \lambda + \alpha x^2 + \beta x^4) = 0$
$x(x^6 - \lambda) = 0$	$x(x^6 - \lambda + \alpha x^2 + \beta x^4) = 0$

The last family of $\mathbb{Z}_2$ symmetric bifurcations we will discuss in this section is perhaps the most interesting and dramatic we have encountered. It differs considerably from the families we have discussed before because the unfolding involves a modal parameter. The mathematical ideas behind moduli are too involved to be dealt with in detail here. Our discussion will have to be confined to a description of the structure of the bifurcations involved with this family, without explaining much of the mathematics behind it.

In our notation we would like to use the symbol H_{11} to describe the simplest member of this family, but, as we will see below, this is not a unique designation because there are three distinct non-degenerate members and only one name. The defining conditions are $a_{10} = a_{01} = 0$ and the normal form is

(13) $\quad x(\varepsilon x^4 + 2m\lambda x^2 + \delta\lambda) = 0$

where $\varepsilon = \text{sign}(a_{20})$, $\delta = \text{sign}(a_{02})$, and $m = a_{11}/(2\sqrt{|a_{20}a_{02}|})$. Non-degeneracy conditions are $a_{20} \neq 0$, $a_{02} \neq 0$ and $m^2 \neq \varepsilon\delta$. The parameter m is the modal parameter, and the fact that it appears in the normal form is the first clue to its difference from the parameters that appeared previously in unfoldings. Usually the coefficients in normal forms are scaled to ± 1, but they cannot be in this case. We note that if $m^2 = \varepsilon\delta$ then the equation can be factored. The other two non-degeneracy conditions will be important in our results; if $a_{20} = 0$ then we will see multiple periodic solutions and if $a_{02} = 0$ we will have interactions between three Hopf points. This is a gross simplification of the true situation, and interested readers are strongly advised to consult the excellent exposition in Golubitsky and Schaeffer's book.

The unfolding of eq. (13) requires two further parameters, giving

(14) $\quad x(\varepsilon x^4 + 2m\lambda x^2 + \delta\lambda + \alpha + 2\beta x^2) = 0.$

There are three distinct bifurcations associated with eq. (14) depending on ε, δ, and the value of m. The first occurs if $\varepsilon\delta = -1$; the transition sets and the diagrams are shown in Fig. 3. Two subcases are apparent in the figure, depending on the sign of m.

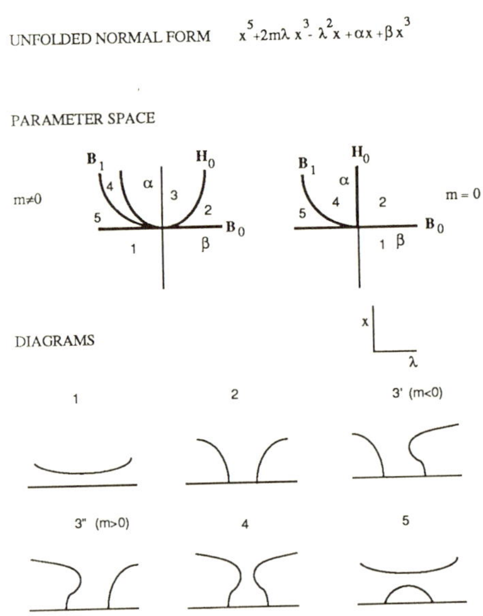

Fig. 3. Unfolded normal form, parameter space, and diagrams for the H_{11} singularity with $\varepsilon=1$ and $\delta=-1$.

The second set of diagrams for eq. (14) corresponds to $\varepsilon\delta = 1$ and $|m| > 1$. The transition set and the diagrams are shown in Fig. 4, assuming $\varepsilon = \delta = 1$ and $m > 1$. The diagrams are exactly the same for $\varepsilon = \delta = -1$ and $m < -1$, but the signs of α and β must be changed. The diagrams for $\varepsilon = \delta = 1$ and $m < -1$ can be obtained by replacing λ with $-\lambda$, and similarly for the remaining case. (We will find such manipulations useful when we apply the theory to our example.) The final bifurcation associated with eq. (14) is for the case $\varepsilon\delta = 1$ and $|m| < 1$. We have found this singularity in our model, but will not present those results here.

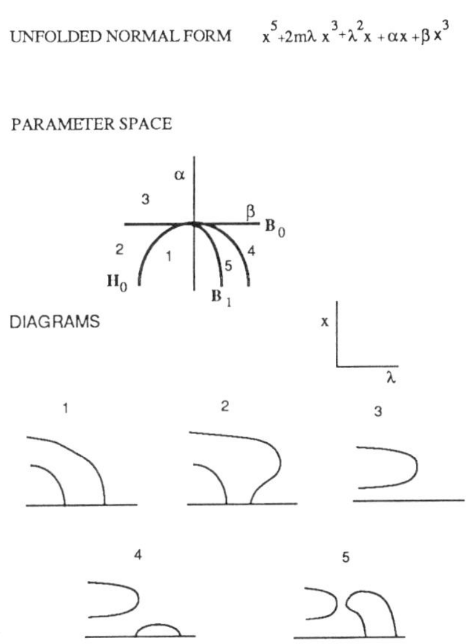

Fig. 4. Unfolded normal form, parameter space, and diagrams for the H_{11} singularity with ε=δ=1 and |m|>1.

Golubitsky and Schaeffer (1985) have developed a useful way to describe this whole family of bifurcations; the moduli bracelet. We want to borrow some of their terminology, so we will briefly discuss their construction. They start by observing that there are four different combinations for the signs of ε and δ; for each combination m can have any real value. A compact represention of this family that also includes the effects of the non-degeneracy conditions is provided by the moduli bracelet, which we have reproduced in Fig. 5.

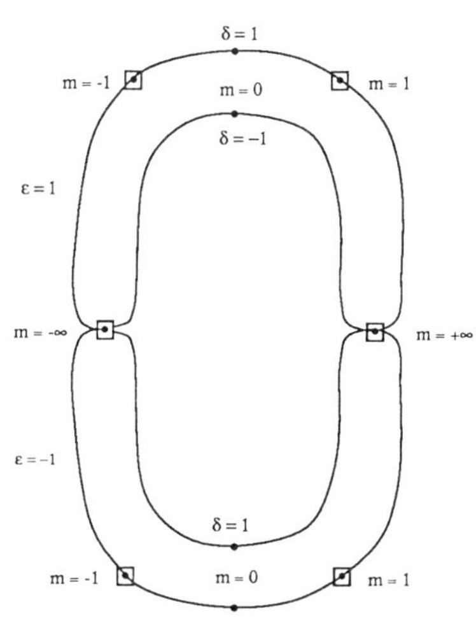

Fig. 5. Moduli bracelet for the unimodal family, showing the different signs and values of m possible in eq. (24). The filled dots represent connector points, identified with boxes, and regular distinguished points (where m=0).

Each of the four combinations is drawn as a semi-circle parametrized by m. Connections are made at $m = \infty$ and $m = -\infty$, corresponding to violations of the conditions $a_{20} \neq 0$ or $a_{02} \neq 0$ and transitions to a different combination of ε and δ. Also identified are the points $m = \pm 1$ on the two branches where $\varepsilon\delta = 1$. These two types of points on the moduli bracelet are termed connector points by G&S because they "connect" two different singularities. Also identified on the bracelet are the points where $m = 0$, since the diagrams (but not the singularities) change there.

The connector points can be thought of as more degenerate singularities. This introduces three more normal forms and their unfoldings into the family: one from $\varepsilon\delta = m^2$ and two more since either $a_{20} = 0$ or $a_{02} = 0$ leads to infinite m. The diagrams associated with these three singularities are much more complicated than any we have seen so far. In Farr (1986) examples of each singularity are described, but in this paper we only present examples of the connector point with $a_{20} = 0$. The connector point with $a_{02} = 0$ will be useful in our results on the F_1 singularities, so we present the normal forms and unfoldings for these two singularities.

If $a_{02} = 0$ at an H_{11} singularity then it is logical to think of the resulting singularity as an interaction of H_{02} and H_{11} singularities. This actually makes sense, as examination of Fig. 6 will show. This figure presents the transition sets and the diagrams for the unfolding of the normal form we will need for our results, and is redrawn from the book of G&S. This normal form is

(15) $\quad x(x^4 - \lambda x^2 - \lambda^3) = 0$

and its unfolding is given by

(16) $\quad x(x^4 - \lambda x^2 - \lambda^3 - \alpha - \beta\lambda - b\lambda^2) = 0.$

The transition sets give a clear picture of the interaction of an H_{02} singularity (the cusp on the $\mathbf{B}_0$ variety) and an H_{11} singularity (the tangential intersection of the $\mathbf{H}_0$, $\mathbf{B}_1$, and $\mathbf{B}_0$ varieties).

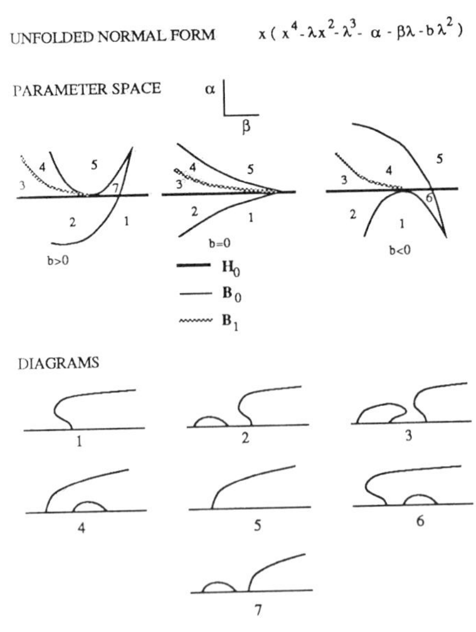

Fig. 6. Unfolded normal form, parameter space, and local diagrams for the connector point with a_{02}=0.

The last singularity of this section is the most complicated one we will encounter. It is one of the two possible normal forms associated with $a_{20} = 0$ at an H_{11} singularity. We have located at least three of these singularities; the two we describe here correspond to the normal form

(17) $x(-x^6 + \lambda x^2 + \lambda^2) = 0$

which has the unfolding

(18) $x(-x^6 + \lambda x^2 + \lambda^2 + \alpha + \beta x^2 + a x^4) = 0.$

We have redrawn the transition set and diagrams from Golubitsky and Schaeffer's book in Fig. 7. There are sixteen diagrams associated with this singularity; finding all of them in an application would be quite a challenge.

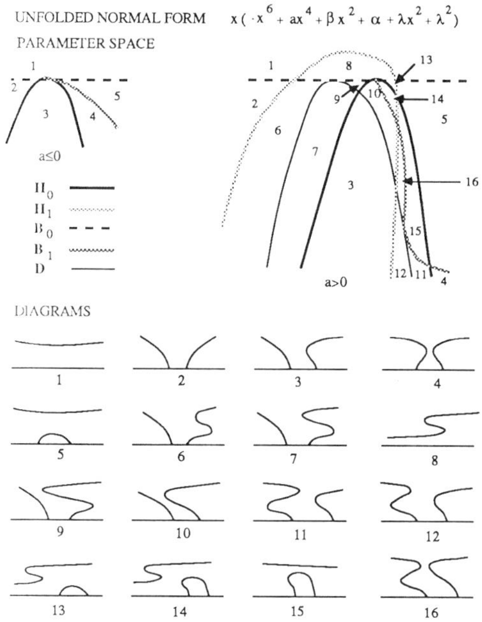

Fig. 7. Unfolded normal form, parameter space, and diagrams for the connector point with $a_{20}=0$.

§3. Steady state stability

In this section we will introduce the equations modelling a continuous stirred tank reactor in which consecutive reactions A→B→C occur. This is a three dimensional model; nonlinearity enters through the temperature dependence of the reaction rates. The equations are given by

(19a) $\dot{u} = 1 - u - u\alpha e^w$

(19b) $\dot{v} = u\alpha e^w - v - v\sigma\alpha e^w$

(19c) $\dot{w} = -(1+\kappa)w + \beta u\alpha e^w + \beta\rho\sigma v\alpha e^w$

We will use α for our bifurcation parameter in this paper. Certain simplifications have been made to obtain these model equations, though they still contain five parameters; details are in Farr (1986) and Farr and Aris (1986). The steady states of our model can be reduced to the single equation involving only w and the parameters

(20) $$w = B \frac{\alpha e^w}{1+\alpha e^w} \left(1 + \rho \frac{\sigma\alpha e^w}{1+\sigma\alpha e^w} \right)$$

where $B = \beta/(1+\kappa)$. The analysis of this equation via singularity theory is straightforward, and has been reported in Jorgensen, et al. (1984).

The next step is to obtain the characteristic equation. The Jacobian matrix is given by

$$(21)\qquad \begin{pmatrix} -(1+\alpha e^{w}) & 0 & -u\alpha e^{w} \\ \alpha e^{w} & -(1+\alpha\sigma e^{w}) & u\alpha e^{w} - v\sigma\alpha e^{w} \\ \beta\alpha e^{w} & \beta\sigma\rho\alpha e^{w} & -(1+\kappa)+\beta\, u\alpha e^{w} + \beta\rho\sigma v\alpha e^{w} \end{pmatrix}$$

As was done in Halbe and Poore (1981), the steady-state relations can be used to express the entries of the Jacobian in terms of u (they actually used 1 - u) and the parameters β,ρ,σ, and κ. We will not do so at this stage, since it is easier to obtain the coefficients of the characteristic equation first and then substitute. We will also be able to use the coefficients in this simpler form to prove quite easily that the G_1 (or triple zero eigenvalue) bifurcation never happens in this system. If we write the characteristic equation as

(22) $-\lambda^3 + P\lambda^2 - Q\lambda + R = 0$

then using the definition $\varepsilon = \alpha e^{w}$ the coefficients P, Q, and R are

(23a) $P = -(1+\kappa) + \beta u\varepsilon + \beta\sigma\rho v\varepsilon - 2 - \varepsilon - \sigma\varepsilon$

(23b) $Q = (1+\kappa)(2+\varepsilon+\sigma\varepsilon) - \beta u\varepsilon(2+\sigma(1+\rho)\varepsilon) - \beta\sigma\rho v\varepsilon(2+\varepsilon) + (1+\varepsilon)(1+\sigma\varepsilon)$

(23c) $R = -(1+\kappa)(1+\varepsilon+\sigma\varepsilon+\sigma\varepsilon^2) + \beta u\varepsilon(1+\sigma(1+\rho)\varepsilon) + \beta\sigma\rho v\varepsilon(1+\varepsilon)$.

To write them all in terms of u and the free parameters, we need only substitute the steady state relations

(24a) $\varepsilon = (1-u)/u$

(24b) $v = u(1-u)/[u(1-\sigma) + \sigma]$

(24c) $w = B(1-u)\{1+\rho\sigma(1-u)/[u(1-\sigma)+\sigma]\}$

(eq. (24c) is not really needed).

It is straightforward to use the characteristic equation to derive conditions for various eigenvalue configurations. By direct substitution, it is easy to show that the condition for a pair of complex conjugate eigenvalues $\pm i\omega$ is

(25) $-PQ + R = 0 \quad , Q > 0$

and that $\omega^2 = Q$ (the reason that $Q > 0$ is required). There are three other eigenvalue configurations of potential interest in this system. These are the F_1 (double zero), the F_2 (zero + imaginary pair), and the G_1 bifurcations. We will show below that the G_1 and F_2 configurations do not occur in this system. The F_1 bifurcation certainly occurs, but there is one minor point we need to resolve. Finding the F_1 configuration from the characteristic equation is not really proper, since it is also necessary that there be only one eigenvector corresponding to the two zero eigenvalues. Fortunately, it is easy to show that this is always the case. If there were two independent eigenvectors with eigenvalue zero, then the rank of the Jacobian matrix would have to be 1 there. The upper left hand corner of the matrix, eq. (21), contains a 2x2 determinant that is never zero; hence the rank is always 2 or 3.

The condition for the G_1 eigenvalue configuration is $P = Q = R = 0$. If we add eq. (23a), (23b), and (23c) the result is

(26) $Q + P + R = -1 - \kappa\sigma\varepsilon^2$

and since the right hand side of eq. (26) is less than -1, the G_1 configuration can never exist in the feasible region, for which κ, σ, and ε must be positive. We can also use eq. (26) to show that the F_2 configuration is not possible, since this requires $R = P = 0$ and $Q > 0$. We rewrite eq. (26) as

(27) $Q = -P - R - 1 - \kappa\sigma\varepsilon^2$

and conclude immediately that if $P = R = 0$ then Q is always less than zero.

Equation (26) implies that the third eigenvalue is always negative at an F_1 singularity. We can see this easily just by setting Q and R equal to zero, and using the fact that the third eigenvalue is given by P. In the context of the center manifold near an F_1 singularity, this result means that the dynamics can be determined from the two dimensional normal form on the center manifold. A similar procedure can be used to show that the third eigenvalue is always negative at a Hopf point as well.

§4. Dynamics results I

The first step in investigating the dynamics is to find a convenient means of describing the steady state behavior. There is a single butterfly singularity at $w = 3$, $\varepsilon = 2+\sqrt{3}$, $B = 3.215$, $\sigma = 0.07180$, $\rho = 0.8666$ (Jorgensen et al., 1984). For $\sigma < 0.0718$, there are two swallowtail points and it is possible to find diagrams with four limit points. In Fig. 8 we show a plot of the hysteresis and double limit varieties for σ=0.01, with B and ρ as parameters. The seven steady state diagrams that correspond to the regions of the central plot are shown around its perimeter.

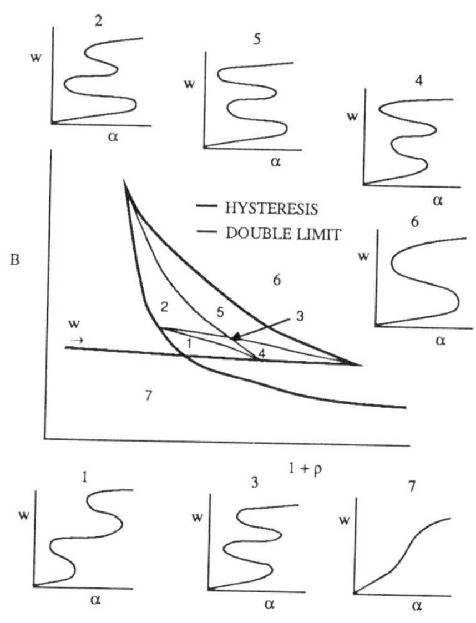

Fig. 8. Plot of hysteresis and double limit varieties for σ=0.01. The seven steady state diagrams are shown around the perimeter of the central diagram, and are identified with numbers corresponding to the open regions.

Fixing a value of ρ and drawing a vertical line we get up to three intersections with each variety. (The number of different lines that can be drawn is 7 for this particular choice of σ, and other choices could lead to different configurations. The value of σ was chosen to match one case treated in Halbe and Poore (1981).) Recalling that we cannot use the combined parameter B in our dynamics investigations and that our β and κ parameters are identical to the B and β parameters used by Uppal, et al. (1974) in their classic study of the CSTR with one reaction, it makes sense to fix ρ and use β and κ as the important parameters in our investigation. Figure 9 demonstrates how the intersections of a ρ=constant line with the hysteresis variety results in straight lines in a $\beta - \kappa$ plot. In what follows we will suppress the double limit lines to simplify the situation, but since we will concentrate our results in the region of unique steady states, this is not a serious defect. The numbering system for the hysteresis lines in Fig. 9 derives from the fact that one can number the limit points in order of increasing w values from 1 to 4 and the two digits identifying the hysteresis lines in Fig. 9 indicate which two limit points come together on that line.

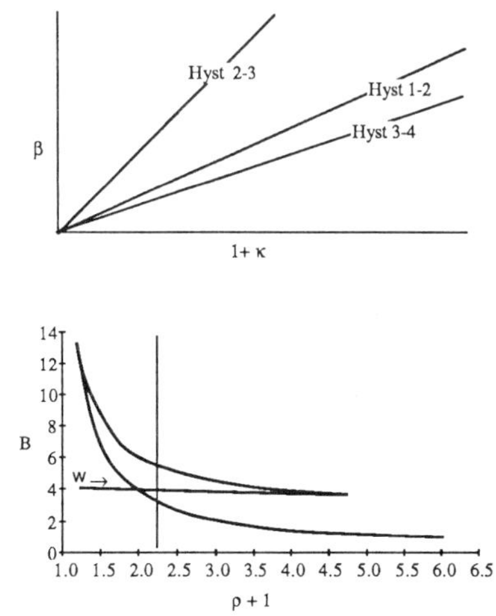

Fig. 9. Bottom - plot of hysteresis variety with σ=0.01. The vertical line corresponds to a particular choice of ρ. Top - $\beta - \kappa$ plot showing how the three intersections with the hysteresis variety become straight lines.

To begin the discussion of dynamics for this system, we first consider Hopf bifurcation points. Previous workers, e.g. Halbe and Poore (1981), had found up to four Hopf points in this system. These four possible Hopf points can occur in various configurations, bringing us to the theme of interactions among the Hopf points and the limit points. The hysteresis lines take care of the limit point - limit point interactions and the Hopf point - limit point interactions result in the F_1 bifurcations mentioned already. The Hopf point - Hopf point interactions fall under the general theory for degenerate Hopf bifurcations described above, but for systems of low dimension (we have worked with 4 - dimensional systems by this method, as well as systems of 2 and 3 dimensions) it is possible to use the characteristic equation to locate degenerate Hopf points of type H_{0m} as follows. Golubitsky and Langford (1981) and earlier Kielhofer (1979) showed that the set of conditions

(28) $a_{01} = a_{02} = \dots = a_{0m} = 0 \ , \quad a_{0m+1} \neq 0$

is equivalent to the set

(29) $\sigma' = \sigma'' = \dots = \sigma^{(m)} = 0 \ , \quad \sigma^{(m+1)} \neq 0$

where $\sigma(\lambda)$ is the real part of the complex conjugate pair, and the derivatives are with respect to λ. In the case of the CSTR, it is easier to express the coefficients of the characteristic equation in terms of the variable u than the bifurcation parameter λ, but the implicit function theorem can be used to show that away from limit points we can always define $u(\lambda)$ as an invertible function locally and this is sufficient to show that conditions (29) are equivalent to

(30) $(-PQ+R)_u = (-PQ+R)_{uu} = \dots = \partial^m(-PQ+R)/\partial u^m = 0, \quad \partial^{m+1}(-PQ+R)/\partial u^{m+1} \neq 0.$

The similarity to singularity theory for steady state multiplicity is clear and gives us a relatively simple way to locate the H_{0m} degeneracy candidates. In the numerical work described below, we used this method to locate simple Hopf points as well as type H_{01} and H_{02} degenerate Hopf points but also calculated the coefficients a_{01} and a_{02} separately. When we come to locating numerically the limits of these singularities as they approach F_1 bifurcation points, having the coefficients of the characteristic equation in terms of u instead of our bifurcation parameter α will be a great help, since derivatives with respect to α are not well behaved there.

The type H_{m0} and H_{11} singularities are found numerically by a two step process. In the first step, Hopf points (or type H_{01} points in the case of H_{11} singularities) are found and the a_{10} and a_{20} coefficients are calculated for several closely spaced values of β and/or κ. These values are then used to find approximations to the derivatives needed for a Newton's method solution for the desired singularity. This method works quite well, and the approximate derivatives are used to relate the parameters of the real problem to those in the unfolded normal form. More details are in Farr (1986).

As an example, Fig. 10 shows the codimension 1 local singularities (in the context of bifurcation diagrams, the F_1 singularity has codimension 1) for $\rho = 1$ and $\sigma = 0.01$ and fairly small values of κ. This particular set of parameters has been used before by Hlavacek et al. (1972), Halbe and Poore (1981), Jorgensen and Aris (1983), and Doedel and Heinemann (1983), but our results cover a lot more ground qualitatively, while giving up some quantitative details. It is convenient to also number the Hopf points from 1 to 4 in order of increasing w value, and this is the source of the designations in Fig. 10. The three hysteresis lines are not shown in the figure, since they are straight lines tangent to the F_1 curve at κ=1(see below). The intersections of the three hysteresis lines and the three H_{01} curves with the F_1 curve correspond to degeneracies in the normal form of the F_1 singularity, but we will not pursue this point in detail. It can be shown (Farr, 1986) that the intersections of the hysteresis lines with the F_1 curve always occur at $\kappa = 1$. This means that the H_{01} 1-2 (i.e. Hopf points 1 and 2 coalesce on this curve) and H_{01} 3-4 curves always intersect the F_1 curve for κ values less than one if the appropriate hysteresis line is present. Conversely, H_{01} 2-3 always intersects the F_1 curve at a value of κ greater than one.

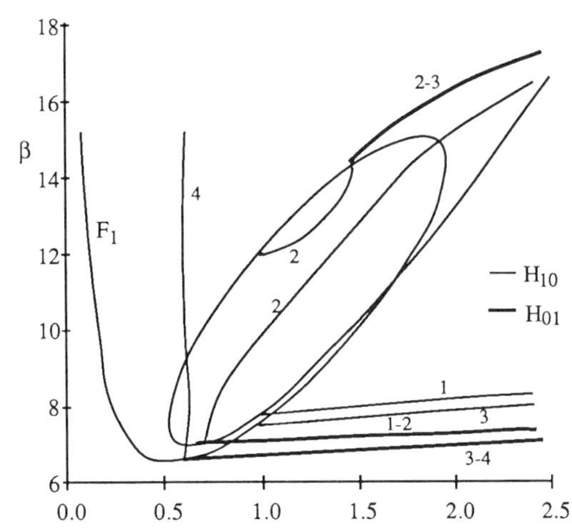

Fig. 10. Codimension 1 local singularities for ρ=1, σ=0.01 and small values of κ. Not shown are the three hysteresis lines, since they are straight lines tangent to the F_1 curve at κ=1.

At larger values of κ the situation is also interesting and we encounter for the first time in this model the H_{20} and H_{11} degenerate Hopf bifurcations. We show in Fig. 11 the H_{01} and H_{10} curves for κ between 5 and 25. The

H_{10} -1 curve is not actually shown in the figure, but it follows the H_{01} 1-2 curve very closely without ever intersecting it. The H_{10} -3 curve, on the other hand, exhibits two turning points, with an H_{20} point occuring near the second limit point , before approaching the H_{01} 2-3 curve and eventually intersecting it tangentially at $\kappa = 38.5$, in an H_{11} bifurcation. This point is not shown in the figure.

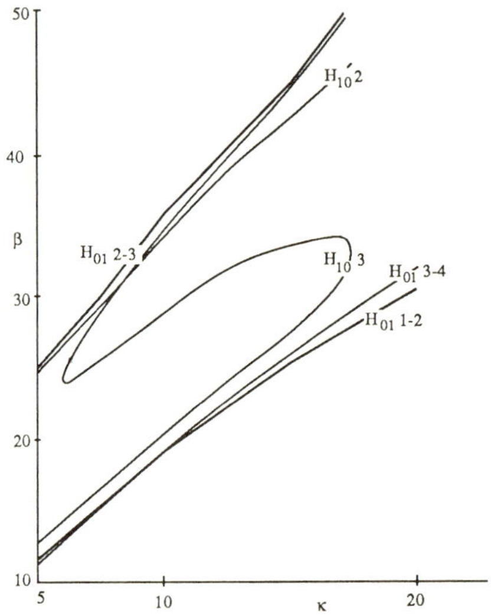

Fig. 11. Codimension 1 degenerate Hopf bifurcations for σ=0.01, ρ=1 and values of κ from 5 to 25.

The next step in our analysis is to compute entire branches of the codimension 2 singularities. In this paper, we restrict the value of ρ to lie between 0.3 and 4 so that these codimension 2 singularities lie in the region of unique steady states. The results are shown in Fig. 12, where we plot both β and κ versus ρ. Several points are distinguished as higher codimension singularities; two are connector points on the moduli bracelet where $a_{20} = 0$ and the third is an example of an H_{30} singularity. One of the H_{20} curves also crosses the H_{11} curve at a point not identified; careful inspection here shows that the values of ρ are not quite the same and so there is not a local singularity associated with these points.

There is quite a lot of information we can associate with the curves in Fig. 12. For each singularity the connected components can be broken up into segments where the local diagrams are qualitatively similar, meaning the coefficients in the normal form have the same sign and the diagrams have the same orientation. This task is easiest for the branch of H_{02} singularities and hardest for the H_{11} singularities since there are more possibilities, depending on the value of the modal parameter as well as the signs of a_{20} and a_{02}. We will discuss each singularity in turn, starting with the easiest. In Farr (1986) these results are extended to large and small values of ρ and used to describe the probable sequence of transitions for our β–κ plots as ρ is changed.

For the values of ρ shown in Fig. 12, the H_{02} singularities always have the same normal form. That is, the sign of a_{10} is always positive and the sign of a_{03} is always negative, so the normal form and diagrams are the same as those of Fig. 1. In our β–κ plots the cusp terminating the H_{01} 2-3 and H_{01} 1-2 curves appears at large values of κ and moves to smaller values as ρ is increased. It is shown in Farr (1986) that there are also H_{02} singularities at small values of ρ (they are off the scale of Fig. 12), but they involve the H_{01} 2-3 curve and the H_{01} 3-4 curve. Thus

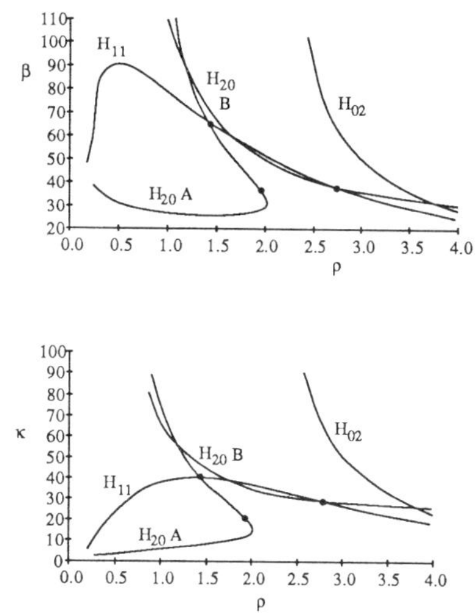

Fig. 12. Codimension 2 degenerate Hopf singularities. The points marked with heavy dots are codimension 3 local singularities.

the H_{02} singularities shown in Fig. 12 must go off to infinity for some intermediate value of ρ so that the two branches of H_{01} singularities are no longer connected. (We checked this by computing the H_{01} singularities out to very large values of κ. A better method would be to use the parameter B in the characteristic equation instead of β and then take the limit for large κ, since all indications are that the value of B remains finite. This we have not yet done.)

There are two distinct continuous branches of H_{20} singularities. They are divided up into segments where the signs in the normal form are constant by the two connector points, an H_{30} singularity, and a limit point. The curve of H_{20} points labeled A in Fig. 12 can be broken up into four segments. The limit point divides it up into two sub-curves; on the lower the H_{20} singularities are all of the same type while on the upper we have an H_{30} singularity and a connector point where $a_{20} = 0$. The other curve of H_{20} singularities, which we label B, is broken by a single connector point where it crosses the H_{11} singularity curve. We might note that these connector points appear on our β–κ plots as an H_{20} point passing through a H_{11} point; this means (among other things) that the Hopf point the H_{20} singularity is associated with changes at one of these connector points.

The information on the H_{20} singularities is summarized in Table 3, where we identify the Hopf point and the signs in the normal form for each segment. Our analysis of the H_{30} singularity is incomplete, since we cannot calculate a_{40}, so we are unable to determine where the regions with four periodic solutions are. The limit point in curve A of H_{20} singularities perhaps needs further explanation. In our β–κ plots we would see two separate H_{20} singularities coalescing. When this happens the $\mathbf{H}_1$ and **D** varieties of the two singularities can either face each other or be opposed. In the second case it is clear that the $\mathbf{H}_1$ and **D** varieties would then detach from the H_{10} singularity curve. If the $\mathbf{H}_1$ and **D** sets for the two H_{20} singularities face each other then as the two singularities coalesce the varieties disappear. An example of this case, including numerical computations of periodic branches, is presented in Planeaux and Jensen (1986). Our calculations indicate that the second case occurs at the limit point in Fig. 12.

Table 3

H_{20} singularities for $0.3 < \rho < 4$.

Normal form: $x(\varepsilon x^6 + \delta\lambda) = 0$, where $\varepsilon = \text{sign}(a_{30})$ and $\delta = \text{sign}(a_{01})$.

Branch	ρ	ε	δ	Hopf point
A upper	0.3-2.03	-1	-1	3
A lower	<1.4	+1	+1	2
	1.4-1.906	+1	-1	3
	1.906-2.03	-1	-1	3
B	< 2.6	+1	+1	2
	2.6-4.0	+1	-1	3

Dividing the curve of H_{11} singularities up into segments is straightforward. We have discussed the moduli bracelet for this family; major changes occur at the connector points and at the distinguished points where $m^2 = \varepsilon\delta$, while more minor changes take place where $m = 0$. These H_{11} singularities are always located on the H_{01} 2-3 curve. The diagrams at any particular point on the H_{11} singularity curve can be identified by determining the values of ε, δ, and m in the normal form, eq. (13). On the other hand, determining the side on which the $\mathbf{B}_1$ variety emerges is simple: the diagram in which a branch of periodic solutions is just about to disappear via an H_{01} bifurcation is always located at smaller κ values than the H_{11} singularity.

We saw above that there are two connector points where $a_{20} = 0$ on the curve of H_{11} singularities. There is also a point on this curve where $m = 0$, so we can divide it up into four segments. Each segment can be associated with a set of diagrams from Fig. 3-4, and this information is summarized in Table 4.

Table 4

H_{11} singularities for $0.3 < \rho < 4$.

Normal form: $x(\varepsilon x^4 + 2mx^2\lambda + \delta\lambda^2) = 0$, where $\varepsilon = \text{sign}(a_{20})$ and $\delta = \text{sign}(a_{02})$.

ρ	ε	δ	m	Diagrams
0.3-1.4	-1	-1	m> 1	Fig. 4 rev.
1.4-2.1	+1	-1	m>0	Fig. 3
2.1-2.6	+1	-1	m<0	Fig. 3
2.6-4	-1	-1	m<-1	Fig. 4

We can get more complete information, however, if we integrate the diagrams for the connector points. The connector point at $\rho = 1.4$ has $a_{30} > 0$, a_{11} and $a_{02} < 0$, so we have to replace λ with $-\lambda$ and take minus the whole equation to put eq. (18) into the right form. This reverses the diagrams and in our $\beta-\kappa$ plot the transition varieties would also be reversed. The other connector point has the same signs for a_{30} and a_{02}, but a_{11} is negative so the diagrams are not reversed. The transition varieties would have their orientation flipped about a vertical axis, just as they did for the other connector point.

Before we end this section, we want to present and describe possible global diagrams near the connector point at $\rho = 1.4$. We chose this singularity because it is removed from the region of multiple steady states. We showed above that the local diagrams can be obtained by flipping those of Fig. 7 about a vertical axis, and that the transition varieties in our β–κ plot were also flipped with respect to those of Fig. 7. To draw the global diagrams, we need information on Hopf points 1 and 4. By our calculations the periodic branch near Hopf point 1 is subcritical and the branch has a limit point. Hopf point 4 is subcritical (i.e. the branch initially takes off to the left from the Hopf point). With this information in mind, we can draw the 16 plausible global digrams shown in Fig. 13-14. Of course we cannot rule out the existence of secondary bifurcations of the periodic orbits, so diagrams actually computed would likely be more complicated.

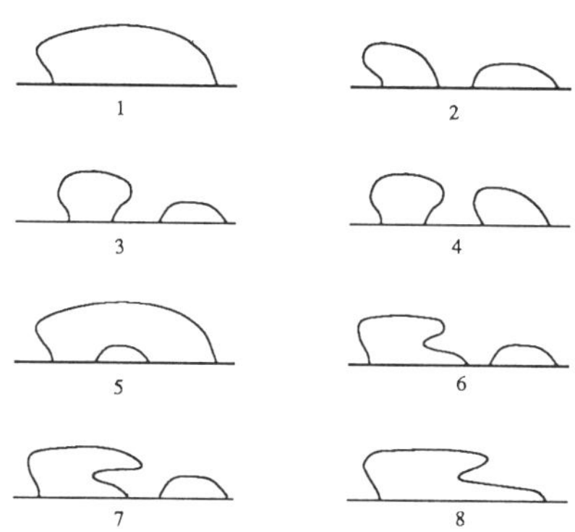

Fig. 13. Global digrams near the connector point at ρ=1.4 corresponding to diagrams 1-8 in Fig. 7.

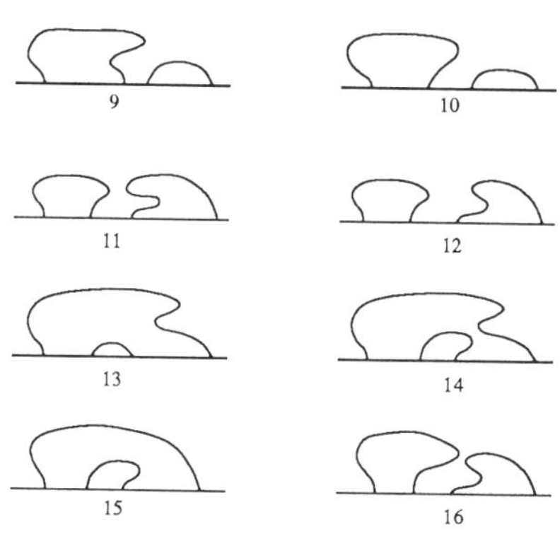

Fig. 14. global diagrams near the connector point at ρ=1.4 corresponding to diagrams 9-16 in Fig. 7.

§5. Dynamics results II

In this section we want to describe the degenerate F_1 bifurcations that occur in our model for $\sigma = 0.01$ and large and small values of ρ. To do so we will find it useful to pay more attention to the F_1 bifurcations. This does not mean that we will try to put our model in the F_1 normal form; we have not found a practical way to do this for this problem. We will be satisfied with being able to locate accurately the equivalents (in codimension) of our codimension

2 degenerate Hopf bifurcations. In this context what we mean by equivalent are the intersection points of H_{01} singularities and hysteresis points with the F_1 curve. Tracking these points will help us to understand the transitions for both large and small values of ρ. We will give evidence below for what we think are degenerate F_1 bifurcations that have not been analyzed yet.

Examination of Fig. 10 shows that when two of the F_1-Hyst singularities approach each other (as they do at each of the two swallowtail points) there is always one F_1-H_{01} singularity trapped between them. Learning how these points interact will help us understand this important transition. The question is which of the F_1-Hyst singularities interacts with the F_1-H_{01} singularity. The answer turns out to be somewhat unexpected; we find that all three points come together simultaneously. Our reasons for this conclusion will be given below.

Locating what we will term the F_1-H_{01} singularities is straightforward, but has to be done numerically. The conditions we used to solve numerically for the H_{01} singularities were

(31) $\quad -PQ + R = (-PQ + R)_u = 0.$

To find the equations for the F_1-H_{01} singularity we add Q = R =0, so the set becomes

(32a) $\quad Q = 0\ ,\ \ R = 0$

(32b) $\quad -P(Q)_u + (R)_u = 0$

since the first condition of eq. (31) is now redundant. We point out that eq. (32cb is obtained from eq. (31) by simply doing the differentiation explicitly and using eq. (32a).

Our results for these degenerate F_1 singularities can be presented by adding the F_1-H_{01} singularities to the plot of the hysteresis variety in Fig. 8. This leaves out κ, but we know that $\kappa = 1$ for the F_1-Hyst singularities and that the values of κ for the F_1-H_{01} singularities behave in a simple manner. In Fig. 15 we show this plot for $\sigma = 0.01$, but schematically so the details are clear. The numbering system for the F_1-Hyst and F_1-H_{01} singularities is derived from our earlier scheme for identifying the hysteresis points and the H_{01} singularities. From this figure we can conclude that for any value of ρ with three hysteresis points, there are three each of the two degenerate F_1 singularities. This precludes any possibility of combined F_1-Hyst-H_{01} bifurcation away from the swallowtail points. We have drawn the figure so that the F_1-H_{01} 1-2 curve passes exactly through the 123 swallowtail point; we have not been able to show this analytically, but our computer results indicate strongly that this is the case. It is straightforward to extend the analysis to find F_1 bifurcations at swallowtail points. They satisfy the conditions for the F_1-Hyst singularities so we know that $\kappa = 1$; details are in Farr (1986). What we did next was to compute the defining conditions for the H_{01} singularity at these points and found that they were zero to within computer accuracy. The same results were also obtained at the 234 swallowtail point. In fact, these same results were found for the whole of both branches of swallowtail points that exist for $\sigma < 0.0718$. At the butterfly point at $\sigma = 0.0718$, the conditions for the H_{02} singularity were also satisfied. This indicates that the whole structure in Fig. 15 collapses at the butterfly point. We have not investigated yet, but we conjecture that for $\sigma > 0.0718$ and both reactions exothermic there is a single hysteresis curve and a single branch of F_1 -H_{01} singularities.

The two apparent crossings in Fig. 15 of the hysteresis curves by the F_1-H_{01} curves away from the swallowtail singularities do not involve any local bifurcations, but are due to the deformation of the F_1 curve near the swallowtail points. We can use the curves in Fig. 15 to sketch the F_1 curves as we vary ρ. To demonstrate this we consider $\rho = 0.5$. The first step is to draw the three hysteresis lines. At $\kappa = 1$ we know that the F_1 curve has tangential intersections with each hysteresis line. The F_1 curve is concave upward at the F_1-Hyst 1-2 and 3-4 points and concave downward at the 2-3 point, so portions near each of these points can be sketched in. The F_1 curve comes down from very large values of β as κ increases from zero and then crosses in order through the F_1-Hyst 3-4 point

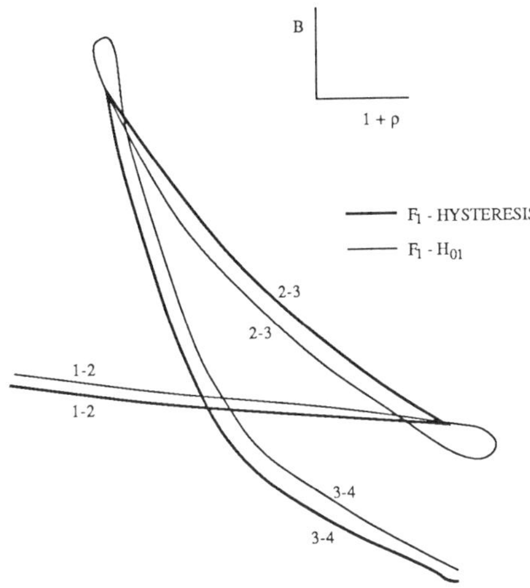

Fig. 15. Diagram of "codimension 2" (codimension 3 by normal forms of vector fields usage) singularities which are interactions of degenerate Hopf bifurcations or hysteresis singularities with double zero eigenvalue singularities.

from left to right, the F_1-Hyst 2-3 point from right to left, and the F_1-Hyst 1-2 point from left to right before going to large values of β and κ. These orientations hold as long as the hysteresis line in question is present. We note in passing that the F_1-H_{01} points can be located easily: the F_1-H_{01} 1-2 and 3-4 occur at κ values less than 1 as long as there are three hysteresis lines. They both pass through 1 at the F_1-swallowtail singularities described above, and then stay above 1 as long as they exist. The κ value for the F_1-H_{01} 2-3 singularity is always above 1.

We give examples of F_1 curves for $\rho = 0.2$ and $\rho = 0.5$ in Fig. 16. There is no loop in the curve for $\rho = 0.2$, since it disappears at the swallowtail point. For smaller values of ρ the upper limit point moves to smaller values of κ and disappears before $\rho = 0$. Moving to large values of ρ, we show F_1 curves for $\rho = 4$ and $\rho = 10$ in Fig. 17. It is not known whether the limit points disappear for even larger values of ρ.

We have not yet described the two limit points that appear in the F_1-H_{01} curves in Fig. 15 where two branches of F_1-H_{01} singularities come together. One appears at $\rho = 0.202$, involving the 2-3 and 3-4 branches and the other is at $\rho = 5.8$ where the 1-2 and 2-3 branches meet. To see what these points mean, we plot the F_1 curve, the H_{01} 2-3 curve, and the H_{01} 3-4 curve for $\rho = 0.2$ and small values of κ in Fig. 18. This indicates that the two H_{01} curves have detached from the F_1 curve and there is now a second branch of H_{02} singularities. The obvious conclusion from Fig. 15 is that the two F_1-H_{01} singularities come together at the instant of detachment in what we would expect to call an F_1-H_{02} singularity. This does not seem to be quite right. It appears that the analog of a_{10} is also zero at this point. Thus this singularity would be more properly described as as an interaction of an F_1 singularity with a connector point

of the H_{11} singularity family, specifically the one with $a_{02} = 0$. This appears to be true for both large and small values of ρ. We draw a schematic diagram of what we believe happens at the instant of detachment for the two cases (i.e. large and small ρ) in Fig. 19.

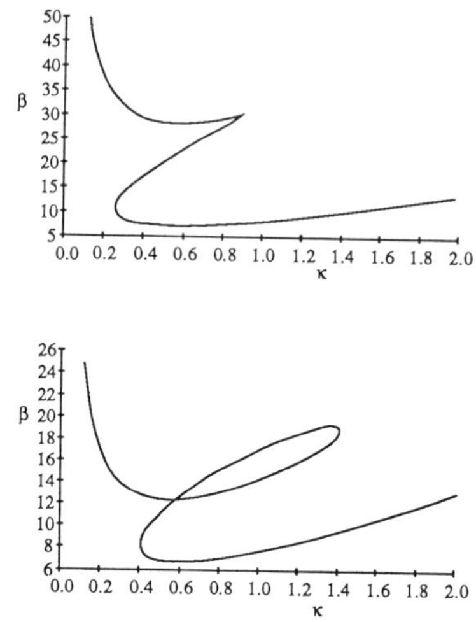

Fig. 16. Double zero eigenvalue curves with σ=0.01. Top - ρ=0.2, only one hysteresis point. Bottom - ρ=0.5 three hysteresis points. In each case, the hysteresis lines would be tangent to the F_1 curve at κ=1.

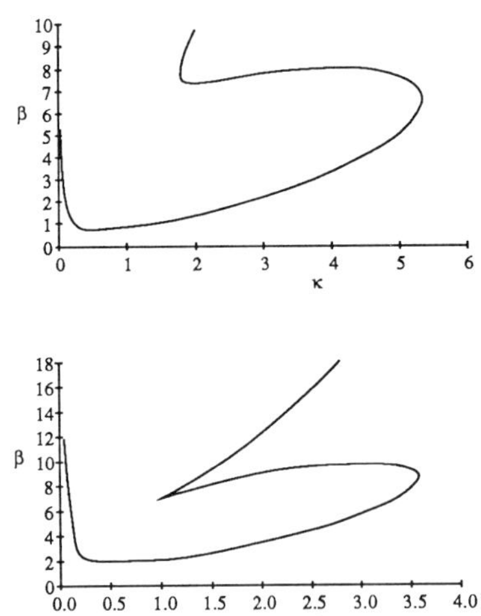

Fig. 17. Double zero eigenvalue curves for σ=0.01. Top - ρ=10, only one hysteresis point. Bottom - ρ=4.0 one hysteresis point

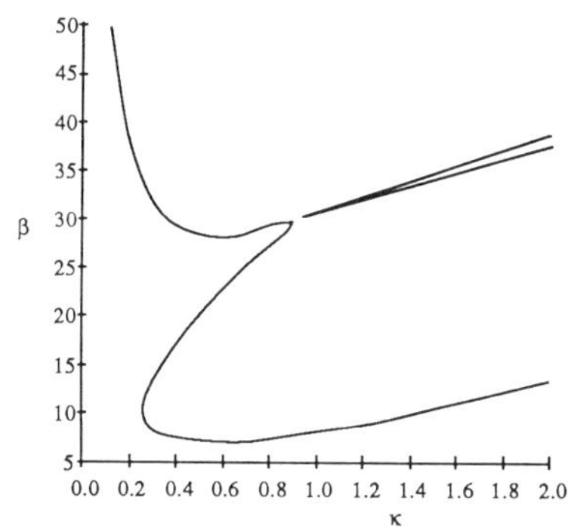

Fig. 18. Partial codimension 1 singularity set for ρ=0.2 showing F_1 curve and H_{01} 2-3 and 3-4 curves. This is just after the H_{01} curves have detached from the F_1 curve.

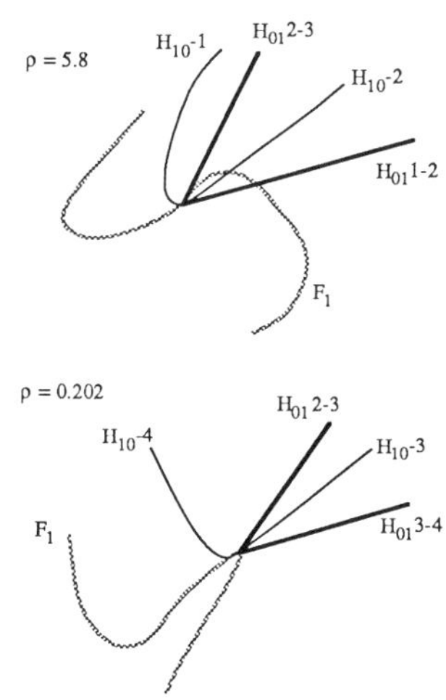

Fig. 19. Schematic diagrams showing the instant the two F_1 - H_{01} singularities come together. Top - ρ=5.8 , the 2-3 and 1-2 curves coalesce. Bottom - ρ=0.202 , the 2-3 and 3-4 points come together. In each case two branches of H_{10} singularities are also involved.

§6. Conclusions

This paper is a brief survey of some of the results on degenerate Hopf bifurcations we have obtained for this system. We feel that this approach is a powerful one, and is most useful when there are several parameters. One advantage the Liapounov-Schmidt, singularity theory approach to Hopf bifurcation has over the normal forms approach is that once a program is written to evaluate the a_{ij} coefficents it can be used for any system with only minor modification. The other advantage is, of course, the power of singularity theory to systematize the investigation of degeneracies.

Quite a lot of further work could be done on this system. The F_1 singularities would be particularly challenging, but on the prosaic side it would be most useful to develop numerical procedures for computing the transition varieties $\mathbf{B}_1$, $\mathbf{H}_1$, and **D**. These could simplify investigations of periodic behavior considerably.

Acknowledgement. During part of this work W. W. Farr was suported by a Kodak fellowship. The authors are also grateful to Jim Planeaux for many stimulating discussions.

References

1. Arnold, V. I., *Geometrical Methods in the Theory of Ordinary Differential Equations*, Springer - Verlag, New York, 1983.

2. Balakotaiah, V. and D. Luss, 1982, Structure of the steady state solutions of lumped-parameter chemically reaction systems, Chem. Engng. Sci., **37**, 1611-1623.

3. Doedel, E. J. and R. F. Heineman, 1983, Numerical computation of periodic solution branches and oscillatory dynamics of the stirred tank reactor with A→B→C reactions, Chem. Engng. Sci., **38**, 1493 - 1499.

4. Farr, W. W. and R. Aris, 1986, Reflections on the multiplicity of steady states of the stirred tank reactor, Chem. Engng. Sci. **41**, 1385-1402.

5. Farr, W. W., 1986, thesis, University of Minnesota.

6. Golubitsky, M and W. F. Langford, 1981, Classification and unfoldings of degenerate Hopf bifurcations, J. Diff. Eq., **41**, 375 - 415.

7. Golubitsky, M. and D. G. Schaeffer, 1979, A theory for imperfect bifurcation via singularity theory, Commun. Pure Appl. Math., **32**, 21-98.

8. Golubitsky, M. and D. G. Schaeffer, *Singularities and Groups in Bifurcation Theory, Vol. 1*, Springer-Verlag, New York, 1985.

9. Halbe, D. C. and A. B. Poore, 1981, Dynamics of the continuous stirred tank reactor with reactions A → B → C, Chem. Engng. J., **21**, 241-253.

10. Hlavacek, V., M. Kubicek, and K. Visnak, 1972, Modelling of chemical reactors - XXVI. Chem. Engng. Sci., **27**, 719-.

11. Jorgensen, D. V. and R. Aris, 1983, On the dynamics of a stirred tank with consecutive reactions, Chem. Engng. Sci., **37**, 45 -.

12. Jorgensen, D.V., W. W. Farr, and R. Aris, 1984, More on the dynamics of the stirred tank with consecutive reactions, Chem. Engng. Sci. **39**, 1741-1752.

13. Kielhofer, H., 1979, Generalized Hopf bifurcation in Hilbert space, Math. Meth. for the Appl. Sci., **1**, 498 - 513.

14. Planeaux, J. B. and K. F. Jensen, 1985, Bifurcation phenomena in CSTR dynamics: a system with extraneous thermal capacitance, Chem. Engng. Sci., **41**, 1497-1523.

15. Uppal, A., W. H. Ray, and A. B. Poore, 1974, On the dynamic behavior of continuous stirred tank reactors, Chem. Engng. Sci., **29**, 967-985.

DEPARTMENT OF CHEMICAL ENGINEERING AND MATERIALS SCIENCE
UNIVERSITY OF MINNESOTA
MINNEAPOLIS, MINNESOTA 55455

Canadian Mathematical Society
Conference Proceedings
Volume 8 (1987)

PERSISTENCE AND BIFURCATION OF INVARIANT TORI

Dietrich Flockerzi

ABSTRACT. We consider a smooth 2-parameter family of ordinary differential equations describing the flow in a neighborhood of an invariant n-dimensional torus.We assume that the normal spectrum contains precisely two different pairs of simple pure imaginary eigenvalues and establish sufficient conditions for the existence of invariant tori of dimension n+1,n+2 and n+3 in certain regions of the parameter plane.

1. Introduction: The Bifurcation Problem

Let us consider a C^∞-smooth m-parameter family of ordinary differential equations describing the flow in a neighborhood of an invariant n-dimensional torus.We assume this family to be of the form

$$\begin{aligned} \dot{x} &= A_o x + X(x,y,z,\mu) = A_o x + O(|x|^2+|z|^2+|\mu|^2), \\ \dot{y} &= \omega + Y(x,y,z,\mu) = \omega + O(|x|+|z|+|\mu|), \\ \dot{z} &= B(y)z + Z(x,y,z,\mu) = B(y)z + O(|x|^2+|z|^2+|\mu|^2) \end{aligned} \tag{1.1}$$

with (x,z,μ) belonging to a neighborhood U of the origin in $\mathbb{R}^k \times \mathbb{R}^\ell \times \mathbb{R}^m$. The variable y is to belong to the standard torus T^n and the right-hand side of (1.1) is assumed to be 2π-periodic in each component of $y \in T^n$. The coordinate system in (1.1) is to be such that

$$X(o,y,o,\mu) \equiv o, \quad Z(o,y,o,\mu) \equiv o \tag{1.2}$$

so that x and z are normal coordinates to the trivial n-torus

$$M^n(\mu) = \{(x,y,z,\mu): x=o, y \in T^n, z=o, (o,o,\mu) \in U\}. \tag{1.3}$$

We have introduced the partitioning of the normal coordinates since we intend to use x for the "critical" and z for the "non-critical" parts of the normal coordinates. We shall assume:

198o Mathematics Subject Classification(1985 Revision).58F14

The spectrum of A_o lies on the imaginary axis. (1.4)

There exist positive constants γ and β such that the fundamental matrix solution $\mathcal{B}(t,s,\eta)$ with $\mathcal{B}(s,s,\eta) = I \in \mathbb{R}^{\ell\times\ell}$ of

$$\dot{z} = B(\omega t+\eta)z, \quad \eta\in T^n, \tag{1.5}$$

satisfies

$$\|\mathcal{B}(t,s,\eta)\| \leq \gamma e^{-\beta(t-s)} \quad \text{for all} \quad t\geq s$$

uniformly in $\eta\in T^n$.

Thus the solutions of the linear equation (1.5) decay exponentially for $t\to\infty$. A particular example of (1.1) is given by the suspension of a nonautonomous system of the form

$$\dot{x} = A_o x + X(x,t,\mu) = A_o x + \mathcal{O}(|x|^2)$$

where X is quasi-periodic in t with n independent frequencies. Systems of the type (1.1) with a nonzero matrix $A_o\in\mathbb{R}^{2x2}$ having eigenvalues $\pm i$ have been considered in [25],[5],[6],[7] and [8]. In the present paper we shall investigate the codimension 2 case of A_o being equal to

$$A_1 = \begin{pmatrix} o & -\beta_1 & o & o \\ \beta_1 & o & o & o \\ o & o & o & -\beta_2 \\ o & o & \beta_2 & o \end{pmatrix}, \quad \beta_1 > \beta_2 > o,\ m = 2. \tag{1.6}$$

The cases with A_o being equal to

$$A_2 = \begin{pmatrix} o & -\beta_o & o \\ \beta_o & o & o \\ o & o & o \end{pmatrix}, \quad \beta_o \neq o, \quad \text{or} \quad A_3 = \begin{pmatrix} o & o \\ o & o \end{pmatrix}$$

are treated in [9] generalizing the results of [15],[21],[22], [24] and [26]. Thus here we shall deal with the interaction of two Hopf bifurcations in the normal part of the invariant n-torus (1.3). We are interested in the existence of invariant tori for (1.1) of dimension n+1,n+2 and n+3 and we shall proceed as follows.

Using invariant manifold theory we reduce the dimension of our problem to n+4 by considering the reduced equation

$$\begin{aligned} \dot{x} &= A_1 x + X(x,y,z,\mu), \quad x\in\mathbb{R}^4, \quad \mu\in\mathbb{R}^2, \\ \dot{y} &= \omega + Y(x,y,z,\mu), \quad y\in T^n, \end{aligned} \tag{1.7}$$

on a center manifold

$$\{(x,y,z,\mu):\ z=S(x,y,\mu),\ |x|+|\mu|<\delta,\ y\in T^n\} \tag{1.8}$$

of (1.1). System (1.7) can be assumed to belong to class C^N for any positive integer N by choosing $\delta=\delta(N)>o$ sufficiently small (see [9]). Since S vanishes for $x=o$ system (1.7) possesses the trivial invariant n-torus $\{x=o\}$. Sufficiently small invariant tori on the above center manifold (1.8) then correspond to invariant tori of the full $(4+n+\ell)$-dimensional system (1.1).

Under suitable nonresonance conditions we simplify system (1.7) by computing "normal forms" for the low-order nonlinearities:

$$\begin{aligned} \dot{x} &= A_1x + \tilde{X}(x,y,\mu) + X_R(x,y,\mu)\ , \\ \dot{y} &= \ \omega\ + \tilde{Y}(x,y,\mu) + Y_R(x,y,\mu)\ . \end{aligned} \tag{1.9}$$

Of course the form of (1.9) depends on the imposed nonresonance conditions. We then consider the truncated systems

$$\begin{aligned} \dot{x} &= A_1x + \tilde{X}(x,y,\mu), \\ \dot{y} &= \ \omega\ + \tilde{Y}(x,y,\mu) \end{aligned} \tag{1.1o}$$

in normal form and investigate the question whether invariant tori of dimension n+1,n+2 or n+3 for the truncated system (1.1o) give rise to invariant tori of the corresponding dimensions for the reduced systems (1.9),(1.7) and hence for the complete system (1.1). At this point we'd like to refer to the work in [1o],[11] and [16] on the interaction of two Hopf bifurcations. There, system (1.1o) without y-variable has been studied. See also [18].

It is known from invariant manifold theory (cf.[3],[9],[14],[23]) that normally hyperbolic invariant manifolds persist under sufficiently small perturbations. Thus we shall proceed in two steps. First we shall establish the existence of invariant tori for the truncated system (1.1o). Then we shall make sure that the invariant tori of (1.1o) in fact lead to invariant tori of (1.9) and hence of (1.1) by keeping the perturbation terms X_R and Y_R sufficiently small. We shall see that either step is connected with certain nonresonance conditions on the frequencies β_1,β_2 and $\omega_1,\dots,\omega_n$.

There are two reasons for not emphasizing the "succession of bifurcations from T^n to T^{n+1} to T^{n+2} to T^{n+3}. Generally, the tori are only known to exist where they are normally hyperbolic and thus not where they are expected to bifurcate. On the other hand, by concentrating on a possible "succession of bifurcations" one unnecessarily restricts his attention to neighborhoods of the expected bifurcation curves. Therefore we prefer to bifurcate the invariant tori of dimension n+1,n+2 or n+3 directly from the underlying trivial n-torus whose existence is known for all small $|\mu|$.

We briefly describe the contents of the following sections. In section 2 we present the normal form result under various nonresonance conditions. The truncations indicated above then lead to planar systems. These are also studied in section 2. The subsequent section is dedicated to the persistence of invariant tori for perturbations of degenerate systems (cf.[4],[23],[9]). The results of this section 3 are then applied in the following sections 4,5 and 6 to establish the existence of invariant tori of dimension n+1,n+2 or n+3 for the complete system (1.1).

Notational Remarks: The notations are mostly self-explanatory. Some shall be introduced now:

$$|m| := |m_1|+\cdots+|m_d| \quad \text{for} \quad m\in\mathbb{Z}^d,$$
$$(k,\omega) := k_1\omega_1+\cdots+k_n\omega_n \quad \text{for} \quad k\in\mathbb{R}^n,\ \omega\in\mathbb{R}^n,$$
$$r*f := (r_1f_1, r_2f_2)^T \quad \text{for} \quad r=(r_1,r_2)^T,\ f=(f_1,f_2)^T,$$
$$e_1 := (1,0)^T,\quad e_2 := (0,1)^T.$$

As long as the dimensions of vectors,matrices etc. are obvious they are not stated explicitly. The $O(.)$- and $o(.)$-statements refer to the case that the argument tends to zero and are supposed to hold uniformly with respect to the remaining variables (unless it is indicated differently).

2. Normal Forms and Truncated Systems

We consider the reduction of system (1.1) on the center manifold (1.8) which we denote for an integer $M>2$ by

$$\begin{aligned}\dot{x} &= \{A_1+\tilde{A}(y,\mu)\}x+X_2(x,y,\mu)+\cdots+X_M(x,y,\mu)+X^*(x,y,\mu),\\ \dot{y} &= \omega+\tilde{\omega}(y,\mu) \quad +Y_2(x,y,\mu)+\cdots+Y_M(x,y,\mu)+Y^*(x,y,\mu).\end{aligned} \tag{2.1}$$

Here, the X_j and Y_j are homogeneous polynomials in x of order j and X^* and Y^* are both of order $O(|x|^{M+1})$.The (4x4)-matrix $\tilde{A}$ and the n-vector $\tilde{\omega}$ are both of order $O(|\mu|)$. As noted in the introduction we may assume system (2.1) to belong to class C^N for a sufficiently large $N\in\mathbb{N}$.

It is our goal to eliminate the toroidal variable y from the leading terms of order $\leq M$ in system (2.1). Hereby we ask for the resulting system in "normal form" to belong to class C^m with $m \geq 2$. Such an elimination shall be achieved by a sequence of transformations familiar from the theory of normal forms (see e.g. [2] and [9]). These transformations lead to linear partial differential equations on a torus which admit solutions under suitable nonresonance conditions on $(\beta,\omega) = (\beta_1,\beta_2,\omega_1,\ldots,\omega_n)$.

<u>Definition 2.1</u>
Suppose there exist positive constants κ_o and κ_1 such that the diophantine condition

$$|\ell_1\beta_1+\ell_2\beta_2+(k,\omega)| \geq \kappa_o(|\ell|+|k|)^{-\kappa_1}$$

holds true for

$$\ell = o \in \mathbb{Z}^2 \quad \text{and} \quad k \in \mathbb{Z}^n-\{o\}$$

and

$$\ell \in \mathbb{Z}^2,\ 1\leq|\ell|\leq j, \quad \text{and} \quad k \in \mathbb{Z}^n.$$

Then we say that the vector (β,ω) satisfies the nonresonance condition NR(j).

<u>Lemma 2.2</u>
Suppose we are given system (2.1) with the above listed properties. If the nonresonance condition NR(2L) and the inequality

$$N \geq m + 4L(n+1+\kappa_1)$$

hold then there exists a smooth change of variables

$$x \to x + h_1(x,y,\mu), \quad y \to y + h_2(x,y,\mu)$$

leading to the following "normal form" for system (2.1) in polar coordinates

$$x_1=r_1\cos\theta_1,\ x_2=r_1\sin\theta_1,\ x_3=r_2\cos\theta_2,\ x_4=r_2\sin\theta_2 : \tag{2.2}$$

$$\dot{r} = r*[\alpha^*(\mu) + \sum_{j=1}^{L-1} R_{2j}(r,\mu)] + R^*(r,\theta,y,\mu), \tag{2.3a}$$

$$\dot{\theta} = \beta + \beta^*(\mu) + \sum_{j=1}^{L-1} \Theta_{2j}(r,\mu) + \Theta^*(r,\theta,y,\mu), \tag{2.3b}$$

$$\dot{y} = \omega + \omega^*(\mu) + \sum_{j=1}^{L} Y_{2j}(r,\mu) + Y^*(r,\theta,y,\mu). \tag{2.3c}$$

The terms on the right-hand-side of (2.3) satisfy the following:
a) α^*,β^* and ω^* are polynomials in μ vanishing for $\mu=o$.

b) R_{2j}, Θ_{2j} and Y_{2j} are homogeneous polynomials in r_1^2, r_2^2 of order j with coefficients that are polynomials in μ.

c) $R^* = O(|r,\mu|^{2L})$ is a C^m-function vanishing for $r=o$.

d) $\Theta_j^* = \hat{\Theta}_j(r,\theta,y,\mu)/r_j$ and $\hat{\Theta}_j = O(|r,\mu|^{2L})$ is a C^m-function $(j=1,2)$.

e) $Y^* = O(|r,\mu|^{2L+1})$ is a C^m-function.

To prove Lemma 2.2 one may follow the ideas presented in [2]. For a detailed proof we refer to [9]. Under the nonresonance condition NR(4) we denote the normal form system (2.3) of system (2.1) by

$$\begin{aligned} \dot{r} &= r*\{A\mu + M(r*r)\} + \tilde{R} , \\ \dot{\theta} &= \beta + \tilde{\beta}\mu + \Theta_2(r,o) + \tilde{\Theta} , \\ \dot{y} &= \omega + \tilde{\omega}\mu + Y_2(r,o) + \tilde{Y} \end{aligned} \tag{2.4}$$

introducing the notations

$$\begin{aligned} \alpha^*(\mu) &= A\mu + O(|\mu|^2), \\ R_2(r,\mu) &= \{M+O(|\mu|)\}(r*r) \quad \text{with} \quad M = \begin{pmatrix} a & b \\ c & d \end{pmatrix} \\ \beta^*(\mu) &= \tilde{\beta}\mu + O(|\mu|^2), \\ \omega^*(\mu) &= \tilde{\omega}\mu + O(|\mu|^2). \end{aligned} \tag{2.5}$$

The remainders are of the form

$$\begin{aligned} \tilde{R}(r,\theta,y,\mu) &= O(|\mu|^2|r| + |\mu||r|^3 + |r,\mu|^4), \quad \tilde{R}(o,\theta,y,\mu) = o, \\ \tilde{\Theta}_j(r,\theta,y,\mu) &= O(|\mu|^2 + |\mu||r|^2 + r_j^{-1}|r,\mu|^4), \quad j=1,2, \\ \tilde{Y}(r,\theta,y,\mu) &= O(|\mu|^2 + |\mu||r|^2 + |r,\mu|^4). \end{aligned} \tag{2.6}$$

As stated above we may assume system (2.4-6) to belong to C^m with $m \geq 2$. To system (2.4-6) we associate the truncated system

$$\begin{aligned} \dot{r} &= r*\{A\mu + M(r*r)\} , \\ \dot{\theta} &= \beta + \tilde{\beta}\mu + \Theta_2(r,o) , \\ \dot{y} &= \omega + \tilde{\omega}\mu + Y_2(r,o) \end{aligned} \tag{2.7}$$

and the planar system

$$\dot{r} = r*\{A\mu + M(r*r)\} . \tag{2.8}$$

Because of the choice of coordinates in (2.2) one expects semipositive equilibria $(\tilde{r}_1,o)$ and $(o,\tilde{r}_2)$ with $\tilde{r}_i > o$ and positive equilibria (r_1,r_2) with $r_i > o$ of system (2.8) to correspond to (n+1)- and (n+2)-dimensional tori of system (2.4) and hence of system (1.1). We shall investigate this correspondance in the sequel. System (2.8) has been extensively studied in [3,1o,11,22].

Under the assumption

$$ae_1^T A \neq (o,o) \tag{2.9}$$

there exists the semi-positive equilibrium

$$(\tilde{r}_1, o) = (\tilde{\gamma}_1(\mu), o) = (\sqrt{-\tfrac{1}{a} e_1^T A\mu}\ , o\ ) \tag{2.10a}$$

of (2.8) in the halfspace

$$H = \{\mu \in \mathbb{R}^2 : ae_1^T A\mu < o\}. \tag{2.10b}$$

Similarily, under the assumption

$$de_2^T A \neq (o,o) \tag{2.11}$$

one has the semi-positive equilibrium

$$(o, \tilde{r}_2) = (o, \tilde{\gamma}_2(\mu)) = (o\ , \sqrt{-\tfrac{1}{d} e_2^T A\mu}\ ) \tag{2.12a}$$

of (2.8) in the halfspace

$$H' = \{\mu \in \mathbb{R}^2 : de_2^T A\mu < o\}. \tag{2.12b}$$

If the conditions

$$\Delta := \det M = ad - bc \neq o\ , \tag{2.13a}$$

$$e_i^T M^{-1} A \neq (o,o),\ i=1,2, \tag{2.13b}$$

hold system (2.8) admits the positive equilibrium

$$(r_1, r_2) = (\gamma_1(\mu), \gamma_2(\mu)),\ \gamma_i(\mu) = \sqrt{-e_i^T M^{-1} A\mu} \quad (i=1,2) \tag{2.14a}$$

in the cone

$$C = \{\mu \in \mathbb{R}^2 : e_i^T M^{-1} A\mu < o\}. \tag{2.14b}$$

In case rank $A = 2$ the conditions (2.9), (2.11) and (2.13) simplify to $a \neq o$, $d \neq o$ and $\Delta \neq o$ respectively. The linearizations of (2.8) at (2.10) and (2.12) are given by

$$\begin{aligned} \dot{r}_1 &= -2e_1^T A\mu r_1 \\ \dot{r}_2 &= (-\tfrac{c}{a}, 1) A\mu r_2 \end{aligned} \quad \text{and} \quad \begin{aligned} \dot{r}_1 &= (1, -\tfrac{b}{d}) A\mu r_1 \\ \dot{r}_2 &= -2e_2^T A\mu r_2 \end{aligned} \tag{2.15}$$

respectively. For

$$\text{(a)} \quad (-\tfrac{c}{a}, 1) A \neq (o,o) \qquad \text{(b)} \quad (1, -\tfrac{b}{d}) A \neq (o,o) \tag{2.16}$$

the above semi-positive equilibria (2.10) and (2.12) undergo a change in stability at the lines

$$\{\mu :\ (-c, a) A\mu = o\} \qquad \text{and} \qquad \{\mu :\ (d, -b) A\mu = o\}$$

respectively (see (2.15)). Under condition (2.13) the positive equilibrium (2.14) bifurcates from the respective semi-positive ones at these lines in μ-space.

For a discussion of the stability properties of (2.14) we shift variables via

$$\rho = r - \gamma(\mu) \quad \text{for} \quad \mu \in C \tag{2.17}$$

to obtain

$$\dot{\rho} = P(\rho,\gamma(\mu)) := C(\gamma(\mu))\rho + F_2(\rho,\gamma(\mu)) + F_3(\rho) \tag{2.18}$$

with

$$C(\gamma) = 2\begin{pmatrix} a\gamma_1^2 & b\gamma_1\gamma_2 \\ c\gamma_1\gamma_2 & d\gamma_2^2 \end{pmatrix} \tag{2.19}$$

and homogeneous polynomials F_2 and F_3 in ρ of order 2 and 3. The following result on the planar system (2.8) and its positive equilibrium (2.14) is easily verified (cf.[9],[1o],[11]).

Lemma 2.3

Under the assumption (2.13) the following is true for $\mu \in C$:

1) The eigenvalues of $C(\gamma(\mu))$ are nonzero and their realparts are of the same (opposite) sign for $\Delta > o$ $(\Delta < o)$. They are pure imaginary if and only if one has

$$\Delta > o \quad \text{and} \quad (a,d)M^{-1}A\mu = o. \tag{2.2o}$$

2) Under condition (2.2o) system (2.18) possesses the integrating factor

$$I(\rho,\gamma(\mu)) = (\rho_1+\gamma_1(\mu))^{m_1}(\rho_2+\gamma_2(\mu))^{m_2},$$
$$m_1 := \frac{2}{\Delta}d(c-a) - 1, \quad m_2 := \frac{2}{\Delta}a(b-d) - 1,$$

so that $\rho=o$ is a center for system (2.18). Moreover, for

$$(a,d)M^{-1}A \neq (o,o) \tag{2.21}$$

system (2.18) is integrable if and only if μ belongs to the "critical halfline"

$$\ell_o := \{\mu \in \mathbb{R}^2 \cap C: (a,d)M^{-1}A\mu = o\}. \tag{2.22}$$

Lemma 2.3 gives a description of the stability properties of the equilibrium (2.14) of (2.8) and reveals a degenerate (i.e. vertical) Hopf bifurcation on ℓ_o (under (2.2o-21)). In section 6 we shall unfold this degeneracy by adjoining fifth order terms to the planar systen (2.8). One then arrives at a positive periodic solution of (2.8) for μ in a one-sided neighborhood of ℓ_o. It is expected to correspond to an invariant (n+3)-torus for the reduced system (2.1) and hence for the complete system (1.1).

3. Persistence of Invariant Tori

For positive integers a and g let us consider a 2-parameter family of ordinary differential equations

$$\begin{aligned} \dot{x} &= A(\nu,\varepsilon)x + \varepsilon^a F(x,\phi,\nu,\varepsilon), \quad x\in U\subset \mathbb{R}^k, \\ \dot{\phi} &= \omega(\nu,\varepsilon) + \varepsilon^g G(x,\phi,\nu,\varepsilon), \quad \phi\in T^n . \end{aligned} \tag{3.1}$$

The "normal coordinate" x is supposed to vary in a neighborhood U of the origin in $\mathbb{R}^k$, the parameters ν and ε in intervals $K:=[\nu_1,\nu_2]$ and $(o,\varepsilon_o]$ respectively. We assume the right-hand side of (3.1) to be continuous in $U\times T^n\times(\nu_1,\nu_2)\times(o,\varepsilon_o)$ and to belong to class $C^r, r\geq 2$, with respect to (x,ϕ). We let K' be the subset of K on which the linear equation

$$\dot{x} = A(\nu,\varepsilon)x \tag{3.2}$$

admits an exponential dichotomy with bound γ and (positive) exponent $\varepsilon^a\alpha(\nu), \nu\in K'$ (see e.g.[9] or [13]).

If $F(o,\phi,\nu,\varepsilon)=o$ then $\{x=o\}$ represents an invariant n-torus for system (3.1). In the following theorem we present sufficient conditions for the existence of an invariant n-torus for (3.1) close to $\{x=o\}$ in case F does not vanish for $x=o$. Thereby we allow $G(o,\phi,\nu,\varepsilon)$ to become unbounded when ν approaches ν_1 or ν_2. We shall model this with the help of the function

$$\sigma:[\nu_1,\nu_2] \to \mathbb{R}, \quad \sigma(\nu) = \min(|\nu-\nu_1|,|\nu-\nu_2|). \tag{3.3}$$

This kind of unboundedness frequently occurs for systems in polar coordinates. A typical example is the above system (2.4-6) since the remainder term in the $\dot{\theta}$-equation contains expressions with r_1 or r_2 as denominator. The following result states in simple but precise terms what parameter constellations lead to an invariant n-torus for the degenerate system (3.1).

Theorem 3.1

Given system (3.1) with the above listed properties let the functions F and G be of the form

$$F(x,\phi,\nu,\varepsilon) = F_1(\phi,\nu,\varepsilon)x + F_2(x,\phi,\nu,\varepsilon) + F_R(\phi,\nu,\varepsilon),$$

$$G(x,\phi,\nu,\varepsilon) = G_1(\nu,\varepsilon)x + G_2(x,\phi,\nu,\varepsilon) + G_R(\phi,\nu,\varepsilon)/\sigma(\nu)$$

with

$$F_1 = O(\varepsilon^b) \quad \text{and} \quad G_1 = O(1) \quad \text{for } \varepsilon\to o,$$

$$F_2 = O(|x|^2) \quad \text{and} \quad G_2 = O(|x|^2) \quad \text{for } x\to o,$$

$$F_R = O(\varepsilon^p) \quad \text{and} \quad G_R = O(\varepsilon^q) \quad \text{for } \varepsilon\to o$$

for positive integers b,p and q.

<u>Claim:</u> If

$$L := \min\{b, \tfrac{p}{2}, g+q-a, \tfrac{g+p-a}{2}, \tfrac{g+b-a+p/2}{2}\} > o \tag{3.4}$$

then there exist positive constants $\bar{\varepsilon}=\bar{\varepsilon}(K)$ and $\kappa=\kappa(K,r)$ with the following properties: If $\varepsilon\in(o,\bar{\varepsilon})$ then there exists a continuous function $x=S(\phi,\nu,\varepsilon)$ for all $\nu\in K'$ and all such ε with

$$\varepsilon^L < \kappa\alpha(\nu), \quad \varepsilon^{g+q-a} < \kappa\alpha(\nu)\sigma(\nu), \quad \varepsilon^{g+p-a} < \kappa\alpha(\nu)\sigma(\nu) \tag{3.5}$$

such that for fixed such ν,ε

$$M := \{(S(\phi,\nu,\varepsilon),\phi,\nu,\varepsilon): \phi\in T^n\} \tag{3.6}$$

represents an invariant n-dimensional torus for system (3.1). The function S is C^{r-2} with respect to ϕ and obeys the estimate

$$S(\phi,\nu,\varepsilon) = O(\varepsilon^p/\alpha(\nu)) \quad \text{for } \varepsilon\to o.$$

Furthermore, if the x-component of a solution $(x,\phi)(t,\nu,\varepsilon)$ of (3.1) has sufficiently small norm for all t then this solution satisfies

$$x(t,\nu,\varepsilon) = S(\phi(t,\nu,\varepsilon),\nu,\varepsilon) \quad \text{for all } t.$$

For fixed $\nu\in K'$ the stability properties of M are determined by the stability properties of the trivial solution of the linear system (3.2).

The restrictions on the size of ε in (3.5) take into account that (i) the normal rate should dominate the tangential rate, (ii) the x- and ϕ-equations in (3.1) are to be weakly coupled and (iii) the unboundedness of G for $\nu\to\nu_1,\nu_2$ is to be compensated. We note that ε necessarily tends to zero when the exponential rate $\alpha(\nu)$ or the "critical denominator $\sigma(\nu)$" approaches zero.

Given the exponents a,b and g condition (3.4) gives a lower bound on the size of p and q. In other words, Theorem 3.1 tells how well one needs to approximate an expected invariant manifold in order to actually prove its existence. Related invariant manifold results for degenerate systems where both the normal and the tangential rate tend to o (with $\varepsilon\to o$) have been established in several places. We'd like to mention the results in [23],[4], [12],[19] and [9].

We give an outline of the proof of Theorem 3.1. In the well-known way (see e.g.[2o,Ch.5]) we extend the right-hand side of (3.1) in a C^∞-fashion to the whole x-space so that the extended and the originally given local system (3.1) coincide for $|x|<\varepsilon^{p/2}$ inside the neighborhood U. By a repeated application of Theorem 2.1 of [9] (generalizing Satz 7.1 in [2o,Ch.5]) one derives the existence of the above manifold (3.6) inside the region where the extended and the local system (3.1) are the same.

4. Existence of (n+1)-Dimensional Invariant Tori

In this section we return to our normal form system (2.4-6) and investigate the question whether system (2.4) possesses (n+1)-dimensional invariant tori in case the planar system (2.8) admits the semi-positive solution (2.1o). Thereby we assume (2.9) and (2.16a), i.e.

$$ae_1^T A \neq (o,o) \quad \text{and} \quad (c,-a)A \neq (o,o). \tag{4.1}$$

The form of the solution (2.1o) leads to the scaling

$$\mu = \varepsilon^2\nu \;,\; r \to \varepsilon r \quad \text{with } \varepsilon>o,\nu\in\mathbb{R}^2 . \tag{4.2}$$

With $f(r_1^2,r_2^2):=\Theta_2(r,o)$ and $g(r_1^2,r_2^2):= Y_2(r,o)$ (see Lemma 2.2) system (2.4) reads in these new coordinates

$$\begin{aligned}
\dot r &= r*\varepsilon^2\{A\nu+M(r*r)\} &&+ O(\varepsilon^3) \;,\; r\in\mathbb{R}^2 ,\\
\dot\theta &= \beta+\varepsilon^2\{\tilde\beta\nu+f(r_1^2,r_2^2)\} &&+ O(\varepsilon^3) \;,\; \theta\in T^2 ,\\
\dot y &= \omega+\varepsilon^2\{\tilde\omega\nu+g(r_1^2,r_2^2)\} &&+ O(\varepsilon^3) \;,\; y\in T^n .
\end{aligned} \tag{4.3}$$

By introducing the notations

$$\begin{aligned}
u &= r_1 - \tilde\gamma_1(\nu) \quad (\text{cf.}(2.1o)) ,\\
v &= r_2(\cos\theta_2,\sin\theta_2)^T ,\\
\hat L_2(\nu) &= (-\tfrac{c}{a},1)A\nu
\end{aligned} \tag{4.4}$$

we obtain

$$\begin{aligned}
\dot u &= \varepsilon^2 U(u,|v|^2,\nu) + O(\varepsilon^3),\\
\dot v &= V(u,|v|^2,\varepsilon^2,\nu)v + O(\varepsilon^3),\\
\dot\theta_1 &= \beta_1 + \varepsilon^2 e_1^T\{\tilde\beta\nu+f(\{u+\tilde\gamma_1(\nu)\}^2,|v|^2)\} + O(\varepsilon^3),\\
\dot y &= \omega + \varepsilon^2\{\tilde\omega\nu+g(\{u+\tilde\gamma_1(\nu)\}^2,|v|^2)\} + O(\varepsilon^3)
\end{aligned} \tag{4.5}$$

where U and $V = \begin{pmatrix} V_1 & -V_2 \\ V_2 & V_1 \end{pmatrix}$ are given by

$$U = -2a\tilde{\gamma}_1^2(\nu)u + 3a\tilde{\gamma}_1(\nu)u^2 + au^3 + b(u+\tilde{\gamma}_1(\nu))|v|^2 ,$$
$$V_1 = \varepsilon^2\{\hat{L}_2(\nu) + 2c\tilde{\gamma}_1(\nu)u + cu^2 + d|v|^2\} , \tag{4.6}$$
$$V_2 = \beta_2 + \varepsilon^2 e_2^T\{ \tilde{\beta}\nu + f(\{u+\tilde{\gamma}_1(\nu)\}^2, |v|^2)\}.$$

We restrict ν to the unit circle and parametrize the closed halfspace $\bar{H}$ of (2.1ob) by

$$\mu = \varepsilon^2\nu, \quad \nu=\nu(\tau)=(\cos\tau,\sin\tau)^T, \varepsilon>o, \quad \tau\in K:=[\underline{\tau},\underline{\tau}+\pi]. \tag{4.7}$$

Then the linear system

$$\dot{u} = -\varepsilon^2 2a\tilde{\gamma}_1^2(\nu(\tau))u, \qquad \dot{v} = V(o,o,\varepsilon^2,\nu(\tau))v \tag{4.8}$$

admits an exponential dichotomy in $K':=K-\{\tau_2\}$ where τ_2 is the unique zero of $\hat{L}_2(\nu(\tau))$ in K (cf. (4.4)). The exponential rate for (4.8) can be taken to be

$$\varepsilon^2\alpha(\tau) := \varepsilon^2\min\{|2a\tilde{\gamma}_1^2(\nu(\tau))|, |\hat{L}_2(\nu(\tau))|\}, \quad \tau\in K'. \tag{4.9}$$

System (4.5-6) at $\nu=\nu(\tau)$ is of the form (3.1). In the notations of section 3 one has

$$a = g = 2, \ b = p = q = 1, \ L = 1/2.$$

Since the $O(\varepsilon^3)$-term in the θ_1-equation contains the "critical denominator" $u+\tilde{\gamma}_1(\nu(\tau))$ we define

$$\sigma(\tau) := \frac{1}{2}\tilde{\gamma}_1(\nu(\tau)), \quad \tau\in K, \tag{4.1o}$$

(cf. (3.3)). As indicated in the outline of the proof of Theorem 3.1 we are interested in the values of u with $|u|<\varepsilon^{p/2}=\varepsilon^{1/2}$. By imposing the condition

$$\varepsilon^{1/2} < \sigma(\tau), \quad \tau\in K, \tag{4.11}$$

we can estimate the critical denominator as follows:

$$u + \tilde{\gamma}_1(\nu(\tau)) > \sigma(\tau). \tag{4.12}$$

By taking (4.9-12) and (3.5) into account an application of Theorem 3.1 leads to

Theorem 4.1

Under the nonresonance condition NR(4) and assumption (2.9) system (2.1) can be reduced to system (4.5-6). In the above setting (including the strengthened version (4.1) of condition (2.9)) there exist positive constants $\bar{\varepsilon}=\bar{\varepsilon}(K)$ and $\kappa=\kappa(K,m)$ with the following properties: If $\varepsilon\in(o,\bar{\varepsilon})$ then there exists a continuous function $(u,v)=S(\theta_1,y,\varepsilon,\tau)$ for $\tau\in K'$ and all such ε with

$$\varepsilon^{1/2} < \kappa\alpha(\tau)$$

so that for fixed such ε,τ

$$M_1^{n+1} := \{(S(\theta_1, y, \varepsilon, \tau), \theta_1, y, \varepsilon, \tau) : (\theta_1, y) \in T^{n+1}\} \qquad (4.13)$$

represents an invariant (n+1)-dimensional torus for system (4.5-6) at $\nu=\nu(\tau)$. The function S belongs to C^{m-2} with respect to (θ_1, y) and is of order $O(\varepsilon/\alpha(\tau)), \varepsilon \to o$.

Of course, the analogous result can be derived for system (2.1) under the assumptions (2.11) and (2.16b). The corresponding invariant (n+1)-torus for the analogue of system (4.5-6) will be denoted by M_2^{n+1}.

Remark 4.2: Starting from the trivial (n+1)-torus $\{(u,v)=(o,o)\}$ of the truncated system, i.e. system (4.5) without the $O(\varepsilon^3)$-terms, Theorem 4.1 establishes the existence of invariant (n+1)-tori for system (4.1) and hence for the full $(4+n+\ell)$-dimensional system (1.1) for $\mu=\varepsilon^2\nu(\tau)$ for all "directions" τ in the halfspace H with the possible exception of the direction τ_2 where the exponential rate for the linear system (4.8) vanishes. Thus Theorem 4.1 shows that the cubic system (2.7) already represents a sufficiently good approximation of the complete system (2.4) in the sense that (n+1)-tori of (2.7) persist under $O(\varepsilon^3)$-perturbations. Under the nonresonance condition NR(5) the O-terms in (4.5) are of order ε^4 which increases the admissible size of ε in Theorem 4.1 to $\varepsilon^1 < \kappa\alpha(\tau), \tau \in K'$. In case rank $A = 2$ the halfline ℓ_2 corresponding to τ_2 is in the interior of H and the (n+1)-torus (4.13) undergoes a change in stability there. This change is due to the fact that the eigenvalues of the v-equation in (4.8) cross the imaginary axis for $\tau=\tau_2$. By considering μ on a circle of radius ε^2 there is a gap $[\tau_2-\tau_\varepsilon, \tau_2+\tau_\varepsilon]$ on which the existence of the invariant (n+1)-tori has not been established.

Remark 4.3: Under the conditions (2.9) and (2.13) the halfline ℓ_2 represents a Hopf bifurcation curve for the truncated (u,v)-system obtained by discarding the $O(\varepsilon^3)$-terms in the (u,v)-part of (4.5). In [9] we have shown that this Hopf bifurcation at ℓ_2 gives rise to invariant (n+2)-tori for system (2.1) in a thin wedge on one side of ℓ_2.

5. Existence of (n+2)-Dimensional Invariant Tori

In the present section we shall show that system (2.4-6), derived under the nonresonance condition NR(4), possesses invariant (n+2)-tori under the assumption

$$\Delta = ad - bc \neq o, \quad e_i^T M^{-1} A \neq (o,o), i=1,2, \tag{5.1a}$$

$$(a,d)M^{-1}A \neq (o,o) \quad \text{in case} \quad \Delta > o \tag{5.1b}$$

(cf. (2.13) and (2.21)). For rank$A = 2$ condition (5.1) simplifies to $\Delta \neq o$ and $ad \neq o$ for $\Delta > o$. The case

$$\Delta > o \quad \text{and} \quad (a,d)M^{-1}A = (o,o),$$

where system (2.18) is integrable in the whole cone C is considered in section 5 of [9]. Under condition (5.1) system (2.18) is integrable at most along the "critical halfline" ℓ_o as shown in Lemma 2.3.

Condition (5.1a) implies the existence of the positive solution $(\gamma_1(\mu), \gamma_2(\mu))$ in the cone C (cf. (2.13-14)). The form of the $\gamma_i(\mu)$ again leads to the scaling (4.2) and to system (4.3). The translation

$$\rho = r - \gamma(\nu), \quad \nu \in C, \tag{5.2}$$

then takes us from system (4.3) to

$$\begin{aligned} \dot{\rho} &= \varepsilon^2 P(\rho, \gamma(\nu)) + O(\varepsilon^3), \\ \dot{\theta} &= \beta + \varepsilon^2 \{ \tilde{\beta}\nu + \Theta_2(\rho + \gamma(\nu), o) \} + O(\varepsilon^3), \\ \dot{y} &= \omega + \varepsilon^2 \{ \tilde{\omega}\nu + Y_2(\rho + \gamma(\nu), o) \} + O(\varepsilon^3) \end{aligned} \tag{5.3}$$

where the polynomial P is given by (2.18) and where the O-term in the $\dot{\theta}$-equation contains the "critical denominators" $\rho_i + \gamma_i(\nu)$, $i=1,2$ (cf. (2.6) and (5.2)).

For an application of Theorem 3.1 we again restrict ν as in (4.7) by introducing the parameter τ via

$$\nu = \nu(\tau) = (\cos\tau, \sin\tau)^T \tag{5.4a}$$

where we assume for convenience that the closed cone $\bar{C}$ is parametrized by ε, τ via

$$\begin{aligned} &\mu = \varepsilon^2 \nu(\tau), \quad \tau \in K := [\tau_1, \tau_2] \quad \text{with} \\ &\gamma_2(\nu(\tau_1)) = o, \quad \gamma_1(\nu(\tau_2)) = o \quad \text{and} \quad \tau_1 < \tau_2 \leq \tau_1 + \pi. \end{aligned} \tag{5.4b}$$

In case the halfline ℓ_o is present in C it corresponds to the unique $\tau^* \in (\tau_1, \tau_2)$ with

$$(a,d)M^{-1}A\nu(\tau^*) = o. \tag{5.5}$$

The set K' of section 3 is then $K' = (\tau_1, \tau_2) - \{\tau^*\}$ and the exponential rate for the linear system

$$\dot{\rho} = \varepsilon^2 C(\gamma(\nu(\tau)))\rho$$

can be chosen to be

$$\varepsilon^2\alpha(\tau) := \frac{1}{2}\varepsilon^2 \min|\mathrm{Re}(\lambda_i(\tau))| \ , \ \tau\in K',$$

where $\lambda_1(\tau),\lambda_2(\tau)$ stand for the eigenvalues of $C(\gamma(\nu(\tau))),\tau\in K$. In the notations of section 3 we have as in the previous section

$$a = g = 2,\ b = p = q = 1,\ L = 1/2$$

so that the "critical denominators" can be bounded from below for

$$|\rho| < \varepsilon^{1/2} < \frac{1}{2}\min\{\gamma_i(\nu(\tau))\} =: \sigma(\tau)$$

by

$$\rho_i+\gamma_i(\nu(\tau)) > \sigma(\tau),\ i=1,2,$$

(cf. (4.9-12)). If we employ the relation $\sigma^2(\tau)=O(\alpha(\tau)),\tau\to\tau_1,\tau_2$, Theorem 3.1 leads to

Theorem 5.1

Under the nonresonance condition NR(4) and condition (5.1) system (2.1) can be reduced to system (5.3). In the above setting there exist positive constants $\bar{\varepsilon}=\bar{\varepsilon}(K)$ and $\kappa=\kappa(K,m)$ with the following properties: If $\varepsilon\in(o,\bar{\varepsilon})$ then there exists a continuous function $\rho=S(\theta,y,\varepsilon,\tau)$ for $\tau\in K'$ and all such ε with

$$\varepsilon^{1/2} < \kappa\alpha(\tau) \tag{5.6}$$

so that for fixed such ε,τ

$$M^{n+2} = \{(S(\theta,y,\varepsilon,\tau),\theta,y,\varepsilon,\tau):\ (\theta,y)\in T^{n+2}\}$$

represents an invariant (n+2)-torus for system (5.3) at $\nu=\nu(\tau)$. The function S is C^{m-2} with respect to (θ,y) and obeys the estimate $O(\varepsilon/\alpha(\tau)),\varepsilon\to o$. Furthermore, if the ρ-component of a solution of (5.3) (at $\nu=\nu(\tau)$) has sufficiently small norm for all t then this solution belongs to M^{n+2} for all t.

Remark 5.2: The manifold M^{n+2} of system (5.3) gives rise to an invariant (n+2)-torus of system (2.1) and hence of system (1.1) for $\mu=\varepsilon^2\nu(\tau)$ for all "directions" τ in the cone C with the possible exception of the direction τ^* along which the truncated planar system (2.18) is integrable. At τ^* the invariant torus M^{n+2} of (5.3) changes from being stable to being unstable due to the eigenvalues $\lambda_1(\tau),\lambda_2(\tau)$ crossing the imaginary axis. We will show that this change in stability may lead to the existence of invariant (n+3)-tori on one side of τ^*. We note that the condition (5.6) forces ε to approach o when τ tends to τ_1,τ_2 or τ^*. Thus, in the direction τ^* where the bifurcation to an (n+3)-torus is to be expected the (n+2)-torus M^{n+2} is not known to exist. We'd like to add that the boundedness property of M^{n+2} stated in Theorem 5.1 allows the continuation of the invariant (n+2)-tori of Remark 4.3.

6. Existence of (n+3)-Dimensional Invariant Tori

We finally investigate system (2.4-6) in the case

$$o > ad > bc\ , \quad \text{rank } A = 2. \tag{6.1}$$

In Lemma 2.3 we have shown that the planar system (2.8) associated to (2.4) is integrable on the halfline ℓ_o in C (see (2.22) and (2.14)). In terms of the parametrization (5.4) ℓ_o corresponds to the unique τ^* determined by (5.5). We shall study system (2.4) in a cusp-like neighborhood of ℓ_o which we parametrize by ε,η via

$$\mu = \varepsilon^2\nu(\tau^*+\varepsilon^2\eta)\ ,\ \varepsilon>o\ ,\ |\eta|<\eta_o\ , \tag{6.2}$$

so that η describes a transversal crossing of ℓ_o given by η=o. An expansion of (6.2) for small $\varepsilon^2\eta$ yields

$$\mu = \varepsilon^2\nu,\quad \nu = \nu^* + \varepsilon^2\eta\nu^{\perp} + \varepsilon^2\eta V(\varepsilon^2\eta) \tag{6.3a}$$

with

$$\nu^* = (\cos\tau^*,\sin\tau^*)^T,\qquad \nu^{\perp} = (-\sin\tau^*,\cos\tau^*)^T\ ,$$
$$V = o(1)\quad \text{for } \varepsilon^2\eta\to o\ . \tag{6.3b}$$

The ε^2-part of the planar system (2.8) along (6.3) in the scaled variables

$$r \to \varepsilon r \tag{6.3c}$$

is then given by

$$\dot{r} = r*\varepsilon^2\ \{A\nu^*+M(r*r)\}\ . \tag{6.4}$$

System (6.4) is integrable. In particular one has

$$\text{trace } C(\gamma(\nu^*)) = (a,d)M^{-1}A\nu^* = o$$

(cf. (2.19), (5.5)). Without loss of generality we assume the ρ-coordinate to be such that

$$C(\gamma(\nu^*)) = \begin{pmatrix} o & -\beta^* \\ \beta^* & o \end{pmatrix},\quad \beta^*=2\sqrt{\Delta}\gamma_1(\nu^*)\gamma_2(\nu^*)\ . \tag{6.5}$$

To unfold the degenerate Hopf bifurcation in (6.4) we assume the nonresonance condition NR(6) to hold. By Lemma 2.2 we then can adjoin fifth order terms to (6.4) that do not depend on the angular variables. System (2.4) is then along (6.3) of the form

$$\dot{r} = r*[\varepsilon^2\{A\nu^*+M(r*r)\}+\varepsilon^4\{R_{2/4}(r,\nu^*)+\eta A\nu^{\perp}\}] + O(\varepsilon^5), \tag{6.6a}$$

$$\dot{\theta} = \beta + \varepsilon^2\Theta_{2/4}(r,\varepsilon^2,\nu) + O(\varepsilon^5), \tag{6.6b}$$

$$\dot{y} = \omega + \varepsilon^2 Y_{2/4}(r,\varepsilon^2,\nu) + O(\varepsilon^5) \tag{6.6c}$$

where the $R_{2/4}, \Theta_{2/4}$ and $Y_{2/4}$ are sums of first and second order homogeneous polynomials in r_1^2, r_2^2. Because of NR(6) these expressions do not depend on (θ,y). By changing coordinates via

$$\rho = r - \Gamma(\varepsilon^2,\eta) \tag{6.7}$$

with $\Gamma(\varepsilon^2,\eta)$ being the positive equilibrium of the system obtained by deleting the O-term in (6.6a) we obtain from (6.6) a radial equation of the form

$$\dot{\rho} = \varepsilon^2 P(\rho,\gamma(\nu^*)) + \varepsilon^4 P_4(\rho,\eta) + O(\varepsilon^5) \tag{6.8}$$

where P is the polynomial from (2.18) (up to the change of coordinates leading to (6.5)). The expression P_4 is a fifth order polynomial in ρ of the form

$$P_4(\rho,\eta) = C_4^*(\eta)\rho + O(|\rho|^2) \tag{6.9}$$

with

$$C_4^*(\eta) = C_{4o} + \eta C_{41}, \tag{6.10a}$$

$$\text{trace } C_4^*(\eta) = \text{trace } C_{4o} - 4\eta(a,d)M^{-1}A\nu^{\perp}. \tag{6.10b}$$

Because of (6.1),(6.3) and (5.5) the η-coefficient in (6.10b) is nonzero so that we may change parameters via

$$\delta = \text{trace } C_4^*(\eta) .$$

Therefore, $\delta=o$ corresponds to the critical curve inside the cusp where the linear system

$$\dot{\rho} = \{\varepsilon^2 C(\gamma(\nu^*)) + \varepsilon^4 C_4(\eta)\}\rho \tag{6.11}$$

has $\rho=o$ as a center. The eigenvalues of the matrix in (6.11) cross the imaginary axis transversally at $\delta=o$ so that one may expect a Hopf bifurcation for the planar system obtained by deleting the O-term in (6.8). In terms of the new parameters ε,δ we denote system (6.6) after the translation (6.7) by

$$\dot{\rho} = \varepsilon^2 P(\rho,\gamma(\nu^*)) + \varepsilon^4 R_4(\rho,\delta) + O(\varepsilon^5) \tag{6.12a}$$

and

$$\dot{\theta} = \beta + \varepsilon^2\Theta_o(\rho,\varepsilon^2,\delta) + O(\varepsilon^5), \tag{6.12b}$$

$$\dot{y} = \omega + \varepsilon^2 Y_o(\rho,\varepsilon^2,\delta) + O(\varepsilon^5). \tag{6.12c}$$

Because of (6.9-1o) R_4 is a fifth order polynomial of the form

$$R_4(\rho,\delta) = C_4(\delta)\rho + O(|\rho|^2), \quad \text{trace } C_4(\delta) = \delta. \tag{6.13}$$

Recalling the integrability property of Lemma 2.3 and the analyticity of the cubic system we may assume without loss of generality that the ε^2-part of the above ρ-equation is in normal form (see e.g.[2]). Thus, introducing polar coordinates

$$\rho_1 = u\cos\psi, \quad \rho_2 = u\sin\psi, \quad u<\rho_o, \tag{6.14}$$

we may assume (6.12) to lead in the coordinates (6.14) to

$$\begin{aligned} \dot{u} &= \varepsilon^4 U(u,\psi,\delta) + O(\varepsilon^5), \\ \dot{\psi} &= \varepsilon^2\Psi(u^2) + \varepsilon^4\tilde{\Psi}(u,\psi,\delta) + O(\varepsilon^5), \\ \dot{\theta} &= \beta + \varepsilon^2\tilde{\Theta}_o(u,\psi,\varepsilon^2,\delta) + O(\varepsilon^5), \\ \dot{y} &= \omega + \varepsilon^2\tilde{Y}_o(u,\psi,\varepsilon^2,\delta) + O(\varepsilon^5) \end{aligned} \tag{6.15}$$

with

$$\Psi(o) = \beta^*, \quad \Psi(u^2) > o \quad \text{for } u<\rho_o \text{ (see (6.5) for } \beta^*).$$

Here, the terms explicitly listed in (6.15) are analytic in (u,ψ). By near-identical transformations familiar from the theory of averaging (cf.[9]) we may remove the ψ-dependence of these terms. E.g.,in a first step, we change coordinates via

$$\begin{aligned} \underline{u} &= u + \varepsilon^2 h(u,\psi,\delta), \\ h(u,\psi,\delta) &= \frac{1}{\Psi(u^2)}\left[\frac{\psi}{2\pi}\int_o^{2\pi} U(u,\tau,\delta)d\tau - \int_o^{\psi} U(u,\tau,\delta)d\tau\right]. \end{aligned} \tag{6.16}$$

This change (6.16) leads to a $\underline{u}$-equation of the form

$$\dot{\underline{u}} = \varepsilon^4 M(\underline{u},\delta) + O(\varepsilon^5), \quad M(u,\delta) = \frac{1}{2\pi}\int_o^{2\pi} U(u,\tau,\delta)d\tau. \tag{6.17a}$$

Because of (6.13) one has

$$M(u,\delta) = u\{\frac{\delta}{2} + M_1(u^2,\delta)\} = \frac{\delta}{2}u + O(u^3). \tag{6.17b}$$

Transformations in the angular variables,similar to (6.16), remove the ψ-dependence in the leading terms of the angular equations so that one arrives at a system of the form

$$\begin{aligned} \dot{u} &= \varepsilon^4 M(u,\delta) + O(\varepsilon^5), \\ \dot{\psi} &= \varepsilon^2\Psi(u^2) + \varepsilon^4\hat{\Psi}(u,\varepsilon^2,\delta) + O(\varepsilon^5), \end{aligned} \tag{6.18a}$$

and

$$\dot{\theta} = \beta + \varepsilon^2\hat{\theta}(u,\varepsilon^2,\delta) + O(\varepsilon^5), \tag{6.18b}$$

$$\dot{y} = \omega + \varepsilon^2\hat{Y}(u,\varepsilon^2,\delta) + O(\varepsilon^5). \tag{6.18c}$$

We note that the coordinates in (6.15) are chosen in such a way that the meanvalue in (6.17) essentially is the familiar Melnikov integral (see [1],[11] and [9]). Simple positive zeros of $M(u,\delta)$ are then expected to give rise to invariant (n+3)-tori for the full system (6.18) and hence for system (6.6) since this is the case for the system obtained by deleting the O-terms in (6.18).

In order to actually prove the existence of such tori for (6.18) we need to check whether the O-terms present a sufficiently small perturbation in the sense of Theorem 3.1. We note that this would be the case if the $O(\varepsilon^5)$-term in the u-equation were of the form

$$\varepsilon^5 U^*(u,\varepsilon,\delta) + O(\varepsilon^7) \tag{6.19}$$

with an angular independent U^*. In the present contribution we restrict ourselves to the simplest case where this is true by strenghtening the above nonresonance condition NR(6) to NR(8). In [9] we have considered system (6.18) under weaker conditions allowing a strong resonance between β_1 and β_2 of the form

$$q\beta_1 + p\beta_2 = o \quad \text{for} \quad |p|+|q| = 7 \text{ or } 8 \ (p,q\in\mathbb{Z}).$$

Under NR(8) the $O(\varepsilon^5)$-term in equation (6.6a) can be written as

$$r{*}\varepsilon^6 R^*(r,\eta) + O(\varepsilon^7).$$

In the obvious way we redo the above steps and arrive at a system (6.18) where the $O(\varepsilon^5)$-term in the u-equation actually is of the form (6.19). To this system we associate the bifurcation equation

$$\frac{\delta}{2} + M_1(u^2,\delta) + \varepsilon^2 U^*(u,\varepsilon,\delta) = o. \tag{6.2o}$$

Let $\zeta(\delta)$ be a simple positive zero of (6.17b) in the domain of M. By the Implicit Function Theorem it determines a unique positive solution $u = \zeta^*(\varepsilon^2,\delta) = \zeta(\delta)+O(\varepsilon^2)$ of (6.2o). If we let

$$\alpha(\delta) := \frac{\partial}{\partial u} M(\zeta(\delta),\delta) \tag{6.21}$$

for such δ belonging to an interval of the form (δ_1,δ_2) then the deviation

$$w = u - \zeta^*(\varepsilon^2,\delta)$$

satisfies a differential equation of the form

$$\dot{w} = \varepsilon^4\{\alpha(\delta)w + W_2(w,\delta) + \varepsilon^2 W_1(w,\delta)\} + O(\varepsilon^7), \tag{6.22a}$$

$$\begin{aligned}
\dot\psi &= \varepsilon^2\Psi(w,\varepsilon,\delta) + O(\varepsilon^5),\\
\dot\theta &= \beta+\varepsilon^2\Theta(w,\varepsilon,\delta) + O(\varepsilon^5),\\
\dot y &= \omega+\varepsilon^2 Y(w,\varepsilon,\delta) + O(\varepsilon^5)
\end{aligned} \tag{6.22b}$$

with

$$\alpha(\delta)\neq o \quad \text{for } \delta\in(\delta_1,\delta_2),\ W_2=O(|w|^2) \quad \text{and } W_1=O(|w|).$$

Thus, the system obtained from (6.22) by deleting the O-terms possesses the invariant (n+3)-torus $\{w=o\}$. Now, system (6.22) is of a form such that Theorem 3.1 is applicable. In the notations of section 3 one has

$$a = 4,\ b = g = 2,\ p = q = 3,\quad L = 1/4 .$$

The "critical denominator" $w+\zeta^*(\varepsilon^2,\delta)$ in the ψ-equation can be bounded away from zero for $|w|<\varepsilon^{p/2}=\varepsilon^{3/2}$ by

$$w + \zeta^*(\varepsilon^2,\delta) > \zeta(\delta)/2$$

provided one has

$$\varepsilon^{3/2} < \kappa_o\zeta(\delta) \tag{6.23}$$

for an appropriate $\kappa_o(\delta_1,\delta_2)$. The "critical denominators" in the θ-equation,given in original coordinates by $\rho_i+\gamma_i(\nu^*)+O(\varepsilon^2)$ for $i=1,2$, can be bounded away from zero by a fixed positive constant for $\delta\in(\delta_1,\delta_2)$ and small ε. Therefore Theorem 3.1 furnishes an invariant (n+3)-torus for (6.22) corresponding to the zero $\zeta(\delta)$. We formulate this result under the additional assumption

$$M_1(u^2,o) = Ku^2 + O(u^4),\quad K\neq o, \tag{6.24}$$

guaranteeing the existence of a simple positive zero $\zeta(\delta)$ of (6.17) for small $|\delta|$ with $\delta K<o$. Because of (6.17) and (6.21) one has

$$\zeta(\delta) = \sqrt{-\delta/2K}\{1+o(1)\},\ \delta\to o, \tag{6.25a}$$

$$\alpha(\delta) = -\delta\{1+o(1)\},\ \delta\to o. \tag{6.25b}$$

We summarize the above derivation in

<u>Theorem 6.1</u>

Under the nonresonance condition NR(8) and the condition (6.1) system (2.4) can be reduced to system (6.22). If the meanvalue $M(u,\delta)$ in (6.17) satisfies condition (6.24) then there exist positive constants $\delta_o,\bar\varepsilon=\bar\varepsilon(\delta_o)$ and $\kappa=\kappa(\delta_o,m)$ with the following properties: If $\varepsilon\in(o,\bar\varepsilon)$ then there exists a continuous function $w=S(\psi,\theta,y,\varepsilon,\delta)$ for $o<|\delta|<\delta_o$, $\delta K<o$ and all such ε with

$$\varepsilon < \kappa|\delta|^4 \tag{6.26}$$

so that for fixed such ε,δ

$$M^{n+3} = \{(S(\psi,\theta,y,\varepsilon,\delta),\psi,\theta,y,\varepsilon,\delta):\ (\psi,\theta,y)\in T^{n+3}\}$$

represents an invariant (n+3)-torus for (6.22). The function S belongs to C^{m-2} with respect to (ψ,θ,y) and is of the order $O(\varepsilon^3/|\delta|)$, $\delta\to o$. The torus M^{n+3} is stable for $K<o$ and unstable for $K>o$.

Remark 6.2

1) The above Theorem describes the analogue to the nondegenerate Hopf bifurcation in the unperturbed ρ-equation (6.8). There are two interpretations of the condition (6.26) determining the parameter set on which the torus M^{n+3} exists. First, for fixed $|\delta|$ with $\delta K<o$, the torus exists on a fixed curve emanating from $\mu=o$ with a quadratic tangency to the critical halfline ℓ_o. The range of $|\mu|=\varepsilon^2$ is given by (6.26). Secondly, for fixed $|\mu|=\varepsilon^2$, (6.26) states how large $|\delta|$ should be in order to get the existence of M^{n+3}. Again, by considering μ on a circle of radius ε^2 there is a gap (o,δ_ε) or $(-\delta_\varepsilon,o)$ on which the (n+3)-torus has not been established.

2) Condition (6.24) just guarantees the existence of a small simple positive zero of the bifurcation equation (6.2o). The analysis preceeding (6.24) shows how to continue the torus M^{n+3} from smaller to larger values of $|\delta|$. As long as the zero remains simple and in the domain in question it corresponds to an invariant (n+3)-torus.

3) There is an alternative way to prove the existence of invariant (n+3)-tori for system (6.6). It is based on the introduction of a moving coordinate system along any periodic solution of the integrable planar system (6.4). In these coordinates one also can derive a system of the form (6.22) so that Theorem 3.1 furnishes (n+3)-dimensional invariant tori for (6.22) (see [1],[9]).

References

1. A.A.Andronov et al.,Theory of Bifurcations of Dynamic Systems on a Plane,Israel Prog.Sci.Transl.,Jerusalem(1971).

2. Y.N.Bibikov,Local Theory of Nonlinear Analytic Ordinary Differential Equations,Lect.Notes in Math.7o2,Springer,Berlin(1979).

3. S.N.Chow&J.K.Hale,Methods in Bifurcation Theory,Grundlehren der

math.Wiss.251,Springer,New York(1982).

4. S.P.Diliberto,"New Results on Periodic Surfaces and the Averaging Principle" in "Differential and Functional Equations", W.A.Harris&Y.Sibuya(eds.),W.A.Benjamin Inc.,New York-Amsterdam (1967),49-87.

5. D.Flockerzi,"Weakly Nonlinear Systems and Bifurcation of Higher Dimensional Tori" in "Equadiff 82,Proceedings",H.W.Knobloch & K.Schmitt(eds.),Lect.Notes in Math.1o17,Springer,Berlin(1983), 185-193.

6. D.Flockerzi,"Generalized Bifurcation of Higher Dimensional Tori",JDE 55.3(1984),346-367.

7. D.Flockerzi,"On the $T^k \to T^{k+1}$ Bifurcation Problem" in "Singularities & Dynamical Systems",S.N.Pnevmatikos(ed.),North-Holland, Amsterdam-London(1985),35-46.

8. D.Flockerzi,"Resonance and Bifurcation of Higher Dimensional Tori",Proc.AMS 94.1(1985),147-157.

9. D.Flockerzi,Invariant Manifolds and Bifurcation of Tori,Habilitationsschrift,Univ.Würzburg(1986).

1o. J.Guckenheimer,"Multiple Bifurcation Problems of Codimension Two",SIAM J.Math.Anal. 15.1(1984),1-49.

11. J.Guckenheimer&P.Holmes,Nonlinear Oscillations,Dynamical Systems and Bifurcations of Vector Fields,Appl.Math.Sci.42,Springer,New York(1983).

12. J.K.Hale,Ordinary Differential Equations,Wiley,New York(1969).

13. D.Henry,Geometric Theory of Semilinear Parabolic Equations, Lect.Notes in Math.84o,Springer,Berlin(1981).

14. M.W.Hirsch,C.C.Pugh&M.Shub,Invariant Manifolds,Lect.Notes in Math.583,Springer,Berlin(1977).

15. P.J.Holmes,"Unfolding a Degenerate Nonlinear Oscillator: A Codimension Two Bifurcation" in "Nonlinear Dynamics",R.H.G.Helleman(ed.),Annals New York Acad.Sci.Vol.357(198o),473-488.

16. G.Iooss&W.F.Langford,"Conjectures on the Routes to Turbulence via Bifurcations" in "Dynamical Systems",R.H.G.Helleman(ed.), Annals New York Acad.Sci.Vol.357(198o),489-5o5.

17. G.Iooss,"Persistance d'un cercle invariant par une application voisine de l'application temps τ d'un champ de vecteurs intégrable",partie I&II,C.R.A.S.Paris,t.296,série I(1983), 27-3o & 113-116.

18. R.Jost&E.Zehnder,"A Generalization of the Hopf Bifurcation Theorem",Helv.Phys.Acta 45(1972),258-276.

19. U.Kirchgraber&E.Stiefel,Methoden der analytischen Störungsrechnung,B.G.Teubner,Stuttgart(1978).

2o. H.W.Knobloch&F.Kappel,Gewöhnliche Differentialgleichungen,B.G. Teubner,Stuttgart(1974).

21. W.F.Langford,"Periodic and Steady-State Mode Interactions Lead to Tori",SIAM J.Appl.Math. 37.1(1979),22-48.

22. W.F.Langford&G.Iooss,"Interactions of Hopf and Pitchfork Bifurcations" in "Bifurcation Problems and Their Numerical Solution", H.D.Mittelmann&H.Weber(eds.),Birkhäuser,Basel(198o),1o3-134.

23. R.J.Sacker,"A New Approach to the Perturbation Theory of Inva-

riant Surfaces",Comm.Pure&Appl.Math.Vol.XVIII(1965),717-732.

24.J.Scheurle&J.Marsden,"Bifurcation to Quasi-Periodic Tori in the Interaction of Steady State and Hopf Bifurcations",SIAM J.Appl. Math. 15.6(1984),1o55-1o74.

25.G.R.Sell,"Bifurcation of Higher Dimensional Tori",ARMA 69.3 (1979),199-23o.

26.F.Spirig,"Sequence of Bifurcations in a Three-Dimensional System Near a Critical Point",J.Appl.Math.Phys.(ZAMP)Vol.34(1983), 259-276.

Acknowledgment: I would like to thank Professor G.Iooss and Professor W.F.Langford for rather helpful conversations. I am grateful to them for letting me have a copy of their unpublished work "On the Interaction of Two Hopf Bifurcations".

Mathematisches Institut
Universität Würzburg
Am Hubland
D-87oo Würzburg
Federal Republic of Germany

Canadian Mathematical Society
Conference Proceedings
Volume 8 (1987)

NONOSCILLATION IN TIME DELAYED LOTKA-VOLTERRA COMPETITION SYSTEMS

K. Gopalsamy[#]

ABSTRACT. Sufficient conditions are derived for the nonoscillation of a time-delayed competition system of the type

$$\dot{x}_i(t) = x_i(t)[b_i - \sum_{j=1}^{n} a_{ij}\, x_j(t-\tau_{ij})] \qquad i = 1,2,\ldots,n.$$

1. INTRODUCTION. The primary purpose of this article is to derive sufficient conditions for a nonlinear autonomous system of delay-differential equations of the type

$$\text{(1.1)} \qquad \frac{dx_i(t)}{dt} = x_i(t)[b_i - \sum_{j=1}^{n} a_{ij}\, x_j(t-\tau_{ij})] \qquad i = 1,2,\ldots,n$$

to be "nonoscillatory about a steady state" where b_i, a_{ij}, τ_{ij} $(i,j = 1,2,\ldots,n)$ are nonnegative constants. Systems of the form (1.1) are known as Lotka-Volterra systems in mathematical ecology. We assume that together with (1.1), we consider continuous initial values of the form $x_i(s) = \varphi_i \in C[-\tau,0]$, $\tau = \max\{\tau_{ij}\ ;\ 1 \le i\ ,\ j \le n\}$ with $\varphi_i(0) > 0$, $\varphi_i(s) \ge 0$, $i = 1,2,3,\ldots,n$. We also suppose (1.1) has a unique steady state $x^* = \{x_1^*,\ldots,x_n^*\}$, $x_i^* > 0$, $i = 1,2,\ldots,n$. Our definition of nonoscillation of (1.1) as follows (see [2]);

DEFINITION. The system (1.1) is said to be nonoscillatory about its positive steady state x^* if (1.1) has at least one solution $x(t) = \{x_1(t),\ldots,x_n(t)\}$ defined on $[-\tau,\infty)$ such that there exists a positive number t^* and the following holds:

$$\text{(1.2)} \qquad [x_i(t) - x_i^*]\frac{dx_i(t)}{dt} \neq 0 \quad \text{for} \quad t \ge t^*\ ;\ i = 1,2,\ldots,n\ .$$

1980 Mathematics Subject Classification (1985 Revision). 34K15, 92A15.

This work was done while the author was a distinguished visitor at the University of Alberta and partially supported by the grant NSERC A-8130.

If we let $x_i(t) \equiv x_i^* + y_i(t)$, $i = 1,2,\ldots,n$ in (1.1) then we have from (1.1),

$$\frac{dy_i(t)}{dt} = -[x_i^* + y_i(t)] \sum_{j=1}^{n} a_{ij}\, y_j(t-\tau_{ij}) \qquad i = 1,2,\ldots,n \tag{1.3}$$

and hence it is sufficient to study the nonoscillation of (1.3) about its trivial solution.

2. NONOSCILLATION IN COMPETITION SYSTEMS. One of the techniques used for showing the existence of a nonoscillatory solution of a differential equation is by means of a suitable fixed point theorem (such as the contraction mapping principle or the Schauder-Tychonoff theorem). In the proof of our result below, we use the following coincidence theorem due to Goebel [1]: GOEBEL'S COINCIDENCE THEOREM: "Let A be an arbitrary set and M be a metric space with a metric ρ. Let S and T be two maps defined on A with values in M. If $S(A) \subset T(A)$, $T(A)$ is a complete subspace of M and if for all $x,y \in A$, $\rho(Sx,Sy) \le \alpha\rho(Tx,Ty)$ where $0 \le \alpha < 1$ then there exists $\tilde{x} \in A$ such that $S\tilde{x} = T\tilde{x}$."

Our main result is formulated as follows:

THEOREM: Assume the following:

(i) There exists a unique steady state $x^* = (x_1^*, x_2^*,\ldots,x_n^*)$ such that $\sum_{j=1}^{n} a_{ij}\, x_j^* = b_i$, $x_i^* > 0$, $i = 1,2,\ldots,n$ and b_i, a_{ij}, τ_{ij} are nonnegative constants for $i,j = 1,2,\ldots,n$.

(ii) Let $e\|A\|\tau \le 1$ where $A = \sum_i \sum_j a_{ij}$, $\tau = \max\{\tau_{ij}\ ;\ i,j = 1,2,\ldots,n\}$.

Then (1.1) is nonoscillatory about $x^* = \{x_1^*, x_2^*,\ldots,x_n^*\}$.

PROOF: First we note that $e\|A\|\tau \le 1$ implies the existence of a positive number μ satisfying $\mu = \|A\|\exp(\mu\tau)$. Define a set E as follows:

$$E = \left[\, z : [-\tau,\infty) \to \mathbb{R}_n \;\middle|\; \begin{array}{l} z \text{ is continuous on } [-\tau,\infty). \\ 0 \le z_j(t) \le e^{\mu\tau}\ ;\ t \ge -\tau\ ;\ j = 1,2,\ldots,n. \\ z_j \text{ is nonincreasing on } [-\tau,\infty) \text{ for} \\ j = 1,2,\ldots,n \text{ and } \sum_{j=1}^{n} z_j(0) \ge 1. \end{array} \right]$$

It can be verified that if $y = \{y_1, y_2,\ldots,y_n\} \in E$, then for $i = 1,2,3\ldots,n$ and $t \ge 0$,

$$\exp[-\mu t] \le \exp\left[-\int_0^t a_{ij}\, y_j(s-\tau_{ij})\,ds\right] \le \exp[\mu\tau].$$

Let p denote a positive number such that $p > \sum_i \sum_j a_{ij}(1 + x_i^*)$ and define two maps T and S as follows:

$$T(y)(t) = \{T_1(y)(t),\dots,T_n(y)(t)\}$$

(2.1) $$T_i(y)(t) = \begin{cases} [y_i(0) + x_i^*]\exp(-pt) & i = 1,2,\dots,n. \quad t \in [-\tau,0] \\ [y_i(t) + x_i^*]\exp(-pt) & t \geq 0 \end{cases}$$

$$S(y)(t) = \{S_1(y)(t),\dots,S_n(y)(t)\}$$

(2.2) $$S_i(y)(t) = \begin{cases} [1+x_i^*]\exp(-pt) \; ; \; t \in [-\tau,0] \\ [1+x_i^*]\exp[-pt - \int_0^t \sum a_{ij}\, y_j(s-\tau_{ij})ds \; ; \; t \geq 0. \end{cases}$$

One can see from (2.1) and (2.2) that S and T map the set E into the set M where

$$M = \left[u : [-\tau,\infty) \to \mathbb{R}_n \;\middle|\; \begin{array}{l} u \text{ is continuous and componentwise} \\ \text{nonnegative, nonincreasing on } [-\tau,\infty) \\ \text{with } \lim_{t\to\infty} \|u(t)\|_{\mathbb{R}_n} = 0. \end{array} \right]$$

The set M can be metrized with the metric ρ where $\rho(u,v) = \sup\{\|u(t) - v(t)\|_{\mathbb{R}_n} \; ; \; t \geq -\tau\}$. $T(E)$ is a complete subspace of the metric space of M; also $S(E) \subset T(E)$. By direct calculation it is possible to show that

$$|S_i(u)(t) - S_i(v)(t)| \leq e^{-pt} \sum_{j=1}^{n} a_{ij}(1+x_i^*)\int_0^t |T_j(u)(s-\tau_{ij}) - T_j(v)(s-\tau_{ij}|e^{ps}ds$$

and hence

(2.3) $$\rho(Su,Sv) \leq \rho(Tu,Tv)(\sum_i \sum_j (1+x_i^*)a_{ij})/p .$$

By Goebel's coincidence theorem there exists a function $u^*(t) = \{u_1^*(t),\dots, u_n^*(t)\}$ in E such that $T(u^*)(t) = S(u^*)(t)$ or equivalently that for $i = 1,2,\dots,n$,

(2.4) $$u_i^*(t) + x_i^* = \begin{cases} (1 + x_i^*) \; ; \; t \in [-\tau,0] \\ (1+x_i^*)\exp[-\int_0^t \sum_j a_{ij}\, u_j^*(s-\tau_{ij})ds \; ; \; t \geq 0 . \end{cases}$$

It follows from (2.4) and the fact $u^* \in E$ that $u^*(t)$, $t \geq -\tau$ is a nonoscillatory solution of (1.3) and this completes the proof.

The following corollaries can be derived from the result of the above Theorem.

COROLLARY 1: The nondelayed competition system

(2.3) $$\frac{dx_i(t)}{dt} = x_i(t)\left[b_i - \sum_{j=1}^{n} a_{ij}\, x_j(t)\right]$$

is nonoscillatory about its positive steady state x^*.

COROLLARY 2. If $(\sum_{j=1}^{n} \sum_{j=1}^{n} a_{ij} x_j^*) e\tau \leq 1$, then (1.1) is nonoscillatory about x^*.

COROLLARY 3. If $ea\tau \leq 1$ then the scalar delay-logistic equation

$$\frac{dy(t)}{dt} = u(t)[a - bu(t-\tau)] \tag{2.6}$$

is nonoscillatory about a/b.

We conclude with the remark that the result of our Theorem provides one of the generalisations to vector systems of the nonoscillation result-discussed for the scalar equation (2.6) by Wright [4] and Kakutani and Markus [3].

BIBLIOGRAPHY

1. K. Goebel. A coincidence theorem. Bull. Acad. Pol. Sci. Ser. Sci. Math. Astronom. Phys. 16(1968), 733 - 735.

2. K. Gopalsamy. Oscillations in linear systems of delay-differential inequalities. Pacific J. Math. (to appear).

3. S. Kakutani and L. Markus. On the nonlinear difference differential equation $y'(t) = [A - By(t-\tau)]y(t)$. Contributions to the theory of nonlinear oscillations, vol. IV. Princeton University Press, 1958.

4. E.M. Wright. A nonlinear difference differential equation. J. Reine Angew. Math. 194(1955), 66 - 87.

School of Mathematical Sciences
The Flinders University of South Australia
BEDFORD PARK, S.A. 5042
AUSTRALIA

Canadian Mathematical Society
Conference Proceedings
Volume 8 (1987)

A CODE FOR HOPF BIFURCATION ANALYSIS OF AUTONOMOUS DELAY-DIFFERENTIAL SYSTEMS

Brian D. Hassard[1]

ABSTRACT. Hopf bifurcation occurs in a delay differential system

$$\dot{x}(t) = f(x(t-t_1(\upsilon)), x(t-t_2(\upsilon)), \dots, x(t-t_m(\upsilon)); \upsilon)$$

$x, f \in \mathbb{R}^n$ when a stationary solution loses linear stability due to a complex conjugate pair of eigenvalues whose real part becomes positive at a critical value of the bifurcation parameter υ. A code has been developed which locates critical values and determines the coefficients in an approximation to the family of periodic solutions which arises. The code also checks the Hopf hypothesis that at a critical value, all eigenvalues except for the pure imaginary pair have negative real parts. This check is based on Gershgorin type bounding for delay differential eigenvalue problems, and on the argument principle.

To illustrate application of the code, two models due to J. M. Mahaffy of a cellular control system with positive and negative feedback are studied. Losses of stability in these models are explained as Hopf bifurcations to families of locally attracting periodic orbits of the delay-differential systems.

1. INTRODUCTION. The purpose of this paper is to describe new features of a code which locates and analyzes Hopf bifurcations for autonomous systems of delay differential equations. This code, BIFDD, generalizes the code BIFOR2 [3] which performs the analogous computations for autonomous systems of ordinary differential equations. The earliest version of BIFDD is outlined in [4]. The current version has the ability to handle systems with delays that may depend upon the bifurcation parameter, and also performs a check of the spectral hypothesis than the eigenvalues of the associated linear stability problem, except for the complex conjugate pair responsible for the bifurcation, all

1980 Mathematics Subject Classification (1985 Revision). 34K20, 34k35, 92A04.

[1]Partially supported by NSF MCS 79-705790 and MCS 81-06657

lie in the negative half plane. The analysis supporting the check is described in this paper: Gershgorin type bounding is used to bound the region in the right half plane where eigenvalues might occur. Then, the number of eigenvalues (counted by multiplicity) with nonnegative real parts in this bounded set is expressed as a contour integral. Two singularities are present on part of the contour along the imaginary axis, so a slight modification of the argument principle is required. The integrand along this portion of the contour is a real-valued function with a continuous extension, but because the function is the computed as the real part of complex-valued function, and a singularity is present in the imaginary part, the quadrature must be done with care so as to avoid excessive accumulation of roundoff error. The range of integration is broken up into pieces, with the choice of subintervals based on an analysis of roundoff and truncation error.

To illustrate the abilities of the present code, we describe application of the code to two cellular control models with linked positive and negative feedback with delays due to J. M. Mahaffy [5]. Hopf bifurcation explains the oscillations in these models, which possibly relate to observed epigenetic oscillations in prokaryotic cells.

We first give the form of the family of periodic solutions for an autonomous system of delay-differential equations, which arises by Hopf bifurcation. Then we outline the four subcomputations performed within BIFDD. We sketch the schemes used to locate critical values of the bifurcation parameter, and to compute the derivatives of the complex conjugate pair of eigenvalues responsible for the bifurcation. Rather than go into the details of the computation of μ_2, τ_2, and β_2, we refer the interested reader to [3] and [4]. We do, however, present the analysis supporting the algorithm to check the hypothesis that $\mathrm{Re}\ \lambda < 0$ for all eigenvalues other than the complex conjugate pair: we give bounds on the region where eigenvalues with nonnegative real parts may lie by a Gershgorin type scheme, and count the number of such eigenvalues by application of the argument principle. Then we describe the Mahaffy systems, and present results of application of BIFDD to these systems. Finally, we discuss several ways in which the code could be improved.

2. THE FORM OF THE PERIODIC SOLUTIONS. Consider the delay-differential system

$$\dot{x}(t) = f(x(t-t_1(\nu)),x(t-t_2(\nu)),\ldots,x(t-t_m(\nu));\nu)$$

where $x, f \in \mathbb{R}^n$, and $t_j(\nu) \geq 0$ for $j=1,\ldots,m$ and for ν real in some interval of interest, f is C^∞ in its arguments jointly, and each $t_j(\nu)$ is also C^∞. Suppose there is a stationary solution (equilibrium) $x_*(\nu)$ for ν in the interval. The stability of the stationary solution is determined by the behavior of solutions of the variational system

$$\dot{y}(t) = \sum_k A_k(\nu)\ y(t-t_k) ,$$

where for $k = 1,\ldots,m$,

$$A_k(\nu) = \partial f/\partial x_k(x_1,x_2,\ldots,x_m;\nu) \Big|_{x_1 = x_2 = \ldots = x_m = x_*(\nu)} .$$

If all roots λ (eigenvalues) of the characteristic equation

$$\det(\sum_k A_k(\nu)\ e^{-\lambda t_k} - \lambda I) = 0$$

have negative real parts, the stationary solution is linearly stable. Suppose that the stationary solution loses linear stability at some point $\nu = \nu_c$ in the interval, due to an isolated complex conjugate pair of eigenvalues

$$\lambda(\nu) = \alpha(\nu) \pm i\ \omega(\nu),$$

such that

$$\alpha(\nu_c) = 0, \qquad \alpha'(\nu_c) \neq 0 , \qquad \omega(\nu_c) > 0 ,$$

and suppose that for all other eigenvalues λ,

$$\text{Re}\ \lambda < 0 .$$

Under these conditions, there exists a family of small amplitude periodic solutions of the system. For each sufficiently small positive ϵ, there is a solution of the form

$$x(t) = p_\epsilon(t) = x_*(\nu_c) + 2\epsilon\ \text{Re}(e^{2\pi i t/T(\epsilon)} v_1) + O(\epsilon^2) ,$$

which occurs for

$$\nu = \nu_c + \mu_2\epsilon^2 + O(\epsilon^2) ,$$

and is orbitally asymptotically stable provided $\beta(\epsilon) < 0$, where

$$\beta(\epsilon) = \beta_2\epsilon^2 + O(\epsilon^4)$$

is a characteristic exponent. The period of p_ϵ is $T(\epsilon)$, where

$$T(\epsilon) = (2\pi/\omega(\nu_c))(1 + \tau_2\epsilon^2 + O(\epsilon^4)) .$$

3. COMPUTATIONS PERFORMED BY BIFDD. The code BIFDD (named for BIfurcation Formulae for Delay-Differential systems) computes all of ν_c, $\omega_0 = \omega(\nu_c)$, $x_*(\nu_c)$, v_1, $\alpha'(\nu_c)$, $\omega'(\nu_c)$, μ_2, τ_2 and β_2. If $\mu_2 \neq 0$, the sign of μ_2 determines the direction of bifurcation, that is, it determines whether the family of bifurcating periodic solutions exists for $\nu > \nu_c$ or for $\nu < \nu_c$.

The user writes a main program which calls BIFDD, and a subroutine which defines the delay-differential system. This subroutine must evaluate the delays

$$t_j = t_j(\nu), \quad j=1,\ldots,m ,$$

the vector function

$$f = f(x_1, x_2, \ldots, x_m; \nu) ,$$

and the Jacobian array consisting of the matrices

$$A_k = \partial f/\partial x_k(x_1, x_2, \ldots, x_m; \nu) , \quad k=1,\ldots,m.$$

The main program must assign initial estimates of ν_c, ω_0, and $x_*(\nu_c)$ before calling BIFDD.

There are four major steps in the computation:

i) location of ν_c

ii) computation of $\alpha'(\nu_c)$ and $\omega'(\nu_c)$

iii) computation of μ_2, τ_2 and β_2

iv) verification of the hypothesis that at $\nu = \nu_c$, all eigenvalues λ other than $\pm i\omega_0$ obey $\mathrm{Re}\,\lambda < 0$.

To locate ν_c, BIFDD solves the pair of nonlinear algebraic equations

$$\mathrm{Re}\,\det(\nu,\omega) = 0,$$

$$\mathrm{Im}\,\det(\nu,\omega) = 0,$$

for the pair (ν_c,ω_0), by means of a Newton iteration with the 2 by 2 Jacobian matrix approximated by one-sided numerical differencing. Here, $\det(\nu,\omega)$ is the determinant of the matrix

$$\sum_k e^{-i\omega t_k(\nu)} A_k(\nu) - i\omega I ,$$

the matrices $A_k(\nu)$ for $k=1,..,m$ are as above, and the stationary solution $x_*(\nu)$ is computed for each different value of ν by solving

$$f(x,x, ..,x; \nu) = 0$$

with Newton's method.

To compute $\alpha'(\nu_c)$ and $\omega'(\nu_c)$, BIFDD uses differencing. Let

$$\lambda_1(\nu) = \alpha(\nu) + i\,\omega(\nu) .$$

This eigenvalue satisfies the equation

$$\det(\lambda_1(\nu); \nu) = 0 ,$$

where now $\det(\lambda; \nu)$ denotes the determinant of the matrix

$$\sum_k e^{-\lambda t_k(\nu)} A_k(\nu) - \lambda I .$$

The derivative

$$\lambda_1'(\nu_c) = \alpha'(\nu_c) + i\,\omega'(\nu_c)$$

is approximated by the symmetric difference quotient

$$(\lambda_1(\nu_c+h) - \lambda_1(\nu_c-h))/2h$$

in which each $\lambda_1(\nu_c \pm h)$ is found by solving the pair of real nonlinear algebraic equations equivalent to the single complex equation $\det(\lambda_1; \nu) = 0$. The value $h = \nu_{ref}\, u^{1/3}$ is used for the increment h, where ν_{ref} is a reference scale for the parameter ν and u is the unit roundoff error characteristic of the machine floating point arithmetic. This value produces a total error (roundoff + truncation) that is $O(u^{2/3})$, so under "usual" circumstances, roughly 2/3 of the computed digits should be correct. Here and throughout BIFDD, when differencing is employed that affects the final values of μ_2, τ_2 and β_2, the schemes and increments chosen are such that the total error is $O(u^{2/3})$, and error estimation is performed based on repeating the differencing with an increment 100 times larger. Note that by design, BIFDD does not require the user to provide analytic expressions for partial derivatives of f of order exceeding 1: this precludes computing $\lambda'(\nu_c)$ by schemes based on implicit differentiation.

The transversality condition $\alpha'(\nu_c) \neq 0$ is checked based on the computed value of $\alpha'(\nu_c)$.

To compute μ_2, τ_2 and β_2, the current version of BIFDD uses the scheme described in [4]: read all t_j's in section 9 of [4] as $t_j(\nu_c)$'s. This scheme follows the theory and algorithm for differential-difference equations described in [3, Ch. 4]. The operator L_μ in the theory is defined by

$$L_\mu \phi = \int_{-r}^{0} d\eta(\theta,\mu)\ \phi(\theta)$$

where in the present context

$$d\eta(\theta,\mu) = \sum_k A_k(\nu_c+\mu)\ \delta(\theta+t_k(\nu_c+\mu))$$

in which $\delta(\theta)$ is the Dirac delta function. The application of the theory is fairly straightforward. For the results of the application in the form of a computational algorithm, see [4]. Note that H. Stech [7] has recently given a nice self-contained derivation of Hopf bifurcation theory for a class of functional differential equations that includes systems of delay-differential equations of the form we study, but by means of Lyapunov-Schmidt theory rather than by Center Manifold reduction.

The next two sections describe the analysis which supports algorithms for verification of the hypothesis that at $\nu = \nu_c$, all eigenvalues λ other than $\pm i\omega_0$ obey $\mathrm{Re}\ \lambda < 0$.

4. BOUNDING THE EIGENVALUES. Consider the delay-differential eigenvalue problem

$$\lambda v = \sum_k e^{-\lambda t_k} A_k v .$$

We shall establish a Gershgorin type result for eigenvalues obeying $\mathrm{Re}\ \lambda \geq \delta$ for some fixed δ, then apply the result to bound $\mathrm{Re}\ \lambda$ for any eigenvalue, and to bound $|\mathrm{Im}\ \lambda|$ for eigenvalues obeying $\mathrm{Re}\ \lambda \geq \delta$. We first rewrite the problem as

$$(\lambda - d_i)v_i = \sum_{k|t_k=0} \sum_{j\neq i} A_k^{ij} v_j + \sum_{k|t_k>0} e^{-\lambda t_k} \sum_j A_k^{ij} v_j \qquad i = 1,\ldots,n$$

where

$$d_i = \sum_{k|t_k=0} A_k^{ii} , \quad i=1,\ldots,n .$$

Then,

$$|\lambda - d_i||v_i| \le r_i(\text{Re } \lambda) \max_j |v_j| , \quad i=1,\ldots,n$$

where, for real γ and for $i=1,\ldots,n$, $r_i(\gamma)$ is defined by

$$r_i(\gamma) = \sum_{k|t_k=0} \sum_{j\neq i} |A_k^{ij}| + \sum_{k|t_k>0} e^{-\gamma t_k} \sum_j |A_k^{ij}| .$$

If we consider eigenvalues λ obeying Re $\lambda \ge \delta$ for some fixed δ, since each $r_i(\gamma)$ is non-increasing,

$$|\lambda - d_i||v_i| \le r_i(\delta) \max_j |v_j|, \quad i=1,\ldots,n .$$

For the index i of a component of v with largest modulus,

$$|\lambda - d_i| \le r_i(\delta)$$

so that all eigenvalues λ with Re $\lambda \ge \delta$ must lie in the union of n disks, where for $i=1,\ldots,n$ the i'th disk is centered at d_i and has radius $r_i(\delta)$. For $m = 1$ and $t_1 = 0$, these are the usual Gershgorin disks.

To bound Re λ, we note that each function $r_j(\gamma)$ is either constant or monotone decreasing: thus for each $j=1,\ldots,n$ there is a unique real solution $\gamma = \gamma_j$ of

$$\gamma = d_j + r_j(\gamma) .$$

Let

$$\gamma_* = \max_j \gamma_j .$$

If Re λ were to exceed γ_* , then for every $j=1,\ldots,n$,

$$\text{Re } \lambda > \gamma_j = d_j + r_j(\gamma_j) \ge d_j + r_j(\gamma_*)$$

which would contradict the previous result that λ must lie in at least one of the disks with centers d_j and radii $r_j(\gamma_*)$. Therefore for any eigenvalue λ,

$$\text{Re } \lambda \le \gamma_* = \max_j \gamma_j .$$

This bound is readily computed: Newton iteration for each γ_j converges monotonically if d_j is used as the initial approximation to γ_j.

Next we bound |Im λ| for eigenvalues with Re $\lambda \ge \delta$. If λ with Re $\lambda \ge \delta$ lies in the disk centered at d of radius r, then

$$|\text{Im } \lambda| \le B(r,d,\delta)$$

where

$$B(r,d,\delta) = \begin{cases} 0 & d \le \delta - r \\ (r^2 - (\delta-d)^2)^{1/2} & \delta - r < d < \delta \\ r & d \ge \delta . \end{cases}$$

Therefore for eigenvalues λ with Re $\lambda \ge \delta$,

$$|\mathrm{Im}\ \lambda| \le \max_j B(r_j(\delta), d_j, \delta) .$$

Again, this bound is readily computed. The same bounding procedures apply to the problem in which all matrices A_k are transposed. For the transposed problem, the eigenvalues are the same but the computed bounds will be in general different. In the present code, bounds are computed both for the original and transposed problems and the lesser bounds are used.

5. COUNTING THE EIGENVALUES WITH NONNEGATIVE REAL PARTS. The argument principle (Ahlfors [1]) counts zeros of an analytic function f(z) which lie inside a given contour C, provided no zeros lie on C. The number of such zeros, counted by multiplicity, is

$$\frac{1}{2\pi i} \int_C \frac{f'(z)}{f(z)}\, dz ,$$

where the contour is traversed counterclockwise.
In the present context, the determinant

$$D(\lambda) = \det(\sum_k e^{-\lambda t_k} A_k - \lambda I)$$

is analytic for all λ and has simple zeros at $\pm i\omega_0$.
Let $B_1 > 0$ be chosen slightly larger than the bound on $|\mathrm{Im}\ \lambda|$ for eigenvalues λ such that Re $\lambda \ge 0$, from the previous section. Let B_2 be chosen slightly more positive than the bound on Re λ from the previous section. All roots with zero real part must belong to the interval $-B_1 < \mathrm{Im}\ \lambda < B_1$ on the imaginary axis. It is a relatively easy matter to check numerically that $\pm i\omega_0$ are the only zeros of $D(\lambda)$ in this interval. We shall proceed under the assumption that this condition is met. The task then becomes one of counting the eigenvalues with positive real parts. If there are any such, they must lie inside the closed contour C consisting of line segments C_1, C_2, C_3, C_4, where

C_1: $\{ z = i s, \quad B_1 \geq s \geq -B_1 \}$,

C_2: $\{ z = s - i B_1, \quad 0 \leq s \leq B_2 \}$,

C_3: $\{ z = B_2 + i s, \quad -B_1 \leq s \leq B_1 \}$,

C_4: $\{ z = s + i B_1, \quad B_2 \geq s \geq 0 \}$.

Now consider the expression

$$\frac{1}{2\pi i} \int_C \frac{D'(\lambda)}{D(\lambda)} \, d\lambda .$$

The function $D'(\lambda)/D(\lambda)$ is continuous along the line segments C_2, C_3 and C_4, and has simple poles at the points $\pm i\omega_0$ on C_1. The expression is well-defined in the sense of a Cauchy principal value integral, and has value 1 greater than the number of eigenvalues with positive real parts, because each of the 2 poles on C_1 contributes 1/2 its residue to the integral. Since the value of the expression is a nonegative integer, the imaginary part vanishes and only the real parts of the integrals along the line segments needs be evaluated. Since C_1 through C_4 are parallel to the axes, the expression may be written

$$\frac{1}{2\pi i} \int_{C_1} \text{Re} \left[\frac{D'(\lambda)}{D(\lambda)} \right] d\lambda + \frac{1}{2\pi} \int_{C_2} \text{Im} \left[\frac{D'(\lambda)}{D(\lambda)} \right] d\lambda +$$
$$+ \frac{1}{2\pi i} \int_{C_3} \text{Re} \left[\frac{D'(\lambda)}{D(\lambda)} \right] d\lambda + \frac{1}{2\pi} \int_{C_4} \text{Im} \left[\frac{D'(\lambda)}{D(\lambda)} \right] d\lambda .$$

By symmetry, the integral along C_4 is the same as along C_2, and the integrals along C_1 and C_3 may be computed by doubling the integrals along the respective top halves. The expression then becomes

$$- \frac{1}{\pi} \int_0^{B_1} \text{Re} \left[\frac{D'(is)}{D(is)} \right] ds + \frac{1}{\pi} \int_0^{B_2} \text{Im} \left[\frac{D'(s-iB_1)}{D(s-iB_1)} \right] ds +$$
$$+ \frac{1}{\pi} \int_0^{B_1} \text{Re} \left[\frac{D'(B_2+is)}{D(B_2+is)} \right] ds .$$

The last two integrals are readily evaluated by standard numerical quadrature algorithms. The first integral requires special care: the integrand $\text{Re}[D'(is)/D(is)]$ is undefined at $s = \omega_0$, and the integral is to be interpreted as a Cauchy principal value integral. By Taylor expansion of $D(\lambda)$ about $\lambda = i\omega_0$,

$$D(\lambda) = a_1(\lambda-i\omega_0) + a_2(\lambda-i\omega_0)^2 + a_3(\lambda-i\omega_0)^3 + \dots ,$$

which implies

$$D'(\lambda)/D(\lambda) = (\lambda - i\omega_0)^{-1} + a_2' + [3a_3' - (a_2')^2](\lambda - i\omega_0) + \dots$$

where $a_2' = a_2/a_1$ and $a_3' = a_3/a_1$. (Since $i\omega_0$ is a simple root, $a_1 \neq 0$.) For $\lambda = is$, and s near ω_0,

$$\mathrm{Re}[D'(is)/D(is)] = \mathrm{Re}(a_2') - (s-\omega_0)\,\mathrm{Im}[3a_3' - (a_2')^2] + \dots .$$

This last expansion shows that the integrand has a continuous extension: at the point $s = \omega_0$, the extended function is defined as $\mathrm{Re}(a_2')$. The Cauchy principal value integral in this case is just the classical integral of the extended function. Despite this mathematical simplicity, however, in the numerical computation the integrand is evaluated as the real part of the complex-valued function $D'(is)/D(is)$, which has a pole. Roundoff error in evaluating $D'(is)/D(is)$ becomes large in relation to $\mathrm{Re}[D'(is)/D(is)]$, for s near ω_0. To minimize this effect, we write

$$\int_0^{B_1} \mathrm{Re}\left[\frac{D'(is)}{D(is)}\right] ds = \int_{\omega_0-h}^{\omega_0+h} \mathrm{Re}\left[\frac{D'(is)}{D(is)}\right] ds + \int_0^{\omega_0-h} \mathrm{Re}\left[\frac{D'(is)}{D(is)}\right] ds + \int_{\omega_0-h}^{B_1} \mathrm{Re}\left[\frac{D'(is)}{D(is)}\right] ds$$

and evaluate the last two parts using numerical quadrature and the first part by means of the approximation

$$\int_{\omega_0-h}^{\omega_0+h} \mathrm{Re}\left[\frac{D'(is)}{D(is)}\right] ds = 2\,h\,\mathrm{Re}\,a_2' + O(h^3) .$$

The values of a_1 and a_2 required are computed by numerical differencing, and the scheme used is such that the total error (truncation + roundoff) present in the value of a_2' is $O(u^{1/2})$, where u is the unit roundoff error of floating point arithmetic on the machine used. The value of h is chosen to minimize the sum of three error estimates, estimates of the roundoff error due to h in the last two integrals and truncation error due to h in the first. This argument leads to the choice $h = \omega_0\, u^{1/3}$.

The attention to roundoff error here is necessary to ensure reliable performance of the quadrature algorithm. The final result only needs to be known with enough accuracy to distinguish between integers, since the object is to count eigenvalues.

6. EXAMPLE APPLICATIONS: THE MAHAFFY MODELS. In [5], J. M. Mahaffy develops models for a cellular control system with positive and negative feedback. He presents two systems of delay-differential equations, which model induction and repression of the lac operon in E. coli. "The lac operon provides an example of a control system where the transcription of the operon is controlled by induction or a positive feedback control and catabolite repression or negative feedback control. These processes are linked through the metabolism of lactose." [5,p.89]. The first model is given by the system

$$\dot{x}_1 = \frac{1}{(1 + (x_5(t-R_1))^{\gamma})} - b_1 x_1(t),$$

$$\dot{x}_2 = \frac{x_1(t)(1 + K_2(x_4(t-R_2))^{\rho})}{(1 + K_3\, x_1(t))(1 + K_4(x_4(t-R_2))^{\rho}) + K_5} - b_2 x_2(t),$$

$$\dot{x}_3 = x_2(t) - b_3 x_3(t),$$

$$\dot{x}_4 = x_3(t-R_3) - b_4 x_3(t) x_4(t),$$

$$\dot{x}_5 = x_3(t)\, x_4(t) - b_5 x_5(t),$$

and the second, simplified model is given by the system

$$\dot{x}_1 = \frac{1}{(1 + (x_5(t-R_1))^{\gamma})} - b_1\, x_1(t),$$

$$\dot{x}_2 = \frac{x_1(t)(1 + K_2(x_4(t-R_2))^{\rho})}{1 + K_4(x_4(t-R_2))^{\rho}} - b_2\, x_2(t),$$

$$\dot{x}_3 = x_2(t) - b_3 x_3(t),$$

$$\dot{x}_4 = x_3(t) - b_4 x_4(t),$$

$$\dot{x}_5 = x_4(t) - b_5 x_5(t).$$

For physical interpretation of the variables and parameters, see [5]. We have changed z's to x's and T's to R's to avoid conflicting notation.

In [6], Mahaffy studies the linear stability of equilibria of these model systems. He varies the delay R_1 in each system, and presents tables which list the critical value of the delay where the eigenvalues cross the imaginary axis, for different combinations of values of the remaining parameters in the systems. He also presents sample trajectories for the second model, for different parameter combinations such that the model has either one or three equilibria.

We applied the code BIFDD to analyze Hopf bifurcations in the two model systems, for the parameter values listed in [6], tables 1 and 2. Our objects were to confirm the values of R_1^c listed, and to extend the results by computing the values of μ_2, τ_2 and β_2 which characterize the Hopf bifurcation. Table 1 of [6] refers to the first model, and for the computations given in this table, the values $R_3 = 0$ and $R_2 = R_1$ are used. Table 2 of [6] refers to the second model, and the value $R_2 = 0$ is used.

For the first model system, for the parameter values

$$b_1 = 1, \quad b_2 = 0.5, \quad b_3 = 0.02, \quad b_4 = b_5 = 1,$$
$$K_2 = 100, \quad K_3 = K_4 = 1, \quad K_4 = 0, \quad \gamma = \rho = 2$$

from line 1 of table 1, we found

$$R_1^c = 57.478, \quad \omega_0 = 0.034167,$$
$$\alpha'(R_1^c) = 0.00012543, \quad \omega'(R_1^c) = -0.00042313,$$
$$x_* = (0.0034012, 0.34235, 17.118, 1, 17.118),$$
$$v_1 = (1, 99.850, 1125.4, 0, 1046.8)$$
$$- (0, 6.8232, 2263.7, 0, 2299.4)\, i,$$
$$\mu_2 = 2.6181 \times 10^6, \quad \tau_2 = 4.2483 \times 10^4, \quad \beta_2 = -656.76.$$

These results show that the loss of stability at R_1^c is indeed a Hopf bifurcation, and since $\mu_2 > 0$, the family of small amplitude periodic orbits exists for values of R_1 slightly greater than the critical value. The individual periodic orbits are locally attracting since $\beta_2 < 0$. Since $\tau_2 > 0$, the period of the solutions increases with R_1.

We reproduced all the critical values listed in table 1 of [6], except for the three cases in which $R_1^c = \infty$ is listed to indicate that the equilibrium is stable for all delays R_1. In these cases, the code failed to locate critical values. For all parameter combinations for which the code succeeded, μ_2 and τ_2 were positive and β_2 negative. In each case for which the check of the spectral hypothesis was performed, the result of the quadrature to count the eigenvalues with nonnegative real parts was 2, to within an expected numerical error. All losses of stability of the equilibrium solution of the first model system produced by increasing the delay R_1, for the parameter combinations given in table 1 of [6], may therefore be explained as Hopf bifurcations to families of locally attracting periodic orbits of the delay-differential system.

For the second model system, for the parameter values

$$b_1 = b_2 = 5, \quad b_3 = 0.2, \quad b_4 = 3.5, \quad b_5 = 1,$$
$$K_2 = 100, \quad K_4 = 1, \quad \gamma = \rho = 2$$

from line 1 of table 2, we found

$$R_1^c = 6.3630, \quad \omega_0 = 0.22535,$$
$$\alpha'(R_1^c) = 0.0099901, \quad \omega'(R_1^c) = -0.020160,$$
$$x_* = (0.072415,\ 0.92915,\ 4.6457,\ 1.3274,\ 1.3274),$$
$$v_1 = (\ 1,\ 13.427,\ 9.9716,\ 1.9078,\ -1.3193)$$
$$-\ (\ 0,\ 7.8996,\ 50.733,\ 14.618,\ 14.321)\ i,$$
$$\mu_2 = 1018.3, \quad \tau_2 = 143.41, \quad \beta_2 = -20.346\ .$$

These results also show Hopf bifurcation to a family of locally attracting periodic orbits. The family exists for values of R_1 slightly greater than R_1^c.

We reproduced (up to differences in rounding) the critical values R_1^c listed in table 2 of [6], for all cases in which there was a unique equilibrium, and for all upper equilibria in the cases in which there were 3 equilibrium solutions. (Equilibria are called upper, middle or lower based on the value of the x_5 component.) For all critical values computed, the associated μ_2 and τ_2 were positive, and β_2 negative. For the cases in which there were three equilibria, we also attempted to locate critical values of R_1 for Hopf bifurcation from the middle and lower equilibria, but found none. By analysis in [6], each of the middle equilibria is unstable, so even if the code had located an apparent Hopf bifurcation, the check of the spectral hypothesis would have revealed the presence of additional eigenvalue(s) with positive real parts. We had hoped to find such a situation in the model, to exercise the code that checks the spectral hypothesis.

For all but one of the lower equilibria given in table 2 of [6], the value $R_1^c = \infty$ is listed. This indicates that the lower equilibrium is stable for all values of R_1, and explains our inability to find critical values. For the lower equilibrium in line 2 of the table, however, the value $R_1^c = 726$ is listed which we were unable to reproduce. We have not yet explained this.

Figure 3 of [6] shows the x_5 component of the second model system for parameter values

$$b_1 = b_2 = 5, \quad b_3 = 0.2, \quad b_4 = 5, \quad b_5 = 1,$$
$$K_2 = 100, \quad K_4 = 1, \quad \gamma = \rho = 2$$

from line 4 of table 2, and for $R_1 = 8$, larger than $R_1^c = 7.13$, the critical value of the delay listed for these parameter values. Two of the trajectories in figure 3 oscillate. By time $t = 200$, the larger amplitude oscillation has peak-to-peak amplitude 1.25 and interval 40 between successive maxima, while the smaller amplitude oscillation has peak-to-peak amplitude 0.1 and time interval 35 between successive maxima. The figure suggests that both larger and smaller oscillations may converge to a periodic orbit with peak-to-peak amplitude of the x_5 component between 0.1 and 1.25, and period between 35 and 40.

We may compare the periodic orbit predicted by BIFDD to these oscillating trajectories. For the same parameter values, we computed

$$R_1^c = 7.1365, \quad \omega_0 = 0.18385,$$
$$\alpha'(R_1^c) = 0.0094234, \quad \omega'(R_1^c) = -0.014792,$$
$$x_* = (0.099033,\ 1.0097,\ 5.0486,\ 1.0097,\ 1.0097),$$
$$v_1 = (\ 1,\ 9.4158,\ -0.17607,\ -0.41270,\ -2.2249) - (\ 0,\ 10.314,\ 51.406,\ 10.266,\ 9.8569)\ i,$$
$$\mu_2 = 937.34, \quad \tau_2 = 164.73, \quad \beta_2 = -17.666.$$

The x_5 component of the periodic solution is approximated by

$$p_\epsilon^5(t) = x_*^5(R_1^c) + 2\epsilon \ \mathrm{Re}\ \{ \exp(2\pi i t/T) v_1^5 \} + O(\epsilon^2)$$

where

$$R_1 = R_1^c + \mu_2\epsilon^2 + O(\epsilon^4)$$

and

$$T = (2\ \pi/\omega_0)(1 + \tau_2\epsilon^2 + O(\epsilon^4)).$$

Neglecting the $O(\epsilon^4)$ terms, we calculate

$$\epsilon \simeq ((R_1 - R_1^c)/\mu_2)^{1/2} = 0.030352$$

from which the peak-to-peak amplitude of $p_\epsilon^5(t)$ is approximately

$$4\epsilon\ |v_1^5| \simeq 1.2268$$

and the period T is approximately 39.362 . The characteristic exponent $\beta(\epsilon)$ is approximately $\beta_2\epsilon^2 = -0.016275$, from which the "half life" characteristic of attraction to the periodic orbit is

$$-\ln(2)/\beta(\epsilon) = 42.589 .$$

These results are all consistent with figure 2 of [6], and suggest that the larger of the two oscillations shown may be closer than the smaller to an attracting periodic orbit.

7. DISCUSSION. The code BIFDD is still being enhanced as the needs become evident. The first version, discussed in [4], allowed only fixed delays, that is, delays independent of the bifurcation parameter. For a class of delay-differential systems with variable delay(s), it is possible to scale time so as obtain equivalent delay-differential systems with fixed delays. This approach was used in [3]. In the general Mahaffy systems, this trick doesn't work, so the enhancement of BIFDD to handle variable delays was a prerequisite to analysis of these systems. Since the computation of μ_2, τ_2 and β_2 involves only the values of the delays at the critical value of the bifurcation parameter, the major changes needed in BIFDD to allow variable delays were in the location of the critical value. The part that computes μ_2, τ_2 and β_2 has remained virtually unchanged since the first version of BIFDD, and has been tested on all of the delay-differential systems in Ch. 4 of [3] for which exact results are known. It has been tested on both fixed and variable delay versions of these systems. It has also been tested by comparison with BIFOR2 results for the centrifugal governor system and the Hodgkin-Huxley system from Ch. 3 of [3]. We therefore feel confident that the computation of μ_2, τ_2 and β_2 is performed reliably.

The portion of the code that bounds Re λ and $|\text{Im } \lambda|$ and then performs contour integration to check the spectral hypothesis is new, and still needs development. For certain systems (including the Mahaffy systems) this computation now takes an excessive amount of time. A partial explanation is that the bounds produced by the Gershgorin type bounding are too large, which causes extra work in the quadratures due to the unnecessarily large intervals of integration. Improving the bounds will therefore lead to a more efficient check of the spectral hypothesis. There is the potential for obtaining sharper bounds by performing a similarity transformation of the Jacobian array before computing the bounds.

If S is any invertible matrix, and for k=1,..,m each matrix A_k is replaced by the similar matrix $A_k' = S^{-1} A_k S$, the eigenvalues are unchanged. Algorithms could be developed that determine the matrices S in (say) the class of diagonal matrices with positive entries, for which the Gershgorin bounds on Re γ and on $|\text{Im } \gamma|$ are minimized. Similar algorithms for balancing individual matrices in preparation for eigenvalue computation are well known.

The algorithm for locating the critical value in BIFDD is specifically designed for locating points of Hopf bifurcation. The code would be more general purpose if the abilities to locate and analyze stationary bifurcations were included. The latter is straightforward; see exercise 11, p. 78 of [3]. (The stationary bifurcation problem is the same for delay-differential systems as for ordinary differential systems.).

At present, the user must supply the initial approximate value of ω_0, which is an additional source of uncertainty in cases where the algorithm to locate a critical value fails to converge. The code BIFOR2 for ordinary differential systems, does not need an initial approximate ω_0 because, after computing the stationary solution for the initial approximate critical value, it invokes the QR algorithm to compute all eigenvalues of the Jacobian matrix at the stationary solution and selects the eigenvalue of interest. While computing the infinity of eigenvalues for the general delay-differential eigenvalue problem is out of the question, it would certainly be feasible to compute the finite set of eigenvalues whose real parts exceed a certain value. A count of the number of such eigenvalues could be constructed by contour integration, and then a root-finding algorithm such as Muller's method with deflation used to compute them.

Hopf bifurcations from unstable stationary solutions of delay-differential systems can also be analyzed by means of BIFDD: we have verified this ability by analyzing the unstable Hopf bifurcations in the Hutchinson-Wright equation [3,p.191] for the critical values $\mu = 2\pi$ and $\mu = 4\pi$. We thank the referee for suggesting this test, which exercises the part of BIFDD that counts eigenvalues with nonnegative real parts. However, the theory of Hopf bifurcations for differential-difference equations described in [3] needs reworking to handle such unstable Hopf bifurcations, a topic which we do not wish to pursue here. The essential observation is that the invariant manifold theory of

Claeyssen [2] still applies when a finite number of eigenvalues with nonnegative real parts are present. Note that Stech's theorem [7] does apply to unstable Hopf bifurcations.

Despite the areas in which it could be improved, the current version of BIFDD can be used effectively, as we hope we have demonstrated with the Mahaffy systems. Copies of BIFDD may be obtained from the author: the best way is via electronic mail. Please address requests for BIFDD to MTHBDH@UBVMSA, on BITNET.

BIBLIOGRAPHY

1. Ahlfors, L. V. Complex Analysis, McGraw-Hill, second edition, 1966.

2. Claeyssen, J. R. "Effect of delays on functional differentia equations," J. Diff. Eqns. 20 (1976) 404-440

3. Hassard, B. D., Kazarinoff, N. D. and Wan, Y-H. Theory and Applications of Hopf bifurcation London Mathematical Society Lecture Note Series 41, Cambridge Univ. Press 1981.

4. Hassard, B. D. "Numerical Hopf Bifurcation Analysis for Dela Differential Systems," in Oscillations in Mathematical Biology , Proceedings, Adelphi University 1982: J.P.E. Hodgson, Ed. Springer-Verlag Lecture Notes in Biomathematics #51, 1983.

5. Mahaffy, J. M. "Cellular Control Models with Linked Positive and Negative Feedback and Delays. I. The Models," J. theor. Biol. (1984) 106, 89-102.

6. Mahaffy, J. M. "Cellular Control Models with Linked Positive and Negative Feedback and Delays. II. Linear Analysis and Local Stability," J. theor. Biol. (1984) 106, 103-118.

7. Stech, H. W. "Hopf Bifurcation Calculations for Functional Differential Equations," J. Math. Anal. Appl. (1985) 109, 472-491.

DEPARTMENT OF MATHEMATICS
STATE UNIVERSITY OF NEW YORK AT BUFFALO
BUFFALO, N.Y. 14214.

Canadian Mathematical Society
Conference Proceedings
Volume 8 (1987)

BIFURCATIONS ASSOCIATED WITH A SIMPLE ZERO AND TWO PAIRS OF PURE IMAGINARY EIGENVALUES

KONCAY HUSEYIN AND PEI YU

ABSTRACT. This paper is concerned with the stability, instability and bifurcation behaviour of a nonlinear autonomous system in the vicinity of a compound critical point which is characterized by a simple zero and two pairs of pure imaginary eigenvalues. The system contains three independent parameters. The static and dynamic bifurcations as well as quasi-periodic motions, resulting from the interactions of the bifurcation modes, and the associated invariant tori are investigated. The analysis is based on a generalized perturbation procedure which employs the *multiple-scale intrinsic harmonic balancing* and embraces the *unification technique* introduced earlier. This more comprehensive perturbation approach leads to explicit asymptotic results concerning stationary solutions, periodic and non-resonant quasi-periodic motions which take place on invariant tori. Also, the stability conditions and the criteria leading to various bifurcations are established.

1. Introduction It is known that the interactions of static and dynamic bifurcation modes in the vicinity of a *compound* critical point of a nonlinear autonomous system may lead to secondary bifurcations and bifurcations into invariant tori associated with quasi-periodic motions. These phenomena have been studied by Langford [1] and others [2,3,4], with particular attention to a critical point at which the Jacobian has a zero and a pair of pure imaginary eigenvalues. It is also known that invariant tori may emerge as a result of bifurcations from periodic motions to quasi-periodic motions as well, without involving any static bifurcations [5]. If the Jacobian of a system has a simple zero and two pairs of pure imaginary eigenvalues at a critical point, the situation becomes more complicated and some new phenomena may be exhibited.

In this paper, a detailed stability and bifurcation analysis of this problem is presented through an original perturbation procedure which yields explicit asymptotic results. The method is based on a *multiple-scale intrinsic harmonic balancing procedure* (which is a generalization of the *intrinsic harmonic balancing* [6,7]) and the *unification technique* introduced earlier [8]. This approach enables one to obtain a set of simplified equivalent differential equations for the analysis of the local behaviour of

0731-1036/87 $1.00 + $.25 per page

the system in the vicinity of the critical point. Thus, based on the simplified equations, incipient bifurcations, secondary bifurcations and bifurcations into periodic as well as non-resonant quasi-periodic motions are explored. Attention is focused on the stability of the solutions and the critical boundaries where various bifurcations take place. In particular, the criteria under which the system exhibits static bifurcations, Hopf bifurcations and bifurcations into 2-dimensional tori are discussed. It is observed that the second-order perturbation equations provide sufficient information for this purpose; however, higher order perturbations are required if one wishes to explore the bifurcations from 2-dimensional tori to 3-dimensional tori. This observation is based on the following facts: first, the rate equations (45) embrace all the second-order terms, in the sense that the contribution of higher order terms in the original equations to the rate equations (45) is also at higher order, as shown in Appendix C. Second, the results such as bifurcation solutions as well as the stability criteria (49), (54), etc. based on the rate equations (45) would not be affected if (45) had higher order terms. All the asymptotic solutions are derived in explicit analytical forms in terms of the convenient system coefficients (derivatives of the vector field). Thus, the theory is directly applicable to specific problems.

Finally, it is noted that — in view of the transformation (2) — the problem considered here is not mathematically the most generic case; nevertheless, the formulation of real systems are often achieved within the framework of this model.

2. Formulation Consider an autonomous system described by

$$(1) \quad \frac{dz^i}{dt} = Z_i(z^j;\, \eta^\beta)$$

where the z^i are the components of the state vector z, and the η^β are certain independent parameters. It is assumed that the functions Z_i are analytic, at least in the region of interest. Attention in this paper is focused on the behaviour of the system in the vicinity of a coincident critical point characterized by a Jacobian which has a simple zero and two pairs of pure imaginary eigenvalues, and all the remaining eigenvalues have negative real parts. For simplicity, therefore, a five dimensional system, involving the critical state variables only, is considered. It is also assumed that the system involves three independent parameters ($\beta = 1, 2, 3$).

Suppose that the system has a single-valued fundamental equilibrium surface, $z^i = f_i(\eta^\beta)$, in the region of interest; system (1) may then be referred to this surface by introducing the transformation

$$(2) \quad z^i = f_i(\eta^\beta) + T_{ij}\, w^j$$

where T_{ij} is chosen appropriately so that the Jacobian matrix at the critical point c — on the fundamental surface — takes the form

$$(3)\quad J = \begin{bmatrix} 0 & 0 & 0 & 0 & 0 \\ 0 & 0 & \omega_{1c} & 0 & 0 \\ 0 & -\omega_{1c} & 0 & 0 & 0 \\ 0 & 0 & 0 & 0 & \omega_{2c} \\ 0 & 0 & 0 & -\omega_{2c} & 0 \end{bmatrix} \qquad \omega_{1c} > 0 \ , \quad \omega_{2c} > 0 \quad \text{and} \quad \omega_{1c} \neq \omega_{2c} \ ,$$

and the transformed system

$$(4)\quad \frac{dw^i}{dt} = W_i(w^j;\, \eta^\beta) \ , \qquad (\ i,\, j = 1,\, 2,\, 3\ 4,\, 5 \quad \text{and} \quad \beta = 1,\, 2,\, 3\ )$$

has the properties

$$(5)\quad W_i(0;\, \eta^\xi) = W_{i\beta}(0;\, \eta^\xi) = W_{i\beta\gamma}(0;\, \eta^\xi) = \ \cdots \ = 0 \ ,$$

where the subscripts on the W_i indicate differentiations with respect to the parameters.

Suppose that in the vicinity of c the Jacobian $\left[W_{ij}(\mu^\beta)\right]$, where $\mu^\beta = \eta^\beta - \eta^\beta_c$, has a real eigenvalue and two distinct pairs of complex conjugate eigenvalues; namely, $\alpha_o(\mu^\beta)$, $\alpha_1(\mu^\beta) \pm \omega_1(\mu^\beta)i$ and $\alpha_2(\mu^\beta) \pm \omega_2(\mu^\beta)i$, so that

$$(6)\quad \alpha_o(0) = \alpha_1(0) = \alpha_2(0) = 0 \ , \quad \omega_1(0) = \omega_{1c} \quad \text{and} \quad \omega_2(0) = \omega_{2c} \ .$$

It is assumed that

$$(7)\quad \det \begin{bmatrix} \dfrac{\partial W_{11}}{\partial \mu^1} & \dfrac{\partial W_{11}}{\partial \mu^2} & \dfrac{\partial W_{11}}{\partial \mu^3} \\ \dfrac{\partial \alpha_1}{\partial \mu^1} & \dfrac{\partial \alpha_1}{\partial \mu^2} & \dfrac{\partial \alpha_1}{\partial \mu^3} \\ \dfrac{\partial \alpha_2}{\partial \mu^1} & \dfrac{\partial \alpha_2}{\partial \mu^2} & \dfrac{\partial \alpha_2}{\partial \mu^3} \end{bmatrix}_c \neq 0 \ ,$$

which represents a generalization of Hopf's transversality condition. It can be shown (see Appendix A) that

$$(8)\quad \frac{\partial \alpha_1}{\partial \mu^\beta} \equiv \alpha_1^\beta = \frac{1}{2}\,(W_{22\beta}+W_{33\beta}) \ , \quad \frac{\partial \omega_1}{\partial \mu^\beta} \equiv \omega_1^\beta = \frac{1}{2}\,(W_{23\beta}-W_{32\beta}) \quad (\beta = 1,\, 2,\, 3);$$

and

$$(9)\quad \frac{\partial \alpha_2}{\partial \mu^\beta} \equiv \alpha_2^\beta = \frac{1}{2}\,(W_{44\beta}+W_{55\beta}) \ , \quad \frac{\partial \omega_2}{\partial \mu^\beta} \equiv \omega_2^\beta = \frac{1}{2}\,(W_{45\beta}-W_{54\beta}) \quad (\beta = 1,\, 2,\, 3).$$

It is clear that, for the analysis of the behaviour of this system, the formulation must be capable of dealing with static, and dynamic bifurcations as well as quasiperiodic bifurcations – preferably in a unified approach. A *unification technique* [8], which is based on the *intrinsic perturbation approach* [6] has recently been applied to the investigation of the bifurcation properties in the vicinity of a coincident critical point (for example, a double zero eigenvalue or a zero and a pair of pure imaginary eigenvalues); and it will be modified here appropriately to cope with the quasi-periodic motions as well.

Suppose that the possible non-trivial solutions (involving stationary, periodic and quasi-periodic solutions) of the system in the vicinity of the critical point may be assumed to be in the parametric form

$$(10) \quad w^i = w^i(\tau_1, \tau_2; \sigma^a), \quad \eta^\beta = \eta^\beta(\sigma^a) \quad \text{and} \quad \omega_k = \omega_k(\sigma^a),$$

where $i = 1, 2, 3, 4, 5$; $\beta = 1, 2, 3$; $\tau_k = \omega_k t$ $(k = 1, 2)$, ω_k are frequencies; and the σ^a $(a = 1, 2, 3)$ are certain unidentified perturbation parameters. Assuming that the ratio ω_1/ω_2 is irrational (i.e. $m_1\omega_1 + m_2\omega_2 \neq 0$ where the m_i are arbitrary integers), one may introduce the two time scale Fourier series

$$(11)\ w^j(\tau_1, \tau_2; \sigma^a) = \sum_{\substack{m=0 \\ (m_1+m_2=m)}}^{M} \Big[p_{j,m_1m_2}(\sigma^a)\cos(m_1\tau_1+m_2\tau_2)+r_{j,m_1m_2}(\sigma^a)\sin(m_1\tau_1+m_2\tau_2)\Big]$$

which provide an approximation for the steady-state solutions in more explicit terms. A complex form of (11) has also been used by Haken [9]. Here, $a = 1, 2, 3$; m_1, m_2 and m are integers and $m \geq 0$ while m_1 and m_2 may be chosen positive or negative; and M is an arbitrary positive integer. It is noted that (11) reduces to the ordinary Fourier series in the case of $m_1 \equiv 0$ or $m_2 \equiv 0$. It is observed that

$$(12) \quad p_{j,\,m_1m_2}(0) = r_{j,\,m_1m_2}(0) = 0 \qquad (\forall\ j, m_1, m_2),$$

since $w^i(\tau_1, \tau_2; 0) = 0$, and $m_1 = m_2 = 0$ leads to a solution $p_{j,00}(\sigma^a) \neq 0$ which represents a post-critical equilibrium (stationary) solution – that is, a static bifurcation.

Substituting the assumed solution (10) into (4) yields the identities

$$(13) \quad \omega_1(\sigma^a)\frac{\partial w^i}{\partial \tau_1} + \omega_2(\sigma^a)\frac{\partial w^i}{\partial \tau_2} \equiv W_i\Big[w^j(\tau_1, \tau_2; \sigma^a); \eta^\beta(\sigma^a)\Big],$$

which are then differentiated with respect to the independent variables σ^a $(a = 1, 2, 3)$ successively to generate a sequence of perturbation equations

$$(14) \quad \omega_1^{,a} w^i_{\tau_1} + \omega_2^{,a} w^i_{\tau_2} + \omega_1 w^{i,a}_{\tau_1} + \omega_2 w^{i,a}_{\tau_2} = W_{ij} w^{j,a} + W_{i\beta}\eta^{\beta,a},$$

and

$$(15) \quad \omega_1^{,ab}w^i_{\tau_1} + \omega_1^{,a}w^{i,b}_{\tau_1} + \omega_1^{,b}w^{i,a}_{\tau_1} + \omega_1 w^{i,ab}_{\tau_1} + \omega_2^{,ab}w^i_{\tau_2} + \omega_2^{,a}w^{i,b}_{\tau_2} + \omega_2^{,b}w^{i,a}_{\tau_2} + \omega_2 w^{i,ab}_{\tau_2}$$
$$= W_{ijk}w^{j,a}w^{k,b} + W_{ij\beta}(w^{j,a}\eta^{\beta,b} + w^{j,b}\eta^{\beta,a}) + W_{ij}w^{j,ab} + W_{i\beta\gamma}\eta^{\beta,a}\eta^{\gamma,b} + W_{i\beta}\eta^{\beta,ab},$$

etc., where subscripts on W_i's denote differentiations with respect to the corresponding variables and/or parameters, and summation convention applies. For clarity, differentiations of variables with respect to the σ^a are indicated by a comma, and the subscripts τ_1 and τ_2 indicate differentiations with respect to τ_1 and τ_2, respectively.

Evaluating the first-order perturbation equation (14) at the critical point c with the aid of (5) results in

$$(16) \quad \omega_{1c}\, w^{i,a}_{\tau_1} + \omega_{2c}\, w^{i,a}_{\tau_2} = W_{ij}\, w^{j,a}$$

where W_{ij} is given by (3). Note that $w^i_{\tau_1} = w^i_{\tau_2} = 0$ due to (12). Substituting the Fourier series (11) into (16) leads to

$$(17) \quad \sum_{\substack{m=0 \\ (m_1+m_2=m)}}^{M} (m_1\omega_{1c}+m_2\omega_{2c}) \Big[r^{i,a}_{i,\,m_1m_2} \cos(m_1\tau_1+m_2\tau_2) - p^{i,a}_{i,\,m_1m_2} \sin(m_1\tau_1+m_2\tau_2) \Big]$$

$$= W_{ij} \sum_{\substack{m=0 \\ (m_1+m_2=m)}}^{M} \Big[p^{i,a}_{j,\,m_1m_2} \cos(m_1\tau_1+m_2\tau_2) + r^{i,a}_{j,\,m_1m_2} \sin(m_1\tau_1+m_2\tau_2) \Big],$$

where again the superscripts on p's and r's, together with commas, indicate differentiations with respect to the σ^a.

Comparing the coefficients of $\cos(m_1\tau_1+m_2\tau_2)$ and $\sin(m_1\tau_1+m_2\tau_2)$ on both sides of (17) leads to certain relationships. It is noted here that this process is facilitated by the assumption that the ratio ω_1/ω_2 is irrational. Indeed, if one assumes that for a set of $(\overline{m}_1, \overline{m}_2)$,

$$m_1\tau_1 + m_2\tau_2 = \overline{m}_1\tau_1 + \overline{m}_2\tau_2,$$

then it follows that

$$\frac{\tau_1}{\tau_2} = \frac{\omega_1}{\omega_2} = \frac{\overline{m}_2 - m_2}{m_1 - \overline{m}_1}$$

is a rational number, contradicting the assumption. Hence, $\overline{m}_1 \equiv m_1$ and $\overline{m}_2 \equiv m_2$, eliminating the possibility of any other combinations.

Thus, (17) yields

$$(18) \quad \begin{cases} (m_1\omega_{1c} + m_2\omega_{2c})\, r^{i,a}_{i,\,m_1m_2} - W_{ij}\, p^{i,a}_{j,\,m_1m_2} = 0, \\ (m_1\omega_{1c} + m_2\omega_{2c})\, p^{i,a}_{i,\,m_1m_2} + W_{ij}\, r^{i,a}_{j,\,m_1m_2} = 0, \end{cases} \quad (\text{ for } i, j = 1, 2, 3, 4, 5\ ).$$

Here for each value of $m = 0, 1, 2, \cdots$, one has to consider all possible choices of m_1 and m_2 satisfying $m_1 + m_2 = m$. For example, when $m = 0$, the possible values of m_1 and m_2 are $(-1, 1)$, $(1, -1)$, $(-2, 2)$, $(2, -2)$, etc., (infinite possibilities). However, it can be shown that $p_{j,m_1m_2}^{;a} = r_{j,m_1m_2}^{;a} = 0$ for $m_1 + m_2 = m \geq 2$ or $|m_i| \geq 2$. Here it is recognized that one may assume

(19) $r_{2,10}(\sigma^a) \equiv 0$ and $r_{4,01}(\sigma^a) \equiv 0$

without loss of generality – since the system under consideration is autonomous. Thus, with this assumption, the first-order perturbation equations yield the following relationships

(20) $p_{2,10}^{;a} = -r_{3,10}^{;a}$, $p_{4,01}^{;a} = -r_{5,01}^{;a}$, $p_{3,10}^{;a} = r_{2,10}^{;a} = 0$ and $p_{5,01}^{;a} = r_{4,01}^{;a} = 0$;

where the subscript notations "j,10" and "j,01", etc. indicate the coefficients of the harmonics $\cos\tau_1$, $\sin\tau_1$ and $\cos\tau_2$, $\sin\tau_2$, etc., respectively.

It is noted that all the other derivatives $p_{j,m_1m_2}^{;a}$ and $r_{j,m_1m_2}^{;a}$ are equal to zero except for the $p_{1,00}^{;a}$ which are not determined at this stage.

Now, evaluating the second-order perturbation equation (15) at the critical point c with the aid of (5) and (20), introducing (11), and setting the coefficients of all the trigonometric functions equal to zero produces a set of non-trivial algebraic equations as follows:

$$(21)\quad W_{i11}p_{1,00}^{;a}p_{1,00}^{;b} + \frac{1}{2}\left[(W_{i22}W_{i33})p_{2,10}^{;a}p_{2,10}^{;b} + (W_{i44}+W_{i55})p_{4,01}^{;a}p_{4,01}^{;b}\right]$$
$$+W_{ij\beta}(p_{1,00}^{;a}\eta^{\beta,b} + p_{1,00}^{;b}\eta^{\beta,a}) + W_{ij}p_{j,00}^{;ab} = 0, \quad \text{for } m_1 = m_2 = 0;$$

$$(22)\quad \begin{cases} (\omega_{1c}-\omega_{2c})r_{i,1-1}^{;ab} = \dfrac{1}{2}(W_{i24}+W_{i35})(p_{2,10}^{;a}p_{4,01}^{;b} + p_{4,01}^{;a}p_{2,10}^{;b}) + W_{ij}p_{j,1-1}^{;ab} , \\ (\omega_{1c}-\omega_{2c})p_{i,1-1}^{;ab} = \dfrac{1}{2}(W_{i34}-W_{i25})(p_{2,10}^{;a}p_{4,01}^{;b} + p_{4,01}^{;a}p_{2,10}^{;b}) - W_{ij}r_{j,1-1}^{;ab} , \end{cases}$$

for $m_1 = 1,\ m_2 = -1$;

(23)

$$\begin{cases} \omega_1^{;a}r_{i,10}^{;b} + \omega_1^{;b}r_{i,10}^{;a} + \omega_{1c}r_{i,10}^{;ab} = W_{i12}(p_{1,00}^{;a}p_{2,10}^{;b} + p_{1,00}^{;b}p_{2,10}^{;a}) + W_{i2\beta}(p_{2,10}^{;a}\eta^{\beta,b} + p_{2,10}^{;b}\eta^{\beta,a}) + W_{ij}p_{j,10}^{;ab}, \\ \omega_1^{;a}p_{i,10}^{;b} + \omega_1^{;b}p_{i,10}^{;a} + \omega_{1c}p_{i,10}^{;ab} = W_{i13}(p_{1,00}^{;a}p_{2,10} + p_{1,00}^{;b}p_{2,10}^{;a}) + W_{i3\beta}(p_{2,10}^{;a}\eta^{\beta,b} + p_{2,10}^{;b}\eta^{\beta,a}) - W_{ij}r_{j,10}^{;ab}, \end{cases}$$

for $m_1 = 1,\ m_2 = 0$;

(24)

$$\begin{cases} \omega_2^{;a}r_{i,01}^{;b} + \omega_2^{;b}r_{i,01}^{;a} + \omega_{2c}r_{i,01}^{;ab} = W_{i14}(p_{1,00}^{;a}p_{4,01}^{;b} + p_{1,00}^{;b}p_{4,01}^{;a}) + W_{i4\beta}(p_{4,01}^{;a}\eta^{\beta,b} + p_{4,01}^{;b}\eta^{\beta,a}) + W_{ij}p_{j,01}^{;ab}, \\ \omega_2^{;a}p_{i,01}^{;b} + \omega_2^{;b}p_{i,01}^{;b} + \omega_{2c}p_{i,01}^{;ab} = W_{i15}(p_{1,00}^{;a}p_{4,01}^{;b} + p_{1,00}^{;b}p_{4,01}^{;a}) + W_{i5\beta}(p_{4,01}^{;a}\eta^{\beta,b} + p_{4,01}^{;b}\eta^{\beta,a}) - W_{ij}r_{j,01}^{;ab}, \end{cases}$$

for $m_1 = 0,\ m_2 = 1$;

$$(25)\quad \begin{cases} 2\,\omega_{1c}\, r_{i,20}^{;ab} = \dfrac{1}{2}\,(\,W_{i22} - W_{i33}\,)\, p_{2,10}^{;a}\, p_{2,10}^{;b} + W_{ij}\, p_{j,20}^{;ab}\,, \\ 2\,\omega_{1c}\, p_{i,20}^{;ab} = W_{i23}\, p_{2,10}^{;a}\, p_{2,10}^{;b} - W_{ij}\, r_{j,20}^{;ab}\,, \end{cases} \quad (\text{for } m_1{=}2,\ m_2{=}0);$$

$$(26)\quad \begin{cases} 2\,\omega_{2c}\, r_{i,02}^{;ab} = \dfrac{1}{2}\,(\,W_{i44} - W_{i55}\,)\, p_{4,01}^{;a}\, p_{4,01}^{;b} + W_{ij}\, p_{j,02}^{;ab}\,, \\ 2\,\omega_{2c}\, p_{i,02}^{;ab} = W_{i45}\, p_{4,01}^{;a}\, p_{4,01}^{;b} - W_{ij}\, r_{j,02}^{;ab}\,, \end{cases} \quad (\text{for } m_1{=}0,\ m_2{=}2);$$

and

$$(27)\quad \begin{cases} (\omega_{1c}+\omega_{2c})r_{i,11}^{;ab} = \dfrac{1}{2}(W_{i24}-W_{i35})\,(p_{2,10}^{;a}p_{4,01}^{;b}+p_{4,01}^{;a}p_{2,10}^{;b}) + W_{ij}p_{j,11}^{;ab}\,, \\ (\omega_{1c}+\omega_{2c})p_{i,11}^{;ab} = \dfrac{1}{2}(W_{i25}+W_{i34})\,(p_{2,10}^{;a}p_{4,01}^{;b}+p_{4,01}^{;a}p_{2,10}^{;b}) - W_{ij}r_{j,11}^{;ab}\,, \end{cases}$$

for $m_1 = m_2 = 1$.

It is noted that the equations (21)–(27) represent a total of 390 equations. It is also observed, however, that if 21 *key derivatives* are determined, the rest of the unknowns may be expressed in terms of these derivatives. Consider, therefore, the following 30 equations out of the above 390 equations:

$$(28)\quad \begin{cases} W_{11\beta}(p_{1,00}^{;a}\eta^{\beta,b} + p_{1,00}^{;b}\eta^{\beta,a}) + W_{111}p_{1,00}^{;a}p_{1,00}^{;b} + l_{11}p_{2,10}^{;a}p_{2,10}^{;b} + l_{12}p_{4,01}^{;a}p_{4,01}^{;b} = 0\,, \\ 2\alpha_1^{\beta}(p_{2,10}^{;a}\eta^{\beta,b} + p_{4,01}^{;b}\eta^{\beta,a}) + 2\alpha_{11}(p_{1,00}^{;a}p_{2,10}^{;b} + p_{1,00}^{;b}p_{2,10}^{;a}) = 0\,, \\ 2\alpha_2^{\beta}(p_{4,01}^{;a}\eta^{\beta,b} + p_{4,01}^{;b}\eta^{\beta,a}) + 2\alpha_{21}(p_{1,00}^{;a}p_{4,01}^{;b} + p_{1,00}^{;b}p_{4,01}^{;a}) = 0\,; \end{cases}$$

and

$$(29)\quad \begin{cases} 2\omega_1^{;a}p_{2,10}^{;b}+2\omega_1^{;b}p_{2,10}^{;a} = 2\omega_1^{\beta}(p_{2,10}^{;a}\eta^{\beta,b}+p_{2,10}^{;b}\eta^{\beta,a}) + 2\omega_{11}(p_{1,00}^{;a}p_{2,10}^{;b}+p_{1,00}^{;b}p_{2,10}^{;a})\,, \\ 2\omega_2^{;a}p_{4,01}^{;b}+2\omega_2^{;b}p_{4,01}^{;a} = 2\omega_2^{\beta}(p_{4,01}^{;a}\eta^{\beta,b}+p_{4,01}^{;b}\eta^{\beta,a}) + 2\omega_{21}(p_{1,00}^{;a}p_{4,01}^{;b}+p_{1,00}^{;b}p_{4,01}^{;a})\,. \end{cases}$$

Here, α_k^{β}, ω_k^{β} $(k = 1, 2)$ are expressed by (8) and (9), respectively, and

$$(30)\quad l_{j1} = \frac{1}{2}\,(\,W_{j22} + W_{j33}\,)\,,\quad l_{j2} = \frac{1}{2}\,(\,W_{j44} + W_{j55}\,)\,,$$

$$(31)\quad \alpha_{1j} = \frac{1}{2}\,(\,W_{22j} + W_{33j}\,)\,,\quad \alpha_{2j} = \frac{1}{2}\,(\,W_{44j} + W_{55j}\,)\,,$$

$$(32)\quad \omega_{1j} = \frac{1}{2}\,(\,W_{23j} - W_{32j}\,)\,,\quad \omega_{2j} = \frac{1}{2}\,(\,W_{45j} - W_{54j}\,)\,,$$

where $j = 1, 2, 3, 4, 5$.

Instead of trying to solve the above equations for the unknown derivatives, the *unification technique* will now be used to derive a set of governing relationships. First, it is noted that each of the above 5 relations involves 6 equations. Multiplying the first equation ($a=b=1$) in each relation by $(\sigma^1)^2/2$, the second equation ($a=1, b=2$) by $\sigma^1\sigma^2$, the third equation ($a=1, b=3$) by $\sigma^1\sigma^3$, the fourth equation ($a=b=2$) by $(\sigma^2)^2/2$, the fifth equation ($a=2, b=3$) by $\sigma^2\sigma^3$ and the last one ($a=b=3$) by $(\sigma^3)^2/2$, and adding them together results in

$$(33)\quad \begin{cases} W_{11\beta}\mu^\beta\, p_{1,00} + \frac{1}{2} W_{111}(p_{1,00})^2 + \frac{1}{2} l_{11}(p_{2,10})^2 + \frac{1}{2} l_{12}(p_{4,01})^2 = 0\,, \\ (\,\alpha_1^\beta \mu^\beta + \alpha_{11}\, p_{1,00}\,)\, p_{2,10} = 0\,, \\ (\,\alpha_2^\beta \mu^\beta + \alpha_{21}\, p_{1,00}\,)\, p_{4,01} = 0\,; \end{cases}$$

and

$$(34)\quad \begin{cases} (\,\omega_1^\beta \mu^\beta + \omega_{11}\, p_{1,00} - \Omega_1\,)\, p_{2,10} = 0\,, \\ (\,\omega_2^\beta \mu^\beta + \omega_{21}\, p_{1,00} - \Omega_2\,)\, p_{4,01} = 0\,; \end{cases}$$

where $\Omega_k = \omega_k - \omega_{kc}$ ($k = 1, 2$).

3. Bifurcation and stability analysis First, it is observed from (33) that

$$(35)\quad p_{1,00} = p_{2,10} = p_{4,01} = 0$$

is a solution which corresponds to the fundamental equilibrium surface ($w^i = 0$).

Static bifurcations from the fundamental equilibrium surface occur for $p_{2,10} = p_{4,01} = 0$ and can readily be determined from the first equation of (33) as

$$(36)\quad p_{1,00} = -\frac{2}{W_{111}} W_{11\beta}\, \mu^\beta\,, \quad p_{2,10} = p_{4,01} = 0\,,$$

assuming that $W_{111} \neq 0$. For the case in which $W_{111} = 0$ (symmetric) higher order terms are required to supplement equation (33).

On the other hand, $p_{2,10} \neq 0$ or $p_{4,01} \neq 0$ signifies the emergence of two distinct families of limit cycles. The *behaviour surface* associated with the first family (Hopf bifurcations, $p_{2,10} \neq 0$, $p_{4,01} = 0$, *with frequency* ω_1) *is given by*

$$(37)\quad \begin{cases} p_{1,00} = -\frac{1}{\alpha_{11}}\, \alpha_1^\beta\, \mu^\beta\,, \quad p_{4,01} = 0\,, \\ (p_{2,10})^2 = \frac{1}{l_{11}\,\alpha_{11}} (\,2\,W_{11\beta} - \frac{W_{111}}{\alpha_{11}}\, \alpha_1^\beta\,)\, \alpha_1^\gamma\, \mu^\beta\, \mu^\gamma\,, \end{cases} \qquad (l_{11}\alpha_{11} \neq 0)\,.$$

The behaviour surface associated with the second family ($p_{2,10} = 0$, $p_{4,01} \neq 0$, with frequency ω_2) is described by

$$
(38)\qquad \begin{cases} p_{1,00} = -\dfrac{1}{\alpha_{21}}\,\alpha_2^{\beta}\,\mu^{\beta} , \quad p_{2,10} = 0 , \\[2ex] (p_{4,01})^2 = \dfrac{1}{l_{12}\,\alpha_{21}}\,\left(2\,W_{11\beta} - \dfrac{W_{111}}{\alpha_{21}}\,\alpha_2^{\beta} \right) \alpha_2^{\gamma}\,\mu^{\beta}\,\mu^{\gamma} , \end{cases} \qquad (l_{12}\alpha_{21} \neq 0) .
$$

The frequencies, corresponding to the bifurcating limit cycles (37) and (38), can be determined from (34) as

$$
(39)\qquad \Omega_1 = \left(\omega_1^{\beta} - \frac{\omega_{11}}{\alpha_{11}}\,\alpha_1^{\beta} \right) \mu^{\beta} , \qquad \text{and}
$$

$$
(40)\qquad \Omega_2 = \left(\omega_2^{\beta} - \frac{\omega_{21}}{\alpha_{21}}\,\alpha_2^{\beta} \right) \mu^{\beta} ,
$$

respectively.

Finally, a family of 2-dimensional tori may bifurcate in the vicinity of the critical point c and is expressed by

$$
(41)\qquad \begin{cases} p_{1,00} = -\dfrac{1}{\alpha_{11}}\,\alpha_1^{\beta}\mu_{\beta} = -\dfrac{1}{\alpha_{21}}\,\alpha_2^{\beta}\,\mu^{\beta} , \\[2ex] l_{11}\,(p_{2,10})^2 + l_{12}\,(p_{4,01})^2 = -\,p_{1,00}\,\left(2\,W_{11\beta}\mu^{\beta} + W_{111}\,p_{1,00} \right) , \end{cases}
$$

with the two frequencies given by (39) and (40).

The asymptotic solutions of the limit cycles (37) or (38), and the 2-dimensional tori (41) may be constructed up to first- or second-order terms. Indeed, setting $\sigma^1 \equiv p_{1,00}$, $\sigma^2 \equiv p_{2,10}$ and $\sigma^3 \equiv p_{4,01}$, and using (20) in (11), one obtains the first-order approximation

$$
(42)\qquad \begin{cases} w^1 = p_{1,00} , \\ w^2 = p_{2,10}\cos\tau_1 , \\ w^3 = -\,p_{2,10}\sin\tau_1 , \\ w^4 = p_{4,01}\cos\tau_2 , \\ w^5 = -\,p_{4,01}\sin\tau_2 . \end{cases}
$$

A second-order approximation takes the form

(43)

$$w^j = \begin{bmatrix} p_{1,00} \\ p_{2,10}\cos\tau_1 \\ -p_{2,10}\sin\tau_1 \\ p_{4,01}\cos\tau_2 \\ -p_{4,01}\sin\tau_2 \end{bmatrix} + \begin{bmatrix} 0 \\ p_{2,00} \\ p_{3,00} \\ p_{4,00} \\ p_{5,00} \end{bmatrix} + p_{j,1-1}\cos(\tau_1-\tau_2) + r_{j,1-1}\sin(\tau_1-\tau_2) + p_{j,20}\cos 2\tau_1 + r_{j,20}\sin 2\tau_1$$

$$+ p_{j,02}\cos 2\tau_2 + r_{j,02}\sin 2\tau_2 + p_{j,11}\cos(\tau_1+\tau_2) + r_{j,11}\sin(\tau_1+\tau_2), \quad (j=1,2,3,4,5),$$

where the amplitudes $p_{j,00}$, $p_{j,1-1}$, $r_{j,02}$, etc. can be determined in terms of $p_{1,00}$, $p_{2,10}$ and $p_{4,01}$, by applying the unification technique to the remaining equations in (21)–(27) which involve all the necessary derivatives. The results – expressed up to second-order terms – are given in Appendix B. In particular, note that

(44)
$$\begin{cases} r_{3,10} = -p_{2,10} + \dfrac{1}{2\omega_{1c}}\Big[(W_{23\beta} + W_{32\beta})\,\mu^\beta + (W_{231} + W_{321})\,p_{1,00} \Big] p_{2,10}, \\ r_{5,01} = -p_{4,01} + \dfrac{1}{2\omega_{2c}}\Big[(W_{45\beta} + W_{54\beta})\,\mu^\beta + (W_{451} + W_{541})\,p_{1,00} \Big] p_{4,01}; \end{cases}$$

which shows that $r_{3,10} = -p_{2,10}$ and $r_{5,01} = -p_{4,01}$ to first order, as indicated by (42).

It is also noted that the perturbation procedure leading to (33) and (34) could have been simplified considerably had one set $\sigma^1 \equiv p_{1,00}$, $\sigma^2 \equiv p_{2,10}$ and $\sigma^3 \equiv p_{4,01}$ in the beginning of the analysis. The reader can recover these equations by adopting such an approach. Nevertheless, the suitability of $p_{1,00}$, $p_{2,10}$ and $p_{4,01}$ for the role of the σ^a was not à priori evident, and a false choice may have led to wrong conclusions.

For simplicity in notation, let $p_{1,00}$, $p_{2,10}$ and $p_{4,01}$ be denoted by y, ρ_1 and ρ_2 respectively. It can be shown, on the basis of (43) (the outline of the proof is given in Appendix C) that the rate equations

(45)
$$\begin{cases} \dfrac{dy}{dt} = W_{11\beta}\,\mu^\beta\, y + \dfrac{1}{2}\,W_{111}\,y^2 + \dfrac{1}{2}\,l_{11}(\rho_1)^2 + \dfrac{1}{2}\,l_{12}(\rho_2)^2, \\ \dfrac{d\rho_1}{dt} = (\,\alpha_1^\beta\,\mu^\beta + \alpha_{11}\,y\,)\,\rho_1, \\ \dfrac{d\rho_2}{dt} = (\,\alpha_2^\beta\,\mu^\beta + \alpha_{21}\,y\,)\,\rho_2; \end{cases}$$

and

$$(46)\quad \begin{cases} \dfrac{d\theta_1}{dt} = \omega_{1c} + \omega_1^\beta \mu^\beta + \omega_{11}\, y\,, \\ \dfrac{d\theta}{dt} = \omega_{2c} + \omega_2^\beta \mu^\beta + \omega_{21}\, y\,; \end{cases}$$

characterize the dynamics in the vicinity of the critical point c.

It is noted that equation (46) can also be inferred from (34) for $p_{2,10} \neq 0$ or $p_{4,01} \neq 0$, and $d\theta_1/\mathrm{dt} = \omega_1$, $d\theta_2/\mathrm{dt} = \omega_2$.

It is also noted that setting $\mu^\beta = 0$ in (45) and (46) results in the so called "normal form" up to second-order terms.

Now, the stability of these solutions may be investigated on the basis of the Jacobian of (45)

$$(47)\quad J(y, \rho_1, \rho_2; \mu^\beta) = \begin{bmatrix} W_{11\beta}\,\mu^\beta + W_{111}\, y & l_{11}\,\rho_1 & l_{12}\,\rho_2 \\ \alpha_{11}\,\rho_1 & \alpha_1^\beta \mu^\beta + \alpha_{11} y & 0 \\ \alpha_{21}\,\rho_2 & 0 & \alpha_2^\beta \mu^\beta + \alpha_{21}\, y \end{bmatrix}.$$

Evaluating the Jacobian on the fundamental equilibrium surface (35) yields

$$(48)\quad J(0, 0, 0; \mu^\beta) = \begin{bmatrix} W_{11\beta}\,\mu^\beta & 0 & 0 \\ 0 & \alpha_1^\beta \mu^\beta & 0 \\ 0 & 0 & \alpha_2^\beta \mu^\beta \end{bmatrix}.$$

It is easy to observe that (48) gives the stability conditions for the fundamental equilibrium surface as

$$(49)\quad W_{11\beta}\,\mu^\beta < 0\,,\quad \alpha_1^\beta \mu^\beta < 0 \quad\text{and}\quad \alpha_2^\beta \mu^\beta < 0\,,$$

which leads to three critical surfaces. One of these is described by

$$(50)\quad S_1:\quad W_{11\beta}\,\mu^\beta = 0\,,\quad (\,\alpha_1^\beta\mu^\beta < 0\,,\ \alpha_2^\beta\mu^\beta < 0\,)$$

which represents a divergence boundary at which static bifurcations (36) occur from the fundamental equilibrium surface (35).

The second critical surface is given by

$$(51)\quad S_2:\quad \alpha_1^\beta \mu^\beta = 0\,,\quad (\,W_{11\beta}\mu^\beta < 0\,,\ \alpha_2^\beta\mu^\beta < 0\,)$$

along which one family of limit cycles associated with (37) and (39) bifurcates from (35).

The critical surface

$$(52)\quad S_3:\quad \alpha_2^\beta \mu^\beta = 0\,,\quad (\,W_{11\beta}\mu^\beta < 0\,,\ \alpha_1^\beta\mu^\beta < 0\,)$$

describes the onset of another family of limit cycles associated with (38) and (40) from (35).

Next, evaluating the Jacobian (47) on the static bifurcation solution (36), one obtains

$$(53)\quad J = \begin{bmatrix} -W_{11\beta}\,\mu^\beta & 0 & 0 \\ 0 & (\,\alpha_1^\beta - \frac{2\,\alpha_{11}}{W_{111}}\,W_{11\beta}\,)\,\mu^\beta & 0 \\ 0 & 0 & (\,\alpha_2^\beta - \frac{2\,\alpha_{21}}{W_{111}}\,W_{11\beta}\,)\,\mu^\beta \end{bmatrix}$$

which givens the stability conditions

$$(54)\quad W_{11\beta}\,\mu^\beta > 0\,,\ (\alpha_1^\beta - \frac{2\alpha_{11}}{W_{111}}\,W_{11\beta})\mu^\beta < 0 \ \text{and}\ (\alpha_2^\beta - \frac{2\alpha_{21}}{W_{111}}W_{11\beta})\mu^\beta < 0\,.$$

These conditions imply two critical surfaces. One of these is given by

$$(55)\quad S_4:\ \ (\,\alpha_1^\beta - \frac{2\,\alpha_{11}}{W_{111}}\,W_{11\beta}\,)\,\mu^\beta = 0 \quad (\,W_{11\beta}\mu^\beta > 0\,,\ (\alpha_2^\beta - \frac{2\alpha_{21}}{W_{111}}W_{11\beta})\mu^\beta < 0\,)\,,$$

along which a secondary Hopf bifurcation appears to occur. However, (55) is, in fact, the intersection of the post-critical equilibrium surface (36) and the bifurcation surface (37), and the family of associated limit cycles is identical to the family represented by (37).

Another critical surface is defined by

$$(56)\quad S_5:\ \ (\,\alpha_2^\beta - \frac{2\,\alpha_{21}}{W_{111}}\,W_{11\beta}\,)\,\mu^\beta = 0 \quad (\,W_{11\beta}\mu^\beta > 0\,,\ (\alpha_1^\beta - \frac{2\alpha_{11}}{W_{111}}W_{11\beta})\mu^\beta < 0\,)\,,$$

which represents a dynamic bifurcation boundary where another secondary Hopf bifurcation appears to occur. But, (56) is again the intersection of (36) and (38). In other words, the secondary Hopf bifurcations from the post-critical equilibrium solution are associated with the incipient family of limit cycles.

To examine the stability of the Hopf bifurcation solution (37), evaluate the Jacobian (47) on this solution to obtain

$$(57)\quad J(y,\,\rho_{1s},\,0;\,\mu^\beta) = \begin{bmatrix} (\,W_{11\beta} - \frac{W_{111}}{\alpha_{11}}\,\alpha_1^\beta\,)\,\mu^\beta & l_{11}\,\rho_1 & 0 \\ \alpha_{11}\,\rho_1 & 0 & 0 \\ 0 & 0 & (\,\alpha_2^\beta - \frac{\alpha_{21}}{\alpha_{11}}\,\alpha_1^\beta\,)\,\mu^\beta \end{bmatrix}.$$

which gives the stability conditions

(58) $(\alpha_2^\beta - \frac{\alpha_{21}}{\alpha_{11}} \alpha_1^\beta)\mu^\beta < 0$, $(W_{11\beta} - \frac{W_{111}}{\alpha_{11}} \alpha_1^\beta)\mu^\beta < 0$ and $l_{11}\, \alpha_{11}(\rho_1)^2 < 0$.

It is noted that the third inequality of (58) implies

(59) $l_{11}\alpha_{11} < 0$.

The conditions (58) imply two critical surfaces. One of these is given by

(60) $S_6 : \quad (W_{11\beta} - \frac{W_{111}}{\alpha_{11}} \alpha_1^\beta)\mu^\beta = 0 \quad ((\alpha_2^\beta - \frac{\alpha_{21}}{\alpha_{11}}\alpha_1^\beta)\mu^\beta < 0 , \ l_{11}\alpha_{11} < 0)$,

along which a secondary Hopf bifurcation from the first bifurcating limit cycles (37), leading to a 2-dimensional torus $(\omega_1, \omega_{2\bar{c}})$, takes place. The second frequency of the torus can be written as

(61) $\omega_{2\bar{c}} = \rho_1 \sqrt{- l_{11}\alpha_{11}} \quad (l_{11}\, \alpha_{11} < 0)$.

On the other hand, the critical surface

(62) $S_7 : \quad (\alpha_2^\beta - \frac{\alpha_{21}}{\alpha_{11}} \alpha_1^\beta) \mu^\beta = 0$, $((W_{11\beta} - \frac{W_{111}}{\alpha_{11}}\alpha_1^\beta)\mu^\beta < 0 , \ l_{11}\alpha_{11} < 0)$,

describes the onset of another secondary Hopf bifurcation from (37), leading to another 2-dimensional torus (ω_1, ω_2). This torus is described by (39)–(41).

Similarly, evaluating the Jacobian on the Hopf bifurcation solution (47) yields

(63) $$J(y, 0, \rho_{2s}; \mu^\beta) = \begin{bmatrix} (W_{11\beta} - \frac{W_{111}}{\alpha_{21}} \alpha_2^\beta) \mu^\beta & 0 & l_{12}\, \rho_2 \\ 0 & (\alpha_1^\beta - \frac{\alpha_{11}}{\alpha_{21}} \alpha_2^\beta) \mu^\beta & 0 \\ \alpha_{21}\, \rho_2 & 0 & 0 \end{bmatrix} .$$

A direct calculation shows that when

(64) $(W_{11\beta} - \frac{W_{111}}{\alpha_{21}} \alpha_2^\beta)\mu^\beta < 0$, $(\alpha_1^\beta - \frac{\alpha_{11}}{\alpha_{21}} \alpha_2^\beta)\mu^\beta < 0$ and $l_{12}\, \alpha_{21}(\rho_2)^2 < 0$,

the Hopf bifurcation solution (38) is stable. It is also observed that

(65) $l_{12}\alpha_{21} < 0$

is a necessary stability condition.

In this case, two critical surfaces can be identified as follows:

(66) $S_8 : \quad (W_{11\beta} - \frac{W_{111}}{\alpha_{21}} \alpha_2^\beta) \mu^\beta = 0$, $((\alpha_1^\beta - \frac{\alpha_{11}}{\alpha_{21}}\alpha_2^\beta)\mu^\beta < 0 , \ l_{12}\alpha_{21} < 0)$,

along which a secondary Hopf bifurcation from the first Hopf bifurcation solution (38) – leading to a 2-dimensional torus – takes place. The second frequency of the torus $(\omega_2, \omega_{1\bar{c}})$ is given by

$$\omega_{1\bar{c}} = \rho_2 \sqrt{-\, l_{12}\alpha_{21}} \quad (l_{12}\, \alpha_{21} < 0) . \tag{67}$$

The critical surface

$$S_9 : \quad (\alpha_1^\beta - \frac{\alpha_{11}}{\alpha_{21}} \alpha_2^\beta) \mu^\beta = 0 \;, \quad ((W_{11\beta} - \frac{W_{111}}{\alpha_{21}} \alpha_2^\beta)\mu^\beta < 0 \;, \; l_{12}\alpha_{21} < 0) \;, \tag{68}$$

represents the onset of another secondary Hopf bifurcation from the first Hopf bifurcation solution (38) – leading to another 2-dimensional torus (ω_2, ω_1). But this family of tori is in fact the same family expressed by (39)–(41).

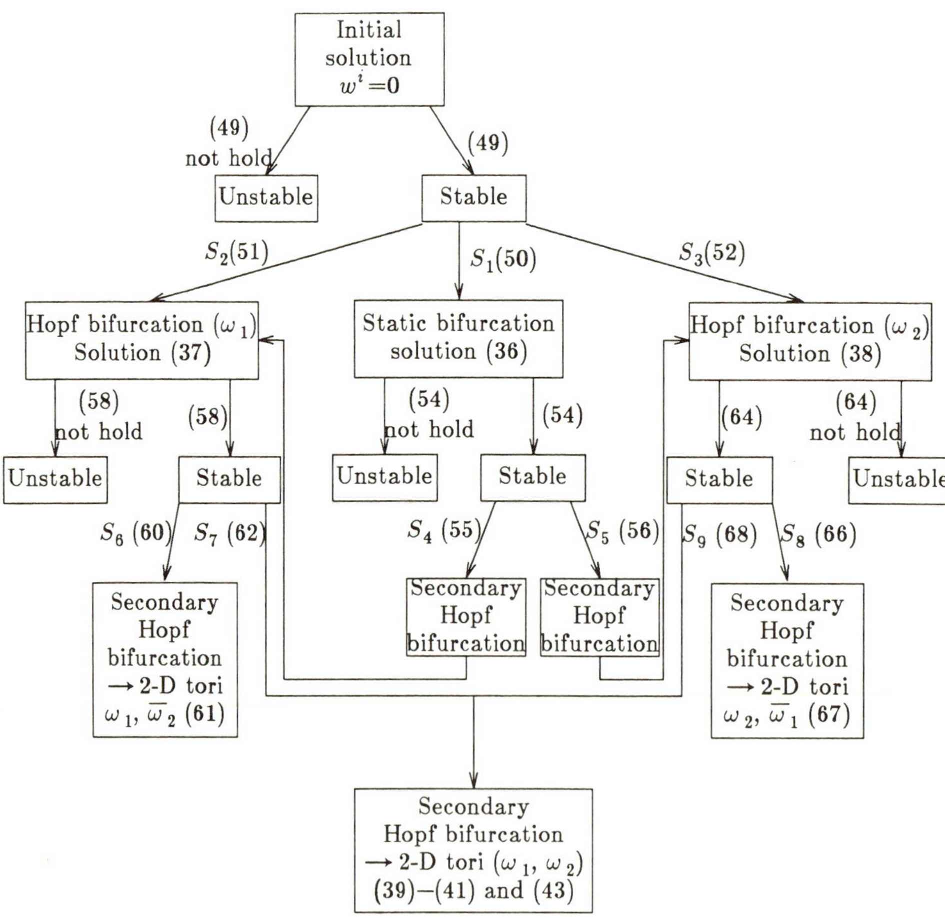

Fig. 1 Bifurcation flow chart for a zero and two pairs of pure imaginary eigenvalues

Finally, evaluating the Jacobian (47) on the quasi-periodic solution (41), one observes that higher order terms are required for a decision concerning stability.

Similarly, higher order terms (i.e. a third order perturbation) are required for exploring bifurcations from 2-dimensional to 3-dimensional tori.

The bifurcation flow chart is sketched in Fig. 1.

No attempt has been made to identify any possible chaotic behaviour.

Appendix A

In this appendix it is shown that the derivatives of the real and imaginary parts of the complex eigenvalues, evaluated at the critical point c, can be expressed in terms of the system coefficients.

In order to obtain the desired relationships, the Jacobian matrix $W(\mu^\beta)=\left[W_{ij}(\mu^\beta)\right]$ is expanded into Taylor series in the vicinity of the compound critical point c,

(A1) $$W_{ij}(\mu^\beta) = W_{ij} + W_{ij\beta}\mu^\beta + \frac{1}{2} W_{ij\beta\gamma}\mu^\beta\mu^\gamma + \cdots .$$

Suppose that the eigenvalues of the Jacobian matrix consist of one real eigenvalue and two pairs of complex conjugate eigenvalues, given by

(A2) $$\lambda_1 = \alpha_o(\mu^\beta) \quad , \quad \lambda_{2,3} = \alpha_1(\mu^\beta) \pm \omega_1(\mu^\beta)i \quad \text{and} \quad \lambda_{4,5} = \alpha_2(\mu^\beta) \pm \omega_2(\mu^\beta)i \ ;$$

and

(A3) $$\alpha_o(0) = \alpha_1(0) = \alpha_2(0) = 0 \quad , \quad \omega_1(0) = \omega_{1c} \quad \text{and} \quad \omega_2(0) = \omega_{2c} .$$

Substituting (A1) into the characteristic equation

(A4) $$\left| W(\mu^\beta) - I\,\lambda(\mu^\beta) \right| = 0 ,$$

one has

(A5)

$$\begin{vmatrix} W_{11\beta}\mu^\beta+..-\lambda & W_{12\beta}\mu^\beta+.. & W_{13\beta}\mu^\beta+.. & W_{14\beta}\mu^\beta+.. & W_{15\beta}\mu^\beta+.. \\ W_{21\beta}\mu^\beta+.. & W_{22\beta}\mu^\beta+..-\lambda & \omega_{1c}+W_{23\beta}\mu^\beta+.. & W_{24\beta}\mu^\beta+.. & W_{25\beta}\mu^\beta+.. \\ W_{31\beta}\mu^\beta+.. & -\omega_{1c}+W_{32\beta}\mu^\beta+.. & W_{33\beta}\mu^\beta+..-\lambda & W_{34\beta}\mu^\beta+.. & W_{35\beta}\mu^\beta+.. \\ W_{41\beta}\mu^\beta+.. & W_{42\beta}\mu^\beta+.. & W_{43\beta}\mu^\beta+.. & W_{44\beta}\mu^\beta+..-\lambda & \omega_{2c}+W_{45\beta}\mu^\beta+.. \\ W_{51}^\beta\mu^\beta+.. & W_{52\beta}\mu^\beta+.. & W_{53\beta}\mu^\beta+.. & -\omega_{2c}+W_{54\beta}\mu^\beta+.. & W_{55\beta}\mu^\beta+..-\lambda \end{vmatrix}=0.$$

Differentiating (A5) with respect to μ^β $(\beta = 1, 2, 3)$ and then evaluating at the critical point c yields

(A6) $$\left[\lambda^2(0)+\omega_{1c}^2\right]\left[\lambda^2(0)+\omega_{2c}^2\right](W_{11\beta}-\lambda^\beta)$$
$$-\lambda(0)\left[\lambda^2(0)+\omega_{2c}^2\right]\left[\omega_{1c}(W_{23\beta}-W_{32\beta})-\lambda(0)(W_{22\beta}+W_{33\beta}-2\lambda^\beta)\right]$$
$$-\lambda(0)\left[\lambda^2(0)+\omega_{1c}^2\right]\left[\omega_{2c}(W_{45\beta}-W_{54\beta})-\lambda(0)(W_{44\beta}+W_{55\beta}-2\lambda^\beta)\right]=0,$$

where $\lambda^\beta = \left.\dfrac{\partial\lambda}{\partial\mu^\beta}\right|_c$.

Substituting $\lambda=\lambda_1\equiv\alpha_o(\mu^\beta)$ into (A6) results in

(A7) $$\alpha_o^\beta = W_{11\beta}.$$

Substituting $\lambda=\lambda_2\equiv\alpha_1(\mu^\beta)+\omega_1(\mu^\beta)i$ into (A6), and replacing $\lambda(0)$ by $\omega_{1c}i$ yields

(A8) $$\alpha_1^\beta=\frac{1}{2}(W_{22\beta}+W_{33\beta}) \quad\text{and}\quad \omega_1^\beta=\frac{1}{2}(W_{23\beta}-W_{32\beta});$$

similarly, substituting $\lambda=\lambda_4\equiv\alpha_2(\mu^\beta)+\omega_2(\mu^\beta)i$ into (A6), and replacing $\lambda(0)$ by $\omega_{2c}i$ gives

(A9) $$\alpha_2^\beta=\frac{1}{2}(W_{44\beta}+W_{55\beta}) \quad\text{and}\quad \omega_2^\beta=\frac{1}{2}(W_{45\beta}-W_{54\beta}).$$

Appendix B

In this appendix, the amplitudes of the harmonics are given in terms of $p_{1,00}$, $p_{2,10}$ and $p_{4,01}$ up to second-order terms as follows:

(B1)

$$p_{i,00}=\pm\frac{1}{\omega_{1c}}\left[W_{j1\beta}\mu^\beta p_{1,00}+\frac{1}{2}W_{j11}(p_{1,00})^2+\frac{1}{4}(W_{j22}+W_{j33})(p_{2,10})^2+\frac{1}{4}(W_{j44}+W_{j55})(p_{4,01})^2\right],$$

for $(i,j)=(2,3),(3,2),(4,5)$ and $(5,4)$;

(B2)

$$p_{1,1-1}=\frac{1}{2(\omega_{1c}-\omega_{2c})}(W_{134}-W_{125})p_{2,10}p_{4,01},\quad r_{1,1-1}=\frac{1}{2(\omega_{1c}-\omega_{2c})}(W_{124}+W_{135})p_{2,10}p_{4,01};$$

(B3)
$$\begin{cases} p_{i,1-1}=\dfrac{1}{2\omega_{2c}(2\omega_{1c}-\omega_{2c})}\left[\pm\omega_{kc}(W_{j24}+W_{j35})-(\omega_{1c}-\omega_{2c})(W_{i34}-W_{i25})\right]p_{2,10}p_{4,01},\\[2ex] r_{i,1-1}=\dfrac{1}{2\omega_{2c}(2\omega_{1c}-\omega_{2c})}\left[\pm\omega_{kc}(W_{j34}-W_{j25})+(\omega_{1c}-\omega_{2c})(W_{i24}+W_{i35})\right]p_{2,10}p_{4,01},\end{cases}$$

for $(i,j)=(2,3),(3,2),(4,5)$ and $(5,4)$;

$$\text{(B4)}\quad \begin{cases} p_{1,10} = \dfrac{1}{\omega_{1c}}(W_{13\beta}\mu^\beta + W_{113}p_{1,00})p_{2,10}\,, \quad r_{1,10} = \dfrac{1}{\omega_{1c}}(W_{12\beta}\mu^\beta + W_{112}p_{1,00})p_{2,10}\,, \\ p_{1,01} = \dfrac{1}{\omega_{2c}}(W_{15\beta}\mu^\beta + W_{151}p_{1,00})p_{4,01}\,, \quad r_{1,01} = \dfrac{1}{\omega_{2c}}(W_{14\beta}\mu^\beta + W_{141}p_{1,00})p_{4,01}\,; \end{cases}$$

$$\text{(B5)}\quad \begin{cases} p_{3,10} = -\dfrac{1}{2\omega_{1c}}\Big[(W_{22\beta} - W_{33\beta})\,\mu^\beta + (W_{221} - W_{331})\,p_{1,00}\Big]\,p_{2,10}\,, \\ r_{3,10} = -\,p_{2,10} + \dfrac{1}{2\omega_{1c}}\Big[(W_{23\beta} + W_{32\beta})\,\mu^\beta + (W_{231} + W_{321})\,p_{1,00}\Big]\,p_{2,10}\,; \end{cases}$$

$$\text{(B6)}\quad \begin{cases} p_{5,01} = -\dfrac{1}{2\omega_{2c}}\Big[(W_{44\beta} - W_{55\beta})\,\mu^\beta + (W_{441} - W_{551})\,p_{1,00}\Big]\,p_{4,01}\,, \\ r_{5,01} = -\,p_{4,01} + \dfrac{1}{2\omega_{2c}}\Big[(W_{45\beta} + W_{54\beta})\,\mu^\beta + (W_{451} + W_{541})\,p_{1,00}\Big]\,p_{4,01}\,; \end{cases}$$

$$\text{(B7)}\quad \begin{cases} p_{i,10} = \dfrac{1}{\omega_{2c}^2 - \omega_{1c}^2}\Big[(\pm\,\omega_{2c}W_{j2\beta} - \omega_{1c}W_{i3\beta})\mu^\beta + (\pm\,\omega_{2c}W_{j21} - \omega_{1c}W_{i31})p_{1,00}\Big]p_{2,10}\,, \\ r_{i,10} = \dfrac{1}{\omega_{1c}^2 - \omega_{2c}^2}\Big[(\pm\,\omega_{2c}W_{j3\beta} + \omega_{1c}W_{i2\beta})\mu^\beta + (\pm\,\omega_{2c}W_{j31} + \omega_{1c}W_{i21})p_{1,00}\Big]p_{2,10}\,, \end{cases}$$

for $(i,\, j) = (4,\, 5)$ or $(5,\, 4)$;

$$\text{(B8)}\quad \begin{cases} p_{i,01} = \dfrac{1}{\omega_{2c}^2 - \omega_{1c}^2}\Big[(\pm\,\omega_{2c}W_{i5\beta} - \omega_{1c}W_{j4\beta})\mu^\beta + (\pm\,\omega_{2c}W_{i51} - \omega_{1c}W_{j41})p_{1,00}\Big]p_{4,01}\,, \\ r_{i,01} = \dfrac{1}{\omega_{1c}^2 - \omega_{2c}^2}\Big[(\pm\,\omega_{2c}W_{i4\beta} + \omega_{1c}W_{j5\beta})\mu^\beta + (\pm\,\omega_{2c}W_{i41} + \omega_{1c}W_{j51})p_{1,00}\Big]p_{4,01}\,, \end{cases}$$

for $(i,\, j) = (2,\, 3)$ or $(3,\, 2)$;

$$\text{(B9)}\quad \begin{cases} p_{1,20} = \dfrac{1}{4\omega_{1c}}\,W_{123}\,(p_{2,10})^2\,, \quad r_{1,20} = \dfrac{1}{4\omega_{1c}}\,(\,W_{122} - W_{133}\,)\,(p_{2,10})^2\,, \\ p_{1,02} = \dfrac{1}{4\omega_{2c}}\,W_{145}\,(p_{4,01})^2\,, \quad r_{1,02} = \dfrac{1}{4\omega_{2c}}\,(\,W_{144} - W_{155}\,)\,(p_{4,01})^2\,; \end{cases}$$

$$\text{(B10)}\quad \begin{cases} p_{i,20} = \dfrac{1}{12\omega_{1c}}\,(\,W_{jjj} - W_{jii} + 4\,W_{i23}\,)\,(p_{2,10})^2\,, \\ r_{i,20} = \dfrac{1}{6\omega_{1c}}\,(\,W_{i22} - W_{i33} \pm W_{j23}\,)\,(p_{2,10})^2\,, \end{cases} \qquad \text{for } (i,\, j) = (2,\, 3) \text{ or } (3,\, 2)\,;$$

(B11)
$$\begin{cases} p_{i,20} = \dfrac{1}{4(4\omega_{1c}^2 - \omega_{2c}^2)} \left[4\omega_{1c} W_{i23} \pm \omega_{2c}(W_{j33} - W_{j22}) \right] (p_{2,10})^2 , \\ r_{i,20} = \dfrac{1}{2(4\omega_{1c}^2 - \omega_{2c}^2)} \left[\pm \omega_{2c} W_{j23} - \omega_{1c}(W_{i33} - W_{i22}) \right] (p_{2,10})^2, \end{cases}$$

for $(i, j) = (4, 5)$ or $(5, 4)$;

(B12)
$$\begin{cases} p_{i,02} = \dfrac{1}{4(4\omega_{2c}^2 - \omega_{1c}^2)} \left[4\omega_{2c} W_{i45} \pm \omega_{1c}(W_{j55} - W_{j44}) \right] (p_{4,01})^2 , \\ r_{i,02} = \dfrac{1}{2(4\omega_{2c}^2 - \omega_{1c}^2)} \left[\pm \omega_{1c} W_{j45} - \omega_{2c}(W_{i55} - W_{i44}) \right] (p_{4,01})^2, \end{cases}$$

for $(i, j) = (2, 3)$ or $(3, 2)$;

(B13)
$$\begin{cases} p_{i,02} = \dfrac{1}{12\omega_{2c}} (W_{jjj} - W_{jii} + 4 W_{i45}) (p_{4,01})^2 , \\ r_{i,02} = \dfrac{1}{6\omega_{2c}} (W_{i44} - W_{i55} \pm W_{j45}) (p_{4,01})^2 , \end{cases} \qquad \text{for } (i, j) = (4, 5) \text{ or } (5, 4) ;$$

(B14)
$$p_{1,11} = \frac{1}{2(\omega_{1c}+\omega_{2c})}(W_{125}+W_{134})p_{2,10}p_{4,01} , \quad r_{1,11} = \frac{1}{2(\omega_{1c}+\omega_{2c})}(W_{124}-W_{135})p_{2,10}p_{4,01} ;$$

(B15)
$$\begin{cases} p_{i,11} = \dfrac{1}{2\omega_c(\omega_{kc}+\omega_{1c}+\omega_{2c})} \Big[\pm \omega_{1c}(W_{j35}-W_{j24})+(\omega_{1c}+\omega_{2c})(W_{i25}+W_{i34})\Big] p_{2,10}p_{4,01} , \\ r_{2,11} = \dfrac{1}{2\omega_c(\omega_{kc}+\omega_{1c}+\omega_{2c})} \Big[\pm \omega_{1c}(W_{j25}+W_{j34})-(\omega_{1c}+\omega_{2c})(W_{i35}-W_{i24})\Big] p_{2,10}p_{4,01} , \end{cases}$$

for $(i, j) = (2, 3), (3, 2), (4, 5)$ and $(5, 4)$.

In the above equations, the "+" sign applies to the cases in which (i, j) is equal to (2, 3) or (4, 5), the "−" sign applies to (3, 2) or (4, 5). In the equations (B3) and (B15), the subscripts denote $k = 1$ for $(i, j) = (2, 3)$ or $(3, 2)$ and $k = 2$ for $(i, j) = (4, 5)$ or $(5, 4)$.

Appendix C

In this appendix, a brief outline of an approach that demonstrates the local equivalence of (45) and (46) to the original system (4) is given. Thus, consider the original system (4) whose Jacobian (evaluated at the critical point c) is in the canonical form (3). The initial equilibrium solution is assumed to be $w^i = 0$.

Expanding (4) into Taylor series around the critical point $(w^j;\ \eta^\beta) = (0;\ \eta_c^\beta)$, one obtains

$$
\text{(C1)}\quad \begin{cases}
\dfrac{dw^1}{dt} = W_{1j\beta}\,\mu^\beta\, w^j + W_{1jk}\, w^j\, w^k + 0(\,|(\mu^\beta,\, w^j)\,|^3)\,, \\
\dfrac{dw^2}{dt} = \omega_{1c} w^3 + W_{2j\beta}\,\mu^\beta\, w^j + W_{2jk}\, w^j\, w^k + 0(\,|(\mu^\beta,\, w^j)\,|^3)\,, \\
\dfrac{dw^3}{dt} = -\,\omega_{1c}\, w^2 + W_{3j\beta}\,\mu^\beta\, w^j + W_{3jk}\, w^j\, w^k + 0(\,|(\mu^\beta,\, w^j)\,|^3)\,, \\
\dfrac{dw^4}{dt} = \omega_{2c}\, w^5 + W_{4j\beta}\,\mu^\beta\, w^j + W_{4jk}\, w^j\, w^k + 0(\,|(\mu^\beta,\, w^j)\,|^3)\,, \\
\dfrac{dw^5}{dt} = -\,\omega_{2c}\, w^4 + W_{5j\beta}\,\mu^\beta\, w^j + W_{5jk}\, w^j\, w^k + 0(\,|(\mu^\beta,\, w^j)\,|^3)\,,
\end{cases}
$$

upon using (3), where $j,\ k = 1,\ 2,\ 3,\ 4,\ 5$ and $\beta = 1,\ 2,\ 3$.

Next, consider perturbating the steady-state solution (43). It is noted from Appendix B that all the coefficients in (43) are functions of $p_{1,00}$, $p_{2,10}$ and $p_{4,01}$. Suppose now that the time-invariant constants $p_{1,00}$, $p_{2,10}$ and $p_{4,01}$ are replaced by time-variant variables y , ρ_1 and ρ_2 , respectively. Furthermore, replacing τ_1 and τ_2 in (43) by θ_1 and θ_2 , respectively, then, the solution (43) takes the form

$$
\text{(C2)}\quad w^i = f_i(y\ ,\ \rho_1\ ,\ \rho_2\ ;\ \theta_1\ ,\ \theta_2\ )\ ,
$$

which may be interpreted as a nonlinear transformation.

Now, substituting (C2) into (C1) and solving for dy/dt , $d\rho_1/\mathrm{dt}$, $d\rho_2/\mathrm{dt}$, $d\theta_1/\mathrm{dt}$ and $d\theta_2/\mathrm{dt}$, one obtains

$$
\text{(C3)}\quad \begin{cases}
\dfrac{dy}{dt} = W_{11\beta}\mu^\beta y + \dfrac{1}{2} W_{111} y^2 + \dfrac{1}{4}(W_{122}+W_{133})(\rho_1)^2 + \dfrac{1}{4}(W_{144}+W_{155})(\rho_2)^2 + 0(\,|(\mu^\beta, w^j)\,|^3), \\
\dfrac{d\rho_1}{dt} = \dfrac{1}{2}\,\rho_1\left[\,(W_{22\beta} + W_{33\beta})\,\mu^\beta + (W_{221} + W_{331})\, y\,\right] + 0(\,|(\mu^\beta,\, w^j)\,|^3)\,, \\
\dfrac{d\rho_2}{dt} = \dfrac{1}{2}\,\rho_2\left[\,(W_{44\beta} + W_{55\beta})\,\mu^\beta + (W_{441} + W_{551})\, y\,\right] + 0(\,|(\mu^\beta,\, w^j)\,|^3)\,;
\end{cases}
$$

and

$$
\text{(C4)}\quad \begin{cases}
\rho_1 \dfrac{d\rho_1}{dt} = \rho_1\left[\omega_{1c} + \dfrac{1}{2}(W_{23\beta} - W_{32\beta})\mu^\beta + \dfrac{1}{2}(W_{231} - W_{321})y\right] + 0(\,|(\mu^\beta, w^j)\,|^3)\,, \\
\rho_2 \dfrac{d\rho_2}{dt} = \rho_2\left[\omega_{2c} + \dfrac{1}{2}(W_{45\beta} - W_{54\beta})\mu^\beta + \dfrac{1}{2}(W_{451} - W_{541})y\right] + 0(\,|(\mu^\beta, w^j)\,|^3)\,;
\end{cases}
$$

after a tedious but straightforward algebra. Truncating the higher order terms in (C3) and (C4), and using (8), (9) and (30)–(32), results in equations (45) and (46), respectively.

REFERENECES

[1] W. F. Langford, *Periodic and steady mode interactions lead to tori,* SIAM J. Appl. Math. (1)**37**(1979), 22–48.

[2] J. Guckenheimer and P. Holmes, *Nonlinear Oscillations, Dynamical Systems, and Bifurcations of Vector Fields,* Springer-Verlag: New York, 1983.

[3] G. Iooss and D. Joseph, *Elementary Stability and Bifurcation Theory,* Springer-Verlag: New York, 1980.

[4] S. N. Chow and J. Hale, *Methods of Bifurcation Theory,* Springer-Verlag: New York, 1982.

[5] J. Guckenheimer, *Multiple bifurcation problems of codimension two,* SIAM J. Math. Anal. (1)**15**(1984), 1–49.

[6] K. Huseyin and A. S. Atadan, *On the analysis of the Hopf bifurcation,* Int. J. Engng. Sci. **21**(1983), 247–262.

[7] K. Huseyin, *Multiple-parameter Stability Theory and Its Applications,* Oxford University Press, 1986.

[8] P. Yu and K. Huseyin, *Static and dynamic bifurcations associated with a double zero eigenvalue,* Dynamic and Stability of Systems **1**(1986), 73–86.

[9] H. Haken, *Advanced Synergetics,* Springer-Verlag: Berlin, Heidelberg, New York, Tokyo, 1983.

Canadian Mathematical Society
Conference Proceedings
Volume 8 (1987)

DYNAMICS AND BIFURCATIONS IN THE MOTION OF TRACTOR-SEMITRAILER VEHICLES

V. Kacani, A. Stribersky, H. Troger, K. Zeman[1]

ABSTRACT. Different types of losses of stability of the straight line steady state motion of a tractor-semitrailer vehicle with and without the control of a driver are studied. The mathematical model is represented by a set of nineteen partly nonlinear ordinary differential equations. The nonlinear stability analysis is done by means of Center Manifold Theory and Singularity Theory. Stability boundaries in parameter space and the corresponding bifurcation diagrams are presented. The mathematical unfolding parameters are interpreted physically.

1. INTRODUCTION. Road safety is to a large amount influenced by the behavior of heavy articulated trucks at high speed. Among them the tractor-semitrailer is a frequently used system which has very interesting dynamical properties. We can explain this assertion easily by considering a simple example. There exists a special state of driving, which is quite common in Europe, but will be relevant in America as well, because it shows that especially stability questions play an important role concerning road safety. The European freeway system reaches quite high elevations, for example, at Brenner pass between Italy and Austria. Consequently long lasting downhill motions must be performed by these heavy trucks. However, in order to save the brakes for emergency braking the driver shifts usually to a low gear and brakes his vehicle by means of the engine. This on the other hand has, basically, two unpleasant consequences concerning the stability behavior of the vehicle. Firstly, the system has braking forces, i.e. tangential tire forces, only at the rear axle of the tractor. Thus the trailer is not braked at all and exerts a compressive force towards the tractor. Obviously this is a configuration prone to instability. Secondly, it is well known from experimental investigations concerning the tire side forces (see Fig. 6), which are the most important physical components concerning stability, that the presence of a tire tangential force has a strong diminishing influence on the tire side force. Hence we have again a destabilizing influence.

1980 Mathematics Subject Classification (1985 Revision) 70K20, 58F14

[1]Supported by the Austrian Fonds zur Förderung der wissenschaftlichen Forschung (P 5519).

We ask now some basic questions in investigating the stability behavior of tractor-semitrailer systems. The first is: at which speed becomes the fundamental state, which is the steady state straight line motion, unstable? This question, of course, can be answered by a linearized analysis, and is given by the stability boundary in parameter space and does not create any problems to answer. Secondly, we ask whether the state of the system, after the loss of stability of the fundamental state had occured, is a nearby stable one or not. This is a very important question, because if such nearby stable states exist, the system, in a practical respect, might behave in a non-critical way. Whereas in cases when nearby stable states do not exist the behavior of the system is critical in so far as strong deviations from the fundamental state might occur. Of course, these types of behavior only can be obtained from a nonlinear mechanical and mathematical model. If we use a nonlinear model we also can answer a third question concerning the influence of imperfections on the behavior of the system.

The mechanical system which we are going to set up describes the motion of a tractor-semitrailer on a flat but possibly also inclined road. Furthermore we simulate the driver by a controller, a fact which seems to be decisive to us in order to make practically meaningful predictions of the driving behavior. To be able to perform a nonlinear stability analysis we use the methods of bifurcation theory as they are developed in /1-4/.

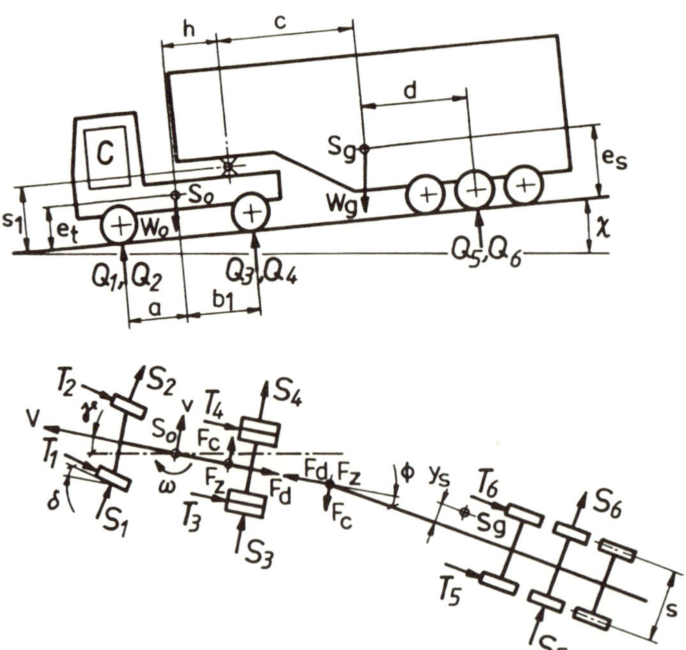

Fig. 1: Mechanical model of the tractor-semi-trailer with controller (C) moving on an inclined road (angle χ) with the three degrees of freedom v, γ and Φ.

2. MECHANICAL MODEL OF THE TRACTOR-SEMITRAILER. We study a mechanical model consisting of two coupled rigid bodies (Fig. 1) having the three degrees of freedom v, γ, ϕ. The inclination of the flat road is given by the angle χ. Aerodynamic forces are neglected. With the notations given in Fig. 1 we can write down the equations of motion from the principles of conservation of momentum and angular momentum, which result in six nonlinear ordinary differential equations for each of the two rigid bodies. We obtain (/5/) for the tractor ($\omega = \dot{\gamma}$)

$$m_0(-v\omega) = -(T_1+T_2)\cos\delta - (S_1+S_2)\sin\delta - (T_3+T_4) - F_d + W_0\sin\chi\cos\gamma$$

$$m_0(\dot{v}+V\omega) = -(T_1+T_2)\sin\delta + (S_1+S_2)\cos\delta + (S_3+S_4) + F_c - W_0\sin\chi\sin\gamma$$

$$0 = -(Q_1+Q_2+Q_3+Q_4) - F_z + W_0\cos\chi \tag{1}$$

$$-I_{x_0z_0}\dot{\omega} = -(S_1+S_2)e_t\cos\delta + (Q_1-Q_2)\ s/2 + F_c(s_1-e_t) - (S_3+S_4)e_t + (Q_3-Q_4)\ s/2 + (T_1+T_2)e_t\sin\delta$$

$$-I_{x_0z_0}\omega^2 = -(S_1+S_2)e_t\sin\delta + (Q_1+Q_2)a - F_zh + F_d(s_1-e_t) - (Q_3+Q_4)b_1 - (T_1+T_2)e_t\cos\delta - (T_3+T_4)e_t$$

$$I_{z_0}\dot{\omega} = (S_2-S_1)(s/2)\sin\delta + (S_1+S_2)a\cos\delta - F_ch - (S_3+S_4)b_1 + (T_2-T_1)(s/2)\cos\delta - (T_1+T_2)a\sin\delta + (T_4-T_3)s/2$$

and the semitrailer

$$m_g[(\dot{\omega}+\ddot{\phi})(c\sin\phi-y_s\cos\phi) + (\omega+\dot{\phi})^2(c\cos\phi + y_s\sin\phi) - v\omega + \omega^2h] = F_d - (T_5+T_6)\cos\phi - (S_5+S_6)\sin\phi + W_g\sin\chi\cos\gamma$$

$$m_g[\dot{v} - \dot{\omega}h + \omega V + (\dot{\omega}+\ddot{\phi})(-c\cos\phi-y_s\sin\phi) + (\omega+\dot{\phi})^2(c\sin\phi-y_s\cos\phi)] = - F_c + (S_5+S_6)\cos\phi - (T_5+T_6)\sin\phi - W_g\sin\chi\sin\gamma$$

$$0 = - (Q_5+Q_6) + F_z + W_g\cos\chi \tag{2}$$

$$-I_{x_g z_g}(\dot{\omega}+\ddot{\phi}) + I_{y_g z_g}(\omega+\dot{\phi})^2 = -(S_5+S_6)e_s + (Q_5-Q_6)s/2 +$$
$$+ (Q_5+Q_6)y_s + (F_c\cos\phi+F_d\sin\phi)(e_s-s_1) - F_z y_s$$

$$-I_{y_g z_g}(\dot{\omega}+\ddot{\phi}) - I_{x_g z_g}(\omega+\dot{\phi})^2 = (F_d\cos\phi-F_c\sin\phi)(e_s-s_1) -$$
$$- (Q_5+Q_6)d - F_z c - (T_5+T_6)e_s$$

$$I_{z_g}(\dot{\omega}+\ddot{\phi}) = (F_d\cos\phi-F_c\sin\phi)y_s - (F_c\cos\phi+F_d\sin\phi)c - (S_5+S_6)d +$$
$$+ (T_6-T_5)(s/2) - (T_5+T_6)y_s \quad .$$

In order to be able to calculate the wheelloads of the tractor, which is a statically indetermined problem, we must introduce the stiffnesses c_F of the front and c_R of the rear suspension of the tractor. Then the following relations

$$Q_1 = Q_{10} + c_F h_1 \;, \quad Q_2 = Q_{20} + c_F h_2 \;,$$
$$Q_3 = Q_{30} + c_R h_3 \;, \quad Q_4 = Q_{40} + c_R h_4 \;,$$

hold where Q_{i0} are the static wheelloads and h_i are the corresponding deflections of the suspension. From the geometric relation (Fig. 2)

$$\frac{1}{2}(h_1 + h_4) = \frac{1}{2}(h_2 + h_3)$$

follows

$$\frac{Q_1-Q_{10}}{c_F} + \frac{Q_4-Q_{40}}{c_R} = \frac{Q_2-Q_{20}}{c_F} + \frac{Q_3-Q_{30}}{c_R} \quad .$$

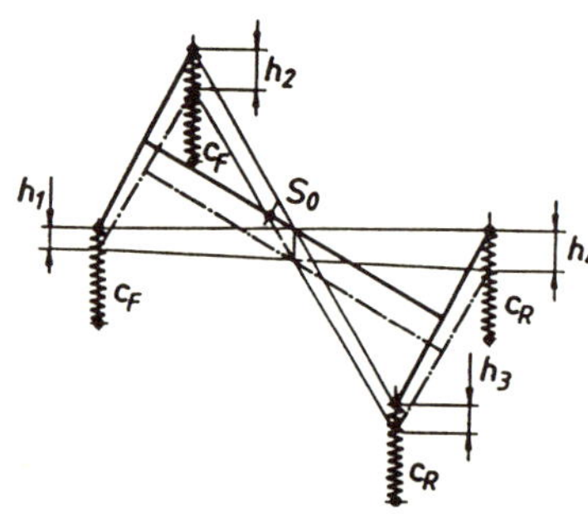

Fig. 2: Deformation of the front and rear suspensions in order to calculate the wheelloads of the tractor.

With $Q_{10} = Q_{20}$ and $Q_{30} = Q_{40}$ we finally obtain

$$(Q_1 - Q_2)c_R = (Q_3 - Q_4)c_F \; . \tag{3}$$

(1), (2), (3) are a set of 13 equations for 13 unknowns v, γ, ϕ, Q_1, ..., Q_6, F_c, F_d, F_z and T_m. Here $Q_1,...,Q_6$ are the wheelloads, F_c, F_d, F_z are the components of the force at the fifth wheel and T_m is the sum of the tangential tire forces at the rear axle of the tractor which is necessary to maintain the prescribed steady state speed of the vehicle. The forces S_i are the tire side forces and they have to be expressed by the variables as it is explained below. The T_i are the tire tangential forces and are given either by T_m or by the braking forces during braking or are due to the rolling resistance of the tires. The latter, however, usually can be neglected.

A model for the controller representing a driver is given in /6/ where also the frequency response function is shown, which has been fitted to experimental data obtained from tests with human drivers. The block diagram of Fig. 3 gives the appropriate controller from which the following relationship

$$\delta = - k_0 \{ \frac{k_5}{s} + [k_1 s^2 (\frac{1}{1+k_2 s})^2 + k_4] \frac{1}{1+k_3 s} \} \gamma \tag{4}$$

between the steering angle δ and the yaw angle γ (Fig. 1) follows. (4) is aquivalent to the following set of linear differential equations

$$\begin{aligned} \dot{\delta} &= - (k_0 k_5 \gamma + \dot{w}) \\ k_2^2 \ddot{u} + 2k_2 \dot{u} + u &= k_0 k_1 \dot{\omega} \\ k_3 \dot{w} + w &= u + k_0 k_4 \gamma \, . \end{aligned} \tag{5}$$

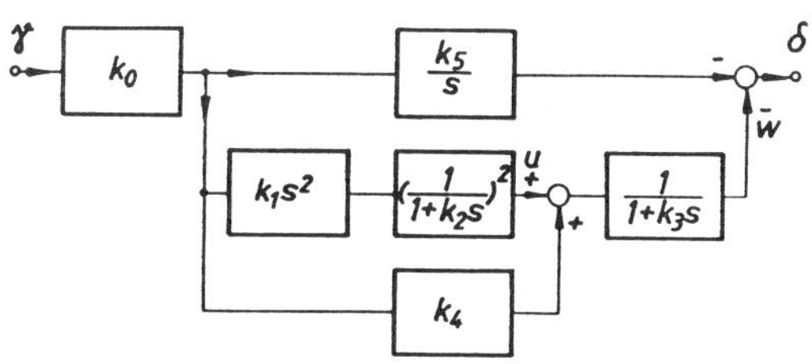

Fig. 3: Block diagram of the active controller simulating a human driver.

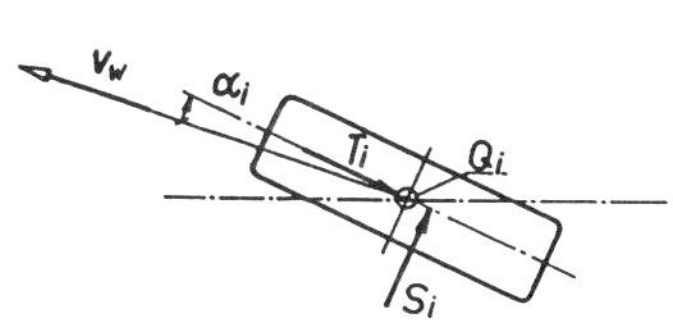

Fig. 4: Tire side force S created by the tire side slip with angle α.

Basically for the validity of the results of a stability analysis in vehicle dynamics is, how good one is able to model the tire side forces which, obviously, are those decisive for maintaining stability. The tire side force S_i depends essentially on three parameters. One is the tire side slip which is expressed by the tire slip angle α_i (Fig. 4). The second is the wheel load Q_i and the third is the tangential tire force T_i. These three influences are taken into consideration in the following relationship

$$\frac{S_i}{P_{ni}} = [(a_{1i}\pi_i + a_{2i}\pi_i^2)\tanh(b_i\alpha_i) + a_{3i}\pi_i\alpha_i] \sqrt{1-(\mu_{Ti}/\mu_{Tmax})^2} \tag{6}$$

where a_{1i}, ... are tire parameters to be obtained from experiments, P_{ni} is a reference wheelload and $\pi_i = Q_i/P_{ni}$. The slip angles α_i are given by

$$\begin{aligned}
\alpha_1 = \alpha_2 &= \delta - \operatorname{arctg} \frac{v + \omega a}{V \pm \omega s/2} \\
\alpha_3 = \alpha_4 &= \quad - \operatorname{arctg} \frac{v - \omega b_1}{V \pm \omega s/2} \\
\alpha_5 = \alpha_6 &= \quad - \operatorname{arctg} \frac{-V\sin\phi + (v - \omega h)\cos\phi - (\omega + \dot{\phi})(c + d)}{V\cos\phi + (v - \omega h)\sin\phi \pm (\omega + \dot{\phi})s/2} \quad .
\end{aligned} \tag{7}$$

In Fig. 5 for $\mu_T = 0$ (explained below) experimental data (dots) are compared with the results given by (6). Up to quite big slip angles α a sufficiently good agreement is given.

The influence of the tire tangential force is given by the square root in (6), where $\mu_T = T_i/Q_i$ is the tangential friction coefficient and $\mu_{Tmax} = T_{max}/Q_i$ is the maximum value of μ_T possible at a given road and for a given tire. Thus μ_{Tmax} is an experimentally obtained number. The whole characteristic field of the tire is given in Fig. 6. This approximation gives practically good results if $\mu_T/\mu_{Tmax} < 0.95$.

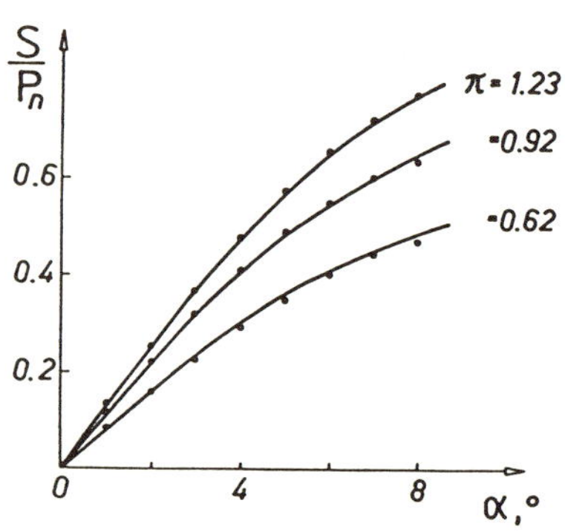

Fig. 5: Tire side force S versus slip angle α (eq. (6)) for different values of the wheelload Q ($\pi = Q/P_n$, P_n being a reference load). Dots mark experimental results.

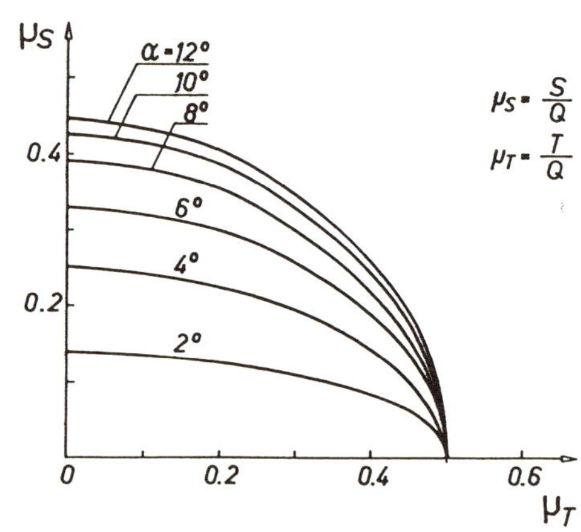

Fig. 6: Influence of the tangential tire force T on the tire side force S.

Taking $\dot{\phi}$ and $\dot{u}$ as new variables we form two vectors

$$x^T = (v,\omega,\phi,\dot{\phi},\gamma,\delta,u,\dot{u},w) \qquad \in \qquad \mathbf{R}^9 \tag{8}$$

$$q^T = (Q_1,Q_2,\ldots,Q_6,F_c,F_d,F_z,T_m) \qquad \in \qquad \mathbf{R}^{10} \ . \tag{9}$$

Adding the trivial equations $\dot{x}_5 = x_2$, $\dot{x}_3 = x_4$ and $\dot{x}_7 = x_8$ we obtain from (1) - (7) the following system of first order autonomous, nonlinear ordinary differential equations

$$W(x)\dot{x} = f(x,q) \tag{10}$$

$$U(x)\dot{x} = t(x,q) \tag{11}$$

where $W(x)$ is a non-degenerate 9×9 matrix (mass-matrix) and $U(x)$ is a 10×9 matrix. From (10) and (11) q must be eliminated. This can be done by calculating

$$\dot{x} = W(x)^{-1}f(x,q) \tag{12}$$

from (10) and inserting this expression into (11) yielding

$$U(x)W(x)^{-1}f(x,q) = t(x,q) \ . \tag{13}$$

$W(x)^{-1}$ can be calculated by series expansions. (13) is an algebraic system from which, according to the Implicit Function Theorem, for values of q close to q_0, we can locally calculate q by series expansions resulting in

$$q = q(x) \ . \tag{14}$$

Inserting (14) into (10) we obtain

$$\dot{x} = G(x,\lambda) \ , \tag{15}$$

which is the desired form of the equations of motion.

In (15) x is given by (8) and

$$\lambda^T = (V,d,\chi) \tag{16}$$

is the parameter vector, where V is the constant driving speed, d is a measure of the loading condition of the trailer designating its center of gravity (Fig. 1) and χ is the inclination angle of the road. It is obvious that the straight line motion of the vehicle, given by the trivial variable vector $x_0 = 0$ results in $G(0,\lambda) = 0$ in (15) for any values of λ, and thus $x_0 = 0$ is the equilibrium position of (15), which we want to study concerning its stability.

3. STABILITY BOUNDARIES AND CENTER MANIFOLD REDUCTION. Proceeding in an engineering style of approach we start with a stable motion of the vehicle, which for an appropriately low speed always will exist. Then we increase the speed quasistatically until we reach a point where the stable steady state loses its stability. This type of approach perfectly fits to the methods of Bifurcation Theory.

Applying these methods we first can reduce the order of system (15), which is nine in our case, considerably to a low dimensional bifurcation system, the dimension of which is equal to the number of eigenvalues with zero realparts and which, as is well known, locally has the same stability properties as the original system. This reduction will be done by means of Centre Manifold Theory (/2/). In order to perform the reduction process we rewrite (15) by expanding the right hand side into a power series with respect to the equilibrium position $x_0 = 0$ which results in $(x,g \in R^9)$

$$\dot{x} = A(\lambda)x + g(x,\lambda) \tag{17}$$

where the components of $g(x,\lambda)$ do not contain linear terms in x and where $A(\lambda)$ is given by

$$A(\lambda) = \left[\frac{\partial G_i}{\partial x_j} \right]_{x=x_0} . \tag{18}$$

Following the ideas in /3/ we select V in (16) as the main parameter. Keeping the two other parameters d and χ in (16) fixed we now vary V quasistatically. I.e. we study the behavior of (17) for increasing but fixed values of V. The stability of $x_0 = 0$ is determined by a theorem of Lyapunov that assures that $x_0 = 0$ is asymptotically stable as long as all eigenvalues of (18) have negative real parts. So, loss of stability can occur only if we reach a value of $V = V_c$ where one or several eigenvalues cross the imaginary axis.

For the uncontrolled motion (steering angle $\delta = 0$) on a flat horizontal road we calculate the stability boundary in a two-dimensional parameter space (V,d) which is shown in Fig. 7. From Fig. 7 we see that depending on the loading condition of the trailer, expressed by the distance d of its mass center (Fig. 1), we obtain generically either a divergence bifurcation or a Hopf bifurcation. However, there exists also a special value $d = d_c$ (Fig. 7) for which a more complicated situation is found, because at loss of stability a zero root and a purely imaginary pair of roots occur simultaneously. This latter case, though it is possible only for a very special loading condition, is yet important for two reasons also from a practical point of view. Firstly, also for nearby systems a coupling between divergence and flutter will be present and secondly, this very special configuration is the one which gives the largest possible stability domain. Thus in optimizing the loading condition of such a vehicle one obtains a multiple critical eigenvalue which is more complicated to analyse than simple generic cases of loss of stability.

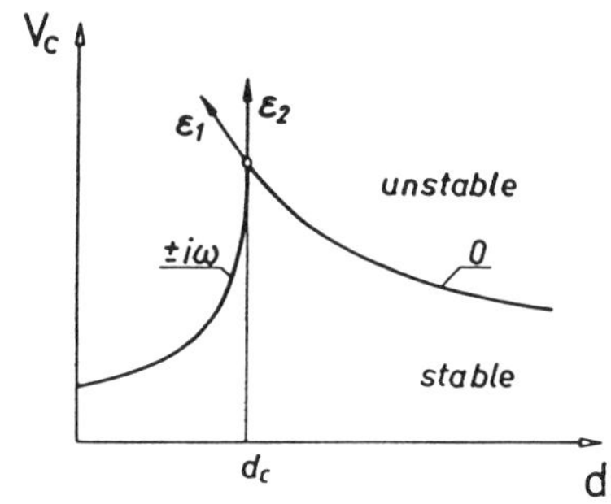

Fig. 7: Stability boundary in V, d space for the model with fixed steering on a horizontal road, showing generically either a zero root or a pair of purely imaginary roots at loss of stability. For $d = d_c$ both cases occur simultaneously.

Including a driver in our mechanical model, we obtain always a Hopf bifurcation at loss of stability no matter whether the vehicle moves on a horizontal or on an inclined road. In Fig. 11 the stability boundary is shown for the vehicle which is braked by the engine, and therefore has tangential tire forces only at the rear axle of the tractor. We also see that for an uphill motion (negative inclination angle χ) no finite domain of stability is obtained. However, for a downhill motion there is a dramatical reduction of the domain of stability. Furthermore the points where the curves end (χ = 3 % and 5 %) mark physically relevant limits, i.e. where the tangential tire forces at the rear axle are larger than their maximum possible values. This means we get locked wheels and no regular type of driving is possible any more.

In order to obtain the bifurcation equations we make a transformation of coordinates to bring (19) into Jordan form. This results in

$$\begin{aligned} \dot{y} &= A_c y + f_c(y, y_n) \\ \dot{y}_n &= A_n y_n + f_n(y, y_n) \end{aligned} \tag{19}$$

where A_c corresponds to one of the cases mentioned above. Thus it is one of the following matrices

$$[0] \quad , \quad \begin{bmatrix} 0 & \omega \\ -\omega & 0 \end{bmatrix} \quad , \quad \begin{bmatrix} 0 & 0 & 0 \\ 0 & 0 & \omega \\ 0 & -\omega & 0 \end{bmatrix} \quad . \tag{20}$$

A_n is a matrix possessing only eigenvalues with negative realpart. y includes the critical variables and y_n the non-critical variables. Center Manifold Theory allows us now to eliminate the non-critical variables y_n from $(19)_1$. By calculating (/2/) $y_n = H(y) = O(|y|^2)$ from $(19)_2$ and introducing this expression in $(19)_1$ we obtain the bifurcation equations

$$\dot{y} = A_c y + f(y,H(y)) \ . \tag{21}$$

4. DISCUSSION OF THE BIFURCATION EQUATIONS. The result of the Center Manifold reduction is a set of one, two or three dimensional bifurcation equations respectively, according (20).

(A) SYSTEM WITHOUT CONTROL. (a) One zero eigenvalue. This case is given if d (Fig. 1) is large. I.e. the center of gravity of the trailer is close to the fifth wheel. We obtain a divergent type of instability. The one-dimensional bifurcation equation is given by

$$\dot{y} = K_1 y^3 + O(y^4) \ . \tag{22}$$

As we are not only interested in the solutions of (22) but also in all solutions possible from the class of equations of third order to which (22) belongs, we embed (22) into a parametrized family. As parameters for this family we use so-called imperfection parameters, which we select with the steering angle δ and with y_s as a measure for a non-symmetrically loaded trailer (Fig. 1). The unfolding is done following the theory presented in /3/ yielding

$$\dot{y} = K_1 y^3 + \lambda_2 y^2 + \lambda_1 y + \lambda_0 \tag{23}$$

where three mathematical parameters λ_0, λ_1, λ_2 are introduced which are related to the physical parameters in the following way (Fig. 1)

$$\lambda_1 \sim V - V_c$$
$$\lambda_0 = p - q \qquad\qquad p = m_1\delta, \quad q = m_2 y_s$$
$$\lambda_2 \sim y_s$$

where m_1 and m_2 are constants. Of course also other parameters could have been selected but nothing qualitatively different would have been obtained.

In Fig. 8 all the qualitatively different ways of behavior are given in the bifurcation diagram. Basically we have an unstable bifurcation point which leads to a critical post-bifurcation behavior, because K_1 in (23) is positiv. From the perfect system (origin of the diagram and drawing 1) we see that small imperfections, either an excentrically loaded trailer ($y_s \neq 0$) or a cornering motion of the vehicle ($\delta \neq 0$) can lead to a strong reduction of the critical speed V_c to V_0 (drawing 2). Furthermore the mode of instability is the socalled jack-knifing which is a very dangerous type of motion. However, it must be mentioned that there also exists a stable solution for $V > V_c$ (drawings 4, 5 and 8, 9) which corresponds to a cornering motion for an excentrically loaded system. A result which one would not expect from practical driving experiences with tractor-semitrailers. Let us try to give a physical interpretation of this state of driving. To be specific we select drawing 4 in Fig. 8. For a steering angle $\delta < 0$ we have at the two different branches stable states. One with $\phi > 0$ and the other with $\phi < 0$ furthermore y_s must be negative (see Fig. 1). The state $\phi > 0$ corresponds to a regular state of driving for a range of speed $0 < V \leq V_0$. The state $\phi < 0$, however, represents a state of driving best characterized by the term "powerslide" from car racing, and exists as stable state only in a very small domain of speed above $V = V_0$.

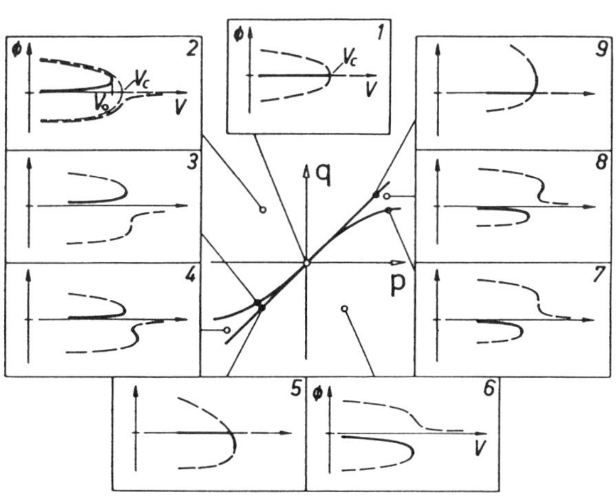

Fig. 8: Bifurcation diagram for the divergent type of instability. V_c is the critical speed of the perfect system.

(b) A pair of purely imaginary roots. This type of loss of stability is found if the location of the center of gravity of the trailer is close to the rear axle of the trailer. We get a Hopf bifurcation described by

$$\dot{r} = K_2 r^3 + \lambda_1 r + O(|r|^5) \tag{24}$$

where K_2 is always positive and $\lambda_1 \sim V-V_c$. In Fig. 9 the amplitude of the unstable limit cycle is shown to be very small, which means that the domain of attraction of the linearly stable steady state motion is very small. Thus if one performs only a linear stability analysis then this could lead to serious mistakes concerning the evaluation of the practical stability of the system.

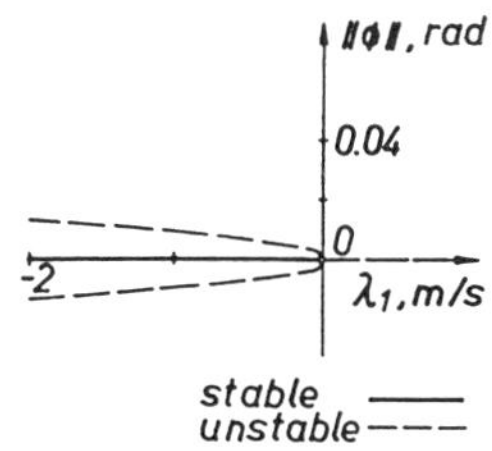

Fig. 9: Amplitude of the unstable limit cycle of the Hopf-bifurcation for the system with fixed steering.

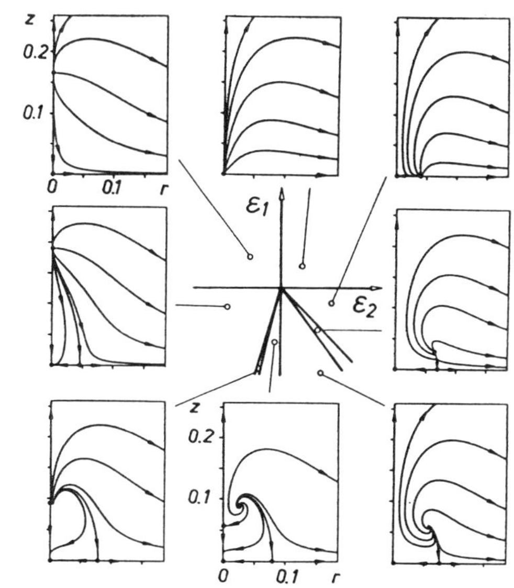

Fig. 10: Bifurcation diagram for the system with fixed steering and one zero root and a pair of purely imaginary roots at loss of stability.

(c) One zero root and a pair of purely imaginary roots. This is the nongeneric case mentioned before. Now we obtain a three-dimensional set of bifurcation equations

$$
\begin{aligned}
\dot{y}_1 &= \qquad\qquad g_1(y_1,y_2,y_3) + \text{h.o.t.} \\
\dot{y}_2 &= \quad \omega y_3 + g_2(y_1,y_2,y_3) + \text{h.o.t.} \\
\dot{y}_3 &= -\omega y_2 \quad + g_3(y_1,y_2,y_3) + \text{h.o.t.}
\end{aligned}
\tag{25}
$$

Introducing polar coordinates and making use of the Averaging Principle /3/ we obtain the following unfolded system of bifurcation equations

$$
\begin{aligned}
\dot{r} &= \varepsilon_1 r + K_3 z^2 r + K_4 r^3 + O(|r|^5 + |z|^5) \\
\dot{z} &= \varepsilon_2 z + K_5 z^3 + K_6 z r^2 + O(|r|^5 + |z|^5)
\end{aligned}
\tag{26}
$$

where r and z are the amplitudes of the oscillatory and the divergent mode respectively. The unfolding parameters ε_1 and ε_2 are linear combinations of the physical parameters $V - V_c$ and $d - d_c$ as it is introduced in Fig. 7 and $K_3,\ldots,K_6$ are constants depending on the given system, which is considered to be symmetrical. In Fig. 10 the bifurcation diagram is given showing all qualitatively possible different ways of behavior. Due to the critical post-bifurcation no stable motions exist after loss of stability of the original motion.

(B) SYSTEM WITH CONTROL. (a) Without delay. Basically we have again a bifurcation equation as it is given in (24). However, there exists a fundamental difference to (24), because now K_2 in (24) changes its sign. I.e. we obtain both critical and non-critical post-bifurcation behaviors of the vehicle (/5/). For three points in Fig. 11 (downhill motion χ = 3 %) the corresponding amplitudes of the limit cycles are given in Fig. 12. It is particularly interesting that just those systems which have the maximum range of stability possess a non-critical post-bifurcational behavior. The explanation for this surprising fact can be given from the frequency response of the controller in our model (/6/). Because just for those frequencies at the stability boundary where we get a non-critical post-bifurcation behavior the controller is very sensitive and able to react strongly.

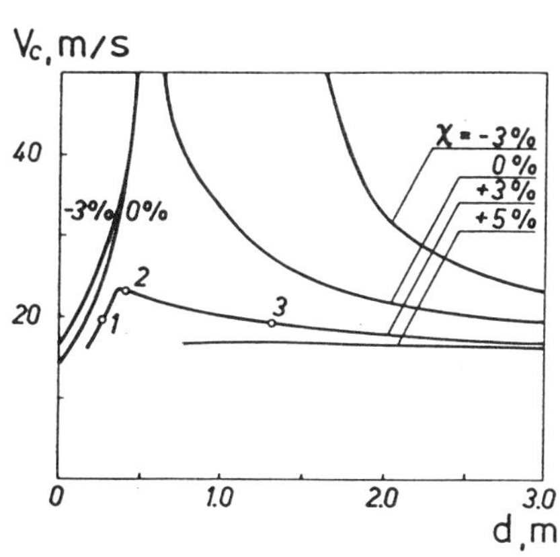

Fig. 11: Stability boundary in V, d space for the actively controlled system having always a pair of purely imaginary roots at loss of stability.

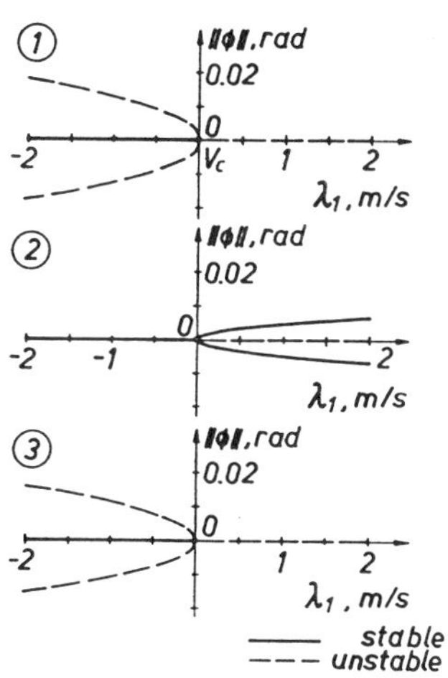

Fig. 12: Amplitudes of stable (2) and unstable (1,3) limit cycles corresponding to the numbers in Fig. 11.

(b) With delay. A recently developed code (/8/) enable us to include also a time delay τ in the controller model. This delay accounts for the influence of the reaction time of the driver. In order to be able to apply this code we had to bring the equations of motion in the form

$$\dot{x} = G(x(t), x(t-\tau); \lambda) \ . \tag{27}$$

To obtain (27) only (4) and (5) had to be changed to

$$\delta = - k_0 e^{-s\tau} \{ \frac{k_5}{s} + [k_1 s(\frac{1}{1+k_2 s})^2 + k_4] \frac{1}{1+k_3 s} \} \gamma \tag{28}$$

$$\begin{aligned} \dot{\delta} &= - (k_0 k_5 \gamma(t-\tau) + \dot{w}) \\ k_2^2 \ddot{u} + 2k_2 \dot{u} + u &= k_0 k_1 \dot{\gamma}(t-\tau) \\ k_3 \dot{w} + w &= u + k_0 k_4 \gamma(t-\tau) \end{aligned} \tag{29}$$

where the time delay τ is introduced. Furthermore a slight change in the controller compared to (4) was necessary for an immediate application of the code in /8/. The results of these calculations are given in Fig. 13, where the influence of the "reaction" time of the driver on the stability boundary is clearly seen. In Fig. 14 the corresponding unstable limit cycles are shown for the loading condition d = 1 in Fig. 13.

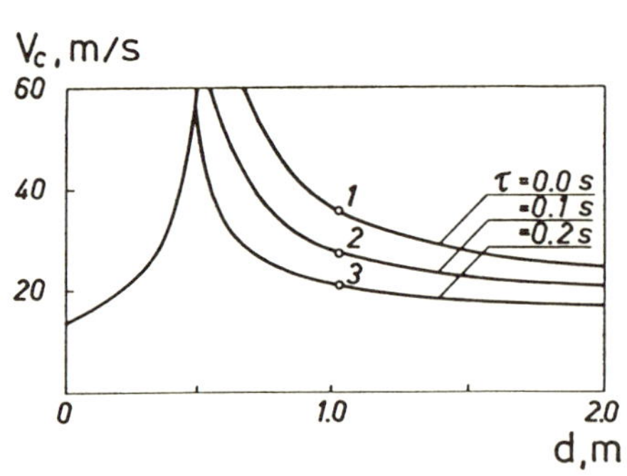

Fig. 13: Stability boundary as in Fig. 11 but for a controller with time delay τ.

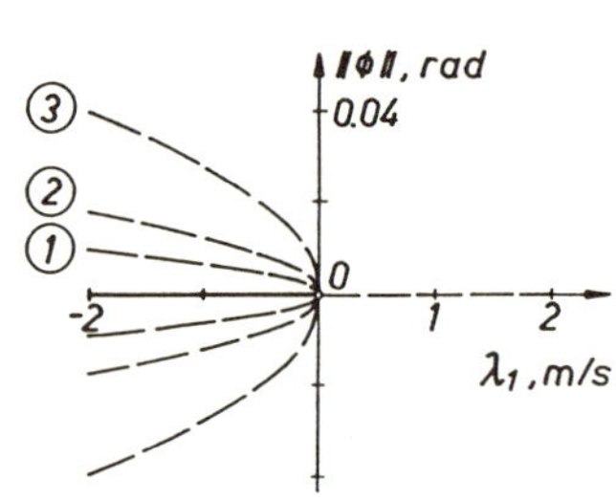

Fig. 14: Amplitudes of the unstable limit cycles corresponding to d = 1 in Fig. 13.

5. CONCLUSIONS. The application of bifurcation theory to the dynamics of heavy road vehicles allows to make predictions on the behavior of the vehicle after loss of stability of the steady state straight motion. Basically two different types of control are considered. The first is with fixed steering and the other with a mathematical model of a human driver. In the first case different types of bifurcations are found including Codimension one and two, whereas for the actively controlled system always a Hopf bifurcation occurs at loss of stability. Whereas for the fixed steering all bifurcations are critical, there exist noncritical bifurcations for the controlled system which is important for the practical operation of these vehicles.

BIBLIOGRAPHY

1. Arnold V.I., Geometrical Methods in the Theory of Ordinary Differential Equations, Springer-Verlag, Berlin, Heidelberg, New York, 1983.
2. Carr J., Applications of Centre Manifold Theory, Appl. Math. Sciences 35, Springer-Verlag, Berlin, Heidelberg, New York, 1981.
3. Golubitsky M., D. Schaeffer, Singularities and Groups in Bifurcation Theory, Appl. Math. Sciences 51, Springer-Verlag, Berlin, Heidelberg, New York, Tokyo, 1985.
4. Guckenheimer J., P. Holmes, Nonlinear Oscillations, Dynamical Systems, and Bifurcations of Vector Fields, Appl. Math. Sciences 42, Springer-Verlag, Berlin, Heidelberg, New York, 1983.
5. Scheidl R., A. Stribersky, H. Troger, K. Zeman, Nonlinear Stability Behavior of a Tractor-Semitrailer in Downhill Motion, O. Nordsträm, ed., Proceedings of the 9th IAVSD Symposium, Linköping (Sweden), Swets and Zeitlinger, 1986.
6. Wallner F., Das Verhalten des Fahrzeuglenkers als regelungstechnisches Problem, Automobilindustrie 1 (1967) 54 - 62.
7. Troger H., K. Zeman, A Nonlinear Analysis of the Generic Types of Loss of Stability of the Steady State Motion of a Tractor-Semitrailer, Vehicle System Dynamics 13 (1984) 161 - 172.
8. Hassard B., A Code for Hopf Bifurcation Analysis of Autonomous Delay-Differential Systems, to appear in these Proceedings.

INSTITUT FÜR MECHANIK
TECHNISCHE UNIVERSITÄT WIEN
KARLSPLATZ 13
A - 1040 VIENNA
AUSTRIA

Canadian Mathematical Society
Conference Proceedings
Volume 8 (1987)

QUASIPERIODIC BIFURCATION OF NONLINEAR VIBRATION SYSTEMS WITH QUASICYCLIC COORDINATES

Lin Rui and Huang Ke-lei

ABSTRACT. This paper studies studies the quasiperiodic oscillation bifurcation phenomena of nonlinear vibration systems having quasicyclic coordinates and unknown displacement functions. The systems considered here are as follows:

$$\dot{X} = \epsilon\ \Phi\ (t,Y,X,\dot{X},\epsilon)$$

$$\dot{Y} = A(u)Y+F(t,X,\dot{X})+\epsilon\bar{F}^*(t,Y,X,\dot{X},\epsilon,u)\quad X\in R^m;\ Y\in R^n;\ t,u\in R$$

where u is the bifurcation parameter and ϵ is the small parameter, ϕ, F, and $\bar{F}^*$ are quasiperiodic functions of t. At the critical point u = 0, A(0) has eigenvalues $\pm\ \omega_1 i,\ldots,\pm\ \omega_r i$ ($2q = r\leq n$) and n-r eigenvalues $\lambda_{r+1},\ldots,\lambda_\eta$ with negative real parts. There exist quasiperiodic bifurcating solutions under certain conditions. We develop KBM averaging methods and use them and perturbation techniques to construct the bifurcating solutions in the nonresonant and resonant cases. At the same time, we obtain two stability theorems for each of the two cases.

1. INTRODUCTION. In recent years there has been a great deal of interest in the bifurcating and chaotic behavior of nonlinear vibration systems [1], [2]. Most of the former research works focus on periodically perturbed systems and their periodic response. But in fact, quasiperiodic behaviors may appear in many nonlinear vibration systems. The difficulties of dealing with quasiperiodic motions are pointed out by Moser [3]. Some developments have been published recently on perturbation problems of almost periodic solutions [4]. The bifurcations which lead to quasiperiodic solutions from equilibrium or periodic trajectories, in autonomous nonlinear vibration systems or nonautonomous ones with periodic excitation, can be studied by means of the general KB averaging method [5].

1980 Mathematics Subject Classification (1985 Revision) 34D10, 58F27.

The aim of this paper is to present an averaging perturbation theory to study the bifurcation phenomena of quasiperiodic oscillations of nonlinear vibration systems, having angular coordinates $x_1,\ldots,x_m$ which describe the rotational motion, in addition to the unknown displacement functions $y_1,\ldots,y_n$. When the small parameter ϵ vanishes, the angular coordinates are functions of time t. The angular coordinates are called quasicyclic coordinates. Nonlinear ordinary differential equations with quasicyclic coordinates frequently occur in studies of rotator vibration problems, in the interactions of the finite power engine and vibration systems with linear, or nonlinear, or parametric excitation, and in other nonlinear vibration problems.

Quasiperiodic functions are defined as the special class of almost periodic functions possessing only a finite basis of frequencies. In other words, we are studying oscillations containing finitely many (rationally independent) frequencies. Such solutions occur naturally if one considers n slightly coupled oscillators.

At first we develop a general averaging perturbation method which can be used to study the bifurcation problems, and then we use singular perturbation techniques to construct the bifurcating solutions in the nonresonant and resonant case. Finally, we establish the averaging variational equations of the bifurcating quasiperiodic solutions and study their stability.

2. STANDARD FORM OF THE EQUATIONS. In this section, we construct an averaging transformation and find orthogonality conditions to obtain the standard form of nonlinear vibration systems with quasicyclic coordinates.

Consider the following equations:

$$\begin{cases} \dfrac{d^2x_i}{dt^2} = \epsilon\phi_i\,(t,y_1,\ldots,y_n,x_1,\ldots,x_m,\dot{x}_1,\ldots,\dot{x}_m,\epsilon) \qquad i = 1,\ldots,m & (2.1) \\ \dfrac{dy_j}{dt} = \sum\limits_{i=1}^{n} a_{ji}\,(\mu)\, y_i + F_j(t,x_1,\ldots,\dot{x}_1,\ldots,\dot{x}_m) & \\ \qquad + \epsilon F_j^*\,(t,y_1,\ldots,y_n,x_1,\ldots,x_m,\dot{x}_1,\ldots,\dot{x}_m,\epsilon,\mu) \quad j = 1,\ldots,n & (2.2) \end{cases}$$

where μ is a bifurcation parameter.

Rewrite equations (2.1) and (2.2) in the form:

$$\begin{cases} \ddot{X} = \epsilon\,\phi\,(t,Y,X,\dot{X},\epsilon) & (2.1a) \\ \dot{Y} = A(\mu)Y + F(t,X,\dot{X}) + \epsilon F^*(t,Y,X,\dot{X},\epsilon,\mu) & (2.2a) \end{cases}$$

where $\phi = (\phi_1,\ldots,\phi_m)^T$, $A(\mu) = (a_{ji}(\mu))_{nxn}$, $F = (F_1,\ldots,F_n)^T$, $F^* = (F_1^*,\ldots,F_n^*)^T$, $X = (x_1,\ldots,x_m)^T$ is the quasicyclic coordinate vector, $Y = (y_1,\ldots,y_n)^T$ the displacement vector, and ϵ is the small parameter. Here ϕ, F^* and F are quasiperiodic functions in t and have period 2π in $x_1,\ldots,x_m$. Also, ϕ, F^* and F are assumed to be analytic functions in ϵ and polynomial in Y and $\dot{X}$.

The elements of ϕ, F^* and F can be expanded into Fourier series:

$$\phi_i = \phi_{i0}(Y,\dot{X},\epsilon) + \Sigma\ [_{ik}(Y,\dot{X},\epsilon)\cos\alpha_k + \phi_{ik}(Y,\dot{X},\epsilon)\sin\alpha_k] \tag{2.3}$$

$$F_j = F_{j0}(X) + \Sigma\ [F_{jk}(X)\cos\alpha_k + F_{jk}(X)\sin\alpha_k] \tag{2.4}$$

$$F_j^* = F_{j0}^*(Y,\dot{X},\epsilon,\mu) + \Sigma\ [F_{jk}^*(Y,X,\epsilon,\mu)\cos\alpha k + \overline{F}_{jk}^*(Y,X,\epsilon,\mu)\sin\alpha_k] \tag{2.5}$$

where $\alpha_k = \overline{\omega}_k t + (K,X)$, $(K,X) = k_1x_1 + \ldots + k_mx_m$, k_i is finite number, $\overline{\omega}_1,\ldots,\overline{\omega}_k$ are incommensurable.

Assume

$$x_i = \Omega_i t + c_i + \epsilon w_i\ (t,X,\Omega,C,\epsilon) \tag{2.6}$$

where $\Omega = (\Omega_1,\ldots,\Omega_m)$, $C = (c_1,\ldots,c_m)$, Ω_i and c_i are constants which will be determined from the following procedure, and w_i is a quasiperiodic function of t with its mean value equal to zero.

System (2.2a) becomes

$$\dot{Y} = A(\mu)Y + F(t,\Omega t + C,\Omega) + \epsilon F_1^*(t,Y,\Omega T + C,\Omega,W,\dot{W},\epsilon,\mu) \tag{2.7}$$

where

$$F_1^* = \frac{\partial F}{\partial X}W\Bigg|_{\substack{X = \Omega t+C\\ \dot{X} = \Omega}} + \frac{\partial F}{\partial \dot{X}}\dot{W}\Bigg|_{\substack{X = \Omega t+C\\ \dot{X} = \Omega}} + F^*(t,Y,\Omega t + C,\Omega,W,\dot{W},\epsilon,\mu)\ . \tag{2.8}$$

Assume that $\mu = 0$ is a critical value of (2.7), and A(0) has imaginary eigenvalues $\pm\omega_1 i,\ldots,\pm\omega_{q1}i$ $(2q_1 = r \le n)$ and n-r eigenvalues $\lambda_{r+1},\ldots,\lambda_n$ with negative real parts. When $\mu < 0$, A(0) has eigenvalues $\bar{\lambda}_1,\ldots,\bar{\lambda}_n$, with $\mathrm{Re}\bar{\lambda}_i < 0$, and when $\mu > 0$, $A(\mu)$ has eignevalues $u_1(\mu) \pm \omega_1(\mu)i,\ldots,u_{q1}(\mu) \pm \omega_{q1}(\mu)i$, $\lambda_{r+1}(\mu)\ldots,\lambda_n(\mu)$ with $\mathrm{Re}\lambda_i^1(\mu)\neq 0$ $(i=1,\ldots,q_1)$, $\mathrm{Re}\lambda_j(\mu)<0$ $(j=r+1,\ldots,n)$. Under certain conditions, the bifurcation of quasiperiodic solutions of (2.7) will occur.

Take $\mu = \epsilon\eta$ and rewrite (2.7) as

$$\dot{Y} = A(0)Y + F(t,\Omega t + C,\Omega) + \epsilon F^*(t,Y,\Omega t + C,\Omega,W,\dot{W},\epsilon,\eta) \tag{2.9}$$

where η is a small parameter, and

$$F^* = D_\mu A(\mu)\Big|_{\mu=0}\ \eta Y +\ldots+ F_1^*(t,Y,\Omega t + C,\Omega,W,\dot{W},\epsilon,\eta) \tag{2.10}$$

The generating system of (2.9) is

$$\dot{Y} = A(0)Y + F(t,\Omega t + C,\Omega) \tag{2.11}$$

LEMMA 1. If the quasiperiodic function $F(t,\Omega t + C,\Omega)$ satisfies the conditions:

$$\lim_{t\to\infty} \frac{1}{t} \int_0^t F(t,\Omega t + C,\Omega)^T \Psi_r(t)\, dt = 0 \qquad r = 1,\dots,n\ , \tag{2.12}$$

then the system (2.11) has a quasiperiodic special solution Y^*. In (2.12), $\Psi_r = (\Psi_{1r},\dots,\Psi_{nr})$ are special solutions of the conjugate equations of the homogeneous part of (2.11), and $\Psi_{ij}(t) = (C_{ij}\cos\omega_j t - D_{ij}\sin\omega_j t)e^{-u_j t}$, $\Psi_{i,q+j} = (C_{ij}\sin\omega j t + D_{ij}\cos\omega_j t)e^{-u_j t}$ $(j=1,\dots,q,q=n/2)$, $u_1 = \dots = u_{q1} = 0$, $\max(u_{q1+1},\dots,u_q) < 0$.

PROOF. Since the system (2.11) is a linear one, this lemma can be proved according to Malkin's book ([6]).

Take the transformation

$$Y = A(t,\eta)[\phi(\theta(t,\eta))E(t)]^T + Y^* \tag{2.13}$$

where $A(t,\eta)$ - $(A_1(t,\eta),\dots,A_q(t,\eta))$, $\theta(t,\eta),\dots,\theta_q(t,\eta))$, $\phi(\theta(t,\eta)) = (\phi_{si}(\theta_i(t,\eta)))_{nxq}$, $E(t) = \mathrm{diag}(e^{u_1 t},\dots,e^{u_q t})$, $\phi_{si} = (P_{si}\cos\theta_i - Q_{si}\sin\theta_i)e^{u_i t}$, $\phi_{s,q+i} = (P_{si}\sin\theta_i + Q_{si}\cos\theta_i)e^{u_i t}$, $A_i(t,\eta)$ and $\theta_i(t,\eta)$ are new variables. when $\epsilon = 0$, $A_i = \text{const}$, ϕ_{si} and $\phi_{s,q+i}$ are special solutions of the homogeneous part of (2.11).

LEMMA 2. Under the above assumptions, the following orthogonality conditions hold:

$$\begin{gathered}\sum_{s=1}^{n} \phi_{si}(\theta_i)\, \Psi_{s,q+j}(\theta_j) = \sum_{s=1}^{n} \phi_{s,q+i}(\theta_i)\, \Psi_{s,q+j}(\theta_j) = 0 \\ \sum_{s=1}^{n} \phi_{si}(\theta_i)\, \Psi_{s,q+j}(\theta_j) = \delta_{ij},\quad \sum_{s=1}^{n} \phi_{s,q+i}(\theta_i)\, \Psi_{sj}(\theta_j) = -\,\delta_{ij}\end{gathered} \tag{2.14}$$

where δ_{ij} is Kronecker's symbol.

PROOF. The solution of the homogeneous part of equation (2.11) can be expressed as

$$\begin{aligned} x_{si}(t) &= (P_{si} + iQ_{si})\, e^{(u_i + i\omega_i)t} \\ x^*_{s,i+q}(t) &= (P_{si} - iQ_{si})e^{(u_i - i\omega_i)t}\end{aligned} \tag{2.15}$$

The conjugate solutions are

$$y_{sj}(t) = (C_{sj} + iD_{sj})e^{(-u_j + i\omega_j)t} \tag{2.16a}$$

They satisfy

$$\sum_{s=1}^{n} x_{si}(t)y_{sj}(t) = \text{const.}$$

When $i \neq j$, we have

$$\sum_{s=1}^{n} (P_{si} + iQ_{si})(C_{sj} + iD_{sj}) = 0$$

$$\text{i.e.} \quad \sum_{s=1}^{n} P_{si}C_{sj} = \sum_{s=1}^{n} P_{si}D_{sj} = \sum_{s=1}^{n} Q_{si}C_{sj} = \sum_{s=1}^{n} Q_{si}D_{sj} = 0. \tag{2.18}$$

When $i = j$, we have

$$\sum_{s=1}^{n} (P_{si}C_{si} - Q_{si}D_{si}) + i\sum_{s=1}^{n} (Q_{si}C_{si} + P_{si}D_{si}) \neq 0 . \tag{2.19}$$

and

$$\sum_{s=1}^{n} (P_{si}C_{si} + Q_{si}D_{si}) + i\sum_{s=1}^{n} (Q_{sisi} - P_{si}D_{si}) \neq 0 . \tag{2.20}$$

Select α as

$$\alpha = -\sum_{s=1}^{n} Q_{si}C_{si} / \sum_{s=1}^{n} Q_{si}D_{si} .$$

Multiplying y_{si} and $y_{s,i+q}$ with $\alpha+i$, we have

$$y_{si} = (C_{si}+iD_{si})(\alpha+i)e^{(-u_i+i\omega_i)t} = (\alpha C_{si}-D_{si})\cos\theta_i - (\alpha D_{si}+C_{si})\sin\theta_i$$

$$+ i[(\alpha C_{si} - D_{si})\sin\theta_i + (\alpha D_{si} + C_{si})\cos\theta_i ,$$

and we denote $C_{si} = \alpha C_{si} - D_{si}$ and $D_{si} = \alpha D_{si} + C_{si}$. (2.21)

By using the relations (2.18), (2.19) and (2.20) and noticing (2.21), we can easily prove (2.14).

Substituting transformation (2.6) and (2.13) into (2.1a) and (2.9) and using the orthogonality conditions (2.14), we obtain the standard forms of (2.1a) and (2.9):

$$\ddot{W} = \epsilon\phi(\bar{\omega}t,\ ut,\ A,\ \theta,\ \Omega t + C,\ \Omega,\ W,\ \dot{W},\ \epsilon,\ \eta) \tag{2.22}$$

$$\begin{cases} \dot{A} = \epsilon R(\bar{\omega}t,\ ut,\ A,\ \theta,\ \Omega t + C,\ \Omega,\ W,\ \dot{W},\ \epsilon,\ \eta) \\ \dot{\theta} = \omega + \epsilon S(\bar{\omega}t,\ ut,\ A,\ \theta,\ \Omega t + C,\ \Omega,\ W,\ \dot{W},\ \epsilon,\ \eta) \end{cases} \tag{2.23}$$

where

$$R_j = \sum_{s=1}^{n} F_s^*(t,A,\theta,\Omega+C,\Omega,W,\dot{W},\epsilon,\eta)\, \Psi_{s,q+j}(\theta_j)e^{-u_j t}$$
$$S_j = \frac{1}{A_j} \sum_{s=1}^{n} F_s^*(t,A,\theta,\Omega t+C,\Omega,W,\dot{W},\epsilon,\eta)\, \Psi_{s,j}(\theta_j)e^{-u_j t} \ . \tag{2.24}$$

When $\epsilon = 0$, ϕ_i, R_j and S_j have Fourier series as follows:

$$\phi_i = \phi_{io}(A,\Omega,\eta,t) + \sum_k [\phi_{ik}(A,\Omega,\eta,t)\cos\alpha_k + \bar{\phi}_{ik}(A,\Omega,\eta,t)\sin\alpha_k]$$
$$R_j = R_{jo}(A,\Omega,\eta,t) + \sum_k [R_{ik}(A,\Omega,\eta,t)\cos\alpha_k + \bar{R}_{ik}(A,\Omega,\eta,t)\sin\alpha_k] \tag{2.25}$$
$$S_j = S_{jo}(A,\Omega,\eta,t) + \sum_k [S_{jk}(A,\Omega,\eta,t)\cos\alpha_k + \bar{S}_{jk}(A,\Omega,\eta,t)\sin\alpha_k]$$

where k is an integer, and

$$\alpha_k = \sum_{i=1}^{q} m_i^k\theta_i + \sum_{i=1}^{p} n_i^k\bar{\omega}_i t + \sum_{i=1}^{m} L_i^k\Omega_i r + \sum_{i=1}^{m} 1_i^k C_i \ .$$

We will study the bifurcation of quasiperiodic solutions and their stability below, both in nonresonant cases and in resonant cases.

3. NONRESONANT CASES. Assume that the frequencies of the generating system $\omega_1,\dots,\omega_{q1}$, the frequencies of driving force $\bar{\omega}_1,\dots,\bar{\omega}_p$ and the quasicyclic frequencies $\Omega_1,\dots,\Omega_m$ satisfy the nonresonant conditions; that is

$$m_1\omega_1 + \dots + m_{q1}\omega_{q1} + n_1\bar{\omega}_1 + \dots + n_p\bar{\omega}_p + \ell_1\Omega_1 + \dots + \ell_m\Omega_m \tag{3.1}$$

may not be a small value in the same order as ϵ only if all $m_1,\dots,m_{q1}$, $n_1,\dots,n_p$, $\ell_1,\dots,\ell_m$ do not vanish.

Take the general KB transformation containing parameter η

$$A_i = \bar{y}_i(\eta) + \epsilon U_i(t,\bar{Y},\theta,\Omega t+C,\Omega,\eta)$$
$$\Theta_i = \omega_i t + \theta_i + \epsilon V_i\ (t,\bar{Y},\theta,\Omega t+c,\Omega,\eta) \tag{3.2}$$

and demand

$$\frac{d\bar{y}_i}{dt} = \epsilon G_i(Y,\Omega,C,\eta) + \epsilon^2 G_i^*(t,\bar{Y},\theta,\Omega t+C,\Omega,W,\epsilon,\eta)$$
$$\frac{d\theta_i}{dt} = \epsilon Z_i(\bar{Y},,C,\eta) + \epsilon^2 Z_i^*(t,\bar{Y},\theta,\Omega t+C,\Omega,W,\epsilon,\eta) \tag{3.3}$$

where $\overline{Y} = (\overline{y}_1(\eta),\ldots,\overline{y}_q(\eta))$, U_i, V_i, $_i^*$ and Z_i^* are periodic functions of t and periodic functions of θ_i and x_j with peirod 2π.

Substituting (3.2) and (3.3) into standard forms (2.22) and (2.23) and comparing the terms with order ϵ, we obtain

$$\frac{d^2W_j}{dt^2} = \phi_{j1}(\overline{Y},\Omega,\eta,t) + \sum_k [\phi_{jk}(\overline{Y},\Omega,\eta,t)\cos\alpha_k + \overline{\phi}_{jk}(\overline{Y},\Omega,\eta,t)\sin\alpha_k]$$

$$G_i + \frac{\partial U_i}{\partial t} = R_{io}(\overline{Y},\Omega,\eta,t) + \sum_k[R_{ik}(\overline{Y},\Omega,\eta,t)\cos\alpha_k + \overline{R}_{ik}(\overline{Y},\Omega,\eta,t)\sin\alpha_k] \tag{3.4}$$

$$Z_i + \frac{\partial V_i}{\partial t} = S_{io}(\overline{Y},\Omega,\eta,t) + \sum_k[S_{ik}(\overline{Y},\Omega,\eta,t)\cos\alpha_k + \overline{S}_{ik}(Y,\overline{\Omega},\eta,t)\sin\alpha_k]$$

Rewrite $R_{io} = R_{io1}(\overline{Y},\Omega,\eta) + R_{io2}(\overline{Y},\Omega,\eta,t)$

$$S_{io} = S_{io1}(\overline{Y},\Omega,\eta) + S_{io2}(\overline{Y},\Omega,\eta,t)$$

$$\phi_{jo} = \phi_{jo1}(\overline{Y},\Omega,\eta) + \phi_{jo2}(\overline{Y},\Omega,\eta,t) .$$

Because the mean value of W is equal to zero, we have

$$\phi_{jo1}(\overline{Y},\Omega,\eta) = 0 \tag{3.5}$$

and

$$G_i = R_{io1}(\overline{Y},\Omega,\eta) \quad , \quad z_i = S_{io1}(\overline{Y},\Omega,\eta) . \tag{3.6}$$

Now, W_j, U_i and V_i can be found by constructing their Fourier series solutions whose coefficients satisfy a set of ordinary differential equations. Omitting the terms smaller than ϵ, we obtain the first approximation equations:

$$\frac{d\bar{y}_i}{dt} = \epsilon G_i(\bar{Y},\Omega,\eta)$$

$$\frac{d\theta_i}{dt} = \epsilon Z_i(\bar{Y},\Omega,\eta) \tag{3.7}$$

$$\phi_{jo1}(\bar{Y},\Omega,\eta) = 0 .$$

If a steady bifurcating solution exists, the terms on the right hand side of eq. (3.7) must vanish, that is the solution can be found from the following algebraic equations:

$$G_i(\bar{Y},\Omega,\eta) = 0 , \qquad \phi_{j01}(\bar{Y},\Omega,\eta) = 0. \tag{3.8}$$

Solving eq. (3.8) together with (3.7), we get the steady bifurcating solutions:

$$\Omega_k = \Omega_k^0(\eta) , \quad \bar{y}_i = \bar{y}_i^0(\eta), \quad \theta_i = \epsilon Z_i(\bar{Y}^0, \Omega^0,\eta)t + \phi_i \tag{3.9}$$
$$k = 1,\dots,m, \quad i = 1,\dots,q ,$$

and a corresponding stability theorem, as follows.

THEOREM 1. There exists steady quasiperiodic bifurcating solutions if and only if

$$\det \left[\frac{\partial(G, \phi_0)}{\partial(\bar{Y}, \Omega)} \right] \Bigg|_{\eta = 0, \ \bar{Y}_0^0, \ \Omega_0^0} \neq 0 \tag{3.10}$$

and their stability is determined by the eigenvalues of the determinant

$$\det \left[J(\frac{\partial(G, \phi_0)}{\partial(\bar{Y}, \Omega)}) - \lambda I \right] \Bigg|_{\eta = 0, \ \bar{Y}^0(\eta), \ \Omega^0(\eta)} = 0 , \tag{3.11}$$

where $G = (G_1,\dots,G_q)$, $\phi_0 = (\phi_{101},\dots,\phi_{m01})$, and $\bar{Y}^0(\eta)$ is the first approximate steady bifurcating solution and $\Omega^0(\eta)$ is the quasicyclic frequency. That is, when all of the real parts of the eigenvalues are negative, the solutions are stable, and when one of the real parts of the eigenvalues is positive, the solutions are unstable. When all of the real parts of the eigenvalues vanish, the higher approximation must be considered.

PROOF. To solve equation (3.8), we use the singular perturbation technique and expand $\bar{Y}$ and Ω as

$$\begin{aligned} \bar{Y}^0(\eta) &= Y_0^0 + \eta Y_1^0 + \eta^2 Y_2^0 + \dots \\ \Omega^0(\eta) &= \Omega_1^0 + \eta\Omega_1^0 + \eta^2\Omega_2^0 + \dots . \end{aligned} \tag{3.12}$$

Substituting (3.12) into (3.8), we have

$$J\left[\frac{\partial(G,\phi_0)}{\partial(\bar{Y},\Omega)}\right] \Bigg|_{\substack{\eta = 0 \\ \bar{Y}_0^0 \\ \Omega_0^0}} \begin{Bmatrix} \bar{Y}_k^0 \\ \Omega_k^0 \end{Bmatrix} = \bar{F}_k(\bar{Y}_1^0,\dots,\bar{Y}_{k-1}^0,\Omega_1^0,\dots,\Omega_{k-1}^0) . \tag{3.13}$$

The necessary and sufficient solvability condition of eq.(3.13) is equivalent to

$$\det\left[J\left(\frac{\partial(G,\ \phi_0)}{\partial(\bar{Y},\ \Omega)}\right)\right]\Bigg|_{\eta\,=\,0,\ \bar{Y}_0^0,\ \Omega_0^0} \neq 0 \quad . \tag{3.14}$$

To study the stability of the solution, we must establish the variational equations for the bifurcating solutions $\bar{y}_i^0(\eta)$, θ_i^0, and $\frac{dx_j^0}{dt}$.

Substitute the perturbation

$$\frac{dx_j}{dt} - \frac{dx_j^0}{dt} = \beta_j \ , \qquad \bar{y}_i - \bar{y}_i^0 = \alpha i \tag{3.15}$$

into the equations

$$\frac{d^2x_j}{dt^2} = \epsilon\ \{\phi_{j01}(\bar{Y},\Omega,\eta) + \phi_{j02}(\bar{Y},\Omega,\eta,t) + \sum_k [\phi_{jk}(\bar{Y},\Omega,\eta,t)\cos\alpha_k$$

$$+ \bar{\phi}_{jk}(\bar{Y},\Omega,\eta,t)\sin\alpha_k]\} + \ldots \tag{3.16a}$$

$$\frac{dy_i}{dt} = \epsilon G_i(\bar{Y},\Omega,\eta) \tag{3.16b}$$

and we obtain the variational equations

$$\frac{d\alpha_i}{dt} = \epsilon\left\{\sum_{j=1}^{q}\frac{\partial G_i}{\partial\bar{y}_j^0}\alpha_j + \sum_{j=1}^{m}\frac{\partial G_i}{\partial\Omega_j^0}\beta_j + \rho_j(\alpha,\beta,\eta)\right\} \tag{3.17a}$$

$$\frac{d\beta_j}{dt} = \epsilon\left\{\sum_{i=1}^{q}\frac{\partial\phi_{j01}}{\partial\bar{y}_i^0}\alpha_i + \sum_{i=1}^{m}\frac{\partial\phi_{j01}}{\partial\Omega_i^0}\beta_i + \rho_j^{(1)}(\alpha,\beta,\eta)\right.$$

$$+ \sum_{i=1}^{q}\frac{\partial\phi_{j02}}{\partial\bar{y}_i^0}\alpha_i + \sum_{i=1}^{m}\frac{\partial\phi_{j02}}{\partial\Omega_i^0}\beta_i + \rho_j^{(2)}(\alpha,\beta,\eta,t)$$

$$+ \sum_{k\neq 0}\left(\sum_{i=1}^{q}\frac{\partial\phi_{jk}}{\partial\bar{y}_i^0}\alpha_i + \sum_{i=1}^{m}\frac{\partial\phi_{jk}}{\partial\Omega_i^0}\beta_i\right)\cos\alpha_k$$

$$\left. + \sum_{k\neq 0}\left(\sum_{i=1}^{q}\frac{\partial\bar{\phi}_{jk}}{\partial\bar{y}_i^0}\alpha_i + \sum_{i=1}^{m}\frac{\partial\bar{\phi}_{jk}}{\partial\Omega_i^0}\beta_i\right)\sin\alpha_k + \rho_j^{(3)}(\alpha,\beta,\eta,t)\right\}$$

$$+ 0(\epsilon^2) \quad . \tag{3.17b}$$

The terms in the right hand side of the equation (3.17b), except the first three terms, consist of the terms

$$C_1(e^{ut})^k \cos[C_2\phi(t)] \tag{3.18}$$

where $u < 0$, C_1 and C_2 are constants, and the mean value of (3.18) is

$$\lim_{t\to\infty} \frac{1}{t}\int_0^t C_1(e^{ut})^k \cos[C_2\phi(t)]dt = 0$$

Averaging eq.(3.17), we get

$$\frac{d\alpha_i}{dt} = \epsilon\left\{\sum_{j=1}^{q} \frac{\partial G_i}{\partial \bar{y}_j^0}\alpha_j + \sum_{j=1}^{m} \frac{\partial G_i}{\partial \Omega_j^0}\beta_j + \rho_i(\alpha,\beta,\eta)\right\} \tag{3.19a}$$

$$\frac{d\beta_j}{dt} = \epsilon\left\{\sum_{i=1}^{q} \frac{\partial \phi_{j01}}{\partial \bar{y}_i^0}\alpha_i + \sum_{i=1}^{m} \frac{\partial \phi_{j01}}{\partial \Omega_i^0}\beta_i + \rho_j^{(1)}(\alpha,\beta,\eta)\right\} \tag{3.19b}$$

From eq.(3.19), we immediately conclude that the stability of the bifurcating solutions depends on the eigenvalues of the characteristic equation of eq.(3.19). So the proof of this theorem is completed.

4. RESONANT CASES. Not only do vibration systems occur in the nonresonant range, but they also often occur in the resonant range. It is necessary and important to study the bifurcation of these systems in resonant cases. Assume that the frequencies of the generating system $\omega_1,\ldots,\omega_{q1}$, the frequencies of the driving force $\bar{\omega}_1,\ldots,\bar{\omega}_p$ and the quasicyclic frequencies $\Omega_1,\ldots,\Omega_m$ satisdy the resonant conditions; that is,

$$D_e = \sum_{i=1}^{q_1} m_i^e\omega_i + \sum_{i=1}^{p} n_i^e\bar{\omega}_i + \sum_{i=1}^{m} l_i^e\Omega_i = 0 \quad e = 1,\ldots,r_0,\ r_0 < q, \tag{4.1}$$

where the sum of the absolute values of m, n and l does not exceed certain value which depends on the nonlinear terms in the primary equations.

If $n_1^e = \ldots = n_p^e = 0$, the system will be autonomous, and if the condition $\Omega_1 = \ldots = \Omega_m$ is required additionally, the problem will turn into the self-synchronization problem. If $n_1 = \ldots = n_p = l_1 = \ldots = l_m = 0$ in the resonant relationships (4.1), the inner resonance will occur.

Take the generic KB transformation containing parameter η for system (2.22) and (2.23) as follows

$$A_i = \bar{y}_i(\eta) + \epsilon U_i(t,\bar{Y},\theta,\Omega t + C,\Omega,\eta) \tag{4.2}$$

$$\theta_i = \omega_i t + \theta_i + \epsilon V_i(t,\bar{Y},\theta,\Omega t + C,\Omega,\eta)$$

and the derivative of the new variables $\bar{y}_i$ and θ_i

$$\frac{d\bar{y}_i}{dt} = \epsilon G_i(\bar{Y},\theta,\Omega,C,\eta) + \epsilon^2 G_i^*(t,\bar{Y},\theta,\Omega t + C,,W,\epsilon,\eta)$$
$$\frac{d\theta_i}{dt} = \epsilon Z_i(\bar{Y},\theta,\Omega mC,\eta) + \epsilon^2 Z_i^*(t,\bar{Y},\theta,\Omega t + C,\Omega,W,\epsilon,\eta) \tag{4.3}$$

where U_i, V_i, G_i^* and Z_i^* are aperiodic functions of t and periodic functions of θ_i and x_j with period 2π.

Substituting (4.2) into (2.22) and (2.23), noticing (4.3) and comparing the terms with order ϵ in both sides of the equations, we get

$$\frac{d^2W_j}{dt^2} = \phi_{j01}(\bar{Y},\Omega,\eta) + \phi_{j02}(\bar{Y},\Omega,\eta,t) + \sum_k [\phi_{jk}\cos\alpha_k + \bar{\phi}_{jk}\sin\alpha_k]$$
$$G_i + \frac{\partial U_i}{\partial t} = R_{i01}(\bar{Y},\Omega,\eta) + R_{i02}(\bar{Y},\Omega,\eta,t) + \sum_k [R_{jk}\cos\alpha_k + \bar{R}_{ik}\sin\alpha_k] \tag{4.4}$$
$$Z_i + \frac{\partial V_i}{\partial t} = S_{i01}(\bar{Y},\Omega,\eta) + S_{i02}(\bar{Y},\Omega,\eta,t) + \sum_k [S_{ik}\cos\alpha_k + \bar{S}_{ik}\sin\alpha_k]$$

where

$$\alpha_k = \left(\sum_{i=1}^{q} m_i^k\omega_i + \sum_{i=1}^{p} n_i^{k}\bar{\omega}_i + \sum_{i=1}^{m} 1_i^k\Omega_i\right)t + \sum_{i=1}^{q} m_i^k\theta_i + \sum_{i=1}^{p} 1_i^k C_i \tag{4.5}$$

Considering the resonant condition (4.1), we see that some α_k are independent of t and the frequencies of these α_k are linear combinations of D_e. That is

$$\sum_{i=1}^{q} m_i^k\omega_i + \sum_{i=1}^{p} n_i^{k}\bar{\omega}_i + \sum_{i=1}^{m} 1_i^k\Omega_i = \sum_{e=1}^{r_0} S_e^k D_e \tag{4.6}$$

where S_e^k is a positive or negative integer.

Introducing symbols

$$\bar{x}e = \sum_{i=1}^{q} m_i^e\theta_i\ , \qquad B_e = \sum_{i=1}^{m} 1_i^e C_i \qquad e = 1,\ldots,r_0\ , \tag{4.7}$$

where $\bar{x}_e$ are non-cyclic coordinates, and assuming that the number of α_k, which is independent of t, is N, we obtain and denote them as

$$\bar{\alpha}_k = \sum_{e=1}^{r_0} S_e^k \bar{x}_e + \sum_{e=1}^{r_0} S_e^k B_e \quad . \tag{4.8}$$

Write

$$\begin{aligned}
\phi_{jk} &= \phi_{jk1}(\bar{Y},\Omega,\eta) + \phi_{jk2}(\bar{Y},\Omega,\eta,t) \\
\bar{\phi}_{jk} &= \bar{\phi}_{jk1}(\bar{Y},\Omega,\eta) + \bar{\phi}_{jk2}(\bar{Y},\Omega,\eta,t) \\
R_{ik} &= R_{ik1}(\bar{Y},\Omega,\eta) + R_{ik2}(\bar{Y},\Omega,\eta,t) \\
\bar{R}_{ik} &= \bar{R}_{ik1}(\bar{Y},\Omega,\eta) + \bar{R}_{ik2}(\bar{Y},\Omega,\eta,t) \\
S_{ik} &= S_{ik1}(Y,\Omega,\eta) + S_{ik2}(\bar{Y},\Omega,\eta,t) \\
\bar{S}_{ik} &= \bar{S}_{ik1}(\bar{Y},\Omega,\eta) + \bar{S}_{ik2}(\bar{Y},\Omega,\eta,t) \quad .
\end{aligned} \tag{4.9}$$

Noticing that G_i, Z_i do not contain t and the mean value of $W(t, \bar{Y}, \theta,\Omega,C,\eta)$ is equal to zero, we have

$$\phi_{j1} = \phi_{jo} + \sum_{k=1}^{N} [\phi_{jk1} \cos(\sum_{e=1}^{r_0} S_e^k \bar{x}_e + \sum_{e=1}^{r_0} S_e^k B_e) + \bar{\phi}_{jk1} \sin(\sum_{e=1}^{r_0} S_e^k \bar{x}_e + \sum_{e=1}^{r_0} S_e^k B_e)]=0 \tag{4.10a}$$

$$G_i = R_{io1} + \sum_{k=1}^{N} [R_{ik1} \cos(\sum_{e=1}^{r_0} S_e^k \bar{x}_e + \sum_{e=1}^{r_0} S_e^k B_e) + \bar{R}_{jk1} \sin(\sum_{e=1}^{r_0} S_e^k \bar{x}_e + \sum_{e=1}^{r_0} S_e^k B_e)] \tag{4.10b}$$

$$Z_i = S_{io1} + \sum_{k=1}^{N} [S_{ik1} \cos(\sum_{e=1}^{r_0} S_e^k + \sum_{e=1}^{r_0} S_e^k B_e) + \bar{S}_{ik1} \sin(\sum_{e=1}^{r_0} S_e^k \bar{x}_e + \sum_{e=1}^{r_0} S_e^k B_e)] \tag{4.10c}$$

Omitting the terms with order higher than $O(\epsilon^2)$ in eq.(4.3), we obtain the first approximation equations

$$\begin{aligned}
\frac{d\bar{y}_i}{dt} &= \epsilon G_i(\bar{Y},\bar{X},\Omega,B,\eta) \qquad i = 1,\ldots,q \\
\frac{d\bar{x}_e}{dt} &= \epsilon \sum_{i=1}^{q} m_i^e \bar{Y},\bar{X},\Omega,B,\eta) = Z^e(\bar{Y},\bar{X},\Omega,B,\eta) \quad e = 1,\ldots,r_0
\end{aligned} \tag{4.11}$$

$$\frac{d\theta_i}{dt} - \epsilon Z_i(\bar{Y},\bar{X},\Omega,B) \quad . \tag{4.12}$$

If the steady bifurcating solution exists, the right hand side of equation (4.11) must vanish. That is

$$\begin{cases}
G_i(\bar{Y},\bar{X},\Omega,B,\eta) = 0 & i = 1,\ldots,q \\
Z^e(\bar{Y},\bar{X},\Omega,B,\eta) = 0 & e = 1,\ldots,r_0 \\
\phi_{j1}(\bar{Y},\bar{X},\Omega,B,\eta) = 0 & j = 1,\ldots,m
\end{cases} \tag{4.13}$$

$$\text{Let } \Omega_j^0(\eta),\ \bar{y}_i^0(\eta),\ \bar{x}_e^0(\eta) \text{ and } \theta_i = \epsilon Z_i t + E_i \tag{4.14}$$

be the solutions of equations (4.13) and (4.12). We can see that the steady bifurcating solutions are connected with quasicyclic phases $c_1,\dots,c_m$.

The constants E_i satisfy

$$\bar{x}_e^0 = \sum_{i=1}^{q} m_i^e E_i \quad . \tag{4.15}$$

THEOREM 2. There exist steady quasiperiodic bifurcating solutions if and only if

$$\det\left[\frac{\partial(G,Z,\phi_1)}{\partial(\bar{Y},\bar{X},\Omega)}\right]\Bigg|_{\eta=0,\ \bar{Y}_0^0,\ \Omega_0^0} \neq 0 \quad . \tag{4.16}$$

The stability of the solutions is determined by the eigenvalues of the determinant

$$\det\left[J\left(\frac{\partial(G,Z,\phi_1)}{\partial(\bar{Y},\bar{X},\Omega)}\right) - \lambda I\right]\Bigg|_{\bar{Y}^0(\eta),\ \bar{X}^0(\eta),\ \Omega^0(\eta)} = 0 \tag{4.17}$$

where $G = (G_1,\dots,G_q)$, $\phi_1 = (\phi_{11},\dots,\phi_{m1})$, $\bar{X} = \text{row}\,(\bar{x}_e)$, and $\bar{Y}^0(\eta)$ is the first approximate steady bifurcating solution, $\bar{X}^0(\eta)$ is the noncyclic coordinate and $\Omega^0(\eta)$ is the quasicyclic frequency. That is, when all of the real parts of the eignevalues are negative, the solutions are stable, and when one of the real parts of the eigenvalues is positive, the solutions are unstable. When all of the real parts of the eigenvalues vanish, higher approximations must be considered.

PROOF. To solve equation (4.13), we use the series expansions

$$\begin{aligned} \bar{Y}^0(\eta) &= \bar{Y}_0^0 + \eta\bar{Y}_1^0 + \eta^2\bar{Y}_2^0 + \dots \\ \bar{X}^0(\eta) &= \bar{X}_0^0 + \eta\bar{X}_1^0 + \eta^2\bar{X}_2^0 + \dots \\ \Omega^0(\eta) &= \Omega_0^0 + \eta\Omega_1^0 + \eta^2\Omega_2^0 + \dots \end{aligned} \tag{4.18}$$

and substitute them into (4.13). We have

$$J\left[\frac{\partial(G,Z,\phi_1)}{\partial(\bar{Y},\bar{X},\Omega)}\right]\Bigg|_{\substack{\eta = 0 \\ \bar{Y}_0^0,\ \bar{X}_0^0,\ \Omega_0^0}} \begin{Bmatrix} \bar{Y}_k^0 \\ \bar{X}_k^0 \\ \Omega_k^0 \end{Bmatrix} = F_k(\bar{Y}_1^0,\dots,\bar{Y}_{k-1}^0,\bar{X}_1^0,\dots,\bar{X}_{k-1}^0,\ \Omega_1^0,\dots,\Omega_{k-1}^0) \tag{4.19}$$

The necessary and sufficient condition for solvability of equation (4.19) is equivalent to

$$\det(J)\Big|_{\eta = 0,\ Y_0^0,\ \bar{X}_0^0,\ \Omega_0^0} \neq 0 \quad . \tag{4.20}$$

To study the stability of the solutions, we substitute the perturbation

$$\dot{x}_j - \dot{x}_j^0 = \beta_j, \quad \bar{y}_i - \bar{y}_i^0 = \alpha_i, \quad \bar{x}_e - \bar{x}_e^0 = \delta_e \tag{4.21}$$

into the equations:

$$\begin{aligned} \frac{d^2 x_j}{dt^2} &= \epsilon\{\phi_{j1}+\phi_{j2}+\sum_{k=1}^{N}(\phi_{jk2}\cos\bar{\alpha}_k+\bar{\phi}_{jk2}\sin\bar{\alpha}_k) + \sum_{k>N}(\phi_{jk}\cos\alpha_k + \bar{\phi}_{jk}\sin\alpha_k)\}+\ldots \\ \frac{d\bar{y}_i}{dt} &= \epsilon G_i(\bar{Y}, \bar{X}, \Omega, B, \eta) \\ \frac{d\bar{x}_e}{dt} &= \epsilon Z^e(\bar{Y}, \bar{X}, \Omega, B, \eta) \end{aligned} \tag{4.22}$$

and make an average of them. We get the averaging variational equations:

$$\begin{aligned} \frac{d\alpha_i}{dt} &= \epsilon\left\{\sum_{j=1}^{q}\frac{\partial G_i}{\partial \bar{y}_j^0}\alpha_j + \sum_{j=1}^{m}\frac{\partial G_i}{\partial \Omega_j^0}\beta_j + \sum_{e=1}^{r_0}\frac{\partial G_i}{\partial \bar{x}_e^0}\delta_e + \rho_i(\alpha,\beta,\delta,\eta)\right\} \\ \frac{d\delta_e}{dt} &= \epsilon\left\{\sum_{j=1}^{q}\frac{\partial Z^e}{\partial \bar{y}_i^0}\alpha_j + \sum_{i=1}^{m}\frac{\partial Z^e}{\partial \Omega_i^0}\beta_j + \sum_{e=1}^{r_0}\frac{\partial Z^e}{\partial \bar{x}_e^0}\delta_e + \rho_e(\alpha,\beta,\delta,\eta)\right\} \\ \frac{d\beta_i}{dt} &= \epsilon\left\{\sum_{i=1}^{q}\frac{\partial \phi_{j1}}{\partial \bar{y}_i^0}\alpha_i + \sum_{i=1}^{m}\frac{\partial \phi_{j1}}{\partial \Omega_i^0}\beta_i + \sum_{e=1}^{r_0}\frac{\partial \phi_{j1}}{\partial \bar{x}_e^0}\delta_e + \rho_j(\alpha,\beta,\delta,\eta)\right\} . \end{aligned} \tag{4.23}$$

It immediately follows that the stability of the bifurcating solutions depends on the eigenvalues of the characteristic equation of equation (4.23). This completes the proof of the theorem.

In conclusion, we remark that the stability condition is related not only to the parameters of the vibration system but also to the quasicyclic phase in the resonant case.

REFERENCES

1. Guckenheimer, J. and Holmes, P.J., Nonlinear oscillations, dynamical systems and bifurcaitons of vector fields, Springer-Verlag, New York, 1983.

2. Lin Rui, Huang Ke-lei and Jin Zisheng, Global bifurcation and chaos in a forced nonlinear oscillator with parametric excitation. Proceedings of 4-th National Conference on Nonlinear Vibration, China, 1986.

3. Moser, J. On the theory of quasiperiodic motions. SIAM Reviews, Vol. 8, No. 2, April, 1966.

4. Lin Zhensheng, The perturbation of almost periodic solution of almost periodic system. China. Ann. of Math., 5B(3), 1984.

5. Lin Rui and Chen Yushu, A new method of solving the Hopf bifurcation in nonlinear vibration systems with many degrees of freedom. Proceedings of ICVPE, Xian, China, 1986.

6. Malkin, J.G., theory of Stability of Motion, U.S. Atomic Energy Commission, Translation AEC-TR-3352 (Translated from Russian original, Moscow, 1956.)

DEPARTMENT OF APPLIED MATHEMATICS AND PHYSICS
BEIJING INSTITUTE OF AERONAUTICS AND ASTRONAUTICS
BEIJING, CHINA

Canadian Mathematical Society
Conference Proceedings
Volume 8 (1987)

BIFURCATION AND STABILITY OF PERIODIC SOLUTIONS OF A CLASS OF NONLINEAR EVOLUTION EQUATION

Lu Qishao

ABSTRACT. The nonlinear evolution equation of the kind:

$$\frac{\partial u}{\partial t} = \alpha \int_{-\infty}^{\infty} d\xi \int_{h}^{u(t,\xi)} g[\xi,\eta;x,u(t,x)]d\eta + \beta u + \omega \equiv F(u;h) \quad (1)$$

is considered, where us is a real periodic function in x with period ℓ for fixed t, $\int_0^\ell |u|^2 dx < \infty$. Equation (1) has a basic solution u = 0. Now we study the spatially periodic static bifurcating solutions of (1) and their stability when the bifurcation parameter h varies. In this paper the necessary condition for bifurcation, the existence of bifurcation, the stability of the equilibrium solutions (i.e. the basic and the bifurcating solutions), and the principle of exchange of stability for the basic solution and the bifurcating solutions are studied. Finally an example of application is given.

1. INTRODUCTION. Bifurcation and stability are of great significance in the study of nonlinear problems. For example, [1], [2] discuss a bifurcation problem for liquid surfaces in hydrostatics, chemical reactions and astrophysics, which is a static bifurcation problem for nonlinear evolution equations with integral operators. In this paper, more general results on bifurcation and stability of periodic solutions of a class of nonlinear evolution equations with integral operators are obtained. At first, the necessary and sufficient conditions for bifurcation are given by means of functional analysis. Next, the stability of equilibrium solutions is analyzed by the method of semigroups of linear operators. Finally, we derive the principle of exchange of stability between the basic and bifurcating solutions.

1980 Mathematics Subject Classificaiton (1985 Revision). 34K20, 58F14.

2. BIFURCATION OF SPATIALLY PERIODIC SOLUTIONS. Let H be a real Hilbert space of all real functions $\phi(x)$ on R with period ℓ, satisfying $\int_0^\ell |\phi|^2 dx < \infty$. The inner product on H is defined by

$$(\phi_1,\phi_2) = \int_0^\ell \phi_1\phi_2 dx\ , \qquad \phi_1,\phi_2 \in H$$

and the norm on H is $\|\phi\| = (\phi,\phi)^{1/2}$.

Now we consider the following nonlinear evolution equation on H:

$$\frac{du}{dt} = G(u;h) - \beta u \equiv F(u;h), \qquad t > 0 \tag{1}$$

where $u \in H$ for fixed $t > 0$, and us is continuous in t for $t \geq 0$ and is continuously differentiable in t for $t > 0$, and G: HXR → H is given by

$$G(u;h) \triangleq \alpha \int_{-\infty}^{\infty} d\xi \int_h^{u(\xi,t)+h} g(\xi - x,\ \eta - u(x,t))d\eta \tag{2}$$

where α, β are real nonzero constants and h is a real parameter. We also suppose that the real function g (ξ,η) satisfies the following conditions:

(1) $g \in L'(R^2)\bigcap C'(R^2)$, $g'_\eta \in L'(R^2)\bigcap C(R^2)$;

(2) $g(\xi,\eta) = 0(e^{-r^2})$ when $r = \sqrt{\xi^2 + \eta^2} \to \infty$; (3)

(3) $g(\xi,\eta) = g(-\xi,\eta)$. (4)

Obviously $F(0;h) = 0$, so $u = 0$ is a basic solution of equation (1). Now we are interested in the static bifurcating periodic solutions in H of the nonlinear evolution equation (1), i.e. the nontrivial solution u bifurcating from $h = \bar{h}$, which is independent of t and satisfies the equilibrium equation

$$F(u;h) = 0, \qquad u \in H\ . \tag{5}$$

From the general theory of bifurcation ([3], [4]) $\bar{h}$ is a bifurcation point of (5) only if the Fréchet derivative of F at (0,h) is singular. Through computation we have the Fréchet derivative at F at (0,h)

$$F_u(0;h)V = (L_h - \beta I)V, \qquad V \in H\ , \tag{6}$$

where the linear integral operator is defined by

$$L_h V = \alpha \int_{-\infty}^{\infty} g(\xi - x, h)\, V(\xi) d\xi \ , \qquad V \in H \ . \tag{7}$$

Therefore the necessary condition for the bifurcation point h is that $\beta \epsilon \sigma(L_h^-)$. To begin with we consider the spectral properties of L_h.

LEMMA 1. L_h is a compact self-adjoint bounded linear operator on H.

PROOF. From the periodicity of the functions in H and the assumption (2), L_h is a bounded linear operator on H.

L_h is a self-adjoint operator on H since its kernel $g(\xi - x, h)$ is real symmetric with respect to ξ and x by the assumption (3).

In order to show that L_h is a compact operator on H, we take a normalized orthogonal basis for H:

$$\{\Psi_\nu\} = \{\Psi_0, \Psi_k^{(1)}, \Psi_k^{(2)}\} \qquad (\text{integers } k \geq 1) \tag{8}$$

where

$$\Psi_0 = \frac{1}{\sqrt{\ell}} \ , \quad \Psi_k^{(1)} = \sqrt{\frac{2}{\ell}} \cos \frac{2\pi k}{\ell} x, \quad \Psi_k^{(2)} = \sqrt{\frac{2}{\ell}} \sin \frac{2\pi k}{\ell} x \ . \tag{9}$$

It leads to the following result:

$$\|L_h\|^2 = \sum_{v} \| L_h \Psi_\nu \|^2 < \infty \tag{10}$$

and then L_h is compact on H.

From Lemma 1 we obtain the properties of the specturm $\sigma(L_h)$ of L_h:

LEMMA 2. The nonzero spectrum of L_h for given $h \in R$ consists of countable discrete real eigenvalues $\lambda_k(h) = 0$, $k \in K$ (a countable set). The null space $N(L_h - \lambda_k I)$ is a finite dimensional subspace. Furthermore, we also have $0 \in \sigma(L_h)$ and it is the only possible accumulation point of the set $\{\lambda_k\}$.

It can also be found that the eigenfunctions of L_h are just the set $\{\psi_\nu\}$ in (8), and the eigenvalues corresponding to Ψ_0, $\Psi_k^{(1)}$ and $\Psi_k^{(2)}$ are

$$\lambda_0(h) = \alpha \int_{-\infty}^{\infty} g(\xi, h) d\xi$$

and

$$\lambda_k(h) - \alpha \int_{-\infty}^{\infty} g(\xi, h) \cos \frac{2\pi k}{\ell} \xi d\xi \qquad (k = 1, 2, \ldots) \tag{11}$$

respectively. Thus the necessary condition for $\bar{h}$ to be a bifurcation point results in the following theorem:

THEOREM 1. Spatially periodic static bifurcaitons from the basic equilibrium solution $u = 0$ at $h = \bar{h}$ may occur only if

$$\lambda_k(\bar{h}) = \beta \qquad (k = 0, 1, 2, \ldots) \tag{12}$$

We denote the values of $\bar{h}$ satisfying (12) by $\bar{h}_k (k = 0,1,2,\ldots)$, and discuss the local existence of the spatially periodic bifurcating solutions at $\bar{h}$ below.

Let $A_0(h) = F_u(0;h)$, $A'_{0h}(h) = f_{uh}(0;h)$ for simplicity. We see that $A_0(\bar{h})_k = L_{\bar{h}_k} - \lambda_k(\bar{h}_k)I$. Now denote $N_k = N(A_0(\bar{h}_k), M_k = R(A_0(\bar{h}_k))$. Since $L_{\bar{h}_k}$ is self-adjoint and compact, we have $M_k = N_k^{\perp}$ and then $H = N_k \oplus M_k$. Define P and Q as the orthogonal projections from H onto N_k and M_k respectively. The dimension of N_k is 1 if $k = 0$, or 2 if $k = 1,2,3,\ldots$.

THEOREM 2. If $PA'_{0h}(\bar{h}_k)\phi \notin M_k$ for any $\phi \in N_k$, then an appropriate parameter ϵ can be chosen so that there exists a nontrivial solution $u = u(\epsilon) \epsilon H$ and $h = h(\epsilon)$ of (5) in a neighborhood of $\epsilon = 0$, which are continuously differentiable in ϵ and satisfy $u(0) = 0$, $h(0) = \bar{h}$, $F(u(\epsilon); h(\epsilon)) = 0$.

In fact, any $u \in H$ can be expressed as

$$u = Pu + Qu = (u,\phi_k)\phi_k + Qu \tag{13}$$

where $\phi_k \in N_k$ with $\|\phi_k\| = 1$, $Qu \in M_k$. Denote $\epsilon = (u,\phi_k)$, and then

$$u = \epsilon(\phi_k + \Psi) \tag{14}$$

where $\Psi = \frac{1}{\epsilon} Qu$. Moreover, we express

$$h = \bar{h}_k + \epsilon\sigma \tag{15}$$

Since $A_0(h) = F_u(0;h)$, the equation (5) can be written as

$$F(u;h) = A_0(h)u + \epsilon^1 B(u;h) = 0 \tag{16}$$

where $B\colon HXR \to H$ and $\|B(u;h)\| = 0(|\epsilon|)$. Evidently $A_0(h)$ is differentiable in h, and it follows that

$$[A_0(\bar{h}_k) + \epsilon\ \sigma A'_{0h}\ (\bar{h}_k + \theta\ \epsilon\ \sigma)]\ (\phi_k + \Psi)$$
$$+ \epsilon\ B(\epsilon(\phi_k + \Psi),\ \bar{h}_k + \epsilon\sigma) = 0, \qquad 0 < \theta < 1 \tag{17}$$

where

$$A'_{0h}(h)V = \alpha \int_{-\infty}^{\infty} g'_\eta(\xi - x,\ h)\ V\ (\xi)d\xi\ . \tag{18}$$

The projection of (17) onto N_k results in

$$\sigma A'_{0h}(\bar{h}_k + \theta\epsilon\sigma)(\phi_k + \Psi) + PB(\epsilon(\phi_k + \Psi),\ \bar{h}_k + \epsilon\sigma) = 0 \tag{19}$$

in consideration that $A_0(\bar{h}_k)\phi_k = 0$, $A_0(\bar{h}_k\Psi\epsilon R(A_0(A_0(\bar{h}_k)) = M_k$ and $M_k = N_k^\perp$. The inner product of (19) with ϕ_k gives

$$\sigma(PA'_{0h}(\bar{h}_k + \theta\epsilon\sigma)(\phi_k + \Psi),\phi_k) + (PB(\epsilon(\phi_k + \Psi),\bar{h}_k + \epsilon\sigma),\phi_k) = 0\ . \tag{20}$$

The projection of (17) onto M_k results in

$$A_0(\bar{h}_k)\Psi + \epsilon\sigma(A_0(\bar{h}_k + \theta\epsilon\sigma)(\phi 2k + \Psi)$$
$$+ \epsilon\ QB(\epsilon(\phi_k + \Psi),\ \bar{h}_k + \epsilon\sigma) = 0 \tag{21}$$

Introduce the pseudoinverse $\hat{T}$: $H \to M_k$ by

$$\hat{T}f = \begin{cases} 0 & (f \in N_k) \\ W \in M_k & (f \in M_k) \end{cases}$$

where $W \in M_k$ is the unique solution of $A_0(\bar{h}_k)V = f$ in M_k if $f \in M_k$. Operating on (21) with $\hat{T}$ we get

$$\Psi + \epsilon\sigma QA_0(\bar{h}_k + \theta\epsilon\sigma)(\phi_k + \Psi)$$
$$+ \epsilon\hat{T}QB(\epsilon(\phi_k + \Psi),\ \bar{h}_k + \epsilon\sigma) = 0 \tag{23}$$

Now (23) and (22) can be combined in an equivalent equation as follows:

$$F(\Psi,\sigma,\epsilon) \overset{\Delta}{=} \begin{pmatrix} \Psi + \epsilon\sigma QA_0(\bar{h}_k + \theta\epsilon\sigma)(\phi_k + \Psi) + \ldots \\ \sigma(PA'_{0h}(\bar{h}_k + \theta\epsilon\sigma)(\phi_k + \Psi),\ \phi_k) + \ldots \end{pmatrix} = 0 \tag{24}$$

Obviously $F(0,\sigma_0,0) = 0$ if we take $\sigma_0 = -(PA'_{0h}(\bar{h}_k)\phi_{ko},\ \phi_{ko})^{-1}$ $(PB(0,\bar{h}_k),\phi_{ko})$, where $\phi_{ko} = \lim_{\epsilon\to 0} \phi_k \in N_k$. Furthermore,

$$F_{(\Psi,\sigma)}(0,\sigma_0,0) = \begin{bmatrix} I & 0 \\ 0 & (PA'_{0h}(\bar{h}_k)\phi_{k0},\phi^{k_0}) \end{bmatrix} \tag{25}$$

$F_{(\Psi,\sigma)}(0,\sigma_0,0)$ is bounded and invertible provided $PA'_{0h}(\bar{h}_k)\phi_{k0}\notin M_k$ for $\phi_{k0}\in N_k$. From the implicit function theorem for a Banach space, there exists $\Psi = \Psi(\epsilon)$, $\sigma = \sigma(\epsilon)$ in a neighbourhood of $\epsilon = 0$ such that $F(\Psi(\epsilon), \sigma(\epsilon),0) = 0$, $\Psi(0) = 0$ $\sigma(0) = \sigma_0$ and $\Psi(\epsilon)$, $\sigma(\epsilon)$ are continuously differentiable in ϵ. When the dimension of N_k is larger than one, we also find ϕ_k from the finite dimensional equation (19) by the technique of Newton's diagram ([4]).

Now take $u = \epsilon(\phi + \Psi)$ and $h = \bar{h}_k + \epsilon\sigma$, obviously they satisfy $F[u(\epsilon; h(\epsilon)] = 0$ in a neighbourhood of $\epsilon = 0$ and $u(0) = 0$, $h(0) = \bar{h}_k$.

We shall consider the stability of the basic equilibrium solution $u = 0$ below. Suppose that there is a initial disturbance $u^0 \in H$ at $t = 0$, we investigate the time-dependent behavior of the solution $u(x,t)$. For small disturbances it can be reduced to a linearized stability problem as follows:

$$\begin{cases} \dfrac{du}{dt} = F_u(0;h)u \equiv A_0(h)u, & t > 0 \\ u\Big|_{t=0} = u_0 \end{cases} \tag{26}$$

Since $A_0(h)$ is a bounded linear operator on H, it is an infinitesimal generator of a uniformly continuous semigroup $T(t;h)$ on H, i.e. $T(t;h)=e^{tA_0(h)}$ and $\lim_{t\to +0} \|T(t;h)-I\| = 0$ for any h. The solution of (26) is

$$u(x,t) = T(t;h)u_0 = e^{tA_0(h)}u_0, \quad t > 0 \tag{27}$$

The time-dependent behavior of $u(x,t)$ is determined by the spectrum $\sigma(A_0(h))$ of $A_0(h)$, which is dependent on h.

THEOREM 3. For given h, if $\mu = \sup\{\mathrm{Re}\Lambda(h): \Lambda(h)\in\sigma(A_0(h))\} < 0$, then the basic equilibrium solution $u = 0$ of (1) is linearly asymptotically stable at h; if $A_0(h)$ has an eigenvalue $\Lambda^*(h)$ with $\mathrm{Re}\Lambda^*(h) > 0$, then the basic equilibrium solution of (1) is unstable.

Actually this theorem can be proved by means of the theory of semigroups of bounded linear operators. Since $\Lambda(h)\in\sigma(A_0(h))$ iff $\lambda(h)\in\sigma(L_n)$ and $\Lambda(h) = \lambda(h)-\beta$ from (6), we see that from lemma 2$\sigma(A_\mu(h))$ includes the eigenvalues

$$\Lambda_k(h) = \alpha \int_{-\infty}^{\infty} g(\xi,h) \cos \frac{2\pi k}{\ell} \xi \, d\xi - \beta \qquad (k = 0,1,2,\ldots) \tag{28}$$

and

$$\hat{\Lambda} = -\beta$$

We conclude immediately that the basic equilibrium solution $u = 0$ can be exponentially stable only if $\beta > 0$.

Finally we show the principle of exchange of stability between the basic and bifurcating equilibrium solutions at $\bar{h}_k$. Let $u(\epsilon) = \epsilon(\phi_k+\Psi)\in H$ be a bifurcating equilibrium solution from $u = 0$ at $h = \bar{h}_k$ (cf. Theorem 2) for sufficiently small $|\epsilon|$. Denote $(W(\epsilon), \Lambda(\epsilon))$ and $(\tilde{W}(\epsilon),\tilde{\Lambda}(\epsilon))$ as the eigen-pairs of $F_u(0,h(\epsilon))$ and $F_u(u(\epsilon); h(\epsilon))$ respectively, i.e.

$$F_u(0; h(\epsilon))W(\epsilon) = \Lambda(\epsilon)W(\epsilon) \tag{30}$$

and

$$F_u(u(\epsilon), h(\epsilon))\tilde{W}(\epsilon = \tilde{\Lambda}(\epsilon)\tilde{W}(\epsilon) \tag{31}$$

holds in a neighbourhood of $\epsilon = 0$ and $\lim\limits_{\epsilon\to 0} W(\epsilon) = \lim\limits_{\epsilon\to 0} \tilde{W}(\epsilon) = \phi_{k0}$, (32)

$\lim \Lambda(\epsilon) = \lim \tilde{\Lambda}(\epsilon) = 0$, where $h(\epsilon) = \bar{h}_k + \epsilon\sigma(\epsilon)$. (33)

We have the following result on the principle of exchange of stability:

THEOREM 4. If $F\colon H\times R \to H$ is continuously Fréchet differentiable in u up to second order, $F_{uu}(0;\bar{h}_k)(\phi,\phi) \notin M_k$ for any $\phi \in N_k$, then under the assumption of Theorem 2, we have

$$\lim_{\epsilon\to 0} \frac{\Lambda(\tilde{\epsilon})}{\Lambda(\epsilon)} = -1 \tag{34}$$

Here we prove this theorem. Taking account of (32) and $u(\epsilon) = \epsilon(\phi_k+\Psi)$, it follows that

$$\epsilon W(\epsilon) = u(\epsilon) + \epsilon p(\epsilon) \tag{35}$$

$$\epsilon\tilde{W}(\epsilon) = u(\epsilon) + \epsilon q(\epsilon) \tag{36}$$

where $p(\epsilon) = 0(\epsilon)$, $q(\epsilon) = 0(\epsilon)$.

On one hand, using the Taylor's expansion of $F(u;h)$ we have

$$\begin{aligned} 0 &= F(u(\epsilon); h(\epsilon)) \\ &= F_u(0;h(\epsilon))u(\epsilon) + \frac{1}{2} F_{uu}(0;h(\epsilon))(u(\epsilon), u(\epsilon)) + 0(\epsilon^2) \end{aligned} \tag{37}$$

From (30) and (35) we get

$$F_u(0;h(\epsilon))(u(\epsilon) + \epsilon p(\epsilon)) = \Lambda(\epsilon)(u(\epsilon) + \epsilon p(\epsilon)) \quad . \tag{38}$$

Substituting (37) into (38) results in

$$\epsilon F_u(0;h(\epsilon))p(\epsilon) - \frac{\epsilon^2}{2} F_{uu}(0;h(\epsilon))(\phi_k + \Psi(\epsilon), \phi_k + \Psi(\epsilon))$$

$$= \epsilon\Lambda(\epsilon)\phi_k + \epsilon\ \Lambda(\epsilon)\ (\Psi(\epsilon) + p(\epsilon)) + 0(\epsilon^2)\ . \tag{39}$$

Considering that $\Lambda(\epsilon) = 0(\epsilon)$, $\Psi(\epsilon) = 0(\epsilon)$, $p(\epsilon) = 0(\epsilon)$, and $h(\epsilon) = \bar{h}_k + \epsilon\sigma(\epsilon)$, we retain terms of the lowest order and obtain

$$\epsilon F_u(0;\bar{h}_k)p(\epsilon) - \frac{\epsilon^2}{2} F_{uu}(0;\bar{h}_k)(\phi_k,\phi_k) = \epsilon\Lambda(\epsilon)\phi_k + 0(\epsilon^2)\ . \tag{40}$$

The inner product of (40) with ϕ_k gives

$$\epsilon\Lambda(\epsilon) = -\frac{\epsilon^2}{2} (F_{uu}(0;\bar{h}_k)(\phi_k,\phi_k),\phi_k) + 0(\epsilon^2) \tag{41}$$

On the other hand, the similar Taylor's expansion for $F(0;h)$ gives

$$\begin{aligned} 0 &= F(0,h(\epsilon)) \\ &= F_u(u(\epsilon);h(\epsilon))(-u(\epsilon)) + \frac{1}{2} F_{uu}(u(\epsilon);\ h(\epsilon))(-u(\epsilon),\ -u(\epsilon)) + 0(\epsilon^2). \end{aligned} \tag{42}$$

By means of (31) and (36), we get the similar result:

$$\epsilon\tilde{\Lambda}(\epsilon) = \frac{\epsilon^2}{2} (F_{uu}(0;\bar{h}_k)(\phi_k,\phi_k),\phi_k) + 0(\epsilon^2) \tag{43}$$

Since $F_{uu}(0,\bar{h}_k)(\phi_k,\phi_k) \notin M_k$ by assumption, $(F_{uu}(0;\bar{h}_k)(\phi_k,\phi_k),\phi_k) \neq 0$ and we finally have

$$\lim_{\epsilon\to 0} \frac{\tilde{\Lambda}(\epsilon)}{\Lambda(\epsilon)} = -1$$

REFERENCES

[1] G. Fusco, An example of bifurcation in hydrostatics, Nonlinear Differential Equations: Invariance, Stability, and Bifurcation (ed. by P. de Mottoni and L. Salvadori), Academic press, 1981, 145-159.

[2] Lu Qishao and Jiang Zhengxin, A bifurcation problem for free surfaces of liquid layers, Proceedings of the International Conference on Nonlinear Mechanics (Shanghai), 1985, 1071-1075.

[3] S.N. Chow and J.K. Hale, Methods of Bifurcation Theory, Springer-Verlag, 1982.

[4] M.G. Crandall and P.H. Rabinowitz, Bifurcation from simple eigenvalues, J. Funct. Anal., 8(1971), 321-340.

[5] M.G. Crandall and P.H. Rabinowitz, Bifurcation, perturbation of simple eigenvalues and linearized stability, Arch. Rat. Mech. Anal., 52(1973), 161-180.

[6] J.T. Schwartz, Nonlinear Functional Analysis, New York Univ., 1965.

[7] H.F. Weinberg, On the stability of bifurcating solutions, Nonlinear Analysis (ed. by L. Cesari and R. Kannan), Academic Press, 1978, 219-233.

[8] A. Pazy, Semigroups of Linear Operators and Applications to Partial Differential Equations, Springer-Verlag, 1983.

DEPARTMENT OF MATHEMATICS
BEIJING INSTITUTE OF AERONAUTICS AND ASTRONAUTICS
BEIJING, CHINA

Canadian Mathematical Society
Conference Proceedings
Volume 8 (1987)

HOMOCLINIC ORBITS AND BERNOULLI BUNDLES IN ALMOST PERIODIC SYSTEMS

Kenneth R. Meyer and George R. Sell

ABSTRACT. This paper announces the natural generalization to an almost periodic system, of the horseshoe invariant set of Smale and the method of detecting transversal homoclinic orbits of Melnikov. We use the construction to define a skew product flow over the hull of an a.p. function. In this way the non-autonomous equations define a dynamical system whose structure reflects the geometry and spectra of the a.p. forcing term.

Within the context of this skew product dynamical system we provide the natural generalization of a hyperbolic orbit and invariant set, the stable and unstable manifolds, transversal homoclinic orbit and the shift automorphism on a symbol space. This last object we call a Bernoulli bundle because it is a fiber bundle with fiber maps which are Bernoulli automorphisms. We then proceed to prove generalizations of the Melnikov theorem for detecting homoclinic orbits, the shadowing lemma, and Smale's theorem on the existence of the horseshoe invariant set.

I. INTRODUCTION. This paper announces the natural generalization to an almost periodic (a.p.) system, of the horseshoe invariant set of Smale(1963) and the method of detecting transversal homoclinic orbits of Melnikov(1963). Detailed proofs will be given in Meyer and Sell(1986b). The results are natural because we use the Miller(1965) and Sell(1967) construction to define a skew product flow over the hull of an a.p. function. In this way the non-autonomous equations define a dynamical system whose structure reflects the geometry and spectra of the a.p. forcing term.

Within the context of this skew product dynamical system we provide the natural generalization of a hyperbolic orbit and invariant set, the stable and unstable manifolds, transversal homoclinic orbit and the shift automorphism on a symbol space. This last object we call a Bernoulli bundle because it is a fiber bundle with fiber maps which are Bernoulli automorphisms. We then proceed to prove generalizations of the Melnikov theorem for detecting homoclinic orbits, the shadowing lemma, and Smale's theorem on the existence of the horseshoe invariant set.

1980 Mathematics Subject Classification: 34C27, 34C35, 54H20, 58F13, 58F15, 58F27, 70K50.

Homoclinic orbits and their implications for autonomous and periodic systems have been investigated since their introduction by Poincaré. The classics in the literature are Poincaré (1892), Birkhoff (1932), Cartwright and Littlewood (1945), Melnikov (1963), and Smale (1963). Our work uses many of the ideas and the method of Palmer(1984). The reader is refered to Chow and Hale(1982) and Guckenheimer and Holmes(1983) for a detailed discussion of and historical remarks on this vast literature.

Recently some work has appeared on almost periodic systems. In Wiggens (1986a,1986b) the method of Melnikov and related results are generalized to cover systems which admit an invariant torus. The detection of a transversal intersection of the stable and unstable manifold is based on a generalized Melnikov function and the asymptotic implications are given. He gives a fairly complete picture in the quasi-periodic case. Scheurle (1986) considers a system of a.p. equations slightly more general than ours, but considers only one equation and not a whole class of equations based on the hull of an a.p. function. He uses the theory of exponential dichotomies as extended by Palmer (1984) to find particular solutions which have a somewhat random form.

II. THE HULL, CROSS SECTIONS AND BERNOULLI BUNDLES. Throughout this paper almost periodic (a.p.) will be in the sense of Bohr (1959). Some authors, e.g. Besicovitch(1932), refer to this as uniformly almost periodic. Let $f: R \to R^n$ (or C^n) be an a.p. function. The spectral theory is based on the fact that the mean

$$M\{f\} = \lim_{T\to\infty} \frac{1}{2T}\int_{-T}^{T} f(s)\, ds$$

exists and for only a countable number of real numbers ω does $M\{f \exp(-i\omega t)\} \neq 0$. This set $\{\omega_k\}$ is called the set of exponents or frequencies of f. We write

$$f(t) \sim \sum_{-\infty}^{\infty} A_k \exp(i\,\omega_k t) \tag{1}$$

where $A_k = M\{f \exp(-i\,\omega_k t)\}$. This series is called the Fourier Series of f.

Consider the real numbers R as a vector space over the rational numbers. The smallest subspace S R which contains $\{\omega_k\}$ is called the <u>modulus</u> of f. In the case S is one dimensional, i.e. $\omega_k = r_k\,\alpha$, where r_k is rational, the function f is said to be <u>limit periodic</u>, because in this case the function is the uniform limit of periodic functions. The example of a limit periodic function which we shall use throughout this paper is

$$\ell(t) = \sum_{0}^{\infty} a_k \, e^{i2\pi(t/2^k)}$$

(or its real part) where the a_k are choosen so that the series converges rapidly. The Nth partial sum of ℓ is 2^N periodic.

If S is a finite module over the integers I, i.e., there exists a finite set $\lambda_1, \ldots, \lambda_l \in R$ such that $\omega_k = \sum \alpha_{ki} \omega_i$ with $\alpha_{ki} \in I$, then f is said to be quasi-periodic. Our standard example of a quasi-periodic function will be

$$q(t) = a_1 e^{i\omega_1 t} + a_2 e^{i\omega_2 t}$$

(or its real part) where ω_1/ω_2 is irrational.

Let $\mathbf{C} = \mathbf{C}(R,R^n)$ (or $= \mathbf{C}(R,C^n)$) be the space of continuous functions from R into R^n (or C^n) with the topology of uniform convergence on compact sets (the compact open topology). Translations define a flow on $\mathbf{C}$ as follows

$$\pi\colon R \times \mathbf{C} \to \mathbf{C}\colon (\tau,f) \to f_\tau$$

where $f_\tau(t) = f(t+\tau)$. For any $f \in \mathbf{C}$ the orbit closure of f is called the hull of f and is denoted by H(f). That is $H(f) = \mathrm{cl}\{g_\tau\colon \tau \in R\}$. If f is a.p. then H(f) is a compact minimal set; each element of H(f) is a.p.; $\pi|H(f)$ is equicontinuous, and H(f) can be given a compact, connected, Abelian group structure. We also use the space **AP** of almost periodic functions with sup norm $\|f\|$. The above results hold in this space also.

If f is a.p. and $f \sim \sum a_k \exp(i\,\omega_k t)$ then $f_\tau \sim \sum a_k \exp(i\,\omega_k(t+\tau))$. If a sequence of translates f_{τ_n} converge, say $f_{\tau_n} \to g$, then one can use the Cantor diagonalization procedure to select a subsequence if necessary such that

$$\tau_n \to \phi_k \pmod{2\pi/\omega_k} \text{ for all } k, \text{ as } n \to \infty$$

Then the Fourier coefficients of f_{τ_n} converge to the Fourier coefficients of g or

$$g \sim \sum A_k \exp(i\,\omega_k\,(t+\phi_k)). \tag{2}$$

Thus if $g \in H(f)$ there are ϕ_k defined $\pmod{2\pi/\omega_k}$ such that (2) holds.

EXAMPLE 1. Consider $q(t) = a_1 \exp(i\,\omega_1 t) + a_2 \exp(i\,\omega_2 t)$ where ω_1/ω_2 is irrational. In this case

$$H(q) = \{a_1 \exp i(\omega_1 t + \alpha_1) + a_2 \exp i(\omega_2 t + \alpha_2)\colon \alpha_i \text{ defined} \pmod{2\pi}\}. \tag{3}$$

The map $h\colon T^2 \to H(q)\colon (\alpha_1,\alpha_2) \to a_1 \exp i(\omega_1 t + \alpha_1) + a_2 \exp i(\omega_2 t + \alpha_2)$ is continuous, 1-1 and onto and thus a homeomorphism of topological spaces. Also it carries the orbits of the dynamical system

$$T^2 \times R \to T^2\colon ((\alpha_1,\alpha_2),t) \to (\alpha_1 + \omega_1 t, \alpha_2 + \omega_2 t)$$

onto the orbits of the translation flow and is an isomorphism of the topological groups. The original function q corresponds to $(0,0) \in T^2$ and the orbit of

q, $\{q_\tau : \tau \in \mathbf{R}\}$ corresponds to the dense line $\{(\omega_1\tau, \omega_2\tau) : \tau \in R\}$ on T^2. See Figure 1.

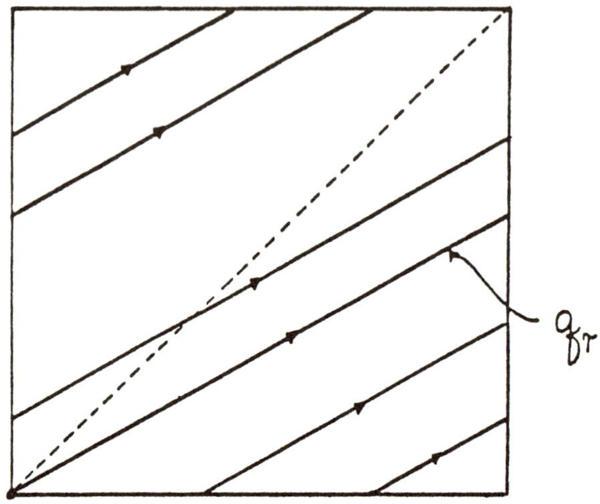

Figure 1: Quasi-periodic hull -- a torus.

EXAMPLE 2: Consider

$$\ell(t) = \sum_0^\infty a_k \exp i\, 2\pi\, (t/2^k) \tag{4}$$

and let τ_n be a sequence such that $\ell_{\tau_n} \to \ell^*$ uniformly where

$$\ell^*(t) = \sum_0^\infty a_k \exp\, (i\, 2\pi\, (t + \phi_k)/2^k). \tag{5}$$

We assume that $\tau_n \to \phi_k \pmod{2^k}$ as $n \to \infty$ and so $\tau_n \to \phi_{k+1} \pmod{2^{k+1}}$ also, or

$$\phi_k = \phi_{k+1} \pmod{2^k}. \tag{6}$$

It is not hard to see that the hull of ℓ is precisely the set of functions ℓ^* as given in (5) where the ϕ_k are defined (mod 2^k) and satisfy (6).
This suggests a coordinate system for $H(\ell)$ as an infinite product

$$\mathop{\times}_{k=0}^{\infty} S^1$$

where $S^1 = \{z \in C^1 : z = \exp i\theta\}$ is the unit circle in the complex plane and the product has the usual product topology. We set $\theta_k = 2\pi\phi_k / 2^k$ so by (6) $\theta_k = 2\theta_{k+1} \pmod{2\pi}$, and we set $z_k = \exp i\theta_k$ so $z_k = z_{k+1}^2$. Then ℓ^* is given the coordinate $\{z_0, z_1, \dots\} \in S$. Since $z_k = z_{k+1}$ the coordinates of ℓ^* are in the inverse limit system

$$S_2 : S^1 \overset{z^2}{\leftarrow} S^1 \overset{z^2}{\leftarrow} S^1 \leftarrow \dots \tag{7}$$

This is the inverse limit of a solenoid, see Hocking and Young(1961).

Let T be a solid torus in a standard embedding in $\mathbf{R}^3$ as given by rotating a meridianal disk $D(0) = \{(x,0,z): (x-10)^2 + z^2 \leq 1\}$ about the z axis as illustrated in Figure 2. Let ϕ_0 be the polar angle in the (x,y) plane

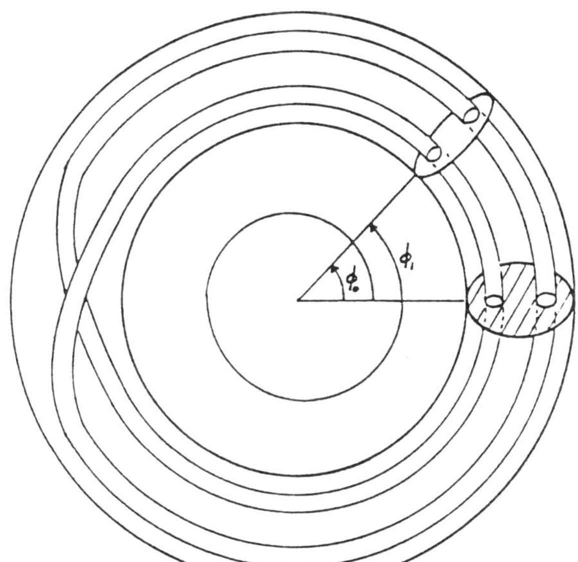

Figure 2: Limit periodic hull -- a solenoid.

normalized so that ϕ_0 is defined (mod 1), and let $D(\phi_0)$ be the image of D(0) after being rotated by ϕ_0. In Figure 2, D(0) is shaded. Let T_1 be a solid torus, lying within the interior of T_0, longitudinally encircling T_0 twice and with medidianal radius 1/4 as illustrated in Figure 2. Let ϕ_1 be an angular variable on T_1 which measures longitudinal displacement in T_1 and is defined (mod 2^1). (See Figure 2). As ϕ_1 traverses [0,2], the meridianal disk $D(\phi_1)$ in T_1 encircles the z-axis twice. Note that T_1 intersects $D(\phi_0)$ in two disks $D(\phi_1')$ and $D(\phi_1'')$ where $\phi_0 \equiv \phi_1' \equiv \phi_1''$ (mod 1).

Continue in this fashion to define T_{k+1} encircling the torus T_k twice with meridianal diameter $1/4^{k+1}$ with longitudinal angle coordinate ϕ_{k+1} defined (mod 2^{k+1}). The 2-solenoid Σ_2 is simply the intersection $\Sigma_2 = \bigcap_{k=0}^{\infty} T_k$ which is a non-empty, connected, compact, one-dimensional subspace of R^3 and so is a "Klosed Kurve" in the sense of Menger. However, Σ_2 is not locally connected and hence cannot be a topological manifold. It is clear from the above construction that $H(\ell)$, the inverse limit system S_2 in (7) and Σ_2 are all homeomorphic.

There is a standard minimal flow on Σ_2 which corresponds to the translation flow restricted to $H(\ell)$. The flow is defined by

$$P_t(\ldots, z_k, \ldots) = (\ldots, z_k \exp(i2\pi t/2^k), \ldots)$$

which corresponds to uniform rotation about the z-axis in the solenoid Σ_2 in Figure 2. The solenoid obtains a continuous Abelian group or structure by component multiplication in the inverse limit representation and this corresponds to the general Abelian group structure on the hull of any a.p. function.

A flow $\pi: X \times R \to X$, X a compact metric space, admits a <u>(global) cross section</u> Z if i) Z is a closed subset of X ii) all trajectories meet Z and iii) there is a continuous function $T: Z \to R$ such that $T(z) > 0$, $\pi(z,T(z)) \in Z$ and $\pi(z,\tau) \notin Z$ for all $z \in Z$ and $0 < \tau < T(z)$. The function T is called the <u>first return time</u>. The <u>Poincaré map</u> (or section map) is the map

$$P: Z \to Z: z \to \pi(z,T(z)).$$

P is a homeomorphism of Z and defines a discrete dynamical system associated with the flow π. Flows which admit global cross sections are precisely supensions of discrete dynamical systems. It should be noted that global cross sections are <u>not</u> unique.

The translation flow on the hull of an almost periodic function always admits a cross section. Let f be a.p. and have a Fourier series as in (1) then $g \in H(f)$ has a Fourier series (3) and

$$g_\tau(t) \sim \sum A_k \exp(i\, \omega_k\, (t + \phi_k + \tau))$$
$$\sim \sum A_k \exp(i\, \omega_k(\phi_k + \tau)) \exp(i\, \omega_k\, t)$$

Thus the Fourier coefficient of g_τ corresponding to the frequency ω_k is

$$s(g_\tau) = M(g_\tau(t)\exp(-i\omega_k t)) = A_k \exp i\, \omega_k(\phi_k + \tau),$$

which has a constantly changing argument as τ varies provided $\omega_k \neq 0$ and $A_k \neq 0$. Thus a cross section to this translation flow is given by

$$Z = \{g \in H(f): \arg s(g) = 0 \text{ or } \phi_k \equiv 0 \pmod{2\pi/\omega_k}\}.$$

In this case the first return time is $T = 2\pi/\omega_k$. Thus cross sections and their first return times are intimately connected with the Fourier spectrum of g.

EXAMPLE 1: $q(t) = a_1 \exp(i\, \omega_1\, t) + a_2 \exp(i\, \omega_2\, t)$ and $q^*(t) \in H(q)$: If $q^*(t) = a_1 \exp i(\omega_1 t + \alpha_1) + a_2 \exp i\,(\omega_2 + \alpha_2)$ so one cross section is $\alpha_1 = 0$ and the first return time is $2\pi/\omega_1$. The angle α_2 is a coordinate on Z so Z is a circle in the torus. The Poincaré map in this coordinate is $P: \alpha_2 \to \alpha_2 + (\omega_2/\omega_1)2\pi$ which is an irrational map of the circle.

EXAMPLE 2: Let $\ell(t)$ and $\ell^*(t)$ be as given previously in (4) and (5). As in the previous example we can define a cross section by requiring that the argument of one of the Fourier coefficients of ℓ^* be zero, or equivalently that $\phi_k \equiv 0 \pmod{2^k}$ for some fixed k. To be specific consider the cross section, Z, defined by $\phi_0 \equiv 0 \pmod 1$ (the shaded disk in Figure 1)), so the first return time is 1. The intersection of this disk and Σ_2 is a Cantor set, and the associated Poincaré map is the classical adding machine of dynamical systems. See Meyer and Sell (1986b).

Let $Q = Q_n = \times_{-\infty}^{\infty} \{1, \ldots, n\}$, i.e., Q is the set of bi-infinite sequences on the symbols 1, 2, ..., n with the usual product topology. So if $q \in Q$ then $q = (\ldots q_{-1}, q_0, q_1 \ldots)$ or more simply written $q = \ldots q_{-1}q_0.q_1 \ldots$ with a decimal point to the right of q_0. Let $A: Q \to Q$ be the shift map or shift automorphism defined by $A(q)_i = q_{i+1}$, i.e., A shifts the "decimal point" one unit to the right. This is the classical Bernoulli dynamical system which has found many applications in contemporary dynamical systems.

Given the flow π on X with section Z, and Poincaré map $P: Z \to Z$ and first return time $T: Z \to R$ we define a Bernoulli product as the suspension of the product of A and P. The product of A and P is simply

$$A*P: Q \times Z \to Q \times Z: (q,z) \to (A(q), P(z)),$$

and so the Bernoulli product flow is the projection of the parallel flow

$$\Psi: (Q \times Z \times R) \times R \to Q \times Z \times R$$

$$: ((q,z,\tau),t) \to (q,z,\tau+t)$$

onto the quotient space

$$Q**Z = (Q \times Z \times R)/\sim$$

where $(q,z,t+\tau(z)) \sim (A(q),P(z),t)$. We called the space Q**Z a Bernoulli bundle.

The shift automorphism, A, has n fixed points $(p_1, \ldots, p_n))$ where $p_i = \ldots iii.iii\ldots$ for $1 \leq i \leq n$. One can see that $\{p_i\}**Z$ is an invariant subset of Q**Z which is globally flow equivalent to the original flow π on X. In other words, the original flow π on X lifts to n copies in Q**Z.

III. THE MELNIKOV METHOD FOR ALMOST PERIODIC SYSTEMS.

The Melnikov method for detecting transverse homoclinic orbits in periodic systems has been discussed by many authors since the landmark paper of Melnikov (1963). Therefore, we shall not develop this theory in great detail nor seek great generality, but simply give an outline the salient features. We follow the more leisure development in Guckenheimer and Holmes (1983) and supplement with results on a.p. differential equations as found in Coppel(1978), Hale(1969), Sell(1978) and Sacker and Sell(1974).

Consider the differential equation

$$\dot{x} = F(x) + \varepsilon f(t) \tag{8}$$

where $x \in R^2$ and $F: R^2 \to R^2$ and $f: R \to R^2$ are smooth. Actually we shall consider (8) as a whole class of differential equations since we shall take f from a subset of **AP**, say the hull of another a.p. function.

While the theory we will describe below extends readily to the more general equation

$$\dot{x} = F(x) + \varepsilon f(x,t),$$

all of the essential features appear in the simpler equation (8). Therefore we shall restrict our attention to (8) in this note. The general case will be treated in a forthcoming paper, Meyer and Sell (1986b).

The unperturbed system (when $\varepsilon = 0$) is assumed to be Hamiltonian so, in particular, the trace of the Jacobian of F is identically zero. Furthermore, the unperturbed system is assumed to have a non-degenerate saddle point at $v_0 \in R^2$ and an orbit $u^0(t)$ homoclinic to v_0, i.e. $u^0(t) \to v_0$ as $t \to \pm \infty$. Thus the stable and unstable manifold of v_0 intersect along the orbit of u^0. See Figure 3.

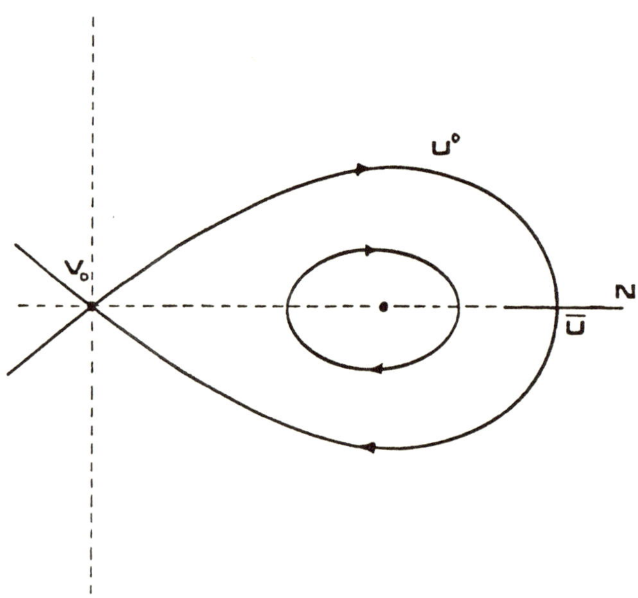

Figure 3: Unperturbed phase portrait.

Let K be a large closed disk in $\mathbf{R}^2$ which contains v_0 and the orbit $u^0(t)$ in its interior, $\varepsilon_0 > 0$ and $\ell_0 > 0$ constants and $L = \{f \in \mathbf{AP} : \|f\| < \ell_0\}$. Let $\phi(t,x_0,f,\varepsilon)$ be the solution of (8) which satisfies $\phi(0,x_0,f,\varepsilon) = x_0$. We consider F as fixed and so suppress the dependence of the solution ϕ on F for simplicity, however f will be taken from a subset of the class of functions L.

In order to understand the underlying geometry of (8) it is useful to recall the concept of a skew product flow. By modifying F outside the large disk K, if necessary, we may assume that the solution $\phi(t,x_0,f,\varepsilon)$ is defined for all t when $|\varepsilon| < \varepsilon_0$ and $f \in L$. Let H be a subset of L invariant under the translation flow. The <u>skew product flow</u> on $R^2 \times H$ is defined by

$$\pi : R \times (R^2 \times H) \to R^2 \times H : (t,x_0,f) \to (\phi(t,x_0,f,\varepsilon),f_t)$$

see Sell(1971) for the verification that π defines a flow and for a background discussion.

When $\varepsilon = 0$ the autonomous system has a non-degenerate saddle point at v_0 and a standard theorem on nonlinear a.p. systems, as found for example in Hale(1969) Chapter 4, applies to (8). For some ε_0, equation (8) has a unique a.p. solution $v(t,f,\varepsilon) = v_0 + O(\varepsilon)$ for $|\varepsilon| < \varepsilon_0$ and $f \in L$ and the $O(\varepsilon)$ is uniform in $t \in \mathbf{R}^1$ and $f \in L$. From the proofs found in Hale(1969), one finds that v satisfies

$$v(t,f_\tau,\varepsilon) = v(t+\tau,f,\varepsilon) = v_\tau(t,f,\varepsilon) \ . \tag{9}$$

Define $x(f,\varepsilon) = v(0,f,\varepsilon)$ so $\phi(t,x(f,\varepsilon),f,\varepsilon) = v(t,f,\varepsilon)$ and define $\mathbf{V} = \{(x(f,\varepsilon),f) \in R^2 \times H : f \in H\}$. The identity (9) shows that $\mathbf{V}$ is an invariant set for the π flow since

$$\phi(t,x(f,\varepsilon),f,\varepsilon) = v(t,f,\varepsilon) = v(0,f_t,\varepsilon) = x(f_t,\varepsilon).$$

The mapping $\Psi : H \to \mathbf{V} : f \to (x(f,\varepsilon),f)$ is a flow equivalence by (9) also. Thus a classical theorem from differential equations says that the skew product flow defined by (8) has, for small ε, an invariant set near $\{v_0\} \times H$, which is flow equivalent to the translation flow on H.

The local stable manifold theorem states that there is a sufficiently small $\delta > 0$ such that

$$W^s_{loc}(f,\varepsilon) = \{x_0 : \|\phi(t,x_0,f,\varepsilon) - v(t,f,\varepsilon)\| < \delta \text{ for } t \geq 0\} \tag{10}$$

is a smooth manifold. In particular, there is a smooth function

$$w^s_{loc} : (-\delta,\delta) \times H \times [0,\varepsilon_0) \to R^2$$

such that

$$W^s_{loc}(f,\varepsilon) = \{w^s_{loc}(\sigma,f,\varepsilon) : \sigma \in (-\delta,\delta)\},$$

and for fixed (f,ε) the map

$$w^s_{loc}(\cdot,f,\varepsilon) : (-\delta,\delta) \to R^2$$

is an embedding of the interval $(-\delta,\delta)$. The function w^s_{loc} is smooth in all its arguments when H is considered as a subset of **AP**, the space of a.p. functions with sup norm. By backward integration w^s_{loc} can be extended to a mapping

$$w^s : R \times H \times [0,\varepsilon_0) \to R^2 ,$$

where $w^s|(-\delta,\delta) \times H \times [0,\varepsilon_0) = w^s_{loc}$, with the property that for fixed (f,ε) the map

$$w^s(\cdot,f,\varepsilon): R \to R^2$$

is an immersion of the line R into R^2. We define the set $W^s(f,\varepsilon)$ by $W^s(f,\varepsilon) = \{w^s(\sigma,f,\varepsilon): \sigma \in R\}$. One then has the characterization:

$$W^s(f,\varepsilon) = \{x_0 : \|\phi(t,x_0,f,\varepsilon)-v(t,f,\varepsilon)\| \to 0 \text{ as } t \to +\infty\}. \tag{11}$$

For the skew product flow we define

$$\mathbf{W}^s(\varepsilon) = \{(W^s(f,\varepsilon),f):f \in H\} .$$

By using (9) one can see that

$$\phi(\tau,W^s(f,\varepsilon),f,\varepsilon) = W^s(f_\tau,\varepsilon).$$

Hence $\mathbf{W}^s(\varepsilon)$ is an invariant set for the skew product flow and is characterized by the formula

$$\mathbf{W}^s(\varepsilon) = \{p\in R^2\times H : \pi(t,p) \to \mathbf{V} \text{ as } t \to +\infty\}. \tag{12}$$

W^s_{loc}, W^s and $\mathbf{W}^s$ are called, respectively, the local stable manifold, the stable manifold and the skew stable manifold. By replacing $t > 0$ by $t < 0$ in (10) and $t \to +\infty$ by $t \to -\infty$ in (11) and (12) one defines the corresponding unstable manifolds $W^u_{loc}, W^u, \mathbf{W}^u$.

Let $p = (x_0,f) \in \mathbf{W}^s(\varepsilon)$. We define the (partial) tangent space by

$$T_p\mathbf{W}^s(\varepsilon) = \text{span}\left\{\frac{\partial w^s}{\partial\sigma}(\sigma_0,f,\varepsilon)\right\}$$

where $x_0 = w^s(\sigma_0,f,\varepsilon)$. That is we consider only the component of the tangent space which lies in the phase plane $\mathbf{R}^2$.

If $p \in \mathbf{W}^s(\varepsilon) \quad \mathbf{W}^u(\varepsilon)$ then we say p is a homoclinic point (homoclinic to $\mathbf{V}$) and $\{\pi(t,p):t \in R\}$ is a homoclinic orbit. Thus $\pi(t,p) \to \mathbf{V}$ as $t \to \pm\infty$. If $p \in \mathbf{W}^s(\varepsilon)\cap \mathbf{W}^u(\varepsilon)$ and $R^2 = T_p\mathbf{W}^s(\varepsilon) + T_p\mathbf{W}^u(\varepsilon)$ then we say $\mathbf{W}^s(\varepsilon)$ and $\mathbf{W}^u(\varepsilon)$ intersect transversally at p. If at each point $p \in \mathbf{W}^s(\varepsilon)\cap \mathbf{W}^u(\varepsilon)$, the sets $\mathbf{W}^s(\varepsilon)$ and $\mathbf{W}^u(\varepsilon)$ intersect transversally then we say $\mathbf{W}^s(\varepsilon)$ and $\mathbf{W}^u(\varepsilon)$ intersect transversally and write $\mathbf{W}^s(\varepsilon) \pitchfork \mathbf{W}^u(\varepsilon)$.

The Melnikov function gives a criteria for the existence of transversal homoclinic orbits. At a point $\overline{u} = u^0(0)$ on the homoclinic orbit of the unperturbed system we construct a normal line n perpendicular to $F(\bar{u})$ and measure the separation of the stable and unstable manifolds on this normal, see Figure 3.

For the perturbed system we let $u_s(0,f,\varepsilon)$ and $u_u(0,f,\varepsilon)$ denote the point on the normal line n where, respectively, the stable and unstable manifolds meet n. We define the separation of the stable and unstable manifold on n to be

$$d(f,\varepsilon) = \| u_s(0,f,\varepsilon) - u_u(0,f,\varepsilon) \| .$$

By a standard argument one finds that

$$d(f,\varepsilon) = \varepsilon \nu M(f) + O(\varepsilon^2)$$

where $\nu = \| F(\bar{u}) \|^{-1}$ is constant, and M is the Melnikov function

$$M(f) = \int_{-\infty}^{\infty} F(u^0(t)) \quad f(t)dt .$$

Define the zero set of M to be

$$Z = \{f \varepsilon H : M(f) = 0\} .$$

We say Z is a <u>simple zero</u> set if

$$\frac{d}{dt} M(f_t)|_{t=0} \neq 0$$

for all $f \varepsilon Z$. If Z is a non-empty simple zero set of M, then Z is a cross section for the restriction of the translation flow to H. See Meyer and Sell(1986b).

If $Z \subset H$ is a cross section for the translation flow π with first return time $T: Z \to R$, then $R^2 \times Z$ is a cross section for the skew product flow Π with the same first return time. Define $\psi(x_0,f,\varepsilon) = \phi(T(f),x_0,f,\varepsilon)$ and $\eta(f) = f_{T(f)}$ so the Poincaré map is

$$\Psi = (\psi,\eta) : R^2 \times Z \to R^2 \times Z : (x_0,f) \to (\psi(x_0,f), \eta(f)),$$

which is a discrete skew product dynamical system.

When $\varepsilon = 0$ the system decouples and the skew product becomes an ordinary product. Thus if f = g of example 1 then Π is the product of the irrational flow on the torus as pictured in Figure 1 and flow pictured in Figure 3. Similarly if $f = \ell$ of example 2, then Π is the product of the solenoidal flow and the flow of Figure 3.

It is easier to visualize the Poincaré map. First consider the case when f = q of Example 1 with $\varepsilon = 0$. The cross section of the translation flow is a circle, the diagonal of Figure 1, and the Poincaré map is an irrational rotation of the circle with return time T. Integrate the autonomous equations when ε = 0 for a time T to obtain a map of the plane as shown in Figure 3. Figure 4

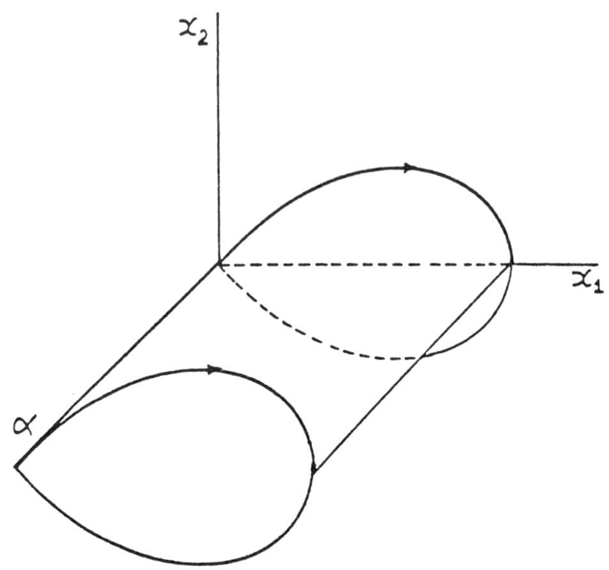

Figure 4: Unperturbed Poincare map.

tries to illustrate the product map. The α axis coming out of the plane of the paper is an angular variable and should be identified (mod 2π) since the space is $R^2 \times S^1$. The map carries a plane perpendicular to the α axis into another such plane. The α axis is an invariant circle for the map, and its stable and unstable manifolds are the products of the loop and a circle.

In the case when $f = \ell$ of example 2 with $\varepsilon = 0$ then one must use some imagination. Think of a Cantor set along the α axis in Figure 4. The Poincaré map is similar to the above except the Cantor set of planes perpendicular to the α-axis are shuffled by the adding machine map.

The main result of this section is:

THEOREM 1: *Let the Melnikov function define a non-trivial, simple zero set* Z H. *Then for each* $f \in Z$ *there is a unique* $\xi(f,\varepsilon) = \bar{u} + O(\varepsilon) \in R^2$ *such that* $(\xi(f,\varepsilon),f) \in W^s(\varepsilon) \cap W^u(\varepsilon)$ *for* $0 < \varepsilon < \varepsilon_0$. *The function* ξ *is continuous. Moreover, if* $\Xi^0 = \{ (\xi(f,\varepsilon),f) : f \in Z \}$, $\Xi^k = \Psi^k(\Xi^0)$ *then* $\Lambda = V \cup \{ \bigcup_{-\infty}^{\infty} \Xi^k \}$ *is a compact invariant set for the Poincaré map* Ψ.

Outline of proof: The separation of the stable and unstable manifold in the normal direction n is

$$d(f,\varepsilon) = \| u_s(0,f,\varepsilon) - u_u(0,f,\varepsilon) \|$$
$$= e\nu\, M(f) + O(\varepsilon)$$

where ν is a non-zero constant. Define a coordinate system near $\overline{u}$ in R^2 as follows: Consider the map $(\alpha,\beta) \to u_s(\alpha,f,\varepsilon) + \beta n$, where n is now a unit normal vector to $F(\overline{u})$. This map takes a neighborhood of the origin in R^2 onto a neighborhood of $\overline{u}$. Note that α is a coordinate along the stable manifold essentially the time parameter), and β is a coordinate in the unit normal direction, see Figure 5. By the fact that n is normal to $F(\overline{u})$, (α,β) constitutes

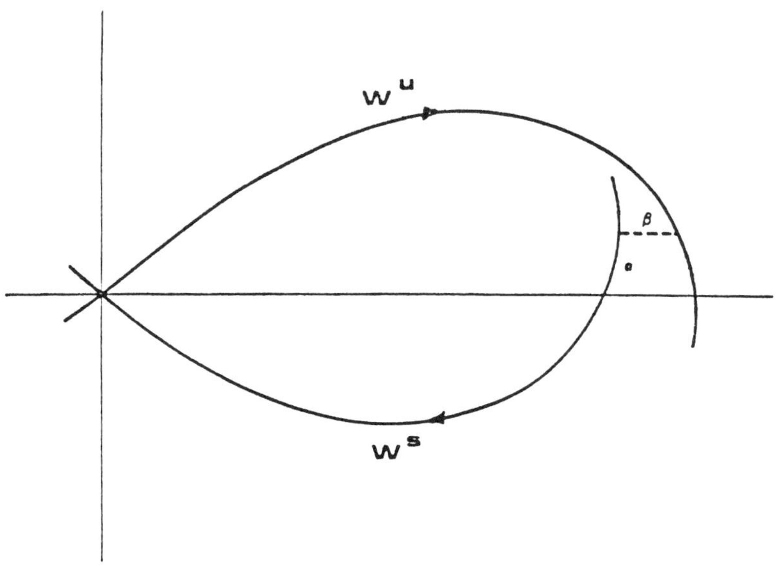

Figure 5: Stable manifold coordinates.

direction, see Figure 5. By the fact that n is normal to $F(\overline{u})$, (α,β) constitutes
a valid coordinate system when ε is small.

In this coordinate system the stable manifold has coordinates $(\alpha,0)$. The unstable manifold

$$u_u(\alpha,f,\varepsilon) = u_s(\alpha,f,\varepsilon) + d(f_\alpha,\alpha)n$$
$$= u_s(\alpha,f,\varepsilon) + \varepsilon\nu\, M(f_\alpha) + O(\varepsilon^2)$$

has coordinates $(\alpha, \varepsilon\nu\, M(f_\alpha) + O(\varepsilon^2))$, and so in these coordinates an intersection of the stable and unstable manifold is obtained when $\varepsilon\nu M(f_\alpha) + O(\varepsilon^2) = 0$. In words $G(\alpha,f,\varepsilon) = 0$ where

$$G(\alpha,f,\varepsilon) = M(f_\alpha) + O(\varepsilon).$$

By assumption

$$G(0,f,0) = M(f) = 0\ ,\ D_1G(0,f,0) \neq 0$$

when $f \varepsilon Z$. Thus by the implicited function theorem we can solve for $\alpha = \overline{\alpha}(f,\varepsilon)$. In these coordinates the intersection of the stable and unstable manifold is $(\alpha, \overline{\alpha}(f,\varepsilon))$, which is just $\xi(f,\varepsilon)$.

Under the Poincaré Ψ map the limit of $(\xi(f,\varepsilon),f)$ lies in **V** under forward and backward iteration since it lies on both the stable and unstable manifolds. Thus Λ is closed and hence compact.

IV. HYPERBOLIC SETS, THE SHADOWING LEMMA AND BERNOULLI BUNDLES. Here we shall concentrate on the Poincaré map considered as a discrete skew product dynamical system. In this section the notation of the last section will be used in a slightly more general setting. Let

$$\Psi = (\psi, \eta): R^n \times Z \to R^n \times Z: (x,f) \to (\psi(x,f), \eta(f))$$

define a discrete skew product dynamical system where Z is an arbitrary compact Hausdorff space. Thus Ψ and η are homeomorphisms. Furthermore assume Ψ and ψ are smooth in their first argument, i.e. $D_1{}^k\psi(x,f)$ is defined and continuous for all $(x,f) \in R^n \times Z$.

A compact, invariant set $\Lambda \subset R^n \times Z$ is called <u>(skew) hyperbolic</u> if there exists a constant μ, $0 < \mu < 1$, and a continuous splitting of R^n into $E_p{}^s + E_p{}^u$, $p \in \Lambda$, such that

$$D_1\psi(p): E_p{}^s \to E_q{}^s \qquad \text{and} \qquad D_1\psi(p): E_p{}^u \to E_q{}^u$$

where $q = \Psi(p)$ and

$$\| D_1\psi(p)(u) \| < \mu \| u \| \quad \text{for } u \in E_p{}^s$$
$$\| D_1\psi^{-1}(p)(v) \| < \mu \| v \| \quad \text{for } v \in E_p{}^u .$$

THEOREM 2: <u>The invariant set Λ of Theorem 1 is skew hyperbolic.</u>

The proof follows by standard results as found in Coppel(1978) and Palmer(1986).

For $\alpha > 0$ a (skew) <u>α-pseudo-orbit</u> for $\Psi|\Lambda$ is a bisequence $\{p_i=(x_i,f_i)\}$, i ranging from $-\infty$ to $+\infty$, $p_i \varepsilon \Lambda$, such that $f_{i+1} = \eta(f_i)$ and $d(\psi(x_i,f_i),x_{i+1}) < \alpha$ for all i. Thus the sequence $\{f_i\}$ is an η-orbit but the components in R^n may jump by as much as α.

One says a Ψ orbit $\{ \Psi^i(y)=(y_i,f_i) \}$ <u>β-shadows</u> an α-pseudo-orbit $\{(x_i,f_i)\}$ provided $d(x_i,y_i) < \beta$ and, of course, $f_{i+1} = \eta(f_i)$. Note that the base orbits are the same. In this context we have:

THEOREM 3 (The Shadowing Lemma): <u>If Λ is a compact, skew hyperbolic invariant set for Ψ: $R^n \times Z \to R^n \times Z$, then for every $\beta > 0$ there is a $\alpha > 0$ such that every α-pseudo-orbit for $\Psi|\Lambda$ is β-shadowed by some Ψ orbit $\{\Psi^i(q)\}$. Moreover, there is a $\beta_0 > 0$ such that if $0 < \beta < \beta_0$ then the Ψ-orbit given above is uniquely deter-</u>

mined by the α-psuedo-orbit.

Proof: A slight variation of the proof given in Meyer and Sell(1986a).

Henceforth, assume that the Melnikov function defines a non-trivial, simple zero set Z H. Let $\Lambda = V \cup \{\bigcup \Xi^k\}$ be the hyperbolic set for the Poincare map $\Psi = (\psi, \eta)$ as given in Theorems 1 and 2. Let A: $Q \to Q$ be the shift automorphism on n symbols as discussed in section II.

THEOREM 4: Under the above assumptions and for small ε, there is an $\iota \in I$ and a compact invariant set Ω $R^2 \times Z$ for the Poincare map Ψ such that $\Psi^\iota|\Omega$ is equivalent to the product map $A\times\eta$: $Q\times Z \to Q\times Z$: $(q,z) \to (A(q), z(\eta))$

Outline of the proof: Assume that $\varepsilon > 0$ is so small that the conclusions of Theorems 1 and 2 hold, so that $\Lambda = V \cup \{\bigcup \Xi^k\}$ is a compact, skew hyperbolic invariant set for the Poincaré map Ψ: $R^2\times Z \to R^2\times Z$. Fix ε and henceforth the ε dependence will not be displayed. Let $\beta = \beta_0$ be as given in the uniqueness part of the shadowing lemma and α corresponding to this β. Assume also the β is so small that $\mathrm{dist}(\Xi^0, \Xi^1) > 4\beta$. Since $\Xi^k \to V$ as $k \to \pm\infty$ there is a $\kappa > 0$ such that $\mathrm{dist}(\Xi^k, V) < \alpha/4$ when $|k| \geq \kappa$. Let N be this $\alpha/2$ neighborhood of **V**.

Our proof is based on Palmer's (1984) construction of an invariant Cantor set for the Poincaré map of a periodic system. His idea is to use the shadowing lemma instead of constructing a horseshoe a lá Smale. Let us relabel some of the Ξ^k as follows: Set $\Xi_0 = \Xi^k$ and $\Xi_\ell = \Psi^{-(2\kappa+1)\ell}(\Xi_0)$ for $1 \leq \ell \leq n$, see Figure 6.

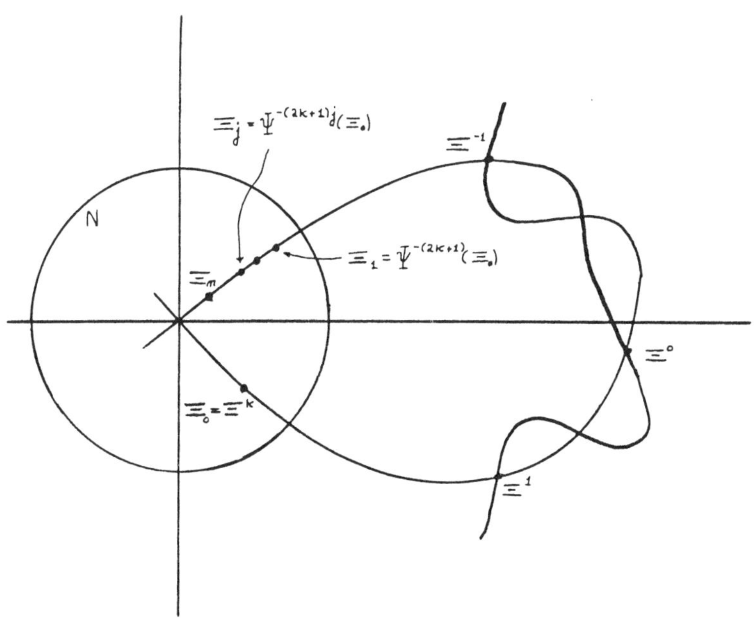

Figure 6: Homoclinic Orbit

Note that Ξ_j, $0 \leq j \leq n$, lie in N. Moreover the Poincaré mapping satisfies $\Psi^{(2\kappa+1)}(\Xi_j) = \Xi_{j-1}$ for $1 \leq j \leq n$. Furthermore for each $z \in Z$ and $0 \leq \ell$, there is a unique point $\xi_\ell(z)$ in R^2 with $(\xi_\ell(z),z) \in \Xi_\ell$, and one has

$$\Psi^{2\kappa+1}(\xi_\ell(z),z) = (\xi_{\ell-1}(\eta^{2\kappa+1}(z)), \eta^{2\kappa+1}(z)),\ 1 \leq \ell \leq n.$$

Given a point $(q,z) \in Q \times Z$ we construct an α-pseudo orbit as follows: We begin with $(\xi_0(z),z)$ in Ξ_0. With $q_0 = j$, $1 \leq j \leq n$, we jump to the point $(\xi_j(\eta^{2\kappa+1}(z)), \eta^{2\kappa+1}(z))$ in Ξ_j. Next we follow the orbit through the latter point by repeated applications of $\Psi^{2\kappa+1}$ until we return to Ξ_0. Since $\Psi^{(2\kappa+1)j}(\Xi_j) = \Xi_0$, this occurs after j applications of $\Psi^{2\kappa+1}$. We have thus returned to the point $(\xi_0(\eta^{(2\kappa+1)j}(z)), \eta^{(2\kappa+1)j}(z))$ in Ξ_0. With $q_1 = \ell$, we then jump to the point $(\zeta_\ell(\eta^{(2\kappa+1)(j+1)}(z)),\eta^{(2\kappa+1)(j+1)}(z))$ in Ξ_ℓ and repeat the above. By using $\Psi^{-(2\kappa+1)}$ with appropriate jumps from Ξ_ℓ to Ξ_0 (when $q_{-n} = \ell$), we construct the α-pseudo orbit for negative time. In this way we obtain an α-pseudo orbit Γ over the trajectory $z^k = \eta^k(z)$ in Z.

Let $\{\Psi^k(P)\}$ be the unique orbit which β-shadows Γ and set $G(q,z) = P$. Thus one has

$$G: Q\times Z \to R^2\times Z.$$

By the shadowing lemma, G is continuous and one-to-one and therefore, since $Q\times Z$ is compact, G is a homeomorphism onto a subset Ω $R^2\times Z$.

It is clear that $\Psi^\iota|\Omega$ where $\iota = 2\kappa+1$ is conjugate to

$$A*\eta: Q\times Z \to Q\times Z: (q,z) \to (A(q),\eta(z)).$$

We thus obtain our final result:

THEOREM 5: *Under the above assumptions the skew product flow for small ε admits an invariant set which is flow equivalent to the flow on a Bernoulli bundle.*

BIBLIOGRAPHY

1. Alekseev, V. M. 1976: 'Symbolic dynamics', 11th *Math. School*, Ed. Mitropolskii and Samoilenko, Kiev.

2. Besicovitch, A.S. 1932: *Almost Periodic Functions*, Cambridge Univ. Press.

3. Birkhoff, G. D. 1932: 'Nouvelles recherches sur les systèms dynamiques', *Mem. Pont. Acad. Soc. Novi. Lyncaei* 1, 85-216.

4. Bohr, H. 1959: *Almost Periodic Functions*, Chelsea Publ., New York.

5. Cartwright, M. L. and J. E. Littlewood 1945: 'On non-linear differential equations of second order: I. The equation $\ddot{y} - k(1-y)y + y = b\lambda k\cos(\lambda t+\alpha)$, k large', *J. London Math. Soc.* 20, 180-189.

6. Chow, S. N. and J. K. Hale 1982: *Methods of Bifurcation Theory*, Springer-Verlag, New York.

7. Chow, S. N., J. K. Hale and J. Mallet-Paret 1980: 'An example of bifurcation to homoclinic orbits', *J. Diff. Eqs.* 37, 351-373.

8. Coppel, W. A. 1978: Dichotomies in Stability Theory, Lecture Notes in Math., 629, Springer-Verlag, New York.

9. Favard, J. 1933: Lecons sur les fonctions presque périodiques, Gauthier-Villars, Paris.

10. Gottschalk, W. H. and G. A. Hedlund 1955: Topological Dynamics, American Math. Soc., Providence, R. I.

11. Guckenheimer, J. and P. Holmes 1983: Nonlinear Oscillations, Dynamical Systems, and Bifurcations of Vector Fields, Springer-Verlag, New York.

12. Hale, J. K. 1969: Ordinary Differential Equations, Wiley-Interscience, New York.

13. Hewitt, E. and K. Ross 1963: Abstract Harmonic Analysis I and II, Springer-Verlag, New York.

14. Hocking, J. G. and G. S. Young 1961: Topology, Addison-Wesley, New York. Holmes, P. 1980: 'Averaging and chaotic motion in forced oscillations', SIAM Appl. Math. 38, 68-80.

15. Hocking, J. G. and G. S. Young 1961: Topology, Addison-Wesley, New York. Holmes, P. 1980: 'Averaging and chaotic motion in forced oscillations', SIAM Appl. Math. 38, 68-80.

16. Markus, L. and K. Meyer 1980: 'Periodic orbits and solenoids in generic Hamiltonian systems', Amer. J. Math. 102, 25-92.

17. Marsden, J. E. 1984: 'Chaos in dynamical systems by the Poincaré-Melnikov-Arnold method', Proc. ARO Workshop on Dynamics, SIAM.

18. Melnikov, V. K. 1963: 'On the stability of the center for time periodic perturbations', Trans. Moscow Math. Soc. 12, 1-57.

19. Meyer, K. and G. Sell 1986a: 'An analytic proof of the shadowing lemma', to appear in Funkcialaj Ekvacioj.

20. Meyer, K. and G. Sell 1986b: 'Melnikov transforms, Bernoulli bundles, and skew product flows', to appear.

21. Miller, R. K. 1965: 'Almost periodic differential equations as dynamical systems with applications to the existence of a.p. solutions', J. Diff. Eqs. 1, 337-345.

22. Nemyckii, V. V. and V. V. Stepanov 1960: Qualitative Theory of Differential Equations, Princeton Univ., Princeton, N. J.

23. Palmer, K. J. 1984: 'Exponential dichotomies and transversal homoclinic points', J. Diff. Eqs. 55, 225-156.

24. Poincaré, H. 1892: Les Méthodes Nouvelles de la Mécanique Céleste III, Gauthier-Villars, Paris.

25. Poinkhoff, H. G. 1973: 'An analytic closing lemma', Proc. Midwest Dyn. Sys. Sem., Northwestern Univ. Press, Evanston, Ill, 128-256.

26. Sacker, R. J. and G. R. Sell 1974: 'Existence of dichotomies and invariant splitting for linear differential systems I', J. Diff. Eqs. 15, 429-58.

27. Sacker, R. J. and G. R. Sell 1977: 'Lifting properties in skew-product flows with applications to differential equations', Memoirs Amer. Math. Soc. 190.

28. Scheurle, J. 1986: 'Chaotic solutions of systems with almost periodic forcing', preprint.

29. Sell, G. R. 1967: 'Nonautonomous differential equations and topological dynamics I, II' Trans. Amer. Math. Soc. 127, 241-62 & 263-83.

30. Sell, G.R. 1971: Topological Dynamics and Ordinary Differential Equations, Van Nostrand, New York.

31. Sell, G. R. 1978: 'The structure of a flow in the vicinity of an almost periodic motion', J. Diff. Eqs. 27(3), 359-93.

32. Smale, S. 1963: 'Diffeomorphisms with infinitly many periodic points', Differential and Combinatorial Topology, S. Chern (ed.), Princeton Univ. Press, 63-80.

33. Wiggens, S. 1986a: 'A generalization ot the method of Melnikov for detecting chaotic invariant sets', preprint.

34. Wiggens, S. 1986b: 'The orbit structure in the neighborhood of a transverse homoclinic torus, preprint.

* * * * * * *

This work was done in part at the Institute for Mathematics and its Applications and at the University of Cincinnati with funds made available by the National Science Foundation.

KENNETH R. MEYER
DEPARTMENT OF MATHEMATICAL SCIENCES
UNIVERSITY OF CINCINNATI
CINCINNATI, OHIO 45221

GEORGE R. SELL
INSTITUTE FOR MATHEMATICS AND ITS APPLICATIONS
UNIVERSITY OF MINNESOTA
MINNEAPOLIS, MINNESOTA 55455

Canadian Mathematical Society
Conference Proceedings
Volume 8 (1987)

PERTURBATIONS OF A SYMMETRIC HOPF BIFURCATION

Wayne Nagata*

ABSTRACT. We consider the effects that perturbations have on the solutions of differential equations near a Hopf bifurcation, when the unperturbed system possesses O(2) symmetry, but the perturbed system possesses less symmetry. There exist invariant two-tori of solutions, and the flow on the tori can be quasiperiodic or phase locked, depending on the remaining symmetry of the perturbed system.

1. INTRODUCTION

The instabilities and transitions in the behaviour of some physical systems can be understood in terms of the bifurcations that occur in differential equations which are used to model the system. It is common to study idealized situations in an effort to obtain tractable mathematical problems, and such idealizations can introduce symmetries to the model differential equations. For example, to study the onset of convection in the presence of a magnetic field, or magnetoconvection, one may consider a two-dimensional horizontal layer of fluid of infinite width and uniform depth, that is heated uniformly from below in the presence of a uniform magnetic field (e.g., Nagata 1986). These idealizations introduce translation and reflection symmetries to the partial differential equations used to model magnetoconvection. To further simplify the problem, one considers only those convective movements that are spatially periodic, with a wavelength fixed at some value. Thus we impose periodic boundary conditions on the differential equations and all translations that differ by an integral number of wavelengths are now identified. In this way the idealizations of the system introduce O(2) symmetry to the mathematical model. Other examples of systems in which

1980 Mathematics Subject Classification (1985 Revision). 58F14, 58F30, 34C30.
*Research supported by NSERC, Canada.

O(2) symmetry appears in the mathematical model due to idealizations made are: the Taylor-Couette flow of a viscous fluid in the gap between two vertical, coaxial, independently rotating cylinders (Chossat and Iooss 1985), chemical reactions in a circular dish (Erneux and Herschkowitz-Kaufman 1977) and lasers with a saturable absorber (Peplowski 1983).

For these systems, under certain conditions there are transitions from stationary to oscillatory behaviour. In the model differential equations these transitions correspond to Hopf bifurcations. However, the bifurcation analysis must take into account the symmetries that are possessed by the model. This has been done in the case of Hopf bifurcations with O(2) symmetry by van Gils (1983), and Golubitsky and Stewart (1985) (see also the references therein).

It is usually hoped that the idealizations made will not qualitatively affect predictions of the behaviour very much, in some physically reasonable sense. However, it is both interesting and useful to see just what effect the removal of symmetry has on a problem, especially if the corresponding idealizations cannot be exactly satisfied by a real system. If it is reasonable to assume that idealizations are "almost" satisfied in a carefully prepared experiment, then one can investigate the behaviour of such a system by studying small symmetry breaking perturbations of symmetric differential equations. In the magnetoconvection problem, one can study the effects of nonuniformities in the externally imposed heating and magnetic field in this way, provided the nonuniformities retain some partial symmetries. It may also be possible to describe convection in some finite containers in terms of a small symmetry breaking perturbation of convection in an infinitely wide layer with periodic boundary conditions.

The papers of van Gils (1983), and van Gils and Mallet-Paret (1986) consider the effect of small perturbations on the branching of solutions near a Hopf bifurcation where the unperturbed equations possess O(2) symmetry. Also, van Gils and Valkering (1985), and Golubitsky and Stewart (1986) have treated the branching of periodic solutions at a Hopf bifurcation with D_n symmetry. In this article, we consider the effect of symmetry breaking perturbations on the flows on bifurcating two-tori which exist near Hopf points. With O(2) symmetry, the flows on the tori are all trivial due to the symmetry, but if the O(2) symmetry is partially broken by a perturbation, it is possible to have phase locked or quasiperiodic flow on the tori.

By the use of center manifold theory, it is possible to reduce the differential equations that model the physical systems described above to a parametrized family of four-dimensional ordinary differential equations which fully describe the dynamics of small-amplitude solutions of the original model equations in a neighbourhood of a Hopf bifurcation with O(2) symmetry. The reduced equations may be written in complexified form as

$$\dot{z} = f(\mu,\nu,z) \tag{1.1}$$

where $z=(z_1, z_2) \in C^2$ are coordinates on a four-dimensional center manifold, $\mu \in R$ is a bifurcation parameter, and $\nu \in R$ is a perturbation parameter. When $\nu = 0$, we assume that the vector field $f(\mu,0,z)$ is equivariant with respect to the following representation of O(2):

$$\begin{aligned} f(\mu,0,\theta z) &= \theta f(\mu,0,z) \text{ for all } \theta \in SO(2) = R/2\pi Z \\ f(\mu,0,\kappa z) &= \kappa f(\mu,0,z), \end{aligned} \tag{1.2}$$

where

$$\theta(z_1, z_2) = (e^{i\theta} z_1, e^{-i\theta} z_2), \quad \kappa(z_1,z_2) = (z_2,z_1).$$

In the applications described above, it is possible to choose coordinates on the center manifold such that the action of O(2) on the reduced equation (1.1) is given by (1.2). We will assume that

$$f(\mu,\nu,0) = 0 \quad \text{for all } \mu,\nu, \tag{1.3}$$

i.e., $z(t) = 0$ is always a solution of (1.1). In the applications the trivial solution $z = 0$ corresponds to a parameterized family of stationary solutions that are invariant under translations and reflections if $\nu = 0$. We consider the physically interesting case where the linearization

$$\dot{z} = Df(\mu,0,0)z$$

of (1.1) about the trivial solution when $\nu = 0$ possesses eigenvalues $\lambda(\mu)$ satisfying $\mathrm{Re}\,\lambda(\mu) < 0$ when $\mu < 0$, and

$$Df(0,0,0) = \begin{bmatrix} i\omega & 0 \\ 0 & i\omega \end{bmatrix}, \quad \omega > 0, \quad \frac{d}{d\mu} \mathrm{Re}\,\lambda(0) > 0 \tag{1.4}$$

when $\mu = 0$. Thus, when ν is fixed at 0, the trivial solution is locally asymptotically stable when $\mu < 0$ and loses stability to an oscillatory mode as μ increases through 0. The critical eigenvalues $\lambda(0) = \pm i\omega$ at $\mu = 0$ are doubled, due to the reflection symmetry κ.

When $\nu \neq 0$, we assume that the vector field $f(\mu,\nu,z)$ is equivariant with respect to a proper subgroup of O(2). We will consider three cases.

<u>Case (a)</u>: $f(\mu,\nu,z)$ is equivariant with respect to SO(2):

$$f(\mu,\nu,\theta z) = \theta f(\mu,\nu,z) \text{ for all } \theta \in SO(2) = R/2\pi Z \tag{1.5a}$$

Case (b): $f(\mu,\nu,z)$ is equivariant with respect to the dihedral group D_n:

$$f(\mu,\nu,\theta_j z) = \theta_j f(\mu,\nu,z) \text{ for } \theta_j \in Z_n = \{2\pi j/n\}_{0 \le j \le n-1} \tag{1.5b}$$
$$f(\mu,\nu,\kappa z) = \kappa f(\mu,\nu,z);$$

Case (c): $f(\mu,\nu,z)$ is equivariant with respect to the cyclic group Z_n:

$$f(\mu,\nu,\theta_j z) = \theta_j f(\mu,\nu,z) \text{ for } \theta_j \in Z_n = \{2\pi j/n\}_{0 \le j \le n-1} \tag{1.5c}$$

Thus when $\nu \neq 0$, the differential equation (1.1) possesses less symmetry than when $\nu = 0$. In this article we describe the behaviour of solutions of (1.1) satisfying (1.2) - (1.4) and either (1.5a), (1.5b) or (1.5c), in certain cases of interest.

2. BRANCHING RESULTS

In this section we summarize results on the branching of solutions of (1.1) for μ,ν near 0. To analyze local bifurcations, we expand in a Taylor series in z to some order k, assuming f is smooth enough. Then by normal form theory (cf. Chow and Hale 1982; Guckenheimer and Holmes 1983), one can construct a finite sequence of polynomial coordinate changes that transform (1.1) into

$$\dot{z} = N_k f(\mu,\nu,z) + O(|z|^{k+1}), \tag{2.1}$$

where the normal form $N_k f(\mu,\nu,\cdot)$ of $f(\mu,\nu,\cdot)$ is a polynomial of degree $\le k$. It is possible to choose the coordinate changes so that all equivariances (1.2), (1.5) are preserved (Ruelle 1973) and thus we may assume that $N_k F$ satisfies (1.2) and (1.5a, b, or c). The normal form possesses an additional symmetry

$$N_k f(\mu,\nu,\xi z) = \xi N_k f(\mu,\nu,z) \text{ for all } \xi \in S^1 = R/2\pi Z, \tag{2.2}$$

where

$$\xi(z_1,z_2) = (e^{i\xi}z_1,\ e^{i\xi}z_2)$$

We use the notation S^1 for the normal form symmetry in order to distinguish it from the SO(2) symmetry in (1.2) or (1.5a), although the two groups are isomorphic. The additional symmetry (2.2) is introduced by the choice of coordinates, and only affects the terms of order $\le k$ in (2.1). The remainder terms $O(|z|^{k+1})$ in (2.1) do <u>not</u> satisfy (2.2), but only possess the symmetries (1.2) and (1.5a, b, or c) of the original vector field in (1.1).

Due to the presence of the additional S^1 symmetry (2.2) in $N_k f$, the analysis of the truncated normal form equation

(2.3) $$\dot{z} = N_k f(\mu,\nu,z)$$

is easier than that of the full equation (2.1). It has been shown, using Liapunov-Schmidt reduction and invariant manifold theory, that the branching diagrams for (2.3) persist for (2.1) when the truncated terms are restored.

By considering the action of all the symmetries (1.2), (1.5a, b, or c), (2.2) on each term in the normal form $N_k f$, one can prove that certain coefficients must vanish, and hence the normal forms must be restricted to certain expressions depending on the symmetries satisfied: see van Gils (1983), Golubitsky and Stewart (1985, 1986), Chossat and Iooss (1985), van Gils and Mallet-Paret (1986).

For the unperturbed equation f possesses the O(2) symmetry (1.2) when $\nu = 0$, and the branching diagrams are determined by the normal form with $k = 4$. For μ near 0, (2.1) has the form

(2.4) $$\begin{aligned} \dot{z}_1 &= z_1(i\omega + \mu a + c_1|z_1|^2 + c_2|z_2|^2) + O(|\mu|^2|z| + |\mu||z|^3 + |z|^5), \\ \dot{z}_2 &= z_2(i\omega + \mu a + c_1|z_2|^2 + c_2|z_1|^2) + O(|\mu|^2|z| + |\mu||z|^3 + |z|^5) \end{aligned}$$

when $k = 4$, and satisfies (1.2) (van Gils 1983), where a, c_1, c_2 are complex constants. In applications, the values of these constants are determined by the model, and usually depend on physically relevant quantities (eg., in magnetoconvection, the constants depend on the viscosity, thermal conductivity, electrical conductivity, magnetic permeability, etc. of the fluid). In the case

(2.5) $$\operatorname{Re} a > 0,\ \operatorname{Re} c_1 < 0,\ \operatorname{Re}(c_1+c_2) < 0,\ \operatorname{Re}(c_1-c_2) < 0$$

one obtains the branching diagram of Figure 1 (van Gils 1983), Golubitsky and Stewart 1985). The trivial solution ϕ loses stability as μ increases through the multiple bifurcation point $\mu = 0$, and branches of unstable "travelling wave" solutions (TW) and stable "standing wave" solutions (SW) bifurcate supercritically. The travelling wave solutions are two isolated periodic orbits of (1.1) that are interchanged by the reflection κ, while the standing wave solutions consist of a smooth family of disjoint periodic orbits which foliate an attracting invariant two-torus. The names of the solutions are motivated by their significance in the applications.

In case (a), if f possesses the SO(2) symmetry (1.5a) when $\nu \neq 0$, then the normal form with $k = 4$ is

$$\begin{aligned} \dot z_1 &= z_1[i\omega + \mu a + \nu b_1 + c_1|z_1|^2 + c_2|z_2|^2] \\ &\qquad + O[(|\mu|+|\nu|)^2|z| + (|\mu|+|\nu|)|z|^3 + |z|^5], \\ \dot z_2 &= z_2[i\omega + \mu a + \nu b_2 + c_1|z_2|^2 + c_2|z_1|^2] \\ &\qquad + O[(|\mu|+|\nu|)^2|z| + (|\mu|+|\nu|)|z|^3 + |z|^5], \end{aligned} \tag{2.6}$$

where ω is real, and a, b_1, b_2, c_1, c_2, are complex constants (van Gils 1983). If the normal form coefficients satisfy (2.5) and

$$\text{(2.7)} \qquad \operatorname{Re} b_1 \neq \operatorname{Re} b_2 \quad , \quad \operatorname{Re}(b_1-b_2)\operatorname{Im}(c_1-c_2) \neq \operatorname{Im}(b_1-b_2)\operatorname{Re}(c_1-c_2)$$

then one obtains the branching diagram of Figure 2 (van Gils 1983). The breaking of the reflection symmetry κ causes the multiple bifurcation point of Figure 1 to be split into two primary (Hopf) bifurcations of periodic travelling wave solutions and a secondary (Neimark-Sacker) bifurcation of an attracting, invariant two-torus. The existence and stability of the torus may be proved using invariant manifold theory (van Gils and Mallet-Paret 1986) or by using the results of Renardy (1982) on bifurcation from rotating waves. In cases other than (2.5), (2.7), the branching diagrams are similar, but the tori are unstable.

In case (b), if f possesses D_n symmetry (1.5b) when $\nu \neq 0$, then the multiple bifurcation point of Figure 1 is not split, but is only shifted along the μ-axis so we may assume that the bifurcation occurs at $\mu = 0$. If

$$n = 3 \text{ or } n \geq 5,$$

then it follows from Golubitsky and Stewart (1986) that (2.1) for $k = 4$ can be written in the form

$$\begin{aligned} \dot z_1 &= z_1[i\omega(\nu) + \mu a(\nu) + c_1(\nu)|z_1|^2 + c_2(\nu)|z_2|^2] \\ &\qquad + O(|\mu|^2 + |\mu||z|^3 + |z|^5), \\ \dot z_2 &= z_2[i\omega(\nu) + \mu a(\nu) + c_1(\nu)|z_2|^2 + c_2(\nu)|z_1|^2] \\ &\qquad + O(|\mu|^2|z| + |\mu||z|^3 + |z|^5), \end{aligned} \tag{2.8}$$

where $\omega(\nu)$ is real, and $a(\nu)$, $c_1(\nu)$, $c_2(\nu)$ are complex. Let us assume

$$\begin{aligned} &\omega(\nu) > 0,\ \operatorname{Re} a(\nu) > 0,\ \operatorname{Re} c_1(\nu) < 0, \\ &\operatorname{Re}[c_1(\nu)+c_2(\nu)] < 0,\ \operatorname{Re}[c_1(\nu)-c_2(\nu)] < 0. \end{aligned} \tag{2.9}$$

Since (2.8), (2.9) for fixed ν is the same as (2.4), (2.5), one can use invariant manifold theory to prove the existence of supercritical bifurcations of two unstable periodic "travelling wave" solutions and of an attracting invariant torus, as in Figure 3. On the other hand,

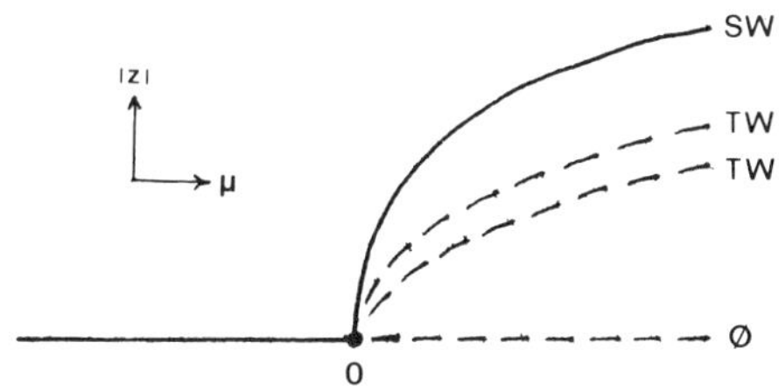

FIG. 1. Branching diagram for the unperturbed equations, Hopf bifurcation with O(2) symmetry. Solid lines represent stable solutions, broken lines represent unstable solutions.

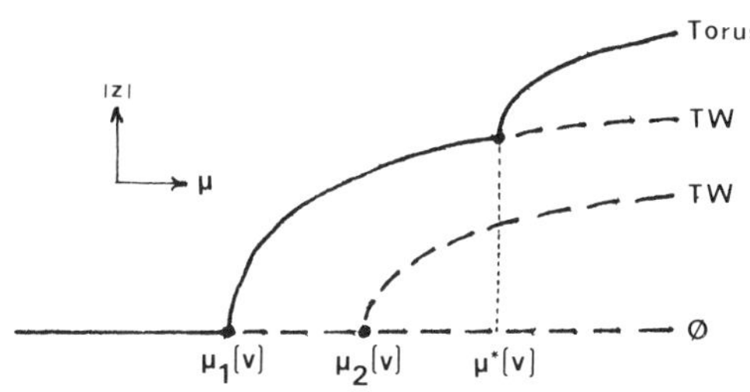

FIG. 2. Branching diagram for $\nu \neq 0$, when reflection symmetry κ is broken.

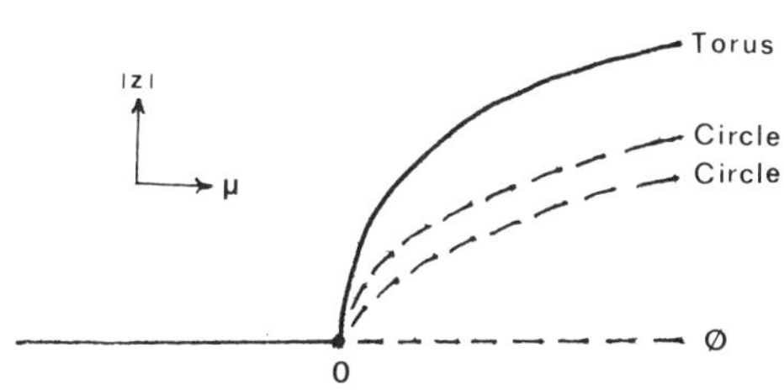

FIG. 3. Branches of invariant manifolds for Hopf bifurcations with D_n symmetry, $n = 3$ or $n \geq 5$.

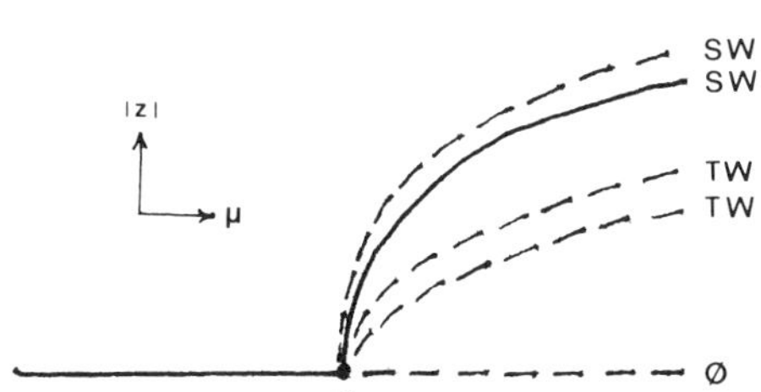

FIG. 4. Branches of periodic solutions for Hopf bifurcations with D_n symmetry, $n = 3$ or $n \geq 5$.

Golubitsky and Stewart (1986) (see also van Gils and Valkering 1985) prove that the branching diagram for periodic solutions is given by Figure 4, where the solid SW branch represents a finite number $m \geq 3$ of stable, periodic orbits, ("standing waves"), the broken SW branch represents m unstable periodic orbits (also "standing waves"), and

$$m = \begin{cases} n, & \text{if } n \text{ is odd}, \\ n/2, & \text{if } n \text{ is even}. \end{cases} \tag{2.10}$$

The stabilities of the standing wave solutions are determined by a higher-order term in the normal form that does not appear in (2.6). In the next section we show that all the standing wave solutions lie on the bifurcating torus.

Finally in case (c), if f possesses Z_n symmetry (1.5c) when $\nu \neq 0$, then provided

$$n = 3 \text{ or } n \geq 5,$$

the normal form with $k = 4$ is the same as (2.6), and the branching diagram in case (2.5), (2.7) is the same as Figure 2. However, as we discuss in the next section, the flow of the torus is different.

3. DYNAMICS ON BIFURCATING TORI

While the normal forms (2.4) and (2.7) are sufficient to determine the branching diagrams for invariant manifolds in Figures 1, 2, and 3, in cases (b) and (c) (D_n and Z_n symmetry) higher-order terms are needed to determine the dynamics on the bifurcating tori. In case (a) (SO(2) symmetry), however, the nature of the flow is not affected by higher-order terms. In this section we describe the flow on the tori near bifurcation in each case. We include a proof that in case (b), when the perturbed ($\nu \neq 0$) equation (1.1) has D_n symmetry (1.5b), the two families of periodic standing wave solutions found by Golubitsky and Stewart (1986) in fact lie on the bifurcating invariant torus if $n = 3$ or $n \geq 5$.

For the unperturbed equation (1.1) the vector field f possesses the full O(2) symmetry, and the bifurcating torus consists of a disjoint union of invariant circles (Schecter 1976). Every trajectory on the torus is a periodic orbit (a standing wave solution), and is invariant under a phase shift of half a period followed by a reflection. Any periodic orbit on the torus can be transformed into another one by a rotation θ (Golubitsky and Stewart 1985). Although normal hyperbolicity will cause the torus to persist for sufficiently small perturbations of the vector field, the flow on the torus for the unperturbed equation is nongeneric and is not structurally stable, and should not be expected to persist for perturbations with less symmetry than O(2).

In fact, for the perturbed equation (1.1) in case (a), where f possesses SO(2) symmetry (1.5b) but the reflection symmetry κ is broken when $\nu \neq 0$, if the normal form (2.6) satisfies (2.5) and (2.7), the flow on the bifurcating torus is quasiperiodic. For a trajectory z(t) on the torus, there is a smooth function $z^*(\mu, \nu, \theta_1, \theta_2)$ such that

$$z(t) = z^*(\mu, \nu, \omega_1 t, \omega_2 t).$$

Furthermore, the two fundamental frequencies ω_1, ω_2, are themselves smooth functions of the parameters μ, ν (Nagata 1986). Although the flow on the torus is no longer as trivial as for the unperturbed equation, it is still not structurally stable due to the remaining SO(2) symmetry. Also, if a smooth one-parameter path in the (μ,ν)-plane crosses the curve of Neimark-Sacker bifurcations $\{(\mu,\nu): \mu = \mu^*(\nu)\}$, the frequency ratios ω_2/ω_1 of the flows on the bifurcating tori will be smooth functions of the curve parameter. This is nongeneric, again due to the remaining SO(2) symmetry.

In case (b) where f has D_n symmetry (1.5b) when $\nu \neq 0$, we have the following result for the flow of the perturbed equation (1.1):

Theorem 1 (Hopf bifurcation with D_n symmetry). Consider the differential equation (1.1), where f possesses D_n symmetry (1.5b) with n = 3 or n ≥ 5, and has normal form (2.8) with coefficients satisfying (2.9). Let m be defined by (2.10). If $f \in C^{2m+1}$, then for $\mu > 0$ there exists an attracting invariant two-torus for (1.1). Furthermore, if an algebraic expression involving one higher-order term in the normal form does not vanish, then on the invariant torus there exist both families of m periodic standing wave solutions whose existence was proved by Golubitsky and Stewart (1986). Except for the m unstable standing wave solutions, every trajectory on the torus approaches asymptotically (as $t \to \infty$) one of the stable standing wave solutions.

The conclusion of this theorem is included in a remark in Golubitsky and Stewart (1986), but without proof. See Figure 5.

Proof of Theorem 1. Assume the hypotheses of Theorem 1: then using the expression given by Golubitsky and Stewart (1986) for an arbitrary $D_n \times S^1$ - symmetric mapping, it follows that the normal form of (2.8) to order k = 2m can be written as

$$\dot{z}_1 = z_1[i\omega(\nu) + \mu a(\nu) + c_1(\nu)|z_1|^2 + c_2(\nu)|z_2|^2 + h(\mu,\nu,|z_1|^2,|z_2|^2)] + d(\nu)\bar{z}_1^{m-1}z_2^m + O(|\mu||z|^{2m-1} + |z|^{2m+1}),$$

(3.1)

$$\dot{z}_2 = z_2[i\omega(\nu) + \mu a(\nu) + c_1(\nu)|z_2|^2 + c_2(\nu)|z_1|^2 + h(\mu,\nu,|z_2|^2,|z_1|^2)] + d(\nu)z_1^m\bar{z}_2^{m-1} + O(|\mu||z|^{2m-1} + |z|^{2m+1}),$$

where $h(\mu,\nu,|z_1|^2,|z_2|^2) = O(|\mu|^2 + |\mu||z|^2 + |z|^4)$, $d(\nu)$ is complex, and $d(0) = 0$. We rewrite (3.1) in rescaled polar coordinates, with

$$z_1 = \epsilon r_1 e^{i\theta_1},\ z_2 = \epsilon r_2 e^{i\theta_2},\ \mu = \epsilon^2\hat{\mu},$$

where $\epsilon > 0$ is a small parameter ($\epsilon \propto |\mu|^{1/2}$), and obtain

$$\dot{r}_1 = \epsilon^2 r_1 R(\nu,\epsilon^2,r_1^2,r_2^2) + \epsilon^{2m-2}r_1^{m-1}r_2^m R_1(\nu,\theta_1-\theta_2) + O(\epsilon^{2m}),$$

$$\dot{r}_2 = \epsilon^2 r_2 R(\nu,\epsilon^2,r_2^2,r_1^2) + \epsilon^{2m-2}r_1^m r_2^{m-1} R_2(\nu,\theta_1-\theta_2) + O(\epsilon^{2m}),$$

(3.2)

$$\dot{\theta}_1 = \omega + \epsilon^2\Theta(\nu,\epsilon^2,r_1^2,r_2^2) + \epsilon^{2m-2}r_1^{m-2}r_2^m\Theta_1(\nu,\theta_1-\theta_2) + O(\epsilon^{2m}),$$

$$\dot{\theta}_2 = \omega + \epsilon^2\Theta(\nu,\epsilon^2,r_2^2,r_1^2) + \epsilon^{2m-2}r_1^m r_2^{m-2}\Theta_2(\nu,\theta_1-\theta_2) + O(\epsilon^{2m}),$$

where

$$R(\nu,0,r_1^2,r_2^2) = \hat{\mu}\ \mathrm{Re}\ a(\nu) + r_1^2\mathrm{Re}\ c_1(\nu) + r_2^2\mathrm{Re}\ c_2(\nu),$$

$$\Theta(\nu,0,r_1^2,r_2^2) = \hat{\mu}\ \mathrm{Im}\ a(\nu) + r_1^2\mathrm{Re}\ c_1(\nu) + r_2^2\mathrm{Re}\ c_2(\nu),$$

$$R_1(\nu,\theta_1-\theta_2) = \mathrm{Re}\ d(\nu)\cos m(\theta_1-\theta_2) + \mathrm{Im}\ d(\nu)\sin m(\theta_1-\theta_2),$$

$$R_2(\nu,\theta_1-\theta_2) = \mathrm{Re}\ d(\nu)\cos m(\theta_1-\theta_2) - \mathrm{Im}\ d(\nu)\sin m(\theta_1-\theta_2),$$

$$\Theta_1(\nu,\theta_1-\theta_2) = \mathrm{Im}\ d(\nu)\cos m(\theta_1-\theta_2) - \mathrm{Re}\ d(\nu)\sin m(\theta_1-\theta_2),$$

$$\Theta_2(\nu,\theta_1-\theta_2) = \mathrm{Im}\ d(\nu)\cos m(\theta_1-\theta_2) + \mathrm{Re}\ d(\nu)\sin m(\theta_1-\theta_2).$$

We first obtain an asymptotic expression for the invariant torus to order ϵ^{2m-4}, in order to subsequently determine the flow on the torus. By (2.9) and the implicit function theorem, for any fixed $\hat{\mu}$ and ν the equation

$$R(\nu,\epsilon^2,X,X) = 0$$

has a unique solution $X^*(\nu,\epsilon^2)$ defined in some neighbourhood of $X = 0$, $\epsilon^2 = 0$. If we fix

$$\hat{\mu} = -\ \mathrm{Re}[c_1(\nu)+c_2(\nu)]/\mathrm{Re}\ a(\nu) > 0,$$

then $X^*(\nu,\epsilon^2) = 1 + O(\epsilon^2)$. Now let

$$r_0(\nu,\epsilon^2) = [X^*(\nu,\epsilon^2)]^{1/2} = 1 + O(\epsilon^2), \tag{3.3}$$

and define new coordinates u_1, u_2 by

$$r_1 = r_0(\nu,\epsilon^2) + \epsilon^{2m-4}u_1, \quad r_2 = r_0(\nu,\epsilon^2) + \epsilon^{2m-4}u_2.$$

Then u_1, u_2 satisfy

$$\begin{aligned} \dot u_1 &= \epsilon^2[2u_1 \operatorname{Re} c_1(\nu) + 2u_2 \operatorname{Re} c_2(\nu) + R_1(\nu,\theta_1-\theta_2)] + O(\epsilon^4), \\ \dot u_2 &= \epsilon^2[2u_1 \operatorname{Re} c_2(\nu) + 2u_2 \operatorname{Re} c_1(\nu) + R_2(\nu,\theta_1-\theta_2)] + O(\epsilon^4). \end{aligned} \tag{3.4}$$

The quantities in the square brackets in (3.4) vanish if

$$\begin{aligned} u_1 = u_1^*(\nu,\theta_1-\theta_2) &= \frac{R_2(\nu,\theta_1-\theta_2) \operatorname{Re} c_2(\nu) - R_1(\nu,\theta_1-\theta_2) \operatorname{Re} c_1(\nu)}{2 \operatorname{Re}[c_1(\nu)-c_2(\nu)] \operatorname{Re}[c_1(\nu)+c_2(\nu)]}. \\ u_2 = u_2^*(\nu,\theta_1-\theta_2) &= \frac{R_1(\nu,\theta_1-\theta_2) \operatorname{Re} c_2(\nu) - R_2(\nu,\theta_1-\theta_2) \operatorname{Re} c_1(\nu)}{2 \operatorname{Re}[c_1(\nu)-c_2(\nu)] \operatorname{Re}[c_1(\nu)+c_2(\nu)]}. \end{aligned} \tag{3.5}$$

Thus if

$$u_1 = u_1^*(\nu,\theta_1-\theta_2) + \rho_1, \quad u_2 = u_2^*(\nu,\theta_1-\theta_2) + \rho_2$$

where u_1^*, u_2^* are as in (3.5), we have the system

$$\begin{aligned} \dot\rho_1 &= 2\epsilon^2[\rho_1 \operatorname{Re} c_1(\nu) + \rho_2 \operatorname{Re} c_2(\nu)] + O(\epsilon^4), \\ \dot\rho_2 &= 2\epsilon^2[\rho_1 \operatorname{Re} c_2(\nu) + \rho_2 \operatorname{Re} c_1(\nu)] + O(\epsilon^4), \\ \dot\theta_1 &= \omega + \epsilon^2\, \Theta(\nu,0,1,1) + O(\epsilon^4) \\ \dot\theta_2 &= \omega + \epsilon^2\, \Theta(\nu,0,1,1) + O(\epsilon^4) \end{aligned} \tag{3.6}$$

Since the eigenvalues $\operatorname{Re}[c_1(\nu) + c_2(\nu)]$, $\operatorname{Re}[c_1(\nu) - c_2(\nu)]$ of the matrix

$$\begin{bmatrix} \operatorname{Re} c_1(\nu) & \operatorname{Re} c_2(\nu) \\ \operatorname{Re} c_2(\nu) & \operatorname{Re} c_1(\nu) \end{bmatrix}$$

are both negative by (2.9), it follows from invariant manifold theory (Chow and Hale 1982, Section 12.5) that (3.6) possesses a smooth attracting invariant torus

$$\rho_1 = \rho_1^*(\nu,\epsilon^2,\theta_1,\theta_2) = O(\epsilon^2), \quad \rho_2 = \rho_2^*(\nu,\epsilon^2,\theta_1,\theta_2) = O(\epsilon^2)$$

for sufficiently small ϵ^2. Hence (3.2) possesses an attracting invariant torus

$$\begin{aligned} r_1 = r_1^*(\nu,\epsilon^2,\theta_1,\theta_2) &= r_o(\nu,\epsilon^2) + \epsilon^{2m-4}u_1^*(\nu,\theta_1-\theta_2) + O(\epsilon^{2m-2}), \\ r_2 = r_2^*(\nu,\epsilon^2,\theta_1,\theta_2) &= r_o(\nu,\epsilon^2) + \epsilon^{2m-4}u_2^*(\nu,\theta_1-\theta_2) + O(\epsilon^{2m-2}), \end{aligned} \tag{3.7}$$

for sufficiently small ϵ^2, where $r_o(\nu,\epsilon^2)$ is given by (3.3), and

$u_j^*(\nu,\theta_1-\theta_2)$, $j = 1,2$, are given by (3.5).

If we now substitute (3.7) into the equations for θ_1 and θ_2 in (3.2), we will be able to determine the flow on the torus represented by (3.7). For convenience we first define new angular variables ϕ and ψ by

$$\phi = \theta_1-\theta_2, \quad \psi = \theta_2. \tag{3.8}$$

Considered as a change of coordinates in R^2, (3.8) is invertible, has unit determinant, and preserves the two-dimensional lattice $(2\pi Z)^2$. Thus (3.8) defines a valid change of coordinates on the two-torus represented by (3.7). From (3.7) and the last two equations of (3.2) we obtain

$$\begin{aligned} \dot\phi &= -2\epsilon^{2m-2} \sin m\phi \, \mathrm{Re}\{[c_1(\nu)-c_2(\nu)]\overline{d(\nu)}\}/ \mathrm{Re}[c_1(\nu)-c_2(\nu)]+O(\epsilon^{2m}) \\ \dot\psi &= \omega + O(\epsilon^2). \end{aligned} \tag{3.9}$$

Thus, if we assume

$$\mathrm{Re}[c_1(\nu)-c_2(\nu)]\overline{d(\nu)} \neq 0, \tag{3.10}$$

then by the implicit function theorem, there exist $2m$ distinct periodic orbits on the torus,

$$\begin{aligned} \phi &= \phi_j(t) = j\pi/m + O(\epsilon^2), \\ \psi &= \psi_j(t) = \psi_j(0) + \omega(\nu)t + O(\epsilon^2), \quad j = 0,1,\dots,2m-1, \end{aligned} \tag{3.11}$$

for ϵ^2 sufficiently small. Furthermore, from (3.2) and (3.9) it follows that the Floquet exponents of the periodic orbits (3.11) are proportional to

$$\begin{aligned} &2\epsilon^2 \,\mathrm{Re}[c_1(\nu)+c_2(\nu)] + O(\epsilon^4), \\ &2\epsilon^2 \,\mathrm{Re}[c_1(\nu)-c_2(\nu)] + O(\epsilon^4), \\ &2m\epsilon^{2m-2}(-1)^{j+1} \,\mathrm{Re}\{[c_1(\nu)-c_2(\nu)]\overline{d(\nu)}\}/\mathrm{Re}[c_1(\nu)-c_2(\nu)] + O(\epsilon^{2m}) \end{aligned} \tag{3.12}$$

for $j = 0,1,\dots,2m-1$ and thus the periodic orbits (3.11) alternate between stable and unstable ones for ϵ^2 sufficiently small. By uniqueness of the periodic orbits, the orbits (3.12) on the torus are the same as the standing wave solutions of Golubitsky and Stewart (1986). Note that the Floquet exponents (3.12) agree with those calculated previously by Golubitsky and Stewart (1986). [ ]

From the proof of Theorem 1, it follows that trajectories on the torus are attracted much more weakly to the stable standing wave solutions than trajectories in C^2 are attracted to the torus, especially if n is large or if $d(\nu) \simeq 0$. If the normal form coefficients have signs other than those in (2.9), then there still exists a branch of invariant tori, with the

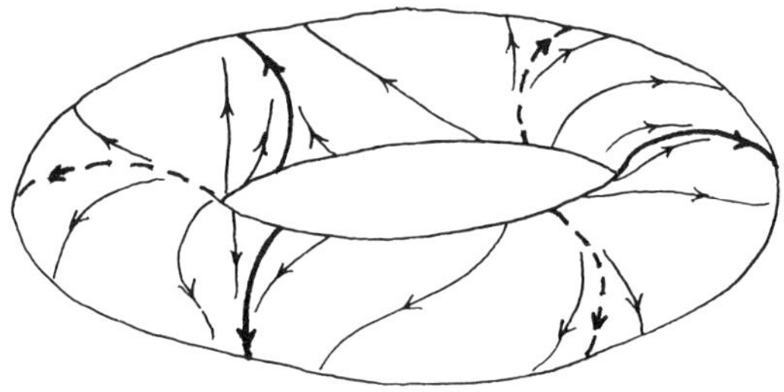

FIG. 5. Schematic representation of invariant torus with m stable periodic oribts and m unstable ones (see Golubitsky and Stewart 1986).

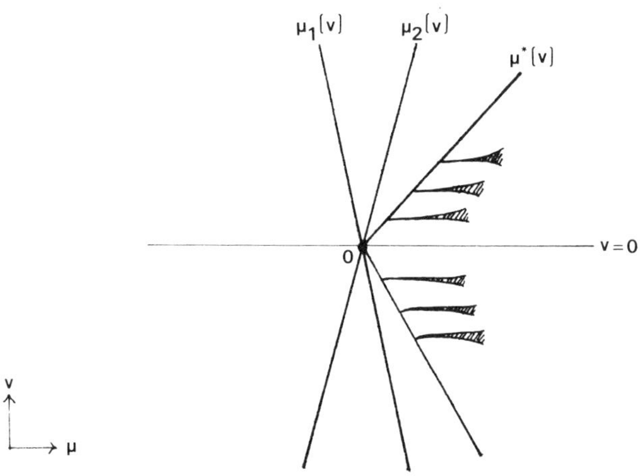

FIG. 6. Arnol'd tongues (shaded regions) for perturbations with Z_n symmetry, $n = 3$ or $n \geq 5$. The curves $\mu = \mu^*(\nu)$ correspond to the secondary bifurcations of invariant tori, as in Fig. 2. Inside the tongues there is phase locking; outside the tongues there is irrational quasiperiodic flow.

flow restricted to each torus satisfying (3.9), but the tori themselves may not be attracting.

Finally, in case (c) where (1.1) possesses Z_n symmetry (1.5c) when $\nu \neq 0$, if $n = 3$ or $n \geq 5$ and f is sufficiently smooth, either quasiperiodic flow or phase locking can occur, depending on the size of the perturbation:

Theorem 2. Consider the differential equation (1.1), where f possesses O(2) symmetry (1.2) when $\nu = 0$, and possesses Z_n symmetry (1.5c), $n = 3$ or $n \geq 5$, when $\nu \neq 0$. Let the normal form of f be given by (2.6) with coefficients satisfying (2.5) and (2.7). If $f \in C^5$, then there exist curves $\{(\mu,\nu): \mu = \mu^*(\nu), \nu > 0\}$ and $\{(\mu,\nu): \mu = \mu^*(\nu)\ \nu < 0\}$ defined for $|\nu|$ sufficiently small, with $\lim_{\nu \to 0} \mu^*(\nu) = 0$, that correspond to secondary Neimark-Sacker bifurcations of attracting two-tori for (1.1). For (μ,ν) belonging to a set of positive two-dimensional Lebesgue measure the flow on the corresponding bifurcating torus is conjugate to a quasiperiodic flow and has an irrational (Poincaré) rotation number. Furthermore, if np/q is sufficiently small, with p,q relatively prime integers and $q \geq 5$, and if $f \in C^{2q+1}$, then generically there is a narrow open set (Arnol'd tongue) in (μ,ν)-space on which the flow on the corresponding bifurcating torus is asymptotically periodic (phase locked) and has rotation number np/q.

A proof of Theorem 2 will appear elsewhere. If we take a generic one-parameter smooth path $(\mu(\sigma),\nu(\sigma))$ that crosses one of the curves of Neimark-Sacker bifurcations into the region $\mu > \mu^*(\nu)$ with increasing curve parameter σ, then a periodic travelling wave solution will lose stability and an attracting invariant torus will bifurcate supercritically. The flow on the torus will be irrational, and hence ergodic, for σ belonging to a set of positive one-dimensional Lebesgue measure, but there will also be phase locked, asymptotically periodic flows for σ belonging to short open intervals. See Figure 6. Except for the restrictions on the possible resonances due to the Z_n symmetry, this behaviour is like that of generic bifrcations of tori from periodic orbits without strong resonances.

4. DISCUSSION

We have described the changes that occur in the nature of flows on the bifurcating invariant tori when a bifurcation originally possessing O(2) symmetry is subjected to perturbation having only SO(2), D_n or Z_n symmetry. If only the reflection symmetry of O(2) is broken by the perturbation, then the flows on the invariant tori are quasiperiodic, with no phase locking (in fact, the flows are "parallel"). If the reflection symmetry is kept but the continuous rotation symmetry is broken, then the flow on each torus locks onto a finite number of periodic standing wave solutions. If both the reflection and the continuous rotation symmetries are broken then in the two-dimensional parameter space there is a set of positive measure for which the corresponding flows on the tori are conjugate to quasiperiodic flows and have irrational rotation numbers, and hence these flows are ergodic. In addition, there are open Arnol'd tongues in the parameter space which correspond to the phase locking of the torus flows onto weakly resonant periodic orbits.

As $n \to \infty$, the flows on the tori with D_n symmetry begin to resemble the flows on the tori with O(2) symmetry, and similarly, the flows on the tori with Z_n symmetry begin to resemble the flows on the tori with SO(2) symmetry. For D_n symmetry, the number of stable standing wave solutions goes to ∞ as $n \to \infty$, while the distances between the stable standing wave solutions approach 0. Thus if n is very large, it would be difficult to distinguish between O(2) and D_n symmetry by observing the flows on bifurcating tori. For Z_n symmetry, the Arnol'd tongues become vanishingly thin and increasingly sparse as $n \to \infty$ so that the dependence of the rotation numbers for the flows on bifurcating tori may appear to depend almost as smoothly on parameters as if there was SO(2) symmetry.

In applications, the perturbations can be due to imperfections or nonuniformities in the externally imposed boundary conditions. For example, in magnetoconvection, there may be ripples in the top and bottom of the fluid layer, or there may be similar nonuniformities in the imposed heating or magnetic field. The analysis suggests that such nonuniformities would have a definite effect on the qualitative nature of the time evolution of the system, although careful measurements may have to be taken to observe the effects. We remark, however, that the foregoing results apply near the bifuraction, for small perturbations. Numerical studies conducted by J. Celetti at the University of Guelph show that phase locking appears to persist for some distance from the bifuraction, for perturbations of moderate size.

REFERENCES

Chossat, P., and Iooss, G. (1985): "Primary and secondary bifurcations in the Couette-Taylor problem," Japan J. Appl. Math. 2, 37-68.

Chow, S.-N., and Hale, J. K. (1982): Methods of Bifurcation Theory, Springer-Verlag, New York.

Erneux, T., and Herschkowitz-Kaufman, M. (1977): "Rotating waves as asymptotic solutions of a model chemical reaction," J. Chem. Phys. 66, 248-250.

van Gils, S. A. (1983): "Hopf bifurcation and symmetry: Travelling and standing waves on the circle," Preprint TW 249/83, Mathematisch Centrum, Amsterdam.

van Gils, S. A., and Mallet-Paret, J. (1986): "Hopf bifurcation and symmetry: Travelling and standing waves on the circle," preprint.

van Gils, S. A., and Valkering, T. (1985): "Hopf bifurcation and symmetry: Standing and travelling waves in a circular chain," Rapportnr. 279, Vrije Universiteit Amsterdam.

Golubitsky, M., and Stewart, I. (1985): "Hopf bifurcation in the presence of symmetry," Arch. Rational Mech. Anal. 87, 107-165.

Golubitsky, M., and Stewart I. (1986): "Hopf bifurcation with dihedral group symmetry: Coupled nonlinear oscillators," in Golubitsky, M., and Guckenheimer, J., eds., Multiparameter Bifurcation Theory, Contemporary Mathematics 56, 131-173.

Guckenheimer, J., and Holmes, P. (1983): Nonlinear Oscillators, Dynamical Systems, and Bifurcations of Vector Fields, Springer-Verlag, New York.

Nagata, W. (1986): "Symmetric Hopf bifurcations and magnetoconvection," in Golubitsky, M., and Guckenheimer, J., eds., Multiparameter Bifurcation Theory, Contemporary Mathematics 56, 237-265.

Peplowski, P. (1983): "Bifurcation of periodic solutions in a laser with saturable absorber," Physica 6D, 364-374.

Renardy, M. (1982): "Bifurcation from rotating waves," Arch. Rational Mech. Anal. 79, 49-84.

Ruelle, D. (1973): "Bifurcations in the presence of a symmetry group," Arch. Rational Mech. Anal. 51, 136-152.

Schecter, S. (1976): "Bifurcations with symmetry," in Marsden, J. E., and McCracken, M., The Hopf Bifurcation and Its Applications, Springer-Verlag, New York.

DEPARTMENT OF MATHEMATICS AND STATISTICS
UNIVERSITY OF GUELPH
GUELPH, ONTARIO, CANADA
N1G 2W1

Canadian Mathematical Society
Conference Proceedings
Volume 8 (1987)

CHAOTIC MOTION OF A FORCED NONLINEAR OSCILLATOR

$\ddot{x} + ax + bx^2 + cx^3 + d = \varepsilon(\rho\cos\nu t + \beta\dot{x})$

N. Sri Namachchivaya and Harry H. Hilton

ABSTRACT. In this paper the behavior of a nonlinear oscillator $\ddot{x} + ax + bx^2 + cx^3 + d = \varepsilon(\rho\cos\nu t + \beta\dot{x})$, under the action of an external forcing function is investigated. Such equations describe the dynamic response of curved panels and shallow arches under small external periodic excitations. It is shown that such a nonlinear system with three dimensions in the extended phase-space can exhibit chaotic oscillations. The method of Melnikov is used to determine the analytical results for the critical parameters at which the dynamic system exhibits horseshoe chaos. The results obtained by the Melnikov technique are compared to those obtained by numerical integration.

1. INTRODUCTION. It is well known that a shell-type structure such as an arch, a curved panel or a shell, may become elastically unstable under a statically applied load. The form of instability in these structures is called "snap-through". The instability phenomenon involves a sudden transition of the structure from one stable equilibrium state to another non-adjacent stable equilibrium configuration. These static problems have been extensively studied in the literature.

When the loading is dynamic, the stability investigation requires the determination of the dynamic response of the structure described by its non-linear equations of motion. Mettler [1] examined the instability of a beam with small initial curvature that is subjected to harmonic load variations and has shown that the phenomenon of dynamic snap-through is somewhat analogous to the jump phenomenon in the theory of non-linear vibrations. Other works by Lock [2], Humphreys [3], [4], and Rehfield [5], report on their studies of the behavior of shallow arches and spherical shells under step and impulsive loadings. Some of the analytical solution techniques employed for periodic loadings are given in Bolotin's text [6].

In such dynamic systems the nonlinear equations of motion are three dimensional (n=3) in the extended phase space. It has been shown that nonlinear deterministic systems with dimension $n \geq 3$ may exhibit under certain conditions a form of complex, irregular behavior called "chaos". A dynamic system undergoing chaos may simultaneously posses a countably infinite number of periodic orbits as well as uncountably many nonperiodic orbits, and is extremely sensitive to variations in the initial conditions. Although a majority of examples in which chaos has been reported in the literature is based on numerical simulations, in certain cases the Melnikov method [7] has yielded some interesting analytical results for the critical parameters at which a dynamic system possesses one specific form of chaos, namely, the Smale horseshoe [8]. The Melnikov technique provides an analytical tool for ascertaining the presence of chaos in certain periodically forced nonlinear systems.

In this paper, the method of Melnikov [7] is applied in order to examine the parameter values at which the dynamic motion of curved panels and shallow arches under periodic excitation becomes chaotic. This method has been previously employed by Holmes [8] to determine the critical parameter values at which a periodically forced Duffing system exhibits chaotic behavior. Modified versions of the Melnikov method have been used by Holmes and Marsden [9,10,11] to examine parametrically excited systems, two degree-of-freedom systems and Arnold diffusion problems in $n \geq 3$ degree-of-freedom mechanical systems under certain configurations and parameter values.

2. FORMULATION OF PHYSICAL PROBLEMS. This section deals with the derivations of equations of motion for some physical systems. Transverse equations of motion of a curved panel and a shallow arch under periodic excitations are considered.

2.1. CURVED PANELS. Consider a shallow cylindrical shell of thickness h, radius R, width b and curvature parameter $k = b^2/Rh$. The shell edges are hinged to fixed supports. At time $t = 0$, the shell is subjected to a uniformly distributed pressure field of intensity $F(t)$, with mean value F_0 (see Figure 1). Under the assumption that the shell is sufficiently flat, the deflection is represented by the fundamental mode whose non-dimensional amplitude is given by $q(t)$. Lagrange's equation of motion for the shell can then be written as

$$\text{(1a)} \qquad \frac{d}{dt}\left(\frac{\partial T}{\partial \dot{q}}\right) + \frac{\partial D}{\partial \dot{q}} - \frac{\partial T}{\partial q} + \frac{\partial V}{\partial q} = Q(t) \ ,$$

where T is the kinetic energy, D the dissipation function, V the potential energy and Q the generalized force corresponding to the time dependent random loading. The expressions for T, D, V and Q(t) are found by the usual methods to be

(1b)
$$T = \frac{bmh^2}{4}\dot{q}^2 , \quad D = \frac{Dh^2\varepsilon}{2}\dot{q}^2 ,$$

$$V = \frac{Eh^5}{b^3(1-\nu^2)}\left[C_1\left(\frac{1}{2}q^2 + \frac{\alpha_o}{3}q^3 + \frac{\beta_o}{4}q^4\right) - \frac{2}{\pi}\lambda q\right] ,$$

$$Q(t) = \frac{2bh}{\pi}\left[F(t) - F_o\right] ,$$

where

$$C_1 = \frac{4k^2}{\pi^2} + \frac{\pi^4}{24} , \quad \alpha_o = -\frac{3}{2}\frac{\pi k}{C_1} ,$$

and

$$\beta_o = \frac{1}{8}\frac{\pi^4}{C_1} , \quad \lambda = \frac{F_o}{E}\left(\frac{b}{h}\right)^4(1-\nu^2) .$$

Here, m is the mass per unit area of the middle surface of the shell, ε is the coefficient of viscous damping, ν and E are Poisson's ratio and Young's modulus respectively. Substituting equations (1b) into equation (1a) gives

(1c)
$$\ddot{q} + 2\zeta\omega\dot{q} + \omega^2(q + \alpha_o q^2 + \beta_o q^3) - \frac{4F_o}{\pi mh} = \frac{4}{\pi mh}\xi(t) ,$$

where $\zeta\omega = \frac{\varepsilon}{m}$, $\omega^2 = \frac{2Eh^3}{b^4 m(1-\nu^2)}C_1 = \bar{\omega}^2 C_1$, and $\xi(t) = F(t) - F_o$.

The introduction of the following quantities

$$q_o = \frac{4k}{\pi^3} , \quad \lambda_o = \frac{32F_o}{m\pi^5\bar{\omega}^2} , \quad \lambda(\tau) = \frac{32\,\xi(t)}{m\pi^5\bar{\omega}^2}$$

$$\tau = \frac{\bar{\omega}\pi^2 t}{2\sqrt{2}} , \quad \beta = \frac{4\sqrt{2}\,\zeta\sqrt{C_1}}{\pi^2}$$

reduces equation (1c) to

(1d)
$$q'' + \beta q' + (1 + 2q_o^2)q - 3q_o q^2 + q^3 - \lambda_o = \lambda(\tau)$$

where the prime denotes differentiation with respect to the new time τ.

2.2. SHALLOW ARCHES. Consider a shallow, pinned, sinusoidal, elastic arch whose undeformed center line is given by $w_o = y_o \sin(\pi x/\ell)$, where x is measured horizontally from the left support as shown in Figure 2. The vertical loading on the arch consists of a distributed static component of intensity $W \sin(\pi x/\ell)$ on which is superimposed a time varying component of intensity $f(t) \sin(\pi x/\ell)$. The time variation of $f(t)$ is assumed to be harmonic of amplitude ρ and frequency ν i.e., $f(t) = \rho \sin(\nu t)$. The arch has constant mass m per unit length, cross-sectional area A, flexural stiffness EI, and cross-sectional radius of gyration $r = (I/A)^{1/2}$. It is assumed that the arch is in equilibrium under the distributed static load of intensity $W \sin(\pi x/\ell)$. At some initial instant, $t = 0$, let the arch be subjected to an additional harmonic load of intensity $f(t) \sin(\pi x/\ell)$. For a sufficiently shallow arch, the deformed center line at any subsequent time t may be described by $w = y(t) \sin(\pi x/\ell)$. The deflection of the crown is then given by $Z(t) = y_o - y(t)$, and the equation of motion relative to the unloaded configuration of the arch is

$$\text{(2a)} \qquad \ddot{Z} + (\beta_o/m)\dot{Z} + \frac{1}{m}\left(\frac{\pi}{\ell}\right)^4 \left\{\left(EI + \frac{1}{2}EAy_o^2\right)Z + \frac{1}{4}EAZ^3 - \frac{3}{4}EAy_oZ^2 - \left(\frac{\ell}{\pi}\right)^4 (W + f(t))\right\} = 0$$

For the following dimensionless quantities

$$\text{(2b)} \qquad q = Z/2r,\ q_o = y_o/2r,\ \lambda_o = \left(\frac{\ell}{\pi}\right)^4 \frac{W}{2rEI},\ \lambda(\tau) = \left(\frac{\ell}{\pi}\right)^4 \frac{f(\tau)}{2rEI}$$

$$\tau = \left(\frac{EI}{m}\right)^{1/2} \left(\frac{\pi}{\ell}\right)^2 t,\ \beta = \beta_o \left(\frac{1}{EIm}\right)^{1/2} \left(\frac{\ell}{\pi}\right)^2 ,$$

equation (2a) reduces to

$$\text{(2c)} \qquad q'' + \beta q' + (1 + 2q_o^2)q - 3q_oq^2 + q^3 - \lambda_o = \lambda(\tau) ,$$

where the prime denotes differentiation with respect to the new time τ.

In this section two physical problems are formulated and the equations of motion are given by equations (1d) and (2c). Furthermore, assuming small damping, i.e., $\beta = \varepsilon\zeta$, $|\varepsilon| \ll 1$, and a small intensity harmonic component $f(t)$ given by $\lambda(\tau) = \varepsilon\rho \cos \omega\tau$, one can in general write the equations of motion for both of these problems as

$$(3) \qquad q'' + (1 + 2q_o^2)\, q - 3q_o q^2 + q^3 - \lambda_o = \varepsilon\,(\rho \cos \omega\tau - \zeta q') \,.$$

Equation (3) will next be examined in order to determine the possible chaotic behavior of the physical systems.

3. UNPERTURBED SYSTEM. The unperturbed motion (for $\varepsilon = 0$ given by equation (3)) is governed by

$$(4) \qquad q'' + (1 + 2q_o^2)q - 3q_o q^2 + q^3 - \lambda_o = 0 \,.$$

For any value of the static load parameter λ_o in equation (4), the possible equilibrium configurations are given by the roots of the cubic equation

$$(5) \qquad q^3 - 3q_o q^2 + (1 + 2q_o^2)q - \lambda_o = 0$$

Equation (5) represents the static load-deflection curve of the arch center line as shown in Figure 3. The maximum and minimum values of the static load parameter λ_o and the corresponding deflection parameters q for which three equilibrium positions can be maintained are:

$$(6) \qquad \lambda_{max} = q_o + 2\,\bar{\alpha}^{\,3/2} \,, \qquad q_{max} = q_o - \bar{\alpha}^{\,1/2} \,,$$

$$(7) \qquad \lambda_{min} = q_o - 2\,\bar{\alpha}^{\,3/2} \,, \qquad q_{min} = q_o + \bar{\alpha}^{1/2} \,,$$

where $\quad \bar{\alpha} = (q_o^2 - 1) / 3 \,.$

The roots of equation (5) can be expressed in the form

$$(8) \qquad q_1 = q_o + 2\bar{\alpha}^{\,1/2} \cos \phi/3 \,, \quad q_{2,3} = q_o - 2\bar{\alpha} \cos (\pi \mp \phi) / 3 \,,$$

where $\quad \cos\phi = \left(\frac{\lambda_o - q_o}{2}\right) (\bar{\alpha})^{-3/2} \,, \quad$ provided $q_o > 1$.

Furthermore, for all three equilibrium configurations corresponding to real roots, the load parameter λ_o must satisfy the inequality $\lambda_{min} < \lambda_o < \lambda_{max}$. Let c represent one of the equilibrium configurations defined by equation (8). By putting $q = c + \eta$, one can write the variational equation about $q = c$ as

$$(9) \qquad \eta'' - \omega_o^2\, \eta + \alpha\eta^2 + \eta^3 = \varepsilon\,(\rho \cos \omega\, \tau - \zeta\eta') \,,$$

where

$$\alpha = 3(c-q_o) ,$$

$$-\omega_o^2 = 3c^2 - 6q_o c + 1 + 2q_o^2 = 3(c-q_{max})(c - q_{min}) .$$

It is obvious that for $q_{min} < c < q_{max}$, $\omega_o^2 > 0$ implies that the linearized variational equation about $\eta = 0$ has $\pm\,\omega_o$ as its eigenvalues. This trivial solution, therefore, corresponds to a saddle point in the phase-plane. The nontrivial solutions of the homogeneous equation corresponding to equation (9), are given by

$$(10) \qquad \eta_{2,3} = -\frac{\alpha}{2} \pm \left[\left(\frac{\alpha}{2}\right)^2 + \omega_o^2\right]^{1/2}$$

and are always stable, provided $\omega_o^2 > 0$. Thus, one has a system whose unperturbed part possesses a homoclinic orbit or separatrix.

The Melnikov technique provides an analytical tool for ascertaining the presence of chaos in certain periodically forced nonlinear systems. This technique provides information about the behavior of the perturbed periodic system based on a calculation of the distance between two branches of the separatrix involving trajectories of the unperturbed system. In the following section some specific forms of the results used in this work are stated without proof. A more complete review of the method of Melnikov may be found, for example, in Guckenheimer and Holmes [12].

4. METHOD OF MELNIKOV. The general theory developed by Holmes [8] and Holmes and Marsden [10] following Melnikov [7] applies to systems of the form (see also Guckenheimer and Holmes [12])

$$(11) \qquad \dot{x} = f(x) + \varepsilon g_\mu(x,t), \qquad x = \binom{u}{v} \in R^2 ,$$

where $f(x) = [\partial H/\partial v, \ -\partial H/\partial u]$ is a Hamiltonian vector field and $g_\mu(\cdot,t) = g_\mu(\cdot,t + T)$ is a vector function, T-periodic in t, and depends upon the parameter μ. The functions f and g_μ are assumed sufficiently smooth in their arguments, and in addition, it is assumed that the unperturbed flow given by $\dot{x} = f(x)$ has a homoclinic orbit $q^o(t)$ to a hyperbolic saddle point whose energy level is h_o, and the interior of which is filled with periodic orbits of periods T_β and energy levels h_β. Furthermore, $T_\beta \to \infty$ monotonically as $\beta \to o$. Under these conditions Melnikov has shown that a function $\Delta_\varepsilon(t_o,t_o)$ can be derived for systems such as (11), so that the value of the integral

$$(12) \qquad \Delta_\varepsilon (t_o, t_o) = \int_{-\infty}^{\infty} f(q^o(t)) \wedge g(q^o(t), t + t_o) dt$$

where $f \wedge g$ denotes the 'wedge' product of the vectors f and g, is a measure of the "distance between the two branches of the separatrix" of the perturbed system. In particular, if $\Delta_\varepsilon (t_o, t_o)$ oscillates and takes positive and negative values (retransversal intersection) then it has been pointed out [8] that such intersections (or homoclinic points) imply the existence of a complex invariant set, i.e., a Smale "horseshoe", in the neighborhood of the saddle point. The perturbed system will then possess orbits of all periods in addition to dense nonperiodic orbits.

5. HAMILTONIAN FORMULATION AND THE MELNIKOV INTEGRAL. In order to apply the method of Melnikov to the present problem, equation (9) is rewritten in the form of a perturbation of a Hamiltonian system (with $\varepsilon = o$) which possesses a "saddle connection". Thus, letting $u = \eta$ and $v = \eta'$ in equation (9) yields

$$u' = v$$

$$(13) \qquad v' = \omega_o^2 u - \alpha u^2 - u^3$$

whose Hamiltonian can be written as

$$(14) \qquad H(u,v) = \frac{1}{2} v^2 - \frac{1}{2} w_o^2 u^2 + \frac{\alpha}{3} u^3 + \frac{1}{4} u^4$$

Since the Hamiltonian is conserved, the level curves H = constant are solutions of equations (13). Furthermore, the value of the Hamiltonian at the saddle point is $H(o,o) = o$, and, in this case, there exist four separatrices leaving and entering the saddle point, which form two loops Γ^0. The interior of Γ^0 is filled with a continuous family of periodic orbits $q^\alpha(t)$ of period T_α. The points of intersection of the saddle loop with the axis $v = o$ can be obtained as

$$(15) \qquad u_{1,2} = -\frac{2}{3}\alpha \pm [(\tfrac{2\alpha}{3})^2 + 2\omega_o^2]^{1/2}, \quad v_{1,2} = o$$

$$u_3 = o \quad , \quad v_3 = o$$

The unperturbed trajectory of the saddle loop at

$$u_\pm^o = -B \pm A, \; v = o \;, \quad \text{where } B = 2\alpha/3, \; A = (2\omega_o^2 + B^2)^{1/2}$$

can be obtained from either equations (13) or equation (14) in the form:

$$u_{\pm}(t-t_o) = \frac{C_{\pm} D e^{-\omega_o(t-t_o)}}{[(C_{\pm} e^{-\omega_o(t-t_o)} + 2B)^2 + D]},$$

(16) $$v_{\pm}(t-t_o) = C_{\pm} D \omega_o \frac{e^{-\omega_o(t-t_o)}[C_{\pm}^2 e^{-2\omega_o(t-t_o)} - 4A^2]}{[(C_{\pm} e^{-\omega_o(t-t_o)} + 2B)^2 + D]^2},$$

where $C_{\pm} = -4 [\frac{\omega_o^2}{B \pm A} + \frac{\alpha}{3}]$ and $D = 8\omega_o^2$.

Equations (16) are parametric equations for the saddle loop Γ^o, whose coordinates tend to $(u_{\pm}^o, 0)$ and $(0, 0)$ as $t \to t_o$ and $t \to \infty$, respectively.

As was discussed in the previous section, in order to compute the Melnikov integral defined by equation (12), the perturbed equations are written in the form:

(17) $$\dot{u} = v ,$$

$$\dot{v} = \omega_o^2 u - \alpha u^2 - u^3 + \varepsilon (\rho \cos \omega t - \zeta v) .$$

It can be shown from equations (17) that the Melnikov integral reduces to

(18) $$\Delta(t_o, t_o) = \int_{-\infty}^{\infty} v(t-t_o) [\zeta v(t-t_o) - \rho \cos \omega t] dt$$

$$= \varepsilon (\zeta I_1 - \rho I_2)$$

where

(19a) $$I_1 = \int_{-\infty}^{\infty} \frac{D^2 \omega_o C^2 e^{-2\tau} (C^2 e^{-2\tau} - 4A^2)^2}{[C^2 e^{-2\tau} + 4BCe^{-\tau} + 4A^2]^4} dt ,$$

(19b) $$I_2 = \int_{-\infty}^{\infty} \frac{DC e^{-\tau}(C^2 e^{-2\tau} - 4A^2) \cos(\Omega\tau + \omega t_o)}{[C^2 e^{-2\tau} + 4BCe^{-\tau} + 4A^2]^2} d\tau ,$$

$\tau = \omega_o (t-t_o)$ and $\Omega = \omega/\omega_o$.

If the integrals defined by equations (19) exist, are finite and are not both identically zero, then one can easily show that transversal zero exists. The first integral, I_1, of equation (19a) can be evaluated as

$$I_1 = D^2\omega_o \{\frac{1}{24 A^2} + 2 B^2 (\frac{8}{\Delta^2} + \frac{1}{3\Delta A^2}) - \frac{64 A^2 B}{\Delta^2\sqrt{\Delta}} (\frac{\pi}{2} - \tan^{-1}(\frac{4B}{\sqrt{\Delta}}))\}, \tag{20}$$

where

$$\Delta = 16 \left(A^2-B^2\right) = 32 \omega_o^2 .$$

Rewriting the second integral, I_2, of equation (19b), as

$$I_2 = \bar{D} \cos\left(\omega t_o\right) \mathrm{Re}\,[J] - \bar{D} \sin\left(\omega t_o\right) \mathrm{Im}\,[J] , \tag{21}$$

where

$$J = \int_{-\infty}^{\infty} \frac{\left(e^{-2\tau} - 4 \bar{A}^2\right) e^{-(1 + i\Omega)\tau}}{\left[e^{-2\tau} + 4 \bar{B} e^{-\tau} + 4 \bar{A}^2\right]^2} d\tau ,$$

$$\bar{A} = A/C, \quad \bar{B} = B/C, \quad \bar{D} = D/C ,$$

one can evaluate the integral J, after some manipulation, to obtain

$$J = \frac{\pi\Omega \sinh(\Omega\Theta) e^{i(\Omega \ell n 2\bar{A} - \pi/4)}}{2\bar{A} \sinh(\Omega\pi) \sin\Theta} , \tag{22}$$

where $\Theta = \cos^{-1}(B/A)$.

Thus, the second integral, I_2, takes the form:

$$I_2 = r \cos\left(\omega t_o + \psi\right) , \tag{23}$$

where
$$r = \frac{D\Omega}{A} \left(\frac{\pi}{2}\right) \frac{\sinh(\Theta\Omega)}{\sinh(\pi\Omega)} \left(\frac{1}{\sin\Theta}\right) ,$$

$$\psi = \frac{\pi}{4} - \Omega \ell n \left(2\bar{A}\right) .$$

It is worth noting that even though the value of ψ in equation (23) depends on C, $\left(\bar{A} = A/C\right)$, the value of r is independent of C. Thus, for both values of V, i.e. $V_{\pm}\left(t-t_o\right)$, the value of r remains the same. The Melnikov function $\Delta_\varepsilon\left(t_o, t_o\right)$ for the homoclinic orbit can now be expressed as

(24) $$\Delta_\varepsilon(t_o,t_o) = \xi I_1 - \rho r \cos(\omega t_o + \psi) .$$

For the right hand side of equation (24) to have a zero, $\tilde{t}_o$, it is necessary to have

(25) $$\left| \frac{\zeta I_1}{r} \right| \leqslant \rho .$$

It is easily verified that the zero, $\tilde{t}_o$, is transverse when the inequality (25) is strict for any nonzero finite frequency ω. When the inequality (25) becomes an equality, the zero $\tilde{t}_o$ is nontransverse and one has to consider higher order ε terms in the evaluation of the Melnikov integral. When ρ is such that inequality (25) is satisfied for a fixed frequency ω, one obtains a complicated phase portrait such as that shown in Figure 4.

6. SPECIAL CASE - DUFFING EQUATION. In this section the special case of equation (10) is considered when the quadratic nonlinearity is identically zero, i.e., $\alpha = 0$. This is realizable when $c = q_o$, which implies $\lambda = q_o$ from equation (5). The equation of motion can then be simplified to

(26) $$\eta'' - \omega_o^2 \eta + \eta^3 = \varepsilon(\rho \cos \omega t - \zeta \eta') ,$$

where $\omega_o^2 = q_o^2 - 1 > 0$ for $q_o > 1$.

Equation (26) is Duffing's equation with negative stiffness, and as remarked earlier, has been extensively studied using Melnikov's method [8, 10]. The parametric equation for the homoclinic orbit based at points $u_\pm^o = \pm\sqrt{2}\,\omega_o$ can be written as

$$u_+(t,t_o) = \sqrt{2}\,\omega_o \operatorname{sech}(\omega_o(t-t_o)) ,$$

$$v_+(t,t_o) = \sqrt{2}\,\omega_o^2 \operatorname{sech}(\omega_o(t-t_o)) \tanh(\omega_o(t-t_o)) ,$$

with $(u_-, v_-) = (-u_+, -v_+)$

It can be shown, by the use of equation (12), that the Melnikov function for the perturbed system (26) is

$$-\Delta(t_o,t_o) = \rho r \sin(wt_o) - \zeta I_2 ,$$

where $r = \sqrt{2}\,\pi\omega \operatorname{sech}\left(\frac{\omega\pi}{2\omega_o}\right)$, $I_2 = \frac{4}{3}\omega_o^3$.

It was shown by Holmes [8] that, for

$$|\rho| > \rho_c = \frac{4}{3}\frac{\zeta}{\sqrt{2}}\frac{\omega_o^3}{\pi w}\cosh\left(\frac{\omega\pi}{2\omega_o}\right) ,$$

transversal intersection of stable and unstable trajectories takes place, giving rise to homoclinic motions. It is worth pointing out that these results can also be deduced from equations (25) by setting $\alpha = 0$ in equations (20), (23) and (24).

7. NUMERICAL RESULTS. The chaotic behavior in this problem is illustrated by numerical simulation. For this purpose, an arch with initial rise $q_o = 1.41$, damping $\zeta = 0.25$ and static load $\lambda_o = 1.475$ was considered. For this configuration and loading, the critical value of ρ from the Melnikov function was obtained to be $\rho_c = 0.21$. Since this analytical value is a lower bound on the amplitude of the forcing term, the amplitudes were always taken to be greater than 0.21 in the numerical simulations. The results of the numerical simulations at $\rho = 0.35 > \rho_c$ are presented in Figures 4, 5 and 6. Figure 4 illustrates the phase plane motion of the trajectories and the corresponding time series for u is given in Figure 5. To gain further insight into the nature of the motion, and to illustrate the foldings of stable and unstable manifolds, an alternative format for presentation of results was considered, namely, a Poincare map. There, the phase plane motion shown in Figure 4 is sampled at discrete time intervals of period $n(2\pi)$ (since the forcing frequency is 1) with n taken from 50 to 1050, and the results are illustrated in Figure 6. In all of these presented results, the initial transient behavior was deleted. It can be seen from the Poincare map that an actual intersection of the stable and unstable manifolds definitely would have taken place at a smaller value of ρ, i.e., $\rho_c \leqslant \rho \leqslant 0.35$.

8. CONCLUSIONS. The effect of small periodic external excitations on the dynamic behavior of shallow arches and of curved panels has been considered. The term associated with the initial rise, q_o, enters the equation of motion as an imperfection. The resulting nonlinear equation possessing quadratic and cubic nonlinearities was theoretically investigated using the Melnikov technique and it was found that the transversal intersection of stable and unstable manifolds takes place at some critical parameter values. It was also shown by numerical simulation that at

parameter values slightly above this critical value, the nonlinear system indeed exhibits chaotic oscillation.

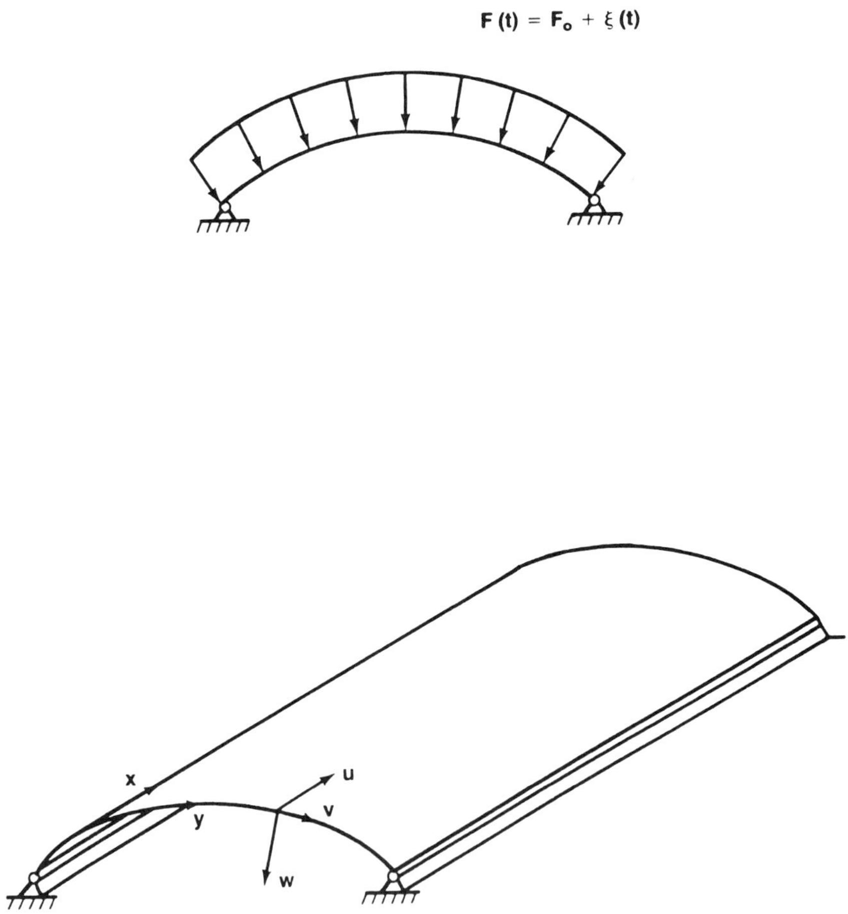

FIGURE 1. A Cylindrical Shell Panel

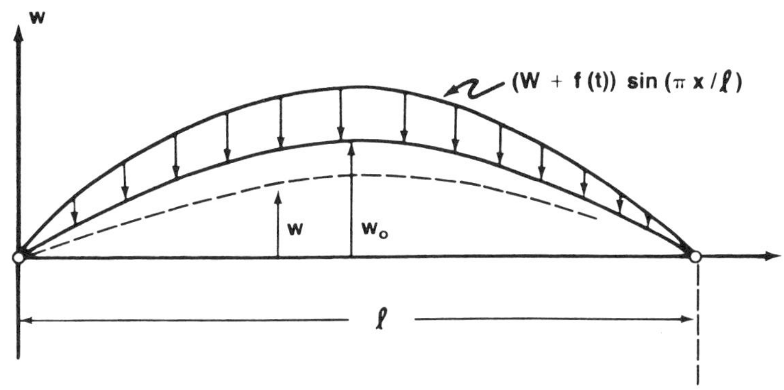

FIGURE 2. Geometry of the Arch

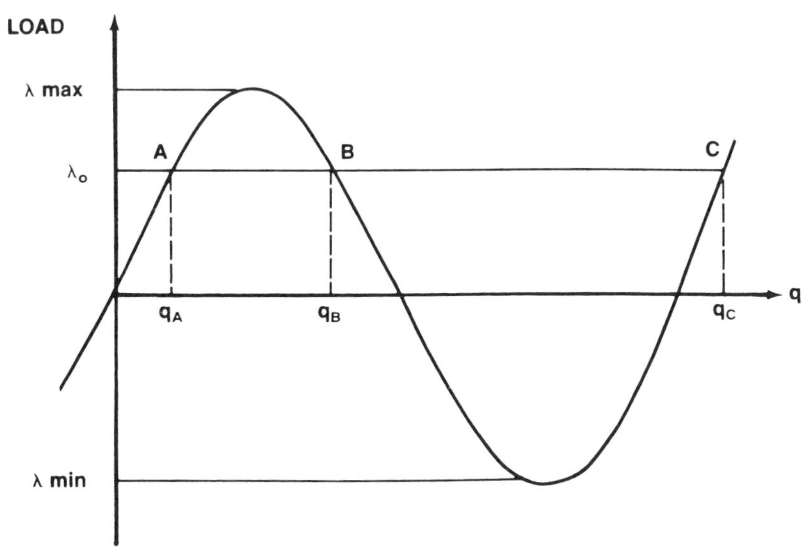

FIGURE 3. Load–Deflection Diagram for the Arch

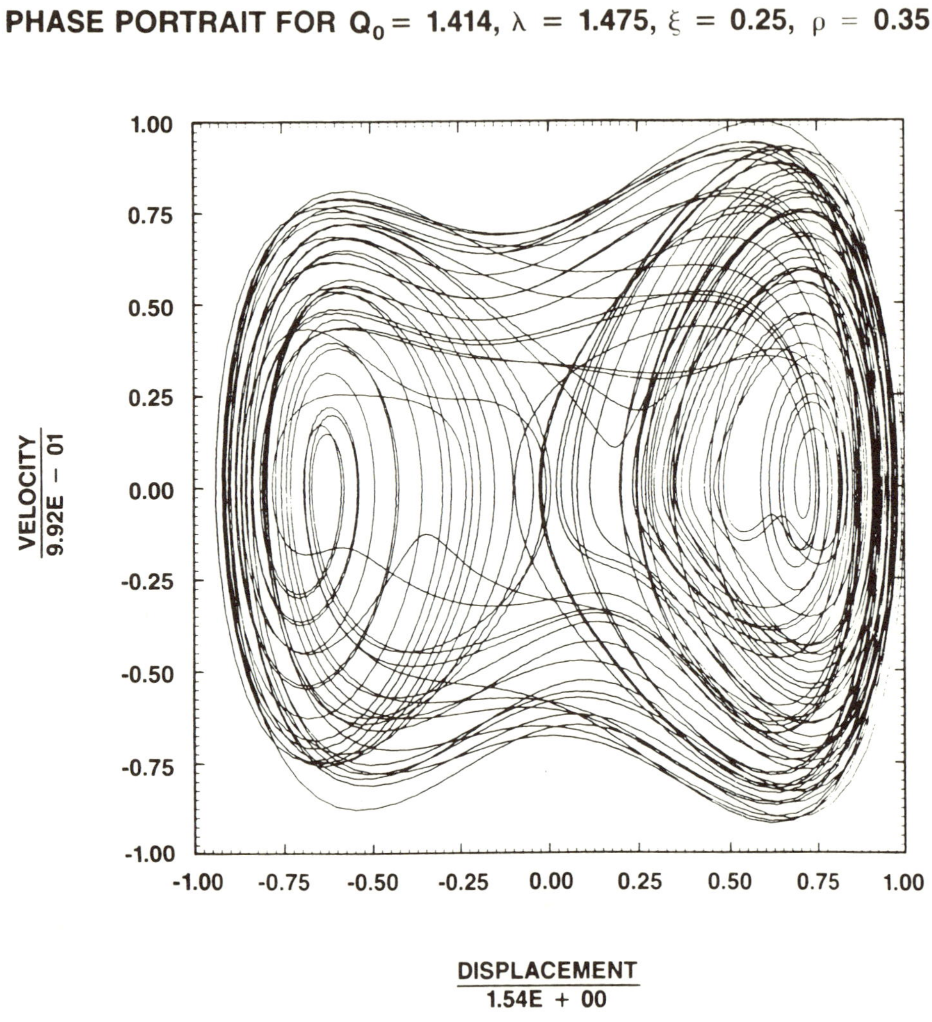

FIGURE 4. Phase-plane motion of the trajectories for $\rho = 0.35$

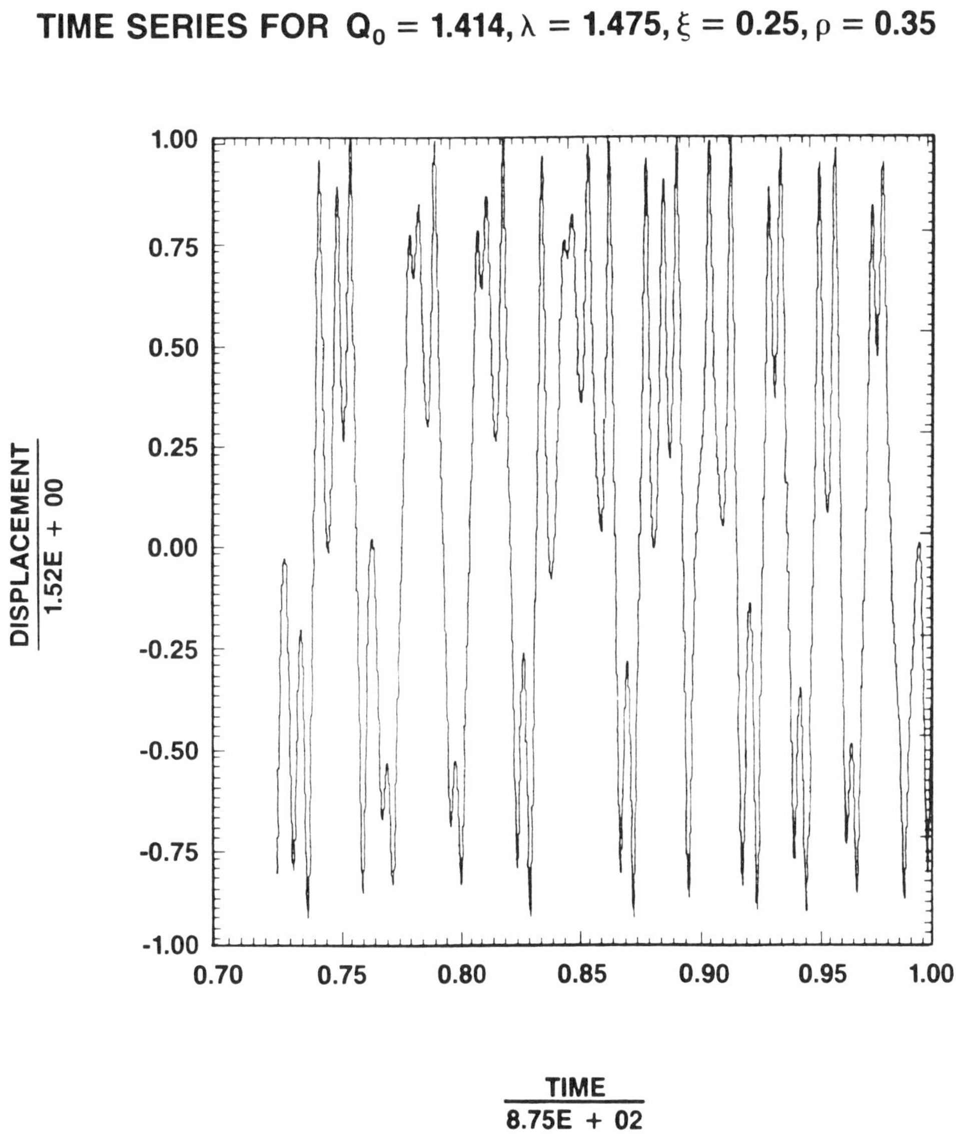

FIGURE 5. Response vs time for $\rho = 0.35$

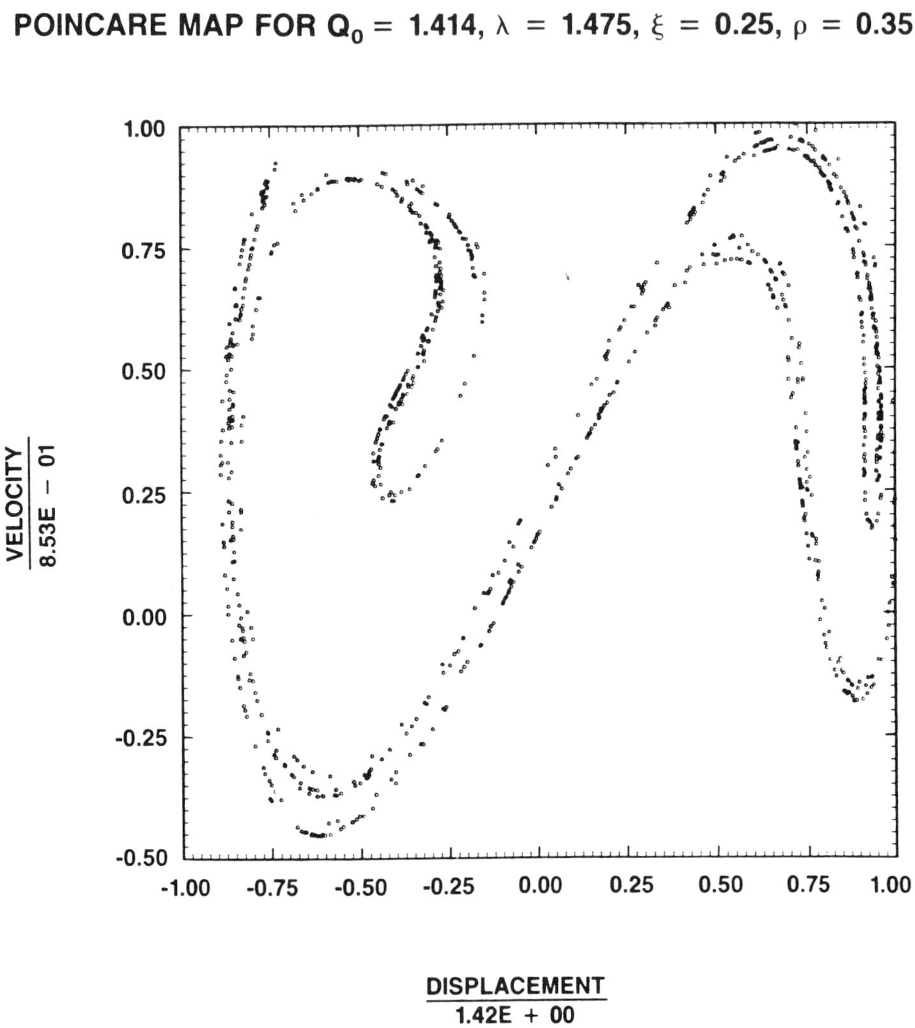

FIGURE 6. Poincare plot from T^{50} to T^{1050} for $\rho = 0.35$

REFERENCES

1. E. Mettler, Handbook of Engineering Mechanics, McGraw-Hill, New York, 1962.

2. M. H. Lock, The snapping of a shallow sinusoidal arch under a step pressure load. Air Force Systems Command, Los Angeles, California, Report No. TDR-469 (5240-10)-16, 1965.

3. J. S. Humphreys, On dynamic snap buckling of shallow arches. AIAA Journal, 4(5), 1966, 878-886.

4. C. S. Hsu, Stability of shallow arches against snap-through under time wise step loads. AIAA Journal, 35(1), 1969, 31-39.

5. L. H. Rehfield, Nonlinear flexural oscillations of shallow arches. AIAA Journal, 12, 1974, 91-93.

6. V. V. Bolotin, Dynamic Stability of Elastic Systems Holden-Day, San Francisco, 1965.

7. V. K. Melnikov, On the stability of the center for time periodic perturbations. Transaction of Moscow Mathematical Society, 12(1), 1963, 1-57.

8. P. Holmes, A nonlinear oscillator with a strange attractor. Philosophical Transaction of Royal Society, A 292, 1979, 419-448.

9. P. Holmes, Averaging and chaotic motions in forced oscillations. SIAM Journal of Applied Mathematics, 88, 1979, 65-80.

10. P. Holmes and J. Marsden, Horseshoes in perturbations of Hamiltonian systems with two degrees of freedom. Communications in Mathematical Physics, 82, 1982, 523-544.

11. P. Holmes and J. Marsden, Melnikov's method and Arnold diffusion for perturbations of integrable Hamiltonian systems. Journal of Mathematical Physics, 23(4), 1982, 669-675.

12. J. Guckenheimer and P. Holmes, Nonlinear Oscillations, Dynamical Systems, and Bifurcations of Vector Fields, Springer-Verlag, New York, 1983.

DEPARTMENT OF AERONAUTICAL AND ASTRONAUTICAL ENGINEERING
UNIVERSITY OF ILLINOIS AT URBANA-CHAMPAIGN
104 SOUTH MATHEWS AVENUE, 102 TRANSPORTATION BUILDING
URBANA, ILLINOIS 61821-2997

Canadian Mathematical Society
Conference Proceedings
Volume 8 (1987)

BOUNDARY LAYER PHENOMENA FOR A DIFFERENTIAL-DELAY EQUATION

Roger D. Nussbaum*
Mathematics Department
Rutgers University
New Brunswick, New Jersey

In this paper we shall present some refinements of results which have been obtained in [3,4,5,6,7] and which deal with the differential-delay equation

$$\varepsilon x'(t) = -x(t) + f(x(t-1)), \ \varepsilon > 0, \qquad (1)_\varepsilon$$

or equivalently (taking $\varepsilon = \lambda^{-1}$)

$$x'(t) = -\lambda x(t) + \lambda f(x(t-1)) \qquad (2)_\lambda$$

These equations have been studied by many authors and are of interest in the theory of dynamical systems and in applications to biology and optics. References to the literature and discussion of the motivation for considering eq. $(1)_\varepsilon$ can be found in [4,5,6,7,10]. Here we shall prove that, under suitable assumptions on f, eq. $(1)_\varepsilon$ has a periodic solution $x_\varepsilon(t)$ for all small ε, and that the number of relative extrema of $x_\varepsilon(t)$ on an interval $[0,\bar{q}]$ (where $\bar{q}$ is the minimal period of $x_\varepsilon(t)$) approaches ∞ as ε approaches 0. Furthermore, there exists $q \in (0,\bar{q})$ such that $x(o) = x(q) = x(\bar{q})$ and $x_\varepsilon(t) \neq 0$ for $t \in (0,\bar{q})$ and $t \neq q$ and the relative extrema referred to above occur in an interval of order ε containing q (the "boundary layer" of the title) or an interval of order ε containing $\bar{q}$.

*Partially supported by NSF Grant DMS 85-03316

We shall need various theorems and definitions from [4,5], and to keep this paper self-contained we shall recall these. The new results (including precise statements of the boundary layer phenomena described above) are given in Theorems 6 and 7, Corollaries 1 and 2 and Proposition 1 below.

Recall that a solution $x: \mathbb{R} \to \mathbb{R}$ of eq.$(2)_\lambda$ is called "slowly oscillating" if any two zeros of $x(t)$ are separated by a distance greater than 1, i.e., $|u-v| > 1$ whenever $x(u) = x(v) = 0$. Following notation in [6], a slowly oscillating solution $x(t)$ of eq. $(2)_\lambda$ will be called a P_N-solution if there exist numbers $z_j, j \in \mathbb{Z}$, such that $z_{j+1} > z_j$ for all $j, (-1)^j x(t) > 0$ on (z_j, z_{j+1}) and $x(t) = (-1)^N x(t+z_N)$ for all t. In [5], P_1-solutions are called S-solutions and P_2-solutions are called slowly oscillating periodic solutions.

Our interest here is in the properties of P_1-solutions and P_2-solutions of eq.$(1)_\varepsilon$. Roughly speaking, if f satisfies a "negative feedback condition" and $f'(0) < -1$, then for $\lambda > 0$ sufficiently large, eq.$(2)_\lambda$ has a P_2-solution. To make this precise, we need to recall some definitions which were introduced in [5] and which we list below as hypotheses H1, H2 and H3.

(H1). <u>The function</u> $f: \mathbb{R} \to \mathbb{R}$ <u>is continuous. There exist positve constants</u> A <u>and</u> B <u>such that</u>

$$f([-B,A]) \subset [-B,A] \quad \underline{\text{and}}$$
$$xf(x) < 0 \quad \underline{\text{for}} \quad x \in [-B,A] \quad \underline{\text{and}} \quad x \neq 0.$$

In this paper we shall reserve the letters A and B to indicate constants as in (H1).

In the statement of the next condition recall that a function f is "monotone decreasing" on an interval J if $x_1 < x_2$ implies $f(x_1) \geq f(x_2)$ for x_1 and x_2 in J and f is "strictly decreasing" on J if $x_1 < x_2$ implies $f(x_1) > f(x_2)$. The function f is "monotone increasing" or "strictly increasing" if $-f$ is monotone decreasing or strictly decreasing respectively.

(H2). *The function* $f: \mathbb{R} \to \mathbb{R}$ *is continuous*, $f(0) = 0$, f *is differentiable at* 0 *and*

$$f'(0) = -k < -1,$$

Furthermore f *is monotone decreasing on some neighborhood of* 0.

(H3). *The function* $f: \mathbb{R} \to \mathbb{R}$ *satisfies* (H1). *Furthermore there exist positive numbers* $a \leq A$ *and* $b \leq B$ *such that for any* $x_0 \in (0,A]$, *if* $x_n = f^n(x_0)$ *denotes the nth iterate of* x_0 *under* f, *then* $\lim_{n\to\infty} x_{2n} = a$ *and* $\lim_{n\to\infty} x_{2n+1} = -b$.

The condition $f(0) = 0$ in (H1)-(H3) is a normalization. In practice, $f(u) = u$ for some u and one is interested in periodic solutions $x(t)$ of $(2)_\lambda$ such that $x(t)-u = y(t)$ is a P_2-solution or P_1-solution of

$$y'(t) = \lambda y(t) + \lambda(f(y(t-1)+u) - u).$$

Verifying that $f(y + u) - u$ satisfies (H1),(H2), (H3) or other conditions may be nontrivial and is discussed in [7].

Before proceeding further we must recall some theorems from [4,5]. If $k > 1$ is as in (H2), define numbers ν_0 and λ_0 by

$$\cos(\nu_0) = -k^{-1}, \quad \nu_0 \in (\tfrac{\pi}{2}, \pi) \quad \text{and}$$
$$\lambda_0 = \nu_0 (\sqrt{k^2-1})^{-1}. \tag{3}$$

THEOREM 1 (See Theorem 1.1 *in* [5]). *Assume that* f *satisfies* (H1) *and* (H2). *Then for each* $\lambda > \lambda_0$, *where* λ_0 *is as in* eq.(3), *equation* $(2)_\lambda$ *has a* P_2-*solution* $x(t)$ *(dependent on* λ) *such that* $x(t) \in (-B,A)$ *for all* t. *If* f *is odd and* $\lambda > \lambda_0$, eq.$(2)_\lambda$ *has a* P_1-*solution solution* $x(t)$ *such that* $x(t) \in (-B,A)$ *for all* t.

We shall always normalize by assuming a P_2-solution $x(t)$ of eq.$(2)_\lambda$ satisfies $x(0) = 0$ and we shall always use letters q and $\bar{q}$ respectively to indicate the first and second zeros of $x(t)$, so $x(t) > 0$ for $0 < t < q$, $x(t) < 0$ for $q < t < \bar{q}$ and $\bar{q}$ is the minimal period of $x(t)$.

THEOREM 2 (See [5]). *Assume that* f *satisfies* (H1) *and* (H2). *There exist positive constants* c_1 *and* c_2 *(independent of* ε*) such that if* $x(t)$ *is a* P_2*-solution of* eq. $(1)_\varepsilon$ *for some* $\varepsilon > 0$, $x(t) \in (-B,A)$ *for all* t, *and* $x(t)$ *has minimal period* $\bar{q}$, *then*

$$2 + c_1\varepsilon \leq \bar{q} \leq 2 + c_2\varepsilon. \tag{4}$$

Suppose that $x(t)$ is a P_2-solution of eq. $(1)_\varepsilon$, $x(0) = 0$, and $x(t)$ has first and second zeros q and $\bar{q}$. If c and d are positive numbers, we say that $x(t)$ satisfies "property M between $-d$ and c" if there exist numbers s_1 and t_1, $0 < s_1 \leq t_1 < q$ and numbers s_2 and t_2, $q < s_2 \leq t_2 < \bar{q}$, such that x is monotone increasing on $[0,s_1]$, $x(t) \geq c$ for $s_1 < t < t_1$, x is monotone decreasing on $[t_1,s_2]$, $x(t) \leq -d$ for $s_2 < t < t_2$ and x is monotone increasing on $[t_2,\bar{q}]$.

Notice that $x(t)$ may satisfy property M between $-d$ and c even if $\sup_t x(t) < c$ or $\inf_t x(t) > -d$. If $\sup_t x(t) < c$, select r_1 so that $0 < r_1 < q$ and $x(r_1) = \sup_t x(t)$ and define $s_1 = r_1 = t_1$. Property M then requires that $x(t)$ be monotonic increasing on $[0,r_1]$ and monotonic decreasing on $[r_1,q]$.

The reason for the introduction of property M is that, as we shall see, P_2-solutions $x(t)$ of eq. $(1)_\varepsilon$ may have numerous local extrema, but these local extrema cannot correspond to values of $x(t)$ near 0.

THEOREM 3 (See [4,5]). *Assume that* f *satisfies* H1 *and* H2. *There exist positive constants* c *and* d, *depending only on* f, *such that if* $x(t)$ *is a* P_2*-solution of* eq. $(1)_\varepsilon$ *for some* $\varepsilon > 0$ *and* $x(t) \in (-B,A)$ *for all* t, *then* $x(t)$ *satisfies property* M *between* $-d$ *and* c. *Furthermore, there exist* $\varepsilon_1 > 0$ *and a constant* D *such that if* $x(t)$ *is a* P_2*-solution of* eq. $(1)_\varepsilon$ *for some* ε *with* $0 < \varepsilon \leq \varepsilon_1$, *then* $\sup_t x(t) > c$, $\inf_t x(t) < -d$ *and*

$$\mu(\{t \in [o,\bar{q}]: -d < x(t) < c\}) \leq D\varepsilon,$$

where μ *denotes Lebesgue measure on* $\mathbb{R}$ *and* $\bar{q}$ *is the minimal period of*

$x(t)$.

If a and b are as (H3) define a piecewise continuous function $x_\infty(t)$ of period 2 by

$$x_\infty(t) = \begin{cases} a & \text{for } 0 \leq t < 1 \\ -b & \text{for } 1 \leq t < 2 \end{cases} \tag{5}$$

and extend by periodicity.

THEOREM 4 (See [4,5]). Suppose that f satisfies (H2) and (H3). If (ε_n) is a sequence of positive reals such that $\lim_{n\to\infty} \varepsilon_n = 0$ and if $x_n(t)$ is a P_2-solution of eq. $(1)_{\varepsilon_n}$ such that $x_n(t) \in (-B,A)$ for all t, then there exists a subsequence x_{n_i} such that $x_{n_i}(t)$ converges uniformly to $x_\infty(t)$ on compact subsets of $\mathbb{R} - \mathbb{Z}$ (i.e., away from the integers).

The previous results suggest that the behaviour of P_2-solutions of eq. $(1)_\varepsilon$ near the integers may be of interest for ε small. To explore this point and to help prove the previous results the so-called "boundary layer equations" or "transition layer equations" were used in [5]. To be more precise, suppose that $x(t)$ is a P_2-solution of eq.$(1)_\varepsilon$ with $x(t) \in (-B,A)$ for all t. If $x(0) = 0$ and q and $\bar{q}$ are the first and second zeros of $x(t)$ and one writes $\bar{q} = 2 + 2r\varepsilon$, $y(t) = x(q - \varepsilon t)$ and $z(t) = x(\frac{1}{2}\bar{q} + q - \varepsilon t)$, one can easily verify that $y(t)$ and $z(t)$ satisfy the equations

$$\begin{aligned} y'(t) &= y(t) - f(z(t-r)) \quad \text{and} \\ z'(t) &= z(t) - f(y(t-r)). \end{aligned} \tag{6$_r$}$$

Note that Theorem 2 implies that there exist positive constants c_1 and c_2, independent of ε such that $c_1 \leq r \leq c_2$.

THEOREM 5 (See [5]). Assume that f satisfies (H1) and (H2). Suppose that (ε_n) is a sequence of positive numbers such that $\lim_{n\to\infty} \varepsilon_n = 0$ and for each ε_n let $x_n(t)$ be a P_2-solution of eq. $(1)_{\varepsilon_n}$ such that $x_n(t)$ has first zero q_n and second zero $\bar{q}_n = 2 + 2\varepsilon_n r_n$. Define $y_n(t)$ and $z_n(t)$ by

$$y_n(t) = x(q_n - \varepsilon_n t) \text{ and}$$

$$z_n(t) = x(\tfrac{1}{2}\bar{q}_n + q_n - \varepsilon_n t).$$

Then there exists a subsequence (n_i) such that $r_{n_i} \to r$, $y_{n_i}(t) \to y(t)$ uniformly on compact subsets of $\mathbb{R}$, $z_{n_i}(t) \to z(t)$ uniformly on compact subsets of $\mathbb{R}$, $y(t)$ and $z(t)$ are C^1 and $(y'_{n_i}(t), z'_{n_i}(t)) \to (y'(t), z'(t))$ uniformly on compact subsets of $\mathbb{R}$. The functions $y(t)$ and $z(t)$ satisfy eq. $(6)_r$. Furthermore, $y(0) = 0$ and $y(t) \neq 0$ for $t \neq 0$ and $z(t) = 0$ for a unique number t. There exists positive numbers c and d and positive numbers t_1 and t_2 such that $y(t) \geq c$ and $z(t) \leq -d$ for $t \geq t_1$ and $y(t) \leq -d$ and $z(t) \geq c$ for $t \leq -t_2$. If f is odd and the functions $x_n(t)$ are P_1-solutions of eq. $(1)_{\varepsilon_n}$ then $y_n(t) = -z_n(t)$ and $y(t)$ satisfies the equation

$$y'(t) = y(t) + f(y(t-r)) \qquad (7)_r$$

Because of Theorem 5 we shall now focus our attention on certain solutions of eq. $(6)_r$ and eq. $(7)_r$. Our results will translate by means of Theorem 5 to results about solutions of eq. $(1)_\varepsilon$ for ε small.

If $y\colon \mathbb{R} \to \mathbb{R}$ and $z\colon \mathbb{R} \to \mathbb{R}$ is a C^1 solution of eq. $(6)_r$ for some $r > 0$ and if f satisfies (H1) and (H2), we shall say that $(y(t),z(t))$ is an "H-solution of eq. $(6)_r$" if $y(t) = 0$ for a unique number t, $z(t) = 0$ for a unique number t, $y(t) \in (-B,A)$ and $z(t) \in (-B,A)$ for all t, and there exist positive numbers c and d such that

$$\liminf_{t\to+\infty} y(t) \geq c \quad\text{and}\quad \limsup_{t\to+\infty} z(t) \leq -d \quad\text{and}$$

$$\limsup_{t\to-\infty} y(t) \leq -d \quad\text{and}\quad \limsup_{t\to-\infty} z(t) \geq c .$$

If $y(t)$ is a C^1 solution of eq. $(7)_r$ and f is odd, then if we define $z(t) = -y(t)$, $(y(t),z(t))$ is a C^1 solution of eq. $(6)_r$. In this case we shall say that $y(t)$ is an H-solution of eq. $(7)_r$ if $(y(t),-y(t))$ is an H-solution of eq. $(6)_r$.

If f satisfies (H2) and (H3) and $(y(t),z(t))$ is an H-solution of eq. $(6)_r$, then it is proved in [5] that $\lim_{t\to+\infty} (y(t),z(t)) = (a,-b)$ and $\lim_{t\to-\infty} (y(t),z(t)) = (-b,a)$. Thus $(y(t),z(t))$ gives a heteroclinic solution of eq. $(6)_r$, connecting the constant solutions $(a,-b)$ and $(-b,a)$. We shall need this result later. In general the terminology "H-solution" is meant to suggest a generalized heteroclinic solution.

Note that even proving the existence of an H-solution of eq. $(6)_r$ (when f satisfies (H1) and (H2)) or of eq. $(7)_r$ (when f is also odd) is highly nontrivial. In fact the only way presently known of proving existence of H-solutions in this generality is by using Theorem 5.

If f satisfies (H2) and (H3) and f is C^1 on a neighborhood of a and $f'(a) > 0$, it is proved in Proposition 4.1 of [5] that if $(y(t),z(t))$ is an H-solution of eq. $(6)_r$, then for every T,

$$\sup_{t\geq T} y(t) > a \quad \text{or} \quad \inf_{t\geq T} z(t) < -b.$$

Our next theorem shows that, under slightly stronger assumptions, either $y(t) - a$ changes sign infinitely often on $[0,\infty]$ or $z(t) + b$ changes sign infinitely often. Because we shall not actually prove that each zero of $y(t) - a$ or of $z(t) + b$ is isolated we must be a little careful in defining the term "changes sign infinitely often." If $w(t)$ is a function defined on an interval $[c,d]$, we shall say that "w changes sign at least N times on $[c,d]$" if there exist numbers $t_j \in [c,d]$ for $1 \leq j \leq N+1$ such that $t_j < t_{j+1}$ for $1 \leq j \leq N$ and $(-1)^j x(t_j) > 0$ for $1 \leq j \leq N+1$ or $(-1)^{j+1} x(t_j) < 0$ for $1 \leq j \leq N+1$. We shall say that "w changes sign exactly N times on $[c,d]$" if w changes sign at least N times on $[c,d]$, but it is not true that w changes sign at least $N+1$ times on $[c,d]$. Finally, we shall say that "w changes sign infinitely often on $[c,d]$" if

w changes sign at least N times on $[c,d]$ for every $N \geq 1$.

THEOREM 6. *Assume that* f *satisfies* (H2) *and* (H3), *that* f *is continuously differentiable on neighborhoods of* a *and* $-b$ *and that* $k_1 = f'(a) > 0$, $k_2 = f'(-b) > 0$ *and* $k_1 k_2 < 1$ (*where* a *and* b *are as in* (H3)). *If* $(y(t),z(t))$ *is an* H-*solution of eq.* $(6)_r$ *for some* $r > 0$, *then either* $y(t) - a$ *or* $z(t) + b$ *changes sign infinitely many times on* $[T,\infty)$ *for every* $T \geq 0$.

PROOF. Suppose the theorem is false, so there exists $T_0 > 0$ such that $y_1(t) = y(t) - a$ satisfies $y_1(t) \geq 0$ for all $t \geq T_0$ or $y_1(t) \leq 0$ for all $t \geq T$. Similarly, by increasing T_0, $z_1(t) = z(t) + b$ satisfies $z_1(t) \geq 0$ for all $t \geq T_0$ or $z_1(t) \leq 0$ for all $t \geq T_0$. Since f satisfies H3, it is proved in [5] that $\lim_{t\to+\infty} y_1(t) = \lim_{t\to+\infty} z_1(t) = 0$.

If one linearizes eq. $(6)_r$ about the rest point $(a,-b)$ and looks for solutions $e^{\zeta t}u$ of the linearized equation ($\zeta \in \mathbb{C}$, $u \in \mathbb{C}^2$), one is led to the characteristic equation

$$\det \begin{pmatrix} \zeta-1 & k_2 e^{-\zeta r} \\ k_1 e^{-\zeta r} & \zeta-1 \end{pmatrix} = 0 . \tag{8}$$

If $k^2 = k_1 k_2$ and $k > 0$, eq. (8) is equivalent to

$$(\zeta - 1 - ke^{-\zeta r})(\zeta - 1 + ke^{-\zeta r}) = 0. \tag{9}$$

Since $k > 0$, the equation

$$\zeta - 1 - ke^{-\zeta r} = 0 \tag{10}$$

always has a positive real root ζ_1 (the expression in eq. (10) is negative for $\zeta = 0$ and positive for large positive ζ). Similarly, because $k < 1$, the equation

$$\zeta - 1 + ke^{-\zeta r} = 0 \tag{11}$$

has a positive real root ζ_2. If $\zeta = \zeta_1$, one easily checks that

$$\begin{pmatrix} \zeta-1 & k_1e^{-\zeta r} \\ k_2e^{-\zeta r} & \zeta-1 \end{pmatrix} \begin{pmatrix} \sqrt{k_1} \\ -\sqrt{k_2} \end{pmatrix} = \begin{pmatrix} 0 \\ 0 \end{pmatrix}, \ \zeta = \zeta_1 \tag{12}$$

(Note that the transpose of the matrix in eq.(8) is used).

If $\zeta = \zeta_2$ one finds that

$$\begin{pmatrix} \zeta-1 & k_1e^{-\zeta r} \\ k_2e^{-\zeta r} & \zeta-1 \end{pmatrix} \begin{pmatrix} \sqrt{k_1} \\ \sqrt{k_2} \end{pmatrix} = \begin{pmatrix} 0 \\ 0 \end{pmatrix}, \ \zeta = \zeta_2. \tag{13}$$

Under our assumptions on $f(x)$ we can write

$$\begin{aligned} f(x) &= a + k_2(x+b) + N_1(x+b) \quad \text{and} \\ f(x) &= -b + k_1(x-a) + N_2(x-a), \end{aligned} \tag{14}$$

where N_1 and N_2 satisfy

$$\lim_{u\to 0} |u^{-1} N_j(u)| = 0 \quad \text{for} \quad j = 1,2. \tag{15}$$

Now we use a standard argument. Let ζ be a solution of eq.(8) such that $\mathrm{Re}(\zeta) > 0$ and let $v = \binom{v_1}{v_2} \in \mathbb{C}^2$ be a nonzero vector such that

$$\begin{pmatrix} \zeta-1 & k_1e^{-\zeta r} \\ k_2e^{-\zeta r} & \zeta-1 \end{pmatrix} \begin{pmatrix} v_1 \\ v_2 \end{pmatrix} = \begin{pmatrix} 0 \\ 0 \end{pmatrix}. \tag{16}$$

Let $\langle u,w\rangle$ denote the standard inner product of vectors u and w. If $T \geq T_0 + r$, consider

$$\int_T^\infty \langle e^{-\zeta(t-T)}v, \begin{pmatrix} y_1'(t) \\ z_1'(t) \end{pmatrix} \rangle \, dt, \tag{17}$$

where $y_1(t)$ and $z_1(t)$ are as before. One can integrate the expression in (17) by parts or by substituting

$$y_1'(t) = y_1(t) - k_2 z_1(t-r) - N_1(z_1(t-r))$$
$$z_1'(t) = z_1(t) - k_1 y_1(t-r) - N_2(y_1(t-r)). \tag{18}$$

If one sets the two expressions one obtains in this way equal and simplifies (using equations (8) and (16)) one obtains

$$-\int_{T-r}^{T} e^{-\zeta(t-T+r)} \langle v, \binom{k_2 z_1(t)}{k_1 y_1(t)}\rangle\, dt + \langle v, \binom{y_1(T)}{z_1(T)}\rangle =$$
$$\int_{T-r}^{\infty} \langle e^{-\zeta(t-T+r)} v, \binom{N_1(z_1(t))}{N_2(y_1(t))}\rangle\, dt \tag{19}$$

The general method involved here is dicussed in more detail in [8], p.301-302. We now consider two cases.

CASE 1. $y_1(t) z_1(t) \leq 0$ for all $t \geq T_0$. In this case choose $\zeta = \zeta_1$, a positive real solution of equation (10) and define $v = \binom{\sqrt{k_1}}{-\sqrt{k_2}}$. We can assume that $y_1(t) \geq T_0$ and $z_1(t) \leq 0$ for $t \geq T_0$ (otherwise use $-v$). Substituting in eq.(19) and using (15) gives (with $\zeta = \zeta_1$)

$$\int_{T-r}^{T} e^{-\zeta(t-T+r)}[\sqrt{k_2}\, k_1 y_1(t) - \sqrt{k_1}\, k_2\, z_1(t)]dt + \sqrt{k_1}\, y_1(T) - \sqrt{k_2}\, z_1(T) \leq$$
$$\int_{T-r}^{\infty} e^{-\zeta(t-T+r)}[\sqrt{k_1}|N_1(z_1(t))| + \sqrt{k_2}\,|N_2\,(y_1(t))|]dt \tag{20}$$

Because $(y_1(t), z_1(t)) \to (0,0)$ as $t \to \infty$, one easily sees (using eq. (15)) that for T large enough

$$0 \leq \int_{T-r}^{T} e^{-\zeta(t-T+r)}[\sqrt{k_1}\, |N_1(z_1(t))| + \sqrt{k_2}\, |N_2(y_1(t))|]dt \leq$$
$$\int_{T-r}^{T} e^{-\zeta(t-T+r)}[k_1\sqrt{k_2}\, y_1(t) - \sqrt{k_1}\, k_2\, z_1(t)]dt. \tag{21}$$

We are using here the fact that $y_1(t) \geq 0$ and $z_1(t) \leq 0$ for $t \geq T_0$. Inequalities (20) and (21) imply that for all large T,

$$\sqrt{k_1}\, y_1(T) - \sqrt{k_2}\, z_1(T) \leq \int_{T}^{\infty} e^{-\zeta(t-T+r)}[\sqrt{k_1}|N_1(z_1(t))|$$
$$+ \sqrt{k_2}|N_2(y_1(t))|]dt \tag{22}$$

If one now chooses T large so that

$$y_1(T)-z_1(T) \geq \frac{1}{2} \sup_{t \geq T}[y_1(t)-z_1(t)]$$

and such that $\sup_{t \geq T} [y_1(t)-z_1(t)]$ is sufficiently small, eq.(15) implies that the left hand side of eq.(22) will actually be strictly greater than the right hand side unless $\sup_{t \geq T}[y_1(t)-z_1(t)] = 0$. In the latter case, $y(t) = a$ and $z(t) = -b$ for all $t \geq T$, so

$$\begin{aligned} &f(y(t-r)) = -b \quad \text{and} \\ &f(z(t-r)) = a \quad \text{for} \quad T \leq t \leq T + r. \end{aligned} \tag{23}$$

Equation (23) implies that

$$\begin{aligned} &f^2(y(t-r)) = a \quad \text{and} \\ &f^2(z(t-r)) = -b \quad \text{for } T \leq t \leq T + r. \end{aligned} \tag{24}$$

Because $0 < (f^2)'(a) < 1$, the inverse function theorem implies that there exists $\delta > 0$ such that $u = a$ is the only solution of

$$f^2(u) = a, \quad |u - a| \leq \delta.$$

Using this fact and the continuity of $y(t-r)$ for $T \leq t \leq T + r$ and the fact that $y(T) = a$, one easily obtains from eq. (24) that $y(t-r) = a$ for $T \leq t \leq T + r$. A similar argument shows that $z(t-r) = -b$ for $T \leq t \leq T + r$, so $y(t) = a$ and $z(t) = -b$ for $t \geq T-r$. Iterating this procedure, one concludes that $y(t) = a$ and $z(t) = -b$ for all t, which contradicts the assumption that $(y(t),z(t))$ is an H-solution of eq. $(6)_r$.

CASE 2. $y_1(t)z_1(t) \geq 0$ for all $t \geq T_0$. In this case, choose $\zeta = \zeta_2$, a positive real solution of equation (11), and $v = (\frac{\sqrt{k_1}}{\sqrt{k_2}})$. We can assume for definiteness that $y_1(t) \geq 0$ and $z_1(t) \geq 0$ for $t \geq T_0$ (otherwise $y_1(t) \leq 0$ and $z_1(t) \leq 0$ for $t \geq T_0$ and use $-v$). Substituting in eq. (19) and using (15) as before gives

$$\int_{T-r}^{T} e^{-\zeta(t-T+r)}[k_1\sqrt{k_2}\, y_1(t) + \sqrt{k_1}\, k_2 z_1(t)]dt + \sqrt{k_1}\, y_1(T) + \sqrt{k_2}\, z_1(T) \leq \int_{T-r}^{\infty} e^{-\zeta(t-T+r)}[\sqrt{k_1}\, |N_1(z_1(t))| + \sqrt{k_2}\, |N_2(y_1(t))|]dt . \tag{23}$$

Essentially the same argument as before (we leave the details to the reader) shows that there exists $T > T_0 + r$ such that $y(t) = a$ and $z(t) = -b$ for all $t \geq T$. As before one derives from this that $y(t) = a$ and $z(t) = -b$ for all t, contradicting the fact that $(y(t),z(t))$ is an H-solution.

Since we have obtained a contradiction by assuming that neither $y(t) - a$ nor $z(t) + b$ changes sign on $[T,\infty)$ for some T, the theorem has been proved. ■

Although we shall not prove this here (for reasons of length), one can prove (under relatively mild assumptions on f) that each zero of $y(t) - a$ is isolated and that $y(t) - a$ changes sign at each of its zeros. More precisely we have the following proposition, whose proof is omitted.

PROPOSITION 1. *Assume that* f *satisfies* H1 *and that* f *is* C^1 *on* $(-B,A)$. *Suppose either that* $f'(x) < 0$ *for all* $x \in (-B,0]$ *or that there exists a number* $\theta \in (-B,0)$ *such that* $f'(x) > 0$ *for* $-B < x < \theta$ *and* $f'(x) < 0$ *for* $\theta < x \leq 0$. *Similarly, assume either that* $f'(x) < 0$ *for all* $x \in [0,A)$ *or that there exists a number* $\psi \in (0,A)$ *such that* $f'(x) < 0$ for $0 \leq x < \psi$ *and* $f'(x) > 0$ *for* $\psi < x < A$. *Assume that* $(y(t),z(t))$ *is an H-solution of* eq. $(6)_r$ *for some* $r > 0$ *and that* $-B < y(t) < A$ *and* $-B < z(t) < A$ *for all* t. *If* $y(\tau) = \alpha \in [0,A)$ *and* $\alpha \neq \psi$, *there exists* $\delta > 0$ *such that* $(t-\alpha)(y(t)-\alpha)$ *is positive for all* $t \in (\tau-\delta,\tau+\delta)$, $t \neq \tau$, *or* $(t-\alpha)(y(t) - \alpha)$ *is negative for all* $t \in (\tau-\delta,\tau+\delta)$, $t \neq \tau$. A *similar statement holds if* $z(\tau) = \beta \in (-B,0]$ *and* $\beta \neq \theta$.

By using Theorem 6 in conjuction with Theorem 5 we easily obtain the following corollary.

COROLLARY 1. *Assume that* f *satisfies* (H2) *and* (H3), *that* f *is continuously differentiable on neighborhoods of* a *and* $-b$, *and that* $f'(a) > 0, f'(-b) > 0$

and $f'(a)\,f'(-b) < 1$, *where* a *and* b *are as in* H3. *Given a positive integer* N, *there exist positive numbers* Θ *and* C *(dependent on* N*) with the following properties: Suppose that* $x(t)$ *is a* P_2*-solution of* eq. $(1)_\varepsilon$ *for some* ε *with* $0 < \varepsilon \leq \Theta$ *and suppose that* $x(0) = 0$ *and* x *has first and second zeros* q *and* $\bar{q}$ *respectively. If* $y(t) = x(q-\varepsilon t)$ *and* $w(t) = x(\bar{q}-\varepsilon t)$, *then either* $y(t) - a$ *has at least* N *sign changes on* $[0,C]$ *or* $w(t) + b$ *has at least* N *sign changes on* $[0,C]$.

PROOF. Suppose not. Then there exists a sequence (ε_j) of positive numbers with $\lim_{j\to\infty} \varepsilon_j = 0$, a sequence (C_j) of positive numbers with $\lim_{j\to\infty} C_j = \infty$ and a sequence $x_j(t)$ of P_2-solutions of eq. $(1)_{\varepsilon_j}$ such that $x_j(0) = 0$, $x_j(t)$ has first and second zeros q_j and $\bar{q}_j$ respectively and if $y_j(t) = x_j(q_j-\varepsilon_j t)$ and $w_j(t) = x_j(\bar{q}_j-\varepsilon_j t)$, $y_j(t) - a$ changes sign fewer than N times on $[0,C_j]$ and $w_j(t) + b$ changes sign fewer than N times on $[0,C_j]$. As usual, define $z_j(t)$ by

$$z_j(t) = x(\tfrac{1}{2}\,\bar{q}_j + q_j - \varepsilon_j t). \tag{24}$$

If we write

$$(\tfrac{1}{2})\bar{q}_j - q_j = d_j\varepsilon_j, \tag{25}$$

Theorem 2 implies that there exists a constant D such that $|d_j| \leq D$ for all j, and the definition of z_j and w_j implies that

$$z_j(t) = w_j(t + d_j)\,.$$

It follows that for j sufficiently large $z_j(t) + b$ changes sign fewer than N times on $[0,C_j-D]$ and of course $y_j(t) - a$ changes sign fewer than N times on $[0,C_j-D]$. Theorem 5 implies that by taking a subsequence we can assume that $(y_j(t),z_j(t))$ converges (uniformly on compact subsets of $\mathbb{R}$) to an H-solution $(y(t),z(t))$ of eq. $(6)_r$ for some $r > 0$. Theorem 5 implies that there exists $C > 0$ such that $y(t) - a$ has at least N sign changes on $[0,C]$ or $z(t) + b$ has at least N sign changes on $[0,C]$. It follows that there exists j_o such that $y_j(t) - a$ or $z_j(t) + b$ has at least N

sign changes on $[0,C]$ for $j \geq j_0$. If j_0 is also chosen so large that $C_j - D > C$ for $j \geq j_0$, we obtain a contradiction.

REMARK 1. For any given convergent subsequence $(y_{j_i}(t), z_{j_i}(t))$ we also obtain from Theorem 5 that the first N oscillations of $y_{j_i}(t)$ about a or of $z_{j_i}(t)$ about $-b$ do not approach zero in amplitude as $\varepsilon_j \to 0$.

Theorem 6 is unsatisfactory in that it gives no information about Gibbs phenomena when $(f^2)'(a) > 1$, i.e., when a is an unstable fixed point of f^2. If f is odd, however, we can obtain some results. We begin with a lemma.

LEMMA 1. *Assume that* f *satisfies* (H1) *and that* f *is odd. Suppose that* $y(t)$ *is an* H-*solution of* eq. $(7)_r$ *and that* $\lim_{t\to\infty} y(t) = a$, *where* $f(a) = -a$ *and* $a \leq A$ (A *as in* (H1)). *If* f *is* C^1 *on a neighborhood of* a, *and* $k_1 = f'(a) > 0$, *then* $y(t) - a$ *changes sign infinitely often on any interval* $[T,\infty)$.

PROOF. Because f is odd we can take $A = B$ in (H1), and it is part of our definition of H-solution that $|y(t)| \leq A$ for all t. Define $z(t) = -y(t)$ so that $(y(t),z(t))$ is an H-solution of eq. $(6)_r$ and $\lim_{t\to\infty}(y(t),z(t)) = (a,-a)$. If the lemma is false, there exists T_0 such that $y(t) - a$ does not change sign on $[T_0,\infty)$. For definiteness, assume that $y(t) - a \geq 0$ for $t \geq T_0$, so $z(t) - b \leq 0$ for $t \geq T_0$ (where $b = -a$). Let ζ be a real, positive root of

$$\zeta - 1 - k_1 e^{-r\zeta} = 0.$$

Such a root exists because $k_1 > 0$. Now exactly the argument used in Case 1 of the proof of Theorem 6 shows that there exists $T \geq T_0 + r$ such that $y(t) = a$ for all $t \geq T$. Define T_1 by

$$T_1 = \inf\{t : y(s) = a \quad \text{for all} \quad s \geq t\}.$$

Because y is an H-solution, $-\infty < T_1 \leq T$. Eq. $(7)_r$ implies that

$$0 = a + f(y(t-r)) \quad \text{for} \quad T_1 \leq t \leq T_1 + r \tag{26}$$

Because $f'(a) > 0$, the inverse function theorem implies that $I + f$ is one-one on a neighborhood of a, and because $y(T) = a$, eq. (26) and a simple continuity argument imply that $y(t-r) = a$ for $T_1 \leq t \leq T_1 + r$, contradicting the definition of T_1. ■

Before proving our next lemma it will be convenient to introduce another set of conditions on f.

(H4) _The function_ $f : \mathbb{R} \to \mathbb{R}$ _is odd_ ($f(-x) = -f(x)$ _for all_ x) _and satisfies_ (H1) _with constants_ A _and_ $B = A$. _In addition there exists a number_ a, $0 < a \leq A$, _such that_ $f(a) = -a$, $|f(x)| > |x|$ _for_ $0 < |x| < a$ _and_ $|f(x)| < |x|$ _for_ $a < |x| \leq A$.

LEMMA 2. _Assume that_ f _satisfies_ (H4) _and that_ $y(t)$ _is an_ H-_solution of eq._ $(7)_r$ _(for some_ $r > 0$) _with_ $|y(t)| \leq A$ _for all_ t. _If_ $y(T) \leq a$ _for some_ T, _then either_ $y(T_1) > a$ _for some_ $T_1 > T$ _or_ $y(t) \leq a$ _for all_ $t \geq T$ _and_ $\lim_{t\to\infty} y(t) = a$. _If_ $y(T) \geq a$ _for some_ T, _then either_ $y(T_1) < a$ _for some_ $T_1 > T$ _or_ $y(t) \geq a$ _for all_ $t \geq T$ _and_ $\lim_{t\to\infty} y(t) = a$.

PROOF. Suppose that $y(t) \leq a$ for all $t \geq T$. We must prove that $\lim_{t\to+\infty} y(t) = a$. We first claim that $\limsup_{t\to\infty} y(t) = a$. If not, there is a number $t_1 \geq T$ and a number $a_1 < a$ such that $y(t) \leq a_1$ for all $t \geq t_1$. In addition, by increasing t_1 and using the definition of an H-solution, there exists $c > 0$ such that $y(t) \geq c$ for all $t \geq t_1$. Assumption (H4) implies that there exists $d > 0$ such that

$$f(u) + u \leq -d \quad \text{for} \quad c \leq u \leq a_1 . \tag{27}$$

Now select any number $t_2 > t_1 + r$ and integrate eq. $(7)_r$ from t_1 to t_2 to obtain

$$\int_{t_1}^{t_2} y'(t)dt = y(t_2) - y(t_1) = \int_{t_1-r}^{t_1} f(y(t))dt + \int_{t_1}^{t_2-r} [f(y(t)) + y(t)]dt + \int_{t_2-r}^{t_2} y(t)dt . \tag{28}$$

The left hand side of eq. (28) is bounded in absolute value by A. On the other hand, by using eq. (27), one sees that the right hand side of eq. (28) is bounded above by

$$2Ar - d(t_2-t_1-r), \text{ so}$$
$$-A \leqslant 2Ar - d(t_2-t_1-r) \tag{29}$$

Since the right hand side of eq. (29) approaches $-\infty$ as $t_2 \to +\infty$, we have obtained a contradiction and therefore must have $\limsup_{t\to+\infty} y(t) = a$.

In order to prove that $\lim_{t\to\infty} y(t) = a$, it now suffices to prove that

$$\liminf_{t\to+\infty} y(t) = a.$$

Suppose not and select numbers c_1 and a_1 such that

$$\liminf_{t\to+\infty} y(t) < c_1 < a_1 < a.$$

We claim that there are sequences of numbers (σ_j) and (τ_j) in $[t_1,\infty)$ for $j \geqslant 1$ such that $\sigma_j < \tau_j < \sigma_{j+1}$ for all $j \geqslant 1$, $\sigma_j \to \infty$ as $j \to \infty$, $y(\sigma_j) = c_1$, $y(\tau_j) = a_1$ and

$$c_1 \leqslant y(t) \leqslant a_1 \quad \text{for } \sigma_j \leqslant t \leqslant \tau_j,\ j \geqslant 1. \tag{30}$$

Assuming the existence of (σ_j) and (τ_j) for the moment, notice that eq. $(7)_r$ implies that $|y'(t)| \leqslant 2A$ for all t. Thus the mean value theorem implies that

$$(y(\tau_j) - y(\sigma_j))(2A)^{-1} = (a_1-c_1)(2A)^{-1} \leqslant \tau_j - \sigma_j. \tag{31}$$

Assumption (H4) implies that there exists a number $\delta > 0$ such that

$$u + f(u) \leq -\delta \quad \text{for} \quad c_1 \leq u \leq a_1. \tag{32}$$

Integrating eq. $(7)_r$ from σ_1 to $\tau_n + r$ gives

$$\int_{\sigma_1}^{\tau_n+r} y'(t)dt = y(\tau_n+r) - y(\sigma_1) = \int_{\sigma_1-r}^{\sigma_1} f(y(t))dt + \int_{\sigma_1}^{\tau_n}[f(y(t)) + y(t)]dt$$
$$+ \int_{\tau_n}^{\tau_n+r} y(t)dt . \tag{33}$$

The left hand side of eq. (33) is bounded below by $-c_1$. However, because $u + f(u) \leq 0$ for $0 < u \leq a$, the right hand side of (33) is bounded above by

$$\int_{\sigma_1-r}^{\sigma_1} f(y(t))dt + \int_{\tau_n}^{\tau_n+r} y(t) + \int_{\sigma_1}^{\tau_n}[f(y(t)) + y(t)] \leq$$
$$2A + \sum_{j=1}^{n} \int_{\sigma_j}^{\tau_j}[f(y(t)) + y(t)]dt . \tag{34}$$

By using equations (31) and (32) we find that

$$2A + \sum_{j=1}^{n} \int_{\sigma_j}^{\tau_j}[f(y(t)) + y(t)]dt \leq 2A - n\delta(a_1-c_1)(2A)^{-1}. \tag{35}$$

Combining these estimates for the right hand side and left hand of eq. (33) we conclude that for all $n \geq 1$,

$$-c \leq y(\tau_n+r) - y(\sigma_1) \leq 2A - n\delta(a_1-c_1)(2A)^{-1}, \tag{36}$$

and for n large enough, eq. (36) gives a contradiction.

We still must prove the existence of (σ_j) and (τ_j). Since we are assuming that $\liminf_{t\to+\infty} y(t) < c_1$, define $\tilde{\sigma}_1$ to be the first number greater than or equal to t_1 such that $y(\tilde{\sigma}_1) = c_1$. Define $\tau_1 = \inf\{t \geq \tilde{\sigma}_1 \mid y(t) = a_1\}$; τ_1 is finite because we know that $\limsup_{t\to\infty} y(t) = a$. Define

$\sigma_1 = \sup\{s \leq \tau_1 \mid y(s) = c_1\}$. In general, if σ_j and τ_j have been defined, define $\tilde{\sigma}_{j+1}$ to be the first number $t \geq \tau_j$ such that $y(t) = c_1$. Then define $\tau_{j+1} = \inf\{t \geq \tilde{\sigma}_{j+1} \mid y(t) = a_1\}$ and define $\sigma_{j+1} = \sup\{s \leq \tau_{j+1} \mid y(s) = c_1\}$. It is easy to check that with these definitions $t_1 \leq \sigma_j < \tau_j < \sigma_{j+1}$ for all $j \geq 1$ and $c_1 \leq x(s) \leq a_1$ for $\sigma_j \leq s \leq \tau_j$. Furthermore, eq. (31) implies that

$$\sigma_n > \sigma_1 + (n-1)(a_1-c_1)(2A)^{-1},$$

so $\sigma_n \to \infty$ as $n \to \infty$.

It remains to prove that if $y(t) \geq a$ for all $t \geq t_1$, then $\lim_{t\to\infty} y(t) = \infty$. However, the proof of this is essentially the same as the previous case if one recalls that

$$u + f(u) > 0 \text{ for } a < u \leq A.$$

Details are left to the reader. ■

THEOREM 7. *Assume that* f *satisfies* (H2) *and* (H4). *If* a *is as in* (H4) *assume that* f *is* C^1 *on an open neighborhood of* a *and* $f'(a) > 0$. *Suppose that* $y(t)$ *is an H-solution of eq.* $(7)_r$ *for some* $r > 0$ *and that* $|y(t)| \leq A$ *for all* t. *Then for any real number* T, $y(t) - a$ *changes sign infinitely often on* $[T,\infty)$. *If* f *is as above but* $f'(a) \leq 0$ *and if* $y(t)$ *is an H-solution of eq.* $(7)_r$ *such that* $y(t) - a$ *does not change sign on some interval* $[T,\infty)$, *then* $\lim_{t\to\infty} y(t) = a$.

PROOF. Assume first that $f'(a) > 0$. If the theorem is false, there exists T such that $y(t) - a$ does not change sign on $[T,\infty)$. Lemma 2 implies that $\lim_{t\to\infty} y(t) = a$, but Lemma 1 then implies that $y(t) - a$ changes sign infinitely often on $[T,\infty)$, a contradiction.

The final statement of the theorem follows immediately from Lemma 2. ■

COROLLARY 2. *Assume that* f *satisfies* (H2) *and* (H4), f *is* C^1 *on a neighborhood of* a (a *as in* (H4)) *and* $f'(a) > 0$. *Given a positive* integer N, there exist positive numbers θ and C (dependent on N) with the following property: Suppose that $x(t)$ is a P_1-solution of eq. $(1)_\varepsilon$ for some ε with $0 < \varepsilon \leq \theta$ and suppose that $x(0) = 0$ and x has first zero q. If $y(t) = x(q-\varepsilon t)$, $y(t)$ has at least N sign changes on $[0,C]$.

PROOF. This follows from Theorem 7 in essentially the same way that Corollary 1 follows from Theorem 6. ∎

It is of interest to apply Theorem 7 and Corollary 2 to a simple example. Suppose that $f(x) = f_\mu(x)$ in eq. $(1)_\varepsilon$, where $f_\mu(x) = -\mu x + x^3$ and $\mu > 0$. It is proved in [7] (and one can prove directly) that f_μ satisfies (H1) if and only if $0 < \mu < (\frac{3\sqrt{3}}{2})$, and it is easy to see that f_μ satisfies (H2) if and only if $\mu > 1$. It is a somewhat harder exercise (carried out in Section 9 of [7]) to prove that f_μ satisfies (H3) if and only if $1 < \mu \leq 2$. The function $-f_\mu(x)$ has the unique positive fixed point $x_\mu = \sqrt{\mu-1}$ and $f_\mu'(x_\mu) = 2\mu - 3$, which is positive if and only if $\mu > \frac{3}{2}$. From this information one concludes that f_μ satisfies (H2) and (H4) if and only if $1 < \mu < (\frac{3\sqrt{3}}{2})$ and that f_μ satisfies the hypotheses of Theorem 7 and Corollary 2 if and only if $(\frac{3}{2}) < \mu < (\frac{3\sqrt{3}}{2})$. Note that the hypotheses of Theorem 6 are only satisfied for $(\frac{3}{2}) < \mu < 2$.

In many applications the function f in eq. $(1)_\varepsilon$ depends on a parameter or parameters, so $f = f_\mu$. Obtaining precise information about the parameter range μ for which the hypotheses of Theorem 6 or Theorem 7 are satisfied may be a difficult question: see [7].

REFERENCES

1. S.-N. Chow and D. Green, Jr., Some results on singular delay-differential equations, in Chaos, Fractals and Dynamics, vol. 1, W. R. Smith and P. Fisher, editors, New York, Marcel Dekker, 1985.

2. S.-N.Chow and J. Mallet-Paret, Singularly perturbed delay-differential equations, in Coupled Nonlinear Oscillators, J. Chandra and A. Scott, editors, North Holland Math. Studies, vol. 80 (1983), 7-12.

3. K.P. Hadeler and J. Tomiuk, Periodic solutions of difference-differential equations, Arch. Rat. Mech. Anal. 65 (1977), 87-95.

4. J. Mallet-Paret and R. D. Nussbaum, Global continuation and complicated trajectories for periodic solutions of a differential-delay equation, Proceedings of Symposia in Pure Mathematics, vol. 45 (1986), 155-167.

5. ________, Global continuation and asymptotic behaviour for periodic solutions of a differential-delay equation, to appear in Annali di Mathematica Pura ed Appl.

6. ________, A bifurcation gap for a singularly perturbed delay equation, Chaotic Dynamics and Fractals, vol. 2, M. Barnsley and S. Demko, eds., Academic Press, New York, 1986, pages 263-287.

7. ________, A differential-delay equation arising in optics and physiology, to be submitted.

8. R. D. Nussbaum, Periodic solutions of nonlinear autonomous functional differential equations, Lecture Notes in Math., vol. 730, Springer-Verlag, 1979, 283-325.

9. ________, Circulant matrices and differential-delay equations, J. Differential Equations 60 (1985), 201-217.

10. ________, The Fixed Point Index and Some Applications, Les Presses de l'Université de Montréal, 1985.

Canadian Mathematical Society
Conference Proceedings
Volume 8 (1987)

ON AN EXAMPLE OF ZIGLIN IN HAMILTONIAN DYNAMICS

David L. Rod[1]

ABSTRACT. This paper uses a theorem of Ziglin to demonstrate the non-integrability of a wide class of classical Hamiltonian systems. The potential, when restricted to an axis, is a polynomial of degree 3 or 4. The geometry of the problem, coupled with the general theory of elliptic functions, allows us to bypass detailed calculations and generalize Ziglin's original example.

INTRODUCTION. In [8] Ziglin gave a theorem which can be used to demonstrate the non-integrability of certain Hamiltonian systems. In Section 4 of [8] he gave an example to which his theorem applied, modulo some additional symmetry requirements. In this paper, we have generalized this result.

Ito in [3] has presented a version of Ziglin's results for analytic Hamiltonian systems with two degrees of freedom which we present as Theorem 1 in Section 1. Sections 2 and 3 then detail the application of Theorem 1 to a wide class of classical (kinetic plus potential energy) Hamiltonians in which the potential, when restricted to an axis, is a polynomial of respective degree 3 and 4. The geometry of the problem, coupled with the general theory of elliptic functions, allows us to bypass the type of detailed calculations with elliptic functions found in [3] and [8].

Theorems 2 and 3 demonstrate when the necessary conditions for integrability given in Theorem 1 are violated, hence no independent second integral exists for the Hamiltonian differential equations. In all the cases where non-integrability of the Hamiltonian system can be demonstrated, the results hold for all but a discrete set of energies and all but isolated parameter ratios of two of the coefficients in the Taylor expansion of the potential. This reflects the generic results on non-integrability of Markus

1980 Mathematics Subject Classification (1985 Revision). 58F05.

[1]Author's research supported in part by the National Sciences and Engineering Research Council of Canada, Grant A8507.

and Meyer [5]. We conclude Section 3 by showing that Theorem 3 applies to the original example of Ziglin which gave rise to the present paper. For further results on the geometrization of the applicability of Theorem 1 to non-integrability conclusions, see [6].

1. ZIGLIN'S THEOREM. In this section, we state Ziglin's Theorem [8]. Ito in [3] has given an exposition of this theorem in the context of Hamiltonian systems with two degrees of freedom which is the case we consider in this paper. Let the complex analytic Hamiltonian

$$H = (1/2)(y_1^2 + y_2^2) + W(x_1,x_2) \tag{1.1}$$

be defined on all of $\mathbb{C}^4$ with coordinates (x_1,x_2,y_1,y_2). The associated Hamiltonian differential equations will then be

$$\begin{cases} \dot{x}_i = (\partial H/\partial y_i) = y_i \\ \dot{y}_i = -(\partial H/\partial x_i) = -(\partial W/\partial x_i) \end{cases} \tag{1.2}$$

for $i = 1,2$, with the time $t \in \mathbb{C}$ and $\cdot = (d/dt)$. The following hypothesis will hold throughout this paper.

HYPOTHESIS 1. (a) $W(x_1,0)$ is a polynomial of degree 3 or 4 in x_1 with $W(0,0) = 0$ and $(\partial W/\partial x_2)(x_1,0) \equiv 0$. When $\deg W(x_1,0) = 4$, the potential W satisfies the additional symmetry condition $W(x_1,x_2) = W(-x_1,x_2)$.

(b) $W(x_1,x_2)$ is an entire real analytic function for $(x_1,x_2) \in \mathbb{R}^2$. (We use the word "entire" to mean that the Taylor expansion has infinite radius of convergence.)

(c) For $x_1 \in \mathbb{R}$ there are real energies $h > 0$ such that $[h-W(x_1,0)] = 0$ has distinct roots, at least two of which are real.

Hypothesis 1 implies that the equations (1.2) admit a non-equilibrium solution

$$P_h(t) = (\pi_h(t),0,\dot{\pi}_h(t),0) \tag{1.3}$$

with positive energy h such that $\pi_h(t)$ is real when $t \in \mathbb{R}$. When $\deg W(x_1,0) = 4$ we will work only with those solutions for which $\pi_h(0) = 0$. From the general theory of elliptic functions (see Chapter 1 of [7]) we know that $\pi_h(t)$ is an elliptic function with two independent periods $\omega_i = \omega_i(h)$, $i = 1,2$. Moreover, when $\deg W(x_1,0) = 3$ this solution will possess one pole of order two in its period parallelogram; when $\deg W(x_1,0) = 4$, there will be two such poles, each of order one. We will restrict to situations in which these poles do not lie on the real t-axis and the period $\omega_1(h)$ is real.

Hypothesis 1 implies that $(\partial^2 W/\partial x_1 \partial x_2)(x_1,0) \equiv 0$. The linearized equations about $P_h(t)$ then decouple and take the form

$$(1.4) \quad \begin{cases} \text{(a)} \ \ddot{\eta}_1 + (\partial^2 W/\partial x_1^2)(\pi_h(t),0)\eta_1 = 0 \\ \text{(b)} \ \ddot{\eta}_2 + (\partial^2 W/\partial x_2^2)(\pi_h(t),0)\eta_2 = 0 \ . \end{cases}$$

Let $\Psi(t;h)$ be a fundamental matrix solution for (1.4b) in a small neighborhood of $t = 0$ with $\Psi(0;h) = I$, the 2×2 identity matrix. Although $\pi_h(t)$ is meromorphic in t, the analytic continuation of the matrix $\Psi(t;h)$ will not in general be single-valued. When $\deg W(x_1,0) = 3$ we construct paths $c_i(h)$, $i = 1,2$, in the complex t-plane that define a period parallelogram for $\pi_h(t)$ and that do not contain any poles of $\pi_h(t)$. Let $c_1(h)$ be the interval $[0,\omega_1(h)]$ on the real t-axis, and let $c_2(h)$ be an appropriate path from 0 to $\omega_2(h)$. When $\deg W(x_1,0) = 4$ we let $c_1(h)$ be the interval $[0,\omega_1(h)/2]$ on the real t-axis. We then choose $c_2(h)$ as a path from 0 to $\omega_2(h)$ such that the two paths $c_i(h)$, $i = 1,2$, and their translates $c_1(h) + \omega_2(h)$ and $c_2(h) + \omega_1(h)/2$, bound half a period parallelogram for $\pi_h(t)$ whose one pole in this region is an interior point. In both cases $\deg W(x_1,0) = 3$ or 4 we let $g_i = g_i(h)$ be the complex analytic continuation of $\Psi(t;h)$ along $c_i(h)$, $i = 1,2$, from 0 to the other endpoint of the path. Since (1.4b) can be expressed as a time-dependent Hamiltonian system, the g_i are (complex) symplectic matrices with $\det(g_i) = 1$; that is $g_i \in SL(2,\mathbb{C})$.

THEOREM 1 (ZIGLIN [8]). Assume Hypothesis 1 and that the Hamiltonian system (1.2) admits an integral F that is meromorphic in a neighborhood in $\mathbb{C}^4$ of the image $\Gamma(h) = \{P_h(t) | t \in \mathbb{C}\}$ of the non-equilibrium solution (1.3). If F is functionally independent of the Hamiltonian H in this neighborhood, and no eigenvalue of the matrix $g_1(h)$ is a root of unity (the non-resonance condition), then the anti-commutator $g_* = g_*(h) = g_2^{-1} g_1^{-1} g_2 g_1 = I$ or $g_1^2(h)$.

REMARK 1. (a) Functional dependence means that the gradients of F and H are dependent vectors at each point where they are defined. Recall that a function meromorphic on $\mathbb{C}^4$ is holomorphic on a dense open subset of $\mathbb{C}^4$.

(b) Assume the system (1.2) admits a non-equilibrium solution $P_h(t)$ that projects to the (x_1,x_2)-plane along some line through the origin other than the x_1-axis. By a linear symplectic change of coordinates, one can rotate this line to the x_1-axis. Hypothesis 1 may then be satisfied for this transformed system.

(c) In our applications, we will be interested in entire real analytic Hamiltonians H on $\mathbb{R}^4$ that extend to complex analytic functions on $\mathbb{C}^4$. The function F may also be initially given as an entire or quotient of two entire functions on $\mathbb{R}^4$. We will use Theorem 1 to conclude for a number of such Hamiltonian systems that an independent second integral F does not exist at energy h by showing that the non-resonance condition holds and that $g_*(h) \neq I$ or $g_1^2(h)$.

(d) We have combined several of Ziglin's results in the statement of Theorem 1. The case $\deg W(x_1,0) = 3$ can be found in Ito [3], whereas the case $\deg W(x_1,0) = 4$ requires additional arguments involving the symmetry condition given in Hypothesis 1(a) for which we must refer to Ziglin [8, Sections 1.5 and 4].

2. APPLICATIONS: $\deg W(x_1,0) = 3$. The following two hypotheses on the potential W will be in force throughout this section.

HYPOTHESIS 2. The potential W of (1.1) is an entire real analytic function of $(x_1,x_2) \in \mathbb{R}^2$ with the following form:

$$W(x_1,x_2) = \sum_{2\leq i+j\leq 3} \gamma_{ij}\, x_1^i\, x_2^j \; + \; \text{h.o.t.}, \tag{2.1}$$

where the γ_{ij} are real constants and the higher order terms (h.o.t.) are given by an analytic function on $\mathbb{C}^2$ whose Taylor expansion about the origin consists of terms of degree ≥ 4, each containing a factor of x_2^3.

HYPOTHESIS 3. The coefficients γ_{ij} in (2.1), in addition to satisfying

$$\gamma_{11} = \gamma_{21} = 0 \quad \text{and} \quad \gamma_{30} \neq 0 \tag{2.2}$$

by Hypothesis 1, also satisfy

$$\gamma_{20} > 0 \;,\; \gamma_{02} < 0 \;, \text{ and } \; \gamma_{12} \neq 0. \tag{2.3}$$

Hypotheses 1-3 imply that there is a solution (1.3) with positive energy and one real period $\omega_1(h)$ in the potential well about the origin. For $t \in \mathbb{R}$ this orbit projects to the real x_1-axis. Let $(0,h^*)$ be the maximal energy interval for which $\pi_h(t)$ has finite real period $\omega_1(h)$. Note that

(1.3) for $t \in \mathbb{R}$ limits to the origin, an equilibrium solution of (1.2), as $h \downarrow 0$. Recall that we have chosen $c_1(h)$ to be the interval $[0,\omega_1(h)]$ in $\mathbb{R}$. Then $\pi_h(t)$ has no poles along $c_1(h)$ and Hypothesis 2 implies that along $\pi_h(t)$ the linearized Poincaré map $g_1(h) = \Psi(\omega_1(h);h) \in SL(2,\mathbb{R})$. We will show for $h \in (0,h^*)$ sufficiently close to 0 that trace $g_1(h) > 2$, hence $\pi_h(t)$ is hyperbolic for $t \in \mathbb{R}$.

The path $c_2(h)$ was chosen so that the second order pole of $\pi_h(t)$ lay interior to the period parallelogram formed by the $c_i(h)$, $i = 1,2$. Let the meromorphic function $\pi_h(t)$ have a local expansion about this pole $t_\infty = t_\infty(h)$,

(2.4) $$\pi_h(t) = A(t - t_\infty)^{-2} + \text{h.o.t.}$$

This orbit satisfies the following equation in the (complex) x_1-plane:

(2.5) $$\ddot{x}_1 + (\partial W/\partial x_1)(x_1,0) = \ddot{x}_1 + (2\gamma_{20}x_1 + 3\gamma_{30}x_1^2) = 0.$$

Thus the coefficient $A \neq 0$ satisfies

(2.6) $$(6A + 3\gamma_{30}A^2) = 0,$$

and hence

(2.7) $$A = -2(\gamma_{30})^{-1}.$$

Now (1.4) becomes

(2.8) $$\begin{cases} \text{(a) } \ddot{\eta}_1 + [2\gamma_{20} + 6\gamma_{30}\pi_h(t)]\eta_1 = 0 \\ \text{(b) } \ddot{\eta}_2 + [2\gamma_{02} + 2\gamma_{12}\pi_h(t)]\eta_2 = 0 . \end{cases}$$

Since $\gamma_{12} \neq 0$ the equation (2.8b) has a regular singular point at $t_\infty = t_\infty(h)$ and takes the following form in a neighborhood of t_∞ (on setting $\gamma = \gamma_{12}(\gamma_{30})^{-1}$):

(2.9) $$\ddot{\eta}_2 + [-4\gamma(t - t_\infty)^{-2} + \text{h.o.t.}]\eta_2 = 0.$$

The eigenvalues of the anti-commutator $g_*(h) = g_2^{-1}g_1^{-1}g_2g_1$ can then be calculated by analytic continuation of a fundamental matrix solution of (2.9) about the pole as

(2.10) $$\lambda_j = \exp(2\pi i\rho_j), \quad j = 1,2,$$

where the ρ_j are the roots of the indicial equation

(2.11) $$\rho(\rho - 1) - 4\gamma = 0$$

of (2.9) (see [1, pp. 230-232]). (The expression (2.10) for the eigenvalues holds even when $(\rho_1-\rho_2)$ = integer.) Note that although the pole $t_\infty(h)$ depends on h, the eigenvalues λ_j and hence trace $g_*(h) = (\lambda_1 + \lambda_2)$ are independent of h. Now recall the notation of Theorem 1.

THEOREM 2. Assume Hypotheses 1-3, and set $\gamma = \gamma_{12}(\gamma_{30})^{-1}$. Then the system (1.2) has no second integral F that is meromorphic in a neighborhood in $\mathbb{C}^4$ of the image $\Gamma(h)$ and functionally independent of the Hamiltonian H in this neighborhood at energies $0 < h < h^*$ when $(1 + 16\gamma) \neq$ (odd integer)^2 (except possibly on a set of measure zero of such energies).

PROOF. In view of Theorem 1, we must show $g_*(h) \neq I$ or $g_1^2(h)$, and that the eigenvalues of $g_1(h)$ are not roots of unity. The solutions to (2.11) have the form

$$(2.12) \qquad \rho_j = (1/2)[1 + (-1)^j(1 + 16\gamma)^{1/2}],$$

and hence

$$(2.13) \qquad \lambda_j = -\exp[(-1)^j\pi i(1 + 16\gamma)^{1/2}], \quad j = 1,2.$$

If $(1 + 16\gamma) \geq 0$ with $(1 + 16\gamma) \neq$ (odd integer)^2, or $(1 + 16\gamma) < 0$, then $\lambda_j \neq 1$, and hence $g_*(h) \neq I$. Note that

$$(2.14) \qquad \text{trace } g_1^2(h) = [\text{trace } g_1(h)]^2 - 2,$$

since $g_1(h) \in SL(2,\mathbb{R})$.

In (2.8b) let

$$(2.15) \qquad Q_h(t) = [2\gamma_{02} + 2\gamma_{12}\pi_h(t)].$$

Then $\sup_{0\leq t\leq\omega_1(h)}|\pi_h(t)| \longrightarrow 0$ as $h \downarrow 0$ implies, with $\gamma_{02} < 0$ by (2.3), that there is a maximal interval $(0,h_0) \subset (0,h^*)$ such that for $t \in [0, \omega_1(h)]$ we have

$$(2.16) \qquad Q_h(t) \leq 0 \quad \text{for} \quad 0 < h < h_0,$$

and $Q_h(t) = 0$ at only a finite number of such t-values. Then by [4, p.14] we have

$$(2.17) \qquad \text{trace } g_1(h) > 2 \quad \text{for} \quad 0 < h < h_0,$$

and the orbit $\pi_h(t)$ for $t \in \mathbb{R}$ is hyperbolic with positive eigenvalues $\neq 1$. Moreover, $g_1^2(h)$ also has positive eigenvalues $\neq 1$. Now, if $(1 + 16\gamma) \geq 0$ we see from (2.13) that the eigenvalues of $g_*(h)$ lie on the unit circle, whereas if $(1 + 16\gamma) < 0$, then these eigenvalues are negative.

In both cases, $g_*(h) \neq g_1^2(h)$, and the theorem is proven on the energy interval $(0,h_0)$, and hence by analyticity on $(0,h^*)$. The excluded energies in $(0,h^*)$ are those for which $g_*(h) = g_1^2(h)$ and those for which $g_1(h)$ has eigenvalues that are roots of unity, the condition $g_*(h) \neq I$ being satisfied when $(1 + 16\gamma) \neq (\text{odd integer})^2$. Q.E.D.

REMARK 2. If F is entire or meromorphic on all of $\mathbb{C}^4$, then one does not need to exclude any energies in the interval $0 < h < h^*$ in Theorem 2.

3. APPLICATIONS: $\deg W(x_1,0) = 4$.

The following two hypotheses will be in force throughout this section.

HYPOTHESIS 4. The potential W of (1.1) is an entire real analytic function of $(x_1, x_2) \in \mathbb{R}^2$ with the following form:

$$W(x_1,x_2) = \sum_{2\leq i+j\leq 4} \delta_{ij} x_1^i x_2^j + \text{h.o.t.}, \tag{3.1}$$

where the δ_{ij} are real constants and the higher order terms (h.o.t.) are given by an analytic function on $\mathbb{C}^2$ whose Taylor expansion about the origin consists of terms of degree ≥ 5, each containing a factor of x_2^3.

HYPOTHESIS 5. The coefficients δ_{ij} in (3.1), in addition to satisfying $\delta_{11} = \delta_{21} = \delta_{12} = \delta_{31} = \delta_{13} = 0$ and $\delta_{40} \neq 0$ by Hypothesis 1, also satisfy either (a) or (b) below:

(a) $\delta_{20} > 0$, and either

(1) $\delta_{02} < 0$ and $\delta_{22} \neq 0$, or

(2) $\delta_{02} = 0$ and $\delta_{22} < 0$,

(b) $\delta_{20} = \delta_{30} = 0$ and $\delta_{40} > 0$, and either

(1) $\delta_{02} < 0$ and $\delta_{22} \neq 0$, or

(2) $\delta_{02} = 0$ and $\delta_{22} < 0$.

Given Hypotheses 1 and 4 and either case (a) or (b) of Hypothesis 5, there will be a solution (1.3) with positive energy and one real period $\omega_1(h)$ in the potential well about the origin. For $t \in \mathbb{R}$, this orbit projects to the real x_1-axis. Let $(0,h^*)$ be the maximal energy interval for which $\pi_h(t)$ has finite real period $\omega_1(h)$. Note that $h^* = \infty$ is possible and that (1.3) for $t \in \mathbb{R}$ limits to the origin, an equilibrium solution of (1.2), as $h \downarrow 0$.

The meromorphic function $\pi_h(t)$ has two poles, each of order one, in its period parallelogram. Let $\pi_h(t)$ have a local expansion about one of these poles $t_\infty = t_\infty(h)$,

$$(3.2)\qquad \pi_h(t) = B(t - t_\infty)^{-1} + \text{h.o.t.}$$

This orbit satisfies the following equation in the (complex) x_1-plane:

$$(3.3)\qquad \ddot{x}_1 + (\partial W/\partial x_1)(x_1,0) = \ddot{x}_1 + \left[2\delta_{20}\, x_1 + 3\delta_{30}\, x_1^2 + 4\delta_{40}\, x_1^3\right] = 0.$$

Thus the coefficient $B \neq 0$ satisfies

$$(3.4)\qquad (2B + 4\delta_{40}\, B^3) = 0,$$

and hence

$$(3.5)\qquad B^2 = -(2\delta_{40})^{-1}.$$

Note that the two solutions for B correspond to the two poles and reflect the fact that the sum of the residues of an elliptic function in its period parallelogram must be zero.

Now (1.4) becomes

$$(3.6)\qquad \begin{cases} \text{(a)}\ \ddot{\eta}_1 + \left[2\delta_{20} + 6\delta_{30}\pi_h(t) + 12\delta_{40}\,\pi_h^2(t)\right]\eta_1 = 0 \\ \text{(b)}\ \ddot{\eta}_2 + \left[2\delta_{02} + 2\delta_{22}\pi_h^2(t)\right]\eta_2 = 0. \end{cases}$$

Thus, (3.6b) has a regular singular point at each of the two poles and takes the following form in a neighborhood of such a pole t_∞ (on setting $\delta = -2\delta_{22}B^2 = \delta_{22}(\delta_{40})^{-1}$):

$$(3.7)\qquad \ddot{\eta}_2 + \left[-\delta(t - t_\infty)^{-2} + \text{h.o.t.}\right]\eta_2 = 0.$$

Recall from Section 1 that we require $\pi_h(0) = 0$. The symmetry in Hypothesis 1(a) implies $\pi_h(t + \omega_1(h)/2) = \pi_h(-t) = -\pi_h(t)$ for $t \in \mathbb{R}$, and hence for $t \in \mathbb{C}$. Thus the coefficients of (3.6b) have real period $\omega_1(h)/2$. As in the proof of Theorem 2, for $h \in (0,h^*)$ sufficiently close to 0 we have (by Hypothesis 4) $g_1(h) = \Psi(\omega_1(h)/2;h) \in SL(2,\mathbb{R})$ with trace $g_1(h) > 2$.

The eigenvalues of the anti-commutator $g_*(h) = g_2^{-1}g_1^{-1}g_2g_1$ can be calculated by analytic continuation in a positive sense of a fundamental matrix solution of (3.7) about one of the poles of $\pi_h(t)$ as

$$(3.8)\qquad \lambda_j = \exp(2\pi i\rho_j),\quad j = 1,2,$$

where the ρ_j are the roots of the indicial equation

(3.9) $\rho(\rho - 1) - \delta = 0.$

Note that although the two poles depend on h, we have trace $g_*(h)$ independent of h. Now recall the notation of Theorem 1.

THEOREM 3. Assume Hypotheses 1, 4, and 5, and set $\delta = \delta_{22}(\delta_{40})^{-1}$. Then the system (1.2) has no second integral F that is meromorphic in a neighborhood in $\mathbb{C}^4$ of the image $\Gamma(h)$ and functionally independent of the Hamiltonian H in this neighborhood at energies $0 < h < h^*$ (except possibly on a set of measure zero of such energies, unless Remark 2 applies) provided
$(1 + 4\delta) \neq (\text{odd integer})^2$.

PROOF. The proof is completely analogous to that of Theorem 2. Q.E.D.

This paper was originally motivated by an example of Ziglin in [8, Section 4]. There he considers the potential

(3.10) $W(x_1,x_2) = (1/2)x_1^2 x_2^2$

which has gradient lines $x_2 = \pm x_1$ (in addition to the two axes). Following Remark 1(b), let T be a rotation of the (x_1,x_2) coordinates through an angle $(\pi/4)$, and call the new coordinates (x_1,x_2) again. Then

(3.11) $(W \circ T)(x_1,x_2) = (1/8)\left[x_1^4 - 2x_1^2 x_2^2 + x_2^4\right],$

with $\delta = \delta_{22}(\delta_{40})^{-1} = -2$. Hence $(1 + 4\delta) < 0$ and Hypotheses 1, 4, and 5(b2) hold for (3.11). Theorem 3 then implies that the system (1.2) with (3.10) is non-integrable, by homogeneity of the potential, at all energies $h > 0$, the flows at any two positive energies being conjugate. This is Ziglin's conclusion in [8, Section 4]. Note that the same conclusion holds for the 4-saddle potential,

(3.12) $W(x_1,x_2) = (1/4)$ Real $(x_1 + ix_2)^4.$

The system (1.2) with (3.12) was shown to be non-integrable at energies $h > 0$ by other means in [2]. For further applications of Theorem 1 to non-integrability conclusions, see [8], [3], [6], and [9].

REFERENCES

[1] G. Birkhoff and G-C. Rota, "Ordinary Differential Equations," third edition, John Wiley and Sons, New York, 1978.

[2] R.C. Churchill, and D.L. Rod, Pathology in dynamical systems II: Applications, J. Differential Equations, 21 (1976), 66-112.

[3] Hidekazu Ito, Non-integrability of Hénon-Heiles system and a theorem of Ziglin, Kodai Math. J., 8 (1985), 120-138.

[4] W. Magnus and S. Winkler, "Hill's Equation," Dover Publications, Inc., New York, 1979.

[5] L. Markus, and K.R. Meyer, Generic Hamiltonian Dynamical Systems are neither Integrable nor Ergodic, Memoirs of the Amer. Math. Soc., No. 144, Amer. Math. Soc., Providence, 1974.

[6] D.L. Rod, On a theorem of Ziglin in Hamiltonian dynamics, to appear.

[7] C.L. Siegel, "Topics in Complex Function Theory," Vol. 1 (Elliptic Functions and Uniformization Theory), Wiley-Interscience, New York, 1969.

[8] S.L. Ziglin, Branching of solutions and nonexistence of first integrals in Hamiltonian mechanics: I and II. Functional Anal. Appl., 16 (1982), 181-189, and 17(1983), 6-17.

[9] Hidekazu Ito, A criterion for non-integrability of Hamiltonian systems with nonhomogeneous potentials, preprint.

DEPARTMENT OF MATHEMATICS AND STATISTICS
THE UNIVERSITY OF CALGARY
CALGARY, ALBERTA
CANADA T2N 1N4

Canadian Mathematical Society
Conference Proceedings
Volume 8 (1987)

SENSITIVE DEPENDENCE, FINITE DECOMPOSITIONS, AND HOMOCLINIC BIFURCATIONS IN A MODEL FOR AN EPIDEMIC

Thomas D. Rogers and Zhou-Cheng Yang

ABSTRACT. A parameterized family of continuous maps on an interval, representing a simple epidemiological model of May, is proved to have sensitive dependence to initial conditions for all relevant values of the parameter. This means that the future values of two nearby initial points will separate widely, implying unpredictability. An additional, contrasting, result is that there exists a finite number of subintervals which attract almost all initial points, the actual number depending on the location of the parameter within a sequence of homoclinic bifurcations. The proofs relay on the recent generalizations of Preston to continuous unimodal mappings of the Guckenheimer–Misiurewicz theory for the dynamics of smooth mappings of the interval. Assuming certain simple growth conditions, suggested by the example, extensions of the results to more general families of maps are also given.

1. INTRODUCTION. In [6] May, and in [7] May and Anderson, derive and study the simplest model for the density dependent regulation of a host population with separated generations by a viral, bacterial or protozoan microparasite which spreads like an epidemic through the generations. Referring to [6], the derivation says that the fraction θ of the population that is infected at epidemic's end is related to the population size N through the Kermack–McKendrick equation

$$1-\theta = \exp\left(-\theta N/N_T\right),$$

where N_T is a threshold density. When $N \leq N_T$ the epidemic does not establish itself and the only possible solution is $\theta = 0$, whereas when $N > N_T$ the epidemic spreads and there is a solution $\theta \neq 0$. Now if in each generation the epidemic ends prior to the event of reproduction and the parasite is lethal, then the fraction of the host which

1980 Mathematics Subject Classification (1985 Revision). Primary 58F40; Secondary 92A17
Research partially supported through grants from National Sciences and Engineering Research Council of Canada.

survives is $1-\theta$. The simplest density dependent difference equation representing the host–parasite is then

$$N_{t+1} = \lambda N_t[1-\theta(N_t)],$$

where $\theta = \theta(N)$ is given implicitly through the Kermack–McKendrick expression and λ is the intrinsic reproductive rate of the host.

Finally, May introduces the dimensionless variable $x = N/N_T$, and his difference equation model becomes

$$x_{n+1} = f_\lambda(x_n) = \begin{cases} \lambda x_n, & 0 \le x_n \le 1 \\ \lambda x_n(1-\theta(x_n)), & x_n \ge 1 \end{cases}$$

where

$$1-\theta(x) = \exp(-x\theta(x)), \qquad x \ge 1.$$

Note that the inverse function to θ, $\varphi : (0,1) \to [1,\infty)$ has the series representation $\varphi(t) = \sum_{n=1}^{\infty} t^{n-1}/n$, $0 < t < 1$, which can be differentiated term by term. Since φ is strictly increasing and differentiable, θ has the same properties.

May's study, and a subsequent one by Rogers, Yang and Yip [10], reveal that the dynamics of the system have certain regularity properties and certain properties which could certainly be described as irregular (unpredictable), depending on the size of the parameter λ; generally the greater the value of λ the more random do the dynamics appear.

Throughout the paper we assume that $\lambda > 1$; for brevity we may occasionally write f in place of f_λ. The point $x = 1$ is always a turning point of the map and the corresponding maximum value is λ. One may verify that the righthand derivative of f_λ at $x = 1$ is $-\lambda$; therefore the map possesses a cusp at its unique turning point $x = 1$ with slope equal to λ in magnitude. The point $x = 0$ is an equilibrium and there is a positive equilibrium point $\bar{x} = \bar{x}(\lambda) = (\lambda \ln \lambda)/(\lambda - 1)$ when $\lambda > 1$. It appears numerically that for small $\lambda > 1$ the solution resembles a noisy two-cycle, while for λ large enough, the iterates appear to densely fill in the attracting interval $[f(\lambda), \lambda)]$. See [10] for a bifurcation diagram. The purpose of this paper is to give a complete topological description of the dynamics of the family.

Our main result states that regardless of the value of $\lambda > 1$, f_λ *has sensitive dependence to initial conditions* — meaning, roughly, that each initial point has neighbors arbitrarily close to it whose corresponding trajectory eventually separates widely from that of the chosen initial point (see Guckenheimer [3]). In proving this, we rely on the property that *all of the periodic orbits of* f_λ *are unstable for all* $\lambda > 1$. The following was proved in [10].

THEOREM 1.1. *For any integer $n \geq 1$ and $\lambda > 1$*

$$\frac{d}{dx}(f_\lambda^n(x)) \text{ exists } \Rightarrow \left|\frac{d}{dx}(f_\lambda^n(x))\right| \geq \frac{f_\lambda^n(x)}{x}$$

and the equality holds if and only if $0 < f^j(x) < 1$ for $j = 0, 1, \ldots, n-1$.

The notation f_λ^n refers to the self-composition function. The proof deals with the implicit form of the function $\theta(x)$, and the reader is referred to [10] for the analytic details. Notice now that if x belongs to an n-cycle of f, $x = f_\lambda^n(x)$, $n \geq 2$, then there are points of the cycle which lie both to the left and to the right of the turning point 1. If no element of the cycle is 1 (so that derivatives exist) we conclude from the theorem

$$\left|\frac{d}{dx} f_\lambda^n(x)\right| > \frac{f_\lambda^n(x)}{x} = 1.$$

This implies the cycle is unstable.

COROLLARY 1.2. *The system $x_{n+1} = f_\lambda(x_n)$ is completely chaotic for all $\lambda > 1$. Specifically if $x \in [0, \infty]$ satisfies $f_\lambda^n(x) = x$, then x is an unstable periodic point.*

Proof. This follows from the theorem as long as 1 does not belong to the cycle, as discussed above. If 1 does belong to the cycle only a slight modification of the proof is needed. Replace the operator $\frac{d}{dx}$ by the operator

$$\left|\frac{d}{dx}\right| f_\lambda(x) = \lim_{h \to 0} \frac{|f_\lambda(x+h) - f_\lambda(x)|}{|h|}.$$

We see that $|\frac{d}{dx}| f_\lambda(x) = |\frac{d}{dx} f_\lambda(x)|$ if $x \neq 1$ and $|\frac{d}{dx}| f_\lambda(x) = \lambda > 1$ if $x = 1$. The operator is continuous and an obvious modification of the chain rule applies. If the turning point $x = 1$ belongs to an n-cycle then $f_\lambda^n(1) = 1$ and it is easy to check that $|\frac{d}{dx}| f_\lambda^n(1) > 1$. Therefore, if h is sufficiently small $|f_\lambda^n(1+h) - 1| > |h|$ and so 1 repells nearby points. We conclude that the system is completely chaotic.

In the next section we will repeatedly use the following fundamental lemma which describes the expansive properties of f and its iterates on intervals which are free of turning points.

FUNDAMENTAL LEMMA. *Let $[a, b]$ be a subinterval of $(0, \lambda]$ such that $f^n = f_\lambda^n$ is monotone on $[a, b]$. Suppose $f^n([a, b]) = [c, d]$. Then $\frac{d}{c} \geq \frac{b}{a}$ and the equality holds if and only if $f^j([a, b]) \subset (0, 1]$, $j = 0, 1, \ldots, n-1$.*

Proof. Since f^n is monotone on $[a, b]$, $1 \notin \operatorname{int}(f^j([a, b]))$ for all $j = 0, 1, \ldots, n-1$. Thus f^n is differentiable on (a, b) and

$$\log \frac{d}{c} = |\log f^n(a) - \log f^n(b)| = \int_a^b |\frac{d}{dx} \log f^n(x)| dx$$

$$= \int_a^b \frac{|\frac{d}{dx} f^n(x)|}{f^n(x)} dx \geq \int_a^b \frac{1}{x}\, dx = \log \frac{b}{a}.$$

The inequality follows from the corresponding inequality for the derivative.

COROLLARY 1.3. *Let $[a,b]$ be a subinterval of $[0,\lambda]$ and let $n \geq 1$. If $f^n([a,b]) \subset [a,b]$ then f^n is not monotone on $[a,b]$.*

Proof. If $f^n([a,b]) \subset [a,b]$ then there exists $0 \leq j \leq n-1$ such that $f^j([a,b]) \not\subset [0,1]$ (otherwise $f^n([a,b]) = [\lambda^n a, \lambda^n b]$ where $\lambda > 1$). Suppose f^n is monotone on $[a,b]$ and $f^n([a,b]) = [c,d]$. Then by the fundamental lemma, $\frac{d}{c} > \frac{b}{a}$. But this implies $[c,d] \not\subset [a,b]$ which contradicts f^n being a self-map of $[a,b]$.

2. CLASSIFICATION OF DYNAMICAL BEHAVIOR. Let $S = S[0,\lambda]$ be the set of continuous self-maps of an interval $[0,\lambda]$ which are *unimodal*, i.e., each $f \in S$ has a unique turning point $\varphi \in [0,\lambda]$ such that f is strictly increasing on $[0,\varphi]$ and strictly decreasing on $(\varphi,\lambda]$. Then if $f \in S$ is sufficiently smooth (negative Schwarzian derivative; see Singer [13]) the combined results of Misiurewicz [8] and Guckenheimer [3] say that f admits only three kinds of limiting behaviour for its iterates. These are roughly described as: (i) f has a stable periodic orbit; (ii) f has no stable periodic orbit and f has sensitive dependence to initial conditions; (iii) f has no stable periodic orbits but has an attracting Cantor set on which f acts like an infinite register shift. In cases (i) and (iii) the points which are attracted form an open dense set (of full Lebesque measure, in fact). The example that we study in this paper requires that we refer to Preston's recent monograph [9] on generalizations of the smooth theory of Guckenheimer and Misiurewicz to the continuous piecewise monotone case, or more specifically, to functions in S which are only continuous. We wish to prove that each of the maps $f = f_\lambda$, $\lambda > 1$, has sensitive dependence to initial conditions, and along the way we need to eliminate a number of other possibilities.

In the following we recall the appropriate definitions (sinks, etc.) as we need them from Preston. These will differ somewhat from the more familiar definitions when f is C', as found in, say, Collet and Eckmann [2].

Let $f \in S$. A *sink* of f is a closed nontrivial interval J such that for some $n \geq 1$, f^n is monotone on J and $f^n(J) \subset J$. This is a useful concept, for if J is a sink then clearly f^m is monotone on J for all $m \geq 0$.

The next result is simply a restatement of Corollary 1.3.

COROLLARY 2.1. *For all $\lambda > 1$ the map $f = f_\lambda$ has no sink.*

Remark. In the terminology of Preston, it follows that f has no traps (a special kind of sink) and further that f is not central (which also requires the existence of a sink).

A special interval which may or may not exist simultaneously with a sink is a homterval. Let $f \in S$ with turning point φ. A nontrivial closed interval J is a *homterval* if $\varphi \notin f^n(J)$ for all $n \geq 0$ and $f^n(J)$ is not contained in any sink for each $n \geq 0$. An important consequence (Proposition 5.8, Preston [9]) is that if J is a homterval then all of the image intervals $f^n(J)$, $n = 0, 1, \ldots$ are disjoint. Furthermore, if J is a nontrivial closed interval and $\varphi \notin f^n(J)$ for all $n \geq 0$, then J is a homterval if and only if no element of J attracts to a periodic orbit of f.

LEMMA 2.2. *For all $\lambda > 1$ the map $f = f_\lambda$ has no homtervals.*

Proof. Suppose $J = [a, b] \subset [0, \lambda]$ is a homterval so that the forward images $f^n(J)$, $n \geq 0$, are disjoint. Then $0 \notin J$ since $J \cap f(J) = \phi$ (0 is a fixed point of f). So there exists a $k \geq 0$ such that $f^k(J) \subset [f(\lambda), \lambda]$. Note that all points in $(0, \lambda]$ map in finite time into the invariant interval $[f(\lambda), \lambda]$. Since $1 \notin f^n(J)$ for all $n \geq 0$, each f^n is monotone on J. Therefore by the fundamental lemma, $|f^n(J)| \geq (\frac{b}{a} - 1) f(\lambda)$ for $n \geq k$, and consequently $\sum_{n=1}^{\infty} |f^n(J)| = \infty$, so that some of the intervals $f^n(J)$ must intersect in $[0, \lambda]$. This is a contradiction. We conclude that f has no homtervals.

Let $f \in S$ with domain $[0, \lambda]$ and let $x \in [0, \lambda]$. Call x *pseudoperiodic* if for each $\varepsilon \geq 0$ there exists $n \geq 1$ such that

(a) $|f^j(x) - f^k(x)| < \varepsilon$ when $j = k \pmod{n}$

(b) x does not lie in between $f^j(x)$ and $f^k(x)$ (inclusive) if $j = k \pmod{n}$ and $j \neq 0 \pmod{n}$.

Note that any periodic point is also pseudoperiodic.

Finally, with Preston, we formally define sensitive dependence. Let $f \in S$ with domain $[0, \lambda]$ and let $\varepsilon > 0$. Define

$$\textstyle\sum_\varepsilon(f) = \Big\{x \in [0, \lambda] : \text{for each } \delta > 0 \text{ there exists } n \geq 0 \text{ such that } |f^k((x - \delta,\, x + \delta))| \geq \varepsilon \text{ for all } k \geq n\Big\}.$$

Say that f has *sensitive dependence to initial conditions* if there is an $\varepsilon > 0$ such that $\sum_\varepsilon(f) = [0, \lambda]$. Preston proves that if f has sensitive dependence, with corresponding $\varepsilon > 0$, then the sets

$$\Omega_\eta(x, f) = \{y \in [0, \lambda] : |f^n(y) - f^n(x)| > \eta \quad \text{for some } n \geq 0\}$$

$$\Omega_\eta^*(x, f) = \{y \in [0, \lambda] : |f^n(y) - f^n(x)| > \eta \quad \text{for infinitely many } n \geq 0\}$$

are, respectively, open dense and residual, provided that $0 < 2\eta < \varepsilon$.

Preston's definition is stronger than that originally given by Guckenheimer, who required only that the measure of $\sum_\varepsilon(f)$ be positive for some $\varepsilon > 0$.

A condition for sensitive dependence is provided by the following theorem.

THEOREM 2.3. *(Preston). Let $f \in S$ with turning point φ satisfying $\varphi < f(\varphi)$, and suppose that 0 is the only fixed point of f in $[0, \varphi]$. Suppose that f has neither sinks nor homtervals. Then, if in addition, φ is not pseudoperiodic, unless it is periodic, f has sensitive dependence to initial conditions.*

Proof. Combine lemmas 5.2 and 5.5 and the major theorem 5.2 of Preston [9].

We may now state our main result.

THEOREM 2.4. *For all $\lambda > 1$ any periodic orbit of the map $f = f_\lambda$ is unstable and f has sensitive dependence to initial conditions.*

Proof. We apply Theorem 2.3. From the above results, all that remains to be shown is that the turning point $\varphi = 1$ of f is not pseudoperiodic, if it is not periodic.

Assume to the contrary that 1 is pseudoperiodic but not periodic.

Let $\{\varepsilon_i\}$ be a sequence of positive numbers decreasing to zero. Then by the definition of pseudoperiodicity, there exists a sequence of integers $\{n_i\}$ such that $|f^j(1) - f^k(1)| < \varepsilon_i$ if $j = k \pmod{n_i}$ and 1 does not lie between $f^j(1)$ and $f^k(1)$ if $j = k \pmod{n_i}$ and $j \neq 0 \pmod{n_i}$.

Now let $n = n_i$ for some fixed i, and let k be any positive integer. Then for any $0 < j < n$, 1 does not lie between $f^j(1)$ and $f^{nk+j}(1)$. So f^n is monotone on $[f^{nk}(1), 1]$ or $[1, f^{nk}(1)]$ according to $f^{nk}(1) < 1$ or $f^{nk}(1) > 1$. (For suppose $[f^{nk}(1), 1]$ is an interval and choose $0 < j < n$ such that $f, \dots, f^j$ are monotone on $[f^{nk}(1), 1]$ but f^{j+1} is not. Then there exists $0 \leq i \leq j$ such that $f^i(x) = 1$ for some $x \in (f^{nk}(1), 1)$. But then $f^{nk+i}(1) < 1 < f^i(1)$ which contradicts the assumed pseudoperiodicity of 1.)

Write $J_k = [f^{nk}(1), 1]$ or $[1, f^{nk}(1)]$ as the case may be. Then $f^n(J_k) = [f^{n(k+1)}(1), f^n(1)]$ or $[f^n(1), f^{n(k+1)}(1)]$ and thus

$$f^n(J_k) \subset J_1 \cup J_{k+1}.$$

Let $J = c\ell \bigcup_{k=1}^{\infty} J_k$. Then $J \subset [1 - \varepsilon_i, 1 + \varepsilon_i]$, because $|f^{nk}(1) - 1| < \varepsilon_i$ by pseudoperiodicity and so $\bigcup_{k=1}^{\infty} J_k \subset [1 - \varepsilon_i, 1 + \varepsilon_i]$.

Observe now that $f^n\left(\bigcup_{k=1}^{\infty} J_k\right) \subset J$ which implies $f^n(J) \subset J$. We have proved that for each $i \geq 0$ there exists a nontrivial closed interval $J \subset [1 - \varepsilon_i, 1 + \varepsilon_i]$ such that $f^{n_i}(J) \subset J$. (Since $1 \varepsilon J_k$ for all k, the union of the J_k's is connected.)

Let $0 < j < n_i = n$. The countable collection of points $f^{nk+j}(1)$, $k = 0, 1, \dots$ is the orbit of $f^j(1)$ under f^n, and 1 can lie between no two of these points. The set $f^j(J)$ is the smallest closed interval containing these points, and thus for each $0 < j < n$ the turning point does not lie in the interior of $f^j(J)$. So if we let $J' = f(J)$ then f^{n-1} is monotone on J' and $f^n(J') \subset J'$. By corollary 1.3, 1 is in the interior of $f^{n-1}(J')$. Thus we can write $f^{n-1}(J') = [c, 1] \cup [1, d]$ and $J' = J_1 \cup J_2$ where J_1 and J_2 are closed subintervals of J' such that $J_1 \cap J_2$ is singleton, $f^{n-1}(J_1) = [c, 1]$ and $f^{n-1}(J_2) = [1, d]$. Also, f^n is monotone on both of J_1 and J_2. Then from the invariance of f^n on J', we have $f^n(J_1) \cup f^n(J_2) \subset J'$ and hence the estimates

$$\int_{J_i} \left|\frac{d}{dx} f^n(x)\right| dx \leq |J'|, \quad i = 1, 2.$$

It follows that

$$\int_{J'} \left|\frac{d}{dx} f^n(x)\right| dx \leq 2|J'|.$$

Fix $\lambda > 1$. When ε_i is sufficiently small, J' does not contain the equilibrium point $\overline{x} = (\lambda \log \lambda)/(\lambda - 1)$. Choose an integer k depending only on λ such that $\lambda^k(f(\lambda) > 1$. Then for any $x \in [f(\lambda), \lambda]$ at least one of the points x, $f(x), \ldots, f^k(x)$, $f^{k+1}(x)$ lies in $[\overline{x}, \lambda]$. Thus among the intervals J', $f(J')$, $f^2(J'), \ldots, f^{n_i - 1}(J')$ at least $\left[\frac{n_i}{k+2}\right]$ of them are contained in $[\overline{x}, \lambda]$. The brackets refer to the least integer function. (The idea is that since $\overline{x}$ can belong to none of the $f^j(J')$, if any point of the array lies in $[\overline{x}, \lambda]$, the corresponding entire interval does.)

Now on the subinterval $[\overline{x}, \lambda]$, f' is continuous and $|f'(x)| > f(x)/x$. Therefore, there is a positive δ such that $|f'(x)| > (1 + \delta) f(x)/x$ for all $x \in [\overline{x}, \lambda]$. Furthermore, using the chain rule we have for $x \in J'$,

$$\left|\frac{d}{dx} f^{n_i}(x)\right| > \frac{f^{n_i}(x)}{x} \cdot (1 + \delta)^{\left[\frac{n_i}{k+2}\right]} \geq \frac{\overline{x}}{\lambda} (1 + \delta)^{\left[\frac{n_i}{k+2}\right]},$$

since $f^{n_i}(x) \in J' \subset [\overline{x}, \lambda]$. This implies the inequality

$$\int_{J'} \left|\frac{d}{dx} f^n(x)\right| dx \geq |J'| \frac{\overline{x}}{\lambda} (1 + \delta)^{\left[\frac{n_i}{k+2}\right]},$$

and consequently,

$$\frac{\overline{x}}{\lambda} (1 + \delta)^{\left[\frac{n_i}{k+2}\right]} \leq 2.$$

This is a contradiction since $n_i \to \infty$ as $i \to \infty$, and hence 1 is not pseudoperiodic unless it is periodic. We have proved that each f_λ, $\lambda > 1$, has sensitive dependence to initial conditions.

3. FINITE SHIFTS AND A SEQUENCE OF HOMOCLINIC BIFURCATIONS. Although there are no infinite register shifts as we have shown, there are *finite shifts*, which furthermore possess arbitrarily large lengths as $\lambda \to 1$. Such shifts were noted in piecewise linear unimodal maps by Milnor and Thurston, and later Jonker and Rand [4] showed generally for continuous unimodal maps that there exists a canonical decomposition of the nonwandering set into at most a countable number of disjoint closed sets on which the map is strongly semiconjugate to a piecewise linear map. The following theorem provides an explicit description in our example of the decomposition of the invariant interval into a finite number of intervals on which the map acts like a shift, along with a description of related homoclinic bifurcations.

THEOREM 3.1. *Consider the family f_λ $\lambda > 1$. Then there exists a sequence λ_i, $i = 0, 1, \ldots$ with the properties*

1. *λ_i decreases to 1 as $i \to \infty$.*
2. *λ_i is the smallest λ such that $f_\lambda^{2^i+1}(1)$ belongs to a 2^{i-1} cycle.*
3. *For $\lambda < \lambda_i$ there exist 2^i disjoint nontrivial closed intervals $J_1, \ldots, J_{2^i}$ such that $J_1 = [f_\lambda^{2^i+1}(1), f_\lambda(1)]$ and $f_\lambda(J_j) = J_{j+1}$ for $j = 1, 2, \ldots, 2^i - 1$ and $f_\lambda(J_{2^i}) = J_1$.*
4. *If $\lambda < \lambda_i$, then for almost all x there exists $n > 0$ such that $f_\lambda^n(x) \in J_j$ for some $1 \le j \le 2^i$.*

Proof. Let $\lambda_0 = +\infty$. We use induction on i. Let $\lambda > 1$ and consider the restriction of f_λ to $[f_\lambda(\lambda), \lambda]$. A calculation shows that $f_\lambda(\lambda) < 1$, f_λ maps $[f_\lambda(\lambda), \lambda]$ onto itself, and $f_\lambda \in S([f_\lambda(\lambda), \lambda])$. Also when λ is large, say $\lambda \ge 4$, then $f_\lambda^2(\lambda)$ is less than the fixed point $\bar{x}(\lambda)$ of f_λ.

Suppose we have found λ_i satisfying (2) and (3) and for $\lambda < \lambda_i$, $f_\lambda^{2^i}$ maps $[f_\lambda^{2^i}(\lambda), \lambda]$ onto itself. Additionally, we suppose $f_\lambda^{2^i} \in S([f_\lambda^{2^i}(\lambda), \lambda])$. Corollary 1.3 says $f_\lambda^{2^i}$ cannot be monotone on this interval.

Write $g = f_\lambda^{2^i}$. The interval becomes $[g(\lambda), \lambda]$. We know that g is differentiable on $(g(\lambda), \varphi) \cup (\varphi, \lambda)$ where φ is its turning point. A calculation shows that when $|g'(x)| < \sqrt{2}$ the value of g at the left endpoint is greater than the fixed point of g (the 2^i-cycle of f). In fact, $|g'(x)| < \sqrt{2}$ when $\lambda < 2^{2^{-(i+1)}}$ since $|f'(x)| \le \lambda$ implies $|g'(x)| \le \lambda^{2^i} < \sqrt{2}$. When λ is close to λ_i, by the inductive assumption (2), $g^2(\lambda)$ is close to $g(\lambda)$, so it is smaller than the fixed point of g. Thus we can find $\lambda_{i+1} \ge 2^{2^{-(i+1)}}$ satisfying (2).

Now suppose that $\lambda < \lambda_{i+1}$. We want to prove that $g(\lambda) < g^3(\lambda) < g^2(\lambda) < \lambda$, and $\varphi < g^3(\lambda)$, $g^4(\lambda) > g^2(\lambda)$.

Since g is not monotone on $[g(\lambda), \lambda]$, $g(g(\lambda)) = g^2(\lambda) < \lambda$. Also we are supposing that $g^2(\lambda) > \bar{x}$, the fixed point of g. We have $g(\lambda) < g^3(\lambda) < g^2(\lambda)$ since g is monotone on $[\bar{x}, \lambda]$. Also $\varphi < q^3(\lambda)$ since otherwise g^2 maps $[g^2(\lambda), \lambda]$ into itself and is monotone there. Finally, let z be the point in $(\varphi, \bar{x})$ such that $g(z) = g(g(\lambda))$. Then g maps $[z, g^2(\lambda)]$ monotonically onto $[g^3(\lambda), g^2(\lambda)]$. By corollary 1.3, $g^3(\lambda) < z$. This and $\varphi < g^3(\lambda)$ together give that $g^3(\lambda) < g^4(\lambda) < \lambda$.

Now the two intervals $I_1 = [g^2(\lambda), \lambda]$ and $I_2 = [g(\lambda), g^3(\lambda)]$ are mapped onto each other by g. This fact and the inductive assumption (3) shows that I_1, $f(I_1), \ldots$, $f^{2^i-1}(I_1)$, $f^{2^i}(I_1) = I_2$, $f(I_2), \ldots, f^{2^i-1}(I_2)$ are disjoint nontrivial closed intervals, and $f(f^{2^i-1}(I_2)) = I_1$. Thus we have proved (3). The map $g^2 = f^{2^{i+1}}$ restricted to $[g^2(\lambda), \lambda]$ belongs to $S([g^2(\lambda), \lambda])$ and maps the interval into itself.

Property (4) is also proved by induction on i. For $i = 0$, let $J_1 = [f_\lambda(\lambda), \lambda]$. Suppose now that (4) is true for $i - 1$. Then for all x except on a countable set there exists a k

such that

$$f_\lambda^k(x) \in [f_\lambda^{2^{i-1}}(\lambda), \lambda].$$

Write $f^{2^{i-1}} = g$ and $f_\lambda^k(x) = y$. We may suppose that y is not $\overline{x} =$ the fixed point of g. This is true for all x except on a countable set (the preimages of $\overline{x}$). Since $\lambda < \lambda_i$, $g^2(\lambda) > \overline{x}$ and the two disjoint subintervals $[g(\lambda), g^3(\lambda)]$ and $[g^2(\lambda), \lambda]$ are mapped onto each other. Note that $\overline{x}$ is the only fixed point of g in $[g(\lambda), \lambda]$. Now consider the sequence g, $g(y)$, $g^2(y), \ldots$. Suppose it is contained in $(g^3(x), g^2(x))$. Then y, $g^2(y)$, $g^4(y), \ldots$ is increasing if $y > \overline{x}$ and decreasing if $y < \overline{x}$ since there is no invariant interval of g in $(g^3(\lambda), g^2(\lambda))$. Thus the sequence y, $g^2(y)$, $g^4(y), \ldots$ converges to a point in $[g^3(x), g^2(x)]$, say to $\overline{y}$. Then $\overline{y}$ is a 2-cycle of g. This is impossible, however, since it would imply the existence of an invariant subinterval of g in $[g^3(\lambda), g^2(\lambda)]$. Therefore, some forward iterate of almost all x under f_λ lies in one of the intervals J_j, $1 \le j \le 2^i$. The theorem is proved.

The results of the paper are subject to immediate generalization to other families of continuous unimodal maps, the main requirement being that the fundamental lemma pertains. Let $f_\lambda : [0, \infty) \to [0, \infty)$ satisfy

(i) $f_\lambda \in S$

(ii) $f_\lambda(1) > 1$ and f_λ has no fixed points in $(0, 1)$

(iii) $\left|\frac{d}{dx} f_\lambda(x)\right| \ge \frac{f_\lambda(x)}{x}$ for $0 < x < 1$ and $\left|\frac{d}{dx} f_\lambda(x)\right| > \frac{f_\lambda(x)}{x}$ for $x > 1$.

THEOREM 3.2. *Let $f_\lambda : [0, \infty) \to [0, \infty)$ be any map satisfying the conditions (i), (ii), (iii). Then f_λ has sensitive dependence for all $\lambda > 1$.*

Proof. Follows immediately from the proof of Theorem 2.4 which relies only on the stated conditions.

The condition (iii) states that f_λ increases faster on $(0, 1)$ and decreases slower on $(1, \infty)$ than the family

$$g_\lambda(x) = \begin{cases} \lambda x, & 0 \le x \le 1 \\ \dfrac{\lambda}{x}, & x > 1. \end{cases}$$

Note that under g_λ all points in $[1, \lambda]$ are 2-cycles except for the fixed point.

We may also state

THEOREM 3.3. *Let $f_\lambda : [0, \infty) \to [0, \infty)$ satisfy (i) – (iii) and in addition the condition: for every $\varepsilon > 0$, there is a $\delta > 0$ such that*

$$\text{(iv)} \qquad \lambda \le 1 + \delta \Rightarrow \left|\frac{d}{dx} f_\lambda(x)\right| \le 1 + \varepsilon, \qquad x \in [f_\lambda^2(\lambda), f_\lambda(\lambda)].$$

Then there exist disjoint closed intervals $J_1, \ldots, J_n$ (n depends on λ) such that

(1) $f_\lambda(J_i) = J_{i+1}$, $i = 1, \ldots, n-1$, and $f_\lambda(J_n) = J_1$;

(2) for almost all x, there is a k such that $f_\lambda^k(x) \in J_i$ for some i. Furthermore, $n \to \infty$ as $\lambda \to 1$ and the endpoints of these intervals are $f_\lambda(1), f_\lambda^2(1), \ldots, f_\lambda^{2n}(1)$.

Proof. The proof is similar to that of Theorem 3.1.

There is an alternate description of the λ_i as a sequence of bifurcations of homoclinic periodic points (Block and Hart [1], Rogers and Marotto [11]). A periodic point p of period n under f is *homoclinic* if $p \in f^n(W^u(p, f^n)\backslash\{p\})$, where $W^u(p, f^n)$ is the unstable manifold of p under f^n. Among other things this characterizes each λ_i as the largest λ such that f_λ does not have $2^i \cdot q$–cycles where q is an odd number. For example, for all $\lambda > \lambda_1 \approx 2.9$ a calculation shows that we have the configuration $f_\lambda^2(1) < 1 < f_\lambda(1)$ and $f_\lambda^5(1) < f_\lambda^4(1)$. Observe that $\lambda = \lambda_1$ marks the first crossing of the critical curves $f_\lambda^3(1)$ and $f_\lambda^4(1)$ beyond $\lambda = 1$, and further that this crossing is transversal. (Recall at $\lambda = \lambda_1$, $f_\lambda^3(1) = \overline{x}(\lambda) =$ fixed point of f_λ.) Then the two noted inequalities imply, by one of the "no division" theorems of Li *et al.* [5], that f_λ has a point of period 6 for all $\lambda > \lambda_1$ and f_λ^2 has points of all periods. Hence f_λ has a homoclinic orbit for all $\lambda > \lambda_1$ (and hence infinitely many homoclinic orbits, of infinitely many periods). By Sharkovsky's theorem [2], f_λ has points of all even periods in the same parameter region. Similar calculations will establish the existence of a homoclinic orbit for f_λ^2 at $\lambda = \lambda_2 \approx 1.6$ (also visible on the bifurcation diagram; see [10]), and so on, in a decreasing sequence of λ_i's. It is worth noting the indicated order (relative position of the critical curves in a neighborhood of the homoclinic bifurcation $\lambda = \lambda_1$ is the same as that of the quadratic family. Compare with [12] in this regard.

Finally, we conjecture that the λ_i converge geometrically with $\lambda_i \geq 2^{2^{-(i+1)}}$ and $\lambda_i - 1 = 0(2^{-(i+1)})$.

References

1. L. Block and D. Hart., *The bifurcation of homoclinic orbits of maps of the interval*, Ergodic Theory and Applications **2** (1982), 131–138.

2. P. Collet and J.-P. Eckmann., "Iterated Maps on the Interval as Dynamical Systems," Birkhauser, Boston, 1981.

3. J. Guckenheimer, *Sensitive dependence on initial conditions for one-dimensional maps*, Communications in Mathematical Physics **70** (1979), 133–160.

4. L. Jonker and D. Rand, *Bifurcations in one dimension*, Inventiones Mathematicae **62** (1981), 347–365.

5. T.Y. Li, M. Misiurewicz, G. Pianigiani, and J.A. Yorke, *No division implies chaos*, Transactions of the American Mathematical Society **273** (1982), 191–199.

6. R.M. May, *Regulation of populations with nonoverlapping generations by microparasites: A purely chaotic system*, Preprint 1982 Department of Biology; Princeton University.

7. R.M. May and R.M. Anderson, *Epidemiology and genetics in the coevolution of parasits and hosts*, Philosophical Transactions of the Royal Society **B219** (1983), 281–313.

8. M. Misiurewicz, *Absolutely continuous measures for certain maps of the interval*, Publ. Math IHES (1980).

9. C. Preston, "Iterates of Maps on an Interval," Springer–Verlag, Berlin–New York, 1983.

10. T.D. Rogers, Z.-C. Yang and L.-W. Yip, *Complete chaos in a simple epidemiological model*, Journal of Mathematical Biology **23** ((1986)), 263–268.

11. T.D. Rogers and F.R. Marotto, *Periodic repellers*, Nonlinear Analysis **7** (1983), 97–100.

12. T.D. Rogers and J.R. Pounder, *Dynamics of a two parameter family of maps of the interval*, Nonlinear Analysis **10** (1986), 415–423.

13. D. Singer, *Stable orbits and bifurcations of maps of the interval*, SIAM Journal of Applied Mathematics **35** (1978), 260–267.

Canadian Mathematical Society
Conference Proceedings
Volume 8 (1987)

THE ENERGY-PERIOD FUNCTION AND PERTURBATIONS OF HAMILTONIAN SYSTEMS IN THE PLANE

Franz Rothe

ABSTRACT. We study perturbations of the Hamiltonian system $\dot{q} = \partial H/\partial p$, $\dot{p} = -\partial H/\partial q$. Using action-angle variables (J,Φ), we consider $\dot{J} = \varepsilon F(J,\Phi,t)$, $\dot{\Phi} = \Omega(J) + \varepsilon G(J,\Phi,t)$ where F and G are invariant under the transformations $t \to t + T_f$ and $\Phi \to \Phi + 2\pi$. We are interested in the case of resonance $\Omega(J_0) = 2\pi n/mT_f$ and look for periodic solutions satisfying $J(t + mT_f) = J(t)$, $\Phi(t + mT_f) = \Phi(t) + 2\pi n$, which lie near to an unperturbed solution $\varepsilon = 0$, $J = J_0$, $\Phi = \alpha + \Omega(J_0)t$ with appropriate phase α. Besides the non-degenerate case $\Omega'(J_0) \neq 0$, well-known from the results of Guckenheimer and Holmes, we consider the degenerate case $\Omega'(J_0) = 0$, $\Omega''(J_0) \neq 0$, which leads to a codimension two bifurcation. In the degenerate case, the bifurcating solutions occur in pairs. The number of pairs of solutions depends on a second Melnikov function involving the angular perturbation G. We give a rigorous proof and a geometric argument, which is not entirely rigorous, but interesting because of its hints about stability. The result is compared with work of Hale and Taboas, where an error is pointed out.

1. INTRODUCTION. The dynamical behavior of Hamiltonian systems in the plane is easily understood. Around equilibrium points of elliptic type, neutrally stable oscillations occur. The behavior of the system is described completely by giving the dependence of the frequency on the amplitude of the oscillations, or, equivalently, the energy-period function.

On the other hand, forced Hamiltonian systems of a single degree of freedom exhibit all the complicated dynamical behavior of systems with many

1985 Mathematics Subject Classification 58F05, 58F22, 58F30, 58F10

Work done while visiting the University of North Carolina at Chapel Hill in the academic year 1985/86 degrees of freedom. Since it is possible - and common practice indeed - to reduce Hamiltonian systems of two degrees of freedom to a periodically forced system of one degree of freedom (see e.g. [1] p.213), one may not expect that the latter systems are more well-behaving than the former ones.

A key tool for the description of a time-periodic perturbation turns out to be the rotation number, which is defined to be the ratio of the natural period of the unforced system and the forcing period. The famous KAM-tori occur for strongly irrational rotation numbers, whereas rational rotation numbers lead to resonances, which can be handled much more elementarily. In either case, perturbation theory is a classical approach. In both cases, the nondegeneracy assumption the unperturbed system has to satisfy if perturbation theory is to apply straightforwardly, turns out to be monotonicity and nonvanishing derivative of the energy-period function. (See e.g. [1] p.185 assumption A3, p. 195 Theorem 4.6.2 and p.219 Theorem 4.8.2)

In this note we are interested in the degenerate case with vanishing derivative of the energy-period function, which is definitely excluded in the book of Guckenheimer and Holmes [1]. We restrict our attention to the more elementary case of resonances (rational rotation numbers). We give a rigorous result about existence and nonexistence of resonant solutions by means of bifurcation theory involving a codimension two bifurcation. Alternatively, an intuitive, but not entirely rigorous, geometric approach is described. This approach gives some information about stability, too.

In a second Theorem we state strict results about linearised stability of the bifurcating solutions.

In contrast to the nondegenerate case of strictly monotone energy-period function, in the degenerate case the periodic solutions appear and disappear in pairs, the number of which depend on the sign of a second Melnikov function. This Melnikov function involves the perturbation of the angular equation, whereas the ordinary Melnikov function deals with the perturbation of the radial or energy equation.

The author wants to thank Prof. Hale for drawing his attention to the paper of Hale and Taboas [2]. They are in fact concerned with the degenerate case, but they use a more abstract setting. The main advantage of our approach via action-angle variables is to make clear the role of the energy-period function $T(E)$. As one expects at first glance, the degenerate case occurs for $T'(E_0) = 0$ and the next nondegeneracy assumption needed is $T''(E_0) \neq 0$. Because this point is stated wrongly in [2] section 6, we take the opportunity to link the correct but not very intuitive result from [2] section 5 to our results.

2. STATEMENT OF THE PROBLEM. We come to a more formal statement. Let $\Omega \subset R^2$ be some open domain with $(0,0) \in \Omega$.
Consider the Hamiltonian system

$$(1) \qquad \dot{q} = \frac{\partial H}{\partial p}, \quad \dot{p} = -\frac{\partial H}{\partial q}$$

with $(q,p) \in \Omega$ and $H \in C^3(\Omega)$.

We suppose that $H(0,0) = 0$ and $H(q,p) > 0$ for all $(q,p) \in \Omega \setminus \{(0,0)\}$. Furthermore, we assume that the center $(0,0) \in \Omega$ is the unique equilibrium in Ω. Let the periodic orbits surrounding the center be given by $(q,p)(t,E)$.

To be precise, we define the map $B: (t,E) \in \mathbb{R} \times [0,\infty) \to (q,p)(t,E) \in \mathbb{R}^2$ by requiring that the restriction of B to fixed E satisfies system (1), whereas the restriction of B to $t = 0$ gives the initial data satisfying $q(0,E) = 0$, $p(0,E) > 0$, $H(q(0,E),p(0,E)) = E$. It is straightforward to check that the map B has determinant of Jacobian equal to one.
The energy-period function $T = T(E)$ is defined via the relations

$$(2) \qquad H(q(t,E),p(t,E)) = E, \quad (q,p)(t+T(E),E) = (q,p)(t,E)$$
$$\text{for all } t \in R, \quad E \in [0,\infty)$$

We are going to study the following perturbation of system (1):

$$(3) \qquad \dot{q} = \frac{\partial H}{\partial p} + \varepsilon f(q,p,t), \quad \dot{p} = -\frac{\partial H}{\partial q} + \varepsilon g(q,p,t).$$

Here ε is a small perturbation parameter and the perturbation has period T_f.

$$(4) \qquad (f,g)(q,p,t+T_f) = (f,g)(q,p,t) \quad \text{for all } q,p,t.$$

The coordinate system appropriate to study equations (3) is given by the action-angle variables (J,Φ) of the unperturbed system (1). In this coordinate system, equations (1) transform into

$$(5) \qquad \dot{J} = 0, \quad \dot{\Phi} = \Omega(J).$$

The change of coordinates $K:(J,\Phi) \to (q,p)$ is a canonical transformation (i.e. the determinant of the Jacobian is equal to one). It can easily be obtained from the energy-period function $T = T(E)$ defined above. Indeed, K is the composition of the maps $A:(J,\Phi) \to (t,E)$ and $B:(t,E) \to (q,p)$, both of which have determinant of Jacobian equal to one. Since the variable Φ is an angle to be understood modulo 2π, the inverse to transformation A is

$$(6) \qquad \Phi = \frac{2\pi}{T(E)}\, t \qquad J = \frac{1}{2\pi}\int_0^E T(E')\,dE' .$$

Comparing (5) and (6), we see that the angular frequency $\Omega = 2\pi/T$ appears in (5) as a function of the action J which is the area of phase space enclosed by the orbit divided by 2π.

The function $\Omega = \Omega(J)$ contains the same information as the energy-period function $T = T(E)$. Differentiating $\Omega(J(E)) = 2\pi/T(E)$ by E yields after some calculations

$$(7) \qquad \frac{d\Omega}{dJ} = - \frac{4\pi^2}{T^3} \frac{dT}{dE} .$$

Now the perturbed system (3) is rewritten in action-angle variable as

$$(8) \qquad \dot{J} = \varepsilon F(J,\Phi,t), \qquad \dot{\Phi} = \Omega(J) + \varepsilon G(J,\Phi,t).$$

By assumption (4) and the nature of action-angle variables, F and G satisfy

$$(9) \qquad (F,G)(J,\Phi,t) = (F,G)(J,\Phi+2\pi,t) = (F,G)(J,\Phi,t+T_f) \quad \text{for all } J,\Phi,t.$$

For some fixed action J_0, let $\Omega_0 = \Omega(J_0)$ and $T_0 = 2\pi/\Omega_0$. We start with the family of unperturbed solutions of (8) given by

$$(10) \qquad \varepsilon = 0, \qquad J = J_0, \qquad \Phi = \alpha + \Omega_0 t$$

with arbitrary initial phase $\alpha \in [0,2\pi)$. The resonance assumption states that the rotation number T_0/T_f (corresponding to the action J_0) is rational.

Hence

$$(11) \qquad T_f = \frac{n}{m} T_0 = \frac{2\pi n}{m\Omega(J_0)} ,$$

where n and m are relatively prime integers. Under the resonance assumption (11), the unperturbed solution (10) satisfies

$$(8b) \qquad J(t+mT_f) = J(t), \qquad \Phi(t+mT_f) = \Phi(t)+2\pi n \quad \text{for all } t.$$

In the following we study the perturbed system (8) together with the boundary condition (8b).

Next define the Melnikov functions which occur in the bifurcation analysis of (8)(8b):

$$(12) \qquad M(\alpha) = \frac{1}{mT_f} \int_0^{mT_f} F(J_0,\alpha+\Omega_0 t,t)dt \quad \text{is the (ordinary) Melnikov function}$$

$$(13) \qquad N(\alpha) = \frac{1}{mT_f} \int_0^{mT_f} G(J_0,\alpha+\Omega_0 t,t)dt \quad \text{is the second Melnikov function}$$

LEMMA. *The Melnikov functions satisfy*

(14) $M(\alpha + 2\pi/m) = M(\alpha)$, $N(\alpha + 2\pi/m) = N(\alpha)$ *for all* $\alpha \in [0,2\pi)$.

PROOF. Since the integrant of (12) is 2π-periodic with aspect to its second argument, we get

(15) $M(\alpha+2\pi) = M(\alpha)$ for all $\alpha \in [0,2\pi)$.

On the other hand, the resonance assumption (11) and (9) imply that

$$M(\alpha+2\pi n/m) = \frac{1}{mT_f} \int_0^{mT_f} F(J_0, \alpha+\Omega_0 (t+T_f), t)dt$$

(16)
$$= \frac{1}{mT_f} \int_0^{mT_f} F(J_0, \alpha+\Omega_0 t, t-T_f)dt$$

$$= M(\alpha)$$

Since n and m are assumed to be relatively prime, there exist integers p and q such that $p+(qn/m) = 1/m$. Hence (15) and (16) imply the assertion (13).

The Melnikov functions have to satisfy the following basic assumptions:

a) $M = M(\alpha)$ has only simple zeros, i.e. $M(\alpha) = 0$ implies $M_\alpha(\alpha) \neq 0$. Since $M(\alpha) = M(\alpha+2\pi/m)$, this assumption implies that there exists a non-negative integer $j \in \mathbb{N}_0$ such that $M(\alpha)$ has exactly $2mj$ zeros $\alpha_\mu \in [0,2\pi)$ and

(17)
$$M(\alpha_\mu) = 0,\ M_\alpha(\alpha_\mu) > 0 \quad \text{for } \mu = 1,3,5 \ldots 2mj-1$$
$$M(\alpha_\mu) = 0,\ M_\alpha(\alpha_\mu) < 0 \quad \text{for } \mu = 2,4,6 \ldots 2mj$$

b) For the degenerate case $\Omega'(J_0) = 0$, which is indeed our main concern in this note, we have to assume that the two Melnikov functions $M(\alpha)$ and $N(\alpha)$ have no common zeros. Hence

(18) $N(\alpha_\mu) \neq 0$ for $\mu = 1,2,3 \ldots 2mj$

3. MAIN RESULTS.

THEOREM I. *Under the resonance assumption* (11) *and assumptions* (17)(18) *from above, we consider the system*

(8) $\dot{J} = \varepsilon F(J,\Phi,t)$, $\dot{\Phi} = \Omega(J) + \varepsilon G(J,\Phi,t)$

(8b) $J(t+mT_f) = J(t)$, $\Phi(t+mT_f) = \Phi(t) + 2\pi n$.

There exists $\varepsilon_0 > 0$ *and a neighbourhood* N *in the* $C[0,mT_f]$*-topology of the family* $\{(J,\Phi) = (J_0,\Omega(J_0)t+\alpha \mid \alpha\varepsilon[0,2\pi)\}$ *of unperturbed solutions with undetermined phase* α, *such that for* $|\varepsilon|<\varepsilon_0$ *all solutions of the perturbed problem* (8)(8b) *in the neighbourhood* N *are exactly the following ones:*

a) in the nondegenerate case $\Omega'(J_0) \neq 0$, *there exist* 2mj *solutions with* $\mu = 1 \ldots 2mj$

(19) $$J(t,\varepsilon) = J_0 - \frac{N(\alpha_\mu)}{\Omega'(J_0)}\varepsilon + J_1^* + O(\varepsilon^2) \quad \text{with} \quad \int_0^{mT} J_1^*(t)dt = 0$$

$$\Phi(t,\varepsilon) = \alpha_\mu + \Omega(J_0)t + O(\varepsilon)$$

b) in the degenerate case $\Omega'(J_0) = 0$, $\Omega''(J_0) \neq 0$, *the number of solutions can vary from* 0 *up to* 4mj. *The solutions occur in pairs. For* $\rho = \pm 1$ *and* $\mu \varepsilon \{1,2 \ldots 2mj \mid N(\alpha_\mu)\Omega''(J_0)\varepsilon < 0\}$, *the solutions are*

(20) $$J(t,\varepsilon) = J_0 + \rho\sqrt{-\frac{2N(\alpha_\mu)}{\Omega''(J_0)}\varepsilon} + O(\varepsilon)$$

$$\Phi(t,\varepsilon) = \alpha_\mu + \Omega(J_0)t + O(\sqrt{\varepsilon})$$

THEOREM II. *For the solutions constructed in* THEOREM I, *the eigenvalues* π_1, π_2 *of the linearised Poincaré map (Floquet multipliers of the linearised variational equations integrated over the time interval* $t \in [0,mT_f]$) *are*

a) in the nondegenerate case $\Omega'(J_0) \neq 0$

(21) $$\pi_{1,2} = 1 \pm mT_f\sqrt{2M_\alpha(\alpha_\mu)\Omega'(J_0)\varepsilon} + O(\varepsilon) \quad \text{for} \quad \mu = 1,2,3\ldots 2mj$$

Hence the fixed points of the Poincaré map (solutions of (8)(8b)) *are elliptic for* $\Omega'(J_0)M_\alpha(\alpha_\mu)\varepsilon < 0$ *and*

hyperbolic for $\Omega'(J_0)M_\alpha(\alpha)\,\varepsilon > 0$.

b) in the degenerate case $\Omega'(J_0) = 0$, $\Omega''(J_0) \neq 0$

(22) $$\pi_{1,2} = 1 \pm mT_f\sqrt{\rho M_\alpha(\alpha_\mu)}\,\sqrt[4]{-2N(\alpha_\mu)\Omega''(J_0)}\,\varepsilon^{3/4} + O(\varepsilon)$$

for $\rho = \pm 1$ *and* $\mu \in \{1,2 \ldots 2mj \mid N(\alpha_\mu)\Omega''(J_0)\varepsilon < 0\}$

Hence the fixed points of the Poincaré map are elliptic for $\rho M_\alpha(\alpha_\mu) < 0$ *and hyperbolic for* $\rho M_\alpha(\alpha_\mu) > 0$.

4. GEOMETRIC APPROACH. Before giving the formal proof of THEOREM I, we describe the geometric argument leading to THEOREM I as well as II. Again, the action-angle variables (J,Φ) are used as coordinate system in the phase-plane $X \ni (q,p)$.

Let $(J,\Phi)(t,\varepsilon,j,\alpha)$ denote the solution of the initial value problem

(8) $\dot{J} = \varepsilon F(J,\Phi,t), \quad \dot{\Phi} = \Omega(J) + \varepsilon G(J,\Phi,t)$

(8i) $J(0,\varepsilon,j,\alpha) = j, \quad \Phi(0,\varepsilon,j,\alpha) = \alpha.$

We integrate (8)(8i) up to time $t = mT_f = nT_0$ and define the Poincaré map

(23) $\Pi_\varepsilon: (j,\alpha) \in X \to (k;\beta) \in X$ by $\Pi_\varepsilon(j,\alpha) = (J,\Phi)(mT_f,\varepsilon,j,\alpha).$

To study the map Π_ε for small ε and action j near to J_0, we define the curve of radial dislocation

(24) $R = \{(j,\alpha) \mid \beta = \alpha + 2\pi n \text{ for } (k,\beta) = \Pi(j,\alpha)\}$

It consists of all points which the Poincaré map turns by the angle $2\pi n$. The difficulties to turn the geometric argument into a rigorous proof arise from the lack of exact knowledge about R. By the resonance assumption (11) and the second equation of system (8), we know that

$$R = \{(j,\alpha) \mid \text{The solution } (J,\Phi)(t,\varepsilon,j,\alpha) \text{ of (8)(8i) satisfies (25)}\}$$

with

(25) $\int_0^{mT_f} \Omega(J) - \Omega(J_0) + \varepsilon G(J,\Phi,t)dt = 0$

Formula (25) leads to some guess about parametrised representations of R, which we mention without proofs.

In the nondegenerate case $\Omega'(J_0) \neq 0$, one gets that for ε small enough

$$R = \{(j(\alpha),\alpha) \mid j(\alpha) = J_0 + O(\varepsilon) \text{ for } \alpha \in [0,2\pi)\}$$

Hence R is a closed Jordan curve in the phase plane enclosing the origin.

In the degenerate case $\Omega'(J_0) = 0$, $\Omega''(J_0) \neq 0$, the structure of R is more complicated. Indeed, R may consist of two branches with distance $O(\sqrt{\varepsilon})$ for some values of α, whereas these branches may disappear both for other values of α. A rigorous proof of this hint would duplicate the independent exact proof of THEOREM I given below.

We consider either the degenerate or the nondegenerate case and assume that some part $R_0 \subset R$ of the radial dislocation curve is given by

(26) $R_0 = \{ (j(\alpha),\alpha) \mid \alpha \in I\}$

where $I \subset \mathbb{R}$ denotes an interval.

The image of R_0 under the Poincaré map Π_ε is given by

(27) $\Pi_\varepsilon(R_0) = \{(k(\alpha),\alpha) \mid \alpha \in A$ and (28) holds $\}$ with

(28) $k(\alpha) - j(\alpha) = \int_0^{mT_f} \dot{J}dt = mT_f\, M(\alpha)\varepsilon + O(\varepsilon^2).$

The solutions of (8)(8b), we look for in THEOREM I, are exactly given by the intersection $R \cap \Pi_\varepsilon(R)$. Neglecting higher order terms, we guess from (28) that

$$R_0 \cap \Pi_\varepsilon(R_0) \simeq \{(k(\alpha),\alpha) \mid \alpha \in A \text{ and } M(\alpha) = 0\} .$$

Thus the zeros of the Melnikov function $M(\alpha)$ give the solutions of (8)(8b). To continue this heuristic discussion, we look at some sketches:

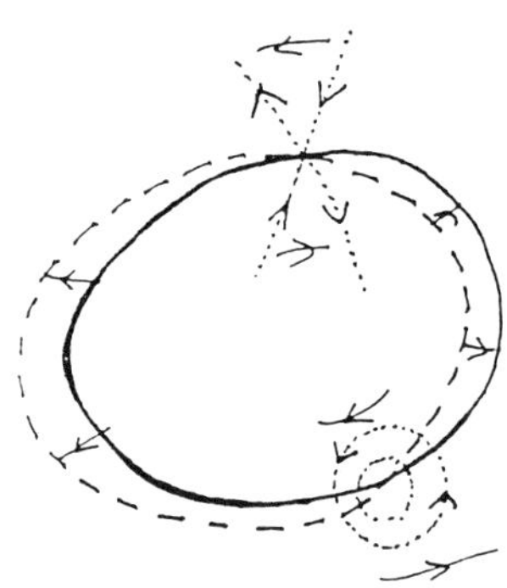

a)α) $T'(E) > 0$, $jm = 1$
There exists one elliptic and one hyperbolic fixed point of Π

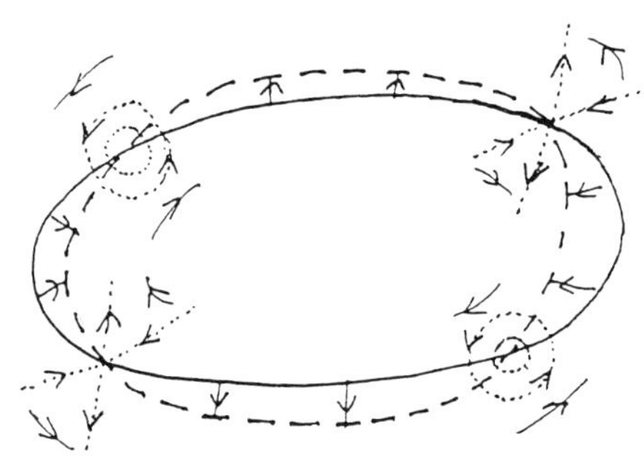

a)β) $T'(E) > 0$, $jm = 2$
There exist two elliptic and two hyperbolic fixed points of Π

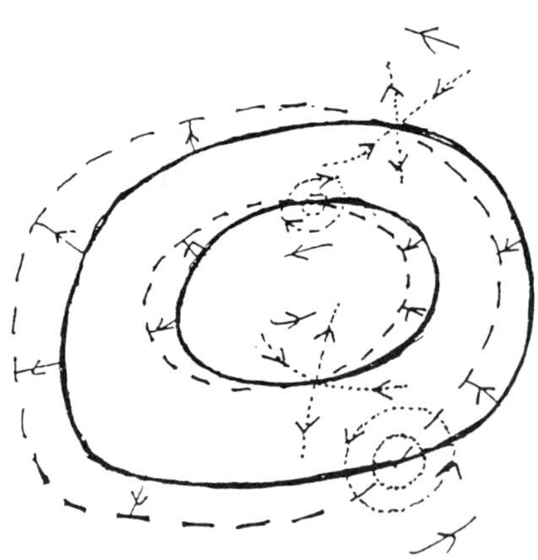

b)α) $T'(E) = 0$, $T''(E) > 0$, $jm = 1$
$N(\Phi) > 0$ for all $\Phi \in [0,2\pi)$
There exist two elliptic and two hyperbolic fixed points

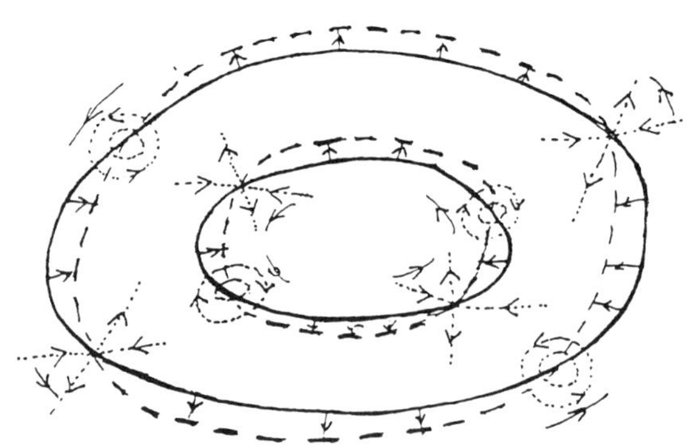

b)β) $T'(E) = 0$, $T''(E) > 0$, $jm = 2$
$N(\Phi) > 0$ for all $\Phi \in [0,2\pi)$
There exist four elliptic and four hyperbolic fixed points

The strong closed curves are the radial dislocation curves R. The strong dotted curves are the images of the radial dislocation curves under the Poincaré map $R' = \Pi(R)$. The thin arrows indicate the action of the Poincaré map. Hence they point e.g. from R to R'. The intersection $R \cap R'$ is the set of fixed points of the Poincaré map. The stability type of these fixed points - either elliptic or hyperbolic - can be deducted from the information on the Poincaré map given by the arrows in the figure. Arrows near elliptic fixed point surround it clockwise or counterclockwise, whereas arrwos near hyperbolic fixed point produce "collisions" which indicate the directions of the stable and unstable manifolds.

The figures a) show the nondegenerate case with $T'(E) > 0$, We have slow oscillations for large amplitude and fast oscillations for small amplitude. Thus the arrows for the action of the Poincaré map point counterclockwise on large circles and clockwise on small circles.

The figures b) show the degenerate case with $T'(E) = 0$, $T''(E) > 0$. Hence both small and large amplitude oscillations are slow leading to counterclockwise arrows.

Furtheron we have chosen the case that $\Omega'(J_0)N(\bar{\Phi})\varepsilon < 0$ for all $\Phi \in [0,2\pi)$ leading to the maximal number $4mj$ of fixed points and a radial dislocation curve consisting of two concentric Jordan curves.

The cases α) and β) correspond to $jm = 1$ - two zeros of the Melnikov function $M(\Phi)$ - and $jm = 2$ - four zeros of the Melnikov function $M(\Phi)$ for $\Phi \in [0,2\pi)$, respectively.

5. PROOF OF THEOREM I. We apply the Lyapunov-Schmidt reduction to problem (8)(8b). The problem can be written formally as $\mathcal{F}(u,\varepsilon) = 0$ with a nonlinear mapping $\mathcal{F}$: $C^1_{per}[0,mT_f]^2 \to C_{per}[0,mT_f]^2$. Both kernel and cokernel of the linearization $D_u\mathcal{F}(0,0)$ are twodimensional and have a basis consisting of constant functions. Let $\mathcal{M}$: $C_{per}[0,mT_f] \to \mathbb{R}$ be the mean value

$$\mathcal{M}x = \frac{1}{mT_f} \int_0^{mT_f} x(t)\, dt.$$

After decomposing by $J_0 = J + \sigma + \varepsilon J_1$, $\Phi = \alpha + \Omega_0 t + \Phi_1$ with $\mathcal{M}J = \mathcal{M}\Phi_1 = 0$, (8)(8b) gets

$$(29) \qquad \begin{aligned} \dot{J}_1 &= F(J_0+\sigma+\varepsilon J_1,\ \alpha+\Omega_0 t+\Phi_1,\ t), \\ \dot{\Phi}_1 &= \Omega(J_0+\sigma+\varepsilon J_1) - \Omega_0 + \varepsilon G(J_0+\sigma+\varepsilon J_1,\ \alpha+\Omega_0 t+\Phi_1,\ t), \end{aligned}$$

$$(30) \qquad \begin{aligned} J_1(t+mT_f) &= J_1(t),\ \mathcal{M}J = 0 \\ \Phi_1(t+mT_f) &= \Phi_1(t),\ \mathcal{M}\Phi_1 = 0 \end{aligned}$$

Now, system (30) is equivalent to $(Id-M)(30)$, $M(30)$. By the implicit function theorem, there exists $\varepsilon_0 > 0$ such that for all $(\alpha,\sigma,\varepsilon) \in [0,2\pi] \times [-\varepsilon_0,\varepsilon_0] \times [-\varepsilon_0,\varepsilon_0]$, $(Id-M)$ (30) has a unique solution $J_1^*(\alpha,\sigma,\varepsilon)$, $\Phi_1^*(\alpha,\sigma,\varepsilon)$ with $J_1 = O(1)$, $\Phi_1 = O(\varepsilon)$. After plugging this solution into equations $M(30)$, we get the bifurcation equations

(31) $$M(\alpha,\sigma,\varepsilon) = 0, \quad N(\alpha,\sigma,\varepsilon) = 0$$

with

(32) $$M(\alpha,\sigma,\varepsilon) \equiv \frac{1}{mT_f} \int_0^{mT_f} F(J_0+\sigma+\varepsilon J_1^*(\alpha,\sigma,\varepsilon),\ \alpha+\Omega_0 t+\Phi_1^*(\alpha,\sigma,\varepsilon),\ t)\ dt$$

(33) $$N(\alpha,\sigma,\varepsilon) \equiv \frac{1}{mT_f} \int_0^{mT_f} [\Omega(J_0+\sigma+\varepsilon J_1^*(\alpha,\sigma,\varepsilon)) - \Omega_0 + $$
$$+ \varepsilon G(J_0+\sigma+\varepsilon J_1^*(\alpha,\sigma,\varepsilon),\ \alpha+\Omega_0 t+\Phi_1^*(\alpha,\sigma,\varepsilon),\ t)]\ dt.$$

Assumption (17) about the simplicity of the zeros of the Melnikov function yields $M(\alpha_\mu,0,0) = 0$ and $M_\alpha(\alpha_\mu,0,0) \neq 0$ for $\mu = 1,2,3 \ldots 2mj$. Hence the first one of the bifurcation equations (27) has exactly $2mj$ solutions

(34) $$\alpha_\mu(\sigma,\varepsilon) = \alpha_\mu + O(|\sigma|+|\varepsilon|) \quad \text{for } \mu = 1,2,3 \ldots 2mj \text{ and } |\sigma|,|\varepsilon| < \varepsilon_1$$

with α_μ given by assumption (17).

Plugging (34) into the second equation of (31) yields the reduced bifurcation equations

(35) $$N_\mu(\sigma,\varepsilon) \equiv N(\alpha_\mu(\sigma,\varepsilon), \sigma,\varepsilon) = 0 \quad \text{for } \mu = 1,2,3 \ldots 2mj$$

Using $J_1^* = O(1)$, $\Phi_1^* = O(\varepsilon)$, (34) and the Taylorexpansion of $\Omega(J)$ (33)(35) imply

(36) $$N_\mu(\sigma,\varepsilon) = \Omega'(J_0)\sigma + \frac{\Omega''(J_0)}{2}[\sigma^2+\varepsilon^2 MJ_1^{*2}] + \varepsilon N(\alpha_\mu) +$$
$$+ O(|\sigma|^3+|\varepsilon|^3) + O(|\varepsilon|(|\sigma|+|\varepsilon|)).$$

We use this expansion to solve the reduced bifurcation equation (35). To this end, we distinguish once more the nondegenerate case $\Omega'(J_0) \neq 0$ and the degenerate case $\Omega'(J_0) = 0$, $\Omega''(J_0) \neq 0$.

a) In the nondegenerate case, each of the bifurcation equations (35) with $\mu \in \{1,2,3 \ldots 2mj\}$ fixed, has exactly one solution

(37) $$\sigma_\mu(\varepsilon) = -\frac{N(\alpha_\mu)}{\Omega'(J_0)}\varepsilon + O(\varepsilon^2) \quad \text{for } \mu = 1,2,3 \ldots 2mj.$$

Plugging these solutions into (34) and using

$$(38)\qquad J = J_0 + \sigma_\mu(\varepsilon) + \varepsilon J_1^*(\alpha_\mu(\sigma_\mu(\varepsilon), \varepsilon), \sigma_\mu(\varepsilon), \varepsilon)$$

$$\Phi = \alpha_\mu(\sigma_\mu(\varepsilon),\varepsilon) + \Omega(J_0)t + \Phi_1^*(\alpha_\mu(\sigma_\mu(\varepsilon),\varepsilon),\sigma_\mu(\varepsilon),\varepsilon)$$

we arrive at part a) of THEOREM 1.

b) In the degenerate case $\Omega'(J_0) = 0$, $\Omega''(J_0) \neq 0$, each of the bifurcation equations (35) with $\mu \in \{1,2,3 \ldots 2mj\}$ fixed has either no or two solutions. We get

$$(39)\qquad \sigma_\mu(\varepsilon) = \rho\sqrt{-\frac{2N(\alpha_\mu)}{\Omega''(J_0)}\varepsilon} + O(\varepsilon) \quad \text{with } \rho = \pm 1 \text{ if } N(\alpha_\mu)\Omega''(J_0)\varepsilon < 0,$$

but no solution otherwise.

Again, plugging (34) and (39) into (38) yields the result stated in THEOREM I.

6. FURTHER DISCUSSION. After this note was nearly finished, the abstract bifurcation result of Hale and Taboas [2], bifurcation near degenerate families, was pointed out to the author. In contrast to Guckenheimer and Holmes [1], Hale and Taboas [2] in fact consider a problem similar to the degenerate case of this note, but they use a more abstract set-up. For example, equation (3.6) from [2] p.30 is equivalent to the system of two bifurcation equations (27) from above.

In order to derive THEOREM I of this note from the result of Hale and Taboas, one has to link their quantities, especially $a(s)$ from assumption (H2) p. 23 to the derivatives to the energy-period function used in this note. This is done in the Proposition below. Thus using Hale and Taboas results from [2] section 1 through 5 and the Proposition below yields an alternative proof of THEOREM I.

Apparently Hale and Taboas were not aware of this Proposition. In fact, [2] section 3 considers a system with constant periods (different from the harmonic oscillator) and states that the hypotheses needed for the bifurcation analysis are satisfied - especially $a(s) \neq 0$ - which turns out to be wrong.

The occurance of this mistake has strengthened the author's opinion that the action-angle variables are the appropriate set-up for the study of perturbations of Hamiltonian systems. In fact, after having introduced the action-angle variables, or equivalently the energy-period function $T(E)$, the assumptions needed for the bifurcation analysis become very lucid: the nondegenerate case is $T'(E_0) \neq 0$ which is well-known from Guckenheimer-Holmes and for the degenerate case $T'(E_0) = 0$, stressed in this note, the next nondegeneracy assumption is indeed simply $T''(E_0) \neq 0$.

To link our note to the results of [2], we consider - similar to section 5 of (2) - the scalar second order differential equation

(40) $\ddot{x} + g(x) = 0$

with $g \in C^3$ and $xg(x) > 0$ for $x \neq 0$.

Let $G(q) = \int_0^q g(x)dx$. Note that (40) is a special case of (1) with $(q,p) = (x,\dot{x})$ and $H(q,p) = G(q) + p^2/2$. Let $x(t,E)$ be the family of solutions of (40) with initial data $x(0,E) = 0$, $x_t(0,E) > 0$ and $H(x(0,E),x_t(0,E)) = E$. Recall definition (2) of the energy-period function.

The following Proposition gives the relation of the energy-period function $T(E)$ to the quantity $a(s)$ appearing in the work of Hale and Taboas [2], hypothesis (H2) and again in section 5, p.35.

PROPOSITION. *Assume that* $T'(E_0) = 0$. *Then*

$$T''(E_0) = - \int_0^{T(E_0)} x_E^3 \, g''(x(t,E)) \, dt,$$

which is equal to $-2a(s)/\nu_1$ *from* [2] p.35.

PROOF. We plug the solution $x(t,E)$ into the differential equation (40) and differentiate twice by the parameter E. Thus we get

(41) $x_{ttEE} + g'(x)x_{EE} = - g''(x)\, x_E^2.$

Multiplying by x_E and integrating yields

$$- \int_0^{T(E)} x_E^3 \, g''(x(t,E)) \, dt = \int_0^{T(E)} x_E[x_{EEtt} + g'(x)x_{EE}] \, dt =$$

(42) $$= [x_E x_{EEt} - x_{Et}x_{EE}]_0^{T(E)} + \int_0^{T(E)} x_{EE}[x_{ttE} + g'(x)x_E] \, dt =$$

$$= [x_E x_{EEt} - x_{Et}x_{EE}]_0^{T(E)} .$$

Here integration by parts has lead to the last vanishing integral.

The definition of the energy-period function is

(43) $x(t+T(E),E) - x(t,E) = 0.$

Differentiation by the parameter E leads to

$$x_E(t+T(E),E) - x_E(t,E) + x_t(t,E)T'(E) = 0$$

and

$$x_{EE}(t+T(E),E) - x_{EE}(t,E) + [x_{Et}(t+T(E),E)+x_{Et}(t,E)]T'(E) + x_t(t,E)T''(E) = 0$$

For the special case $T'(E_0) = 0$, we get

$$x_E(t+T(E_0),E_0) - x_E(t,E_0) = 0$$

$$x_{EE}(t+T(E_0),E_0) - x_{EE}(t,E_0) = - x_t(t,E_0)T''(E_0)$$

and hence

$$(44) \qquad [x_E x_{EEt} - x_{Et}x_{EE}]_0^{T(E_0)} = \begin{vmatrix} x_t & x_E \\ x_{tt} & x_{Et} \end{vmatrix}(t,E_0)T''(E_0) = T''(E_0).$$

Here we used that the determinant in (44) is identically 1, since it is independent of time t and equal to 1 for $t = 0$. Combining (42) and (44) proves the Proposition.

Sufficient criteria for monotonicity of the energy-period function and for excluding the degenerate case $T'(E) = 0$ have been worked out by many authors quite recently. Rothe [3] states the bulk of the known results and gives some relations between them.

PROOF OF THEOREM II. The linearised variational equation corresponding to one fixed bifurcating solution $(J,\Phi)(t,\varepsilon)$ of system (8)(8b) from Theorem I is:

$$(45) \qquad \frac{dP}{dt} = A(t,\varepsilon)P(t,\varepsilon), \quad P(0,\varepsilon) = I.$$

Here P and A are 2×2-matrices, I is the identity matrix and

$$(46) \qquad A(t,\varepsilon) = \begin{pmatrix} 0 & 0 \\ \Omega'(J) & 0 \end{pmatrix} + \begin{pmatrix} F_J & F_\Phi \\ G_J & G_\Phi \end{pmatrix}$$

with the solution $(J,\Phi)(t,\varepsilon)$ from Theorem I plugged in. We look for the eigenvalues $\pi_{1,2}(\varepsilon)$ of $P(mT_f,\varepsilon)$, which are exactly the eigenvalues of the linearised Poincaré map Π_ε.

We use the following decompositions of A and P:

$$(47) \qquad A(t,\varepsilon) = A_0(\varepsilon) + B(t,\varepsilon), \quad P(t,\varepsilon) = e^{A_0(\varepsilon)t}(I + Q(t,\varepsilon)) \quad \text{with}$$

$$A_0(\varepsilon) = \begin{pmatrix} 0 & 0 \\ r(\varepsilon) & 0 \end{pmatrix} \quad \text{and} \quad e^{A_0(\varepsilon)t} = \begin{pmatrix} 1 & 0 \\ r(\varepsilon)t & 1 \end{pmatrix}$$

The choice of $r(\varepsilon)$ is determined by

$$(48) \qquad r(\varepsilon) = \begin{cases} \Omega'(J_0) & \text{nondegenerate case } \Omega'(J_0) \neq 0 \\ \rho\sqrt{-2N(\alpha_u)\Omega''(J_0)\varepsilon} & \text{degenerate case } \Omega'(J_0) = 0,\ \Omega''(J_0) \neq 0 \end{cases}$$

Using the Taylor expansion $\Omega'(J) = \Omega'(J_0)(J-J_0) + O(J-J_0)^2$ and formulas (19)(20) from THEOREM I, we see that the bifurcating solution $(J,\Phi)(t,\varepsilon)$ satisfies

(49) $$\Omega'(J(t,\varepsilon)) = r(\varepsilon) + O(\varepsilon) \quad \text{and} \quad B(t,\varepsilon) = O(1)$$
for small ε uniformly for $t \in [0,mT_f]$.

By the substitution (47), the variational equation (45) is transformed into

(50) $$\frac{dQ}{dt} = e^{-A_0(\varepsilon)t} B(t,\varepsilon)\, e^{A_0(\varepsilon)t} (I + \varepsilon Q(t,\varepsilon))$$
$$Q(0,\varepsilon) = 0.$$

Neglecting the term εQ on the right-hand side of (50) and recalling (47) leads us to the definition

$$\tilde{P}(t,\varepsilon) = e^{A_0(\varepsilon)t}(I + \varepsilon\tilde{Q}(t,\varepsilon)) =$$
$$= e^{A_0(\varepsilon)t} + \varepsilon e^{A_0(\varepsilon)t} \int_0^t e^{-A_0(\varepsilon)\tau} B(\tau,\varepsilon) e^{A_0(\varepsilon)\tau}\, d\tau$$

Comparison with (50) and neglecting terms via (49) leads to

(51) $$Q(t,\varepsilon) = \tilde{Q}(t,\varepsilon) + O(\varepsilon), \quad P(t,\varepsilon) = \tilde{P}(t,\varepsilon) + O(\varepsilon^2),$$

which assures that $\tilde{P}(t,\varepsilon)$ is a useful approximation of $P(t,\varepsilon)$.
For the calculation of the eigenvalues $\pi_{1,2}(\varepsilon)$ of $P(mT_f,\varepsilon)$, we need the trace and determinant of the matrix.
Since $\mathrm{Tr}\,(e^{-A_0 t} e^{-A_0\tau} B e^{A_0\tau}) = \mathrm{Tr}\,(e^{-A_0 t} B)$, the calculation of the trace from (51) yields

(52) $$\mathrm{Tr}\, P(mT_f,\varepsilon) = 2 + \varepsilon\int_0^{mT_f} \mathrm{Tr}[e^{A_0(\varepsilon)mT_f} B(t,\varepsilon)]\,dt + O(\varepsilon^2) =$$

$$= 2 + \varepsilon\int_0^{mT_f} \mathrm{Tr}B(t,\varepsilon)\,dt + \varepsilon r(\varepsilon) mT_f \int_0^{mT_f} B_{12}(t,\varepsilon)\,dt + O(\varepsilon^2)$$

$$= 2 + \varepsilon s(\varepsilon) + \varepsilon m(\varepsilon) + O(\varepsilon^2) \quad \text{with}$$

(53) $$s(\varepsilon) = \int_0^{mT_f} (F_J + G_\Phi)(\tau,\varepsilon)\,d\tau \quad \text{and} \quad m(\varepsilon) = r(\varepsilon) mT_f \int_0^{mT_f} F_\Phi(\tau,\varepsilon)\,d\tau.$$

By the definition (48) of $r(\varepsilon)$, the definition (13) of the Melnikov function M and (19)(20) from THEOREM I, we get

$$m(\varepsilon) = m_0(\varepsilon) + O(\varepsilon) \quad \text{with}$$

$$(54)\qquad m_0(\varepsilon) = \begin{cases} (mT_f)^2\Omega'(J_0)M_\alpha(\alpha_\mu) & \text{nondegenerate case } \Omega'(J_0) \neq 0 \\ (mT_f)^2\ \rho\ \sqrt{-2N(\alpha_\mu)\Omega''(J_0)\varepsilon}\ M_\alpha(\alpha_\mu) & \text{degenerate case } \Omega'(J_0) = 0 \end{cases}$$

Now (52)(53)(54) yield

$$(55)\qquad \operatorname{Tr} P(mT_f,\varepsilon) = 2 + \varepsilon s(\varepsilon) + \varepsilon m_0(\varepsilon) + O(\varepsilon^2)$$

The determinant of $P(mT_f,\varepsilon)$ is easy to get:

$$(56)\qquad \det P(mT_f,\varepsilon) = \exp \int_0^{mT_f} A(t,\varepsilon)\, dt = \exp(\varepsilon s)$$

Finally, the eigenvalues $\pi_{1,2}(\varepsilon)$ are calculated from (55)(56):

$$\pi_{1,2}(\varepsilon) = \operatorname{Tr}/2 \pm \sqrt{-\det + \operatorname{Tr}^2/4}$$

$$= 1 + \varepsilon(s(\varepsilon) + m_0(\varepsilon))/2 + O(\varepsilon^2) \pm$$

$$\pm \sqrt{(1 + \varepsilon(s(\varepsilon) + m_0(\varepsilon))/2)^2 - e^{\varepsilon s} + O(\varepsilon^2)}$$

$$(57)\qquad = 1 \pm \sqrt{\varepsilon m_0(\varepsilon)} + O(\varepsilon)$$

since $m_0(\varepsilon) \neq 0$ for $\varepsilon > 0$ by assumption (17).

Now THEOREM II follows from (54) and (57).

BIBLIOGRAPHY

1. John Guckenheimer and Philip Holmes, Nonlinear Oscillations, Dynamical Systems and Bifurcations of Vector Fields, Springer-Verlag New York Berlin 1984

2. J.K.Hale and P.Z.Taboas, Bifurcation near degenerate families, Applicable Analysis, 11 (1980) 21-37

3. F.Rothe, Periods of oscillation, nondegeneracy and specific heat of Hamiltonian systems in the plane, Proceedings of the International Conference on Differential Equations and Mathematical Physics, Birmingham 1986

Fakultät für Informatik, Institut für Mathematik
Universität der Bundeswehr München
Werner Heisenberg Weg 39, D8014 Neubiberg,
Federal Republic of Germany

Canadian Mathematical Society
Conference Proceedings
Volume 8 (1987)

BIFURCATION METHODS IN QUADRATIC SYSTEMS

Christiane Rousseau

ABSTRACT. In this paper we present an exhaustive treatment of all the bifurcation methods that have been employed to solve Hilbert's 16th problem for quadratic systems, i.e. determine the maximum number and position of limit cycles for a quadratic system. These methods involve in particular an extensive use of the Hopf bifurcation, but we consider also limit circles rising from a homoclinic loop bifurcation, from a loop through a saddle-node, and through bifurcations involving the points at infinity.

1. INTRODUCTION. The second part of Hilbert's 16th problem is to determine the number and position of limit cycles for a polynomial system :

$$(1.1) \qquad \dot{x} = P(x,y) \qquad \dot{y} = Q(x,y)$$

where P and Q are polynomials of degree n. The problem is still open for $n = 2$. Petrovskii and Landis had "proved" that the answer was three [1], but Novikov found a mistake in the crucial lemmas and their papers were withdrawn. Petrovskii and Landis used in their "proof" methods of complex algebraic geometry and it was mentioned by Arnold [2] that their powerful methods have never been fully exploited. There were several attempts to prove that the maximal number of limit cycles was three, until 1979, when there were given independantly two examples of quadratic systems with four limit cycles [3], [4]. In both cases the position of limit cycles is three cycles around a singular point and a unique one around the other singular point : we call this arrangement of limit cycles position (3,1).

It was shown [5] that a quadratic system cannot have limit cycles around more than two points and that a limit cycle contains at most one singular point. This limits us to the consideration of at most two limit points of the system.

There were never found examples of systems with four limit cycles in position (2,2), and it was shown by Ye Yanqian [6] that all limiting positions (i.e. all simultaneous bifurcations) that could give birth to four limit cycles in that position, through Hopf or homoclinic loop bifurcations is impossible in a quadratic system. From this Ye Yanqian conjectures that the position (2,2) is impossible. In his limiting positions he includes integrable systems, and homoclinic and heteroclinic loops through points at infinity. In the case of position (3,1) the limiting case through Hopf bifurcation at weak foci of finite order is impossible. It is possible

---

1980 Mathematics Subject Classification (1985 Revision). 34-06, 34C05.

Supported by NSERC and the ministry of Education of Québec

through a 4-parameter bifurcation from an integrable system with the four parameters λ, δ ,ε , c

$$(1.2) \qquad \dot{x} = \lambda x - y + ax^2 + c(5+\delta)xy + y^2 \qquad \dot{y} = x + cx^2 + (3a + 5 + \varepsilon + \delta(a + 1)) .$$

Such a bifurcation from an integrable system is of infinite codimension in planar analytic systems. Here the restriction to quadratic systems makes it of finite codimension. In opposition we call a bifurcation finite iff it is a bifurcation of finite codimension in planar analytic systems. when we speak of the Hopf bifurcation, we always mean it is of finite codimension inside analytic systems.

It is shown in [7] that the position (3,1) for limit cycles can be obtained by successive Hopf bifurcations. We make the following conjecture :

CONJECTURE 1.1. Any simultaneous finite bifurcations leading to four limit cycles is impossible in quadratic systems.

This comes from the following consideration : In the 12-dimensional manifold of quadratic vector fields, the submanifold of integrable systems in the neighborhood of a singular point or of a homoclinic loop [8] is of codimension 3. Therefore any bifurcation of codimension 4 leading to the simultaneous birth of four cycles should lead to an integrable system. This would explain also why people believed for so long that the maximal number of limit cycles in quadratic systems is three.

The principal tool for the birth of limit cycles is the Hopf bifurcation and this was the first tool to be deeply studied . It started by the work of Bautin [9] who proved that no more than three limit cycles can rise from a Hopf bifurcation of a singular point in a quadratic system. The maximal number of limit cycles that can rise in simultaneous Hopf bifurcations are positions (3,0) or (1,1) ([7] or [10]). The examples of quadratic systems with four limit cycles used mixed methods : a Hopf bifurcation at one singular point, and a study of singular points at infinity, which led to the consideration of large Poincaré-Bendixson domains, containing large limit cycles (too large to be seen at numerical simulations). These methods are well described in [11] and [12]. The examples in [3] and [4] concerned systems with a unique saddle-point at infinity. The method was then generalized to systems with three singular points at infinity [13]. In [7] , together with D. Schlomiuk, the author introduced a new idea, namely that the known systems with four limit cycles could be constructed by a succession of two Hopf bifurcations, i.e. by local methods only. This was shown on the example of Shi Songling. We now show it for all known systems with four limit cycles.

In this paper we study the effect of saddle-node, transcritical or pitchfork bifurcations of singular points at infinity on the number of limit cycles. (Note that these are the only bifurcations that can happen at infinity for quadratic systems, since singular points at infinity are given by a cubic equation).

The problem of the birth of limit cycles from a homoclinic loop has been observed more recently by Ye Yanqian [6] and the author [8]. In [8.] there is an example of a quadratic system with two limit cycles rising in a Hopf bifurcation of order 2 and dying in a homoclinic loop of order 2. This leads to the existence of a triangular region in the two-dimensional parameter space in which there are two limit cycles. The sides of the triangle correspond to the Hopf bifurcation, the homoclinic loop bifurcation, and the saddle-node bifurcation for cycles (fusion of two limit cycles before they disappear). The more general problem in codimension 3 was considered, and some general results on the impossibility of some codimension 3 and higher bifurcations were obtained. Recently Guckenheimer and Rand showed that no more than two limit cycles can rise from a homoclinic loop, by perturbation of a hamiltonian system (such a bifurcation is not a finite one in the sense we defined earlier).

We present here some results on the birth of limit cycles from a saddle-node loop. These are the first results on this bifurcation inside quadratic systems. In particular we analyse a quadratic system depending on three parameters which has a very rich structure : saddle-node loop together with Hopf bifurcations, connection of a saddle-node with a saddle point at infinity, bifurcation of critical points at infinity.

The presentation of the paper is as follows. In Sec. 2 we present the results on the Hopf bifurcations in quadratic systems. In Sec. 3 we present results on homoclinic loop bifurcation, and simultaneous Hopf and homoclinic loop bifurcations. Sec. 4 presents the effect of bifurcations of singular points at infinity. Sec. 5 contains the results on saddle-node loops.

2. HOPF BIGURCATIONS IN QUADRATIC SYSTEMS.

We first recall briefly the notation and result of the generalized Hopf bifurcation theorem. We suppose that the system :

(2.1) $\dot{x} = P(x,y) \qquad \dot{y} = Q(x,y)$

has a singular point at the origin, and that the linearized system at the origin has two pure imaginary eigenvalues $\pm i\omega$. Then the system can be brought around the origin to Poincaré normal form [7]:

(2.2) $\dot{r} = c_1 r^3 + c_2 r^5 + \ldots + c_k r^{2k+1} + O(r^{2k+3}) \qquad \dot{\theta} = 1$

DEFINITION. We say that the system has a Hopf bifurcation of order $k = 1$ at the origin if $c_1 \neq 0$, and a bifurcation of order $k > 1$ if $c_1 = \ldots = c_{k-1} = 0$ and $c_k \neq 0$. We also say that the origin is a weak focus of order k.

THEOREM. [7] If (2.1) has a Hopf bifurcation of order k at the origin, then
1) any perturbation of (2.1) will have at most k limit cycles in a neighborhood of the origin.
2) for any $0 \leq i \leq k$, there exists a perturbation of (2.1), which has exactly i limit cycles. For $i = k$ the perturbation is constructed in the following way : we add parameters $c_1, \ldots, c_{k-1}$ as in (2.2) and a term $c_0 r$ in (2.2) in such a way that $c_0, \ldots, c_k$ have alternate sign and $c_0 \ll c_1 \ll \ldots \ll c_k$.

THEOREM. (Bautin [9]) A quadratic system cannot have more than three limit cycles rising from a Hopf bifurcation at a singular point.

The proof goes the following way : if $c_1 = c_2 = c_3 = 0$ then $c_k = 0$, for any $k > 3$. In other words we can verify that the c_i's are polynomials in the coefficients of (2.1) (in finite number). By Hilbert's basis theorem the ideal generated by the c_i's is finitely generated. Bautin's theorem states that the ideal is generated by c_1, c_2, c_3.

To obtain systems with a great number of cycles one can try to have Hopf bifurcations simultaneously around two singular points. For this purpose we reduce (2.1) to a particularly simple form. We can suppose that (0,0) and (0,1) are two singular points of our system, and that the linear part around the origin is in the form: $\dot{x} = -\omega y$, $\dot{y} = \omega x$. A change of scale in time allows us to bring our system to the form :

(2.3) $\dot{x} = -y + ax^2 + bxy + y^2$ $\dot{y} = x + cx^2 + dxy$

THEOREM. ([7] and [10]): By Hopf bifurcations around (0,0) and (0,1), we can only obtain the positions (1,0), (1,1), (2,0), (3,0).

The proof uses the characterization of a quadratic integrable system in the neighborhood of a singular point [7], namely (2.3) is integrable in the neighborhood of the origin iff one of the following conditions is satisfied :

I $a + 1 = c = 0$

(2.4) II $b(a + 1) = c(d + 2a)$ and $c(d + 2a)^3 + b(d - a)(d + 2a)^2 - b^3 (d + 2a) - b^3 = 0$

III $d - 3a - 5 = -b + 5c = a + 2(c^2 + 1) = 0$

The system (2.3) depends on 4 parameters and the submanifold of integrable systems given by (2.4) is of codimension 2. Therefore we expect to have generically two more degeneracies before arriving at an integrable system. Since (2.3) leads at least to position (1,0), we could expect (1,1), (2,0), (2,1), (3,0). However (2,1) cannot arrive with two simultaneous Hopf bifurcations. We will have a similar result with simultaneous Hopf and homoclinic loop bifurcations.

The system of Shi Songling [3] is the following :

(2.5) $\dot{x} = \lambda x - y - 10x^2 + (5 + \delta)xy + y^2$ $\dot{y} = x + x^2 + (-25 + 8\varepsilon - 9\delta)xy$

with $\delta = -10^{-13}$, $\varepsilon = -10^{-52}$, $\lambda = -10^{-200}$. We consider rather the system:

(2.6) $\dot{x} = -y - 10x^2 + \beta xy + y^2$ $\dot{y} = x + x^2 - 25xy$

The system (2.6) has a Hopf bifurcation at $\beta = 0$ around the point (0,1) and a limit cycle for $\beta > 0$. We can check that the limit cycle stays until $\beta = 5$. (There exists no singular point other than the origin and a unique saddle point at infinity, which gives us a Poincaré-Bendixson domain around (0,1) for $0 < \beta < 5$). For $\beta = 5$ we have a Hopf bifurcation of order 3 around the point (0,1). Since the cycle around (0,1) is structurally stable it remains for small values of λ, δ, ε (cf. [7] for details).

Shi Songling [3] does not consider the concept of successive Hopf bifurcations and limits himself to consider the Poincaré-Bendixson domain around (0,1). Later in [11]-[12] he considers general conditions for a system to have small perturbations with four limit cycles : the conditions require that the system has the same topological type as in (2.5) (cf. Fig.1), namely:

1) Only two singular points one of which is a focus, and the other a weak focus of order 3 (i.e. the system is having there a Hopf bifurcation of order 3).

2) A change of coordinates and scaling of time brings the system to:

(2.7) $\dot{x} = -y + ax^2 + 5cxy + y^2$ $\dot{y} = x + cx^2 + (3a + 5)xy$

3) A unique singular point at infinity of saddle type. For $c > 0$ the singular point is located above the line $1 + cx + (3a + 5)y = 0$.

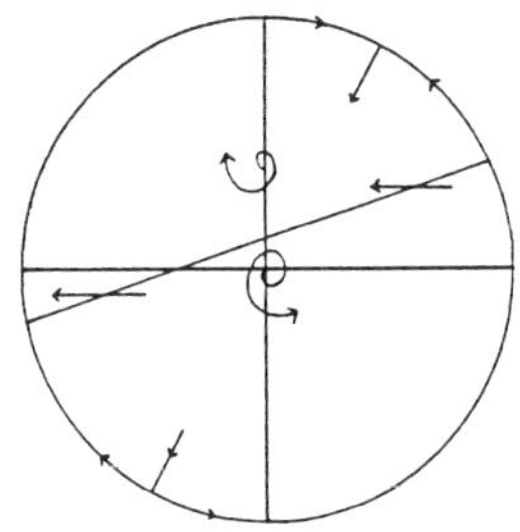

Figure 1. Topological type of a system giving four limit cycles after bifurcation in case 1.

Later in [13], one finds similar conditions, but for a system with three singular points at infinity. The line $1 + cx + (3a + 5)y = 0$ defines two hemispheres, each with three singular points on the equator, but not of the same stability type with respect to the interior of the domain. The authors then construct a hyperbola without contact (Fig.2 a)), in order to apply the same ideas as before. They work in a rotated system of coordinates in which one of the singular point at infinity is on the y-axis. If we rotate the system back, in order to have (0,0) and (0,1) as singular points, we obtain Fig.2 b). We now show that, in the two cases the limit cycle obtained around (0,1) can be viewed as born in a previous Hopf bifurcation.

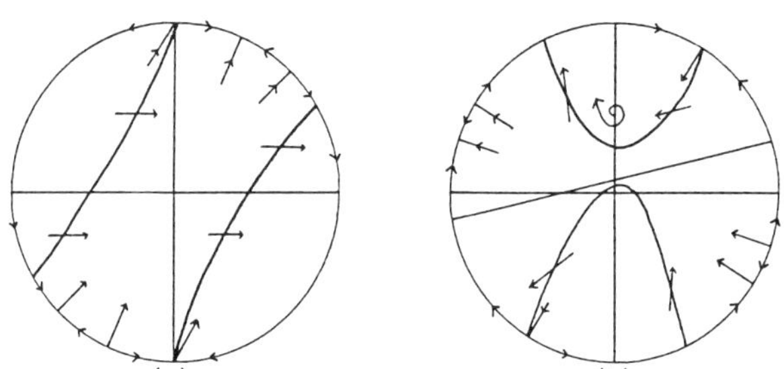

(a) (b)

Figure 2. Topological type of a system giving four limit cycles after bifurcation in case II. a) corresponds to a system with a singular point at infinity on the y-axis and b) to the equivalent system with the singular point (0,1) .

THEOREM. All known cases of systems with four limit cycles can be brought to the form:

(2.8) $\dot{x} = \lambda x - y + ax^2 + (\beta + \delta)xy + y^2 \qquad \dot{x} = x + cx^2 + (3a + 5 + \varepsilon + \delta(a + 1)/c)xy$

The four cycles appear in the following way. First for $\lambda = \varepsilon = \delta = \beta = 0$ we have a Hopf bifurcation at the point (0,1). This bifurcation gives a limit cycle for β with same sign as c. Then a small perturbation with $\lambda, \varepsilon, \delta$ gives three more cycles around the origin.

Proof : The known examples are of three kinds. All involve systems with exactly two singular points, which we can suppose to be (0,0) and (0,1). In case I and II (0,0) is a weak focus of order 3, so the system can be brought to the form [11]-[12]:

(2.9) $\dot{x} = -y + ax^2 + 5cxy + y^2 \qquad \dot{y} = x + cx^2 + (3a + 5)xy$

In case I and III we have a unique saddle point at infinity [11]-[12], and a line without contact. In Case II we have three singular points at infinity and a hyperbola without contact [13]. In case III [4], the origin is a weak focus of order 2, so the Hopf bifurcation at the origin gives two cycles at (0,0) and consideration of the system at infinity gives two Poincaré-Bendixson, so two large limit cycles, one around each singular point.

Case I and II : They are perturbations of (2.9) in suitable coordinates. We restrict ourselves to the case $c > 0$. The case $c < 0$ follows from the change of coordinates $x \mapsto -x$, $t \mapsto -t$. For $\lambda = \varepsilon = \delta = \beta = 0$, (2.8) has a Hopf bifurcation at the point (0,1) for $a < -2$. The Hopf bifurcation is nondegenerate and gives an attractive limit cycle for $\beta > 0$. We want to show that the cycle remains until $\beta = 5c$. We first show that there is no other point than the origin for $\beta \in [0,5c]$. Such singular points would be given by:

(2.10) $$x = -1/c - y(3a+5)/c$$

$$y^2[a(3a+5)^2 - \beta c(3a+5) + c^2] + y\,[2a(3a+5) - \beta c - c^2] + a = 0$$

The discriminant of the second equation is $\Delta = c^2(-12a^2 - 24a + (\beta + c)^2)$.
$\Delta < 0$ for $a < -1 - \sqrt{1+(\beta+c)^2/12}$, which is automatically satisfied for $\beta \in [0,5c]$ if $\Delta < 0$ at $\beta = 5c$, i.e. $a < -1 - \sqrt{1+3c^2}$. Therefore there is no other singular points for $\beta \in [0,5c]$ if there is none at $\beta = 5c$.

The limit cycle can have no interaction with the origin since we have the line without contact $1 + cx + (3a+5)y = 0$ on which $\dot{x} < 0$.

We now consider the behaviour at infinity. If we take $z = 1/x$ and $u = y/x$ we get the following system (once multiplied by $1/z$):

(2.11) $$\dot{u} = -u^3 - \beta u^2 + (2a+5)u + c + u^2 z + z \qquad \dot{z} = -az - \beta uz - u^2 z + uz^2$$

Singular points are given by $z = 0$ and :

(2.12) $$-u^3 - \beta u^2 + (2a+5)u + c = 0$$

(2.12) has a unique positive real root u_0 and two roots u_1 and u_2 with negative real parts (either real or complex) When u_1 and u_2 are real we suppose $u_2 < u_1$. The Jacobian matrix for $z = 0$ has eigenvalues $\lambda_1 = -3u^2 - 2\beta u + (2a+5)$ in the u-direction and $\lambda_2 = -a - \beta u - u^2$ in the z-direction. (2.12) gives that $u\lambda_2 = -(3a+5)u - c$. Therefore $\lambda_2 > 0$ for u_1 and u_2 whenever they exist, and for u_0 precisely when $u_0 > -c/(3a+5)$.
We analyse how u_0, u_1 and u_2 vary when $\beta \in [0,5c]$. We get $u_0', u_2' < 0$, $u_1' > 0$.
So as β goes from $5c$ to 0, u_0 moves up and u_1 and u_2 move towards each other. Therefore, if the points are situated as in Fig.1 (resp. Fig.2 b)) relatively to the line (resp. hyperbola) for $\beta = 5c$, they remain so for $\beta \in [0,5c]$. We only need to show that the upper half of the hyperbola, which is without contact at $\beta = 5c$, remains without contact for $\beta \in [0,5c]$. This comes from the consideration of the perturbation $((\beta - 5c)xy, 0)$ of the vector field at $\beta = 5c$. The perturbed vector field is given in dotted lines on Fig. 3. In both quadrants this gives a vector with greater slope.

Case III : The system of Chen and Wang [4] is the following:

(2.13) $$\dot{x} = -y - \delta_2 x - 3x^2 + (1 - \delta_1)xy + y^2 \qquad \dot{y} = x + 2/9\,x^2 - 3xy$$

This example can be viewed as equivalent to :

(2.14) $$\dot{x} = \lambda x - y + ax^2 + (\beta + \delta)xy + y^2 \qquad \dot{y} = x + cx^2 + (3a + 5 + \varepsilon + \delta(a+1)/c)\,xy$$

with :

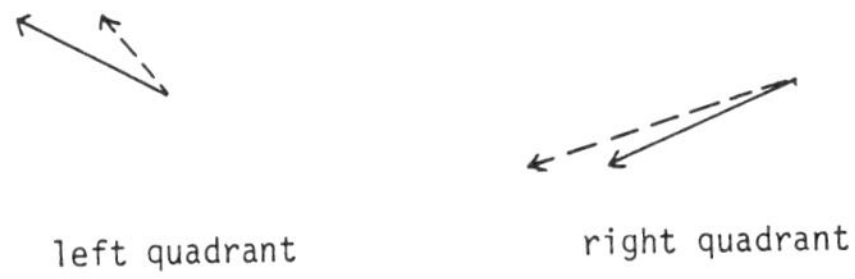

Figure 3. Perturbed vector field (in dotted line) for Figure 2.

(2.15) $\lambda = \delta_2$, $a = -3$, $c = 2/9$, $\beta = 10/9$, $\delta = -1/9$, $\varepsilon = \delta_1$

The difference with the previous case is that here δ is chosen as big as -1/9, instead of being arbitrarily small. The difference between (2.14) and (2.15) lies in the position of ε. It is easily checked that these two positions are equivalent.

REMARK. In [11]-[12] we find a condition that guarantees that u_0 is above the line $1 + cx + (3a + 5)y = 0$. This condition is unnecessary. In fact from (2.12) we get that $\lambda_1 < 0$ for u_0 , u_2 , and $\lambda_1 > 0$ for u_1 . Therefore u_1 and u_2 are respectively a node and a saddle point. From the Poincaré index theorem it follows that u_0 is a saddle. Then $\lambda_2 > 0$ for u_0, and u_0 is above the line.

3. HOMOCLINIC LOOP BIFURCATIONS IN QUADRATIC SYSTEMS.

We present here the results obtained on birth of limit cycles from homoclinic loop bifurcations (HLB) in quadratic systems. The study of HLB in quadratic systems is more recent than the study of the Hopf bifurcation. We start by recalling the known numerical examples, before giving the general results.

The following example was given by Bogdanov [14] :

(3.1) $\dot{x} = y \qquad \dot{y} = -1 - \mu y + x^2 + xy$

In this system a limit cycle rises from the point (-1,0) in a Hopf bifurcation at $\mu = -1$, and grows bigger monotonically in μ, until it dies in a nondegenerate HLB starting at the saddle point (0,1), for $\mu = -5/7$ (cf. Fig.4).

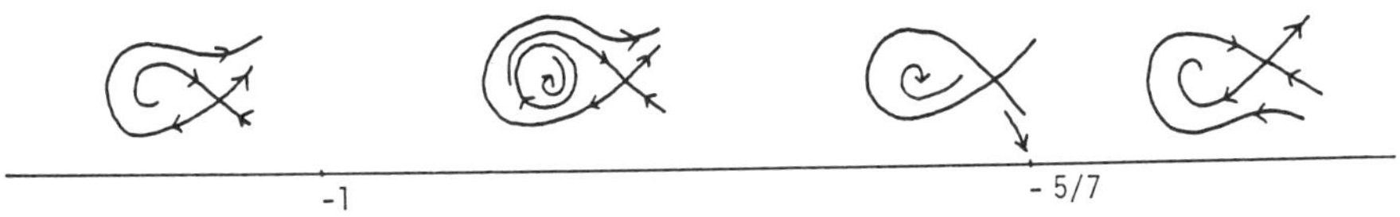

Figure 4. Bifurcation in the system of Bogdanov (3.1)

REMARK. The technique used here to study the HLB is the method of Melnikov. In quadratic sytems it is generally not possible to bring a system close to a hamiltonian or integrable system, and the only way to prove that one has homoclinic loops is to try to follow the global behaviour of invariant stable and unstable manifolds of a saddle point

The following system is an example in which two limit cycles rise from a HLB of order 2 (cf. [8]) :

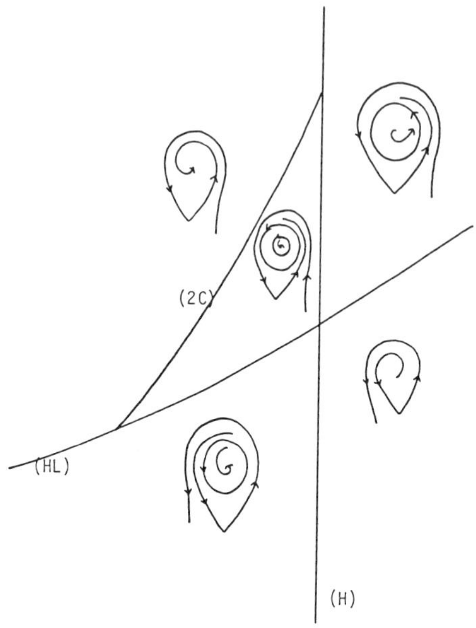

Figure 5. Bifurcation diagram for system (3.2)

$$(3.2) \qquad \dot{x} = (\lambda - \delta - 1)x + y - x^2 + (1 + \partial)xy - y^2 \qquad \dot{y} = 2x + 2/3x^2 - xy$$

There is a triangular region in the parameter space (cf. Fig.5) for which this system has two limit cycles, bounded by three codimension 1 curves: a curve of Hopf bifurcation of order 1, a curve of HLB of order 1 and a curve of saddle-node for cycles. The three vertices of the triangular region correspond to: a Hopf bifurcation of order 2 (at $\lambda = \delta = 0$), a HLB of order 2 (approximatly at $\lambda = \delta + 1 \simeq -0.0651$), and a Hopf bifurcation of order 1 simultaneous to a HLB of order 1 (approximatly at $\lambda = 0, \delta \simeq -0.4$).

In [8] we also study the following system :

$$(3.3) \qquad \dot{x} = (\lambda - \delta - 5)x + y - x^2 + (5 + \partial)xy - y^2$$
$$\dot{y} = (9 - \varepsilon + 2\delta)x + x^2 + (-8 + \varepsilon - 2\delta)xy .$$

It has a Hopf bifurcation of order 2 together with a HLB of order 1 at $\lambda = \varepsilon = 0, \delta \simeq 1.785$. The system also has a Hopf bifurcation of order 3 for $\lambda = \delta = \varepsilon = 0$. The system has three cycles in a region $0 << \lambda << \varepsilon << \delta$.

The examples (3.1) and (3.2) are an illustration of the principle : *Limit cycles which appear from a singular point in a Hopf bifurcation die in a HLB* . This principle appears to be rather generic in many dynamical systems. We would then expect that there exists a tetraedral regioin the parameter space of (3.3) in which there would exist three limit cycles. However one of the expected vertex does not exist. This follows from the following result :

PROPOSITION. ([8] or [10]) It is impossible to have a Hopf bifurcation of order one together with a HLB of order ≥ 2.

In [8] we find that in simultaneous Hopf bifurcation of order k and HLB of order 1 conjecture 1.1 lies on the result of [15] that there cannot exist a limit cycle around a weak focus of order 3. It would follow if we could show that the submanifolds corresponding to Hopf bifurcation of order 3 and to homoclinic loop are transversal inside the manifold of quadratic systems.

We now give an example of a system which has simultaneously two homoclinic loops. It is shown in [16] that two homoclinic loops are necessarily disjoint in a quadratic system.

THEOREM. The system :

$$(3.6) \qquad \dot{x} = -\lambda x - y + 2x^2 + 2\lambda xy + y^2 \qquad \dot{y} = x - (1 - \lambda)x^2 - 2xy$$

has two limit cycles, one around (0,0) and the other around (0,1) for $\lambda > 0$ enough small. For $\lambda = \lambda_0 < 1$ the two cycles disappear simultaneously in homoclinic loop bifurcations.

Proof : The system has simultaneous Hopf bifurcations around (0,0) and (0,1), at $\lambda = 0$, and limit cycles around the two points for $\lambda > 0$ small.

The system is symmetric under the symmetry $x \mapsto -x$, $y \mapsto 1-y$, $t \mapsto -t$. So the two cycles, around (0,0) and around (0,1) have the same behaviour. We show that the cycles have disappeared before $\lambda = 1$ (numerical simulation gives $\lambda \in [0.05,0.1]$). At $\lambda = 1$ the system can be brought to :

$$(3.7) \qquad \dot{x} = \lambda x - y + ax^2 + bxy + y^2 \qquad \dot{y} = x + xy$$

with $\lambda < 0$, $b < 0$, $-1 \leq c \leq 0$ and we can apply the result of [17] to conclude that the system has at most one limit cycle. Because of the symmetry it has none. We also want to show that there is no double heteroclinic loop at $\lambda = 1$, containing the two saddle points :

$p_{1,2} = (x_{1,2} = \pm(5\lambda^2 - 6\lambda + 9)^{-\frac{1}{2}}, y_{1,2} = \frac{1}{2}(1 - (1-\lambda)\, x_{1,2})$ and the invariant line $y = \frac{1}{2}$. Elsewhere for $\lambda < 1$, the line $y = \frac{1}{2}$ is without contact except at $x = 0$ and is always crossed downwards. Since we have a unique singular point at infinity which is a node, the only possibility for thee limit cycles to have died is two simultaneous homoclinic loops, one starting from p_1 and going around (0,0), and the other starting from p_2 and going around (0,1). (Fig.6).

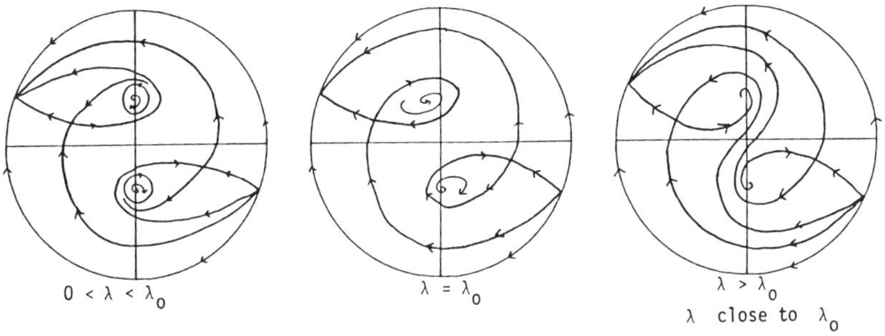

Figure 6. Simultaneous homoclinic loop bifurcation for system (3.6). The horizontal line is the line $y = \frac{1}{2}$.

There is no heteroclinic connection with p_1 and p_2: both the unstable manifold W^u of p_2 and the stable manifold W^s of p_1 start with negative slope inside the ellipse $\dot{x} = 0$. Once outside the ellipse we can check that symmetrical points with respect to the y-axis have same vertical speed, and that the point to the right has a smaller horizontal speed. Consequently the slope of W^s is greater than that of W^u and there is no connection between the two manifolds.

One can check that the trace of p_1 (resp. p_2) is positive (resp. negative), which corresponds to the fact that a repulsive (resp. attractive) cycle disappears in the homoclinic loop from p_1 (resp. p_2).

4. BIFUCATIONS WITH SINGULAR POINTS AT INFINITY.

The equation defining singular points at infinity is a cubic equation, so we have a possibility of one or three singular points at infinity.

We now give an example of a system with two cycles rising simultaneously around two different singular points in Hopf bifurcations and dying in a double heteroclinic loop at infinity.

THEOREM. The following system depending on the parameter λ:

(4.1) $\quad \dot{x} = -\lambda x - y - x^2 + 3/2\, \lambda xy + y^2 \qquad \dot{y} = x + (1 - \lambda)x^2 - 2xy$

has a simultaneous Hopf bifurcation at $\lambda = 0$ at the points (0,0) and (0,1). It has two limit cycles for $\lambda \in (0,1)$. The two cycles disappear simultaneously at $\lambda = 1$ in a double heteroclinic loop at infinity.

Proof : The Hopf bifurcation is of order 1 for the two singular points, and the limit cycles appear for $\lambda > 0$. For $\lambda > 0$ the origin is attractive and (0,1) is repulsive. Therefore the cycle around (0,0) is repulsive and the one around (0,1) attractive. The system has no other singular points than (0,0) and (0,1). The equation for singular points at infinity is :

(4.2) $\quad -u^3 - 3/2\, \lambda u^2 - u + (1 - \lambda) = 0 \qquad z = 0$

We have a unique singular point at infinity $u = u_0$, $z = 0$ for λ in an open set containing [0,1]. The singular point is a saddle. Moreover $u_0 > (1 - \lambda)/2$ for $\lambda < 1$ and $u_0 < (1 - \lambda)/2$ for $\lambda > 1$. This shows that the limit cycles remain until $\lambda = 1$ (cf. Fig.7). At $\lambda = 1$ a bifurcation occurs (Fig.7 c)) which correspond to a double loop at infinity. It occurs for $u_0 = 0$, i.e. the singular point is on the invariant line $y = \frac{1}{2}$. The fact that there is no limit cycles at $\lambda = 1$ follows [6] , which implies that the limit cycles have disappeared for $\lambda > 1$.

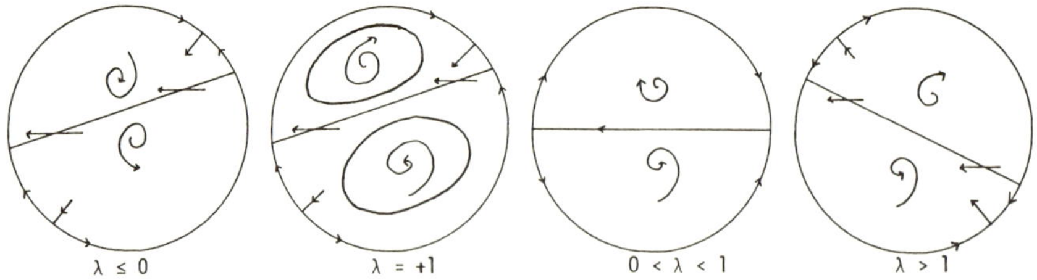

Figure 7. Bifurcation diagram for system (4.1).

REMARK. Two cycles appearing simultaneously around two different singular points in Hopf bifurcations are necessarily one attractive, the other repulsive, since the system before perturbation has the form :

(4.3) $\quad \dot{x} = -y + ax^2 + y^2 \qquad \dot{y} = x + cx^2 + dxy \qquad \text{with } 1 + d < 0$

If instead of the system (4.1), we had taken the system :

(4.4) $\quad \dot{x} = -\lambda x - y - x^2 + 2\lambda xy + y^2 \qquad \dot{y} = x + (1 - \lambda)x^2 - 2xy .$

We get the same behaviour as in Fig.7, except that at $\lambda = 1$ a double singular point at infinity appears at $u = -1$. In this case there is a symmetrical situation between (4.4) and the system obtained in localizing (4.4) at (0,1), through the change $x \mapsto -x$, $y \mapsto -y$, $t \mapsto -t$.

THEOREM. The system

(4.5) $\dot{x} = -\lambda x - y - x^2 + a\lambda xy + y^2 \qquad \dot{y} = x + (1-\lambda)x^2 - 2xy \qquad$ with $a > 1$

has two cycles around (0,0) and (0,1) appearing for $\lambda > 0$ in simultaneous Hopf bifurcations at $\lambda = 0$. There exists $a_0 \in (2,32]$ such that for $a > a_0$ the system has a bifurcation of heteroclinic loop at infinity (Fig.8) which makes the limit cycle around (0,1) disappear for some $\lambda < 1$.

Proof : The points at infinity are given by the equation :

(4.6) $-u^3 - au^2 - u + (1-\lambda) = 0$

For $a > 2$ the system has three singular points at infinity for values of $\lambda < 1$. We then necessarily have a positive one and two negative ones. At $\lambda = 1$ we have as before the invariant line $y = \frac{1}{2}$. It is easily checked that there are no finite singular points other than (0,0) and (0,1), for $\lambda > 0$. For $a \geq 32$, there always exists some $\lambda \in (0,1)$ such that the divergence line $-\lambda-4x + a\lambda y = 0$ is parallel to the line without contact $1 + (1-\lambda)x - 2y = 0$, and lies below that line, so the upper domain can have no limit cycles. The only possibility for the limit cycle is to have died in a heteroclinic loop at infinity as in Fig.8 . For $a = 32$ we get $\lambda = \frac{1}{2}$. The behaviour in Fig.8 may of course have happened already before $a = 32$, if we have $a > 2$ (in order to have three limit points at infinity for $\lambda < 1$).

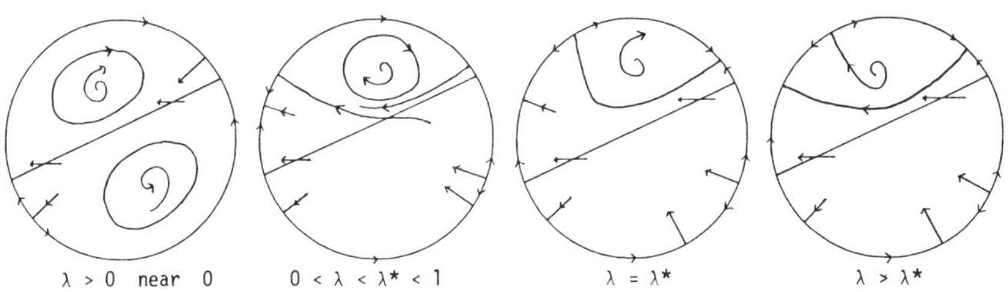

Figure 8. Bifurcation diagram for system (4.5), with $a > a_0$.

5. SADDLE NODE LOOP BIFURCATION.

In this bifurcation we are concerned with systems with at least three singular points : a double one, the saddle-node and another one (which is a focus or a center, since a singular point inside a limit cycle is necessarily a focus or a center [18]). In the general case we have to deal with systems with four finite singular points. In view of conjecture 1.1 we first prove :

THEOREM. There exists no quadratic system with a Hopf bifurcation of order 3 together with a loop through a saddle-node.

Proof : Around (0,1) the system has the form [11] :

(5.1) $\dot{X} = -Y + aX^2 + 5cXY - Y^2 \qquad \dot{Y} = X + cX^2 + dXY \qquad$ with $d = 3a - 5$

The system localized at the origin becomes :

(5.2) $\dot{x} = -5cx + y + ax^2 + 5cxy - y^2 \qquad \dot{y} = x(1-d) + cx^2 + dxy$

For (0,0) to be a saddle-node we need $d = 1 \Longrightarrow a = 2$. So we consider :

(5.3) $\dot{x} = -5cx + y + 2x^2 + 5cxy - y^2 \qquad \dot{y} = cx^2 + xy$

It is clear from Fig.9 in the case $c < 0$ that a loop starting from (0,0) tangent to the hyperbola should go up, but the region $\dot{y} < 0$ leaves no place to go up. The case $c > 0$ follows from the case $c < 0$ through the change of variables $x \mapsto -x$, $t \mapsto -t$.

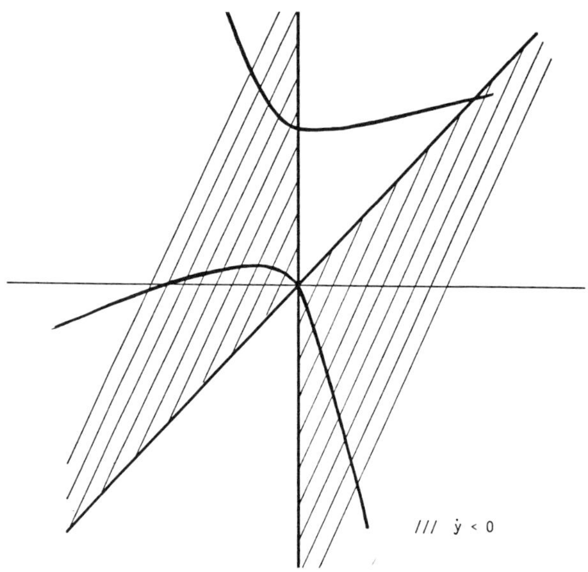

Figure 9. Impossibility of a saddle loop around (0,1) in system (5.2)

The following system gives an example of a loop through a saddle-node in a quadratic system :

(5.4) $\dot{x} = (\lambda + 5)x + y + ax^2 - 5xy - y^2 \qquad \dot{y} = \varepsilon + x^2 + xy$

We study the phase portrait of (5.4) for $\varepsilon = 0$ for various values of λ and a. This system has a very rich structure. In all regions where we have a saddle-node loop we get a limit cycle once the saddle-node has disappeared, i.e. for $\varepsilon < 0$ close to the plane $\varepsilon = 0$.

i) **Singular points of (5.4) for $\varepsilon = 0$:**

The system has the singular points (0,0), (0,1) and $p = (x^*, -x^*)$ with $x^* = -(\lambda - 4)/(4 + a)$. The point (0,0) is a saddle-node with center manifold tangent to the vector $(-1, \lambda + 5)$. For $\lambda > -5$ its unstable manifold is tangent to the x-axis. The point (0,1) goes through a Hopf bifurcation of order 2 for $\lambda = 0$, $a = 4/7$, and of order 1 for $\lambda = 0$ elsewhere (see curve (H) in Fig.10). $\mathrm{Re}(c_2) > 0$, so there are two cycles in a tongue $\lambda > 0$, $a > 4/7$. This tongue is limited on one side by the curve (H) and on the other side by the curve (2C). The point p is located above the x-axis for $\lambda, a > -4$. It becomes a saddle for

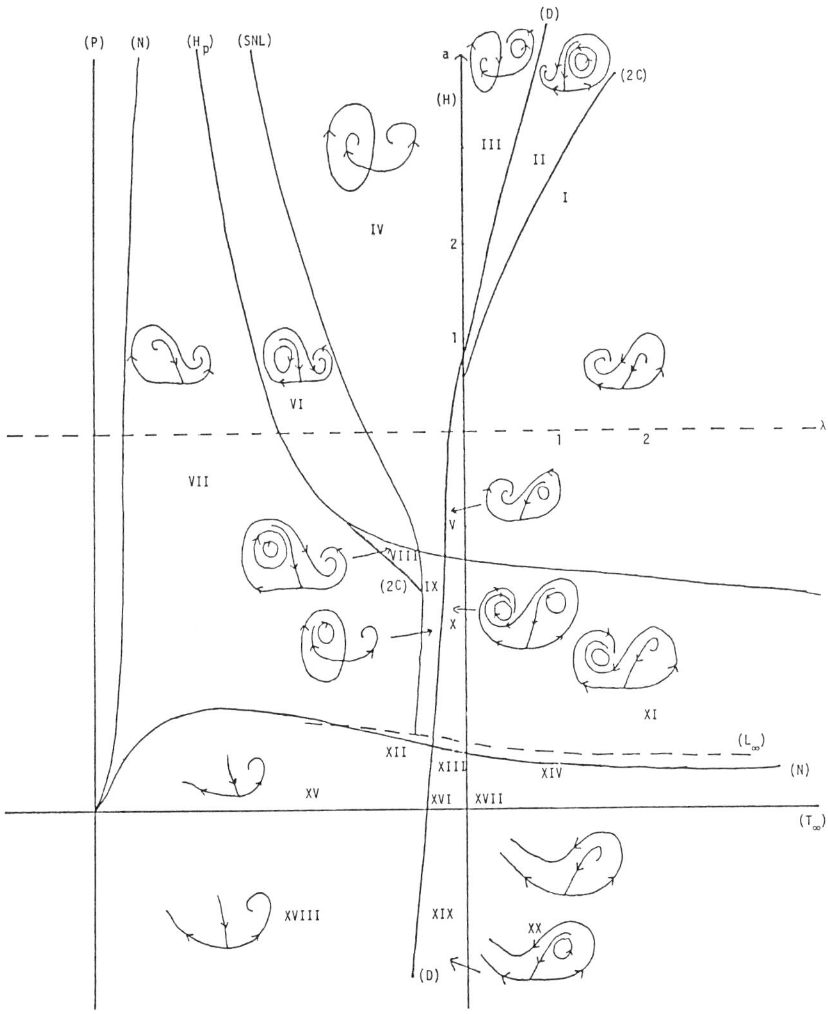

Figure 10. Bifurcation diagram for system (5.4).

$a < -4$. It undergoes a Hopf bifurcation on the branch of the hyperbola $a\lambda + 3a + 2\lambda + 4 > 0$ located above the line $a = -4$ (curve (H_p) in Fig.10). We have there :
$Re(c_1) = (4a^2 + 19a + 16)/8x^*$ and $Re(c_2) = 0$ for $a = (-19 + \sqrt{105})/8$. For this a:
$Re(c_2) = 11\,(13 + \sqrt{105})^{1/2}\ (73\sqrt{105} + 1045)/768\sqrt{2} > 0$.
There is a pitchfork for $\lambda = -4$, since $p = (0,0)$ (curve (P) in Fig.10). At $\lambda = a = -4$ the system is very degenerate: $\dot{x}$ and $\dot{y}$ have the common factor $x + y$.

ii) **Points at infinity :** They are given by

(5.5) $\quad u^3 + 5u^2 + u(1 - a) + 1 = 0$ and $z = 0$

We have a unique point at infinity P_1 for $a \in (a^*, +\infty)$, with $a^* \in (-4,-3)$. P_1 corresponds to a value of $u < -1$. For $a > -4$, P_1 is a saddle point. At a^* there is a saddle node leading to two new points P_1, P_2, with u-coordinate in the interval [-1,0]. The one with smallest u-coordinate undergoes a transcritical bifurcation (curve (T_∞) in Fig. 10) with P_1 at $a = -4$ (cf. Fig.11). Therefore all limit cycles that existed around p must have disappeared before $a = -4$.

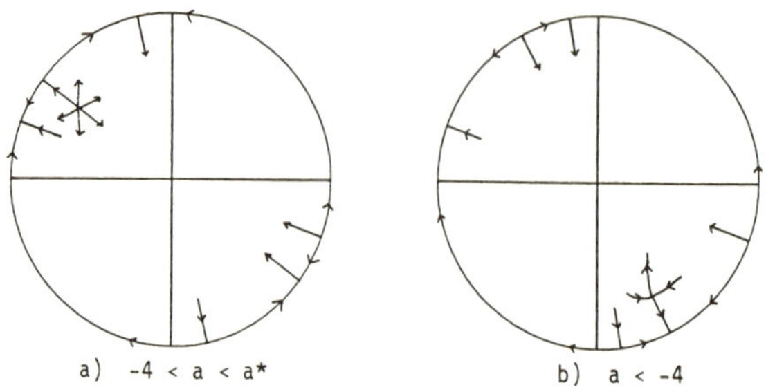

Figure 11. Bifurcation (T_∞) in system (5.4).

iii) **Invariant manifolds of (0,0) :** The two branches of the unstable manifold are called W_-^u (on the left) and W_+^u (on the right). The incoming branch of the center manifold is called W_-^0 . There is an infinity of outgoing branches. In the case $a = \lambda = 0$ we can show that they are as in Fig.12, since the hyperbola $\dot{x} = 0$ reduces to two lines. For the other values of a and λ we obtain numerically the behaviour given in Fig.10

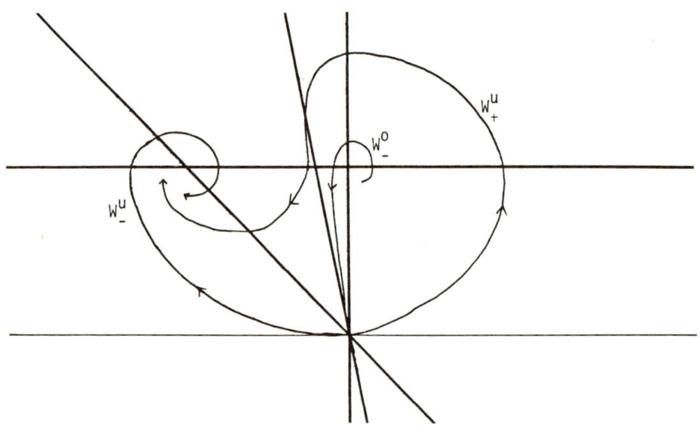

Figure 12. Relative position of W_+^u, W_-^u and W_-^0 in system (5.4) for $\lambda = a = 0$

The bifurcation curve (D) between regions II and III (resp. V to IV, X to IX) "should" indeed be a double curve since the transition from II to III (resp. V to IV, X to IX) requires two steps :

- First a loop formed by the intersection of W_+^u with W_-^0 .
- Then a bifurcation at infinity, with the unstable manifold of P_1 coinciding with W_-^0 in order to have W_-^0 on the left.

Between the two curves we have the phase portrait in Fig.13, but this has not been seen numerically, nor the two limit cycles that should result around (0,1) below that region, for $\varepsilon < 0$.

The curve (SNL) in Fig.10 corresponds to the intersection of W_-^0 with W_-^u .
It is clear numerically that the curve (D) crosses (T_∞) approximatly at $\lambda = -0.45$. We conjecture that the curve (SLN) and (D) cross (T_∞) at the same place since all limit cycles around p must have disappeared before (T_∞). Note that (SNL) is hard to follow numerically when we approach (T_∞) since p goes to infinity.

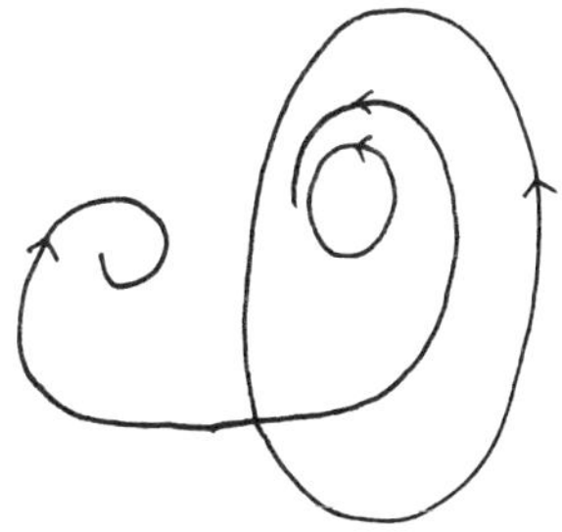

Figure 13. Phase portrait of system (5.4) between the two branches of the "double curve" (D) .

The point p becomes a node on the curve (N) (see Fig.10) given by :

(5.6) $\Delta = a^2 (\lambda + 3)^2 - a\ (12\lambda + 40) - 12\lambda^2 - 112\lambda - 240 = 0$

Therefore all limit cycles around p must have disappeared before we reach (N), since a limit cycle cannot contain a node [18]. The two branches of (N) meet at $\lambda = a = -4$. One of the branch ensures us that p is a node when the pitchfork happens. The point (0,1) also becomes a node for $|\lambda| > 2$ but this implies no additional bifurcation. The existence of (N) implies the existence of a curve where the limit cycle around p dies in region IX and X and XI. The only possibility is bifurcations involving the points at infinity (note that the curve (N) is situated below $a = a^*$, where the points at infinity P_2 and P_3 appear. This could for example be realized by a sequence of two bifurcations : The first one would be the intersection of W^u_- with the stable manifold of P_2 (Fig.14 b)). The second one would be a heteroclinic loop through P_2 and P_3 (Fig.14 c)) containing only the singular point p. We draw in Fig.10 a hypothetical curve (L_∞) corresponding to the effect of the two bifurcations. We limit the drawing of the curve to the region of interest.

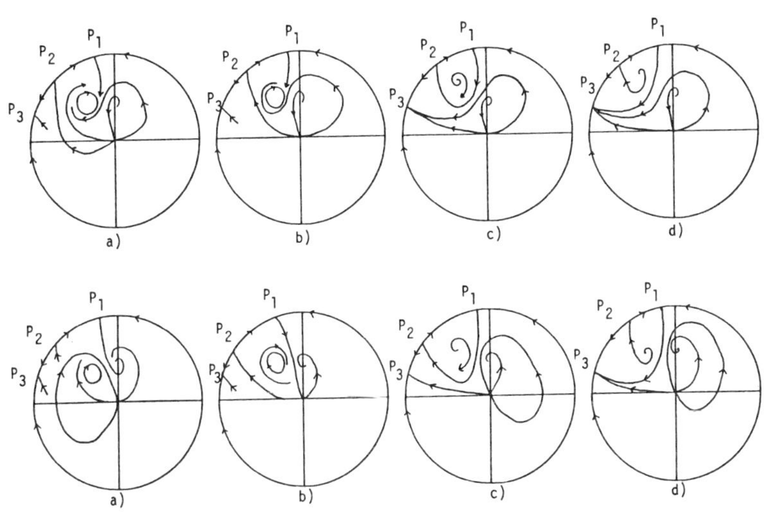

Figure 14. Sequence of bifurcations rerpresented by curve (L_∞) in Fig. 10. The upper pictures correspond to the passage from X to XIII. (The passage from XI to XIV uis obtained by replacing the repulsive (0,1) by a repulsive limit cycle). The lower pictures represent the passage from IX to XII. a) represents the phase portrait in region X (resp. XI, IX), b) and c) represent the two bifurcations and d) represents the phase portraits in region XIII (resp. XIV, XII).

iv) **Behaviour for $\varepsilon \neq 0$**: For $\varepsilon > 0$ the system (5.4) has two singular points, a saddle and a repulsive node instead of the singular point at the origin. The curves where (5.4) has a connection between W_-^u and W_-^0, or between (0,0) and one of the singular points at infinity, for $\varepsilon = 0$, are the boundaries of surfaces where (5.4) has saddle connections or saddle loops for $\varepsilon > 0$. For $\varepsilon < 0$ close to 0, the system has a limit cycle below each region of the plane $\varepsilon = 0$ where the system has a loop through the saddle-node.

ACKNOWLEDGEMENTS. We are greatful to Philip Holmes and Dana Schlomiuk for helpful discussions.

BIBLIOGRAPHY

1. Petrovskii I. and Landis E., On the number of limit cycles of the equation $dy/dx = P(x,y)/Q(x,y)$, where P and Q are polynomials of the second degree , Amer. Math. Soc. Trans. Ser. 2, 10 , (1958), 177-221.

2. Arnold V. I., Real algebraic geometry (the 16th Hilbert problem), in proceedings in Pure Mathemetics, 28 ,Amer. Math. Soc., (1976), 50-51.

3. Shi Songling, A concrete example of the existence of four limit cycles for plane quadratic systems, Sci. Sinica, 23 , (1980), 153-158.

4. Chen L. S. and Wang M. S., The relative position and number of limit cycles of a quadratic differential system, Acta Math. Sinica, 22 , (1979), 751-758.

5. Tung Chin-Chu, Positions of limit cycles of the system :
$dx/dt = \sum_{0 \le i+k \le 2} a_{ik} x^i y^k$, $dy/dt = \sum_{0 \le i+k \le 2} b_{ik} x^i y^k$,
Sci. Sinica, 8 , (1959),151-171, or Chinese J. Math., 8 , (1966), 854-874.

6. Ye Yanqian, On the impossibility of (2,2) distribution of limit cycles of any real quadratic system, Jour. of Nanjing Univ., (1985), 161-182.

7. Rousseau C. and Schlomiuk D., Generalized Hopf bifurcation and application to planar quadratic systems, submitted to Ann. Math. Pol. , (1985).

8. Rousseau C., Example of a quadratic system with two limit cycles rising in a homoclinic loop bifurcation, to appear in the J. of Diff. Eqs. (1986).

9. Bautin N. N., on the number of limit cycles which appear with the variation of coefficients from an equilibrium position of focus or center type, Mat. Sb., 30 , (1952), 181-196; Amer. Math. Soc. Trans. No 100 (1954).

10. Ye Yanqian, An important property of the weak focus of a quadratic differential system, Chinese Ann. Math A, 4 , n^0 1, (1983), 65-69.

11. Shi Songling, A method of constructing cycles without contact around a weak focus, J. of Diff. Eq., 41 , (1981), 301-312.

12. Qin Yuanxun, Shi Songling and Cai Suilin, On limit cycles of planar quadratic system, Sci. Sinica series A, 25 , (1982), 41-50.

13. Qin Yuanxun, Suo Guangjian and Du Xingfu, On limit cycles of planar quadratic system, Sci. Sinica series A, 26 ,(1983),1025,1033.

14. Bogdanov R. I., Bifurcation of the limit cycle of a family of plane vector fields, and Versal deformations of a singularity of a vector field on the plane in the case of zero eigenvalues, Sel. Math. Sov., 1 , (1981), 373-421, (Russian original in Trudy Sem. Petrovskii,1976).

15. Li Cheng Zhi, Around a weak focus of order 3 of any RQDS there can exist no limit cycles (to appear in Chin. Ann. Math.)

16. Dueng Jen-Zhu, Dong Jin-Zhu and Tung Chin-Chu, The structure of the separatrix cycles of a quadratic system, Acta Math. Sinica , 12 , (1962), 251-257.

17. Cherkas L. A. and Zhilevich L. I., Some criteria for the absence of limit cycles and for the existence of a single limit cycle, Differential equations, 6 (1970), 1170-1178.

18. Coppel W. A., A survey of quadratic sytems, J. Diff. Eqs, 2 , (1966), 293-304.

DÉPARTEMENT DE MATHÉMATIQUES ET DE STATISTIQUE
and CENTRE DE RECHERCHES MATHÉMATIQUES
UNIVERSITÉ DE MONTRÉAL
C.P. 6128, Succ. A
MONTRÉAL, QUÉ., H3C 3J7, CANADA

Canadian Mathematical Society
Conference Proceedings
Volume 8 (1987)

ASYMPTOTIC PROPERTIES OF ARNOLD TONGUES

Jürgen Scheurle

ABSTRACT. A periodically perturbed Hopf bifurcation problem is considered, where the perturbation is rapidly oscillating. Bifurcation of periodic solutions is analyzed. Particular attention is paid to estimating the "thickness" of the so-called Arnold tongues in the parameter space. We conclude from our results that for very large perturbation frequencies the behavior of the bifurcating solutions is "quasiperiodic" and frequency locking is practically not observable close to the bifurcation point.

1. INTRODUCTION. Consider a dynamical system with two scalar angular variables, i.e., a system on a two-dimensional torus. It is well known that generically there are open regions in the parameter space where the so-called rotation number is constant and rational. In these regions the flow is asymptotically periodic. This phenomenon is called phase locking. Near a curve of Neimark-Sacker bifurcations in a two-dimensional parameter space these regions are wedge-like domains with tip on the bifurcation curve, called Arnold tongues.

In this contribution we consider a periodically perturbed Hopf bifurcation problem, where the frequency of the perturbation grows to infinity as the size of the perturbation shrinks to zero. We analyze the asymptotic behavior of the Arnold tongues when the bifurcation curve is reached. In particular, we give upper bounds for their widths. Estimates for Arnold tongues, in the case where the frequency of the periodic perturbation remains fixed, have appeared in the article of Takens [12] and in the book of Arnold [2]. It turns out that the Arnold tongues get thinner and thinner and the corresponding rotation numbers tend to zero as the frequency of the perturbation tends to infinity. So, asymptotically there is no phase locking in this case. Also, the bifurcating periodic solutions behave like quasiperiodic solutions in that they cover the torus more or less densely.

1980 Mathematics Subject Classification (1985 Revision), 58F14, 58F22.

Such a situation also occurs in the unfolding of a codimension two singularity which has been considered in Scheurle and Marsden [11] and Scheurle [10]. There it has been shown that the bifurcating tori are quasiperiodic along continuous curves in the parameter space which meet the bifurcation curve transversally and fill a set of positive measure. So in a way the present results supplement this work. For other unfoldings where the situation described above occurs, see e.g.Iooss and Langford [6]. Our methods immediately carry over to handle these problems, also.

The plan of the papers is as follows. In Section 2 we state the basic equation and provide a certain normal form. Our later analysis, in particular the estimates, is heavily based on this normal form. In Section 3 we use Sacker's theorem [9] to prove existence of a smooth curve of Neimark-Sacker bifurcations through the origin in the parameter space (a similar proof has been sketched in the book of Chow and Hale [3]). In Section 4 periodic solutions on the bifurcating tori are constructed and estimates for the Arnold tongues are derived. Finally, Section 5 is devoted to our conclusion.

2. BASIC EQUATION AND A NORMAL FORM. We consider the following non-autonomous differential system

$$(2.1) \qquad \dot{x} = g(\lambda, x) + \mu h(\mu, \lambda, x, t/\mu) \qquad (x \in \mathbb{R}^2, \ t \in \mathbb{R}),$$

where as usual the dot denotes the derivative with respect to time t, and λ and $\mu \neq 0$ are real parameters. We assume that the maps $g: \mathbb{R} \times \mathbb{R}^2 \to \mathbb{R}^2$ and $h: \mathbb{R} \times \mathbb{R} \times \mathbb{R}^2 \times \mathbb{R} \to \mathbb{R}^2$ are infinitely often differentiable in some neighborhood of $\lambda = 0$, $\mu = 0$, $x = 0$ and $t \in \mathbb{R}$. We also assume that

$$(\mathrm{H1}) \qquad \begin{aligned} g(\lambda, 0) &= 0 \\ h(\mu, \lambda, x, \tau + 1) &= h(\mu, \lambda, x, \tau) \end{aligned}$$

holds for all values of μ, λ, x and τ, i.e. h is 1-periodic with respect to τ. Furthermore, for the Jacobian of g at x = 0 we make the assumption

(H2) $D_x g(\lambda, 0)$ has the eigenvalues $\alpha(\lambda) \pm i\beta(\lambda)$ with $\alpha(0) = 0$ and $\beta(0) > 0$.

Next we provide a certain normal form for this problem. It is based on a certain averaging procedure with respect to t. The main purpose of this procedure is to eliminate the t-dependence of the lower-order terms in the Taylor expansion with respect to x of the right-hand side of equation (2.1). But we also simplify the remaining t-independent terms as much as possible. It turns out that the order of the terms up to which this normal form can be achieved increases to infinity as the value of the parameter μ tends to zero. Our later analysis of the problem heavily relies upon this fact.

THEOREM 2.1. *(Normal form) Let the maps* g *and* h *have the above properties, and let* N *be an arbitrary odd, positive integer. Then there exist positive constants* μ_0, λ_0, ρ_0 *and a coordinate transformation*

$$(\rho, \theta) \mapsto x = T(\mu, \lambda, \rho, \theta, t/\mu)$$

in some neighborhood of $x = 0$, *such that for* $0 < |\mu| \le \mu_0/N$, $|\lambda| \le \lambda_0$, $\rho \in [0, \rho_0]$, $\theta \in [0, 1]$ *and* $t \in \mathbb{R}$, *in the new coordinates equation* (2.1) *reads as follows:*

$$(2.2)\qquad \begin{aligned} \dot{\rho} &= \xi\rho + \gamma_1\rho^3 + \gamma_3\rho^5 + \ldots + \gamma_{N-4}\rho^{N-2} + f_1(\mu, \lambda, \rho, \theta, t/\mu) \\ \dot{\theta} &= \zeta + \gamma_2\rho^2 + \gamma_4\rho^4 + \ldots + \gamma_{N-3}\rho^{N-3} + f_2(\mu, \lambda, \rho, \theta, t/\mu). \end{aligned}$$

Here the coefficients ξ, ζ *and* γ_k *are* C^∞ *functions of* μ *and* λ; *the functions* T, f_1 *and* f_2 *are also of class* C^∞ *and* 1*-periodic in each of their last two arguments. Moreover, we have*

$$(2.3)\qquad \begin{gathered} T(0, \lambda, \rho, \theta, t/\mu) = 0, \quad \frac{\partial}{\partial\theta} T(\mu, \lambda, 0, \theta, t/\mu) = 0, \\ \xi(0, \lambda) = \alpha(\lambda), \quad \zeta(0, \lambda) = \beta(\lambda), \end{gathered}$$

and the estimates

$$(2.4)\qquad |f_k(\mu, \lambda, \rho, \theta, t/\mu)| \le c_k\rho^{N+1-k} \quad (k = 1, 2)$$

hold with some constants c_k *which depend on* N *only.*

PROOF: Rescaling the time by setting

$$(2.5)\qquad \tau = t/\mu$$

and considering the suspended system

$$(2.6)\qquad \begin{aligned} \dot{x} &= \mu\{g(\lambda, x) + \mu h(\mu, \lambda, x, \tau)\} \\ \dot{\tau} &= 1 \end{aligned}$$

the proof follows exactly the same lines as the proofs of Lemmas 3.1, 3.2 and 3.3 in [10]. We therefore omit it here. ∎

We consider equation (2.2) to be the normal form of the original equation (2.1). Obviously the lower-order terms of the right-hand side are not only independent of t but also of θ. As a first consequence we read off from this normal form that $\rho = 0$, $\theta = \zeta t + \text{const.}$ is a solution for all parameter values. According to (2.3), the corresponding solution of (2.1) is μ-periodic and close to $x = 0$ for small values of $|\mu|$; actually it approaches $x = 0$ as $|\mu|$ tends to 0. So it can be considered as the continuation with respect to μ of the equilibrium point $x = 0$ of the system $\dot{x} = g(\lambda, x)$. Subsequently we shall refer to it as the *basic solution* of (2.1).

3. NEIMARK-SACKER BIFURCATION. In this section we study the bifurcation of certain toroidal integral manifolds of dimension 2 from the basic solution. Such a bifurcation is generally called a Neimark-Sacker bifurcation. But we

also can speak of a perturbed Hopf bifurcation, since it reduces to a standard Hopf bifurcation when the terms f_1 and f_2 are neglected in our normal form. Actually, then it even reduces to a bifurcation of the solution $\rho = 0$ of the simple algebraic equation

$$\xi\rho + \gamma_1\rho^3 + \dots + \gamma_{N-4}\rho^{N-2} = 0 \tag{3.1}$$

since the θ-equation decouples from the ρ-equation. In this sense, nontrivial solutions of (3.1) are approximations for the desired integral manifolds of (2.1).

We now make the additional non-degeneracy assumptions

$$\text{(H3)} \qquad \gamma_1(0,\ 0) \neq 0, \quad \kappa = \alpha'(0)/\gamma_1(0,\ 0) \neq 0 \qquad (\alpha' = \tfrac{d}{d\lambda}\alpha).$$

In view of $\alpha'(0) \neq 0$ and (2.3) the implicit function theorem implies that the basic solution $\rho = 0$ is singular along a C^∞ curve $\mathcal{C}$ in the parameter space given by

$$\mathcal{C}: \lambda = \lambda^*(\mu),$$

where $\lambda^*(\mu)$ is the unique solution of the equation $\xi = 0$ with respect to λ for small $|\mu|$, and $\lambda^*(0) = 0$. Moreover, it follows that there are nontrivial solutions of (3.1) which bifurcate from the basic solution along the curve $\mathcal{C}$. The condition $\gamma_1(0,\ 0) \neq 0$ implies that they bifurcate either to the right or to the left of $\mathcal{C}$ depending on the sign of the product κ in (H3). Indeed, set

$$\lambda = \lambda^*(\mu) + \Lambda \tag{3.2}$$

and blow up the singularity along $\mathcal{C}$ by setting

$$\rho = |\Lambda|^{1/2}\tilde{\rho}. \tag{3.3}$$

Then the implicit function theorem yields a C^∞ family of solutions

$$\tilde{\rho} = \rho^*(\mu,\ \Lambda),$$

where $|\Lambda|$ is sufficiently small and $\Lambda > 0$ (< 0, respectively) when the sign of κ is negative (positive, respectively), and

$$\rho^*(0,\ 0) = |\kappa|^{1/2}. \tag{3.4}$$

In order to continue these solutions to integral manifolds for the full system (2.2) we introduce a new variable $R \in [0,\ 1]$ by setting

$$\rho = |\Lambda|^{1/2}(\rho^*(\mu,\ \Lambda) + |\Lambda|^{(N-4)/2}R), \tag{3.5}$$

which leads to the following system of equations:

$$\begin{aligned} \dot{R} &= \Omega(\mu,\ \Lambda)R + F_1(\mu,\ \Lambda,\ R,\ \theta,\ t/\mu) \\ \dot{\theta} &= \omega(\mu,\ \Lambda) + F_2(\mu,\ \Lambda,\ R,\ \theta,\ t/\mu) \end{aligned} \qquad (\Lambda \gtrless 0), \tag{3.6}$$

where

$$\Omega(\mu,\ \Lambda) = \xi + \sum_{k=1}^{(N-3)/2} (2k + 1)\gamma_{2k-1}|\Lambda|^k\rho^*(\mu,\ \Lambda)^{2k},$$

$$\omega(\mu, \Lambda) = \zeta + \sum_{k=1}^{(N-3)/2} \gamma_{2k}|\Lambda|^k \rho^*(\mu, \Lambda)^{2k},$$

the functions F_1 and F_2 are of class C^∞ and 1-periodic in each of their last two arguments. Moreover, for $N \geq 5$ they fulfill the following estimates

$$(3.7)\qquad \begin{aligned} |F_1(\mu, \Lambda, R, \theta, t/\mu)| &\leq c_3|\Lambda|^{3/2} \\ |F_2(\mu, \Lambda, R, \theta, t/\mu)| &\leq c_4|\Lambda|^{(N-2)/2} \end{aligned}$$

with some constants c_3 and c_4 depending on N only.

Now we proceed as in the proof of Theorem 2.1. After rescaling the time in (3.6) according to (2.5), we consider the suspended system with the additional equation $\dot{\tau} = 1$. Then Sacker's theorem [9] applies to prove the following theorem, because $|F_k(\mu, \Lambda, R, \theta, \tau)/\xi| \leq c_5|\Lambda|^{1/2}$ $(k = 1, 2)$, where c_5 is another constant.

THEOREM 3.1 *(Existence of a 2-dimensional toroidal integral manifold) Let* (H1) *and* (H2) *hold, let* (H3) *hold with* $\kappa < 0$ ($\kappa > 0$, *respectively), and let* N *be an arbitrary integer greater than or equal to* 5. *Then, given any* $m \in \mathbb{N}$, *there exist a positive constant* $\Lambda_0 = \Lambda_0(N, m)$ *and a positive constant* μ_0 *which is independent of* N *and* m, *such that for* $0 < |\mu| \leq \mu_0/N$ *and* $\Lambda \in [0, \Lambda_0]$ ($\Lambda \in [-\Lambda_0, 0]$, *respectively) system* (3.6) *possesses an integral manifold* $\mathcal{T} = \mathcal{T}(\mu, \Lambda)$ *given by the equation*

$$(3.8)\qquad \mathcal{T}:\ R = R(\mu, \Lambda, \theta, \tau),$$

where the function on the right-hand side is of class C^m *and 1-periodic in* θ *and* τ; *moreover, the function itself and all its derivatives with respect to* θ *and* τ *depend continuously on* μ *and* Λ, *and* $R(\mu, 0, \theta, \tau) = 0$.

It follows that the motion on $\mathcal{T}$ is described by the equation

$$(3.9)\qquad \dot{\theta} = \mu\{\omega(\mu, \Lambda) + G(\mu, \Lambda, \theta, \tau)\},$$

where G is obtained by plugging the representation (3.8) for $\mathcal{T}$ into F_2. Hence, G has the same properties as $R(\mu, \Lambda, \theta, \tau)$ in Theorem 3.1. Moreover, we have the estimate

$$(3.10)\qquad |G(\mu, \Lambda, \theta, \tau)| \leq c_4|\Lambda|^{(N-2)/2}.$$

Thus we have reduced the dimension of the original problem (2.1) by one. Equation (3.9) will be studied in the next section.

REMARK 3.2. It also follows from (3.6) that as $t \to \infty$, $\mathcal{T}$ is attractive for nearby solutions if the sign of $\alpha'(0)\Lambda$ is negative and locally repelling if this sign is positive.

4. PERIODIC SOLUTIONS AND ARNOLD TONGUES. Obviously, every solution of (3.9) being of the form

(4.1) $$\theta = \frac{p}{q}\tau + \tilde{\Theta} \quad (p,\ q \in \mathbf{Z} \text{ with no common prime factor})$$

where $\tilde{\Theta}$ is a q-periodic function of τ, represents a μq-periodic solution of (2.1). If we think of $\mathcal{T}$ as a 2-torus, then this solution winds around $\mathcal{T}$ p times in the θ-direction and q times in the τ-direction during one period. Therefore, we speak of a p:q *resonance*. So, although solutions of this kind may have the same period for different pairs (p, q), their qualitative behavior differs.

In order to find such solutions we set

(4.2) $$\tilde{\Theta} = \theta_0 + \Theta,$$

where θ_0 is an arbitrary fixed real number, and look for q-periodic functions $\Theta = \Theta(\tau)$, such that

(4.3) $$[\Theta]_0 = \frac{1}{q}\int_0^q \Theta(\tau)d\tau = 0$$

and the equation

(4.4) $$\dot{\Theta} = \eta + \mu G(\mu,\ \Lambda,\ \frac{p}{q}\tau + \theta_0 + \Theta,\ \tau)$$

is satisfied for some $\eta \in \mathbb{R}$.

LEMMA 4.1. *Let the integers* N, m, p *and* q *be given and let all previous assumptions be satisfied. Assume that* $|\Lambda|$ *is sufficiently small depending on* N, m *and* μq. *Then there is a unique* $\eta \in \mathbb{R}$ *such that* (4.4) *has a q-periodic solution* $\Theta(\tau)$ *of class* C^m *which satisfies* (4.3). *Moreover,* η *and* $\Theta(\tau)$ *together with all its derivatives depend continuously on* μ, Λ *and* θ_0, $\Theta \equiv 0$ *for* $\Lambda = 0$, *and for* $\eta = \eta(\mu, \Lambda, \theta_0)$ *the estimate*

(4.5) $$|\eta(\mu,\ \Lambda,\ \theta_0)| \leq \mu c_4|\Lambda|^{(N-2)/2}$$

holds.

PROOF: We use a kind of Liapunov-Schmidt method to prove this lemma. In fact, we split equation (4.4) into two equivalent equations as follows

$$\eta = -\mu\mathcal{P}G(\mu,\ \Lambda,\ \frac{p}{q}\cdot + \theta_0 + \Theta(\cdot),\ \cdot)$$

$(4.6)_{1,2}$

$$\Theta(\tau) = \mu\int_0^\tau [\mathrm{id} - \mathcal{P}]G(\mu,\ \Lambda,\ \frac{p}{q}\sigma + \theta_0 + \Theta(\sigma),\ \sigma)d\sigma$$

for $\eta \in \mathbb{R}$ and $\Theta \in \mathcal{X}$, wherefore we introduce the Banach space

$$\mathcal{X} = \{\Theta \in C^0(\mathbb{R})\,|\,\Theta \text{ is q-periodic and } [\Theta]_0 = 0\}$$

equipped with the usual sup-norm, and the projection operator

$$\mathcal{P}:\ C^0(\mathbb{R}) \to \mathbb{R}, \quad \Theta \mapsto [\Theta]_0.$$

It is easily seen that the right-hand side of the second equation defines a Lipschitz-continuous map in $\mathcal{X}$, because G is continuously differentiable with respect to θ. Since $G(\mu,\ 0,\ \theta,\ \tau) = 0$, the corresponding Lipschitz constant is less than one if $|\Lambda|$ is sufficiently small. (For fixed Λ, the Lipschitz constant is proportional to the product μq.) Also, this map depends continuously on μ, Λ and θ_0. Hence, by the contraction mapping principle, $(4.6)_2$ has a unique solution $\Theta \in \mathcal{X}$, which depends continuously on the parameters. Furthermore, $\Theta = 0$ for $\Lambda = 0$. The additional smoothness properties of Θ which are claimed in the lemma follow by a bootstrapping argument. Finally, η is determined by $(4.6)_1$ as a continuous function of μ, Λ and θ_0. The estimate (4.5) is a consequence of (3.10). ∎

In order to be able to formulate the following corollary we note that because of $\omega(0, 0) = \beta(0) > 0$, the equation

$$\mu\omega(\mu, \Lambda) = \delta \tag{4.7}$$

can be solved uniquely for μ in terms of Λ and δ near $\Lambda = 0$ and $\delta = 0$. The solution which is of class C^∞ in both variables will be denoted by $\mu_\delta(\Lambda)$. Note that the bound for $|\delta|$ does not depend on N.

COROLLARY 4.2. (*Existence of* p:q *resonant periodic solutions*) *Let* (H1), (H2) *and* (H3) *with* $\kappa < 0$ ($\kappa > 0$, *respectively*) *hold. Let the constant* μ_0 *be chosen according to Theorem* 3.1. *Furthermore, let integers* $N \geq 5$, $m \geq 1$ *and a pair of integers* p, q *with no common prime factor be given, such that the inequalities*

$$(4.8)_{1,2} \qquad \begin{aligned} &|\mu^0_{p/q}| < \mu_0/N \qquad\qquad (\mu^0_{p/q} = \mu_{p/q}(0)) \\ &\omega(\mu^0_{p/q}, 0) + \mu^0_{p/q}\frac{\partial\omega}{\partial\mu}(\mu^0_{p/q}, 0) > \epsilon > 0 \end{aligned}$$

are satisfied. Then there exists a positive constant $\Lambda_0 = \Lambda_0(N, m, p, q)$ *with the following property. For every* $\Lambda \in [0, \Lambda_0]$ ($\Lambda \in [-\Lambda_0, 0]$, *respectively*) *and for every* $\theta_0 \in [0, 1)$ *there exist parameter values*

$$\mu = \mu^*_{p/q}(\Lambda, \theta_0),$$

such that (2.1) *has a* p:q *resonant periodic solution of class* C^m. *For these parameter values the estimate*

$$|\mu^*_{p/q}(\Lambda, \theta_0) - \mu_{p/q}(\Lambda)| \leq c_6|\Lambda|^{(N-2)/2} \tag{4.9}$$

holds with some constant c_6 *depending on* N *only. For other parameter values with* $0 < |\mu| < \mu_0/N$ *and* $|\Lambda| \leq \Lambda_0$ *no* p:q *resonant periodic solution exists.*

PROOF: Obviously, every solution of (4.4) given by Lemma 4.1 defines a solution of type (4.1), (4.2) for equation (3.9) and vice versa, provided that

$$\mu\omega(\mu, \Lambda) - p/q = \eta(\mu, \Lambda, \theta_0) \tag{4.10}$$

holds. Hence, in view of the uniqueness of η it remains to solve this equation for μ and to show the estimate (4.9) for its solutions. To this end we set

$$\mu = \mu_{p/q}(\Lambda) + |\Lambda|^{(N-2)/2}\tilde{\mu} \qquad (\tilde{\mu} \in \mathbb{R}). \tag{4.11}$$

Substituting this expression for μ into (4.10) and using (4.5), (4.7) and $(4.8)_2$ we end up with a fixed point problem

$$\tilde{\mu} = M(\tilde{\mu}, \Lambda, \theta_0) \tag{4.12}$$

for $\tilde{\mu}$, where M is a continuous real-valued map of $\tilde{\mu}$. It maps some bounded interval around $\tilde{\mu} = 0$ into itself, if $|\Lambda|$ is sufficiently small. Therefore all solutions of (4.12) are contained in this interval, and Brouwer's fixed point theorem applies to prove existence. Consequently, (4.9) follows from (4.11). Finally we have to make sure that $0 < |\mu| < \mu_0/N$ holds for the solutions of (4.10). In view of $(4.8)_1$ and (4.11) this leads to another restriction for Λ_0.■

Let us call the region in the (μ, λ)-space which is described by the parameter values determined by Corollary 4.2, the p:q *resonance zone*. Generically, this zone is a wedge-like region with tip on the bifurcation curve, called an *Arnold tongue*, and inside this whole zone all solutions on $\mathcal{T}$ behave like a p:q resonant periodic solution, at least asymptotically for large $|t|$. This phenomenon is usually called *phase locking*; the ratio p/q corresponds to the *Poincaré rotation number* for a 2-torus (cf. [1, 2, 3, 4, 5]).

REMARK 4.3. There is some evidence that the constants c_k (k=1,2,3,4,5,6) can be chosen independently of N, if the maps g and h in (2.1) are analytic in x (cf. [7, 8]).

5. CONCLUSION. It follows from (4.7) that the estimate

$$c_7|p/q| \le |\mu^0_{p/q}| \le c_8|p/q| \tag{5.1}$$

holds where the constants c_7 and c_8 do not depend on N, since $\omega(\mu, 0) = \varsigma(\mu, \lambda^*(\mu))$. Actually, $\mu^0_{p/q}$ is "close" to $\beta(0)p/q$. Consequently, the ratio p/q tends to 0 as the tips of the Arnold tongues approach the origin in the (μ, λ)-space. Moreover, we can choose N proportional to q/p in order to satisfy assumption $(4.8)_1$ of Corollary 4.2. Therefore, by (4.9) the Arnold tongues get thinner and thinner near the bifurcation curves $\mathbb{C}$, close to the origin.

Also, recall from Section 4 that an p:q resonant periodic solution winds around the torus $\mathcal{T}$ p times in the θ-direction and q times in the τ-direction before it comes back to the starting point. So, as $|q|$ tends to infinity, these solutions look more and more like quasiperiodic solutions on $\mathcal{T}$ which would never come back to an earlier point and thus cover the torus densely. As a matter of fact, between the Arnold tongues there are continuous curves in the parameter space along which the motion on $\mathcal{T}$ is quasiperiodic. These curves meet $\mathbb{C}$ transversally, where the relative measure of such intersection points tends to one as the origin is approached. A proof of this result requires ideas from KAM theory (cf. [10, 11]).

Our conclusion from these results is that for values of μ and λ near zero, the behavior of the solutions of (2.1) on the integral manifold $\mathcal{T}$ is "quasiperiodic". Also, since the Arnold tongues are very thin near $\mathbb{C}$, close to the origin of the parameter space, the phenomenon of phase locking is generically not observable there. This means, on a generic path through this region phase locking occurs only in very small intervals.

BIBLIOGRAPHY

1. V. I. Arnold, "Small denominators I. Mappings of the circumference into itself", AMS Translations, Ser. 2, 46 (1965), 213-284.
2. V. I. Arnold, Geometrical Methods in the Theory of Ordinary Differential Equations, Springer, New York, 1983.
3. S. N. Chow and J. K. Hale, Methods of Bifurcation Theory, Springer, New York, 1982.
4. J. Guckenheimer and P. Holmes, "Nonlinear oscillations, dynamical systems and bifurcation of vector fields", Springer, New York, 1983.
5. R. M. Herman, "Mesure de Lebesgue et nombre de rotation", Springer Lecture Notes in Math. 597 (1977), 271-293.
6. G. Iooss and W. F. Langford, "Conjectures on routes to turbulence via bifurcations", Ann. of the New York Acad. Sci. 357 (1980), 489-505.
7. A. I. Neishtadt, "The separation of motions in systems with rapidly oscillating phase", PMM U.S.S.R. 18 (2) (1984), 133-139.
8. N. N. Nekhorosev, "Exponential estimate of the time of stability of Hamiltonian systems close to integrable II", Tr. Seminara in I.G.Petrovskogo, Iss. 5, Izd. MGU, 1979.
9. R. J. Sacker, "A new approach to the perturbation theory of invariant surfaces", Comm. Pure Appl. Math. 18 (1965), 717-732.
10. J. Scheurle, "Successive bifurcations in the interaction of steady state and Hopf bifurcation", Cont. Math. 28 (1984), 419-433.
11. J. Scheurle and J. E. Marsden, "Bifurcation to quasiperiodic tori in the interaction of steady state and Hopf bifurcations", SIAM J. Math. Anal. 15 (6) (1984), 1055-1074.
12. F. Takens, "Forced oscillations and bifurcations", Comm. Math. Inst. Rijksuniversiteit Utrecht 3 (1974), 1-59.

DEPARTMENT OF MATHEMATICS
COLORADO STATE UNIVERSITY
FORT COLLINS, CO 80523

Canadian Mathematical Society
Conference Proceedings
Volume 8 (1987)

BIFURCATIONS OF POLYNOMIAL VECTOR FIELDS IN THE PLANE

J.Sotomayor[1] and R. Paterlini

ABSTRACT. The simplest patterns of qualitative change on planar polynomial vector fields whose coefficients depend smoothly on a real parameter are studied in this paper.

1. INTRODUCTION. This article is concerned with the description of the simplest patterns according to which vector fields of the form $X=\sum(P_k\partial/\partial x+Q_k\partial/\partial y)$, $k=0,1,\ldots,n$, where $P_k=\sum a_{ij}x^iy^j$, $Q=\sum b_{ij}x^iy^j$,$i+j=k$, called polynomial vector fields of degree n, change qualitatively - bifurcate - under small perturbations of their $N=(n+1)(n+2)$ coefficients a_{ij}, b_{ij}.

The compact bifurcations, which are the qualitative changes that occur on bounded regions of the plane, have been studied in [A-L,S_1,S_2,T] for the general case in which X is a smooth vector field. The simplest,but most common, patterns through which singular points and periodic orbits, located on compact regions, disappear or emerge due to small deformations of the vector field, have been described in these works. Nothing of essentially new will be added here to the knowledge of compact bifurcations for the case of polynomial vector fields.

The main concern of this paper will be the study of the simplest bifurcations at infinity; these are the patterns of qualitative change of polynomial vector fields that can be observed only by their global consideration on the whole plane and not by their restrictions to any compact region.

How do singularities and periodic orbits tend to (or spring from) infinity, overflowing (or inflowing to) arbitrarily large compact sets of the plane, as the coefficients a_{ij} , b_{ij} change smoothly? This is a typical question , whose answer, in a context explained below, will be attempted here.

Denote by $\mathfrak{X}_n$ the vector space of polynomial vector fields of degree n in the plane, identified with the space R^N , $N=(n+1)(n+2)$, of their coefficients. The bifurcation set $\mathfrak{X}^1{}_n$, where bifurcations occur, is defined by exclusion from $\mathfrak{X}_n$ of the set Σ_n of structurally stable vector fields, at which qualitative changes are not

Mathematics Subject Classification (1985 Revision). 34C25, 58F14
[1]Supported by CNPq,Brazil, and J.S. Guggenheim Memorial Foundation.

produced by small perturbations in $\mathfrak{X}_n$. It has been proved in [S_2] that $\mathfrak{X}^1{}_n$ is contained in the union of countably many analytic submanifolds of $\mathfrak{X}_n$, of codimension greater than or equal to one. In particular, $\mathfrak{X}^1{}_n$ has null Lebesgue measure. The point set topological properties of Σ_n have been studied in [Pu,Sa]. The characterization of this set however has not yet been achieved ; see problem (2.14). For this reason, here we consider instead the class Σ_n of regularly stable polynomial vector fields which has a characterization in terms of conditions reviewed in (2.13).

Theorems 1,2,and 3 of this paper distinguish, through conditions which violate in the mildest possible way those for Σ_n, an analytic hypersurface $\Sigma^1{}_n$, injectively immersed in $\mathfrak{X}_n$. The simplest bifurcations, compact and at infinity, as illustrated geometrically in (2.17),occur by the effect of small perturbations transverse to $\Sigma^1{}_n$. Furthermore, the elements in $\Sigma^1{}_n$ remain regularly stable under small perturbations along $\mathfrak{X}^1{}_n$, in a sense made precise in (2.12).

One of the important problems left open in this paper is to prove that this stability property along $\mathfrak{X}^1{}_n$ actually characterizes $\Sigma^1{}_n$.

Another equally relevant problem is to prove that the bifurcations of $\Sigma^1{}_n$ are generic. That is, to show that $\mathfrak{X}^2{}_n=\mathfrak{X}^1{}_n-\Sigma^1{}_n$ is contained in a countable union of submanifolds of codimension greater than or equal to two in $\mathfrak{X}_n$. By Thom's Transversality Theorem [Th], this would imply that, generically, bifurcations along one-parameter families ,i.e. curves, in $\mathfrak{X}_n$ occur only at points of transversal crossing with $\Sigma^1{}_n$.

The extension of these properties, valid for smooth vector fields in the plane and sphere [S_1,S_2],to $\mathfrak{X}_n$, n≥2, constitute the fundamental problems of qualitative bifurcation theory for planar polynomial vector fields.

2. THE MAIN RESULTS. After reviewing some preliminary definitions, most of which are based on [A-L,S_1], and introducing a few others, the main results of this work will be stated. Instrumental to this end is the *Poincaré Compactification* , $\mathcal{P}$, on the space $\mathfrak{X}_n$. Recall from [P,S_2] that $\mathcal{P}(X)$ is the analytic vector field tangent to the sphere $S^2=\{x^2+y^2+z^2=1\}$ as well as to the equator $S^1= S^2\cap\{z=0\}$, whose restriction to the northern hemisphere $S^2{}_+=S^2\cap\{z\geq 0\}$, is given by $\mathcal{P}(X)(x,y,z)=z^{n-1}P_*(X)(x,y,z)$, where $P:R^2\to S^2{}_+$ is the central projection: $P(u,v)=(u,v,1)/(u+v+1)^{1/2}$.

It will be useful to introduce polar coordinates (θ,ρ), 2π-periodic in θ, defined by the covering map from $R\times(-1,1)$ onto $S^2-\{(0,0,\pm1)\}$, given by $(\theta,\rho)\to(x,y,z)=(1+\rho^2)^{-1/2}(\cos\theta,\sin\theta,\rho)$. The expression for $\mathcal{P}(X)$ in these coordinates is

(2.1) $$(1+\rho^2)^{(1-n)/2}[(\Sigma\rho^i A_{n-i}(\theta))\partial/\partial\theta-\rho(\Sigma\rho^i R_{n-i}(\theta))\partial/\partial\rho],\ i=0,1,\ldots,n,$$ where

(2.2) $$A_k(\theta)=-P_k(\cos\theta,\sin\theta)\sin\theta+Q_k(\cos\theta,\sin\theta)\cos\theta,$$
$$R_k(\theta)= P_k(\cos\theta,\sin\theta)\cos\theta+Q_k(\cos\theta,\sin\theta)\sin\theta.$$

Here, $X=P\partial/\partial x+Q\partial/\partial y$ and $P=\sum P_k$, $Q=\sum Q_k$, $k=0,1,\ldots,n$, where P_k, Q_k are homogeneous polynomials of degree k.

The expression for $\mathbb{P}(X)$ in the coordinates on S^2-S^1 defined by the central projection P onto $S^2{}_+$ is given by

(2.3) $(u^2+v^2+1)^{(1-n)/2}X(u,v)$.

Therefore, a singularity of $\mathbb{P}(X)$ at $p\in S^2-S^1$, called *finite*, is of the same type except for a reparametrization, as that of X at $P^{-1}(p)$.

Recall that a singularity is *hyperbolic* if the eigenvalues of its Jacobian matrix have non-vanishing real parts. The singularities of $\mathbb{P}(X)$ on S^1, called singularities of X at *infinity*, are given by $A_n(\theta)=0$. At these points the Jacobian matrix of $\mathbb{P}(X)$ is

$$(2.4) \qquad \begin{bmatrix} A'_n(\theta) & A_{n-1}(\theta) \\ 0 & -R_n(\theta) \end{bmatrix}$$

Therefore, hyperbolicity for singularities at infinity is equivalent to $A'_n(\theta)\neq 0$ and $R_n(\theta)\neq 0$.

(2.5) Recall [A-L,S_1] that a non-hyperbolic singular point p is called a *saddle-node* of a (C^2) vector field X if its derivative at p has only one eigenvalue equal to 0. Furthermore, calling v the null eigenvector and w the other, it holds that the second derivative $D^2X_p(v,v) = av+bw$ satisfies $a\neq 0$.

For polynomial vector fields, there are two types of saddle-node singularities at infinity, say at $(\theta,0)$:

a) *transversal saddle-node*, provided $A'_n(\theta)\neq 0$, $R_n(\theta)=0$, and $(R'_nA_{n-1}-R_{n-1}A'_n)(\theta)\neq 0$.

b) *tangential saddle-node*, if $A'_n(\theta)=0$, $A''_n(\theta)\neq 0$ and $R_n(\theta)\neq 0$.

A direct computation of second derivatives in the direction of the eigenvectors $\partial/\partial\theta$ and $-A_{n-1}\partial/\partial\theta+(A'_n+R_n)\partial/\partial\rho$ of (2.4) shows that these points are in fact saddle-nodes with null eigenvector transversal to $S^1=\{\rho=0\}$ in case a) and tangent to it in case b). See Lemma (3.1).

(2.6) A singular point whose eigenvalues have non-vanishing imaginary parts is called a *focus*. Take a segment L through a focus p of X. The orbits of X define a return map π which is of the same class as X on a neighborhood L_0 of p. The focus is *hyperbolic* if $\pi'(p)\neq 1$ and it is a *double* or *composed focus* if $\pi'(p)=1$ and $\pi^{(3)}(p)\neq 0$ [A-L,S_1]. Notice that by (2.4) $\mathbb{P}(X)$ cannot have foci at infinity.

(2.7) For reference purposes, a finite saddle-node, a pair of opposite infinity saddle-nodes and a double focus will be called *non-hyperbolic simple* singularities. The local phase portraits of these singularities are well known [A-L,S_1]. See Fig.1. Recall however, for future reference, that a saddle point has four hyperbolic sectors, bounded by four *separatrices*. A saddle-node has a parabolic sector bounded by the two *strong separatrices*, tangent to the non-zero eigenspace, and two hyperbolic sectors,each bounded by a strong separatrix and separated from each other by the *weak separatrix* , tangent to the null eigenspace.

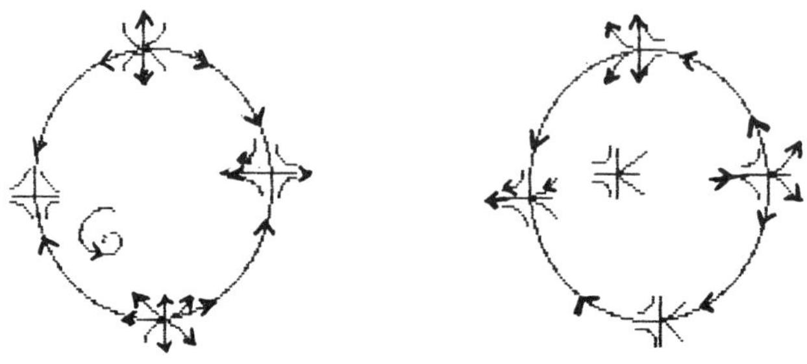

Figure 1. Singular Points

(2.8) An orbit γ of $\mathbb{P}(X)$ which is a separatrix for two different singular points or twice a separatrix of the same singularity, will be called a *double separatrix* or , when the singularities involved are saddle points,a *saddle connection* [A-L,S_1].

Five essentially different types of saddle connections γ in S^2-S^1, called *simple* described in a),...,e) below, will occur in this paper for polynomial vector fields.

a) *finite saddle connection*, joining two different finite saddles.

b) *loop* ,joining a finite saddle p to itself, with $\sigma(p)=\mathrm{trace}(DX(p))\neq 0$.The sign of σ determines if the loop is an attractor (negative) or a repellor (positive) [A-L,S_1].

c) *mixed saddle connection* , joining a finite to an infinite saddle.

d) *non-symmetric saddle connection*, which joins two infinite saddles $p_i=\{\theta_i, \mathrm{mod} 2\pi\}$, $i=1,2$, with $\theta_1\neq\theta_2$, modπ, so that the arc $[\theta_1,\theta_2]$ is free from singular points, and $\sigma_1(\theta_1,\theta_2)=A'_n(\theta_1)R_n(\theta_2)-A'_n(\theta_2)R_n(\theta_1)\neq 0$.

The sign of σ_1 determines if the *graph* or *separatrix cycle* Γ of $\mathbb{P}(X)$ with corners p_1, p_2 and edges are γ, going from p_1 to p_2, and $[\theta_1 ,\theta_2]$, oriented from θ_2 to θ_1, is an attractor (positive) or repellor (negative). This follows from a criterion given by Dulac [D,S_3]. See also [R_1].

e) *symmetric saddle connections* , joining two infinite symmetric saddles, going from $p_2=\{\theta_2$, mod $2\pi\}$ to $p_1=\{\theta_1=\theta_2-\pi$, mod$\pi\}$ so that the arc $\alpha=[\theta_1,\theta_2]$ is free from singularities (and $A_n(\theta)>0$ there) and

$$\sigma_2(\pm) = \int_\gamma \mathrm{div}X \pm (A'_n(\theta_1)/R_n(\theta_1))\int_\alpha(-R_n(\theta)/A_n(\theta))d\theta \neq 0 .$$

The signs of $\sigma_2(\pm)$ determine the attracting or repelling character of the two singular cycles $\Gamma(\pm)$ of $\mathbb{P}(X)$, with corners p_1 and p_2 , bounded by γ and $\pm\alpha$. In section 5 it is shown after [D,R$_2$,Pe] that the return maps $\pi(\pm)$ associated to these singular cycles are differentiable. The exponentials of $\sigma_2(\pm)$ turn out to be the derivatives of these maps at points of γ. See (5.1).

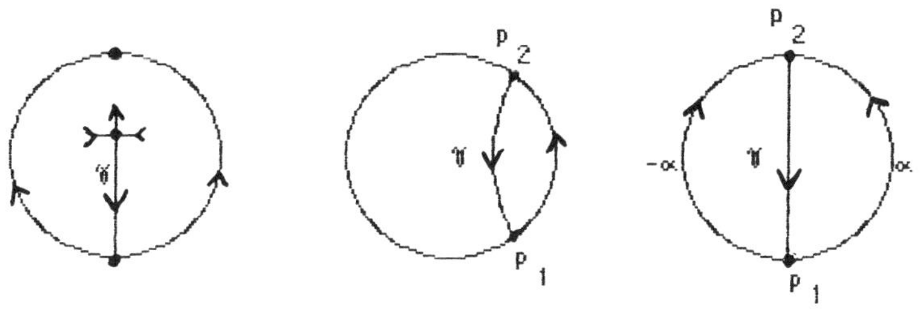

c d e

Figure 2 . Saddle Connections

(2.9) To a periodic orbit γ of a vector field and a segment L transversal to it at a point p, it is associated a return map π, called Poincaré map , defined on a neighborhood L_0 of p into L , which has p as fixed point an is of the same class as the vector field. The periodic orbit γ is called *hyperbolic* if $\pi'_p \neq 1$. It is called of *multiplicity k* if $\pi_p^1 = 1$, $\pi_p^2 = 0, \ldots, \pi_p^{(k-1)} = 0$, and $\pi_p^{(k)} \neq 0$. If k=2, γ is called *double*, if k=3 it is called *triple*.

A classical formula of Poincaré establishes that for a planar vector field X, $\log(\pi_p^1) = \int_\gamma \operatorname{div} X$. From (2.1) it follows that S^1 is a periodic orbit of $\mathbb{P}(X)$ if and only if $A_n(\theta)$ has no real zeroes, which happens only if n is odd. In this case , X is said to have a *periodic orbit at infinity* .

For purpose of future reference, a finite double or a triple periodic orbit at infinite will be called *simple non-hyperbolic periodic orbit* .

The condition for the hyperbolicity for periodic orbits at infinity is given by

$$\text{(2.10)} \qquad \int_0^{2\pi} R_n(\theta) A_n^{-1}(\theta)\, d\theta \neq 0.$$

This follows from Poincaré's formula and (2.1). See also (4.1) where equivalent conditions are established.

(2.11) Notice that the multiplicity of an orbit at infinity cannot be even,due to the invariance of $\mathbb{P}(X)$ with respect to the antipodal map on S^2.

(2.12) Let $\mathcal{C}$ be a subset of $\mathfrak{X}_n$. A vector field X in $\mathcal{C}$ is said to be *structurally stable along* $\mathcal{C}$ provided for every smooth curve ξ: $(R,0) \to (\mathcal{C}, X)$, there is a family of homeomorphisms h: $(R,0) \to (\mathrm{Hom}(S^2,S^1), \mathrm{Id})$ such that for small λ, $h(\lambda)$ maps orbits of $\mathbb{P}(X)$ onto those of $\mathbb{P}(\xi(\lambda))$, preserving S^1. If it also happens that that $h(\lambda)$ restricted to every periodic orbit of X is continuously differentiable in λ, it is said that X is *regularly structurally stable along* $\mathcal{C}$.

(2.13) THEOREM [S_2] A vector field X in $\mathfrak{X}n$ is regularly structurally stable (along $\mathfrak{X}_n$) if and only if it belongs to the set $\Sigma_n = \Sigma_n(1) \cap \Sigma_n(2) \cap \Sigma_n(3)$, where $\Sigma_n(i)$ is defined by the property i = 1,2,3 below

1) All the singularities of $\mathbb{P}(X)$ are hyperbolic.

2) All periodic orbits of $\mathbb{P}(X)$ are hyperbolic.

3) All double separatrices of $\mathbb{P}(X)$ are contained in S^1.

Furthermore, the regular bifurcation set, $\mathfrak{X}^1_n = \mathfrak{X}_n - \Sigma_n$, is contained in a countable union of analytic submanifolds of codimension ≥ 1.

(2.14) Problem [S_2]. Can a non hyperbolic attracting or repelling periodic orbit occur in a (non- regularly) structurally stable polynomial vector field?

This is the only problem left open in order to characterize structurally stable polynomial vector fields as the class Σ_n described in (2.13).

(2.15) Σ^1_n is defined as the class of vector fields in $\mathfrak{X}^1_n = \mathfrak{X}_n - \Sigma_n$ for which conditions 1,2,3 of (2.13) are violated in the mildest possible way. This idea has its origin in [A-L]; see also [S_1,T].

It is natural to express Σ^1_n as the union of three essentially different subsets, $\Sigma^1_n(i)$ of $\mathfrak{X}^1_n(i) = \mathfrak{X}_n - \Sigma^1_n(i)$, i = 1,2,3.

1) $\Sigma^1_n(1)$ is the set of X in $\Sigma_n(2) \cap \Sigma_n(3)$ for which $\mathbb{P}(X)$ has only non-hyperbolic simple singularity (2.7).

2) $\Sigma^1_n(2)$ is the set of X in $\Sigma_n(1) \cap \Sigma_n(2)$ for which $\mathbb{P}(X)$ has only a non-hyperbolic periodic orbit, which is simple (2.9).

3) $\Sigma^1_n(3)$ is the set of X in $\Sigma_n(1) \cap \Sigma_n(2)$ for which $\mathbb{P}(X)$ has only one saddle connection which is simple (2.8).

THEOREM 1. Let n ≥ 2.

a) Every X in $\Sigma^1_n(1)$ is regularly stable along $\mathfrak{X}^1_n$.

b) Also, $\Sigma^1_n(1)$ is an embedded analytic submanifold of codimension one in $\mathfrak{X}_n$.

c) Furthermore, every X in $\Sigma^1_n(1)$ has a neighborhood U where an analytic function F: $(U,X) \to (R,0)$ with $dF_X \neq 0$ and $F^{-1}(0) \subset \Sigma^1_n(1)$ can be defined. Call $U_+ = F^{-1}(0,+\infty)$ and $U_- = F^{-1}(-\infty,0)$. Then U_+ and U_- can be taken in Σ_n.

THEOREM 2. Let n ≥ 2.

a) Every X in $\Sigma^1_n(2)$ is regularly stable along $\mathfrak{X}^1_n$.

b) $\Sigma^1_n(2)$ is an embedded analytic submanifold of codimension one in $\mathfrak{X}_n$.

c) Furthermore, every X in $\Sigma^1_n(2)$ has a neighborhood U in $\mathfrak{X}_n$ where an analytic function F: $(U,X) \to (R,0)$ with $dF_X \neq 0$ and $F^{-1}(0) \subset \Sigma^1_n(2)$ can be defined. Let U_+ and U_- be as in Theorem 1. Then U_- can be taken in Σ_n. U_+ can also be taken in Σ_n, unless X has a semi-stable periodic orbit which is the α and ω-limit set of saddle separatrices of $\mathbb{P}(X)$; in this case, for U sufficiently small, $U_+ \subset \Sigma_n \cup \mathfrak{X}^1_n(3)$ with $U_+ \cap \mathfrak{X}^1_n(3)$ non empty and contained in $\Sigma_n(1) \cap \Sigma_n(2)$. See Fig. 5 and also (2.13).

THEOREM 3. Let $n \geq 2$.

a) Every X in $\Sigma^1{}_n(3)$ is regularly stable along $X^1{}_n$.

b) Also, $\Sigma^1{}_n(3)$ is an injectively immersed analytic submanifold of codimension one in X_n.

c) Furthermore, every X in $\Sigma^1{}_n(3)$ has a neighborhood U in X_n where an analytic function $F: (U,X) \to (R,0)$ with $dF_X \neq 0$ and $F^{-1}(0) \subset \Sigma^1{}_n(3)$ can be defined. Let U_+ and U_- be as in Theorem 1. Then U_- and U_+ can be taken in Σ_n unless

c_1) the saddle connection in $S^2 - S^1$ is contained in the limit set of some separatrix of $\mathbb{P}(X)$. That is, in Figures 7, 14, 15 and 16 the dashed separatrices are present.

In this case, for U small, $U_- \subset \Sigma_n \cup X^1{}_n(3)$ with $U_- \cap X^1{}_n(3)$ non empty and contained in $\Sigma_n(1) \cap \Sigma_n(2)$. Also, $U_+ \subset \Sigma_n$, unless

c_2) the saddle connection in $S^2 - S^1$ is symmetric and the two singular cycles of $\mathbb{P}(X)$ which it defines are the limit sets of separatrices of X. That is, in Figure 15, the two dashed separatrices actually exist. In this case, $U_+ \subset \Sigma_n \cup X^1{}_n(3)$ with $U_+ \cap X^1{}_n(3)$ non empty, contained in $\Sigma^1{}_n(1) \cap \Sigma^1{}_n(2)$. See (2.13).

Theorems 1,2 and 3 will be proved in sections 3,4 and 5, respectively.(2.16)

(2.16) Problems. The main problems left open in this work will be stated now.

Problem 1 (Characterization). Does $\Sigma^1{}_n$ coincide with the set of all vector fields of $X^1{}_n$ which are regularly stable along $X^1{}_n$?

Problem 2 (Genericity). Is $X^2{}_n = X^1{}_n - \Sigma^1{}_n$ contained in the union of countably many submanifolds of codimension ≥ 2 of X_n?

(2.17) Geometric description of simple bifurcations. The concise statements of Theorems 1, 2 and 3, while providing a synthesis of the common qualitative properties of bifurcations from $\Sigma^1{}_n$, do not give an idea of the diversity of possibilities for individual cases. A geometric illustration follows below.

(2.17.1) Compact bifurcations [A-L, S_1]

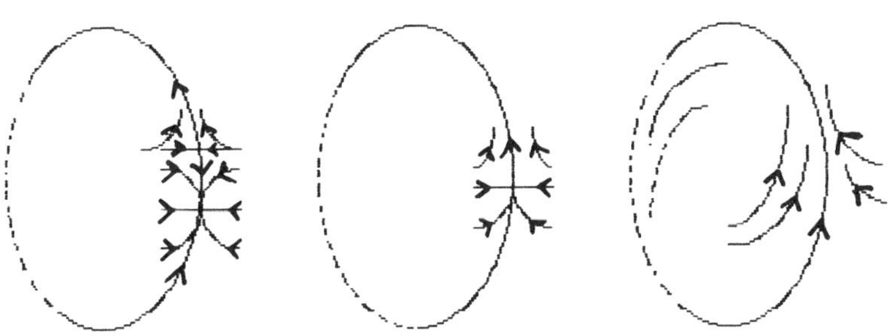

F<0 F=0 F>0

Figure 3. Bifurcation of saddle-node and saddle-node loop (dashed)

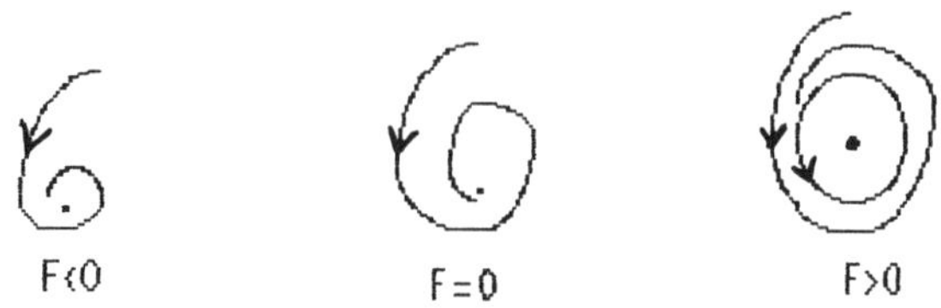

Figure 4. Andronov-Hopf bifurcation of double focus.

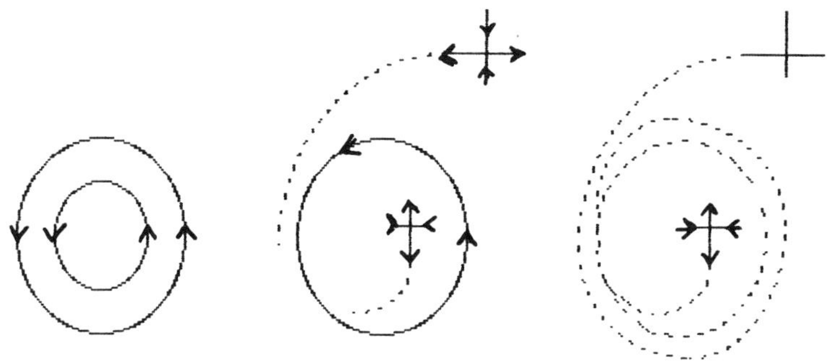

Figure 5. Semi-stable orbit bifurcation and possible saddle connection (dashed).

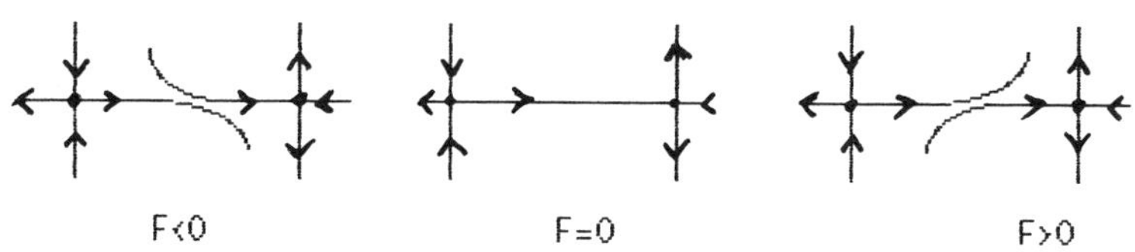

Figure 6. Saddle connection.

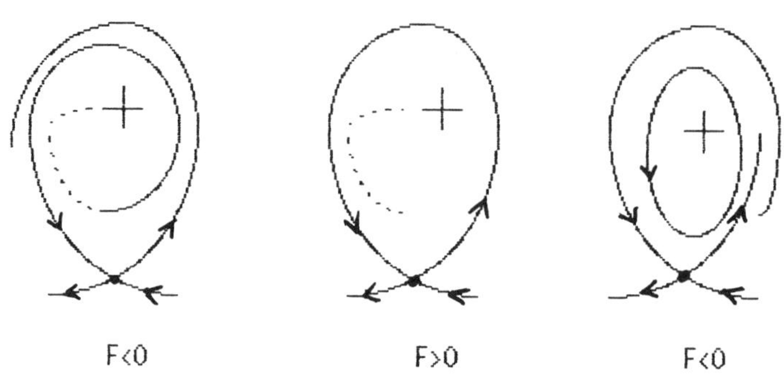

Figure 7. Saddle loop bifurcation with possible saddle connection (dashed)

(2.17.2) Bifurcations at Infinity.

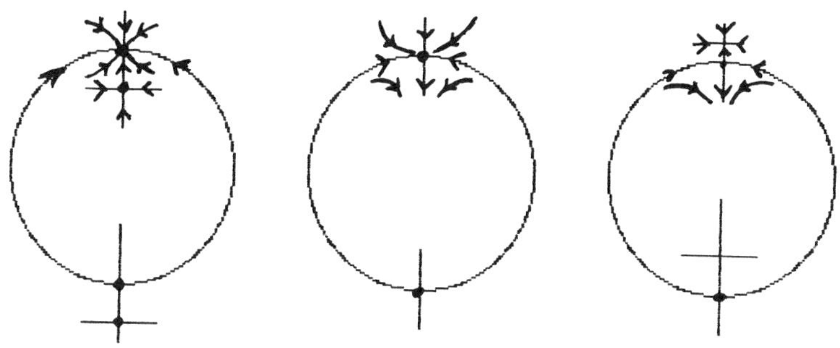

F<0 F=0 F>0

Figure 8. Bifurcation of transversal saddle-node.

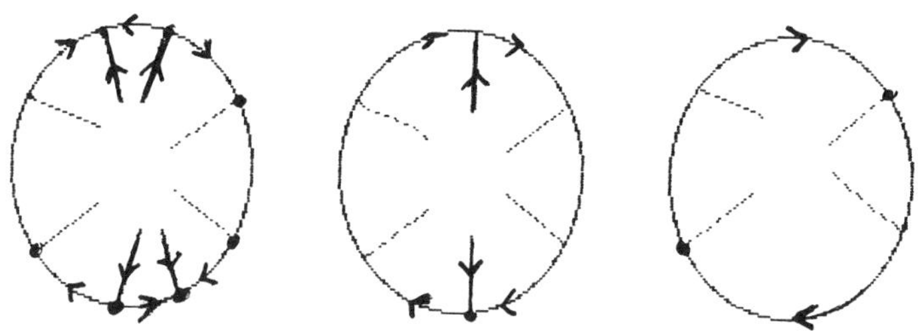

F>0 F=0 F<0

Figure 9. Bifurcation of tangential saddle-node (periodic orbit at infinity, when singularities with dashes separatrices are absent)

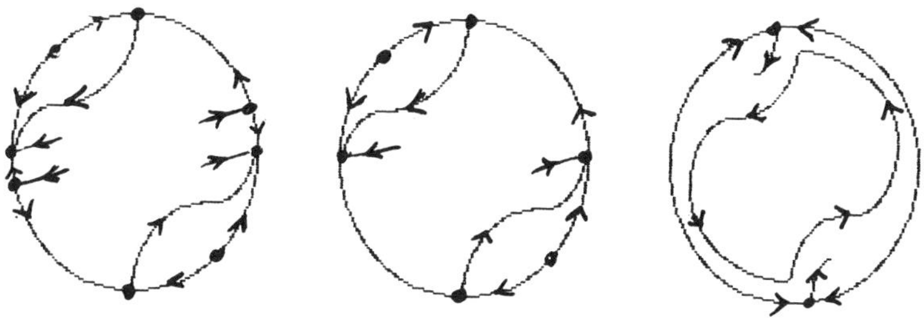

F<0 F=0 F>0

Figure 10. Bifurcation of tangential saddle node: one periodic orbit.

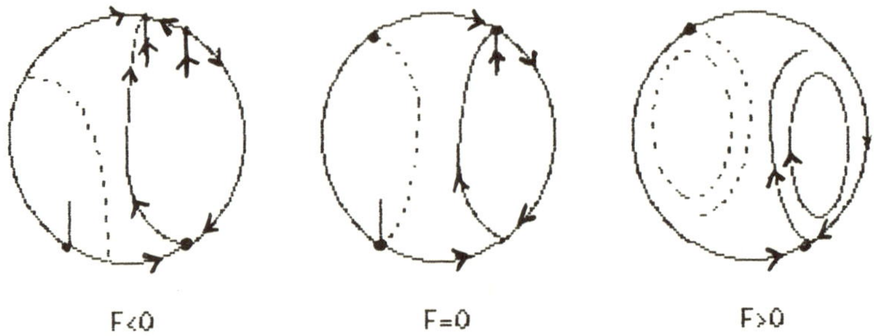

Figure 11. Bifurcation of tangential saddle node: one or two periodic orbits.

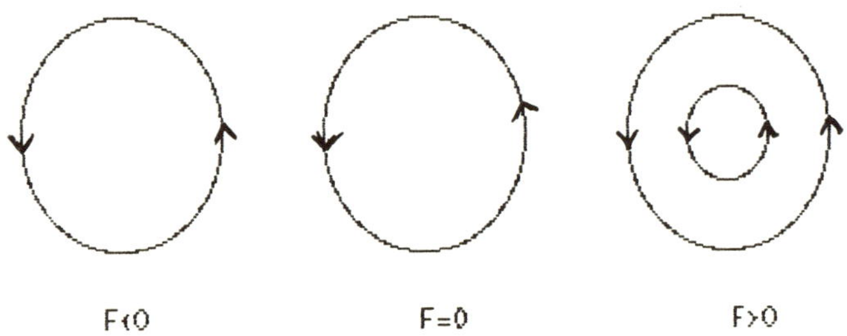

Figure 12. Bifurcation of periodic orbit at infinity.

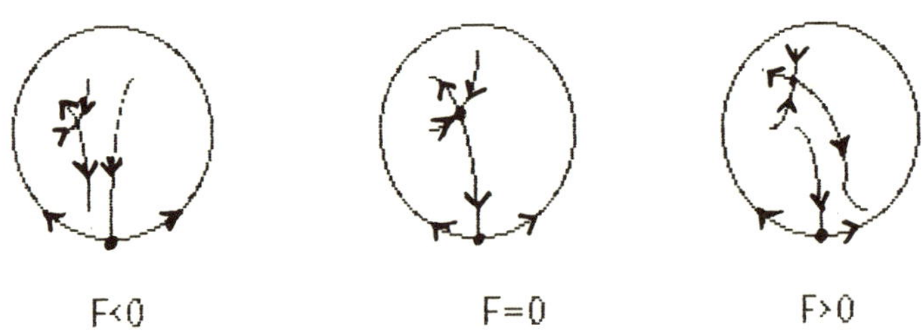

Figure 13. Bifurcation of mixed saddle connection (finite and infinite)

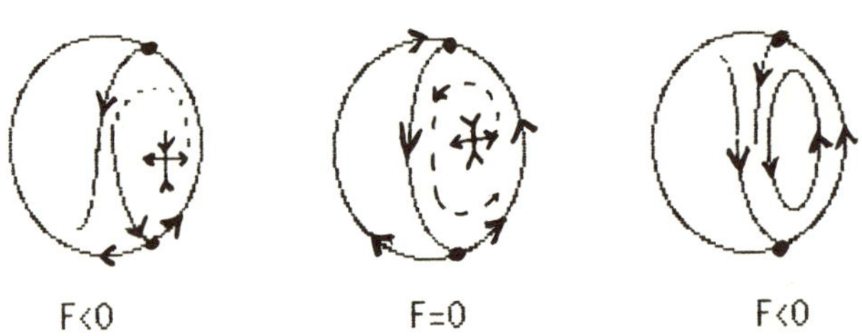

Figure 14. Bifurcation of non symmetric infinite saddle connection with possible mixed connection (dashed).

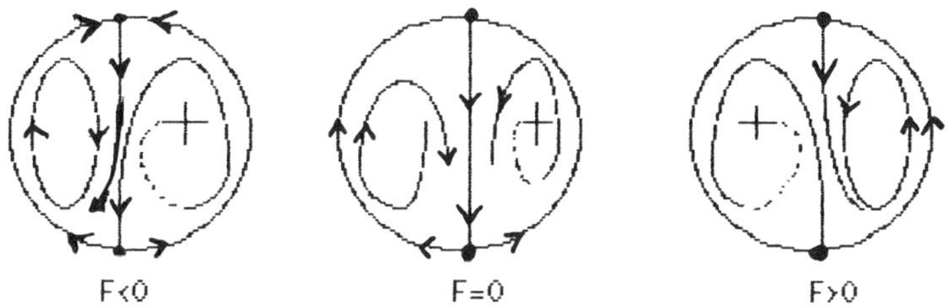

Figure 15. Bifurcation of symmetric infinite saddle connection, first case, with possible mixed saddle connections (dashed).

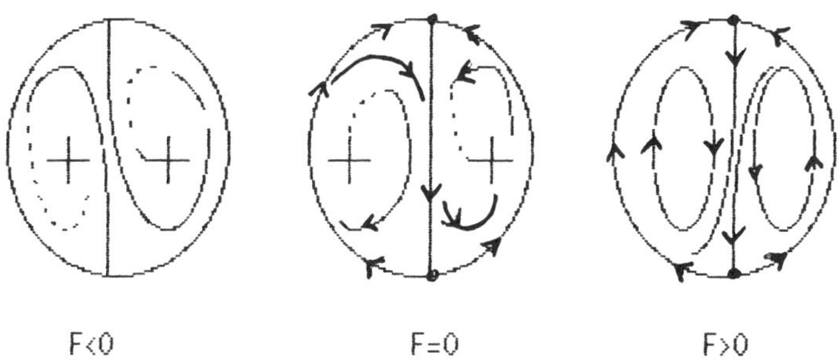

Figure 16. Bifurcation of symmetric infinite saddle connection, second case, with possible mixed saddle connections (dashed).

3. PROOF OF THEOREM 1. We begin with the following.

(3.1)<u>LEMMA. The singularities a) and b) described in (2.5) are saddle-nodes. Every singularity of $\mathbb{P}(X)$ at infinity which is a saddle-node is either of type a) or b).</u>

Proof. Let $Y=A\,\partial/\partial\theta+R\partial/\partial\rho$ be the expression for the components of $\mathbb{P}(X)$ in polar coordinates (2.1). Let $(\theta,0)$ be a singularity at infinity. Then

$$(\partial^2/\partial\theta^2)A(\theta,0)=A''_n(\theta),\quad (\partial^2/\partial\rho\partial\theta)A(\theta,0)=A'_{n-1}(\theta)$$
$$(\partial^2/\partial\rho^2)A(\theta,0)=(1-n)A_n(\theta)+2A_{n-2}(\theta)\ ,\ (\partial^2/\partial\theta^2)R(\theta,0)=0,$$
$$(\partial^2/\partial\rho\partial\theta)R(\theta,0)=-R'_n(\theta),\quad (\partial^2/\partial\rho^2)R(\theta,0)=-2R_{n-1}(\theta)\,.$$

In case a) (transversal saddle-node), the eigenvectors of $DY(\theta,0)$ are $v_0=-A_{n-1}(\theta)\partial/\partial\theta+A'_n(\theta)\partial/\partial\rho$ and $v_1=\partial/\partial\theta$ corresponding respectively to the eigenvalues 0 and $A'_n(\theta)$. The second derivative of Y at $(\theta,0)$ satisfies $D^2Y(\theta,0).v_0{}^2=2(R'_nA_{n-1}-R_{n-1}A'_n)(\theta)v_0+a_1v_1$, for some a_1. Therefore $(\theta,0)$ is a saddle-node if and only if $(R'_nA_{n-1}-R_{n-1}A'_n)(\theta)\neq 0$.

In case b) (tangencial saddle-node),, the eigenvectors of $DY(\theta,0)$ are $v_0=\partial/\partial\theta$ and $v_1=-A_{n-1}(\theta)\partial/\partial\theta+R_n(\theta)\partial/\partial\rho$ corresponding respectively to the eigenvalues 0 and $-R_n(\theta)$. We have $D^2Y(\theta,0)=A''_n(\theta)v_0$. Therefore $(\theta,0)$ is a saddle-node if and only if $A''_n(\theta)\neq 0$. ∎

(3.2) $\sum^1{}_n(1)$ can be naturally partitioned into nine essentially different subsets defined as follows:

a) The saddle-node is finite. There are two subcases ,see Fig.3 :

a_1) The weak separatrix is disjoint from parabolic sector.

a_2) The weak separatrix is contained in the interior of the parabolic sector.

b) There is a double (or composed) focus.See Fig.4.

c) There is a pair of transversal saddle-nodes at infinity.

d) There is a pair of tangencial saddle-nodes at infinity.

Four subcases of d) are possible:

d_1) There is no singular cycle containing the saddle-nodes. See Fig.9.

d_2) There is only one singular cycle containing one of the sadde-nodes. See Fig.11, ignoring dashed orbits.

d_3) There are two singular cycles, each containing one of the saddle-nodes. See Fig.11, considering also dashed orbits.

d_4) There is only one singular cycle containing both saddle-nodes. Two subcases of d_4) are possible:

d_{4i}) The cycle is contained in S^1. See Fig.9.

d_{4f}) The cycle contains two saddle separatrices in S^2-S^1. See Fig.10.

Denote by $\sum^1{}_n(1,j)$, $j=a,a_1,\ldots,d_{4f}$ the different parts of $\sum^1{}_n(1)$. The cases $\sum^1{}_n(1,a)$ and $\sum^1{}_n(1,b)$ have been studied in [A-L, S_1] in the smooth case. The proofs given there apply to the present case, with no essential modification. This is also valid for $\sum^1{}_n(1,d)$ except for part c) of Theorem 1, cases d_2, d_3 and d_{4f}. These cases need an explicit justification for the fact that the periodic orbits that bifurcate from the graphs, when the saddle-node is eliminated (for F>0), are hyperbolic. We deal with these cases in (3.3) and (3.4) . In (3.5) we study $\sum^1{}_n(1,c)$, where a bifurcation pattern different from those analyzed in [A-L,S_1] appears . See Fig8.

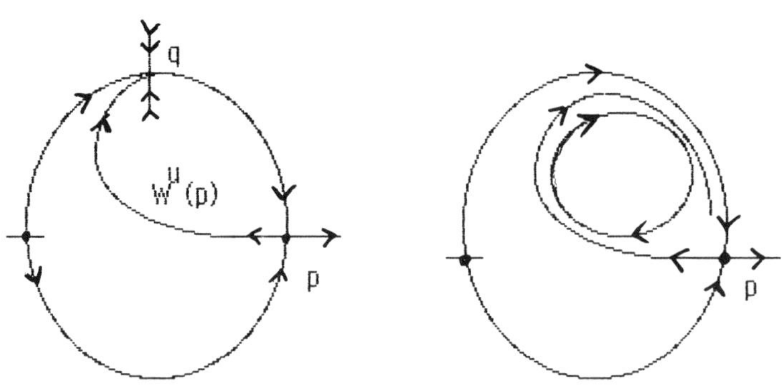

Figure 17

(3.3) Let X_0 in X_n have a saddle point p and a tangencial saddle-node q at infinity whose separatrices form a graph such as in Fig.17,left,that is:

(i) The weak separatrix of the tangencial saddle-node q, denoted here $W^u(q)$, coincides with a stable separatrix of p.

(ii) One unstable saddle separatrix $W^u(p)$ tends to q through the interior of the nodal sector $W^s(q)$ of the saddle-node q, which is an attractor (that is $R_n(q)>0$, (2.4)).

<u>(3.4) PROPOSITION. Take X_0 as in (3.3). There are neighborhoods $\mathcal{U}(X_0)$ of X_0 and $U(q)$ in S^1 such that if X in $\mathcal{U}(X_0)$ has no singular point on U(q), then X has a unique attracting hyperbolic periodic orbit whose central projection is near the graph of $\mathcal{P}(X_0)$.</u>

Proof. By means of an affine transformation of the plane we can suppose that p, q, $W^u(q)$ are in $U_1=\{(y_1,y_2,y_3)\in S^2;\ y_1>0\}$ and $q=(1,0,0)$. Taking the chart $\varphi_1:U_1\to R^2$ given by $z_1=y_2/y_1$, $z_2=y_3/y_1$, we translate the situation to the plane (z_1,z_2) as shown in Fig 18. Let $Z_0=(\varphi_1)_*\mathcal{P}(X_0)$. Let X in $\mathfrak{X}_n$ be close to X_0 with $\mathcal{P}(X)$ having no singularity near q. Put $Z=(\varphi_1)_*\mathcal{P}(X)$. The Poincaré return map of Z maps the segment I (see Fig 19) into itself and has (at least) a fixed point. To conclude, let us prove that it is a hyperbolic attractor. This will imply the uniqueness.

Let φ be a C^1 linearization of Z around p=p(Z), as in Fig 20, [Ha] ; p=p(Z) is the saddle point of Z and we can take φ depending continuously on Z. Let $\varepsilon>0$ be small enough and $J_3=\{(x,\varepsilon);0\le x\le\varepsilon\}$ and $J_4=\{(\varepsilon,y);0\le y\le\varepsilon\}$. Write $I_3=\varphi^{-1}(J_3)$ and $I_4=\varphi^{-1}(J_4)$.

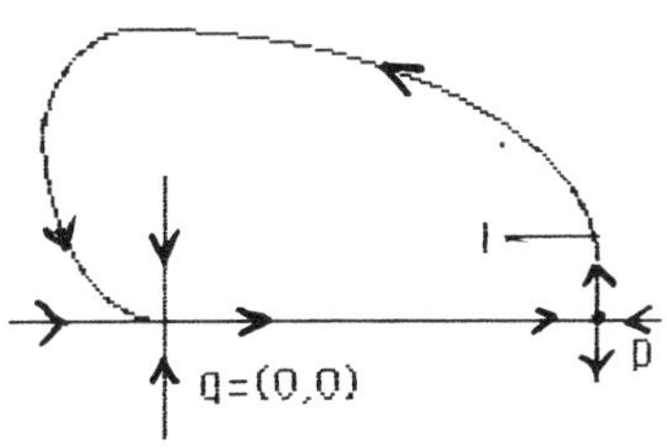

Figure 18. The vector field Z_0.

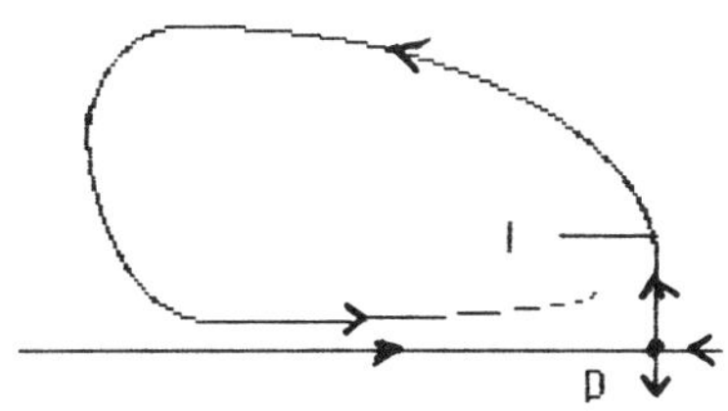

Figure 19. The vector field Z

If $\tau_3:J_3\to J_4$ is the Poincaré map of φ_*Z defined on J_3, then $\tau_3(x)=\varepsilon^{(1+\mu/\lambda)}x^{-\mu/\lambda}$ for all x in J_3, where $\mu<0<\lambda$ are the eigenvalues of the saddle at p(Z). Let ψ be the smooth change of coordinates around (0,0) constructed in the following way. Let $\mathcal{V}$ be a sufficiently small neighborhood of X_0 in $\mathfrak{X}_n$ and let V be a small neighborhood of (0,0) in the plane (z_1,z_2). Consider the vector field $Z:V\times\mathcal{V}\to R^2\times\mathfrak{X}_n$ defined by $Z(z_1,z_2,X)=((\varphi_1)_*\mathcal{P}(X)(z_1,z_2),0)$. The chart ψ takes this field into its Takens normal form [Ta, p.144]: $A(z_1,X)\partial/\partial z_1-B(z_1,X)z_2\partial/\partial z_2$ where, $A(0,X_0)=0$, $A(z_1,X)>0$ e $B(z_1,X)>0$ for $|z_1|\le\alpha$ sufficiently small and for all X close enough to X_0. By the Weierstrass-Malgrange Preparation Theorem [W] we can write: $A/B=Q[z_1{}^2+z_1R_1(X)+R_0(X)]$, where $Q=Q(z_1,X)>0$, $R_0(X_0)=0$ and $R_1(X_0)=0$. Notice that $F(X)=R_0(X)-(1/4)R_1{}^2(X)$ defines the analytic function required by Theorem 1. Let $0<a\le Q(z_1,X)\le b$ for all $|z_1|\le\alpha$ and X in $\mathcal{V}$. It is easy to show that $DF(X_0)\neq0$. Therefore ψ is a smooth change of variables which verifies

$$(\psi_*Z)(z_1,z_2)=B(z_1,Z)[Q(z_1,Z)(z_1{}^2+z_1R_1(Z)+R_0(Z))\partial/\partial z_1-z_2\partial/\partial z_2].$$

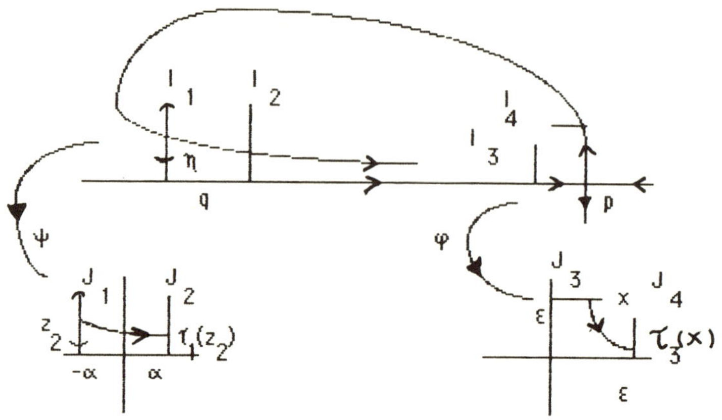

Figure 20.

Let $J_1=\{(-\alpha,z_2); 0<\eta\le z_2<\alpha\}$ and $J_2=\{(\alpha,z_2); 0\le z_2<\alpha\}$. Here ,$\eta>0$ is taken with the following property: for X close enough to X_0 with $F(X)>0$, the image of Poincaré map π_4 from I_4 is contained in $I_1=\psi^{-1}(J_1)$.

Write $I_2=\psi^{-1}(J_2)$. The Poincaré map τ_1: $J_1\to J_2$ for $\psi * Z$ is given by $\tau_1(z_2)=z_2\exp M(Z)$, where

$$M(Z)=-\int_{-\alpha}^{\alpha}[Q(z_1,Z)(z_1^2+z_1R_1(Z)+R_0(Z)]^{-1}dz_1$$

Notice that since $z_1^2+z_1R_1(Z)+R_0(Z)=F(Z)+(z_1+1/2R_1(Z))^2$, integration leads to the inequalities

$$\eta\exp[-ma^{-1}F^{-1/2}] \le \tau_1(z_2) \le z_2\exp[-mb^{-1}F^{-1/2}],$$

$$\exp[-ma^{-1}F^{-1/2}] \le \tau'_1=\exp M(Z) \le \exp[-mb^{-1}F^{-1/2}],$$

where $m = 2\mathrm{arctg}[\alpha F^{-1/2}] < 2\pi$.

Let π_2:$I_2\to I_3$ be the Poincaré map. The return map π of X of J_1 into itself is given by $\pi(z_2)=(\pi_4\varphi^{-1})\tau_3(\varphi\pi_2\psi^{-1})\tau_1(z_2)$. Put $h=\tau_1(z_2)$, $L=(\varphi\pi_2\psi^{-1})|J_2$ and $x=L(h)$. Then, $\pi'(z_2) = D(\pi_4\varphi^{-1})(\tau_3(x)).D\tau_3(x).DL(h).D\tau_1(z_2)$.

We can write $L(h)= d(X)h(1+r(X,h))$, where r tends to 0 with h and d is bounded , uniformly on a neighborhood of X_0. Then for all z_2 in J_1,

$$|\pi'(z_2) | \le c|\tau'_3(x).\tau'_1(z_2)| \le c|x^{-\mu/\lambda-1}\tau'_1(z_2)| \le c_2|[\tau_1(z_2)]^{-\mu/\lambda-1}\tau'_1(z_2)|,$$

where c_2 is a constant uniform for all X near X_0 , with $F(X)>0$.

Write $\nu=-\mu/\lambda-1$. The inequalities above for τ_1 and τ'_1 will be used now.

If $\nu\ge 0$, $|\pi'(z_2)|\le c_2|z_2|^{\nu}\exp(-\mu/\lambda\, M(X))$, for all z_2 in J_1 , which tends to 0 as $F(X)$ goes to 0 .

If $\nu<0$, $|\pi'(z_2)|\le c_2\eta^{\nu}\exp(-\mu/\lambda M(X))$, which tends to 0 ,uniformly in J_1, as $F(X)$ goes to 0. ■

Proposition (3.4) gives directly the proof of Theorem 1,part c,for cases d_2 and d_3. Also , it provides the essential step , which must be applied twice for the conclusion in case d_{4f}. See Fig 10.

(3.5) Now we will give the proof of Theorem 1 in case $\Sigma^1{}_n(1,c)$. By (2.1), the singularities on and near infinity, for Y near X, are given by

$$\Sigma\rho^i A_{n-i}(Y,\theta)=0 \quad \text{and} \quad -\rho(\Sigma\rho^i R_{n-i}(Y,\theta))=0,\ i=0,\dots,n.$$

The conditions for transversal saddle-node of $\mathcal{P}(X)$ at $(\theta_0,0)$ in (2.5,a) imply that the first of these equations has an analytic solution of the form $\theta=\theta(Y,\rho)$. The finite singularities of $\mathcal{P}(Y)$ near $(\theta_0,0)$ are given by substituting $\theta=\theta(Y,\rho)$ into the second equation divided by ρ. This equation has an implicit analytic solution of the form $\rho=F(Y)$, which provides the function F required in Theorem 1, part b. This conclusion is possible by the Implicit Function Theorem whose application is legitimate due to the last two conditions in (2.5,a). Part a follows from [A-L, S_1] and part c is similar to the analysis of the saddle-node bifurcation in [S_1]. ∎

4. PROOF OF THEOREM 2. Recall from (2.15,2) that that the set $\Sigma^1{}_n(2)$ is partitioned into $\Sigma^1{}_n(2,f)$ and $\Sigma^1{}_n(2,i)$, according to the simple non-hyperbolic periodic orbit be double or semi-stable and finite (case f) or triple and infinite (case i). There is nothing new to add here to the study in [A-L,S_1,S_2] for the proof of Theorem 2 in case f. Notice only that there are two subcases of f to consider: f_1 (resp. f_2) if the semi-stable periodic orbit is (resp. is not) simultaneously the α and ω-limit set of saddle separatrices. It is only in case f_1 that U_+ (Theorem 2,c) has non-empty intersection with $\mathfrak{X}^1{}_n(3)$.

Now we deal with $\Sigma^1{}_n(2,i)$.

(4.1)LEMMA. Let X in $\mathfrak{X}_n$, n odd, have a periodic orbit at infinity. Then the periodic orbit is hyperbolic if and only if one of the following equivalent conditions is satisfied:

$$\int_0^{2\pi} R_n(\theta)A_n^{-1}(\theta)\,d\theta \neq 0,$$

$$\int_{-\infty}^{\infty} [P_n(1,s)+sQ_n(1,s)][(1+s^2)(Q_n(1,s)-s\,P_n(1,s))]^{-1}ds \neq 0,$$

$$\int_{-\infty}^{\infty} P_n(1,s)(Q_n(1,s)-s\,P_n(1,s))^{-1}ds \neq 0,$$

Where the integral from $-\infty$ to ∞ denotes the symmetric improper integral which is the limit as t tends to ∞ of the corresponding integral from -t to t. See [L, pag.241].

Proof. With the notation introduced in (2.1) and (2.2), write

$G(\theta,\rho)= -\rho(\Sigma\rho^i R_{n-i}(\theta))/\,\Sigma\rho^i A_{n-i}(\theta)$, $i=0,1,\dots,n$.

The Poincaré map of the periodic orbit at infinity satisfies:

$\pi'(0)=\exp\int_0^{2\pi}\partial G(\theta,0)/\partial\rho = \exp\int_0^{2\pi}-R_n(\theta)A_n^{-1}(\theta)\,d\theta$,

provides the positive orientation of the periodic orbit coincides with the trigonometric sense on S^1,i.e., if $A_n(\theta)>0$. Otherwise, when $A_n(\theta)<0$,

$\pi'(0)=\exp\int_0^{2\pi}R_n(\theta)A_n^{-1}(\theta)\,d\theta$. This proves the first part of the Lemma.

Since n is odd, $R_n(\theta)A_n^{-1}(\theta)$ has period π. Therefore

$$\int_0^{2\pi} R_n(\theta)A_n^{-1}(\theta)\,d\theta = 2\int_{-\pi/2}^{\pi/2} R_n(\theta)A_n^{-1}(\theta)\,d\theta =$$

$$2\int_{-\infty}^{\infty} [P_n(1,s)+sQ_n(1,s)][(1+s^2)(Q_n(1,s)-s\,P_n(1,s))]^{-1}ds\text{ , where } s=tg\theta.$$

Furthermore, we have

$$\int_{-\infty}^{\infty} [P_n(1,s)+sQ_n(1,s)][(1+s^2)(Q_n(1,s)-s\,P_n(1,s))]^{-1}ds\ -$$

$$\int_{-\infty}^{\infty} P_n(1,s)(Q_n(1,s)-s\,P_n(1,s))^{-1}ds = \int_{-\infty}^{\infty}(s/1+s^2)ds = 0. \qquad ■$$

(4.2)<u>LEMMA</u>. <u>The set $\mathfrak{X}^1_n(2,i)$ of vector fields in $\mathfrak{X}_n$ which have a non-hyperbolic periodic at infinity is an analytic embedded submanifold of codimension one .</u>

Proof. The function $F(X)=\int_0^{2\pi}R_n(X,\theta)A_n^{-1}(X,\theta)\,d\theta$ is defined on the open set U of $\mathfrak{X}_n$ on which $A_n(X,\theta)=-P_n(X,\cos\theta,\sin\theta)\sin\theta+Q_n(X,\cos\theta,\sin\theta)\cos\theta$ does not have real roots. Here P_n ,Q_n are the n degree homogeneous components of X. Certainly F=0 defines $\mathfrak{X}^1_n(2,i)$. Direct calculation shows that X is a regular value of 0. In fact,

$$DF_X(Y)=\int_0^{2\pi}\ [Q_n(X,\theta)P_n(Y,\theta)-P_n(X,\theta)Q_n(Y,\theta)][A_n(X,\theta)]^{-2}d\theta.$$

This shows that $DF_X\neq 0$ everywhere. ■

Last lemma proves part b or Theorem 2 for $\Sigma^1_n(2,i)$ since this set is open in $\mathfrak{X}^1_n(2,i)$. Notice also that parts a and c follow from the results in [S_1], applied to the projective plane P^2. In fact every vector field in $\Sigma^1_n(2,i)$ projects into a vector field in P^2 with a one sided non-hyperbolic triple periodic orbit, which is the case studied in [S_1] .

(4.3)<u>REMARK</u>. By computing the integral in (4.1) using the last expression, it can be proved that there is a non-trivial real polynomial function which vanishes on $\mathfrak{X}^1_n(2,i)$. This algebraic property does not hold for $\mathfrak{X}^1_n(2,f)$ as proved in [I].

(4.4)<u>PROPOSITION</u>. <u>$\mathfrak{X}^2_n(2,i)= \mathfrak{X}^1_n(2,i)-\Sigma^1_n(2,i)$ is ,for $n\geq 3$ odd, is an codimension one analytic submanifold of $\mathfrak{X}^1_n(2,i)$.</u>

Proof. Define H(X) on $\mathfrak{X}^1_n(2,i)$ to be $\pi_X^{(3)}(p)$, where π_X is th return map of the periodic orbit at infinity of $\mathcal{P}(X)$ and p=(1,0,0). Clearly H is analytic and $H^{-1}(0)=\mathfrak{X}^2_n(2,i)$. Let us prove that dH_X ,for all X in $\mathfrak{X}^2_n(2,i)$, does not vanish on vectors tangent to $\mathfrak{X}^1_n(2,i)$.

In fact, let $\rho=\rho(\theta,u)$ be the solution of the differential equation $d\rho/d\theta=G(\theta,\rho)$ as in the proof (4.1), with initial condition $\rho(0,u)=u$. The derivatives $\rho_1=\partial\rho/\partial u$, $\rho_2=\partial^2\rho/\partial u^2$ and $\rho_3=\partial^3\rho/\partial u^2$, at u=0 , as functions or θ, are solutions of the following differential equations:

$\rho'_1 = a(\theta)\rho_1,\ \rho_1(0)=1,\quad \rho'_2 = a(\theta)\rho_2+b(\theta),\ \rho_2(0)=0,$

$\rho'_3 = a(\theta)\rho_3+c(\theta),\ \rho_3(0)=0,$ where $a(\theta)=-R_n(\theta)A_n^{-1}(\theta),$

$b(\theta)=(\partial^2G(\theta,0)/\partial\rho^2)(\rho_1(\theta))^2$ and

$c(\theta)=(\partial^3G(\theta,0)/\partial\rho^3)(\rho_1(\theta))^3+(\partial^2G(\theta,0)/\partial\rho^2)\rho_1(\theta)\rho_2(\theta)$.

Direct computation gives:

$\partial^2 G(\theta,0)/\partial\rho^2 = -2[-R_n(\theta)A_{n-1}(\theta)+R_{n-1}(\theta)A_n(\theta)]\,(A_n(\theta))^{-2}$,

$\partial^3 G(\theta,0)/\partial\rho^3 = -6[(-R_nA_nA_{n-2}+R_n(A_{n-1})^2-R_{n-1}A_{n-1}A_n+R_{n-2}(A_n)^2)(\theta)](A_n(\theta))^{-2}$

We have $\pi_X^{(3)}(\rho) = \rho_3(2\pi)$. Therefore, H(X) can be expressed as an integral of a function involving only R_i, A_i, $i = n, n-1, n-2$.

Since $a(\theta)$ and $\partial^2 G(\theta,0)/\partial\rho^2$ depend only on the homogeneous parts P_i, Q_i, $i=n,,n-1$, and $\partial^3 G(\theta,0)/\partial\rho^3$ depends linearly on P_{n-2} and Q_{n-2}, an easy calculation with a vector field Y with vanishing homogeneous parts of degrees n and n-1 and therefore in the tangent space to $X^1_n(2,i)$, gives:

$$dH_X(Y) = 6\int_0^{2\pi}\exp(-2\int_0^{\theta}R_n(\tau)A_n^{-1}(\tau)d\tau)[R_n(\theta)A_{n-2}(Y,\theta) - A_n(\theta)R_{n-2}(Y,\theta)]\,(A_n(\theta))^{-2}d\theta.$$

Taking, for example, $Y=(x^2+y^2)^{k-1}(x\partial/\partial x+y\partial/\partial y)$, where $n=2k+1$, gives a non vanishing expression for $dH_X(Y)$. ∎

5. PROOF OF THEOREM 3. Recall that $\Sigma^1_n(3)$ is the set of vector fields in $\Sigma_n(1)\cap\Sigma_n(2)$ which have only one saddle connection in S^2-S^1 which is simple (see (2.8)). The proof of the fact that that $\Sigma^1_n(3)$ is an injectively immersed analytic submanifold of codimension one in X_n appears in [S_2; S_3]. Therefore, part b of Theorem 3 holds.

Let $\Sigma^1_n(3,j)$, j=a,b,...,e, be the the set of vector fields in $\Sigma^1_n(3)$ whose connection is of type j, defined in (2.8). There is nothing new to add here to to the analysis carried out in [A-L; S_1] for the proof of parts a and c of Theorem 3, in cases j= a, b,c, and d.

The periodic orbit that bifurcates from the separatrix cycle in case j=d, as illustrated in in Fig 14, is hyperbolic. In fact, the return map π associated to the singular cycle is defined on a segment J in S^2 transversal to γ at a point 0. It follows from [D,R_1,Pe] that, with the notation in (2.8,d),

$$s_1 = A'_n(\theta_1)R_n(\theta_2)/A'_n(\theta_2)R_n(\theta_1) = \lim(\log\pi(x)/\log x),\ x\to 0.$$

The assertion in (2.8,d) concerning the sign of σ_1 and the attracting properties of the singular cycle follows from this limit. The hyperbolicity of the bifurcating periodic orbit, which is attracting (resp. repelling) if $s_1>1$ (resp. <1) follows from the analysis in [R_2, Pe].

The case j=e is essentially different. In fact, in this case n must even and, due to the symmetry of the saddles at the corners of the singular cycles $\Gamma(\pm)$ of (2.8,e), always $s_1=1$. The proof of the assertions concerning the meaning of the the integral expressions $\sigma_2(\pm)$ as the logarithms of the derivatives of the Poincaré maps $\pi(\pm)$ associated to the singular cycles will be given now. The hyperbolicity, unicity, and attracting-repelling character of the bifurcating periodic orbits follow also from [R_2, Pe], with hypotheses on the signs of $\pi'(\pm)-1$ ($\sigma_2(\pm)$, by 5.1). Bifurcations illustrated in Figs.15 and 16 show the two possibilities: same and opposite signs, respectively.

(5.1) PROPOSITION. With the notation introduced in (2.8,e) and Fig.2e,

(a) $\pi'(\pm) = \exp[\int_\gamma \operatorname{div} X \pm (A'_n(\theta_1)/R_n(\theta_1)) \int_\alpha (-R_n(\theta)/A_n(\theta))d\theta]$,

where $\pi(\pm)$ denotes the return map of the singular cycle $\Gamma(\pm)$, relative to a section at a point of γ. The integrals in this formula are the symmetric improper integrals:

$$\int_\alpha = \lim_{\epsilon\to 0} \int_{-\pi/2+\epsilon}^{\pi/2-\epsilon}; \quad \int_\gamma = \lim_{\epsilon\to 0} \int_{A+\epsilon}^{B-\epsilon} \text{ or } \lim_{t\to\infty} \int_{-t}^{t},$$ according to (A,B), finite, or R be the maximal interval of definition for γ.

Proof. Call Y the expression of $\mathcal{P}(X)$ in polar coordinates introduced in (2.1). By means of a rotation of the plane, we can assume that the symmetric infinite saddles $p_1=\theta_1$ and $p_2 = \theta_1+\pi$, corners of the singular cycle, are given by $p_1=(-\pi/2,0)$ and $p_2=(\pi/2,0)$ in the (θ,ρ) plane. See Fig 21. Let L_1 be a C^1 linearization [Ha], valid on a neighborhood V_1 of p_1, such that $L_1(p_1)=(0,0)$, $L_1(\{\rho=0\})$ is contained in $\{y=0\}$ and $L_1(W^s(p_1))$ is contained in $\{x=0\}$. Let $L_2 = L_1 \circ A^{-1}$, where $A:(\theta,\rho) \to (\theta+\pi,-\rho)$, from $V_2 = A(V_1)$ to the plane. We have that $(L_1)_*Y(L_2(\theta,\rho)) = -(L_2)_*Y(L_2(\theta,\rho))$. Then L_2 is a C^1 linearization of Y around $(\pi/2,0)$.

For δ sufficiently small, let $J_1=(0,\delta)\times\{\delta\}$, $J_2=\{\delta\}\times(0,\delta)$, $J_5=\{-\delta\}\times(-\delta,0)$, $J_6=(-\delta,0)\times\{-\delta\}$. Write $I_i=(L_1)^{-1}(J_i)$, $i=1,2$ and $I_i=(L_2)^{-1}(J_i)$, $i=5,6$. Put $(\alpha',0)=(L_1)^{-1}(\delta,0)$ and $(\beta',0)=(L_2)^{-1}(-\delta,0)$. Let $I_3=\{\alpha'\}\times(0,\delta')$ and $I_4=\{\beta'\}\times(0,\delta'')$, for δ' and δ'' sufficiently small. Put $J_3=L_1(I_3)$ and $J_4=L_2(I_4)$. Let $\mu=-R_n(-\pi/2)<0$ and $\lambda=A'_n(-\pi/2)>0$ be the eigenvalues of $DY(p_1)$ (see (2.4)), and write $\nu=1+\mu/\lambda$.

With the notation introduced in Fig.21 for the Poincaré maps, we have, for $x=L_1(s)$

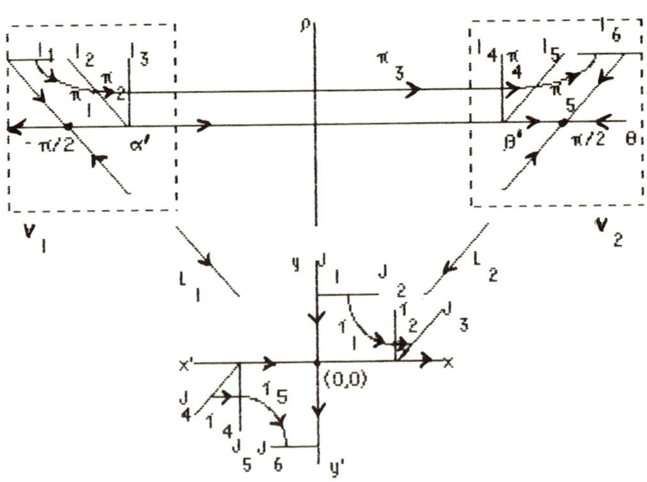

Figure 21

and $y=\delta^\nu x^{-\mu/\lambda}$, that $\pi_5\pi_4\pi_3\pi_2\pi_1(s) = \pi_5\pi_4\pi_3\pi_2(L_1)^{-1}\tau_1(x) = \pi_5\pi_4\pi_3\pi_2(L_1)^{-1}(y)$. Write $\rho_0=\pi_2((L_1^{-1}(y)) = a_1y(1+A_1(y))$, where $a_1=D(\pi_2L_1^{-1})(0)$ and $\lim A_1(y)=0$, as y tends to 0. Write $\rho_1 = \pi_3(\rho_0) = a_2\, \rho_0(1+A_2(\rho_0))$, with $\lim A_2(\rho_0)=0$, as ρ_0 tends to 0, and

(b) $\quad a_2 = \exp[-\int_{\alpha'}^{\beta'} R_n(\theta)/A_n(\theta)\, d\theta\,]$.

We have, $L_2\pi_4(\rho_1) = a_3\, \rho_1(1+A_3(\rho_1))$, where $a_3 = D(L_2\pi)(0)$ and $\lim A_3(\rho_1)=0$, as ρ_1 tends to 0. Put $y' = L_2\pi_4(\rho_1)$, then $x' = \tau_5(y') = \delta^{1+\lambda/\mu}(y')^{-\lambda/\mu}$ can be written as $x' = (a_1a_2a_3)^{-\lambda/\mu}x(1+B(x))$, where $\lim B(x)=0$, as x tends to 0.

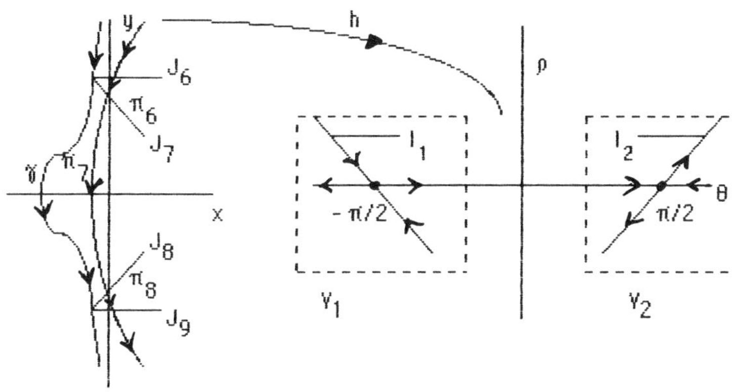

Figure 22

Consider now Fig. 22. We can suppose that the finite separatrix $\gamma(t)$ is assymptotic to the y-axis. Denote by h the chart $(\theta,\rho) \to (\rho^{-1}\cos\theta, \rho^{-1}\sin\theta)$. Put $J_6 = h^{-1}(l_6)$ and $J_1 = h^{-1}(l_1)$. Let J_7 and J_8 be line segments normal to γ as illustrated in Fig. 22.

For all $x' = \tau_5(y') \in J_6$ let $x'' = \pi_6 h^{-1} L_2(x') = a_4 x'(1+A_4(x'))$, where $a4 = D(\pi_6 h^{-1} L_2^{-1})(0)$ and $\lim A_4(x')=0$, as x' tends to 0. Also, $\hat{x} = \pi_7(x'') = a_5 x''(1+A_5(x''))$, with $\lim A_5(x'')=0$, as x'' tends to 0, and

(c) $\quad a_5 = |X(\gamma(a))|(|X(\gamma(b))|^{-1})\exp[-\int_a^b \operatorname{div}X(\gamma(t)dt]$,

where $\gamma(a) \in J_7$ and $\gamma(b) \in J_8$.

Let $\pi = \pi(+)$ be the return map on J_1. Then $\pi(x) = L_1 h\pi_8(\hat{x}) = a_8\hat{x}(1+A_5(\hat{x}))$, with $\lim A_5(\hat{x})=0$, as $\hat{x}$ tends to 0. We have

$$\pi(x) = a_2{}^{-\lambda/\mu}\, a_5\,(a_1a_3)^{-\lambda/\mu} a_4 a_6\, x(1+C(x)), \text{ where } \lim C(x)=0, \text{ as x tends to 0.}$$

Therefore, for $\delta > 0$, small, we get that

(d) $\quad \pi'(0) = a_2{}^{-\lambda/\mu}\, a_5\,(a_1a_3)^{-\lambda/\mu} a_4 a_6.$

We claim that $\lim a_1a_3 = 1$ and $\lim a_4a_6 = 1$, as δ tends to 0. Let us prove the first assertion. Parametrize l_2 as the graph of a fuction $\theta = \alpha' + \hat{\theta}(\rho)$ of class C^1. We have that $D\pi_2(0) = 1$. Similarly, $D\pi_4(0) = 1$. Let P_2 be the second projection of R^2, then $\lim a_1 = P_2 D(L_1)^{-1}{}_{(0,0)}(0,1)$ and $\lim (a_3)^{-1} = P_2(DA)(DL_1{}^{-1})_{(0,0)}(0,-1)$, as δ tends to

0. Therefore, lim a_1 = lim $(a_3)^{-1}$ and lim $a_1a_3=1$, as δ tends to 0. The proof of the second assertion is similar. Also it holds that lim $(|X(\gamma(a))|(|X(\gamma(b))|^{-1})=1$.

Therefore, taking the limit in (d), as δ tends to 0, we obtain the expression in (a) for the case $\Gamma(+)$. In case $\Gamma(-)$, the integral in (b) converges to $\int_{[-\alpha]} R_n/A_n = \int_{\alpha-\pi} R_n/A_n = -\int_{\alpha} R_n/A_n$, since the integrand changes sign when translated by π. ∎

6. FINAL REMARKS. (6.1) It is possible to show that the submanifolds $\Sigma^1{}_n(i)$, $i=1,2,3$ are non empty for n appropriate, but large. However it is not obvious how to construct explicit examples when n=2,3. Concerning this point, for n=2, see [B]. Interesting particular situations of bifurcations, for n=2, have been studied from different points of view by Cherkas [Ch], Keith and Rand [K-R] and Ye Yan Quian [Y], among others.

(6.2) A very particular version of Problems 1 and 2 can be solved presently. Call $\mathbb{X}^1{}_n(1)$ the set of X in $\mathfrak{X}^1{}_n$ for which $\mathbb{P}(X)$ has a pair of non-hyperbolic singular points with vanishing Jacobian determinant and, if on S^1, with vanishing normal to S^1 eigenvalue. It can be proved using the methods in [S_2] that Problems 1 and 2, reformulated by changing $\mathfrak{X}^1{}_n$ by $\mathbb{X}^1{}_n(1)$, $\Sigma^1{}_n$ by $\mathbb{S}^1{}_n(1)=\Sigma^1{}_n(1)\cap\mathbb{X}^1{}_n(1)$ and $\mathfrak{X}^2{}_n$ by $\mathbb{X}^2{}_n=\mathbb{X}^1{}_n(1)-\mathbb{S}^1{}_n(1)$, have affirmative answers. The essential fact that enables to use the method of [S_2] here is that $\mathbb{X}^1{}_n(1)$ is invariant under rotations of vector fields.

BIBLIOGRAPHY

[A-L] Andronov, A., Leontovich, E. et al. Theory of Bifurcations of Dynamical Systems on the Plane. IPST, Jerusalem, 1971.

[B] Bamón, R., Solution of Dulac's Problem for Quadratic Vector Fields. Informes de Matematica 038/85, IMPA. To appear in Pub. Math. IHES.

[Ch] Cherkas, L., Bifurcation of Limit Cycles of a Quadratic Vector Field with Variation of Parameter Rotating the Field. Diff. Eq.,17;11 (1981), 1265-1276.

[D] Dulac,H., Sur les Cycles Limites. Bull. Soc.Mat. de France, 51 (1923).

[Ha] Hartman, Ph., On Local Homeomorphisms of Euclidean Spaces. Bol. Soc. Mat. Mexicana, 5 (1960) 220-241.

[I] Iliashenko, Ju., Non Algebraic Character of the Manifolds of Differential Equations with Rational Right Hand Sides and with Multiple Limit Cycles. Math. USSR Sbornik, 12 (1970) 453-456.

[K-R] Keith, W ; Rand, R., Dynamics of a System Exhibiting the Global Bifurcation of a Limit Cucle at Infinity,Int. Jour. Non-Linear Mechanics, 20 (1985) 325-338.

[L] Lang, S., Analysis I, Addison-Wesley, Reading, Mass, (1971).

[P] Poincaré, H., Memoire sur les courbes définies par une équation differentielle. J. Mathematiques , 7 (1881), 375-422.

[Pe] Pereira, J. M., Bifurcações dos Campos Vetoriais Equivariantes, Tese, IMPA, (1981).

[Pu] Pugh, C., Structural Stability and Hilbert's 16th Problem, Lecture Notes in Mathematics 468, Springer Verlag, (1975).

[R_1] Reyn, J. W., A Stabiliy Criterion for Separatrix Polygons in the Plane, Nieuw Archief woor Wiskunde, (3), XXVII, (1979) 238-254.

[R_2] Reyn, J. W., Generation of Limit Cycles from Separatrix Polygons in the Phase Plane, Springer Lecture Notes in Mathematics, Springer Verlag, 810(1980)177 -201.

[S_1] Sotomayor, J., Generic One Parameter Families of Vector Fields on Two Dimensional Manifolds, Publ. Math. IHES , 43 (1974).

[S_2] Sotomayor, J., On Stable Planar Polynomial Vector Fields, Rev. Mat. Iberoamericana, 2 (1985) 15-23.

[S_3] Sotomayor, J., Curvas Definidas por Equações Diferenciais no Plano, IMPA, Rio de Janeiro, (1981).

[Sa] dos Santos, G., Classification of Generic Quadratic Vector Fields with no Limit Cycles, Lecture Notes in Mathematics, Springer Verlag, 597, (1977) 605-640.

[T] Teixeira,M., Generic Bifurcation on Manifolds with Boundary, Jour. Diff. Eq., 25;1(1977).

[Ta] Takens, F., Partially Hyperbolic Fixed Points, Topology,10 (1971), 133-147.

[Th] Thom, R., Quelques Propriétés Globales des Varietés Differentiables, Comm. Math. Helv. 28 (1954),17-86.

[W] Wall,C., Introduction to the Preparation Theorem, Lecture Notes in Mathematics, Springer Verlag ,192 (1971).

[Y] Ye Yan Quian et al., Theory of Limit Cycles, Translation of Math. Monographs, AMS, 66 (1986).

Instituto de Matemática Pura e Aplicada
Estrada Dona Castorina, 110
CEP 22.460, Rio de Janeiro, RJ
Brazil

Universidade Federal de São Carlos
Caixa Postal 676
CEP 13.560, São Carlos, SP
Brazil

Canadian Mathematical Society
Conference Proceedings
Volume 8 (1987)

EXAMPLES OF BIFURCATION WITHOUT MIXED MODE SOLUTIONS

Lawrence Turyn

ABSTRACT. We consider nonlinear problems, with parameter ϵ, which have $\mathbb{Z}_2 \oplus \mathbb{Z}_2$ symmetry for all ϵ and $O(2)$ symmetry for $\epsilon = 0$. Under certain conditions, solutions of a nonlinear problem will not mix the eigenvectors of the linearization. We present examples including (i) a boundary value problem for a partial differential equation with perturbed domain, and (ii) existence of periodic solutions of ordinary differential equations and integral equations.

1. Introduction In recent years many authors have shown how symmetry can be used to analyze solutions of nonlinear problems. In particular, several monographs [1, 2, 3] have shown how symmetry in differential equations is inherited by the bifurcation equations. In a previous paper [4] we established an abstract result about symmetry breaking from $O(2)$ symmetry to $\mathbb{Z}_2 \oplus \mathbb{Z}_2$ symmetry and discussed an example of a nonlinear Mathieu equation. The purpose of this note is to present some other examples.

The first example, suggested by Prof. Golubitsky, is a Dirichlet problem on a parameter dependent domain: A circle is perturbed into an ellipse. By a change of variables the perturbation appears in the differential equation.

The next two examples involve existence of periodic solutions. The second example perturbs a system of two linear oscillators in resonance, and the third example is an integral equation.

As in [4], we consider a problem

(1.1) $\qquad M(x; \nu, \epsilon) = 0 \in Z, \; x \in X, \; \nu \in \Lambda, \; \epsilon \in \mathbb{R},$

where X, Λ, Z are real Banach spaces. Assume that $x = 0$ is a solution of (1.1) for all ν, ϵ. Assume M is C^∞ smooth and $L_0 = D_x M(0;0,0)$ is a Fredholm operator with two-dimensional kernel $\mathcal{N} = \text{sp}\,\{x_1, x_2\}$ and two-codimensional range $\mathcal{R}$. Define closed proper subspaces $X_i \subset X$, $i = 1, 2$, by

1980 Mathematics Subject Classification 47H15, 35J65, 34C25, 45B05

(1.2) $X_1 = k_2^{\perp}$, $X_2 = k_1^{\perp}$

where bounded linear functionals k_1, k_2 satisfy

(1.3) $k_i(x_j) = \delta_{ij}$, the usual Kronecker symbol.

We think of x as the "solution" and ν, ϵ as "parameters". A <u>mixed mode solution</u> of (1.1) is an $x \neq 0$ for which $M(x;\nu,\epsilon) = 0$ for some particular ν and some particular $\epsilon \neq 0$ and $x \notin X_i$ for $i = 1,2$. The "modes" we have in mind are $x \in X_1$ versus $x \in X_2$. Under appropriate hypotheses there are no mixed mode solutions. Because those hypotheses include enough symmetry in $M(\cdot;\nu,0)$ to allow the existence of a family of non-trivial solutions $x \notin X_i$ for $i = 1,2$ for some ν, we exclude $\epsilon = 0$ from the definition of mixed mode solutions.

For a fixed neighborhood $0 \in W \subset X$ one can define local solution sections $S(\nu,\epsilon) = \{x \in W : M(x;\nu,\epsilon) = 0\}$, $S_i(\nu,\epsilon) = \{x \in W \cap X_i : M(x;\nu,\epsilon) = 0\}$ for $i = 1,2$, and $S_3(\nu) = \{x \in W : M(x;\nu,0) = 0\}$. Define corresponding <u>local bifurcation sets</u>: $\mathcal{B}$ consists of (ν,ϵ) where S has cardinality which is not locally constant, $\mathcal{B}_1$, $\mathcal{B}_2$ consist of (ν,ϵ), with $\epsilon \neq 0$, where S_1, S_2 has cardinality which is not locally constant, and let $\mathcal{B}_3 = \{\nu,0) : S_3(\nu) \neq \{0\}\}$. Under appropriate hypotheses we will establish the disjoint decomposition $\mathcal{B} = \mathcal{B}_1 \cup \mathcal{B}_2 \cup \mathcal{B}_3$.

<u>Symmetry of bifurcation equations</u>. A <u>compact group</u> is a group C which is a compact topological space for which the map $(c_1,c_2) \mapsto c_1c_2^{-1}$ is continuous. Given a compact group C and bounded linear operators $K(c)$ on a Banach space Y for all $c \in C$, we say (C,K) is a <u>representation of C over Y</u> if the map $c \mapsto K(c)$ is a group homeomorphism which is strongly continuous, i.e. $c \mapsto K(c)y$ is continuous for all fixed $y \in Y$. An operator $N : X \times \mathcal{M} \to Z$ on Banach spaces $X,\mathcal{M},Z$ is <u>equivariant</u> under $(C,K,\tilde{K})$ if (C,K), $(C,\tilde{K})$ are group representations of C over X,Z,resp., and $N(\kappa x,\mu)=\tilde{\kappa}\, N(x,\mu)$ whenever $\kappa = K(c)$, $\tilde{\kappa} = \tilde{K}(c)$ for some $c \in C$. In the applications in this paper, $\tilde{\Gamma}$ and Γ differ only in the technical sense of acting on spaces $Z \subset X$. Our principal assumptions are

(H1) $M : X \times (\Lambda \times \mathbb{R}) \to Z$ is equivariant under $(G, \Gamma, \tilde{\Gamma})$, where $G = \mathbb{Z}_2 \oplus \mathbb{Z}_2 = \{e,s_1\} \oplus \{e, s_2\}$ is a subgroup of $G_3 = O(2)$

(H2) $M(x;\nu,\epsilon)|_{\epsilon=0}$ is equivariant under $(G_3,\Gamma_3, \tilde{\Gamma}_3)$, where $G_3 = O(2)$

(H3) There exist projections $P : X \to \mathcal{N}$, $Q : Z \to \mathcal{R}$ such that

$\gamma P = P\gamma$, $\tilde{\gamma}Q = Q\tilde{\gamma}$ whenever $\gamma = \Gamma(s)$, $\tilde{\gamma} = \tilde{\Gamma}(s)$ for some $s \in G_3$

(H4) With $Px = k_1(x)x_1 + k_2(x)x_2$ and $(I - Q)z = \ell_1(z)z_1 + \ell_2(z)z_2$:

$K(s_1)y_j = (-1)^j y_j$, $1 \leq j \leq 2$, $K(s_2) = -K(s_1)$ on $sp\{y_1, y_2\}$

$K(\theta)\, y_1 = \cos 2\pi\theta\, y_1 - \sin 2\pi\theta\, y_2$

$K(\theta)y_2 = \sin 2\pi\theta y_1 + \cos 2\pi\theta\, y_2$, for $0 \leq \theta < 1$,

where $K(s) = \Gamma(s)$, $y_j = x_j$, $j = 1,2$ (resp. $K(s) = \tilde{\Gamma}(s)$, $y_j = z_j$, $j = 1,2$). Note that $\kappa(s_2) = \kappa(s_1)\,\kappa(1/2)$ on sp $\{y_1, y_2\}$.

(H5) $\ell_1(Cx_1) \neq \ell_2(Cx_2)$, where $C = D_\epsilon D_x M(0;0,0)$

(H6) For j=1,2, there is an integer n with $\ell_j\, (D_x^n M(0;0,0)x_j) \neq 0$

Theorem 1: Under hypotheses (H1) - (H5) there are no mixed mode solutions of $0 = M(x;\nu,\epsilon)$, hence follows $S(\nu,\epsilon) = S_1(\nu,\epsilon) \cup S_2(\nu,\epsilon)$, a disjoint union, for $\epsilon \neq 0$.

Theorem 2: Under hypotheses (H1) - (H6) follows the decomposition $\mathcal{B} = \mathcal{B}_1 \cup \mathcal{B}_2 \cup \mathcal{B}_3$, a disjoint union.

The proofs are found in [4].

An anonymous referee has remarked that instead of the definitions of X_1, X_2 given in (1.2) one could define $X_1 = \{x \in X \,|\, \Gamma(s_2)x = x\}$, $X_2 = \{x \in X \,|\, \Gamma(s_1)x = x\}$. This is more elegant from the point of view of isotropy subspaces, as espoused in [1,2,3]. The equivalence of the different definitions, locally near $x = 0$, follows from the equivariant implicit function theorem used in the method of Liapunov-Schmidt; see, for example, [3, Lemma 3.3.5].

A reader interested in this paper or [4] may wish to read Prof. Langford's paper in these proceedings, where bifurcation without mixed mode solutions in a Mathieu equation appears as a special case of a grander scheme.

2. A Partial Differential Equation with Perturbed Domain

Let $E(\epsilon)$ be the elliptical region $x^2 + (1 + \epsilon)\, y^2 < 1$ and $\partial E(\epsilon)$ its boundary. Consider the Dirichlet problem

$$\begin{aligned} v_{xx} + v_{yy} + \lambda v + g(v) &= 0 \text{ in } E(\epsilon) \\ v &= 0 \text{ on } \partial E(\epsilon) \end{aligned} \tag{2.1}$$

where subscripts will denote partial derivatives. After change of variables $\tilde{x} = x$, $\tilde{y} = (1 + \epsilon)^{1/2} y$, $u(\tilde{x},\tilde{y}) = v(\tilde{x}, (1+\epsilon)^{-1/2}\tilde{y})$, (2.1) is equivalent to

$$\begin{aligned} u_{\tilde{x}\tilde{x}} + (1+\epsilon)u_{\tilde{y}\tilde{y}} + \lambda\, u + g(u) &= 0 \quad \text{in } D \\ u &= 0 \quad \text{on } \partial D \end{aligned} \tag{2.2}$$

where D is the unit disk and ∂D its boundary. With polar coordinates $\tilde{x} = r \cos\theta$, $\tilde{y} = r \sin\theta$, $w(r,\theta) = u(r\cos\theta, r\sin\theta)$, (2.2) becomes

(2.3) $Lw + \lambda w + \epsilon Cw + g(w) = 0,\ 0 \le \theta \le 2\pi,\ 0 < r < 1$

$w(1, \theta) = 0,\ 0 \le \theta \le 2\pi$

(2.4) $w(r, \theta)$ is bounded as $r \to 0+$

$w(r, .)$ is 2π-periodic,

where $Lw = w_{rr} + \frac{1}{r} w_r + \frac{1}{r^2} w_{\theta\theta}$ and

$$Cw = \sin^2\theta\, w_{rr} + \frac{1}{r}\cos^2\theta\, w_r + \frac{1}{r}\sin 2\theta\, w_{r\theta} - \frac{1}{r^2}\sin 2\theta\, w_\theta + \frac{1}{r^2}\cos^2\theta\, w_{\theta\theta}.$$

Linearization: Assume $\lambda_0 = \mu^2$, μ positive, and $J_1(\mu) = 0$, where J_1 is the first Bessel function of the first kind. Then $w_1 = \cos\theta\, J_1(\mu r)$, $w_2 = \sin\theta\, J_1(\mu r)$ satisfy $Lw + \lambda_0 w = 0$ and (2.4).

The nonlinear problem (2.3), (2.4) can be posed in the Banach spaces $X = \{w \in W^{2,2}(D)$: w satisfies (2.4)$\}$ and $Z = L^2(D)$, where $W^{2,2}$ denotes a Sobolev space. Then (2.3), (2.4) has for all ϵ the symmetries $\theta \mapsto \pi - \theta$ and $\theta \mapsto -\theta$ required for (H1) of section 1, and (2.3), (2.4) has for $\epsilon = 0$ the symmetries $\theta \mapsto \theta + \alpha$, all real α, required for (H2). With w_1, w_2 replacing x_1, x_2 and $z_1 = w_1$, $z_2 = w_2$, (H3) and (H4) are satisfied: note $w_1(r,\pi-\theta) = -w_1(r,\theta)$, $w_2(r,\pi-\theta) = w_2(r,\theta)$, $w_1(r,-\theta) = w_1(r,\theta)$, and $w_2(r,-\theta) = -w_2(r,\theta)$. The "double eigenvalue splitting" hypothesis (H5) is verified: Given two functions $f(r,\theta)$, $g(r,\theta)$, one denotes $\langle f,g\rangle = \int_0^{2\pi}\int_0^1 f(r,\theta)\, g(r,\theta)\, r\, dr\, d\theta$ and notes that $\ell_j(z) = \langle w_j, z\rangle$ for j=1,2. Since

$$\langle Cw_1,w_1\rangle - \langle Cw_2,w_2\rangle = \frac{\pi}{2}\int_0^1 \left(\frac{1}{r} + \mu r\right) J_1^2(\mu r)\,dr > 0,$$

we see that (H5) is true. So, with the assumptions on λ_0 and g we see that (2.1) exhibits the phenomenon of bifurcation without mixed mode solutions.

3. A System of Ordinary Differential Equations

Consider

$$\ddot{x} = -x - 2x\,(x^2+y^2) + \epsilon(x^2+y^2)x - \nu x.$$
(3.1)
$$\ddot{y} = -y - 2y\,(x^2+y^2) + 2\epsilon(x^2+y^2)y - \nu y$$

For all ϵ, (3.1) has the symmetries $(x,y) \mapsto (-x,y)$, $(x,y) \mapsto (x,-y)$. For $\epsilon = 0$, (3.1) has the symmetry $(x,y) \mapsto (\cos\theta\, x - \sin\theta\, y,\ \sin\theta\, x + \cos\theta\, y)$ for all θ; for $\epsilon = 0$ there is additional symmetry given by $t \mapsto t + \alpha$, $0 \le \alpha \le 2\pi$, which is irrelevant to the discussion below. Denote $\hat{\mathcal{P}}^k = \{x \in C^k(\mathbb{R},\mathbb{R}): x(t+2\pi) = x(t) = -x(-t)$ for all $t\}$, $X = \hat{\mathcal{P}}^2 \times \hat{\mathcal{P}}^2$, and $Z = \hat{\mathcal{P}}^0 \times \hat{\mathcal{P}}^0$. Then the linearization of (3.1) has, for $\epsilon = 0$, solution set $\mathcal{N} \subseteq X$ given by

$$\mathcal{N} = \mathrm{sp}\{(\sin t,\ 0),\ (0,\ \sin t)\}.$$

One can generalize (H5) to $C = D_\epsilon D_x^j\, M(0;0,0)$, e.g. $j=3$. Here, $C(x,y) = ((x^2+y^2)x,\ 2(x^2+y^2)y)$ and

$$\ell_1(C(\sin t,0)) = \int_0^{2\pi} \sin t \sin^3 t\, dt = \frac{3}{8},\quad \ell_2(C(0,\ \sin t)) = 2\int_0^{2\pi} \sin t \sin^3 t\, dt = \frac{3}{4}$$

so (3.1) exhibits the phenomenon of bifurcation without mixed mode solutions.

4. Perturbation of a Fredholm Integral Equation

For a fixed finite T consider

$$(4.1)\quad u(x) = \lambda\int_0^T K(x-t)u(t)dt + \epsilon\int_0^T L(x-t)u(t)dt + (G(u))(x).$$

Denote $\mathcal{P}_T = \{u\in C(\mathbb{R},\mathbb{R}):\ u(z + T) = u(z) \text{ for all } z\}$ and assume

(4.2) $K \in \mathcal{P}_T$ and $K(z) = K(-z)$ for all z

(4.3) $L \in C(\mathbb{R},\ \mathbb{R})$ and $L(-z) = L(z-T)$ for all z

(4.4) $G : \mathcal{P}_T \to \mathcal{P}_T$ is continuously differentiable, $G(-u) = -G(u)$, and $G(u(-\cdot)) = G(u(\cdot))$.

Assume that λ_0 is such that

$$(4.5)\qquad u(x) = \lambda_0\int_0^T K(x-t)u(t)dt$$

has a two-dimensional solution set in $\mathcal{P}_T$. Then that set is invariant under O(2): If u(x) is in $\mathcal{P}_T$ and solves (4.5) then so does $u(x+\theta)$, for all θ, as well as $u(-x)$. The first assertion follows from

$$u(x+\theta) = \lambda_0\int_0^T K(x+\theta-t)u(t)dt = \lambda_0\int_{-\theta}^{T-\theta} K(x-\tau)u(\tau+\theta)d\tau = \lambda_0\int_0^T K(x-\tau)u(\tau+\theta)d\tau$$

because of (4.2) and the fact that the integral of a periodic function over an interval of a period is independent of the interval. The second assertion follows from

$$u(-x) = \lambda_0\int_0^T K(-x-t)u(t)dt = \lambda_0\int_0^T K(x+t)u(t)dt = \lambda_0\int_0^T K(-T+x+t)u(t)dt$$

$$= \lambda_0\int_T^0 K(x-\tau)u(-\tau)(-d\tau) = \lambda_0\int_0^T K(x-\tau)u(-\tau)d\tau,$$

again using (4.2).

In addition to assuming that (4.5) has two-dimensional solution set $\mathcal{N} \subseteq \mathcal{P}_T$, assume

(4.6) If $0 \neq u$ solves (4.5) then there is a θ such that $v(\cdot) = u(\cdot + \theta)$ is linearly independent of u.

By (4.2), K is a Hilbert-Schmidt kernel, hence by [3, Theorem 2.6.19] there are $u_1,\ u_2 \in \mathcal{P}_T$ and positive integer k such that for $0 \le \theta \le T$

$$u_1(\cdot + \theta) = \cos(2\pi k\theta/T)\, u_1 - \sin(2\pi k\theta/T)\, u_2$$
(4.7)
$$u_2(\cdot + \theta) = \sin(2\pi k\theta/T)\, u_1 + \cos(2\pi k\theta/T)\, u_2.$$

We note in passing that [1, Lemma VII, 2.5] generalizes to $k \geq 2$ in (4.7). That lemma is crucial to the proofs of theorems 1, 2 found in section 1.

By (4.2), (4.3), (4.4), and a calculation involving L which is similar to that which showed u(-x) solves (4.5) if u(x) does, one can see that the solutions of (4.1) are invariant under the symmetries of $u \mapsto -u$ and $x \mapsto -x$, for all ϵ, required for (H1) of section 1. Further, (4.1) has the symmetry $x \mapsto x+\theta$, all real θ, for $\epsilon = 0$, required for (H2). With u_1, u_2 replacing x_1, x_2 and $z_1 = u_1$, $z_2 = u_2$, (H3) and (H4) are satisfied: without loss of generality, u_1, u_2 are such that $u_1(-x) = -u_1(x)$, $u_2(-x) = u_2(x)$; if not, by the invariance of $\mathcal{N}$ under $u \mapsto -u$, $x \mapsto -x$, one could choose linear combinations of u_1, u_2 and then relabel them as u_1, u_2 to have the desired properties. The "double eigenvalue splitting" hypothesis reduces to

$$\int_0^T u_1(x) \int_0^T L(x-t)\, u_1(t)dt\, dx \neq \int_0^T u_2(x) \int_0^T L(x-t)u_2(t)dt\, dx.$$

BIBLIOGRAPHY

1. M. Golubitsky and D. G. Schaeffer, Singlarities and Groups in Bifurcation Theory, Volume I, Springer, New York, 1985.
2. D. H. Sattinger, Group Theoretic Methods in Bifurcation Theory, Lecture Notes in Mathematics 762, Springer-Verlag, Heidelberg, 1979.
3. A. Vanderbauwhede, Local Bifurcation and Symmetry, Pitman, Boston, 1982.
4. L. Turyn, Bifurcation without mixed mode solutions, in Multiparameter Bifurcation Theory. Contemporary Mathematics 56 (M. Golubitsky and J. Guckenheimer, eds.) American Mathematical Society, Providence, 1986.

Department of Mathematics and Statistics
Wright State University
Dayton, Ohio 45435

Canadian Mathematical Society
Conference Proceedings
Volume 8 (1987)

SECONDARY BIFURCATIONS OF PERIODIC SOLUTIONS IN AUTONOMOUS SYSTEMS

A. Vanderbauwhede

ABSTRACT. We describe a new approach to the problem of bifurcation of periodic solutions near a nonconstant periodic solution of an autonomous system. Instead of the usual Poincaré map we use a Liapunov-Schmidt reduction near an orbit of solutions in a space of periodic mappings. We are particularly interested in symmetric systems. The advantage of the method is that the resulting bifurcation problem has the full symmetry of the critical periodic solution at the bifurcation point. For such symmetric systems we recover some recent results of B. Fiedler [2].

1. INTRODUCTION. Consider an autonomous system

$$(1.1)_\lambda \qquad \dot{x} = f(x,\lambda) ,$$

with $x \in \mathbb{R}^n$, $\lambda \in \mathbb{R}^k$, and $f : \mathbb{R}^n \times \mathbb{R}^k \to \mathbb{R}^n$ sufficiently smooth. Suppose that for some $\lambda_0 \in \mathbb{R}^k$ the equation $(1.1)_{\lambda_0}$ has a nontrivial periodic solution $\tilde{x}_0(t)$, with minimal period $T_0>0$. We want to discuss the following problem : what bifurcations will $(1.1)_\lambda$ undergo near the orbit $C_0 := \{\tilde{x}_0(t) \mid t \in \mathbb{R}\}$ as we change the parameter λ near λ_0? More in particular we will be interested in the bifurcation of periodic solutions of $(1.1)_\lambda$ near C_0.

The classical way to attack this problem is via a Poincaré map. We take a section S at a point $\tilde{x}_0(t_0) \in C_0$, such that S is transversal to C_0 and with $C_0 \cap S = \{\tilde{x}_0(t_0)\}$. Then we define the Poincaré map $\Pi : S \times \mathbb{R}^k \to S$ as the first return map on S under the flow of $(1.1)_\lambda$. Periodic solutions of $(1.1)_\lambda$ near C_0 and with minimal period near T_0 correspond then to fixed points of Π; i.e. we obtain such periodic solutions by solving the equation

$$(1.2) \qquad \Pi(x,\lambda) = x .$$

For subharmonic solutions we have to solve fixed point equations of the form

$$(1.3) \qquad \Pi^m(x,\lambda) = x ,$$

for some m>1, and with $\Pi^m : S \times \mathbb{R}^k \to S$ defined recursively by $\Pi^1 = \Pi$,

---

1980 Mathematics Subject Classification : 58F14.

$\Pi^{m+1}(x,\lambda) = \Pi(\Pi^m(x,\lambda),\lambda)$. Methods to handle (1.2) and (1.3) are well established; for a survey, see e.g. Iooss [4].

Now suppose that (1.1) is symmetric; by this we mean that f is equivariant with respect to some closed subgroup Γ of $O(n)$:

$$(1.4) \qquad f(\gamma x,\lambda) = \gamma f(x,\lambda) \quad , \quad \forall \gamma \in \Gamma .$$

An immediate consequence is that if $\tilde{x}(t)$ is a solution of (1.1) , then so is $\gamma\tilde{x}(t)$, for each $\gamma \in \Gamma$; in particular, each periodic solution may generate a whole family of periodic solutions. We can describe the symmetry of the periodic solution $\tilde{x}_0(t)$ by means of two subgroups of Γ, defined respectively by

$$(1.5) \qquad K_0 := \{\gamma \in \Gamma \mid \gamma\tilde{x}_0(t)=\tilde{x}_0(t),\ \forall t \in \mathbb{R}\}$$

and

$$(1.6) \qquad H_0 := \{\gamma \in \Gamma \mid \gamma(C_0)=C_0\} .$$

Remark that, by uniqueness of solutions, a symmetry-operator $\gamma \in \Gamma$ will belong to H_0 if and only if $\gamma\tilde{x}_0(t_0) \in C_0$ for some $t_0 \in \mathbb{R}$. We call H_0 the *orbital symmetry* and K_0 the *spatial symmetry* of $\tilde{x}_0(t)$. There exists a group-homomorphism $\theta_0 : H_0 \to S^1 := \mathbb{R}/\mathbb{Z}$ such that

$$(1.7) \qquad \gamma\tilde{x}_0(t) = \tilde{x}_0(t+\theta_0(\gamma)T_0) \quad , \quad \forall \gamma \in H_0,\ \forall t \in \mathbb{R} .$$

We have $K_0 = \ker \theta_0$; hence K_0 is a normal subgroup of H_0, and either $H_0/K_0 \cong \mathbb{Z}_m := m^{-1}\mathbb{Z}/\mathbb{Z}$ for some $m \geq 1$, or $H_0/K_0 \cong S^1$. Now we would like to see these symmetries of $\tilde{x}_0(t)$ reflected in some way in the associated Poincaré map Π. This is quite easy for the spatial symmetry : one can choose the section S such that $\gamma(S) = S$ for all $\gamma \in K_0$, from which it follows easily that Π is K_0-equivariant. For the orbital symmetry things are not so simple. B. Fiedler [2] has shown that if $\mathbb{Z}_m \subset H_0/K_0$ for some $m>1$ then we can write $\Pi = \gamma^m\Pi_0^m$ for some $\gamma \in H_0$ and with $\Pi_0 : S \times \mathbb{R}^k \to S$ a K_0-equivariant "partial" Poincaré map. However, in case $H_0/K_0 \cong S^1$ this does not account for the full orbital symmetry H_0.

In this paper we present an attempt to incorporate the full symmetry of $\tilde{x}_0(t)$ in the bifurcation analysis for periodic solutions near C_0. We do not use the Poincaré map, but instead work in a space of periodic functions and make an equivariant Liapunov-Schmidt reduction near an orbit of solutions, as explained in chapter 8 of [7]. We give the details of this reduction in section 2; in order to present the main ideas in as simple a context as possible we consider at that stage general systems without any particular symmetry. In section 3 we discuss briefly a few generic bifurcations which appear in such systems. In section 4 we show how symmetries can be incorporated in the theory, and how a systematic study of the symmetry groups involved leads (in a more general context) to the elementary bifurcations obtained by Fiedler [2] via the Poincaré map.

2. REDUCTION TO AN EQUIVALENT PROBLEM. The problem of finding all periodic solutions of the system $(1.1)_\lambda$ is equivalent to that of finding for all $\sigma>0$ all 1-periodic solutions of the equation

$$\sigma\dot{x} = f(x,\lambda) \ . \tag{2.1}$$

Indeed, if $x_0(t)$ is a 1-periodic solution of (2.1) for some $(\eta,\sigma) \in \mathbb{R}\times(0,\infty)$ then $\tilde{x}_0(t) := x_0(\sigma t)$ is a $(1/\sigma)$-periodic solution of $(1.1)_\lambda$; also the converse holds. Now let Z (resp. X) be the Banach space of all C^0 (resp. C^1) 1-periodic mappings $z : \mathbb{R} \to \mathbb{R}^n$, endowed with the C^0 (resp. C^1) supremum norm. The problem is then equivalent to that of solving the equation

$$M(x,\lambda,\sigma) = 0 \tag{2.2}$$

for $(x,\lambda,\sigma) \in X\times\mathbb{R}^k\times(0,\infty)$, with $M : X\times\mathbb{R}^{k+1} \to Z$ defined by

$$M(x,\lambda,\sigma)(t) := -\sigma\dot{x}(t) + f(x(t),\lambda) \quad , \quad \forall t \in \mathbb{R} \ . \tag{2.3}$$

An important property of M is its S^1-equivariance :

$$M(T_\phi x,\lambda,\sigma) = T_\phi M(x,\lambda,\sigma) \quad , \quad \forall\phi \in S^1 \ , \tag{2.4}$$

where, for each $\phi \in S^1$, the operator $T_\phi \in \mathcal{L}(Z)$ is defined by

$$(T_\phi z)(t) := z(t-\phi) \quad , \quad \forall t \in \mathbb{R} \ , \ \forall z \in \mathbb{Z} \ . \tag{2.5}$$

Consequently, if

$$M(x_0,\lambda_0.\sigma_0) = 0 \ , \tag{2.6}$$

then (2.2) has for $(\lambda,\sigma) = (\lambda_0,\sigma_0)$ a whole orbit of solutions, given by

$$C := \{T_\phi x_0 \mid \phi\in S^1\} \ . \tag{2.7}$$

Let us return for a moment to our original problem : we want to describe, for all λ near λ_0, all periodic solutions of $(1.1)_\lambda$ near the orbit C_0 of a periodic solution $\tilde{x}_0(t)$ of $(1.1)_{\lambda_0}$ with minimal period $T_0 = 1/\sigma_0 > 0$. Excluding periodic solutions whose minimal period blows up to infinity as one approaches the bifurcation point, we can restrict to periodic solutions with minimal period near NT_0, for some $N\geqq 1$; the time scaling needed to make such solutions 1-periodic will depend on N, and hence we will also have to consider different scalings of $\tilde{x}_0$. Formally we define for each $N\geqq 1$ a function $x_N \in X$ by $x_N(t) := \tilde{x}_0(N\sigma_0^{-1}t)$; then $(x_N,\lambda_0,N^{-1}\sigma_0)$ is a solution of (2.2). Our problem is then to determine , for each $N\geqq 1$, all solutions $(x,\lambda,\sigma) \in X\times\mathbb{R}\times(0,\infty)$ of (2.2) near $\{(T_\phi x_N,\lambda_0,N^{-1}\sigma_0) \mid \phi\in S^1\}$. We will do this separately for each $N\geqq 1$, writing (x_0,λ_0,σ_0) for $(x_N,\lambda_0,N^{-1}\sigma_0)$.

So assume (2.6), and define $C \subset X$ by (2.7). The first step is to describe a neighborhood of C in X. The tangent space to C at x_0 is spanned by $\dot{x}_0$; hence also $\dot{x}_0 \in \ker L_0$, where $L_0 := D_x M(x_0,\lambda_0,\sigma_0)$. We define a projection $P \in \mathcal{L}(Z)$ by

$$Pz := \langle z,\dot{x}_0\rangle\langle\dot{x}_0,\dot{x}_0\rangle^{-1}\dot{x}_0 \quad , \quad \forall z \in Z \ , \tag{2.8}$$

where

$$(2.9)\qquad \langle z_1,z_2\rangle := \int_0^1 (z_1(t),z_2(t))dt \quad , \quad \forall z_1,z_2 \in Z .$$

Remark that P commutes with T_ϕ for each $\phi \in \Sigma_0 := \{\phi \in S^1 | T_\phi x_0 = x_0\}$; since x_0 is a nonconstant periodic solution we will have $\Sigma_0 = \mathbb{Z}_N$ for some $N \geq 1$. Then we can apply theorem 8.2.5 of [7] to obtain the following :

LEMMA 1. Let $Y := N(P) = \{z \in Z | \langle \dot{x}_0, z\rangle = 0\}$. Then there exists a neighborhood Ω of the origin in $Y \cap X$ such that

$$(2.10)\qquad \{T_\phi(x_0+y) \mid \phi \in S^1,\ y \in \Omega\}$$

forms an open neighborhood of C in X.

As an immediate consequence we can replace x by $T_\phi(x_0+y)$ in the equation (2.2); since M is S^1-equivariant and T_ϕ an isomorphism on Z, the resulting equation is equivalent to

$$(2.11)\qquad \hat{M}(y,\lambda,\sigma) := M(x_0+y,\lambda,\sigma) = 0 .$$

We have $\hat{M}(0,\lambda_0,\sigma_0) = 0$, $\hat{M} : (Y \cap X) \times \mathbb{R}^{k+1} \to Z$ is smooth and Σ_0-equivariant, and we want to find all solutions $(y,\lambda,\sigma) \in \Omega \times \mathbb{R}^{k+1}$ of (2.11) near $(0,\lambda_0,\sigma_0)$.

Next we rewrite (2.11) as the system

$$(2.12a)\qquad P\hat{M}(y,\lambda,\sigma) = 0 ,$$
$$(2.12b)\qquad (I-P)\hat{M}(y,\lambda,\sigma) = 0 .$$

Since $PD_\sigma\hat{M}(0,\lambda_0,\sigma_0) = -\dot{x}_0 \neq 0$ we can solve (2.12a) by the implicit function theorem for $\sigma = \tilde{\sigma}(y,\lambda)$; bringing this solution in (2.12b) we obtain the reduced equation :

$$(2.13)\qquad \tilde{M}(y,\lambda) := (I-P)\hat{M}(y,\lambda,\tilde{\sigma}(y,\lambda)) = 0 .$$

The mapping $\tilde{M} : (Y \cap X) \times \mathbb{R}^k \to Y$ is defined and smooth near $(0,\lambda_0)$, and is Σ_0-equivariant. Also $\tilde{M}(0,\lambda_0) = 0$, and it is a little exercise in Banach space calculus to prove the following.

LEMMA 2. The operator $\tilde{L}_0 := D_y\tilde{M}(0,\lambda_0) \in \mathcal{L}(X \cap Y, Y)$ is a Fredholm operator with index 0; also $\text{Im } \tilde{L}_0 = (I-P) \text{ Im } L_0$. If $\dot{x}_0 \notin R(L_0)$ then $\ker \tilde{L}_0 = (I-P) \ker L_0$ and $\dim \ker \tilde{L}_0 = \dim \ker L_0 - 1$. If $\dot{x}_0 \in R(L_0)$ then $\ker \tilde{L}_0 = \text{span}\{y_0\} \oplus (I-P)\ker L_0$, where $y_0 \in Y \cap X$ is such that $L_0 y_0 = \dot{x}_0$; in this case $\dim \ker \tilde{L}_0 = \dim \ker L_0$.

It follows that the equation (2.13) is in a form on which we can apply the whole machinery of equivariant bifurcation theory via the Liapunov-Schmidt reduction (see e.g. [3] and [7]). The basic group is $\Sigma_0 = \mathbb{Z}_N$. By lemma 2 we can relate $\ker \tilde{L}_0$ to $\ker L_0$, which in turn can be determined from a study of the characteristic multipliers and corresponding eigenvectors of the periodic solution $\tilde{x}_0(t)$ of $(1.1)_{\lambda_0}$. In the next section we briefly describe some of the elementary bifurcations which one can expect from such analysis; we will give more details in a forthcoming paper.

3. ELEMENTARY BIFURCATIONS IN THE GENERAL CASE. In one-parameter problems (i.e. when k=1) one expects that generically one has $\dim \ker \tilde{L}_0 = 1$ at possible bifurcation points (see e.g. M. Medved [6] or Alligood, Mallet-Paret and Yorke [1]). The assumption $\dim \ker \tilde{L}_0 = 1$ leads to a scalar bifurcation equation and, under appropriate and generically satisfied transversality conditions, to two different bifurcation pictures, depending on how the Σ_0-symmetry acts on $\ker \tilde{L}_0$.

Suppose first that $\ker \tilde{L}_0 = \text{span}\{u_0\}$ and that $T_\phi u_0 = u_0$ for all $\phi \in \Sigma_0 = \mathbb{Z}_N$. Then all bifurcating solutions will also be Σ_0-symmetric, i.e. they will have minimal period 1/N; we can then use this information to perform a time rescaling which reduces us to the case N=1. Generically one then finds a turning point, corresponding to two periodic solutions of $(1.1)_\lambda$ with approximately the same period which coalesce and disappear as λ crosses λ_0. We remark that at such a turning point one has generically that $\dot{x}_0 \in R(L_0)$ (see lemma 2).

The second case is when $T_\phi u_0 = -u_0$ for some $\phi \in \Sigma_0$. This is only possible when N is even; all bifurcating solutions will then have a minimal period equal to either 1/N or 2/N, and by a time rescaling we can assume that N=2, with $T_{1/2}u_0 = -u_0$. The resulting picture is that of a pitchfork bifurcation : the solutions along the primary branch (containing x_0) have minimal period 1/2; while those along the bifurcating branches have minimal period 1. So we find here the well known period-doubling bifurcation.

In two-parameter problems (i.e. with k=2) and when $N \geq 3$ we may generically have $\dim \ker \tilde{L}_0 = 2$ at bifurcation points, with the action of $\mathbb{Z}_N$ on $\ker \tilde{L}_0$ equivalent to the standard action of $\mathbb{Z}_M$ on $\mathbb{R}^2$, for some divisor $M \geq 3$ of N. Again a time rescaling allows us to consider only the case M=N. We have then a two-dimensional $\mathbb{Z}_N$-equivariant problem, describing bifurcation of subharmonic solutions; in particular one finds so-called Arnol'd tongues in the two-dimensional parameter space, corresponding to those parameter values for which a subharmonic solution exists (compare e.g. with [8] and [9]).

4. ELEMENTARY BIFURCATIONS IN SYMMETRIC SYSTEMS. Now we return to the symmetric case which formed the main reason for introducing our new approach. We assume that there is a compact subgroup Γ of O(n) such that (1.4) holds. Then the mapping M defined by (2.3) is $\Gamma \times S^1$-equivariant, with $\Gamma \times S^1 = \{(\gamma,\phi) \mid \gamma \in \Gamma,\ \phi \in S^1\}$ acting on Z by

$$(4.1) \qquad (\gamma,\phi).z(t) := \gamma z(t-\phi) \quad , \quad \forall t \in \mathbb{R} .$$

So if (2.6) holds then (2.2) has for $(\lambda,\sigma) = (\lambda_0,\sigma_0)$ an orbit of solutions given by

$$(4.2) \qquad \tilde{C} := \{(\gamma,\phi).x_0 \mid (\gamma,\phi) \in \Gamma \times S^1\} .$$

We will assume that we start with a solution (x_0,λ_0,σ_0) for which $\tilde{C}$ is one-dimensional. Since $\tilde{C}$ is compact and $x \in \tilde{C}$ implies $\{T_\phi x \mid \phi \in S^1\} \subset \tilde{C}$ this means that there exist a finite number of elements $\gamma_j \in \Gamma$ $(j=1,\dots,r)$, with $\gamma_1 = Id$, such that

$$\tilde{C} = \bigcup_{j=1}^{r} \gamma_j(C) , \tag{4.3}$$

with C given by (2.7). We can then still do the reduction of section 2 near C; in fact, acting with γ_j on the solutions near C will give us also the solutions near each of the other components of $\tilde{C}$. Moreover, if we put

$$\Sigma_0 := \{(\gamma,\phi) \in \Gamma\times S^1 \mid (\gamma,\phi).x_0 = x_0\} , \tag{4.4}$$

then it is easily seen that P is Σ_0-equivariant, and hence also the reduced problem (2.13) will be Σ_0-equivariant.

Now let us consider the simplest possible case for bifurcation, namely when k=1, $\dim N(\tilde{L}_0) = 1$ and $N(\tilde{L}_0) = \mathrm{span}\{u_0\}$, $u_0 \in Y\setminus\{0\}$. Let

$$\Sigma := \{(\gamma,\phi) \in \Sigma_0 \mid (\gamma,\phi).u_0 = u_0\} . \tag{4.5}$$

As in section 3 we have to consider two cases : $\Sigma = \Sigma_0$ and $\Sigma \neq \Sigma_0$. The case $\Sigma = \Sigma_0$ leads (generically) to a turning point, corresponding to two periodic solutions with the same symmetry and approximately the same period coalescing and disappearing as λ crosses λ_0.

The interesting case is the one for which $\Sigma \neq \Sigma_0$. Then Σ will be a normal subgroup of Σ_0, with $\Sigma_0/\Sigma \cong \mathbb{Z}_2$; this leads generically to a pitchfork bifurcation, with two branches of Σ-symmetric solutions bifurcating at x_0 from a primary branch of Σ_0-symmetric solutions. Depending on Σ_0 and Σ this will correspond to different types of symmetry-breaking bifurcations, as we will explain now. We start by analyzing more closely the structure of Σ_0.

Since x_0 is assumed to be a non-constant periodic solution we will have that

$$\{\phi \in S^1 \mid (Id,\phi) \in \Sigma_0\} = \mathbb{Z}_N \tag{4.6}$$

for some $N \geq 1$; this means that 1/N is the minimal period of x_0. Next we define two subgroups of Γ by

$$H_0 := \{\gamma \in \Gamma \mid \exists \phi \in S^1 \text{ such that } (\gamma,\phi) \in \Sigma_0\} \tag{4.7}$$

and

$$K_0 := \{\gamma \in \Gamma \mid (\gamma,0) \in \Sigma_0\} ; \tag{4.8}$$

H_0 and K_0 are respectively the orbital and the spatial symmetry of x_0 (and $\tilde{x}_0$), as introduced in section 1. Then there exists a group homomorphism

$$\theta_0 : H_0 \to S^1/\mathbb{Z}_N = \{\phi + \mathbb{Z}_N \mid \phi \in S^1\} \tag{4.9}$$

such that

(4.10) $\Sigma_0 = \{(\gamma,\phi) \mid \gamma \in H_0,\ \phi \in \theta_0(\gamma)\}$;

moreover $K_0 = \ker \theta_0$ is a normal subgroup of H_0, and we have either $H_0/K_0 \cong \mathbb{Z}_m$ for some $m \geq 1$, or $H_0/K_0 \cong S^1$; in this last case we call x_0 a rotating wave solution.

Assuming that Σ is not the whole Σ_0 there are several possibilities for the structure of Σ as related to the structure of Σ_0. We consider separately each of the possible cases.

CASE 1 : $\{Id\} \times \mathbb{Z}_N \subset \Sigma$.

The bifurcating solutions have the same minimal period $1/N$ as x_0; by a time rescaling we may then assume that $N=1$, so that Σ_0 and Σ have the form

(4.11) $\Sigma_0 = \{(\gamma,\theta_0(\gamma)) \mid \gamma \in H_0\}$, $\Sigma = \{(\gamma,\theta_0(\gamma)) \mid \gamma \in H\}$,

with $\theta_0 : H_0 \to S^1$ a group homomorphism, and H a normal subgroup of H_0 such that $H_0/H \cong \mathbb{Z}_2$. There are two subcases :

(a) $K_0 \subset H$: the bifurcating solutions have the same spatial symmetry as the solutions on the primary branch. We have a pitchfork bifurcation with breaking of the orbital symmetry and conservation of spatial symmetry and minimal period. Note that since H/K_0 must be a normal subgroup of H_0/K_0 with $(H_0/K_0)/(H/K_0) \cong \mathbb{Z}_2$ this case is only possible if $H_0/K_0 \cong \mathbb{Z}_m$ with m even.

(b) $K_0 \not\subset H$: in this case we have a pitchfork bifurcation with breaking of the spatial symmetry; indeed, the spatial symmetry of the bifurcating solutions is given by $K := K_0 \cap H \neq K_0$. Remark however that $H/K \cong H_0/K_0$. We have also that $K_0/K \cong \mathbb{Z}_2$, so that this case can only appear if K_0 has a normal subgroup K such that $K_0/K \cong \mathbb{Z}_2$.

CASE 2 : $\{Id\} \times \mathbb{Z}_N \not\subset \Sigma$.

This case is only possible if N is even. The bifurcating solutions will have minimal period $2/N$, so that we have period-doubling; a time rescaling allows us to put $N=2$. The isotropy Σ of the bifurcating solutions has then the form

(4.12) $\Sigma = \{(\gamma,\theta(\gamma)) \mid \gamma \in H_0\}$

with $\theta : H_0 \to S^1$ a group homomorphism such that $\theta(\gamma) \in \theta_0(\gamma)$ for each $\gamma \in H_0$. We see that the solutions along the bifurcating branch keep the same orbital symmetry as the solutions along the primary branch. As for the spatial symmetry there are again two cases :

(a) $K := \ker \theta = K_0$: the bifurcating solutions keep also the same spatial symmetry. This case is only possible if $H_0/K_0 \cong \mathbb{Z}_m$ with m odd. Indeed, suppose that $\mathbb{Z}_m \subset H_0/K_0$ for some $m \geq 1$; then there is some $\gamma \in H_0$ such that $\theta_0(\gamma) = \{\frac{1}{2m}, \frac{1}{2m} + \frac{1}{2}\}$. It follows that $\theta(\gamma) = \frac{1}{2m} + \frac{k}{2}$, with $k=0$ or 1. Since $\gamma^m \in K_0 = K$ we must have that $\frac{1}{2} + \frac{mk}{2} = 0 \pmod 1$, i.e. $mk = 1 \pmod 2$. So necessarily m is odd and $k=1$.

(b) $K \neq K_0$: then $K_0/K \cong \mathbb{Z}_2$, and we have period-doubling with breaking of the spatial symmetry. This case is only possible if K_0 has a normal subgroup K such that $K_0/K \cong \mathbb{Z}_2$.

We conclude that if $H_0/K_0 \cong \mathbb{Z}_m$ with m even, or if $H_0/K_0 \cong S^1$, and if at the same time K_0 has no normal subgroup K with $K_0/K \cong \mathbb{Z}_2$, then a simple period doubling as described here cannot occur.

We finish with some remarks on the relation of the foregoing with the results of Fiedler [2], in particular with table 4.2 of [2] which gives a list of generic bifurcations of periodic orbits in one-parameter $\mathbb{Z}_q$-equivariant systems. Using Fiedler's terminology, case 1a above corresponds to a flip pitchfork (also discussed to some extent by Klič [5]), case 1b to a flip-flop pitchfork, and case 2a to flip doubling. As for case 2b, it corresponds to both flop doubling and flip-flop doubling, two distinct cases in [2] which can in fact be shown to be equivalent.

The foregoing shows that our approach allows to recover Fiedler's results in a more general context, since we did not restrict to cyclic groups; also we do not need a separate approach to discuss bifurcation near rotating wave solutions. We hope to expand on the method in a later paper and to apply the same technique also for a discussion of subharmonic bifurcation in symmetric systems.

ACKNOWLEDGEMENTS. We thank Bernold Fiedler, Marty Golubitsky and Gerard Iooss for a number of discussions and remarks on topics connected to this paper, and Bill Langford for the kind invitation to attend the meeting.

BIBLIOGRAPHY

1. K.T. Alligood, J. Mallet-Paret and J.A. Yorke, "An index for the continuation of relatively isolated sets of periodic orbits", in : J. Palis (Ed.), Geometric Dynamics, Lect. Notes Math. 1007, Springer-Verlag, New York, 1983,1-21.
2. B. Fiedler, "Global Hopf bifurcation in reaction diffusion systems with symmetry", Proc. Workshop on Dynamics of Infinite Dimensional Systems, Lisbon, 1986, to appear.
3. M. Golubitsky and D.G. Schaeffer, Singularities and Groups in Bifurcation Theory I, Springer-Verlag, New York, 1985.
4. G. Iooss, Bifurcation of Maps and Applications, Mathematics Studies, Vol. 36, North-Holland, Amsterdam, 1979.
5. A. Klič, "Bifurcations in the systems with involutory symmetry", preprint 1985.
6. M. Medved, "Generic properties of parametrized vector fields I,II", Czech. Math. J. 25 (1975), 375-388 and 26 (1976), 71-83.
7. A. Vanderbauwhede, Local bifurcation and symmetry, Research Notes in Math., Vol. 75, Pitman, London, 1982.
8. A. Vanderbauwhede, "Bifurcation of subharmonic solutions : a generic approach", in : I.W. Knowles and R.T. Lewis (Eds.), Differential Equations, Mathematics Studies, Vol. 92, North-Holland, Amsterdam, 1984, 545-552.

9. A. Vanderbauwhede, "Bifurcation of subharmonic solutions in time-reversible systems", Z. Angew. Math. Phys., 37 (1986), 455-478.

INSTITUUT VOOR THEORETISCHE MECHANICA
RIJKSUNIVERSITEIT GENT
KRIJGSLAAN 281
B-9000 GENT, BELGIUM

Canadian Mathematical Society
Conference Proceedings
Volume 8 (1987)

THE NUMBER OF ORIENTATION REVERSING CYCLES IN THE QUADRATIC MAP

A. WEISS AND T.D. ROGERS

ABSTRACT. A formula is derived for the number of bifurcations of stable periodic orbits in the quadratic family of maps $x \to x^2 - a$. A bijection with the set of transitive unimodal permutations makes the combinatorial structure of the cyles transparent. Furthermore, the twisted shift operator defined on a binary symbol space is introduced as the natural device for studying the dynamics of unimodal maps of the interval.

INTRODUCTION. Let Δ_n denote the set of permutations δ of $I_n = \{1, 2, \ldots, n\}$ which act transitively on I_n and which are unimodal, i.e., there exists $m = m_\delta$ so that

i) whenever $k < \ell \leq m$ then $\delta k > \delta \ell$.

ii) whenever $m \leq k < \ell$ then $\delta k < \delta \ell$.

THEOREM. *Δ_n has cardinality*

$$\frac{1}{n} \sum_{d/n, d \text{ odd}} \mu(d) 2^{\frac{n}{d}-1}$$

where μ is the Möbius function of number theory.

The problem of counting Δ_n was posed in Rogers [6] in the context of the study of iterated maps on an interval of the real line as a dynamical system. In fact the formula in the theorem was conjectured on the basis of a comparison of empirical counts of Δ_n with the numbers in the table on page 11 of Myrberg [5], which concerns "negative" cycles of the map $x \to x^2 - 2$. There is bijection behind this which is sketched in §5.

The proof of the theorem occupies §§1–4 and is entirely elementary and self-contained. Nevertheless, the constructions involved come from the same source as the problem.

First, and most important, the map π_δ in §3 is essentially the invariant co-ordinate of Milnor and Thurston [4], which is always non-decreasing. So half of Lemma 3 holds in greater generality which, in particular, suggests that the theorem does also.

1980 Mathematics Subject Classification (1985 Revision). Primary 58F20; Secondary 54H14.
Supported through grants from NSERC.

Second, the introduced twisted shift map F on a space of sequences is nearly a map of an interval (see §5), and, because of the commutative square in §3, can be viewed as a "universal unimodal map". Moreover, iii) of §1 says that invariant coordinates and natural coordinates coincide for F. Even though all of these properties are trivial they do suggest that F provides a better model for iterated maps on an interval than the ordinary shift operator which is usually employed.

Finally, cycles of F have a signature. This is what makes the other half of Lemma 3 (i.e., π_δ is injective) work, yet still remain somewhat mysterious. There are other combinatorial settings in which the formula of the theorem occurs (see [3]), however we emphasize there are no natural bijections which relate the various applications, and the ubiquity of the formula merely reflects the usefulness of the Möbius function in counting situations.

§1. THE TWISTED SHIFT OPERATOR F. Let E denote the set of sequences $\varepsilon = (\varepsilon_1, \varepsilon_2, \dots)$ with $\varepsilon_i = +1$ or -1 for all i. Order E lexicographically, i.e., $\varepsilon < \eta$ means that the smallest t with $\varepsilon_t \neq \eta_t$ has $\varepsilon_t = -1 < 1 = \eta_t$.

Define $A : E \to \{-1, 1\}$ by

$$A(\varepsilon) = \varepsilon_1.$$

This auxiliary function is compatible with the ordering as follows:

i) if $A(\eta) = 1$ and $\varepsilon > \eta$ then $A(\varepsilon) = 1$

ii) if $A(\eta) = -1$ and $\varepsilon < \eta$ then $A(\varepsilon) = -1$

as is easy to verify.

Our main interest is in the map $F : E \to E$ defined by

$$F(\varepsilon)_i = \varepsilon_1 \varepsilon_{i+1} \qquad (= A(\varepsilon)\varepsilon_{i+1}).$$

The kth iterate F^k of F satisfies

$$F^k(\varepsilon)_i = \varepsilon_k \varepsilon_{i+k}$$

by induction on $k \geq 1$.

Properties of F are conveniently expressed via A. The ones we need are

iii) $\prod_{i=0}^{j-1} A(F^i \varepsilon) = \varepsilon_j$ for $j \geq 1$

and all ε, and the order properties

iv) F is increasing on the domain $A(\varepsilon) = +1$

v) F is decreasing on the domain $A(\varepsilon) = -1$

(where "increasing" means <u>strictly</u> increasing, always) which are also clear once one observes that if $\varepsilon < \eta$ and $A(\varepsilon) = A(\eta)$ then the least t with $\varepsilon_t \neq \eta_t$ has $t \geq 2$.

From the formula for F^k one immediately deduces

vi) $F^d(\varepsilon) = \varepsilon$ if and only if $\varepsilon_{i+d} = \varepsilon_d \varepsilon_i$ for all i,

and from this a simple calculation shows

vii) if $F^d(\varepsilon) = \varepsilon$ then $F(\varepsilon)_d = \varepsilon_d$.

We are now ready to study cycles of F. By an <u>n-cycle</u> of F we mean a subset C of E with n elements so that F maps C back into C and indeed F acts transitively on C. Write $E(n)$ for the set of n-cycles of F.

If C is an n-cycle of F and ε is in C then $F^n(\varepsilon) = \varepsilon$ so by vii) and transitivity of the action of F on C it follows that ε_n is independent of the choice of ε in C. Call this value the <u>signature</u> of C. Write $E_-(n)$ for the set of n-cycles of F which have signature -1.

LEMMA 1. *$E_-(n)$ has cardinality*

$$\frac{1}{n} \sum_{d/n, d \text{ odd}} \mu(d)\, 2^{\frac{n}{d}-1}.$$

Proof: Set $E_n = \{\varepsilon \in E : F^n(\varepsilon) = \varepsilon\}$. By construction $F(E_n) \subseteq E_n$, and so, by $F(F^{n-1}(\varepsilon)) = \varepsilon$, F maps E_n onto E_n. However, by vi) elements of E_n are determined by their first n coordinates, so E_n is finite, and F is a permutation of E_n. Therefore, we can decompose E_n into a disjoint union of cycles of F. In fact

$$E_n = \dot{\bigcup_{d/n}} \dot{\bigcup_{C \in E(n/d)}} C$$

since F^n = identity on E_n implies that the cycles contained in E are precisely those whose order divides n.

Consider now $E_n^- = \{\varepsilon \in E_n : \varepsilon_n = -1\}$. By vii) we have $F(E_n^-) \subseteq E_n^-$ so by last paragraph F is also a permutation of E_n^-. To determine the cycle decomposition of $E_n^- \subseteq E_n$ we need only decide which of the cycles in the decomposition of E_n are in fact contained in E_n^-. We <u>claim</u>:

$$E_n^- = \dot{\bigcup_{d/n, d \text{ odd}}} \dot{\bigcup_{C \in E_-(n/d)}} C.$$

Let C be a cycle in E_n, of order n/d say, with d dividing n, and let $\varepsilon \in C$. We must show that $\varepsilon_n = -1$ if and only if d is odd and $\varepsilon_{\frac{n}{d}} = -1$.

By $F^{n/d}(\varepsilon) = \varepsilon$ and vi) we have $\varepsilon_{i+\frac{n}{d}} = \varepsilon_{\frac{n}{d}} \varepsilon_i$ for all i. Then $\varepsilon_{i+k\cdot\frac{n}{d}} = \left(\varepsilon_{\frac{n}{d}}\right)^k \varepsilon_i$ for all i, by induction on $k \geq 1$. Setting $k = d$ we get $\varepsilon_{i+n} = \left(\varepsilon_{\frac{n}{d}}\right)^d \varepsilon_i$ for all i. On the other hand, by $F^n(\varepsilon) = \varepsilon$ and vi) we also have $\varepsilon_{i+n} = \varepsilon_n \varepsilon_i$ for all i. Comparing, we deduce $\varepsilon_n = \left(\varepsilon_{\frac{n}{d}}\right)^d$, which, by the last paragraph, settles the claim.

By vi) we see that E_n^- has 2^{n-1} elements, so setting $e_-(n) = \text{card } E_-(n)$, the claim gives the relation

$$2^{n-1} = \sum_{d/n, d \text{ odd}} \frac{n}{d} e_-(\frac{n}{d})$$

from which the lemma follows by Möbius inversion.

To prove the theorem it now suffices to construct maps $\Phi : E_-(n) \to \Delta_n$ and $\Psi : \Delta_n \to E_-(n)$ so that $\Psi \circ \Phi$ and $\Phi \circ \Psi$ are the identity map on $E_-(n)$ and Δ_n respectively.

§2. **CONSTRUCTION OF Φ.** Let $C \in E_-(n)$. Since C has $n < \infty$ elements the ordering on E induces the linear ordering on C hence there is a unique <u>increasing</u> bijection

$$\lambda = \lambda_C : I_n \to C.$$

Since C is a cycle of F we can use commutativity of the square

$$\begin{array}{ccc} I_n & \xrightarrow{\lambda} & C \\ \delta \downarrow & & \downarrow F \\ I_n & \xrightarrow{\lambda} & C \end{array}$$

to define a mapping δ. We want to set $\Phi(C) = \delta$.

We must check Φ is well-defined, i.e., $\delta \in \Delta_n$. Since F acts transitively on C the commutative square ensures the same of δ. It remains to check that δ is unimodal which we delay until we have

LEMMA 2. *Notation as above determine $m = m_\delta \in I_n$ by $\delta(m) = 1$. Then*

i) $(A \circ \lambda)(k) = +1$ *for* $k > m$

ii) $(A \circ \lambda)(k) = -1$ *for* $k < m$

iii) $\prod_{k=1}^{n} (A \circ \lambda)(k) = -1$

iv) $\prod_{i=0}^{j-1} (A \circ \lambda)(\delta^i k) = \lambda(k)_j$ *for all* $j \geq 1$.

Proof: Set $\lambda(m) = \eta \in C$. By the commutative square with λ increasing we must have

v) $F(\eta) = \min\{F(\varepsilon) : \varepsilon \in C\}$

and i), ii) will follow if we can show for all $\varepsilon \in C$ that

$$A(\varepsilon) = \begin{cases} 1 & \varepsilon > \eta \\ -1 & \varepsilon < \eta. \end{cases}$$

Suppose first that $A(\eta) = 1$. If $\varepsilon > \eta$ then $A(\varepsilon) = 1$ follows by i) of §1. If $\varepsilon < \eta$ we show $A(\varepsilon) = 1$ is impossible: for then iv) of §1 implies $F(\varepsilon) < F(\eta)$ contrary to v).

Similarly suppose $A(\eta) = -1$. If $\varepsilon < \eta$ then $A(\varepsilon) = -1$ by ii) of §1. If $\varepsilon > \eta$ we show $A(\varepsilon) = -1$ is impossible: for then v) of §1 implies $F(\varepsilon) < F(\eta)$ again contrary to v).

Turning now to iv) we first check

$$\lambda\delta^j = F^j\lambda \qquad \text{for all } j \geq 1$$

by induction on j, based on the commutative square. Thus

$$\prod_{i=0}^{j-1}(A \circ \lambda)(\delta^i k) = \prod_{i=0}^{j-1} A(\lambda\delta^i(k)) = \prod_{i=0}^{j-1} A(F^i(\lambda k)) = \lambda(k)_j$$

by iii) of §1. Finally iii) is a consequence of iv) and transitivity of δ since then

$$\prod_{k=1}^{n}(A \circ \lambda)(k) = \prod_{j=0}^{n-1}(A \circ \lambda)(\delta^j h) = \lambda(h)_n$$

for any $h \in I_n$, and this is indeed -1 because C has signature -1 by hypothesis. The lemma is proved.

Finally, we check δ is unimodal by verifying

i′) whenever $k < \ell < m$ then $\delta k > \delta\ell$

ii′) whenever $m < k < \ell$ then $\delta k < \delta\ell$

with $m = m_\delta$ given by Lemma 2. Since δm is the smallest element of I_n this is equivalent with the original definition.

So suppose $k < \ell < m$. Then $\lambda k < \lambda\ell$ since λ is increasing. And by ii) of Lemma 2 we can apply v) of §1 to deduce $F(\lambda k) > F(\lambda\ell)$. By commutativity of the square this means $\lambda(\delta k) > \lambda(\delta\ell)$ and finally by λ increasing we have $\delta k > \delta\ell$ which is i′). The verification of ii′) is entirely analogous.

§3. CONSTRUCTION OF Ψ. Let $\delta \in \Delta_n$. Define $\alpha = \alpha_\delta : I_n \to \{-1, 1\}$ by

i) $\alpha(k) = +1, \quad k > m_\delta$

ii) $\alpha(k) = -1, \quad k < m_\delta$

iii) $\prod_{k=1}^{n} \alpha(k) = -1$

where $m = m_\delta$ is defined by unimodality of δ (and necessarily has $\delta m = 1$) and where $\alpha(m)$ is defined by iii). Now define $\pi = \pi_\delta : I_n \to E$ by setting

$$\pi(k)_j = \prod_{i=0}^{j-1} \alpha(\delta^i k), \qquad j \geq 1.$$

We want to put $\Psi(\delta) = \pi_\delta(I_n)$.

This construction forces again a commutative square

$$\begin{array}{ccc} I_n & \xrightarrow{\pi} & E \\ \delta\downarrow & & \downarrow F \\ I_n & \xrightarrow{\pi} & E \end{array}$$

which is a trite computation. We delay checking that Ψ is well-defined, i.e., $\pi_\delta(I_n) \in E_-(n)$, until we have the crucial

LEMMA 3. *$\pi_\delta : I_n \to E$ is increasing.*

Proof: Define $S : I_n \times I_n \to \{-1, 0, 1\}$ by

$$S(\ell, k) = \begin{cases} +1, & \ell > k \\ -1, & \ell < k \\ 0, & \ell = k \end{cases}$$

From the construction of $\alpha = \alpha_\delta$ one verifies

iv) if $\alpha(\ell) = \alpha(k)$ then $S(\delta\ell, \delta k) = \alpha(k) S(\ell, k)$

v) if $\alpha(\ell) \neq \alpha(k)$ then $S(\ell, k) = \alpha(\ell)$,

by exhausting cases in iv) using $\delta \in \Delta_n$.

Claim: Let $r \geq 1$ and suppose k, ℓ satisfy $\alpha(\delta^i k) = \alpha(\delta^i \ell)$ for $0 \leq i < r$. Then $S(\delta^r \ell, \delta^r k) = \pi(k)_r S(\ell, k)$.

For iv) and induction on j yields

$$S(\delta^j \ell, \delta^j k) = \left(\prod_{i=0}^{j-1} \alpha(\delta^i k) \right) S(\ell, k)$$

for $1 \leq j \leq r$. The claim follows by setting $j = r$.

Since δ acts transitively on I_n we have

$$\pi(k)_n = \prod_{i=0}^{n-1} \alpha(\delta^i k) = \prod_{\ell=1}^{n} \alpha(\ell)$$

and so iii) gives

vi) $\pi(k)_n = -1$ for all k in I_n.

We first deduce that π is injective. Suppose $\pi(k) = \pi(\ell)$. Then the definition of π shows that we can take $r = n$ in the Claim so have

$$S(\ell, k) = S(\delta^n \ell, \delta^n k) = \pi(k)_n S(\ell, k) = -S(\ell, k)$$

by vi). Thus $S(\ell, k) = 0$, i.e., $\ell = k$ as required.

Finally, we show π is increasing. Suppose $k < \ell$. By the last paragraph there is a smallest t with $\pi(k)_t \neq \pi(\ell)_t$. It suffices to show $\pi(k)_t = 1$, by definition of lexicographic order.

By definition of π and minimality of t we have

vii) $\alpha(\delta^i k) = \alpha(\delta^i \ell)$ for $0 \leq i < t-1$

while $\alpha(\delta^{t-1}k) \neq \alpha(\delta^{t-1}\ell)$ which by v) implies

viii) $S(\delta^{t-1}\ell, \delta^{t-1}k) = \alpha(\delta^{t-1}\ell)$

and of course $k < \ell$ says

ix) $S(\ell, k) = +1$.

Suppose $t = 1$: combining viii) and ix) gives immediately $\alpha(\ell) = S(\ell, k) = +1$ hence $\pi(\ell)_1 = 1$, as required.

Suppose $t \geq 2$: by vii) we can apply the Claim with $r = t-1$ together with ix) to conclude $S(\delta^{t-1}\ell, \delta^{t-1}k) = -\pi(k)_{t-1}$. By viii) we deduce $\alpha(\delta^{t-1}\ell) = \pi(k)_{t-1}$. Combining this with the definition of π then gives

$$\pi(\ell)_t = \pi(\ell)_{t-1}\alpha(\delta^{t-1}\ell) = \pi(\ell)_{t-1}\pi(k)_{t-1} = 1$$

and the lemma is proved.

Finally, to check $\pi(I_n) \in E_-(n)$ we only need that π is injective. This says $\pi(I_n)$ has n elements and then the commutative square forces $\pi(I_n)$ to be an n-cycle of F. That $\pi(I_n)$ has signature -1 is just vi) above.

§4. PROOF OF THE THEOREM. It remains only to show that the consequences of the constructions in §2 and §3, which were not needed to show that the maps are well-defined, are indeed enough to conclude that $\Phi \circ \Psi$ and $\Psi \circ \Phi$ are identity maps.

Let $\delta \in \Delta_n$. Then $C = \Psi(\delta)$ is in $E_-(n)$ so

$$\begin{array}{ccc} I_n & \xrightarrow[\pi_\delta]{} & C \\ {\scriptstyle\delta}\big\downarrow & & \big\downarrow{\scriptstyle F} \\ I_n & \xrightarrow[\pi_\delta]{} & C \end{array}$$

commutes. To compute $\Phi(C)$ one takes the increasing bijection $\lambda_C : I_n \to C$, which by Lemma 3 can only be π_δ, and then $\Phi(C)$ is defined by the commutativity of just this square which is already satisfied by δ. Thus $\Phi(C) = \delta$, i.e., $\Phi \circ \Psi$ is identity on Δ_n.

Let $C \in E_-(n)$. Then $\delta = \Phi(C)$ in Δ_n is defined via a bijection $\lambda_C : I_n \to C$. Now compute $\Psi(\delta)$: first, we need α_δ which is defined by i), ii), iii) of §3. However, by i), ii), iii) of Lemma 2 we see that λ_C has the property that $A \circ \lambda_C = \alpha_\delta$.

Next we compute π_δ:

$$\pi_\delta(k)_j = \prod_{i=0}^{j-1} \alpha_\delta(\delta^i k) = \prod_{i=0}^{j-1} (A \circ \lambda_C)(\delta^i k) = \lambda_C(k)_j$$

by iv) of Lemma 2. Thus $\pi_\delta = \lambda_C$ and so $\Psi(\delta) = \pi_\delta(I_n) = \lambda_C(I_n) = C$, i.e., $\Psi \circ \Phi$ is the identity on $E_-(n)$.

The theorem is completely proved.

§5. VISUALIZING F. Define $\rho : E \to [-1,1]$ by $\rho(\varepsilon) = \sum_{k=1}^{\infty} \varepsilon_k 2^{-k}$. Denote by E_* that which remains of E after we throw out all eventually constant sequences and by $[-1,1]_*$ what remains of $[-1,1]$ when we throw out all rational numbers with 2-power denominator. Then it is easy to see that $\rho : E_* \to [-1,1]_*$ is an increasing bijection.

Define $f : [-1,1] \to [-1,1]$ by $f(x) = 2|x| - 1$. Then

$$\begin{array}{ccc} E_* & \xrightarrow[\rho]{} & [-1,1]_* \\ F\big\downarrow & & \big\downarrow f \\ E_* & \xrightarrow[\rho]{} & [-1,1]_* \end{array}$$

commutes. In particular the order properties of F are transparent for f. Moreover, the n-cycles of F actually lie in E_*, by vi) of §1, so also must appear for f. Indeed if ε satisfies vi) of §1 then

$$\rho(\varepsilon) = \sum_{i=1}^{n} \varepsilon_i 2^{n-i} / (2^n - \varepsilon_n)$$

by summing the geometric series; thus $\rho(\varepsilon)$ is an odd rational number (i.e., with odd numerator and denominator) and in fact $\rho(\varepsilon) \equiv 1 \mod 4$. It is again easy to verify that precisely these elements of $[-1,1]_*$ occur in cycles of f.

Myrberg [5] considers the map $x \to x^2 - 2$ of $[-2,2]$ which corresponds to f by the change of coordinates $x \to 2\sin(\frac{\pi}{2} x)$, and signature -1 corresponds to "negative multiplier", i.e., an orientation reversing cycle. (Although Myrberg states a number of theorems related to the present work, there are no complete proofs; refer, for example, to his Hypothesis H.)

In the language of bifurcation theory, it follows from the recent work of Douady and Hubbard [1], that the kneading invariant of critical point of the quadratic family $x \to x^2 - \mu$ is monotone. As a consequence of this, the formula stated in our theorem also counts the number of bifurcations of stable periodic orbits of period n as well as the number of distinct superstable cycles in this family.

Apparently the results of [1] also serve to verify that the generic flip and fold bifurcations as described by Guckenheimer [2] are the only bifurcation of periodic orbits that occur in the quadratic family. As noted by Smale, if K_n = the number of periodic orbits of period n for the map $x^2 - 2$ then $K_n = \frac{1}{n} \sum_{d/n} \mu(d) 2^{\frac{n}{d}}$. Let S_n = the number of cycles of period n which were ever stable in the family $x^2 - \mu$. Then from Guckenheimer [2], S_n satisfies the recursion $S_{2n} = \dfrac{K_{2n} + S_n}{2}$, $S_{2n+1} = \dfrac{K_{2n+1}}{2}$, $n = 1, 2, \ldots$, where $S_1 = S_2 = 1$. It is easily checked that the formula of the theorem in the introduction is the unique solution to this initial value problem. This fact alone, however, ignores the basic combinatorial structure of the stable cycles, which is that of Δ_n.

REFERENCES

1. Douady, A. and J.H. Hubbard. "Intération des polynomes quadratiques complexes," C.R. Acad. Sci. Paris, 294 (1982), 123–126.

2. Guckenheimer, J., "On the Bifurcations of Maps of the Interval," Inventiones Math., 39 (1977), 165–178.

3. Miller, R.L. "Necklaces, symmetries and self-reciprocal polynomials," Discrete Mathematics, 22 (1978), 25–33.

4. Milnor, J. and Thurston, W. "On iterated maps of the interval, I, II," Preprint, Princeton (1977).

5. Myrberg, P.J. "Iteration der reelen polynome zweiten grades III," Ann. Acad. Scient. Fennicae A.I., 336/3 (1963), 3–18.

6. Rogers, T.D. "Chaos in systems in population biology," pp. 91–146 in Progress in Theoretical Biology, Volume 6, Academic Press, 1981.

DEPARTMENT OF MATHEMATICS
UNIVERSITY OF ALBERTA
EDMONTON, ALBERTA
CANADA T6G 2G1

BCDEFGHIJ-898